Student's Solution:

to accompany

College Algebra

John W. Coburn
St. Louis Community College at Florissant Valley

Written by
Rosemary M. Karr, Ph.D.
and
Lesley A. Seale

Boston Burr Ridge, IL Dubuque, IA Madison, WI New York San Francisco St. Louis
Bangkok Bogotá Caracas Kuala Lumpur Lisbon London Madrid Mexico City
Milan Montreal New Delhi Santiago Seoul Singapore Sydney Taipei Toronto

The McGraw·Hill Companies

Student's Solutions Manual to accompany
COLLEGE ALGEBRA
JOHN W. COBURN

Published by McGraw-Hill Higher Education, an imprint of The McGraw-Hill Companies, Inc., 1221 Avenue of the Americas, New York, NY 10020.

Recycled/acid free paper
This book is printed on recycled, acid-free paper containing 10% postconsumer waste.

1 2 3 4 5 6 7 8 9 0 QPD/QPD 0 9 8 7 6

ISBN-13: 978-0-07-291761-1
ISBN-10: 0-07-291761-X

www.mhhe.com

Chapter R: A Review of Basic Concepts and Skills

Chapter 1: Equations and Inequalities

Chapter 2: Functions and Graphs

Chapter 3: Operations on Functions and Analyzing Graphs

Chapter 4: Polynomial and Rational Functions

Chapter 5: Exponential and Logarithmic Functions

Chapter 6: Systems of Equations and Inequalities

Chapter 7: Conic Sections and Nonlinear Systems

Chapter 8: Systems of Equations and Inequalities

R.1 Exercises

1. Subset of; element of

3. Positive; negative; 7; -7; principal

5. Order of operations requires multiplication before addition.

7. a. $\{1, 2, 3, 4, 5\}$
   b. $\{\ \}$

9. True

11. True

13. True

15. $\frac{4}{3} = 1.\overline{3}$

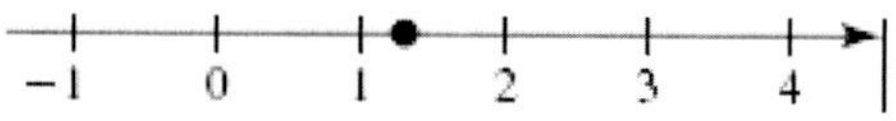

17. $2\frac{5}{9} = 2.\overline{5}$

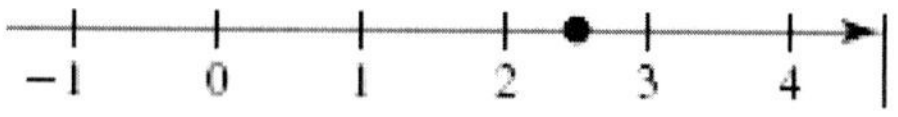

19. $\sqrt{7} \approx 2.65$

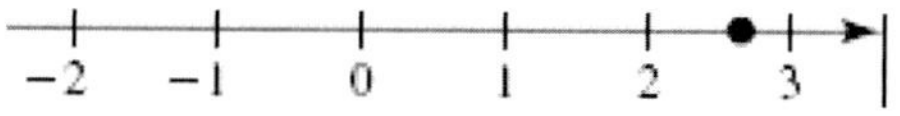

21. $\sqrt{3} \approx 1.73$

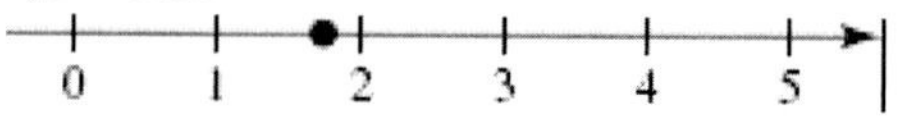

23. a. i) $\{8, 7, 6\}$
   ii) $\{8, 7, 6\}$
   iii) $\{-1, 8, 7, 6\}$
   iv) $\left\{-1,\ 8,\ 0.75,\ \frac{9}{2},\ 5.\overline{6},\ 7,\ \frac{3}{5},\ 6\right\}$
   v) $\{\ \}$
   vi) $\left\{-1,\ 8,\ 0.75,\ \frac{9}{2},\ 5.\overline{6},\ 7,\ \frac{3}{5},\ 6\right\}$

   b. $\left\{-1,\ \frac{3}{5},\ 0.75,\ \frac{9}{2},\ 5.\overline{6},\ 6,\ 7,\ 8\right\}$

   c.

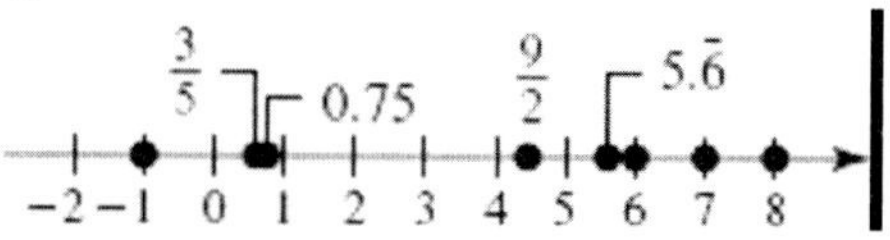

25. a. i) $\{\sqrt{49},\ 2,\ 6,\ 4\}$
   ii) $\{\sqrt{49},\ 2,\ 6,\ 0,\ 4\}$
   iii) $\{-5,\ \sqrt{49},\ 2,\ -3,\ 6,\ -1,\ 0,\ 4\}$
   iv) $\{-5,\ \sqrt{49},\ 2,\ -3,\ 6,\ -1,\ 0,\ 4\}$
   v) $\{\sqrt{3},\ \pi\}$
   vi)
   $\{-5,\ \sqrt{49},\ 2,\ -3,\ 6,\ -1,\ \sqrt{3},\ 0,\ 4,\ \pi\}$

   b. $\{-5,\ -3,\ -1,\ 0,\ \sqrt{3},\ 2,\ \pi,\ 4,\ 6,\ \sqrt{49}\}$

   c.

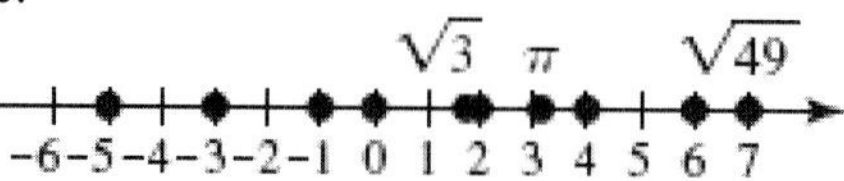

27. False; not all real numbers are irrational.

29. False; not all rational numbers are integers.

31. False; $\sqrt{25} = 5$ is not irrational.

33. c; IV

35. a; VI

37. d; III

39. Let *a* represent Kylie's age: $a \geq 6$.

41. Let *n* represent the number of incorrect words: $n \leq 2$.

43. $|-2.75| = 2.75$

45. $-|-4| = -4$

47. $\left|\frac{1}{2}\right| = \frac{1}{2}$

49. $\left|-\frac{3}{4}\right| = \frac{3}{4}$

51. $|-7.5 - 2.5| = |-10| = 10$;
   $|2.5 - (-7.5)| = |10| = 10$

53. Negative

55. -*n*

57. -8 and 2

59. Undefined; since $12 \div 0 = k$, implies $k \cdot 0 = 12$.

61. Undefined, since $7 \div 0 = k$, implies $k \cdot 0 = 7$.

63. a. Positive
b. Negative
c. Negative
d. Negative

65. $-\sqrt{\frac{121}{36}} = -\frac{11}{6}$

67. $\sqrt[3]{-8} = -2$

69. $9^2 = 81$ is closest.

71. -24 – (-31) = -24 + 31 = 7

73. 7.045 – 9.23 = -2.185

75. $4\frac{5}{6} + \left(-\frac{1}{2}\right) = 4\frac{5}{6} + \left(-\frac{3}{6}\right) = 4\frac{2}{6} = 4\frac{1}{3}$

77. $\left(-\frac{2}{3}\right)\left(3\frac{5}{8}\right) = \left(-\frac{2}{3}\right)\left(\frac{29}{8}\right) = -\frac{58}{24} = -\frac{29}{12}$

or $-2\frac{5}{12}$

79. (12)(-3)(0) = 0

81. $-60 \div 12 = -5$

83. $\frac{4}{5} \div (-8) = \frac{4}{5} \cdot \left(-\frac{1}{8}\right) = -\frac{4}{40} = -\frac{1}{10}$

85. $-\frac{2}{3} \div \frac{16}{21} = -\frac{2}{3} \cdot \frac{21}{16} = -\frac{42}{48} = -\frac{7}{8}$

87. $-3^2 + 15 - |5-15| - \sqrt{169}$
$= -9 + 15 - |-10| - 13$
$= -9 + 15 - 10 - 13$
$= 6 - 10 - 13$
$= -4 - 13$
$= -17$

89. $\sqrt{\frac{9}{16}} - \frac{3}{5} \cdot \left(\frac{5}{3}\right)^2$
$= \frac{3}{4} - \frac{3}{5} \cdot \frac{25}{9}$
$= \frac{3}{4} - \frac{75}{45}$
$= \frac{3}{4} - \frac{5}{3}$
$= \frac{9}{12} - \frac{20}{12}$
$= -\frac{11}{12}$

91. $\frac{4(-7) - 6^2}{6 - \sqrt{49}} = \frac{-28 - 36}{6 - 7} = \frac{-64}{-1} = 64$

93. $2475\left(1 + \frac{0.06}{4}\right)^{4 \cdot 10} = 4489.70$

95. $D = \frac{d \cdot n}{n+2} = \frac{5 \cdot 12}{12+2} = \frac{60}{14} \approx 4.3$ cm

97. 50 + (-3)(6) = 50 – 18 = 32° F

99. 134 – (-45) = 134 + 45 = 179° F

101. $3\frac{1}{7} = \frac{22}{7} \approx 3.14285$;
$\frac{355}{113} \approx 3.14159$;
$\frac{62832}{20000} = 3.1416$;
$\sqrt{10} \approx 3.1623$;
$\pi \approx 3.141592654$;
Tsu-Ch'ung-chih: $\frac{355}{113}$

103. Negative

R.2 Exercises

1. Constant

3. Coefficient

5. -5 + 5 = 0; $-5 \cdot \left(-\frac{1}{5}\right) = 1$

7. Two; 3 and -5

9. Two; 2 and $\frac{1}{4}$

11. Three; -2, 1 and -5

13. One; -1

15. $n-7$

17. $n+4$

19. $(n-5)^2$

21. $2n-13$

23. n^2+2n

25. $\frac{2}{3}n-5$

27. $3(n+5)-7$

29. Let w represent the width. Then $2w$ represents twice the width and $2w-3$ represents three meters less than twice the width.

31. Let b represent the speed of the bus. Then $b+15$ represents 15 mph more than the speed of the bus.

33. $4x-2y$; $4(2)-2(-3)=8+6=14$

35. $-2x^2+3y^2; -2(2)^2+3(-3)^2$
$=-2(4)+3(9)=-8+27=19$

37. $2y^2+5y-3$; $2(-3)^2+5(-3)-3$
$=2(9)-15-3=18-15-3=0$

39. $-2(3y+1)$;
$-2(3(-3)+1)=-2(-9+1)=-2(-8)=16$

41. $3x^2y$; $3(2)^2(-3)=3(4)(-3)=-36$

43. $(-3x)^2-4xy-y^2$;
$(-3\cdot 2)^2-4(2)(-3)-(-3)^2$
$=(-6)^2-8(-3)-9=36+24-9=51$

45. $\frac{1}{2}x-\frac{1}{3}y$; $\frac{1}{2}(2)-\frac{1}{3}(-3)=1+1=2$

47. $(3x-2y)^2$;
$(3\cdot 2-2(-3))^2=(6+6)^2=12^2=144$

49. $\frac{12y+5}{-3x+1}$; $\frac{-12(-3)+5}{-3(2)+1}=\frac{36+5}{-6+1}=\frac{-41}{5}$

51. $\sqrt{-12y}\cdot 4$;
$\sqrt{-12(-3)}\cdot 4=\sqrt{36}\cdot 4=6\cdot 4=24$

53. x^2-3x-4

x	Output
-3	$(-3)^2-3(-3)-4=14$
-2	$(-2)^2-3(-2)-4=6$
-1	$(-1)^2-3(-1)-4=0$
0	$(0)^2-2(0)-3=-3$
1	$(1)^2-3(1)-4=-6$
2	$(2)^2-3(2)-4=-6$
3	$(3)^2-3(3)-4=-4$

-1 has an output of 0.

55. $-3(1-x)-6$

x	Output
-3	$-3(1-(-3))-6=-18$
-2	$-3(1-(-2))-6=-15$
-1	$-3(1-(-1))-6=-12$
0	$-3(1-(0))-6=-9$
1	$-3(1-(1))-6=-6$
2	$-3(1-(2))-6=-3$
3	$-3(1-(3))-6=0$

3 has an output of 0.

57. x^3-6x+4

x	Output
-3	$(-3)^3-6(-3)+4=-5$
-2	$(-2)^3-6(-2)+4=8$

-1	$(-1)^3-6(-1)+4=9$
0	$(0)^3-6(0)+4=4$
1	$(1)^3-6(1)+4=-1$
2	$(2)^3-6(2)+4=0$
3	$(3)^3-6(3)+4=13$

2 has an output of 0.

59. a. $-5+7=7+(-5)=2$
    b. $-2+n=n+(-2)$
    c. $-4.2+a+13.6=a+(-4.2)+13.6$
    $=a+9.4$
    d. $7+x-7=x+7-7=x$

61. a. $x+(-3.2)+\underline{3.2}=x$
    b. $n-\frac{5}{6}+\underline{\frac{5}{6}}=n$

63. $-5(x-2.6)=-5x+13$

65. $\frac{2}{3}\left(-\frac{1}{5}p+9\right)=-\frac{2}{15}p+6$

67. $3a+(-5a)=3a-5a=-2a$

69. $\frac{2}{3}x+\frac{3}{4}x=\frac{8}{12}x+\frac{9}{12}x=\frac{17}{12}x$

71. $3(a^2+3a)-(5a^2+7a)$
    $=3a^2+9a-5a^2-7a$
    $=-2a^2+2a$

73. $x^2-(3x-5x^2)=x^2-3x+5x^2=6x^2-3x$

75. $(3a+2b-5c)-(a-b-7c)$
    $=3a+2b-5c-a+b+7c=2a+3b+2c$

77. $\frac{3}{5}(5n-4)+\frac{5}{8}(n+16)$
    $=3n-\frac{12}{5}+\frac{5}{8}n+10$
    $=\frac{24}{8}n-\frac{12}{5}+\frac{5}{8}n+\frac{50}{5}$
    $=\frac{29}{8}n+\frac{38}{5}$

79. $(3a^2-5a+7)+2(2a^2-4a-6)$
    $=3a^2-5a+7+4a^2-8a-12$
    $=7a^2-13a-5$

81. $-4b+7b-9b=3b-9b=-6b$

83. $13g-4h+4g+13h=17g+9h$

85. $5x+12x^2-8x-3x^2=9x^2-3x$

87. $6.3y-11.9x-7.2y+0.5x=-0.9y-11.4x$

89. $R=\frac{kL}{d^2}$;
    $R=\frac{(0.000025)(90)}{(0.015)^2}=\frac{0.00225}{0.000225}=10\text{ ohms}$

91. Let j represent the speed of the jet.
    $\frac{1}{2}j$

93. Let w represent the width.
    $2w+3$

95. Let c represent the cost of the 1978 stamp.
    $c+22$; 15 + 22 = 37¢

97. Let t represent the number of hours of labor.
    $25t+43.50$; 25(1.5) + 43.50 = \$81.00

99. a. positive odd integer

R.3 Exercises

1. Power

3. $20x$; 0

5. a. has addition of unlike terms. b. is multiplication.

7. $(6p^2q)(p^3q^3)=6p^5q^4$

9. $(-3.2a^2b^2)(5a^3b)=-16a^5b^3$

11. $\frac{2}{3}x^6y\cdot 21xy^6=14x^7y^7$

13. $(6pq^2)^3=6^3(p)^3(q^2)^3=216p^3q^6$

15. $\left(3.2hk^2\right)^3 = (3.2)^3(h)^3\left(k^2\right)^3 = 32.768h^3k^6$

17. $\left(\dfrac{p}{2q}\right)^2 = \dfrac{(p)^2}{(2)^2(q)^2} = \dfrac{p^2}{4q^2}$

19. $\left(-0.7c^4\right)^2\left(10c^3d^2\right)^2$
$= (-0.7)^2\left(c^4\right)^2(10)^2\left(c^3\right)^2\left(d^2\right)^2$
$= 0.49c^8 \cdot 100c^6d^4$
$= 49c^{14}d^4$

21. $\left(\dfrac{3}{4}x^3y\right)^2 = \left(\dfrac{3}{4}\right)^2\left(x^3\right)^2(y)^2 = \dfrac{9}{16}x^6y^2$

23. $\left(-\dfrac{3}{8}x\right)^2\left(16xy^2\right)$
$= \left(-\dfrac{3}{8}\right)^2(x)^2\left(16xy^2\right)$
$= \dfrac{9}{64}x^2 \cdot 16xy^2 = \dfrac{9}{4}x^3y^2$

25. a. $V = S^3$;
$V = \left(3x^2\right)^3$
$V = 3^3\left(x^2\right)^3$
$V = 27x^6$

b. $V = 27x^6$;
$V = 27(2)^6$
$V = 27(64)$
$V = 1728 \text{ units}^3$

27. $\dfrac{-6w^5}{-2w^2} = 3w^3$

29. $\dfrac{-12a^3b^5}{4a^2b^4} = -3ab$

31. $\left(\dfrac{2}{3}\right)^{-3} = \dfrac{2^{-3}}{3^{-3}} = \dfrac{3^3}{2^3} = \dfrac{27}{8}$

33. $\dfrac{2}{h^{-3}} = 2h^3$

35. $\left(\dfrac{2p^4}{q^3}\right)^2 = \dfrac{2^2\left(p^4\right)^2}{\left(q^3\right)^2} = \dfrac{4p^8}{q^6}$

37. $\left(\dfrac{0.2x^2}{0.3y^3}\right)^3 = \dfrac{(0.2)^3\left(x^2\right)^3}{(0.3)^3\left(y^3\right)^3}$
$= \dfrac{0.008x^6}{0.027y^9} = \dfrac{8x^6}{27y^9}$

39. $\left(\dfrac{5m^2n^3}{2r^4}\right)^2 = \dfrac{(5)^2\left(m^2\right)^2\left(n^3\right)^2}{(2)^2\left(r^4\right)^2} = \dfrac{25m^4n^6}{4r^8}$

41. $\left(\dfrac{5p^2q^3r^4}{-2pq^2r^4}\right)^2 = \left(\dfrac{5pq}{-2}\right)^2$
$= \dfrac{5^2p^2q^2}{(-2)^2} = \dfrac{25p^2q^2}{4}$

43. $\dfrac{9p^6q^4}{-12p^4q^6} = \dfrac{3p^2}{-4q^2}$

45. $\dfrac{20h^{-2}}{12h^5} = \dfrac{5}{3}h^{-2-5} = \dfrac{5}{3}h^{-7} = \dfrac{5}{3h^7}$

47. $\dfrac{\left(a^2\right)^3}{a^4 \cdot a^5} = \dfrac{a^6}{a^9} = \dfrac{1}{a^3}$

49. $\left(\dfrac{a^{-3}b}{c^{-2}}\right)^{-4} = \dfrac{a^{12}b^{-4}}{c^8} = \dfrac{a^{12}}{b^4c^8}$

51. $\dfrac{-6\left(2x^{-3}\right)^2}{10x^{-2}} = \dfrac{-6\left(4x^{-6}\right)}{10x^{-2}} = \dfrac{-24x^{-6}}{10x^{-2}}$
$= \dfrac{-12x^{-6-(-2)}}{5} = \dfrac{-12x^{-6+2}}{5} = \dfrac{-12x^{-4}}{5} = \dfrac{-12}{5x^4}$

53. $\dfrac{14a^{-3}bc^0}{-7\left(3a^2b^{-2}c\right)^3} = \dfrac{14a^{-3}bc^0}{-7\left(27a^6b^{-6}c^3\right)}$

$$= \frac{14a^{-3}bc^0}{-189a^6b^{-6}c^3} = -\frac{2a^{-3-6}b^{1-(-6)}c^{0-3}}{27}$$
$$= \frac{-2a^{-9}b^{1+6}c^{-3}}{27} = \frac{-2b^7}{27a^9c^3}$$

55. $4^0 + 5^0 = 1 + 1 = 2$

57. $2^{-1} + 5^{-1} = \frac{1}{2} + \frac{1}{5} = \frac{5}{10} + \frac{2}{10} = \frac{7}{10}$

59. $3^0 + 3^{-1} + 3^{-2} = 1 + \frac{1}{3} + \frac{1}{3^2}$
$= \frac{9}{9} + \frac{3}{9} + \frac{1}{9} = \frac{13}{9}$

61. $-5x^0 + (-5x)^0 = -5(1) + 1 = -5 + 1 = -4$

63. 14,500,000,000 = $1.45\text{x}10^{10}$

65. $4.8\text{x}10^9$ = \$4,800,000,000

67. $\frac{465{,}000{,}000}{17{,}500} = \frac{4.65\text{x}10^8}{1.75\text{x}10^4} = \frac{4.65}{1.75}\text{x}\frac{10^8}{10^4}$
$= 2.657142857\text{x}10^4 = 26571.42857$
≈ 26571.4 *hours*
$\frac{26571.4}{24} = 1107.141667 \approx 1107.1$ *days*

69. $-35w^3 + 2w^2 + (-12w) + 14$
Polynomial; None of these; Degree 3

71. $5n^{-2} + 4n + \sqrt{17}$
Non-polynomial because exponents are not whole numbers; NA; NA

73. $p^3 - \frac{2}{5}$
Polynomial; Binomial; Degree 3

75. $7w + 8.2 - w^3 - 3w^2$
$= -w^3 - 3w^2 + 7w + 8.2$
Lead coefficient: -1

77. $c^3 + 6 + 2c^2 - 3c = c^3 + 2c^2 - 3c + 6$
Lead coefficient: 1

79. $12 - \frac{2}{3}x^2 = -\frac{2}{3}x^2 + 12$
Lead coefficient: $-\frac{2}{3}$

81. $(3p^3 - 4p^2 + 2p - 7) + (p^2 - 2p - 5)$
$= 3p^3 - 4p^2 + 2p - 7 + p^2 - 2p - 5$
$= 3p^3 - 3p^2 - 12$

83. $(5.75b^2 + 2.6b - 1.9) + (2.1b^2 - 3.2b)$
$= 5.75b^2 + 2.6b - 1.9 + 2.1b^2 - 3.2b$
$= 7.85b^2 - 0.6b - 1.9$

85. $\left(\frac{3}{4}x^2 - 5x + 2\right) - \left(\frac{1}{2}x^2 + 3x - 4\right)$
$= \frac{3}{4}x^2 - 5x + 2 - \frac{1}{2}x^2 - 3x + 4$
$= \frac{3}{4}x^2 - 5x + 2 - \frac{2}{4}x^2 - 3x + 4$
$= \frac{1}{4}x^2 - 8x + 6$

87.
$$\begin{array}{rrrrrrr} q^6 + & 2q^5 + & q^4 + & 2q^3 & & & \\ -(& q^5 + & 2q^4 + & & q^2 + & 2q) & \\ \hline q^6 + & q^5 - & q^4 + & 2q^3 - & q^2 - & 2q & \end{array}$$

89. $-3x(x^2 - x - 6) = -3x^3 + 3x^2 + 18x$

91. $(3r - 5)(r - 2)$
$= 3r^2 - 6r - 5r + 10 = 3r^2 - 11r + 10$

93. $(x - 3)(x^2 + 3x + 9)$
$= x^3 + 3x^2 + 9x - 3x^2 - 9x - 27 = x^3 - 27$

95. $(b^2 - 3b - 28)(b + 2)$
$= b^3 + 2b^2 - 3b^2 - 6b - 28b - 56$
$= b^3 - b^2 - 34b - 56$

97. $(7v - 4)(3v - 5)$
$= 21v^2 - 35v - 12v + 20$
$= 21v^2 - 47v + 20$

99. $(3-m)(3+m)$
$=9+3m-3m-m^2$
$=9-m^2$

101. $(p-2.5)(p+3.6)$
$=p^2+3.6p-2.5p-9$
$=p^2+1.1p-9$

103. $\left(x+\frac{1}{2}\right)\left(x+\frac{1}{4}\right)$
$=x^2+\frac{1}{4}x+\frac{1}{2}x+\frac{1}{8}$
$=x^2+\frac{1}{4}x+\frac{2}{4}x+\frac{1}{8}$
$=x^2+\frac{3}{4}x+\frac{1}{8}$

105. $\left(m+\frac{3}{4}\right)\left(m-\frac{3}{4}\right)$
$=m^2-\frac{3}{4}m+\frac{3}{4}m-\frac{9}{16}=m^2-\frac{9}{16}$

107. $(3x-2y)(2x+5y)$
$=6x^2+15xy-4xy-10y^2$
$=6x^2+11xy-10y^2$

109. $(4c+d)(3c+5d)$
$=12c^2+20cd+3cd+5d^2$
$=12c^2+23cd+5d^2$

111. $(2x^2+5)(x^2-3)$
$=2x^4-6x^2+5x^2-15$
$=2x^4-x^2-15$

113. $4m-3$; Conjugate: $4m+3$
$(4m-3)(4m+3)$
$=16m^2+12m-12m-9=16m^2-9$

115. $7x-10$; Conjugate: $7x+10$
$(7x-10)(7x+10)$
$=49x^2+70x-70x-100=49x^2-100$

117. $6+5k$; Conjugate: $6-5k$
$(6+5k)(6-5k)$
$=36-30k+30k-25k^2=36-25k^2$

119. ab^2+c ; Conjugate: ab^2-c
$(ab^2+c)(ab^2-c)$
$=a^2b^4-ab^2c+ab^2c-c^2=a^2b^4-c^2$

121. $(x+4)^2=x^2+2(4\cdot x)+16=x^2+8x+16$

123. $(4g+3)^2$
$=16g^2+2(4g\cdot 3)+9=16g^2+24g+9$

125. $(4p-3q)^2$
$=16p^2-2(4p\cdot 3q)+9q^2$
$=16p^2-24pq+9q^2$

127. $(2m+3n)^2$
$=4m^2+2(2m\cdot 3n)+9n^2$
$=4m^2+12mn+9n^2$

129. $(x-3)(y+2)=xy+2x-3y-6$

131. $(k-5)(k+6)(k+2)$
$=(k-5)(k^2+2k+6k+12)$
$=(k-5)(k^2+8k+12)$
$=k^3+8k^2+12k-5k^2-40k-60$
$=k^3+3k^2-28k-60$

133. $M=0.5t^4+3t^3-97t^2+348t$

a. $M=0.5(2)^4+3(2)^3-97(2)^2+348(2)$
$M=0.5(16)+3(8)-97(4)+696$
$M=8+24-388+696$
$M=340$ mg

b. $M=0.5(3)^4+3(3)^3-97(3)^2+348(3)$
$M=0.5(81)+3(27)-97(9)+1044$
$M=40.5+81-873+1044$
$M=292.5$ mg

c. Less, the amount is decreasing.

d. Using the grapher, $y=0$ when $t=5$. The drug will wear off after 5 hours.

135. $F=\frac{kPQ}{d^2}$
$F=kPQd^{-2}$

137. $\frac{5}{x^3}+\frac{3}{x^2}+\frac{2}{x^1}+4=5x^{-3}+3x^{-2}+2x^{-1}+4$

139. $R=(20-1x)(200+20x)$

$R=4000+400x-200x-20x^2$

$R=4000+200x-20x^2$

Let x represent the number of \$1 decreases.

x	$R(x)$
1	4180
2	4320
3	4420
4	4480
5	4500
6	4480
7	4420
8	4320
9	4180
10	4000

Using the table, maximum revenue occurs at $x = 5$. Thus, $20 - 1x = 20 - 5 = 15$

The most revenue will be earned when the price is \$15.

141. $(3x^2+kx+1)-(kx^2+5x-7)+(2x^2-4x-k)$

$=-x^2-3x+2$

$3x^2+kx+1-kx^2-5x+7+2x^2-4x-k$

$=-x^2-3x+2$

$(kx-kx^2-k)+5x^2+8-9x=-x^2-3x+2$

$k(x-x^2-1)=-6x^2+6x-6$

$k=\frac{-6(x^2-x+1)}{x-x^2-1}$

$k=\frac{-6(x^2-x+1)}{-(x^2-x+1)}$

$k=6$

R.4 Exercises

1. Product

3. Binomial; Conjugate

5. Answers will vary;

$4x^2-36=4(x^2-9)=4(x+3)(x-3)$

7. a. $-17x^2+51=-17(x^2-3)$

b. $21b^3-14b^2+56b=7b(3b^2-2b+8)$

c. $-3a^4+9a^2-6a^3$

$=-3a^2(a^2-3+2a)$

$=-3a^2(a^2+2a-3)$

9. a. $2a(a+2)+3(a+2)=(a+2)(2a+3)$

b. $(b^2+3)3b+(b^2+3)2=(b^2+3)(3b+2)$

c. $4m(n+7)-11(n+7)=(n+7)(4m-11)$

11. a. $9q^3+6q^2+15q+10$

$=(9q^3+6q^2)+(15q+10)$

$=3q^2(3q+2)+5(3q+2)$

$=(3q+2)(3q^2+5)$

b. $h^5-12h^4-3h+36$

$=(h^5-12h^4)-(3h-36)$

$=h^4(h-12)-3(h-12)$

$=(h-12)(h^4-3)$

c. $k^5-7k^3-5k^2+35$

$=(k^5-7k^3)-(5k^2-35)$

$=k^3(k^2-7)-5(k^2-7)$

$=(k^2-7)(k^3-5)$

13. a. $b^2-5b-14=(b-7)(b+2)$

b. $a^2-4a-45=(a-9)(a+5)$

c. $n^2-9n+20=(n-4)(n-5)$

15. a. $3p^2-13p-10=(3p+2)(p-5)$

b. $4q^2+7q-15=(4q-5)(q+3)$

c. $10u^2-19u-15=(5u+3)(2u-5)$

17. a. $4s^2-25$

$=(2s)^2-(5)^2$

$=(2s+5)(2s-5)$

b. $9x^2-49$

$=(3x)^2-(7)^2$

$=(3x+7)(3x-7)$

c. $50x^2-72$

$=2(25x^2-36)$

$=2[(5x)^2-(6)^2]$

$=2(5x+6)(5x-6)$

d. $121h^2 - 144$
$= (11h)^2 - (12)^2$
$= (11h + 12)(11h - 12)$

19. a. $a^2 - 6a + 9 = (a-3)^2$
b. $b^2 + 10b + 25 = (b+5)^2$
c. $4m^2 - 20m + 25 = (2m-5)^2$
d. $9n^2 - 42n + 49 = (3n-7)^2$

21. a. $8p^3 - 27 = (2p)^3 - (3)^3$
$= (2p-3)(4p^2 + 6p + 9)$
b. $m^3 + \frac{1}{8} = (m)^3 + \left(\frac{1}{2}\right)^3$
$= \left(m + \frac{1}{2}\right)\left(m^2 - \frac{1}{2}m + \frac{1}{4}\right)$
c. $g^3 - 0.027 = (g)^3 - (0.3)^3$
$= (g - 0.3)(g^2 + 0.3g + 0.09)$
d. $-2t^4 + 54t = -2t(t^3 - 27)$
$= -2t\left[(t)^3 - (3)^3\right]$
$= -2t(t-3)(t^2 + 3t + 9)$

23. a. $9 + x^4 - 10x^2$
$= x^4 - 10x^2 + 9$
Let u represent x^2
$= u^2 - 10u + 9$
$= (u-9)(u-1)$
$= (x^2 - 9)(x^2 - 1)$
$= (x+3)(x-3)(x+1)(x-1)$
b. $13x^2 + x^4 + 36$
$= x^4 + 13x^2 + 36$
Let u represent x^2
$= u^2 + 13u + 36$
$= (u+9)(u+4)$
$= (x^2 + 9)(x^2 + 4)$
c. $-8 + x^6 - 7x^3$
$= x^6 - 7x^3 - 8$
Let u represent x^3
$= u^2 - 7u - 8$
$= (u-8)(u+1)$
$= (x^3 - 8)(x^3 + 1)$
$= (x-2)(x^2 + 2x + 4)(x+1)(x^2 - x + 1)$

25. a. $n^2 - 1 = (n+1)(n-1)$
b. $n^3 - 1 = (n-1)(n^2 + n + 1)$
c. $n^3 + 1 = (n+1)(n^2 - n + 1)$
d. $28x^3 - 7x = 7x(4x^2 - 1)$
$= 7x(2x+1)(2x-1)$

27. $a^2 + 7a + 10 = (a+5)(a+2)$

29. $x^2 - 12x + 20 = (x-2)(x-10)$

31. $64 - 9m^2 = (8)^2 - (3m)^2 = (8+3m)(8-3m)$

33. $-9r + r^2 + 18 = r^2 - 9r + 18$
$= (r-3)(r-6)$

35. $2h^2 + 7h + 6 = (2h+3)(h+2)$

37. $9k^2 - 24k + 16 = (3k-4)^2$

39. $2x^2 - 13x - 21$ Prime

41. $12m^2 - 40m + 4m^3 = 4m^3 + 12m^2 - 40m$
$= 4m(m^2 + 3m - 10) = 4m(m+5)(m-2)$

43. $a^2 - 7a - 60 = (a+5)(a-12)$

45. $8x^3 - 125 = (2x)^3 - (5)^3$
$= (2x-5)(4x^2 + 10x + 25)$

47. $4m^2 - 19m + 12 = (m-4)(4m-3)$

49. $x^3 - 5x^2 - 9x + 45$
$= (x^3 - 5x^2) - (9x - 45)$
$= x^2(x-5) - 9(x-5)$
$= (x-5)(x^2 - 9)$
$= (x-5)(x+3)(x-3)$

51. a. prime polynomial:
h. $x^2 + 9$
b. standard trinomial a = 1:

e. $x^2 - 3x - 10$

c. perfect square trinomial:

c. $x^2 - 10x + 25$

d. difference of cubes:

f. $8s^3 - 125t^3$

e. binomial square:

b. $(x+3)^2$

f. sum of cubes:

a. $x^3 + 27$

g. binomial conjugates:

i. $(x-7)$ and $(x+7)$

h. difference of squares:

d. $x^2 - 144$

i. standard trinomial $a \neq 1$:

g. $2x^2 - x - 3$

53. $H = x^3 + 2x^2 + 5x - 9$
$$H = (x^3 + 2x^2 + 5x) - 9$$
$$H = x(x^2 + 2x + 5) - 9$$
$$H = x[(x^2 + 2x) + 5] - 9$$
$$H = x[x(x+2) + 5] - 9;$$
$$H(-3) = -3[-3(-3+2) + 5] - 9$$
$$- -3[-3(-1) + 5] - 9$$
$$= -3[3+5] - 9$$
$$= -3(8) - 9$$
$$= -24 - 9$$
$$= -33$$

55. $S = 2\pi r^2 + \pi r^2 h$
$$S = \pi r^2(2+h);$$
$$= \pi(2.5)^2(2+10)$$
$$= \pi(2.5)^2(12)$$
$$\approx 235.62\ cm^2$$

57. $V = x^3 + 8x^2 + 15x$
$$V = x(x^2 + 8x + 15)$$
$$= x(x+3)(x+5)$$

a. If the height is x inches then the width is $x + 3$ which would make the width 3 inches more than the height.

b. If the height is x inches then the length is $x + 5$ which would make the width 5 inches more than the height.

c. $V = 2(2+3)(2+5) = 2(5)(7) = 70\,ft^3$

59. $L = L_0\sqrt{1 - \left(\frac{v}{c}\right)^2}$
$$L = L_0\sqrt{\left(1+\frac{v}{c}\right)\left(1-\frac{v}{c}\right)}$$
$$L = 12\sqrt{(1+0.75)(1-0.75)}$$
$$L = 12\sqrt{(1.75)(0.25)}s$$
$$L = 12\sqrt{0.4375} = 12\sqrt{0.0625(7)}$$
$$= 12(0.25)\sqrt{7} \approx 7.94 \text{ inches}$$

61. a. $\frac{1}{2}x^4 + \frac{1}{8}x^3 - \frac{3}{4}x^2 + 4$
$$= \frac{1}{8}(4x^4 + x^3 - 6x^2 + 32)$$

b. $\frac{2}{3}b^5 - \frac{1}{6}b^3 + \frac{4}{9}b^2 - 1$
$$= \frac{1}{18}(12b^5 - 3b^3 + 8b^2 - 18)$$

63. $192x^3 - 164x^2 - 270x$
$$= 2x(96x^2 - 82x - 135)$$
$$= 2x(16x-27)(6x+5)$$

R.5 Exercises

1. 1; -1

3. Common factor

5. False; $x - (x + 1) = x - x - 1 = -1$

7. a. $\frac{a-7}{-3a+21} = \frac{a-7}{-3(a-7)} = -\frac{1}{3}$

b. $\frac{2x+6}{4x^2-8x} = \frac{2(x+3)}{4x(x-2)} = \frac{x+3}{2x(x-2)}$

9. a. $\frac{x^2-5x-14}{x^2+6x-7} = \frac{(x-7)(x+2)}{(x+7)(x-1)}$

Simplified

b. $\frac{a^2+3a-28}{a^2-49} = \frac{(a+7)(a-4)}{(a+7)(a-7)} = \frac{a-4}{a-7}$

11. a. $\frac{x-7}{7-x} = \frac{-(7-x)}{7-x} = -1$

b. $\frac{5-x}{x-5}=\frac{-(x-5)}{x-5}=-1$

13. a. $\frac{-12a^3b^5}{4a^2b^{-4}}=-3a^{3-2}b^{5-(-4)}=-3ab^9$

b. $\frac{7x+21}{63}=\frac{7(x+3)}{63}=\frac{x+3}{9}$

c. $\frac{y^2-9}{3-y}=\frac{(y+3)(y-3)}{-(y-3)}=-1(y+3)$

d. $\frac{m^3n-m^3}{m^4-m^4n}=\frac{m^3(n-1)}{m^4(1-n)}$

$=\frac{m^3(n-1)}{-m^4(n-1)}=\frac{-1}{m}$

15. a. $\frac{2n^3+n^2-3n}{n^3-n^2}=\frac{n\left(2n^2+n-3\right)}{n^2(n-1)}$

$=\frac{n(2n+3)(n-1)}{n^2(n-1)}=\frac{2n+3}{n}$

b. $\frac{6x^2+x-15}{4x^2-9}=\frac{(2x-3)(3x+5)}{(2x-3)(2x+3)}$

$=\frac{3x+5}{2x+3}$

c. $\frac{x^3+8}{x^2-2x+4}=\frac{(x+2)\left(x^2-2x+4\right)}{x^2-2x+4}$

$=x+2$

d. $\frac{mn^2+n^2-4m-4}{mn+n+2m+2}$

$=\frac{\left(mn^2+n^2\right)-(4m+4)}{(mn+n)+(2m+2)}$

$=\frac{n^2(m+1)-4(m+1)}{n(m+1)+2(m+1)}$

$=\frac{(m+1)\left(n^2-4\right)}{(m+1)(n+2)}$

$=\frac{(m+1)(n+2)(n-2)}{(m+1)(n+2)}$

$=n-2$

17. $\frac{a^2-4a+4}{a^2-9}\cdot\frac{a^2-2a-3}{a^2-4}$

$=\frac{(a-2)(a-2)}{(a+3)(a-3)}\cdot\frac{(a-3)(a+1)}{(a+2)(a-2)}$

$=\frac{(a-2)(a+1)}{(a+3)(a+2)}$

19. $\frac{x^2-7x-18}{x^2-6x-27}\cdot\frac{2x^2+7x+3}{2x^2+5x+2}$

$=\frac{(x-9)(x+2)}{(x-9)(x+3)}\cdot\frac{(2x+1)(x+3)}{(2x+1)(x+2)}=1$

21. $\frac{p^3-64}{p^3-p^2}\div\frac{p^2+4p+16}{p^2-5p+4}$

$=\frac{p^3-64}{p^3-p^2}\cdot\frac{p^2-5p+4}{p^2+4p+16}$

$=\frac{(p-4)\left(p^2+4p+16\right)}{p^2(p-1)}\cdot\frac{(p-4)(p-1)}{\left(p^2+4p+16\right)}$

$=\frac{(p-4)^2}{p^2}$

23. $\frac{3x-9}{4x+12}\div\frac{3-x}{5x+15}=\frac{3x-9}{4x+12}\cdot\frac{5x+15}{3-x}$

$=\frac{3(x-3)}{4(x+3)}\cdot\frac{5(x+3)}{-(x-3)}=\frac{-15}{4}$

25. $\frac{a^2+a}{a^2-3a}\cdot\frac{3a-9}{2a+2}=\frac{a(a+1)}{a(a-3)}\cdot\frac{3(a-3)}{2(a+1)}=\frac{3}{2}$

27. $\frac{8}{a^2-25}\cdot\left(a^2-2a-35\right)$

$=\frac{8}{(a+5)(a-5)}\cdot(a-7)(a+5)=\frac{8(a-7)}{a-5}$

29. $\frac{xy-3x+2y-6}{x^2-3x-10}\div\frac{xy-3x}{xy-5y}$

$=\frac{(xy-3x)+(2y-6)}{x^2-3x-10}\cdot\frac{xy-5y}{xy-3x}$

$=\frac{x(y-3)+2(y-3)}{(x-5)(x+2)}\cdot\frac{y(x-5)}{x(y-3)}$

$=\frac{(y-3)(x+2)}{(x-5)(x+2)}\cdot\frac{y(x-5)}{x(y-3)}$

$=\frac{y}{x}$

31. $\frac{m^2+2m-8}{m^2-2m}\div\frac{m^2-16}{m^2}$

$$=\frac{m^2+2m-8}{m^2-2m}\cdot\frac{m^2}{m^2-16}$$
$$=\frac{(m+4)(m-2)}{m(m-2)}\cdot\frac{m^2}{(m+4)(m-4)}$$
$$=\frac{m}{m-4}$$

33. $\frac{y+3}{3y^2+9y}\cdot\frac{y^2+7y+12}{y^2-16}\div\frac{y^2+4y}{y^2-4y}$
$$=\frac{y+3}{3y(y+3)}\cdot\frac{(y+3)(y+4)}{(y+4)(y-4)}\cdot\frac{y^2-4y}{y^2+4y}$$
$$=\frac{y+3}{3y(y+3)}\cdot\frac{(y+3)(y+4)}{(y+4)(y-4)}\cdot\frac{y(y-4)}{y(y+4)}$$
$$=\frac{y+3}{3y(y+4)}$$

35. $\frac{x^2-0.49}{x^2+0.5x-0.14}\div\frac{x^2-0.10x+0.21}{x^2-0.09}$
$$=\frac{x^2-0.49}{x^2+0.5x-0.14}\cdot\frac{x^2-0.09}{x^2-0.10x+0.21}$$
$$=\frac{(x+0.7)(x-0.7)}{(x+0.7)(x-0.2)}\cdot\frac{(x+0.3)(x-0.3)}{(x-0.3)(x-0.7)}$$
$$=\frac{x+0.3}{x-0.2}$$

37. $\frac{3a^3-24a^2-12a+96}{a^2-11a+24}\div\frac{6a^2-24}{3a^3-81}$
$$=\frac{3(a^3-8a^2-4a+32)}{a^2-11a+24}\cdot\frac{3a^3-81}{6a^2-24}$$
$$=\frac{3[(a^3-8a^2)-(4a-32)]}{(a-8)(a-3)}\cdot\frac{3(a^3-27)}{6(a^2-4)}$$
$$=\frac{3[a^2(a-8)-4(a-8)]}{(a-8)(a-3)}\cdot\frac{3(a-3)(a^2+3a+9)}{6(a+2)(a-2)}$$
$$=\frac{3(a-8)(a^2-4)}{(a-8)(a-3)}\cdot\frac{3(a-3)(a^2+3a+9)}{6(a+2)(a-2)}$$
$$=\frac{3(a-8)(a+2)(a-2)}{(a-8)(a-3)}\cdot\frac{3(a-3)(a^2+3a+9)}{6(a+2)(a-2)}$$
$$=\frac{3(a^2+3a+9)}{2}$$

39. $\frac{4n^2-1}{12n^2-5n-3}\cdot\frac{6n^2+5n+1}{2n^2+n}\cdot\frac{12n^2-17n+6}{6n^2-7n+2}$
$$=\frac{(2n+1)(2n-1)}{(4n-3)(3n+1)}\cdot\frac{(3n+1)(2n+1)}{n(2n+1)}\cdot\frac{(4n-3)(3n-2)}{(3n-2)(2n-1)}$$
$$=\frac{2n+1}{n}$$

41. $\frac{3}{8x^2}+\frac{5}{2x}=\frac{3}{8x^2}+\frac{20x}{8x^2}=\frac{3+20x}{8x^2}$

43. $\frac{7}{4x^2y^3}-\frac{1}{8xy^4}=\frac{7(2y)}{8x^2y^4}-\frac{x}{8x^2y^4}=\frac{14y-x}{8x^2y^4}$

45. $\frac{4p}{p^2-36}-\frac{2}{p-6}$
$$=\frac{4p}{(p-6)(p+6)}-\frac{2}{p-6}$$
$$=\frac{4p}{(p-6)(p+6)}-\frac{2(p+6)}{(p-6)(p+6)}$$
$$=\frac{4p-2p-12}{(p-6)(p+6)}$$
$$=\frac{2p-12}{(p-6)(p+6)}$$
$$=\frac{2(p-6)}{(p-6)(p+6)}$$
$$=\frac{2}{p+6}$$

47. $\frac{m}{m^2-16}+\frac{4}{4-m}$
$$=\frac{m}{(m+4)(m-4)}-\frac{4}{m-4}$$
$$=\frac{m}{(m+4)(m-4)}-\frac{4(m+4)}{(m+4)(m-4)}$$
$$=\frac{m-4m-16}{(m+4)(m-4)}$$
$$=\frac{-3m-16}{(m+4)(m-4)}$$

49. $\frac{2}{m-7}-5=\frac{2}{m-7}-\frac{5(m-7)}{m-7}$
$$=\frac{2-5m+35}{m-7}=\frac{-5m+37}{m-7}$$

51. $\dfrac{y+1}{y^2+y-30}-\dfrac{2}{y+6}$

$=\dfrac{y+1}{(y+6)(y-5)}-\dfrac{2(y-5)}{(y+6)(y-5)}$

$=\dfrac{y+1-2y+10}{(y+6)(y-5)}=\dfrac{-y+11}{(y+6)(y-5)}$

53. $\dfrac{1}{a+4}+\dfrac{a}{a^2-a-20}$

$=\dfrac{1}{a+4}+\dfrac{a}{(a+4)(a-5)}$

$=\dfrac{a-5}{(a+4)(a-5)}+\dfrac{a}{(a+4)(a-5)}$

$=\dfrac{a-5+a}{(a+4)(a-5)}$

$=\dfrac{2a-5}{(a+4)(a-5)}$

55. $\dfrac{3y-4}{y^2+2y+1}-\dfrac{2y-5}{y^2+2y+1}$

$=\dfrac{3y-4-(2y-5)}{y^2+2y+1}$

$=\dfrac{3y-4-2y+5}{y^2+2y+1}$

$=\dfrac{y+1}{(y+1)(y+1)}$

$=\dfrac{1}{y+1}$

57. $\dfrac{2}{m^2-9}+\dfrac{m-5}{m^2+6m+9}$

$=\dfrac{2}{(m+3)(m-3)}+\dfrac{m-5}{(m+3)^2}$

$=\dfrac{2(m+3)+(m-5)(m-3)}{(m+3)^2(m-3)}$

$=\dfrac{2m+6+m^2-8m+15}{(m+3)^2(m-3)}$

$=\dfrac{m^2-6m+21}{(m+3)^2(m-3)}$

59. $\dfrac{y+2}{5y^2+11y+2}+\dfrac{5}{y^2+y-6}$

$=\dfrac{y+2}{(5y+1)(y+2)}+\dfrac{5}{(y+3)(y-2)}$

$=\dfrac{1}{5y+1}+\dfrac{5}{(y+3)(y-2)}$

$=\dfrac{(y+3)(y-2)+5(5y+1)}{(5y+1)(y+3)(y-2)}$

$=\dfrac{y^2+y-6+25y+5}{(5y+1)(y+3)(y-2)}$

$=\dfrac{y^2+26y-1}{(5y+1)(y+3)(y-2)}$

61. a. $p^{-2}-5p^{-1}=\dfrac{1}{p^2}-\dfrac{5}{p};\dfrac{1-5p}{p^2}$

b. $x^{-2}+2x^{-3}=\dfrac{1}{x^2}+\dfrac{2}{x^3};\dfrac{x+2}{x^3}$

63. $\dfrac{\frac{5}{a}-\frac{1}{4}}{\frac{25}{a^2}-\frac{1}{16}}=\dfrac{\left(\frac{5}{a}-\frac{1}{4}\right)16a^2}{\left(\frac{25}{a^2}-\frac{1}{16}\right)16a^2}=\dfrac{80a-4a^2}{400-a^2}$

$=\dfrac{4a(20-a)}{(20+a)(20-a)}=\dfrac{4a}{20+a}$

65. $\dfrac{p+\frac{1}{p-2}}{1+\frac{1}{p-2}}=\dfrac{\left(p+\frac{1}{p-2}\right)(p-2)}{\left(1+\frac{1}{p-2}\right)(p-2)}$

$=\dfrac{p(p-2)+1}{p-2+1}=\dfrac{p^2-2p+1}{p-1}$

$=\dfrac{(p-1)(p-1)}{p-1}$

$=p-1$

67. $\dfrac{\frac{2}{3-x}+\frac{3}{x-3}}{\frac{4}{x}+\frac{5}{x-3}}=\dfrac{\frac{-2}{x-3}+\frac{3}{x-3}}{\frac{4}{x}+\frac{5}{x-3}}$

$=\dfrac{\left(\frac{-2}{x-3}+\frac{3}{x-3}\right)x(x-3)}{\left(\frac{4}{x}+\frac{5}{x-3}\right)x(x-3)}$

$=\dfrac{-2x+3x}{4(x-3)+5x}=\dfrac{x}{4x-12+5x}=\dfrac{x}{9x-12}$

69. $\dfrac{\dfrac{2}{y^2-y-20}}{\dfrac{3}{y+4}-\dfrac{4}{y-5}} = \dfrac{\dfrac{2}{(y-5)(y+4)}}{\dfrac{3}{y+4}-\dfrac{4}{y-5}}$

$$= \frac{\left(\dfrac{2}{(y-5)(y+4)}\right)(y-5)(y+4)}{\left(\dfrac{3}{y+4}-\dfrac{4}{y-5}\right)(y-5)(y+4)}$$

$$= \frac{2}{3(y-5)-4(y+4)}$$

$$= \frac{2}{3y-15-4y-16}$$

$$= \frac{2}{-y-31}$$

71. a. $\dfrac{1+3m^{-1}}{1-3m^{-1}} = \dfrac{1+\dfrac{3}{m}}{1-\dfrac{3}{m}}$

$$= \frac{\left(1+\dfrac{3}{m}\right)m}{\left(1-\dfrac{3}{m}\right)m}$$

$$= \frac{m+3}{m-3}$$

b. $\dfrac{1+2x^{-2}}{1-2x^{-2}} = \dfrac{1+\dfrac{2}{x^2}}{1-\dfrac{2}{x^2}}$

$$= \frac{\left(1+\dfrac{2}{x^2}\right)x^2}{\left(1-\dfrac{2}{x^2}\right)x^2}$$

$$= \frac{x^2+2}{x^2-2}$$

73. $C = \dfrac{450P}{100-P}$

P	$\dfrac{450P}{100-P}$
40	$\dfrac{450(40)}{100-40} = 300$
60	$\dfrac{450(60)}{100-60} = 675$
80	$\dfrac{450(80)}{100-80} = 1800$
90	$\dfrac{450(90)}{100-90} = 4050$
93	$\dfrac{450(93)}{100-93} = 5979$
95	$\dfrac{450(95)}{100-95} = 8550$
98	$\dfrac{450(98)}{100-98} = 22050$
100	$\dfrac{450(100)}{100-100} = \mathit{error}$

a. $C = \dfrac{450(40)}{100-40} = \$300$ million;

$C = \dfrac{450(85)}{100-85} = \$2550$ million

b. It would require many resources.

c. No

75. $\dfrac{1}{f_1} + \dfrac{1}{f_2}$

$$\frac{f_2+f_1}{f_1 f_2}$$

77. $\dfrac{\dfrac{a}{x+h}-\dfrac{a}{x}}{h}$

$$= \frac{\left(\dfrac{a}{x+h}-\dfrac{a}{x}\right)x(x+h)}{hx(x+h)}$$

$$= \frac{ax-a(x+h)}{hx(x+h)}$$

$$= \frac{ax-ax-ah}{hx(x+h)}$$

$$= \frac{-ah}{hx(x+h)}$$

$$= \frac{-a}{x(x+h)}$$

79. $\dfrac{\dfrac{1}{2(x+h)^2}-\dfrac{1}{2x^2}}{h}$

$$= \frac{\left(\frac{1}{2(x+h)^2} - \frac{1}{2x^2}\right)(x+h)^2(2x^2)}{h(x+h)^2(2x^2)}$$

$$= \frac{x^2 - (x+h)^2}{h(x+h)^2(2x^2)}$$

$$= \frac{x^2 - x^2 - 2xh - h^2}{h(x+h)^2(2x^2)}$$

$$= \frac{-2xh - h^2}{h(x+h)^2(2x^2)}$$

$$= \frac{-h(2x+h)}{h(x+h)^2(2x^2)}$$

$$= \frac{-(2x+h)}{2x^2(x+h)^2}$$

81. $P = \dfrac{50(7d^2+10)}{d^3+50}$

d	$P = \dfrac{50(7d^2+10)}{d^3+50}$
0	$P = \dfrac{50(7(0)^2+10)}{(0)^3+50} = 10$
1	$P = \dfrac{50(7(1)^2+10)}{(1)^3+50} = 16.67$
2	$P = \dfrac{50(7(2)^2+10)}{(2)^3+50} = 32.76$
3	$P = \dfrac{50(7(3)^2+10)}{(3)^3+50} = 47.40$
4	$P = \dfrac{50(7(4)^2+10)}{(4)^3+50} = 53.51$
5	$P = \dfrac{50(7(5)^2+10)}{(5)^3+50} = 52.86$
6	$P = \dfrac{50(7(6)^2+10)}{(6)^3+50} = 49.25$
7	$P = \dfrac{50(7(7)^2+10)}{(7)^3+50} = 44.91$
8	$P = \dfrac{50(7(8)^2+10)}{(8)^3+50} = 40.75$
9	$P = \dfrac{50(7(9)^2+10)}{(9)^3+50} = 37.04$
10	$P = \dfrac{50(7(10)^2+10)}{(10)^3+50} = 33.81$

Price rises rapidly for first four days, then begins a gradual decrease.
Yes, on the 35th day of trading.

83.

t	$N = \dfrac{60t-120}{t}$
3	$N = \dfrac{60(3)-120}{3} = 20$
4	$N = \dfrac{60(4)-120}{4} = 30$
5	$N = \dfrac{60(5)-120}{5} = 36$
6	$N = \dfrac{60(6)-120}{6} = 40$
7	$N = \dfrac{60(7)-120}{7} = 42.9$
8	$N = \dfrac{60(8)-120}{8} = 45$

$t = 8$ weeks

85. (b); $20 \cdot n \div 10 \cdot n$
All the others equal 2.

87. $\dfrac{1}{\frac{5}{2}+\frac{4}{3}} = \dfrac{1(6)}{\left(\frac{5}{2}+\frac{4}{3}\right)(6)} = \dfrac{6}{15+8} = \dfrac{6}{23}$

The reciprocal of the sum of their reciprocals is $\dfrac{6}{23}$.

$\dfrac{1}{\frac{b}{a}+\frac{d}{c}} = \dfrac{1(ac)}{\left(\frac{b}{a}+\frac{d}{c}\right)(ac)} = \dfrac{ac}{bc+ad}$

R.6 Exercises

1. Even

3. $\left(16^{\frac{1}{4}}\right)^3$

5. Answers will vary.

7. $\sqrt{x^2}$

a. $\sqrt{(9)^2} = \sqrt{81} = 9$

b. $\sqrt{(-10)^2} = \sqrt{100} = 10$

9. a. $\sqrt{49p^2} = \sqrt{(7p)^2} = 7|p|$

b. $\sqrt{(x-3)^2} = |x-3|$

c. $\sqrt{81m^4} = \sqrt{(9m^2)^2} = 9m^2$

d. $\sqrt{x^2-6x+9} = \sqrt{(x-3)^2} = |x-3|$

11. a. $\sqrt[3]{64} = \sqrt[3]{(4)^3} = 4$

b. $\sqrt[3]{-125x^3} = \sqrt[3]{(-5x)^3} = -5x$

c. $\sqrt[3]{216z^{12}} = \sqrt[3]{(6z^4)^3} = 6z^4$

d. $\sqrt[3]{\frac{v^3}{-8}} = \sqrt[3]{\left(\frac{v}{-2}\right)^3} = \frac{v}{-2}$

13. a. $\sqrt[6]{64} = \sqrt[6]{(2)^6} = 2$

b. $\sqrt[6]{-64}$ Not a real number

c. $\sqrt[5]{243x^{10}} = \sqrt[5]{(3x^2)^5} = 3x^2$

d. $\sqrt[5]{-243x^5} = \sqrt[5]{(-3x)^5} = -3x$

e. $\sqrt[5]{(k-3)^5} = k-3$

f. $\sqrt[6]{(h+2)^6} = |h+2|$

15. a. $\sqrt[3]{-125} = \sqrt[3]{(-5)^3} = -5$

b. $-\sqrt[4]{81n^{12}} = -\sqrt[4]{(3n^3)^4} = -3|n^3|$

c. $\sqrt{-36}$ Not a real number

d. $\sqrt{\frac{49v^{10}}{36}} = \sqrt{\left(\frac{7v^5}{6}\right)^2} = \frac{7|v^5|}{6}$

17. a. $8^{\frac{2}{3}} = (8^{1/3})^2 = \sqrt[3]{8}^2 = 2^2 = 4$

b. $\left(\frac{16}{25}\right)^{\frac{3}{2}} = \left[\left(\frac{16}{25}\right)^{\frac{1}{2}}\right]^3$

$= \sqrt{\left(\frac{16}{25}\right)}^3 = \left(\frac{4}{5}\right)^3 = \frac{64}{125}$

c. $\left(\frac{4}{25}\right)^{-\frac{3}{2}} = \left(\frac{25}{4}\right)^{\frac{3}{2}} = \left[\left(\frac{25}{4}\right)^{\frac{1}{2}}\right]^3$

$= \sqrt{\left(\frac{25}{4}\right)}^3 = \left(\frac{5}{2}\right)^3 = \frac{125}{8}$

d. $\left(\frac{-27p^6}{8q^3}\right)^{\frac{2}{3}} = \left[\left(\frac{-27p^6}{8q^3}\right)^{\frac{1}{3}}\right]^2$

$= \sqrt[3]{\left(\frac{-27p^6}{8q^3}\right)}^2 = \left(\frac{-3p^2}{2q}\right)^2 = \frac{9p^4}{4q^2}$

19. a. $-144^{\frac{3}{2}} = -\left[(144)^{\frac{1}{2}}\right]^3 = -\sqrt{144}^3$

$= -(12)^3 = -1728$

b. $\left(-\frac{4}{25}\right)^{\frac{3}{2}} = \left[\left(-\frac{4}{25}\right)^{\frac{1}{2}}\right]^3 = \sqrt{\left(-\frac{4}{25}\right)}^3$

Not a real number

c. $(-27)^{\frac{-2}{3}} = \left(-\frac{1}{27}\right)^{\frac{2}{3}} = \left[\left(-\frac{1}{27}\right)^{\frac{1}{3}}\right]^2$

$= \sqrt[3]{\left(-\frac{1}{27}\right)}^2 = \left(-\frac{1}{3}\right)^2 = \frac{1}{9}$

d. $-\left(\frac{27x^3}{64}\right)^{-\frac{4}{3}} = -\left(\frac{64}{27x^3}\right)^{\frac{4}{3}}$

$$= -\left[\left(\frac{64}{27x^3}\right)^{\frac{1}{3}}\right]^4 = -\sqrt[3]{\left(\frac{64}{27x^3}\right)}^{4}$$

$$= -\left(\frac{4}{3x}\right)^4 = \frac{-256}{81x^4}$$

21. a. $\left(2n^2 p^{\frac{-2}{5}}\right)^5 = 32n^{10}p^{-2} = \frac{32n^{10}}{p^2}$

b. $\left(\frac{8y^{\frac{3}{4}}}{64y^{\frac{3}{2}}}\right)^{\frac{1}{3}} = \frac{8^{\frac{1}{3}}y^{\frac{1}{4}}}{64^{\frac{1}{3}}y^{\frac{1}{2}}}$

$= \frac{\sqrt[3]{8}y^{\frac{1}{4}-\frac{1}{2}}}{\sqrt[3]{64}} = \frac{2y^{\frac{-1}{4}}}{4} = \frac{1}{2y^{\frac{1}{4}}}$

23. a. $\sqrt{18m^2} = \sqrt{(3m)^2 \cdot 2} = 3m\sqrt{2}$

b. $-2\sqrt[3]{-125p^3q^7} = -2\sqrt[3]{\left(-5pq^2\right)^3 \cdot q}$

$= -2\left(-5pq^2\right)\sqrt[3]{q} = 10pq^2\sqrt[3]{q}$

c. $\frac{3}{8}\sqrt[3]{64m^3n^5} = \frac{3}{8}\sqrt[3]{(4mn)^3n^2}$

$= \frac{3}{8}(4mn)\sqrt[3]{n^2} = \frac{3}{2}mn\sqrt[3]{n^2}$

d. $\sqrt{32p^3q^6} = \sqrt{\left(16p^2q^6\right)\cdot 2p}$

$\sqrt{\left(4pq^3\right)^2 \cdot 2p} = 4pq^3\sqrt{2p}$

e. $\frac{-6+\sqrt{28}}{2} = \frac{-6+\sqrt{4\cdot 7}}{2} = \frac{-6+2\sqrt{7}}{2}$

$= \frac{-6}{2} + \frac{2\sqrt{7}}{2} = -3+\sqrt{7}$

f. $\frac{27-\sqrt{72}}{6} = \frac{27-\sqrt{36\cdot 2}}{6} = \frac{27-6\sqrt{2}}{6}$

$= \frac{27}{6} - \frac{6\sqrt{2}}{6} = \frac{9}{2} - \sqrt{2}$

25. a. $2.5\sqrt{18a}\sqrt{2a^3} = 2.5\sqrt{36a^4}$

$= 2.5\sqrt{\left(6a^2\right)^2} = 2.5\left(6a^2\right) = 15a^2$

b. $-\frac{2}{3}\sqrt{3b}\sqrt{12b^2} = -\frac{2}{3}\sqrt{36b^2}\sqrt{b}$

$= -\frac{2}{3}(6b)\sqrt{b} = -4b\sqrt{b}$

c. $\sqrt{\frac{x^3y}{3}}\sqrt{\frac{4x^5y}{12y}} = \sqrt{\frac{4x^8y^2}{36y}} = \sqrt{\frac{x^8y}{9}}$

$= \sqrt{\frac{x^8}{9}}\sqrt{y} = \frac{x^4\sqrt{y}}{3}$

d. $\sqrt[3]{9v^2u}\,\sqrt[3]{3u^5v^2} = \sqrt[3]{27v^4u^6}$

$= \sqrt[3]{27u^6v^3}\sqrt[3]{v} = \sqrt[3]{\left(3u^2v\right)^3}\sqrt[3]{v} = 3u^2v\,\sqrt[3]{v}$

27. a. $\frac{\sqrt{8m^5}}{\sqrt{2m}} = \sqrt{\frac{8m^5}{2m}} = \sqrt{4m^4} = 2m^2$

b. $\frac{\sqrt[3]{108n^4}}{\sqrt[3]{4n}} = \sqrt[3]{\frac{108n^4}{4n}} = \sqrt[3]{27n^3} = 3n$

c. $\sqrt{\frac{45}{16x^2}} = \frac{\sqrt{9\cdot 5}}{\sqrt{16x^2}} = \frac{3\sqrt{5}}{4x}$

d. $12\sqrt[3]{\frac{81}{8z^9}} = 12\frac{\sqrt[3]{81}}{\sqrt[3]{8z^9}} = 12\frac{\sqrt[3]{27\cdot 3}}{2z^3}$

$= \frac{12(3)\sqrt[3]{3}}{2z^3} = \frac{18\sqrt[3]{3}}{z^3}$

29. a. $\sqrt[5]{32x^{10}y^{15}} = 2x^2y^3$

b. $x\sqrt[4]{x^5} = x\sqrt[4]{x^4\cdot x}$

$= x\cdot x\sqrt[4]{x} = x^2\sqrt[4]{x}$

c. $\sqrt[4]{\sqrt[3]{b}} = \sqrt[4]{b^{\frac{1}{3}}} = \left((b)^{\frac{1}{3}}\right)^{\frac{1}{4}}$

$= b^{\frac{1}{12}} = \sqrt[12]{b}$

d. $\frac{\sqrt[3]{6}}{\sqrt{6}} = \frac{6^{\frac{1}{3}}}{6^{\frac{1}{2}}} = 6^{\frac{1}{3}-\frac{1}{2}} = 6^{-\frac{1}{6}} = \frac{1}{\sqrt[6]{6}}$

31. a. $12\sqrt{72} - 9\sqrt{98}$

$= 12\cdot 6\sqrt{2} - 9\cdot 7\sqrt{2}$

$= 72\sqrt{2} - 63\sqrt{2}$

$=9\sqrt{2}$

b. $8\sqrt{48}-3\sqrt{108}$
$=8\cdot 4\sqrt{3}-3\cdot 6\sqrt{3}$
$=32\sqrt{3}-18\sqrt{3}$
$=14\sqrt{3}$

c. $7\sqrt{18m}-\sqrt{50m}$
$=7\cdot 3\sqrt{2m}-5\sqrt{2m}$
$=21\sqrt{2m}-5\sqrt{2m}$
$=16\sqrt{2m}$

d. $2\sqrt{28p}-3\sqrt{63p}$
$=2\cdot 2\sqrt{7p}-3\cdot 3\sqrt{7p}$
$=4\sqrt{7p}-9\sqrt{7p}$
$=-5\sqrt{7p}$

33. a. $3x\sqrt[3]{54x}-5\sqrt[3]{16x^4}$
$=3x\cdot 3\sqrt[3]{2x}-5\cdot 2x\sqrt[3]{2x}$
$=9x\sqrt[3]{2x}-10x\sqrt[3]{2x}$
$=-x\sqrt[3]{2x}$

b. $\sqrt{4}+\sqrt{3x}-\sqrt{12x}+\sqrt{45}$
$=2+\sqrt{3x}-2\sqrt{3x}+3\sqrt{5}$
$=2-\sqrt{3x}+3\sqrt{5}$

c. $\sqrt{72x^3}+\sqrt{50}-\sqrt{7x}+\sqrt{27}$
$=6x\sqrt{2x}+5\sqrt{2}-\sqrt{7x}+3\sqrt{3}$

35. a. $(7\sqrt{2})^2=49\cdot 2=98$

b. $\sqrt{3}(\sqrt{5}+\sqrt{7})=\sqrt{15}+\sqrt{21}$

c. $(n+\sqrt{5})(n-\sqrt{5})=n^2-5$

d. $(6-\sqrt{3})^2=36-12\sqrt{3}+3=39-12\sqrt{3}$

37. a. $(3+2\sqrt{7})(3-2\sqrt{7})=9-4(7)$
$=9-28=-19$

b. $(\sqrt{5}-\sqrt{14})(\sqrt{2}+\sqrt{13})$
$=\sqrt{10}+\sqrt{65}-\sqrt{28}-\sqrt{182}$
$=\sqrt{10}+\sqrt{65}-2\sqrt{7}-\sqrt{182}$

c. $(2\sqrt{2}+6\sqrt{6})(3\sqrt{10}+\sqrt{7})$
$=6\sqrt{20}+2\sqrt{14}+18\sqrt{60}+6\sqrt{42}$
$=12\sqrt{5}+2\sqrt{14}+36\sqrt{15}+6\sqrt{42}$

39. $x^2-4x+1=0$

a. $(2+\sqrt{3})^2-4(2+\sqrt{3})+1=0$
$4+4\sqrt{3}+3-8-4\sqrt{3}+1=0$
$0=0$

b. $(2-\sqrt{3})^2-4(2-\sqrt{3})+1=0$
$4-4\sqrt{3}+3-8+4\sqrt{3}+1=0$
$0=0$

41. $x^2+2x-9=0$

a. $(-1+\sqrt{10})^2+2(-1+\sqrt{10})-9=0$
$1-2\sqrt{10}+10-2+2\sqrt{10}-9=0$
$0=0$

b. $(-1-\sqrt{10})^2+2(-1-\sqrt{10})-9=0$
$1+2\sqrt{10}+10-2-2\sqrt{10}-9=0$
$0=0$

43. a. $\frac{3}{\sqrt{12}}=\frac{3}{2\sqrt{3}}\cdot\frac{\sqrt{3}}{\sqrt{3}}=\frac{3\sqrt{3}}{6}=\frac{\sqrt{3}}{2}$

b. $\sqrt{\frac{20}{27x^3}}=\frac{2\sqrt{5}}{3x\sqrt{3x}}\cdot\frac{\sqrt{3x}}{\sqrt{3x}}=\frac{2\sqrt{15x}}{9x^2}$

c. $\sqrt{\frac{27}{50b}}=\frac{3\sqrt{3}}{5\sqrt{2b}}\cdot\frac{\sqrt{2b}}{\sqrt{2b}}=\frac{3\sqrt{6b}}{10b}$

d. $\sqrt[3]{\frac{1}{4p}}=\frac{1}{\sqrt[3]{4p}}\cdot\frac{\sqrt[3]{2p^2}}{\sqrt[3]{2p^2}}=\frac{\sqrt[3]{2p^2}}{2p}$

45. a. $\frac{8}{3+\sqrt{11}}\cdot\frac{3-\sqrt{11}}{3-\sqrt{11}}$
$=\frac{8(3-\sqrt{11})}{9-11}=\frac{8(3-\sqrt{11})}{-2}$
$=-4(3-\sqrt{11})=-12+4\sqrt{11}\approx 1.27$

b. $\frac{6}{\sqrt{x}-\sqrt{2}}\cdot\frac{\sqrt{x}+\sqrt{2}}{\sqrt{x}+\sqrt{2}}$
$=\frac{6\sqrt{x}+6\sqrt{2}}{x-2}$

47. a. $\frac{\sqrt{10}-3}{\sqrt{3}+\sqrt{2}}\cdot\frac{\sqrt{3}-\sqrt{2}}{\sqrt{3}-\sqrt{2}}$
$=\frac{\sqrt{30}-\sqrt{20}-3\sqrt{3}+3\sqrt{2}}{3-2}$
$=\sqrt{30}-2\sqrt{5}-3\sqrt{3}+3\sqrt{2}$

≈ 0.05

b. $\dfrac{7+\sqrt{6}}{3-3\sqrt{2}} \cdot \dfrac{3+3\sqrt{2}}{3+3\sqrt{2}}$

$= \dfrac{21+21\sqrt{2}+3\sqrt{6}+3\sqrt{12}}{9-18}$

$= \dfrac{21+21\sqrt{2}+3\sqrt{6}+6\sqrt{3}}{-9}$

$= \dfrac{7+7\sqrt{2}+\sqrt{6}+2\sqrt{3}}{-3}$

≈ -7.60

49. $L = 1.13(W)^{\frac{1}{3}}$

$L = 1.13(400)^{\frac{1}{3}} \approx 8.33$ feet

51. $c^2 = a^2 + b^2$

$c^2 = 9^2 + (21.5)^2$

$c^2 = 81 + 462.25$

$c^2 = 543.25$

$c \approx 23.3;$

$23.3 + 0.3 + 0.3 = 23.9$ m

53. $\dfrac{\sqrt{x+2}-\sqrt{x}}{2} \cdot \dfrac{\sqrt{x+2}+\sqrt{x}}{\sqrt{x+2}+\sqrt{x}}$

$= \dfrac{x+2-x}{2\sqrt{x+2}+2\sqrt{x}}$

$= \dfrac{2}{2\sqrt{x+2}+2\sqrt{x}} = \dfrac{1}{\sqrt{x+2}+\sqrt{x}}$

55. $T = 0.407R^{\frac{3}{2}}$

a. $T = 0.407(93)^{\frac{3}{2}} \approx 365.02$ days

b. $T = 0.407(142)^{\frac{3}{2}} \approx 688.69$ days

c. $T = 0.407(36)^{\frac{3}{2}} \approx 87.91$ days

57. $V = 2\sqrt{6L}$

a. $V = 2\sqrt{6(54)} = 2\sqrt{324} = 36$ mph

b. $V = 2\sqrt{6(90)} = 2\sqrt{540} \approx 46.5$ mph

59. $S = \pi r\sqrt{r^2 + h^2}$;

$S = \pi(6)\sqrt{6^2 + 10^2}$

$S = 6\pi\sqrt{136}$

$S = 12\pi\sqrt{34}$

$S \approx 219.82\ \text{m}^2$

61. a. $x^2 - 5$

$x^2 - \left(\sqrt{5}\right)^2$

$\left(x+\sqrt{5}\right)\left(x-\sqrt{5}\right)$

b. $n^2 - 19$

$n^2 - \left(\sqrt{19}\right)^2$

$\left(n+\sqrt{19}\right)\left(n-\sqrt{19}\right)$

63. Because $m^4 \geq 0$ *for* $m \in \mathrm{R}$.

65. $$\left(\left(\left(\left(3^{\frac{5}{6}}\right)^{\frac{3}{2}}\right)^{\frac{4}{5}}\right)^{\frac{3}{4}}\right)^{\frac{2}{5}}\Bigg)^{\frac{10}{3}}$$

$$= \left(\left(3^{\frac{5}{4}}\right)^{\frac{3}{5}}\right)^{\frac{4}{3}} = \left(3^{\frac{3}{4}}\right)^{\frac{4}{3}} = 3$$

Chapter R Practice Test

1. a. True
   b. True
   c. False; $\sqrt{2}$ cannot be expressed as a ratio of integers, denominator $\neq 0$.
   d. True

2. a. $\sqrt{121} = 11$
   b. $\sqrt[3]{-125} = -5$
   c. $\sqrt{-36}$ Not a real number
   d. $\sqrt{400} = 20$

3. a. $\frac{7}{8}-\left(-\frac{1}{4}\right)=\frac{7}{8}+\frac{2}{8}=\frac{9}{8}$

b. $-\frac{1}{3}-\frac{5}{6}=-\frac{2}{6}-\frac{5}{6}=-\frac{7}{6}$

c. $-0.7+1.2=0.5$

d. $13.+(-5.9)=1.3-5.9=-4.6$

4. a. $(-4)\left(-2\frac{1}{3}\right)=(-4)\left(-\frac{7}{3}\right)=\frac{28}{3}$

b. $(-0.6)(-1.5)=0.9$

c. $\frac{-2.8}{-0.7}=4$

d. $4.2\div(-0.6)=-7$

5. $2000\left(1+\frac{0.08}{12}\right)^{12\cdot 10}\approx 4439.28$

6. a. $0\div 6=0$

b. $6\div 0$ Undefined

7. a. $-2v^2+6v+5$

Terms: 3

Coefficients: -2, 6, 5

b. $\frac{c+2}{3}+c$

Terms: 2

Coefficients: $\frac{1}{3}$, 1

8. a. $2x-3y^2$

$=2(-0.5)-3(-2)^2$

$=-1-12$

$=-13$

b. $\sqrt{2}-x\left(4-x^2\right)+\frac{y}{x}$

$=\sqrt{2}-(-0.5)\left(4-(-0.5)^2\right)+\frac{-2}{-0.5}$

$=\sqrt{2}-(-0.5)(3.75)+4$

$=\sqrt{2}+1.875+4$

≈ 7.29

9. a. $x^3-(2x-9)$

b. $2n-3\left(\frac{n}{2}\right)^2$

10. a. Let r represent Earth's radius. Then $11r-119$ represents Jupiter's radius.

b. Let e represent this year's earnings. Then $4e+1.2$ represents last year's earnings.

11. a. $8v^2+4v-7+v^2-v=9v^2+3v-7$

b. $-4(3b-2)+5b$

$=-12b+8+5b=-7b+8$

c. $4x-\left(x-2x^2\right)+x(3-x)$

$=4x-x+2x^2+3x-x^2$

$=6x+x^2$

12. a. $9x^2-16=(3x+4)(3x-4)$

b. $4v^3-12v^2+9v$

$=v\left(4v^2-12v+9\right)$

$=v(2v-3)^2$

c. $x^3+5x^2-9x-45$

$=\left(x^3+5x^2\right)-(9x+45)$

$=x^2(x+5)-9(x+5)$

$=(x+5)\left(x^2-9\right)$

$=(x+5)(x+3)(x-3)$

13. a. $\frac{5}{b^{-3}}=5b^3$

b. $\left(-2a^3\right)^2\left(a^2b^4\right)^3$

$=\left(4a^6\right)\left(a^6b^{12}\right)$

$=4a^{12}b^{12}$

c. $\left(\frac{m^2}{2n}\right)^3=\frac{m^6}{8n^3}$

d. $\left(\frac{5p^2q^3r^4}{-2pq^2r^4}\right)^2$

$=\left(\frac{5pq}{-2}\right)^2$

$=\frac{25}{4}p^2q^2$

14. a. $\frac{-12a^3b^5}{3a^2b^4}=-4ab$

b. $\left(3.2\text{x}10^{-17}\right)\text{x}\left(2.0\text{x}10^{15}\right)$

$= (3.2\text{x}2)(10^{-17}\text{x}10^{15})$

$= 6.4\text{x}10^{-2} = 0.064$

c. $\left(\frac{a^{-3}b}{c^{-2}}\right)^{-4} = \frac{a^{12}b^{-4}}{c^8} = \frac{a^{12}}{b^4c^8}$

d. $-7x^0 + (-7x)^0 = -7(1) + 1 = -6$

15. a. $(3x^2 + 5y)(3x^2 - 5y) = 9x^4 - 25y^2$

b. $(2a + 3b)^2 = 4a^2 + 12ab + 9b^2$

16. a. $(-5a^3 + 4a^2 - 3) + (7a^4 + 4a^2 - 3a - 15)$

$= -5a^3 + 4a^2 - 3 + 7a^4 + 4a^2 - 3a - 15$

$= 7a^4 - 5a^3 + 8a^2 - 3a - 18$

b. $(2x^2 + 4x - 9) - (7x^4 - 2x^2 - x - 9)$

$= 2x^2 + 4x - 9 - 7x^4 + 2x^2 + x + 9$

$= -7x^4 + 4x^2 + 5x$

17. a. $\frac{x-5}{5-x} = \frac{-(5-x)}{5-x} = -1$

b. $\frac{4-n^2}{n^2 - 4n + 4}$

$= \frac{(2+n)(2-n)}{(n-2)(n-2)}$

$= \frac{(2+n)(2-n)}{-(2-n)(n-2)}$

$= \frac{2+n}{-(n-2)}$

$= \frac{2+n}{2-n}$

c. $\frac{x^3 - 27}{x^2 + 3x + 9}$

$= \frac{(x-3)(x^2 + 3x + 9)}{x^2 + 3x + 9} = x - 3$

d. $\frac{3x^2 - 13x - 10}{9x^2 - 4}$

$= \frac{(3x+2)(x-5)}{(3x+2)(3x-2)}$

$\frac{x-5}{3x-2}$

e. $\frac{x^2 - 25}{3x^2 - 11x - 4} \div \frac{x^2 + x - 20}{x^2 - 8x + 16}$

$= \frac{x^2 - 25}{3x^2 - 11x - 4} \cdot \frac{x^2 - 8x + 16}{x^2 + x - 20}$

$= \frac{(x+5)(x-5)}{(3x+1)(x-4)} \cdot \frac{(x-4)^2}{(x+5)(x-4)}$

$= \frac{x-5}{3x+1}$

f. $\frac{m+3}{m^2 + m - 12} - \frac{2}{5(m+4)}$

$= \frac{m+3}{(m+4)(m-3)} - \frac{2}{5(m+4)}$

$= \frac{5(m+3) - 2(m-3)}{5(m+4)(m-3)}$

$= \frac{5m + 15 - 2m + 6}{5(m+4)(m-3)}$

$= \frac{3m + 21}{5(m+4)(m-3)}$

$= \frac{3(m+7)}{5(m+4)(m-3)}$

18. a. $\sqrt{(x+11)^2} = |x + 11|$

b. $\sqrt[3]{\frac{-8}{27v^3}} = \frac{-2}{3v}$

c. $\left(\frac{25}{16}\right)^{\frac{-3}{2}} = \left(\frac{16}{25}\right)^{\frac{3}{2}}$

$= \sqrt{\frac{16}{25}}^{\,3} = \left(\frac{4}{5}\right)^3 = \frac{64}{125}$

d. $\frac{-4 + \sqrt{32}}{8} = \frac{-4 + 4\sqrt{2}}{8} = \frac{-1}{2} + \frac{\sqrt{2}}{2}$

e. $7\sqrt{40} - \sqrt{90} = 14\sqrt{10} - 3\sqrt{10} = 11\sqrt{10}$

f. $(x + \sqrt{5})(x - \sqrt{5}) = x^2 - 5$

g. $\sqrt{\frac{2}{5x}} = \frac{\sqrt{2}}{\sqrt{5x}} \cdot \frac{\sqrt{5x}}{\sqrt{5x}} = \frac{\sqrt{10x}}{5x}$

h. $\frac{8}{\sqrt{6} - \sqrt{2}} \cdot \frac{\sqrt{6} + \sqrt{2}}{\sqrt{6} + \sqrt{2}}$

$= \frac{8\sqrt{6} + 8\sqrt{2}}{6 - 2}$

$= \frac{8\sqrt{6} + 8\sqrt{2}}{4}$

$= 2\sqrt{6} + 2\sqrt{2}$

$= 2(\sqrt{6} + \sqrt{2})$

19. $R = (30 - 0.5x)(40 + x)$

$R = 1200 + 10x - 0.5x^2$

a.

x	$R = 1200 + 10x - 0.5x^2$
1	$R = 1200 + 10(1) - 0.5(1)^2 = 1209.50$
2	$R = 1200 + 10(2) - 0.5(2)^2 = 1218$
3	$R = 1200 + 10(3) - 0.5(3)^2 = 1225.50$
4	$R = 1200 + 10(4) - 0.5(4)^2 = 1232$
5	$R = 1200 + 10(5) - 0.5(5)^2 = 1237.50$
6	$R = 1200 + 10(6) - 0.5(6)^2 = 1242$
7	$R = 1200 + 10(7) - 0.5(7)^2 = 1245.50$
8	$R = 1200 + 10(8) - 0.5(8)^2 = 1248$
9	$R = 1200 + 10(9) - 0.5(9)^2 = 1249.50$
10	$R = 1200 + 10(10) - 0.5(10)^2 = 1250$
11	$R = 1200 + 10(11) - 0.5(11)^2 = 1249.50$

Ten decreases of 0.50 or \$5.00

b. Maximum revenue is \$1250.

20. Diagonal of the bottom face:

$a^2 + b^2 = c^2$

$32^2 + 24^2 = c^2$

$1024 + 576 = c^2$

$1600 = c^2$

$40 = c$

Diagonal of the rectangular prism:

$a^2 + b^2 = c^2$

$40^2 + 42^2 = c^2$

$1600 + 1764 = c^2$

$3364 = c^2$

$58 = c$

The diagonal of the rectangular prism is 58 cm.

1.1 Technology Highlight

1. Less; 0.554 is less than 0.56.

2. Answers will vary.

1.1 Exercises

1. Identity, unknown

3. Literal, two

5. Answers will vary.

7. $-2x+7=60$, linear

9. $7+9d=5$, linear

11. $2xy-3=5$, non-linear, two variables are multiplied together.

13. $-2(3y+5)=-7+4y-12$
$$\begin{aligned} -6y-10&=-19+4y\\ -10y&=-9\\ y&=\frac{-9}{-10}\\ y&=\frac{9}{10} \end{aligned}$$

15. $8-(3n+5)=-5+2(n+1)$
$$\begin{aligned} 8-3n-5&=-5+2n+2\\ 3-3n&=-3+2n\\ -5n&=-6\\ n&=\frac{6}{5} \end{aligned}$$

17. $2(3m+5)=5-2(m-1)$
$$\begin{aligned} 6m+10&=5-2m+2\\ 6m+10&=7-2m\\ 8m&=-3\\ m&=\frac{-3}{8} \end{aligned}$$

19. $\frac{1}{2}x+5=\frac{1}{3}x+7$
$$\begin{aligned} 6\left(\frac{1}{2}x+5\right)&=\left(\frac{1}{3}x+7\right)6\\ 3x+30&=2x+42\\ x&=12 \end{aligned}$$

21. $15=-6-\frac{3y}{8}$
$$\begin{aligned} 21&=-\frac{3y}{8}\\ \left(-\frac{8}{3}\right)21&=-\frac{3y}{8}\left(-\frac{8}{3}\right)\\ -56&=y \end{aligned}$$

23. $\frac{n}{2}+\frac{n}{5}=\frac{2}{3}$
$$\begin{aligned} 30\left(\frac{n}{2}+\frac{n}{5}\right)&=30\left(\frac{2}{3}\right)\\ 15n+6n&=20\\ 21n&=20\\ n&=\frac{20}{21} \end{aligned}$$

25. $\frac{2}{3}(m+6)=\frac{-1}{2}$
$$\begin{aligned} \frac{2}{3}m+4&=\frac{-1}{2}\\ 6\left(\frac{2}{3}m+4\right)&=6\left(\frac{-1}{2}\right)\\ 4m+24&=-3\\ 4m&=-27\\ m&=-\frac{27}{4} \end{aligned}$$

27. $0.2(2.4-3.8x)-5.4=0$
$$\begin{aligned} 0.48-0.76x-5.4&=0\\ -4.92-0.76x&=0\\ -0.76x&=4.92\\ x&=\frac{4.92}{-0.76}\\ x&=-\frac{123}{19} \end{aligned}$$

29. $-5-(3n+4)=-8-2n$
$$\begin{aligned} -5-3n-4&=-8-2n\\ -9-3n&=-8-2n\\ -n&=1\\ n&=-1 \end{aligned}$$

31. $-3(4z+5)=-15z-20+3z$
$$\begin{aligned} -12z-15&=-15z-20+3z\\ -12z-15&=-12z-20\\ -15&\neq-20 \end{aligned}$$
Contradiction

33. $8-8(3n+5)=-5+6(1+n)$
$$\begin{aligned} 8-24n-40&=-5+6+6n\\ -24n-32&=1+6n\\ -30n&=33\\ n&=-1.1 \end{aligned}$$
Conditional; $n = -1.1$

35. $-4(4x+5)=-6-2(8x+7)$
$$-16x-20=-6-16x-14$$
$$-16x-20=-20-16x$$
$$0=0$$
Identity

37. $I=PRT$
$$\frac{I}{PT}=R$$

39. $C=2\pi r$
$$\frac{C}{2\pi}=r$$

41. $W=I^2R$
$$\frac{W}{I^2}=R$$

43. $V=\frac{4}{3}\pi r^2h$
$$\frac{3}{4}V=\pi r^2h$$
$$\frac{3V}{4\pi r^2}=h$$

45. $\frac{A}{6}=s^2$
$$A=6s^2$$

47. $S=B+\frac{1}{2}PS$
$$S-B=\frac{1}{2}PS$$
$$2(S-B)=PS$$
$$\frac{2(S-B)}{S}=P$$

49. $Ax+By=C$
$$By=-Ax+C$$
$$y=\frac{-Ax}{B}+\frac{C}{B}$$

51. $\frac{5}{6}x+\frac{3}{8}y=2$
$$\frac{3}{8}y=-\frac{5}{6}x+2$$
$$\left(\frac{8}{3}\right)\left(\frac{3}{8}y\right)=\left(\frac{8}{3}\right)\left(-\frac{5}{6}x+2\right)$$
$$y=-\frac{20}{9}x+\frac{16}{3}$$

53. $y-3=\frac{-4}{5}(x+10)$
$$y-3=\frac{-4}{5}x-8$$
$$y=\frac{-4}{5}x-5$$

55. $3x+2=-19$
$$a=3, b=2, c=-19$$
$$x=\frac{-19-2}{3}$$
$$x=-7$$

57. $-6x+1=33$
$$a=-6, b=1, c=33$$
$$x=\frac{33-1}{-6}$$
$$x=-\frac{16}{3}$$

59. $2x-13=-27$
$$a=2, b=-13, c=-27$$
$$x=\frac{-27-(-13)}{2}$$
$$x=-7$$

61. $SA=2\pi r^2+2\pi rh$
$$1256=2(3.14)(8)^2+2(3.14)(8)h$$
$$1256=401.92+50.24h$$
$$854.08=50.24h$$
$$17=h$$
$$h=17\text{cm}$$

63. Let x represent the length of the second descent.
$$2x+198=1218$$
$$2x=1218-198$$
$$2x=1020$$
$$x=510$$
The second spelunker descended 510 feet.

65. Let L represent the length of the package.
$$2(14+12)+L=108$$
$$2(26)+L=108$$
$$52+L=108$$
$$L=56$$
The package can be up to 56 inches long.

67. Let L represent the length of the Shimotsui bridge.

$$\begin{aligned}364+2L&=6532\\2L&=6168\\L&=3084\end{aligned}$$

The Shimotsui bridge is 3084 feet long.

69. Let x represent the first consecutive even integer.
Let $x+2$ represent the second consecutive even integer.
$$\begin{aligned}2x+x+2&=146\\3x+2&=146\\3x&=144\\x&=48\end{aligned}$$
The first integer is 48.
The second integer is 50.

71. Let x represent the first consecutive odd integer.
Let $x+2$ represent the second consecutive odd integer.
$$\begin{aligned}7x&=5(x+2)\\7x&=5x+10\\2x&=10\\x&=5\end{aligned}$$
The first integer is 5.
The second integer is 7.

73. Let t represent the number of hours when Chris overtakes Belinda.
$$D_B=D_C$$
$$\begin{aligned}30(t+2)&=45t\\30t+60&=45t\\60&=15t\\4&=t\end{aligned}$$
4 hours after 11am is 3pm.

75. Let t represent the number of hours Jeff was driving in the construction zone.
$$D_1+D_2=72$$
$$\begin{aligned}30t+60(1.5-t)&=72\\30t+90-60t&=72\\-30t+90&=72\\-30t&=-18\\t&=0.6\text{ hour}\end{aligned}$$
$0.6(60)=36$ minutes.

77. 2 quarts + 2 quarts = 4 quarts;
$$2(1.00)+2(0.00)=2$$
$$\frac{2}{4}=50\%\text{ juice}$$
4 quart mixture of 50% orange juice.

79. 8 lbs + 8 lbs = 16 lbs;
$$8(2.50)+8(1.10)=28.8$$
$$\frac{\$28.80}{16}=\$1.80\text{ per pound}$$
Sixteen pound mixture at a cost of $1.80 per pound.

81. Let x represent the gallons of pure anti-freeze.
$$\begin{aligned}1.00(x)+0.20(10)&=0.50(x+10)\\x+2&=0.50x+5\\0.5x&=3\\x&=6\end{aligned}$$
6 gallons of pure anti-freeze.

83. Let x represent the pounds of walnuts.
$$\begin{aligned}0.84x+1.20(20)&=1.04(x+20)\\0.84x+24&=1.04x+20.8\\-0.2x&=-3.2\\x&=16\end{aligned}$$
16 pounds of walnuts.

85. 1.4 million

87. Let x represent the number of ounces of 15% acid solution.
Let y represent the number of ounces of 20% acid solution.
Let z represent the number of ounces of 35% acid solution.
Let w represent the number of ounces of 45% acid solution.
There are 6 possible combinations of two solutions that must be checked.

Case 1: $\begin{cases}x+y=200\\15x+20y=200(29)\end{cases}$

$C(x,y)=120x+180y$;

Case 2: $\begin{cases}x+z=200\\15x+35z=200(29)\end{cases}$

$C(x,z)=120x+280z$;

Case 3: $\begin{cases}x+w=200\\15x+45w=200(29)\end{cases}$

$C(x,w)=120x+359w$;

Case 4: $\begin{cases}y+z=200\\20y+35z=200(29)\end{cases}$

$C(y,z)=180y+280z$;

Case 5: $\begin{cases}w+z=200\\45w+35z=200(29)\end{cases}$

$C(z,w)=280z+359w$;

Case 6: $\begin{cases}y+zw=200\\20y+45w=200(29)\end{cases}$

$C(y,w)=180y+359w$;

Using a grapher to solve Case 1 gives the point of intersection (-360,560). Since you cannot have a negative number of ounces, this case is impossible.

Using a grapher to solve Case 2 gives the point of intersection (60,140). Then $C(60,140)=120(60)+280(140)=\$46{,}400$.

Using a grapher to solve Case 3 gives the point of intersection $\left(106\frac{2}{3},93\frac{1}{3}\right)$. Then

$$C\left(106\frac{2}{3},93\frac{1}{3}\right)=120\left(106\frac{2}{3}\right)+359\left(93\frac{1}{3}\right)$$

$\approx \$46{,}306.67$

Using a grapher to solve Case 4 gives the point of intersection (80,120). Then $C(80,120)=180(80)+280(120)=\$48{,}000$.

Using a grapher to solve Case 5 gives the point of intersection (320,-120). Since you cannot have a negative number of ounces, this case is impossible.

Using a grapher to solve Case 6 gives the point of intersection $(72,128)$. Then

$C(72,128)=120(72)+359(128)$

$=\$54{,}592$

Case 3 minimizes the cost. Thus, $106\frac{2}{3}$ ounces of the 15% acid solution and $93\frac{1}{3}$ ounces of the 45% acid should be used.

89. $P+Q+S=40$

$P+R+U=34$
$S+T+U=30$

$Q+R=26$
$Q+T=23$
$R+T=19;$

$Q+R=26$
$-Q-T=-23$
$R-T=3;$

$R-T=3$
$R+T=19$
$2R=22$
$R=11;$

$Q+R=26$
$Q+11=26$
$Q=15;$

$Q+T=23$
$15+T=23$
$T=8;$

$P+R+U=34$
$P+11+U=34$
$P+U=23;$

$P+Q+S=40$
$P+15+S=40$
$P+S=25;$

$P+U=23$
$-P-S=-25$
$U-S=-2;$

$S+8+U=30$
$S+U=22;$

$U-S=-2$
$S+U=22$
$2U=20$
$U=10;$

$S+U=22$
$S+10=22$
$S=12;$

$P+Q+S=40$
$P+15+12=40$
$P=13;$

$P+Q+R+S+T+U$
$=13+15+11+12+8+10=69$

91. $-3v^3+v^2-\frac{v}{3}+7$

$-3,1,-\frac{1}{3},7$

93. $\frac{6}{7}\cdot 5\cdot 21=\frac{6}{7}\cdot 21\cdot 5$

Commutative property for multiplication since the order of the 5 and 21 is reversed.

95. $2x-3x^2+5x-9+x^3=x^3-3x^2+7x-9$

1.2 Technology Highlight

1. $\{0\,0\,0\,0\,1\}$; only 4 satisfies $2<L_1$ and $L_1<16$
   $\{1\,1\,1\,1\,1\}$; all elements satisfy $2<L_1$ or $L_1<16$
   $\{1\,1\,1\,0\,0\}$; only -4,-2,0 satisfy $2>L_1$ and $L_1<16$

2. $\{0\,1\,1\,1\,1\,1\,1\,0\}$; not included
   $\{1\,1\,1\,1\,1\,1\,1\,1\}$; not included
   $\{0\,0\,0\,0\,0\,0\,0\,0\}$; not included

1.2 Exercises

1. Set, interval

3. Intersection, union

5. Answers will vary.

7. $w\ge 45$

9. $250<T<450$

11. $y < 3$

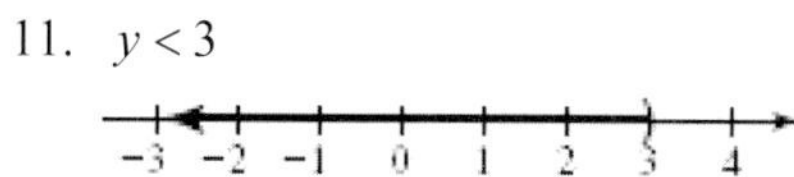

13. $m \le 5$

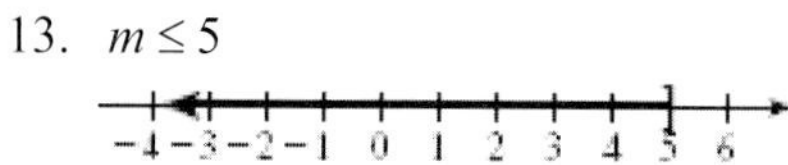

15. $x \ne 1$

17. $5 > x > 2$

19. $5a - 11 \ge 2a - 5$
$3a \ge 6$
$a \ge 2$
$\{a | a \ge 2\}$, Interval notation: $a \in [2, \infty)$

21. $2(n+3) - 4 \le 5n - 1$
$2n + 6 - 4 \le 5n - 1$
$2n + 2 \le 5n - 1$
$-3n \le -3$
$n \ge 1$
$\{n | n \ge 1\}$, Interval notation: $n \in [1, \infty)$

23. $\frac{3x}{8} + \frac{x}{4} < -4$
$8\left(\frac{3x}{8} + \frac{x}{4}\right) < 8(-4)$
$3x + 2x < -32$
$5x < -32$
$x < -\frac{32}{5}$
$\left\{x \middle| x < \frac{-32}{5}\right\}$,
Interval notation: $x \in \left(-\infty, \frac{-32}{5}\right)$

25. $1 < \frac{1}{2}y < \frac{5}{2}$
$2(1) < 2\left(\frac{1}{2}y\right) < 2\left(\frac{5}{2}\right)$
$2 < y < 5$
$\{y | 2 < y < 5\}$, Interval notation: $y \in (2, 5)$

27. $-3 < 2m - 7 \le 5$
$4 < 2m \le 12$
$2 < m \le 6$
$\{m | 2 < m \le 6\}$, Interval notation: $m \in (2, 6]$

29. $3 > 3(2m - 1) - 5 \ge 0$
$3 > 6m - 3 - 5 \ge 0$
$3 > 6m - 8 \ge 0$
$11 > 6m \ge 8$
$\frac{11}{6} > m \ge \frac{8}{6}$
$\frac{11}{6} > m \ge \frac{4}{3}$
$\left\{m \middle| \frac{11}{6} > m \ge \frac{4}{3}\right\}$,
Interval notation: $m \in \left[\frac{4}{3}, \frac{11}{6}\right)$

31. $\{x | x \ge -2\}; [-2, \infty)$

33. $\{x | -2 \le x \le 1\}; [-2, 1]$

35. $A \cap B = \{2\}$
$A \cup B = \{-3, -2, -1, 0, 1, 2, 3, 4, 6, 8\}$

37. $A \cap D = \{\ \}$
$A \cup D = \{-3, -2, -1, 0, 1, 2, 3, 4, 5, 6, 7\}$

39. $B \cap D = \{4, 6\}$
$B \cup D = \{2, 4, 5, 6, 7, 8\}$

41. $x < 5$ and $x \ge -2$
$[-2, 5)$

43. $x < -2$ or $x > 1$

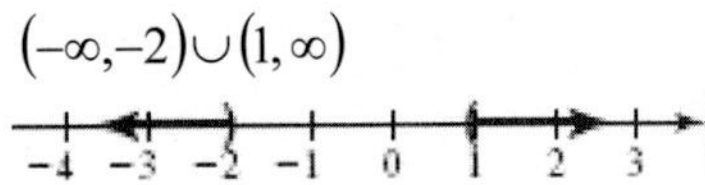

45. $x \geq 3$ and $x \leq 1$
no solution

47. $4(x-1) \leq 20$ or $x+6>9$
$4x-4 \leq 20$ or $x>3$
$4x \leq 24$
$x \leq 6$ or $x>3$
$x \in (-\infty, \infty)$

49. $-2x-7 \leq 3$ and $2x \leq 0$
$-2x \leq 10$ and $x \leq 0$
$x \geq -5$ and $x \leq 0$
$x \in [-5,0]$

51. $\frac{3}{5}x+\frac{1}{2}>\frac{3}{10}$ and $-4x>1$
$10\left(\frac{3}{5}x+\frac{1}{2}\right)>\left(\frac{3}{10}\right)10$ and $-4x>1$
$6x+5>3$ and $x<-\frac{1}{4}$
$6x>-2$ and $x<-\frac{1}{4}$
$x>-\frac{1}{3}$ and $x<-\frac{1}{4}$
$x \in \left(\frac{-1}{3}, \frac{-1}{4}\right)$

53. $\frac{3x}{8}+\frac{x}{4}<-4$ or $x+1>-5$
$8\left(\frac{3x}{8}+\frac{x}{4}\right)<8(-4)$ or $x>-6$
$3x+2x<-32$ or $x>-6$
$5x<-32$ or $x>-6$
$x<-\frac{32}{5}$ or $x>-6$
$x \in (-\infty, \infty)$

55. $-3 \leq 2x+5<7$
$-8 \leq 2x<2$
$-4 \leq x<1$
$x \in [-4,1)$

57. $-0.5 \leq 0.3-x \leq 1.7$
$-0.8 \leq -x \leq 1.4$
$0.8 \geq x \geq -1.4$
$x \in [-1.4,0.8]$

59. $-7<-\frac{3}{4}x-1 \leq 11$
$-6<-\frac{3}{4}x \leq 12$
$\left(-\frac{4}{3}\right)(-6)>\left(-\frac{4}{3}\right)\left(-\frac{3}{4}x\right) \geq \left(-\frac{4}{3}\right)12$
$8>x \geq -16$
$x \in [-16,8)$

61. $\frac{12}{m}$
$m \neq 0$
$m \in (-\infty,0) \cup (0,\infty)$

63. $\frac{5}{y+7}$
$y+7 \neq 0$
$y \neq -7$
$y \in (-\infty,-7) \cup (-7,\infty)$

65. $\frac{a+5}{6a-3}$
$6a-3 \neq 0$
$6a \neq 3$
$a \neq \frac{1}{2}$
$a \in \left(-\infty, \frac{1}{2}\right) \cup \left(\frac{1}{2}, \infty\right)$

67. $\frac{15}{3x-12}$
$3x-12 \neq 0$
$3x \neq 12$
$x \neq 4$

$x \in (-\infty, 4) \cup (4, \infty)$

69. $\sqrt{x-2}$
$x - 2 \geq 0$
$x \geq 2$
$x \in [2, \infty)$

71. $\sqrt{3n-12}$
$3n - 12 \geq 0$
$3n \geq 12$
$n \geq 4$
$n \in [4, \infty)$

73. $\sqrt{b - \frac{4}{3}}$
$b - \frac{4}{3} \geq 0$
$b \geq \frac{4}{3}$
$b \in \left[\frac{4}{3}, \infty\right)$

75. $\sqrt{8-4y}$
$8 - 4y \geq 0$
$-4y \geq -8$
$y \leq 2$
$y \in (-\infty, 2]$

77. $<$

79. $<$

81. $<$

83. $>$

85. $BMI \geq \frac{704W}{H^2}$
$27 \geq \frac{704W}{(68)^2}$
$27 \geq \frac{704W}{4624}$
$\left(\frac{4624}{704}\right)27 \geq \left(\frac{4624}{704}\right)\left(\frac{704W}{4624}\right)$
$177.34 \geq W$
Weight could be 177.34 pounds or less.

87. $\frac{68 + 75 + x}{3} \geq 80$
$68 + 75 + x \geq 240$
$143 + x \geq 240$
$x \geq 97\%$

89. $F = \frac{9}{5}C + 32;$
$45 = \frac{9}{5}C + 32$
$13 = \frac{9}{5}C$
$7.2° = C;$
$85 = \frac{9}{5}C + 32$
$53 = \frac{9}{5}C$
$29.4° = C;$
$7.2°C < T < 29.4°C$

91. $\frac{1125 + 850 + 625 + 400 + b}{5} \geq 1000$
$1125 + 850 + 625 + 400 + b \geq 5000$
$3000 + b \geq 5000$
$b \geq \$2000$

93. $\frac{1}{2}(12)h \geq 48$
$6h \geq 48$
$h \geq 8$ inches

95. Alaska: $-80°F \leq T \leq 100°F;$
Hawaii: $12°F \leq T \leq 100°F;$
Alaska;
Answers will vary.

97. $x - 2 \geq 4$ or $x - 2 \leq -4$
$x \geq 6$ or $x \leq -2$
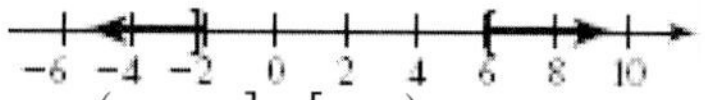

$x \in (-\infty, -2] \cup [6, \infty)$

99. $2n - 8$

101. $V = lwh + \frac{1}{2}\pi r^2 h$
$V = 12(5)(7) + \frac{1}{2}\left(\pi(3.5)^2\right)(5)$
$V = 420 + 30.625\pi \text{ cm}^3$

103. $-4(x-7)-3=2x+1$
$$-4x+28-3=2x+1$$
$$-4x+25=2x+1$$
$$-6x=-24$$
$$x=4$$

1.3 Technology Highlight

1. 14.37892522

2. 27

3. $\frac{49}{25}$

4. $\frac{-1}{32}$

1.3 Exercises

1. Excluded

3. Extraneous

5. Answers will vary.

7. $x^2-15=2x$
$$x^2-2x-15=0$$
$$(x-5)(x+3)=0$$
$$x-5=0 \text{ or } x+3=0$$
$$x=5 \quad \text{or } x=-3$$

9. $m^2=8m-16$
$$m^2-8m+16=0$$
$$(m-4)(m-4)=0$$
$$m-4=0$$
$$m=4$$

11. $5p^2-10p=0$
$$5p(p-2)=0$$
$$5p=0 \text{ or } p-2=0$$
$$p=0 \text{ or } p=2$$

13. $-14h^2=7h$
$$-14h^2-7h=0$$
$$-7h(2h+1)=0$$
$$-7h=0 \text{ or } 2h+1=0$$
$$h=0 \quad \text{or } 2h=-1$$
$$h=0 \quad \text{or } h=\frac{-1}{2}$$

15. $a^2-17=-8$
$$a^2-9=0$$
$$(a+3)(a-3)=0$$
$$a+3=0 \text{ or } a-3=0$$
$$a=-3 \quad \text{or } a=3$$

17. $g^2+18g+70=-11$
$$g^2+18g+81=0$$
$$(g+9)(g+9)=0$$
$$g+9=0$$
$$g=-9$$

19. $m^3+5m^2-9m-45=0$
$$m^2(m+5)-9(m+5)=0$$
$$(m+5)(m^2-9)=0$$
$$(m+5)(m+3)(m-3)=0$$
$$m+5=0 \text{ or } m+3=0 \text{ or } m-3=0$$
$$m=-5 \quad \text{or } m=-3 \quad \text{or } m=3$$

21. $(c-12)c-15=30$
$$c^2-12c-15=30$$
$$c^2-12c-45=0$$
$$(c-15)(c+3)=0$$
$$c-15=0 \text{ or } c+3=0$$
$$c=15 \quad \text{or } c=-3$$

23. $9+(r-5)r=33$
$$9+r^2-5r=33$$
$$r^2-5r-24=0$$
$$(r-8)(r+3)=0$$
$$r-8=0 \text{ or } r+3=0$$
$$r=8 \quad \text{or } r=-3$$

25. $(t+4)(t+7)=54$
$$t^2+11t+28=54$$
$$t^2+11t-26=0$$
$$(t+13)(t-2)=0$$
$$t+13=0 \text{ or } t-2=0$$
$$t=-13 \quad \text{or } t=2$$

27. $2x^2-4x-30=0$
$$2(x^2-2x-15)=0$$
$$2(x-5)(x+3)=0$$
$$x-5=0 \text{ or } x+3=0$$
$$x=5 \quad \text{or } x=-3$$

29. $2w^2-5w=3$
$2w^2-5w-3=0$
$(2w+1)(w-3)=0$
$2w+1=0$ or $w-3=0$
$2w=-1$ or $w=3$
$w=-\frac{1}{2}$ or $w=3$

31. $\left(x^2-3x\right)^2-14\left(x^2-3x\right)+40=0$
Let $u=x^2-3x$;
$u^2=\left(x^2-3x\right)^2$;
$u^2-14u+40=0$
$(u-4)(u-10)=0$
$u-4=0$ or $u-10=0$
$u=4$ or $u=10$
$x^2-3x=4$ or $x^2-3x=10$
$x^2-3x-4=0$ or $x^2-3x-10=0$
$(x-4)(x+1)=0$ or $(x-5)(x+2)=0$
$x-4=0$ or $x+1=0$ or $x-5=0$ or $x+2=0$
$x=4$ or $x=-1$ or $x=5$ or $x=-2$

33. $\frac{2}{x}+\frac{1}{x+1}=\frac{5}{x^2+x}$
$\frac{2}{x}+\frac{1}{x+1}=\frac{5}{x(x+1)}$
$x(x+1)\left[\frac{2}{x}+\frac{1}{x+1}=\frac{5}{x(x+1)}\right]$
$2x+2+x=5$
$3x=3$
$x=1$

35. $\frac{4}{a+2}=\frac{3}{a-1}$
$(a+2)(a-1)\left[\frac{4}{a+2}=\frac{3}{a-1}\right]$
$4a-4=3a+6$
$a=10$

37. $\frac{1}{3y}-\frac{1}{4y}=\frac{1}{y^2}$
$12y^2\left[\frac{1}{3y}-\frac{1}{4y}=\frac{1}{y^2}\right]$
$4y-3y=12$
$y=12$

39. $x+\frac{14}{x-7}=1+\frac{2x}{x-7}$
$(x-7)\left[x+\frac{14}{x-7}=1+\frac{2x}{x-7}\right]$
$x(x-7)+14=1(x-7)+2x$
$x^2-7x+14=x-7+2x$
$x^2-7x+14=3x-7$
$x^2-10x+21=0$
$(x-7)(x-3)=0$
$x-7=0$ or $x-3=0$
$x=7$ or $x=3$
x=3, x=7 is extraneous

41. $\frac{6}{n+3}+\frac{20}{n^2+n-6}=\frac{5}{n-2}$
$\frac{6}{n+3}+\frac{20}{(n+3)(n-2)}=\frac{5}{n-2}$
$(n+3)(n-2)\left[\frac{6}{n+3}+\frac{20}{(n+3)(n-2)}=\frac{5}{n-2}\right]$
$6(n-2)+20=5(n+3)$
$6n-12+20=5n+15$
$6n+8=5n+15$
$n=7$

43. $\frac{a}{2a+1}-\frac{a^2+5}{2a^2-5a-3}=\frac{2}{a-3}$
$\frac{a}{2a+1}-\frac{a^2+5}{(2a+1)(a-3)}=\frac{2}{a-3}$
$(2a+1)(a-3)\left[\frac{a}{2a+1}-\frac{a^2+5}{(2a+1)(a-3)}=\frac{2}{a-3}\right]$
$a(a-3)-\left(a^2+5\right)=2(2a+1)$
$a^2-3a-a^2-5=4a+2$
$-3a-5=4a+2$
$-7a=7$
$a=-1$

45. $\frac{1}{f}=\frac{1}{f_1}+\frac{1}{f_2}$
$f\,f_1f_2\left[\frac{1}{f}=\frac{1}{f_1}+\frac{1}{f_2}\right]$
$f_1f_2=f\,f_2+f\,f_1$
$f_1f_2=f(f_2+f_1)$

$$\frac{f_1 f_2}{f_1 + f_2} = f$$

47. $I = \dfrac{E}{R+r}$

$(R+r)\left[I = \dfrac{E}{R+r}\right]$

$IR + Ir = E$

$Ir = E - IR$

$r = \dfrac{E - IR}{I}$

49. $V = \dfrac{1}{3}\pi r^2 h$

$3V = \pi r^2 h$

$\dfrac{3V}{\pi r^2} = h$

51. $V = \dfrac{4}{3}\pi r^3$

$3V = 4\pi r^3$

$\dfrac{3V}{4\pi} = r^3$

53. a. $-3\sqrt{3x-5} = -9$

$\sqrt{3x-5} = 3$

$3x - 5 = 9$

$3x = 14$

$x = \dfrac{14}{3}$

b. $-11 = \dfrac{\sqrt{3x-4}}{-2} - 10$

$-1 = \dfrac{\sqrt{3x-4}}{-2}$

$2 = \sqrt{3x-4}$

$4 = 3x - 4$

$8 = 3x$

$\dfrac{8}{3} = x$

c. $2\sqrt{m-4} = \sqrt{3m+24}$

$4(m-4) = 3m + 24$

$4m - 16 = 3m + 24$

$m = 40$

55. a. $2 = \sqrt[3]{3m-1}$

$8 = 3m - 1$

$9 = 3m$

$3 = m$

b. $2\sqrt[3]{7-3x} - 3 = -7$

$2\sqrt[3]{7-3x} = -4$

$\sqrt[3]{7-3x} = -2$

$7 - 3x = -8$

$-3x = -15$

$x = 5$

c. $\dfrac{\sqrt[3]{2m+3}}{-5} + 2 = 3$

$\dfrac{\sqrt[3]{2m+3}}{-5} = 1$

$\sqrt[3]{2m+3} = -5$

$2m + 3 = -125$

$2m = -128$

$m = -64$

d. $\sqrt[3]{2x-9} = \sqrt[3]{3x+7}$

$2x - 9 = 3x + 7$

$-x = 16$

$x = -16$

57. a. $\sqrt{x-2} - \sqrt{2x} = -2$

$\sqrt{x-2} = \sqrt{2x} - 2$

$\left(\sqrt{x-2}\right)^2 = \left(\sqrt{2x} - 2\right)^2$

$x - 2 = 2x - 4\sqrt{2x} + 4$

$-x - 6 = -4\sqrt{2x}$

$(-x-6)^2 = \left(-4\sqrt{2x}\right)^2$

$x^2 + 12x + 36 = 16(2x)$

$x^2 + 12x + 36 = 32x$

$x^2 - 20x + 36 = 0$

$(x-18)(x-2) = 0$

$x - 18 = 0$ or $x - 2 = 0$

$x = 18$ or $x = 2$

b. $\sqrt{x+5} + \sqrt{x-10} = 5$

$\sqrt{x+5} = 5 - \sqrt{x-10}$

$\left(\sqrt{x+5}\right)^2 = \left(5 - \sqrt{x-10}\right)^2$

$x + 5 = 25 - 10\sqrt{x-10} + x - 10$

$-10 = -10\sqrt{x-10}$

$1 = \sqrt{x-10}$

$1 = x - 10$

$11 = x$

59. $x^{\frac{3}{5}}+17=9$

$x^{\frac{3}{5}}=-8$

$\left(x^{\frac{3}{5}}\right)^{\frac{5}{3}}=(-8)^{\frac{5}{3}}$

$x=-32$

61. $0.\bar{3}x^{\frac{5}{2}}-39=42$

$\frac{1}{3}x^{\frac{5}{2}}=81$

$x^{\frac{5}{2}}=243$

$\left(x^{\frac{5}{2}}\right)^{\frac{2}{5}}=(243)^{\frac{2}{5}}$

$x=9$

63. $8x^{-\frac{3}{2}}-17=-\frac{11}{8}$

$8x^{\frac{-3}{2}}=\frac{125}{8}$

$x^{\frac{-3}{2}}=\frac{125}{64}$

$\left(x^{\frac{-3}{2}}\right)^{-\frac{2}{3}}=\left(\frac{125}{64}\right)^{-\frac{2}{3}}$

$x=\frac{16}{25}$

65. $x^{\frac{2}{3}}-2x^{\frac{1}{3}}-15=0$

Let $u=x^{\frac{1}{3}}$

$u^2=x^{\frac{2}{3}};$

$u^2-2u-15=0$

$(u-5)(u+3)=0$

$u-5=0$ or $u+3=0$

$u=5$ or $u=-3$

$x^{\frac{1}{3}}=5$ or $x^{\frac{1}{3}}=-3$

$\left(x^{\frac{1}{3}}\right)^3=(5)^3$ or $\left(x^{\frac{1}{3}}\right)^3=(-3)^3$

$x=125$ or $x=-27$

67. $x^3-9x^{\frac{3}{2}}+8=0$

Let $u=x^{\frac{3}{2}}$

$u^2=x^3;$

$u^2-9u+8=0$

$(u-1)(u-8)=0$

$u-1=0$ or $u-8=0$

$u=1$ or $u=8$

$x^{\frac{3}{2}}=1$ or $x^{\frac{3}{2}}=8$

$\left(x^{\frac{3}{2}}\right)^{\frac{2}{3}}=(1)^{\frac{2}{3}}$ or $\left(x^{\frac{2}{3}}\right)^{\frac{2}{3}}=(8)^{\frac{2}{3}}$

$x=1$ or $x=4$

69. $(x^2+x)^2-8(x^2+x)+12=0$

Let $u=x^2+x$

$u^2=(x^2+x)^2;$

$u^2-8u+12=0$

$(u-6)(u-2)=0$

$u-6=0$ or $u-2=0$

$u=6$ or $u=2$

$x^2+x=6$ or $x^2+x=2$

$x^2+x-6=0$ or $x^2+x-2=0$

$(x+3)(x-2)=0$ or $(x+2)(x-1)=0$

$x+3=0$ or $x-2=0$ or $x+2=0$ or $x-1=0$

$x=-3$ or $x=2$ or $x=-2$ or $x=1$

71. $x^{-2}-3x^{-1}-4=0$

Let $u=x^{-1}$

$u^2=x^{-2};$

$u^2-3u-4=0$

$(u-4)(u+1)=0$

$u-4=0$ or $u+1=0$

$u=4$ or $u=-1$

$x^{-1}=4$ or $x^{-1}=-1$

$(x^{-1})^{-1}=(4)^{-1}$ or $(x^{-1})^{-1}=(-1)^{-1}$

$x=\frac{1}{4}$ or $x=-1$

73. $x^{-4} - 13x^{-2} + 36 = 0$

Let $u = x^{-2}$

$u^2 = x^{-4};$

$u^2 - 13u + 36 = 0$

$(u-9)(u-4) = 0$

$u - 9 = 0$ or $u - 4 = 0$

$u = 9$ or $u = 4$

$x^{-2} = 9$ or $x^{-2} = 4$

$\frac{1}{x^2} = 9$ or $\frac{1}{x^2} = 4$

$x^2 = \frac{1}{9}$ or $x^2 = \frac{1}{4}$

$x = \pm\frac{1}{3}$ or $x = \pm\frac{1}{2}$

75. $x + 4 = 7\sqrt{x+4}$

Let $u = (x+4)^{\frac{1}{2}}$

$u^2 = x + 4;$

$u^2 = 7u$

$u^2 - 7u = 0$

$u(u-7) = 0$

$u = 0$ or $u - 7 = 0$

$u = 0$ or $u = 7$

$\sqrt{x+4} = 0$ or $\sqrt{x+4} = 7$

$x + 4 = 0$ or $x + 4 = 49$

$x = -4$ or $x = 45$

77. $2\sqrt{x+10} + 8 = 3(x+10)$

Let $u = (x+10)^{\frac{1}{2}}$

$u^2 = x + 10;$

$2u + 8 = 3u^2$

$0 = 3u^2 - 2u - 8$

$0 = (3u+4)(u-2)$

$3u + 4 = 0$ or $u - 2 = 0$

$3u = -4$ or $u = 2$

$u = -\frac{4}{3}$ or $u = 2$

$\sqrt{x+10} = -\frac{4}{3}$ or $\sqrt{x+10} = 2$

$x + 10 = \frac{16}{9}$ or $x + 10 = 4$

$x = -\frac{74}{9}$ or $x = -6$

$x = -6$, $x = -\frac{74}{9}$ is extraneous

79. $S = \pi r\sqrt{r^2 + h^2}$

$S = \pi(6)\sqrt{(6)^2 + (10)^2}$

$S = 6\pi\sqrt{36 + 100}$

$S = 6\pi\sqrt{136}$

$S = 12\pi\sqrt{34}\ \text{m}^2$

81. Let x represent the first consecutive odd integer.
Let $x + 2$ represent the second consecutive odd integer.
Let $x + 4$ represent the third consecutive odd integer.

$x(x+4) = 9(x+2) - 4$

$x^2 + 4x = 9x + 18 - 4$

$x^2 + 4x = 9x + 14$

$x^2 - 5x - 14 = 0$

$(x-7)(x+2) = 0$

$x - 7 = 0$ or $x + 2 = 0$

$x = 7$ or $x = -2$

The first integer is 7.
The second integer is 9.
The third integer is 11.

83. $\frac{3+n}{4-n} = -8$

$(4-n)\frac{3+n}{4-n} = -8$

$3 + n = -8(4-n)$

$3 + n = -32 + 8n$

$-7n = -35$

$n = 5$

85. Let w represent the width.

$w(w+2) = 143$

$w^2 + 2w = 143$

$w^2 + 2w - 143 = 0$

$(w-11)(w+13) = 0$

$w - 11 = 0$ or $w + 13 = 0$

$w = 11$ or $w = -13$

11inches by 13 inches

87. $\frac{a}{9}=\frac{8}{12}$

$12a=72$

$a=6\text{ cm}$

89. Let x represent the number of decreases in price.

$(70-2x)(15+3x)=2250$

$1050+210x-30x-6x^2=2250$

$6x^2-180x+1200=0$

$x^2-30x+200=0$

$(x-10)(x-20)=0$

$x-10=0$ or $x-20=0$

$x=10$ or $x=20$

10, \$2 decreases results in a price of \$50 and a sale of 45 shoes.
20, \$2 decreases results in a price of \$30 and a sale of 75 shoes.

91. $h=-16t^2+vt+k$

$h=-16t^2+176t-480$

a. $h=-16(4)^2+176(4)-480$

$h=-32$ feet

b. $0=-16t^2+176t$

$0=-16t(t-11)$

$-16t=0$ or $t-11=0$

$t=0$ or $t=11$

The pebble returns after 11 seconds.

c. $h=-16(5)^2+176(5)-480=0;$

$h=-16(6)^2+176(6)-480=0;$

The pebble is at the canyon's rim.

93. Let x represent the time to fill the pool if both pipes are open.

$\frac{1}{20}+\frac{1}{28}=\frac{1}{x}$

$560x\left[\frac{1}{20}+\frac{1}{28}=\frac{1}{x}\right]$

$28x+20x=560$

$48x=560$

$x=11\frac{2}{3}$ hours

11 hr 40 min

95. Let t represent the time to travel from Bloomingdale to Chicago.

$D_{going}=D_{return}$

$70t=50\left(t+\frac{4}{5}\right)$

$70t=50t+40$

$20t=40$

$t=2$ hours

Distance = 70(2)=140 miles.

97. $C=\frac{92P}{100-P}$

$100=\frac{92P}{100-P}$

$(100-P)\left[100=\frac{92P}{100-P}\right]$

$10000-100P=92P$

$10000=192P$

$52.1\%=P$

99. $T=0.407R^{\frac{3}{2}}$

a. $88=0.407R^{\frac{3}{2}}$

$216.22\approx R^{\frac{3}{2}}$

$(216.22)^{\frac{2}{3}}\approx\left(R^{\frac{3}{2}}\right)^{\frac{2}{3}}$

$R\approx 36$ million miles

b. $225=0.407R^{\frac{3}{2}}$

$552.83\approx R^{\frac{3}{2}}$

$(552.83)^{\frac{2}{3}}\approx\left(R^{\frac{3}{2}}\right)^{\frac{2}{3}}$

$R\approx 67$ million miles

c. $365=0.407R^{\frac{3}{2}}$

$896.81\approx R^{\frac{3}{2}}$

$(896.81)^{\frac{2}{3}}\approx\left(R^{\frac{3}{2}}\right)^{\frac{2}{3}}$

$R\approx 93$ million miles

d. $687=0.407R^{\frac{3}{2}}$

$1687.96\approx R^{\frac{3}{2}}$

$$(1687.96)^{\frac{2}{3}} \approx \left(R^{\frac{3}{2}}\right)^{\frac{2}{3}}$$

$R \approx 142$ million miles

e. $4333 = 0.407R^{\frac{3}{2}}$

$10646.19 \approx R^{\frac{3}{2}}$

$$(10646.19)^{\frac{2}{3}} \approx \left(R^{\frac{3}{2}}\right)^{\frac{2}{3}}$$

$R \approx 484$ million miles

f. $10759 = 0.407R^{\frac{3}{2}}$

$26434.89 \approx R^{\frac{3}{2}}$

$$(26434.89)^{\frac{2}{3}} \approx \left(R^{\frac{3}{2}}\right)^{\frac{2}{3}}$$

$R \approx 887$ million miles

101. The constant "3" was not multiplied by the LCM;

$3 - \frac{8}{x+3} = \frac{1}{x}$

$3x(x+3) - 8x = x + 3$

$3x^2 + 9x - 8x = x + 3$

$3x^2 + x = x + 3$

$3x^2 = 3$

$x^2 = 1$

$x = \pm 1$

103. Answers will vary.

105. a. $4x^2 - 23x + 15 = 0$

$4x^2 - 20x - 3x + 15 = 0$

$4x(x-5) - 3(x-5) = 0$

$(x-5)(4x-3) = 0$

$x - 5 = 0$ or $4x - 3 = 0$

$x = 5$ or $4x = 3$

$x = 5$ or $x = \frac{3}{4}$

b. $3x^2 + 23x + 14 = 0$

$3x^2 + 21x + 2x + 14 = 0$

$3x(x+7) + 2(x+7) = 0$

$(x+7)(3x+2) = 0$

$x + 7 = 0$ or $3x + 2 = 0$

$x = -7$ or $3x = -2$

$x = -7$ or $x = -\frac{2}{3}$

107. Let t represent the time traveled.

$250t + 325t = 980$

$575t = 980$

$t \approx 1.7$ hours

109. Let x represent the first consecutive odd integer.
Let $x + 2$ represent the second consecutive odd integer.
Let $x + 4$ represent the third consecutive odd integer.

$2x - 3(x+4) = x + 2$

$2x - 3x - 12 = x + 2$

$-x - 12 = x + 2$

$-2x = 14$

$x = -7$

The first integer is -7.
The second integer is -5.
The third integer is -3.

111. $2^{-1} + (2x)^0 + 2x^0$

$= \frac{1}{2} + 1 + 2(1)$

$= \frac{1}{2} + 1 + 2$

$= 3\frac{1}{2}$

Chapter 1 Mid-Chapter Check

1. a. $9x^2 - 4 = 0$

$(3x+2)(3x-2) = 0$

$3x + 2 = 0$ or $3x - 2 = 0$

$3x = -2$ or $3x = 2$

$x = \frac{-2}{3}$ or $x = \frac{2}{3}$

b. $4x^2 - 12x + 9 = 0$

$(2x-3)(2x-3) = 0$

$2x - 3 = 0$

$2x = 3$

$x = \frac{3}{2}$

c. $3x^3 - 9x^2 - 30x = 0$

$3x(x^2 - 3x - 10) = 0$

$3x(x-5)(x+2) = 0$

$3x=0$ or $x-5=0$ or $x+2=0$
$x=0$ or $x=5$ or $x=-2$

d. $x^2-\frac{1}{9}=0$

$\left(x+\frac{1}{3}\right)\left(x-\frac{1}{3}\right)=0$

$x+\frac{1}{3}=0$ or $x-\frac{1}{3}=0$

$x=-\frac{1}{3}$ or $x=\frac{1}{3}$

e. $x^2+13x-48=0$
$(x+16)(x-3)=0$
$x+16=0$ or $x-3=0$
$x=-16$ or $x=3$

f. $4x^2-15x+9=0$
$(4x-3)(x-3)=0$
$4x-3=0$ or $x-3=0$
$4x=3$ or $x=3$

$x=\frac{3}{4}$ or $x=3$

g. $4x^3-8x^2-5x+10=0$
$4x^2(x-2)-5(x-2)=0$
$(x-2)(4x^2-5)=0$
$x-2=0$ or $4x^2-5=0$
$x=2$ or $4x^2=5$

$x=2$ or $x^2=\frac{5}{4}$

$x=2$ or $x=\pm\frac{\sqrt{5}}{2}$

h. $2n^2-n=3$
$2n^2-n-3=0$
$(2n-3)(n+1)=0$
$2n-3=0$ or $n+1=0$
$2n=3$ or $n=-1$

$n=\frac{3}{2}$ or $n=-1$

2. $\frac{5}{x+2}+\frac{3}{x}=2$

$x(x+2)\left[\frac{5}{x+2}+\frac{3}{x}=2\right]$

$5x+3(x+2)=2x(x+2)$
$5x+3x+6=2x^2+4x$
$8x+6=2x^2+4x$
$0=2x^2-4x-6$
$0=x^2-2x-3$
$0=(x-3)(x+1)$
$x-3=0$ or $x+1=0$
$x=3$ or $x=-1$

3. $\frac{-2}{x^2-4}+\frac{1}{3x-6}=\frac{3}{x+2}$

$\frac{-2}{(x+2)(x-2)}+\frac{1}{3(x-2)}=\frac{3}{x+2}$

$3(x+2)(x-2)\left[\frac{-2}{(x+2)(x-2)}+\frac{1}{3(x-2)}=\frac{3}{x+2}\right]$

$-6+1(x+2)=9(x-2)$
$-6+x+2=9x-18$
$-4+x=9x-18$
$-8x=-14$

$x=\frac{7}{4}$

4. $x+3=\sqrt{x^2-3}$

$(x+3)^2=\left(\sqrt{x^2-3}\right)^2$

$x^2+6x+9=x^2-3$
$6x+9=-3$
$6x=-12$
$x=-2$

5. $\sqrt{x-5}+1=\sqrt{x}$

$\left(\sqrt{x-5}+1\right)^2=\left(\sqrt{x}\right)^2$

$x-5+2\sqrt{x-5}+1=x$

$x-4+2\sqrt{x-5}=x$

$2\sqrt{x-5}=4$

$\sqrt{x-5}=2$

$\left(\sqrt{x-5}\right)^2=(2)^2$

$x-5=4$
$x=9$

6. $H=-16t^2+v_0t$

$H+16t^2=v_0t$

$\frac{H+16t^2}{t}=v_0$

7. $S=2\pi x^2+\pi x^2y$

$S=x^2(2\pi+\pi y)$

$\frac{S}{2\pi+\pi y}=x^2$

$$\sqrt{\frac{S}{2\pi + \pi y}} = x$$

$$\sqrt{\frac{S}{\pi(2+y)}} = x$$

8. a. $-5x+16 \le 11$ or $3x+2 \le -4$
$-5x \le -5$ or $3x \le -6$
$x \ge 1$ or $x \le -2$

–4 –3 –2 –1 0 1 2 3 4

b. $\frac{1}{2} < \frac{1}{12}x - \frac{5}{6} \le \frac{3}{4}$
$\frac{4}{3} < \frac{1}{12}x \le \frac{19}{12}$
$16 < x \le 19$

0 15 16 17 18 19 20

9. Let n represent the number.
$\frac{n}{n-3} = \frac{1}{4}n$
$4(n-3)\left[\frac{n}{n-3} - \frac{1}{4}n\right]$
$4n = n(n-3)$
$4n = n^2 - 3n$
$0 = n^2 - 7n$
$0 = n(n-7)$
$n = 0$ or $n - 7 = 0$
$n = 0$ or $n = 7$

10. a. $H = -16t^2 + 96t$
$H = -16(1)^2 + 96(1)$
At $t = 1$, $H = 80$ feet

b. $140 = -16t^2 + 96t$
$140 = -16t^2 + 96t$
$16t^2 - 96t + 140 = 0$
$4(4t^2 - 24t + 35) = 0c$
$4(2t-5)(2t-7) = 0$
$2t - 5 = 0$ or $2t - 7 = 0$
$t = \frac{5}{2}$ sec or $t = \frac{7}{2}$ sec
At $H = 140$, $t = \frac{5}{2}$ sec or $t = \frac{7}{2}$ sec

Reinforcing Basic Concepts

1. $x^2 + 4x = 0$
$x(x+4) = 0$
$x = 0$ or $x + 4 = 0$
$x = 0$ or $x = -4$

2. $x^2 + 7x = 0$
$x(x+7) = 0$
$x = 0$ or $x + 7 = 0$
$x = 0$ or $x = -7$

3. $x^2 - 5x = 0$
$x(x-5) = 0$
$x = 0$ or $x - 5 = 0$
$x = 0$ or $x = 5$

4. $x^2 - 2x = 0$
$x(x-2) = 0$
$x = 0$ or $x - 2 = 0$
$x = 0$ or $x = 2$

5. $x^2 + \frac{1}{2}x = 0$
$x\left(x + \frac{1}{2}\right) = 0$
$x = 0$ or $x + \frac{1}{2} = 0$
$x = 0$ or $x = -\frac{1}{2}$

6. $x^2 + \frac{2}{5}x = 0$
$x\left(x + \frac{2}{5}\right) = 0$
$x = 0$ or $x + \frac{2}{5} = 0$
$x = 0$ or $x = -\frac{2}{5}$

7. $x^2 - \frac{2}{3}x = 0$
$x\left(x - \frac{2}{3}\right) = 0$
$x = 0$ or $x - \frac{2}{3} = 0$
$x = 0$ or $x = \frac{2}{3}$

8. $x^2 - \frac{5}{6}x = 0$
$x\left(x - \frac{5}{6}\right) = 0$

$x = 0$ or $x - \frac{5}{6} = 0$
$x = 0$ or $x = \frac{5}{6}$

9. $x^2 - 81 = 0$
$(x+9)(x-9) = 0$
$x+9=0$ or $x-9=0$
$x=-9$ or $x=9$

10. $x^2 - 121 = 0$
$(x+11)(x-11) = 0$
$x+11=0$ or $x-11=0$
$x=-11$ or $x=11$

11. $x^2 - 7 = 0$
$(x+\sqrt{7})(x-\sqrt{7}) = 0$
$x+\sqrt{7}=0$ or $x-\sqrt{7}=0$
$x=-\sqrt{7}$ or $x=\sqrt{7}$

12. $x^2 - 31 = 0$
$(x+\sqrt{31})(x-\sqrt{31}) = 0$
$x+\sqrt{31}=0$ or $x-\sqrt{31}=0$
$x=-\sqrt{31}$ or $x=\sqrt{31}$

13. $x^2 - 49 = 0$
$(x+7)(x-7) = 0$
$x+7=0$ or $x-7=0$
$x=-7$ or $x=7$

14. $x^2 - 13 = 0$
$(x+\sqrt{13})(x-\sqrt{13}) = 0$
$x+\sqrt{13}=0$ or $x-\sqrt{13}=0$
$x=-\sqrt{13}$ or $x=\sqrt{13}$

15. $x^2 - 21 = 0$
$(x+\sqrt{21})(x-\sqrt{21}) = 0$
$x+\sqrt{21}=0$ or $x-\sqrt{21}=0$
$x=-\sqrt{21}$ or $x=\sqrt{21}$

16. $x^2 - 16 = 0$
$(x+4)(x-4) = 0$
$x+4=0$ or $x-4=0$
$x=-4$ or $x=4$

17. $x^2 + 4x - 45 = 0$
$(x+9)(x-5) = 0$
$x+9=0$ or $x-5=0$
$x=-9$ or $x=5$

18. $x^2 - 13x + 36 = 0$
$(x-9)(x-4) = 0$
$x-9=0$ or $x-4=0$
$x=9$ or $x=4$

19. $x^2 + 10x + 16 = 0$
$(x+8)(x+2) = 0$
$x+8=0$ or $x+2=0$
$x=-8$ or $x=-2$

20. $x^2 - 7x - 44 = 0$
$(x-11)(x+4) = 0$
$x-11=0$ or $x+4=0$
$x=11$ or $x=-4$

21. $x^2 + 6x - 16 = 0$
$(x+8)(x-2) = 0$
$x+8=0$ or $x-2=0$
$x=-8$ or $x=2$

22. $x^2 - 20x + 51 = 0$
$(x-17)(x-3) = 0$
$x-17=0$ or $x-3=0$
$x=17$ or $x=3$

23. $x^2 + 8x + 7 = 0$
$(x+1)(x+7) = 0$
$x+1=0$ or $x+7=0$
$x=-1$ or $x=-7$

24. $x^2 - 6x - 27 = 0$
$(x-9)(x+3) = 0$
$x-9=0$ or $x+3=0$
$x=9$ or $x=-3$

1.4 Technology Highlight

1. GC $3-3.464101615i$;
Manual $3-2i\sqrt{3}$

2. GC $23+43i$;
Manual $-12+15i+28i-35i^2 = 23+43i$
Check: $\frac{23+43i}{4-5i} = -3+7i$

3. GC $0.2+0.6i$;

Manual $\dfrac{2i(3-i)}{(3+i)(3-i)}=\dfrac{6i-2i^2}{10}=0.2+0.6i$

Check: $(0.2+0.6i)(3+i)$

$=0.6+0.2i+1.8i+0.6i^2=2i$

1.4 Exercises

1. $3-2i$

3. $2,\ 3\sqrt{2}$

5. b is correct.

7. a. $\sqrt{-16}=4i$
   b. $\sqrt{-49}=7i$
   c. $\sqrt{27}=\sqrt{9(3)}=3\sqrt{3}$
   d. $\sqrt{72}=\sqrt{36(2)}=6\sqrt{2}$

9. a. $-\sqrt{-18}=-\sqrt{-1(9)(2)}=-3i\sqrt{2}$
   b. $-\sqrt{-50}=-\sqrt{-1(25)(2)}=-5i\sqrt{2}$
   c. $3\sqrt{-25}=3(5i)=15i$
   d. $2\sqrt{-9}=2(3i)=6i$

11. a. $\sqrt{-19}=i\sqrt{19}$
    b. $\sqrt{-31}=i\sqrt{31}$
    c. $\sqrt{\dfrac{-12}{25}}=\dfrac{\sqrt{-1(4)3}}{\sqrt{25}}=\dfrac{2\sqrt{3}}{5}i$
    d. $\sqrt{\dfrac{-9}{32}}=\dfrac{\sqrt{-9}}{\sqrt{32}}\cdot\dfrac{\sqrt{2}}{\sqrt{2}}=\dfrac{3\sqrt{2}}{\sqrt{64}}i=\dfrac{3\sqrt{2}}{8}i$

13. a. $\dfrac{2+\sqrt{-4}}{2}=\dfrac{2+2i}{2}=1+i$
    $a=1, b=1$
    b. $\dfrac{6+\sqrt{-27}}{3}=\dfrac{6+3i\sqrt{3}}{3}=2+\sqrt{3}\,i$
    $a=2, b=\sqrt{3}$

15. a. $\dfrac{8+\sqrt{-16}}{2}=\dfrac{8+4i}{2}=4+2i$
    $a=4, b=2$
    b. $\dfrac{10-\sqrt{-50}}{5}=\dfrac{10-5i\sqrt{2}}{5}=2-\sqrt{2}\,i$
    $a=2, b=-\sqrt{2}$

17. a. $5=5+0i$
    $a=5, b=0$
    b. $3i=0+3i$
    $a=0, b=3$

19. a. $2\sqrt{-81}=2(9i)=0+18i$
    $a=0, b=18$
    b. $\dfrac{\sqrt{-32}}{8}=\dfrac{4\sqrt{2}}{8}i=0+\dfrac{\sqrt{2}}{2}i$
    $a=0, b=\dfrac{\sqrt{2}}{2}$

21. a. $4+\sqrt{-50}=4+5\sqrt{2}\,i$
    $a=4, b=5\sqrt{2}$
    b. $-5+\sqrt{-27}=-5+3\sqrt{3}\,i$
    $a=-5, b=3\sqrt{3}$

23. a. $\dfrac{14+\sqrt{-98}}{8}=\dfrac{14+7i\sqrt{2}}{8}=\dfrac{7}{4}+\dfrac{7\sqrt{2}}{8}i$
    $a=\dfrac{7}{4}, b=\dfrac{7\sqrt{2}}{8}$
    b. $\dfrac{5+\sqrt{-250}}{10}=\dfrac{5+5i\sqrt{10}}{10}=\dfrac{1}{2}+\dfrac{\sqrt{10}}{2}i$
    $a=\dfrac{1}{2}, b=\dfrac{\sqrt{10}}{2}$

25. a. $(12-\sqrt{-4})+(7+\sqrt{-9})$
    $=(12-2i)+(7+3i)$
    $=19+i$
    b. $(3+\sqrt{-25})+(-1-\sqrt{-81})$
    $=(3+5i)+(-1-9i)$
    $=2-4i$
    c. $(11+\sqrt{-108})-(2-\sqrt{-48})$
    $=11+\sqrt{-108}-2+\sqrt{-48}$
    $=11+\sqrt{-1(36)(3)}-2+\sqrt{-1(16)(3)}$
    $=9+6\sqrt{3}\,i+4\sqrt{3}\,i$
    $=9+10\sqrt{3}\,i$

27. a. $(2+3i)+(-5-i)$
    $=2+3i-5-i$
    $=-3+2i$
    b. $(5-2i)+(3+2i)$
    $=5-2i+3+2i$
    $=8$
    c. $(6-5i)-(4+3i)$

$= 6 - 5i - 4 - 3i$
$= 2 - 8i$

29. a. $(3.7 + 6.1i) - (1 + 5.9i)$
$= 3.7 + 6.1i - 1 - 5.9i$
$= 2.7 + 0.2i$

b. $\left(8 + \frac{3}{4}i\right) - \left(-7 + \frac{2}{3}i\right)$
$= 8 + \frac{3}{4}i + 7 - \frac{2}{3}i$
$= 15 + \frac{1}{12}i$

c. $\left(-6 - \frac{5}{8}i\right) + \left(4 + \frac{1}{2}i\right)$
$= -6 - \frac{5}{8}i + 4 + \frac{1}{2}i$
$= -2 - \frac{1}{8}i$

31. a. $5i \cdot (-3i)$
$= -15i^2$
$= 15$

b. $4i \cdot (-4i)$
$= -16i^2$
$= 16$

33. a. $-7i(5 - 3i)$
$= -35i + 21i^2$
$= -21 - 35i$

b. $6i(-3 + 7i)$
$= -18i + 42i^2$
$= -42 - 18i$

35. a. $(-3 + 2i)(2 + 3i)$
$= -6 - 9i + 4i + 6i^2$
$= -12 - 5i$

b. $(3 + 2i)(1 + i)$
$= 3 + 3i + 2i + 2i^2$
$= 1 + 5i$

37. a. conjugate $4 - 5i$
$= (4 + 5i)(4 - 5i)$
$= 16 - 20i + 20i - 25i^2$
$= 16 + 25$
$= 41$

b. conjugate $3 + i\sqrt{2}$
$= (3 + i\sqrt{2})(3 - i\sqrt{2})$
$= 9 - 3\sqrt{2}i + 3\sqrt{2}i - 2i^2$
$= 9 + 2$
$= 11$

39. a. conjugate $-7i$
$(7i)(-7i)$
$= -49i^2$
$= 49$

b. conjugate $\frac{1}{2} + \frac{2}{3}i$
$\left(\frac{1}{2} + \frac{2}{3}i\right)\left(\frac{1}{2} - \frac{2}{3}i\right)$
$= \frac{1}{4} - \frac{1}{3}i + \frac{1}{3}i - \frac{4}{9}i^2$
$= \frac{25}{36}$

41. a. $(4 - 5i)(4 + 5i)$
$= 16 + 20i - 20i - 25i^2$
$= 41$

b. $(7 - 5i)(7 + 5i)$
$= 49 + 35i - 35i - 25i^2$
$= 74$

43. a. $(3 - i\sqrt{2})(3 + i\sqrt{2})$
$= 9 + 3i\sqrt{2} - 3i\sqrt{2} - 2i^2$
$= 11$

b. $\left(\frac{1}{6} + \frac{2}{3}i\right)\left(\frac{1}{6} - \frac{2}{3}i\right)$
$= \frac{1}{36} - \frac{1}{9}i + \frac{1}{9}i - \frac{4}{9}i^2$
$= \frac{17}{36}$

45. a. $(2 + 3i)^2$
$= (2 + 3i)(2 + 3i)$
$= 4 + 6i + 6i + 9i^2$
$= -5 + 12i$

b. $(3 - 4i)^2$
$= (3 - 4i)(3 - 4i)$
$= 9 - 12i - 12i + 16i^2$
$= -7 - 24i$

47. a. $(-2 + 5i)^2$
$= (-2 + 5i)(-2 + 5i)$
$= 4 - 10i - 10i + 25i^2$
$= -21 - 20i$

b. $(3 + i\sqrt{2})^2$
$= (3 + i\sqrt{2})(3 + i\sqrt{2})$
$= 9 + 3i\sqrt{2} + 3i\sqrt{2} + 2i^2$
$= 7 + 6\sqrt{2}\ i$

49. $x^2+36=0, x=-6;$

$(-6)^2+36=0$
$36+36=0$
$72\neq 0$ no

51. $x^2+49=0, x=-7i;$

$(-7i)^2+49=0$
$49i^2+49=0$
$-49+49=0$
$0=0$ yes

53. $(x-3)^2=-9, x=3-3i;$

$(3-3i-3)^2=-9$
$(-3i)^2=-9$
$9i^2=-9$
$-9=-9$ yes

55. $x^2-2x+5=0, x=1-2i;$

$(1-2i)^2-2(1-2i)+5=0$
$1-4i+4i^2-2+4i+5=0$
$4+4i^2=0$
$4-4=0$
$0=0$ yes

57. $x^2-4x+9=0, x=2+i\sqrt{5};$

$\left(2+i\sqrt{5}\right)^2-4\left(2+i\sqrt{5}\right)+9=0$
$4+4i\sqrt{5}+5i^2-8-4i\sqrt{5}+9=0$
$5+5i^2=0$
$5-5=0$
$0=0$ yes

59. $x^2-2x+17=0, x=1+4i, 1-4i;$

$(1+4i)^2-2(1+4i)+17=0$
$1+8i+16i^2-2-8i+17=0$
$16+16i^2=0$
$16-16=0$
$0=0$
$1+4i$ is a solution.
$(1-4i)^2-2(1-4i)+17=0$
$1-8i+16i^2-2+8i+17=0$
$16+16i^2=0$
$16-16=0$
$0=0$
$1-4i$ is a solution.

61. a. $i^{48}=\left(i^4\right)^{12}=(1)^{12}=1$

b. $i^{26}=\left(i^4\right)^6 i^2=(1)^6(-1)=-1$

c. $i^{39}=\left(i^4\right)^9 i^3=(1)^9(-i)=-i$

d. $i^{53}=\left(i^4\right)^{13} i^1=(1)^{13}(i)=i$

63. a. $\dfrac{-2}{\sqrt{-49}}=\dfrac{-2}{7i}\cdot\dfrac{i}{i}=\dfrac{-2i}{7i^2}=\dfrac{2}{7}i$

b. $\dfrac{4}{\sqrt{-25}}=\dfrac{4}{5i}\cdot\dfrac{i}{i}=\dfrac{4i}{5i^2}=\dfrac{-4}{5}i$

65. a. $\dfrac{7}{3+2i}\cdot\dfrac{3-2i}{3-2i}=\dfrac{21-14i}{9-4i^2}=\dfrac{21-14i}{13}$

$=\dfrac{21}{13}-\dfrac{14}{13}i$

b. $\dfrac{-5}{2-3i}\cdot\dfrac{2+3i}{2+3i}=\dfrac{-10-15i}{4-9i^2}=\dfrac{-10-15i}{13}$

$=\dfrac{-10}{13}-\dfrac{15}{13}i$

67. a. $\dfrac{3+4i}{4i}\cdot\dfrac{i}{i}=\dfrac{3i+4i^2}{4i^2}=\dfrac{-4+3i}{-4}=1-\dfrac{3}{4}i$

b. $\dfrac{2-3i}{3i}\cdot\dfrac{i}{i}=\dfrac{2i-3i^2}{3i^2}=\dfrac{3+2i}{-3}=-1-\dfrac{2}{3}i$

69. $|a+bi|=\sqrt{a^2+b^2}$

a. $|2+3i|=\sqrt{(2)^2+(3)^2}=\sqrt{13}$

b. $|4-5i|=\sqrt{(4)^2+(-5)^2}=\sqrt{41}$

c. $\left|3+\sqrt{2}\,i\right|=\sqrt{(3)^2+\left(\sqrt{2}\right)^2}=\sqrt{11}$

71. $5+\sqrt{15}\,i+5-\sqrt{15}\,i=10$
$10=10$
verified;
$\left(5+\sqrt{15}\,i\right)\left(5-\sqrt{15}\,i\right)=40$
$25-5\sqrt{15}\,i+5\sqrt{15}\,i-15i^2=40$
$40=40$
verified

73. $Z=R+iX_L-iX_C$
$Z=7+i(6)-i(11)=7-5i\ \Omega$

75. $V=IZ$

$V=(3-2i)(5+5i)$
$V=15+15i-10i-10i^2$
$V=25+5i$ volts

77. $Z=\dfrac{Z_1Z_2}{Z_1+Z_2}$

$Z=\dfrac{(1+2i)(3-2i)}{1+2i+3-2i}$

$Z=\dfrac{3-2i+6i-4i^2}{4}$

$Z=\dfrac{7+4i}{4}$

$Z=\dfrac{7}{4}+i\ \Omega$

79. $x^2+9=x^2-(-9)=(x-3i)(x+3i)$

81. $(A+B)^3=A^3+3A^2B+3AB^2+B^3$

$\left(-\dfrac{1}{2}+\dfrac{\sqrt{3}}{2}i\right)^3=\left(-\dfrac{1}{2}\right)^3+3\left(-\dfrac{1}{2}\right)^2\left(\dfrac{\sqrt{3}}{2}i\right)$

$+3\left(-\dfrac{1}{2}\right)\left(\dfrac{\sqrt{3}}{2}i\right)^2+\left(\dfrac{\sqrt{3}}{2}i\right)^3$

$=-\dfrac{1}{8}+\left(\dfrac{3\sqrt{3}}{8}i\right)-\left(\dfrac{3}{2}\right)\left(\dfrac{3}{4}i^2\right)+\left(\dfrac{3\sqrt{3}}{8}i^3\right)$

$=-\dfrac{1}{8}+\dfrac{3\sqrt{3}}{8}i+\dfrac{9}{8}-\dfrac{3\sqrt{3}}{8}i=1$

83. $\sqrt{z}=\dfrac{\sqrt{2}}{2}\left(\sqrt{|z|+a}\right)\pm i\left(\sqrt{|z|-a}\right)$

a. $\sqrt{-7+24i}=$

$=\dfrac{\sqrt{2}}{2}\left(\sqrt{|-7+24i|-7}\pm i\sqrt{|-7+24i|+7}\right)$

$=\dfrac{\sqrt{2}}{2}\left(\sqrt{\sqrt{(-7)^2+24^2}-7}\pm i\sqrt{\sqrt{(-7)^2+24^2}+7}\right)$

$=\dfrac{\sqrt{2}}{2}\left(\sqrt{18}+i\sqrt{32}\right)$

$=\dfrac{\sqrt{36}}{2}\pm\dfrac{i\sqrt{64}}{2}$

$=3\pm 4i$

$=3+4i$

b. $\sqrt{5-12i}=$

$=\dfrac{\sqrt{2}}{2}\left(\sqrt{|5-12i|+5}\pm i\sqrt{|5-12i|-5}\right)$

$=\dfrac{\sqrt{2}}{2}\left(\sqrt{\sqrt{5^2+(-12)^2}+5}\pm i\sqrt{\sqrt{5^2+(-12)^2}-5}\right)$

$=\dfrac{\sqrt{2}}{2}\left(\sqrt{18}\pm i\sqrt{8}\right)$

$=\dfrac{\sqrt{36}}{2}\pm\dfrac{\sqrt{16}}{2}i$

$=3-2i$

c. $\sqrt{4+3i}=$

$=\dfrac{\sqrt{2}}{2}\left(\sqrt{|4+3i|+4}\pm i\sqrt{|4+3i|-4}\right)$

$=\dfrac{\sqrt{2}}{2}\left(\sqrt{\sqrt{4^2+3^2}+4}\pm i\sqrt{\sqrt{4^2+3^2}-4}\right)$

$=\dfrac{\sqrt{2}}{2}\left(\sqrt{9}\pm i\sqrt{1}\right)$

$=\dfrac{\sqrt{2}}{2}(3\pm i)$

$=\dfrac{3\sqrt{2}}{2}\pm\dfrac{\sqrt{2}}{2}i$

$=\dfrac{3\sqrt{2}}{2}+\dfrac{\sqrt{2}}{2}i$

85. a. Six is not a rational number-False
b. The rational numbers are a subset of the reals-True
c. 103 is an element of the set $\{3,4,5,\ldots\}$-True
d. The real numbers are not a subset of the complex-False

87. $\dfrac{x^2-4x+4}{x^2+3x-10}\cdot\dfrac{x^2-25}{x^2-10x+25}$

$=\dfrac{(x-2)(x-2)}{(x+5)(x-2)}\cdot\dfrac{(x+5)(x-5)}{(x-5)(x-5)}$

$=\dfrac{(x-2)}{(x-5)}$

$x\neq -5, 2, 5$

89. a. x^4-16
$=(x^2+4)(x^2-4)$
$=(x^2+4)(x+2)(x-2)$

b. n^3-27
$=(n-3)(n^2+3n+9)$

c. $x^3 - x^2 - x + 1$
$= x^2(x-1) - (x-1)$
$= (x-1)(x^2-1)$
$= (x-1)(x+1)(x-1)$
$= (x-1)^2(x+1)$

d. $4n^2m - 12nm^2 + 9m^3$
$= m(4n^2 - 12nm + 9m^2)$
$= m(2n-3m)^2$

1.5 Technology Highlight

1. D is a perfect square
2. D = 0
3. No
4. Yes, $2 \pm \sqrt{2}, -2 \pm \sqrt{3}$
   or approximately 3.41, 0.59, –0.27, –3.73

1.5 Exercises

1. Descending, 0

3. Quadratic, 1

5. The square root property is easier.
$4x^2 - 5 = 0$
$4x^2 = 5$
$x^2 = \frac{5}{4}$
$x = \pm\frac{\sqrt{5}}{2}$

7. $2x - 15 - x^2 = 0$
$-x^2 + 2x - 15 = 0$
Quadratic; $a = -1, b = 2, c = -15$

9. $\frac{2}{3}x - 7 = 0$
not quadratic

11. $\frac{1}{4}x^2 = 6x$
$\frac{1}{4}x^2 - 6x = 0$
Quadratic; $a = \frac{1}{4}, b = -6, c = 0$

13. $2x^2 + 7 = 0$
Quadratic; $a = 2, b = 0, c = 7$

15. $-3x^2 + 9x - 5 + 2x^3 = 0$
Not quadratic

17. $(x-1)^2 + (x-1) + 4 = 9$
$x^2 - 2x + 1 + x - 1 - 5 = 0$
$x^2 - x - 5 = 0$
Quadratic; $a = 1, b = -1, c = -5$

19. $m^2 = 16$
$m = \pm 4$

21. $y^2 - 28 = 0$
$y^2 = 28$
$y = \pm\sqrt{28}$
$y = \pm 2\sqrt{7} \approx \pm 5.29$

23. $p^2 + 36 = 0$
$p^2 = -36$
$p = \pm\sqrt{-36}$
No real solutions

25. $x^2 = \frac{21}{16}$
$x = \pm\frac{\sqrt{21}}{4} \approx \pm 1.15$

27. $(n-3)^2 = 36$
$n - 3 = \pm 6$
$n = 6 + 3$ or $n = -6 + 3$
$n = 9$ or $n = -3$

29. $(w+5)^2 = 3$
$w + 5 = \pm\sqrt{3}$
$w = -5 \pm \sqrt{3}$
$w \approx -3.27$ or $w \approx -6.73$

31. $(x-3)^2 + 7 = 2$
$(x-3)^2 = -5$
$x - 3 = \pm\sqrt{-5}$
No real solutions

33. $(m-2)^2 = \frac{18}{49}$

$m-2 = \pm\frac{\sqrt{18}}{7}$

$m = 2 \pm \frac{3\sqrt{2}}{7}$

$m \approx 2.61$ or $m \approx 1.39$

35. $x^2 + 6x + \underline{9}$

$(x+3)^2$

37. $n^2 + 3n + \underline{\frac{9}{4}}$

$\left(n + \frac{3}{2}\right)^2$

39. $p^2 + \frac{2}{3}p + \underline{\frac{1}{9}}$

$\left(p + \frac{1}{3}\right)^2$

41. $x^2 + 6x = -5$

$x^2 + 6x + 9 = -5 + 9$

$(x+3)^2 = 4$

$x + 3 = \pm 2$

$x = -3 \pm 2$

$x = -1$ or $x = -5$

43. $p^2 - 6p + 3 = 0$

$p^2 - 6p = -3$

$p^2 - 6p + 9 = -3 + 9$

$(p-3)^2 = 6$

$p - 3 = \pm\sqrt{6}$

$p = 3 \pm \sqrt{6}$

$p \approx 5.45$ or $p \approx 0.55$

45. $p^2 + 6p = -4$

$p^2 + 6p + 9 = -4 + 9$

$(p+3)^2 = 5$

$p + 3 = \pm\sqrt{5}$

$p = -3 \pm \sqrt{5}$

$p \approx -0.76$ or $p \approx -5.24$

47. $m^2 + 3m = 1$

$m^2 + 3m + \frac{9}{4} = 1 + \frac{9}{4}$

$\left(m + \frac{3}{2}\right)^2 = \frac{13}{4}$

$m + \frac{3}{2} = \pm\frac{\sqrt{13}}{2}$

$m = -\frac{3}{2} \pm \frac{\sqrt{13}}{2}$

$m \approx 0.30$ or $m \approx -3.30$

49. $n^2 = 5n + 5$

$n^2 - 5n = 5$

$n^2 - 5n + \frac{25}{4} = 5 + \frac{25}{4}$

$\left(n - \frac{5}{2}\right)^2 = \frac{45}{4}$

$n - \frac{5}{2} = \pm\frac{\sqrt{45}}{2}$

$n = \frac{5}{2} \pm \frac{3\sqrt{5}}{2}$

$n \approx 5.85$ or $n \approx -0.85$

51. $2x^2 = -7x + 4$

$2x^2 + 7x = 4$

$x^2 + \frac{7}{2}x = 2$

$x^2 + \frac{7}{2}x + \frac{49}{16} = 2 + \frac{49}{16}$

$\left(x + \frac{7}{4}\right)^2 = \frac{81}{16}$

$x + \frac{7}{4} = \pm\frac{9}{4}$

$x = -\frac{7}{4} \pm \frac{9}{4}$

$x = \frac{1}{2}$ or $x = -4$

53. $2n^2 - 3n - 9 = 0$

$2n^2 - 3n = 9$

$n^2 - \frac{3}{2}n = \frac{9}{2}$

$n^2 - \frac{3}{2}n + \frac{9}{16} = \frac{9}{2} + \frac{9}{16}$

$\left(n - \frac{3}{4}\right)^2 = \frac{81}{16}$

$$n-\frac{3}{4}=\pm\frac{9}{4}$$
$$n=\frac{3}{4}\pm\frac{9}{4}$$
$$n=3 \text{ or } n=-\frac{3}{2}$$

55. $4p^2-3p-2=0$
$$4p^2-3p=2$$
$$p^2-\frac{3}{4}p=\frac{1}{2}$$
$$p^2-\frac{3}{4}p+\frac{9}{64}=\frac{1}{2}+\frac{9}{64}$$
$$\left(p-\frac{3}{8}\right)^2=\frac{41}{64}$$
$$p-\frac{3}{8}=\pm\frac{\sqrt{41}}{8}$$
$$p=\frac{3}{8}\pm\frac{\sqrt{41}}{8}$$
$$p\approx 1.18 \text{ or } p\approx -0.43$$

57. $\frac{m}{2}=\frac{7}{2}-\frac{2}{m}$
$$2m\left[\frac{m}{2}=\frac{7}{2}-\frac{2}{m}\right]$$
$$m^2=7m-4$$
$$m^2-7m=-4$$
$$m^2-7m+\frac{49}{4}=-4+\frac{49}{4}$$
$$\left(m-\frac{7}{2}\right)^2=\frac{33}{4}$$
$$m-\frac{7}{2}=\pm\frac{\sqrt{33}}{2}$$
$$m=\frac{7}{2}\pm\frac{\sqrt{33}}{2}$$
$$m\approx 6.37 \text{ or } m\approx 0.63$$

59. $x^2-3x=18$
$$x^2-3x-18=0$$
$$(x-6)(x+3)=0$$
$$x-6=0 \text{ or } x+3=0$$
$$x=6 \quad \text{or } x=-3$$

61. $4m^2-25=0$
$$4m^2=25$$
$$m^2=\frac{25}{4}$$
$$m=\pm\frac{5}{2}$$

63. $4n^2-8n-1=0$
$$a=4, b=-8, c=-1$$
$$n=\frac{-(-8)\pm\sqrt{(-8)^2-4(4)(-1)}}{2(4)}$$
$$n=\frac{8\pm\sqrt{80}}{8}$$
$$n=\frac{8\pm 4\sqrt{5}}{8}$$
$$n=\frac{2\pm\sqrt{5}}{2}$$
$$n\approx 2.12 \text{ or } n\approx -0.12$$

65. $6w^2-w=2$
$$6w^2-w-2=0$$
$$(2w+1)(3w-2)=0$$
$$2w+1=0 \text{ or } 3w-2=0$$
$$2w=-1 \quad \text{or } 3w=2$$
$$w=-\frac{1}{2} \quad \text{or } w=\frac{2}{3}$$

67. $4m^2=12m-15$
$$4m^2-12m+15=0$$
$$a=4, b=-12, c=15$$
$$m=\frac{-(-12)\pm\sqrt{(-12)^2-4(4)(15)}}{2(4)}$$
$$m=\frac{12\pm\sqrt{-96}}{8}$$
$$m=\frac{12\pm 4\sqrt{6}\,i}{8}$$
$$m=\frac{3\pm\sqrt{6}\,i}{2}$$
$$m=\frac{3}{2}\pm\frac{\sqrt{6}}{2}i$$
$$m\approx 1.5\pm 1.22i$$

69. $4n^2-9=0$
$$4n^2=9$$
$$n^2=\frac{9}{4}$$

$n = \pm\frac{3}{2}$

71. $5w^2 = 6w + 8$

$5w^2 - 6w - 8 = 0$
$(5w+4)(w-2) = 0$
$5w + 4 = 0$ or $w - 2 = 0$
$5w = -4$ or $w = 2$
$w = -\frac{4}{5}$ or $w = 2$

73. $3a^2 - a + 2 = 0$
$a = 3, b = -1, c = 2$

$a = \frac{-(-1) \pm \sqrt{(-1)^2 - 4(3)(2)}}{2(3)}$

$a = \frac{1 \pm \sqrt{-23}}{6}$

$a = \frac{1 \pm \sqrt{23}\,i}{6}$

$a = \frac{1}{6} \pm \frac{\sqrt{23}}{6}i$

$a \approx 0.1\overline{6} \pm 0.80i$

75. $5p^2 = 6p + 3$

$5p^2 - 6p - 3 = 0$
$a = 5, b = -6, c = -3$

$p = \frac{-(-6) \pm \sqrt{(-6)^2 - 4(5)(-3)}}{2(5)}$

$p = \frac{6 \pm \sqrt{96}}{10}$

$p = \frac{6 \pm 4\sqrt{6}}{10}$

$p = \frac{3 \pm 2\sqrt{6}}{5}$

$p = \frac{3}{5} \pm \frac{2\sqrt{6}}{5}$

$p \approx 1.58$ or $p \approx -0.38$

77. $5w^2 - w = 1$

$5w^2 - w - 1 = 0$
$a = 5, b = -1, c = -1$

$w = \frac{-(-1) \pm \sqrt{(-1)^2 - 4(5)(-1)}}{2(5)}$

$w = \frac{1 \pm \sqrt{21}}{10}$

$w = \frac{1}{10} \pm \frac{\sqrt{21}}{10}$

$w \approx 0.56$ or $w \approx -0.36$

79. $2a^2 + 5 = 3a$

$2a^2 - 3a + 5 = 0$
$a = 2, b = -3, c = 5$

$a = \frac{-(-3) \pm \sqrt{(-3)^2 - 4(2)(5)}}{2(2)}$

$a = \frac{3 \pm \sqrt{-31}}{4}$

$a = \frac{3 \pm \sqrt{31}\,i}{4}$

$a = \frac{3}{4} \pm \frac{\sqrt{31}}{4}i$

$a \approx 0.75 \pm 1.39i$

81. $2p^2 - 4p + 11 = 0$
$a = 2, b = -4, c = 11$

$p = \frac{-(-4) \pm \sqrt{(-4)^2 - 4(2)(11)}}{2(2)}$

$p = \frac{4 \pm \sqrt{-72}}{4}$

$p = \frac{4 \pm 6\sqrt{2}\,i}{4}$

$p = 1 \pm \frac{3\sqrt{2}}{2}i$

$p = 1 \pm 2.12i$

83. $w^2 + \frac{2}{3}w = \frac{1}{9}$

$9\left[w^2 + \frac{2}{3}w = \frac{1}{9}\right]$

$9w^2 + 6w = 1$
$9w^2 + 6w - 1 = 0$
$a = 9, b = 6, c = -1$

$w = \frac{-(6) \pm \sqrt{(6)^2 - 4(9)(-1)}}{2(9)}$

$w = \frac{-6 \pm \sqrt{72}}{18}$

$w = \frac{-6 \pm 6\sqrt{2}}{18}$

$w = \frac{-1 \pm \sqrt{2}}{3}$

$w = \frac{-1}{3} \pm \frac{\sqrt{2}}{3}$
$w \approx 0.14$ or $w \approx -0.80$

85. $0.2a^2 + 1.2a + 0.9 = 0$
$a = 0.2, b = 1.2, c = 0.9$
$a = \frac{-(1.2) \pm \sqrt{(1.2)^2 - 4(0.2)(0.9)}}{2(0.2)}$
$a = \frac{-1.2 \pm \sqrt{0.72}}{0.4}$
$a = \frac{-1.2 \pm 0.6\sqrt{2}}{0.4}$
$a = \frac{-6 \pm 3\sqrt{2}}{2}$
$a = \frac{-6}{2} \pm \frac{3\sqrt{2}}{2}$
$a \approx -0.88$ or $a \approx -5.12$

87. $\frac{2}{7}p^2 - 3 = \frac{8}{21}p$
$21\left[\frac{2}{7}p^2 - 3 = \frac{8}{21}p\right]$
$6p^2 - 63 = 8p$
$6p^2 - 8p - 63 = 0$
$a = 6, b = -8, c = -63$
$p = \frac{-(-8) \pm \sqrt{(-8)^2 - 4(6)(-63)}}{2(6)}$
$p = \frac{8 \pm \sqrt{1576}}{12}$
$p = \frac{8 \pm 2\sqrt{394}}{12}$
$p = \frac{4 \pm \sqrt{394}}{6}$
$p = \frac{4}{6} \pm \frac{\sqrt{394}}{6}$
$p \approx 3.97$ or $p \approx -2.64$

89. $w + 3 - \frac{10}{w} = 0$
$w\left[w + 3 - \frac{10}{w} = 0\right]$
$w^2 + 3w - 10 = 0$
$(w-2)(w+5) = 0$
$w - 2 = 0$ or $w + 5 = 0$
$w = 2$ or $w = -5$

91. $a - \frac{7}{2a} = \frac{1}{2}$
$2a\left[a - \frac{7}{2a} = \frac{1}{2}\right]$
$2a^2 - 7 = a$
$2a^2 - a - 7 = 0$
$a = 2, b = -1, c = -7$
$a = \frac{-(-1) \pm \sqrt{(-1)^2 - 4(2)(-7)}}{2(2)}$
$a = \frac{1 \pm \sqrt{57}}{4}$
$a = \frac{1}{4} \pm \frac{\sqrt{57}}{4}$
$a \approx 2.14$ or $a \approx -1.64$

93. $n - \frac{4}{n-3} = 0$
$(n-3)\left[n - \frac{4}{n-3} = 0\right]$
$n^2 - 3n - 4 = 0$
$(n-4)(n+1) = 0$
$n - 4 = 0$ or $n + 1 = 0$
$n = 4$ or $n = -1$

95. $-3x^2 + 2x + 1 = 0$
$a = -3, b = 2, c = 1$
$(2)^2 - 4(-3)(1) = 16$
two rational solutions

97. $-4x + x^2 + 13 = 0$
$x^2 - 4x + 13 = 0$
$a = 1, b = -4, c = 13$
$(-4)^2 - 4(1)(13) = -36$
two complex solutions

99. $15x^2 - x - 6 = 0$
$a = 15, b = -1, c = -6$
$(-1)^2 - 4(15)(-6) = 361$
two rational solutions

101. $-4x^2 + 6x - 5 = 0$
$a = -4, b = 6, c = -5$
$(6)^2 - 4(-4)(-5) = -44$
two complex solutions

103. $2x^2 + 8 = -9x$

$2x^2 + 9x + 8 = 0$
$a = 2, b = 9, c = 8$

$(9)^2 - 4(2)(8) = 17$
two irrational solutions

105. $4x^2 + 12x = -9$

$4x^2 + 12x + 9 = 0$
$a = 4, b = 12, c = 9$

$(12)^2 - 4(4)(9) = 0$
one repeated solution

107. $-6x + 2x^2 + 5 = 0$

$2x^2 - 6x + 5 = 0$
$a = 2, b = -6, c = 5$

$$x = \frac{-(-6) \pm \sqrt{(-6)^2 - 4(2)(5)}}{2(2)}$$

$$x = \frac{6 \pm \sqrt{-4}}{4}$$

$$x = \frac{6 \pm 2i}{4}$$

$$x = \frac{3}{2} \pm \frac{1}{2}i$$

109. $5x^2 + 5 = -5x$

$5x^2 + 5x + 5 = 0$
$a = 5, b = 5, c = 5$

$$x = \frac{-(5) \pm \sqrt{(5)^2 - 4(5)(5)}}{2(5)}$$

$$x = \frac{-5 \pm \sqrt{-75}}{10}$$

$$x = \frac{-5 \pm 5\sqrt{3}\,i}{10}$$

$$x = -\frac{1}{2} \pm \frac{\sqrt{3}}{2}i$$

111. $-2x^2 = -5x + 11$

$0 = 2x^2 - 5x + 11$
$a = 2, b = -5, c = 11$

$$x = \frac{-(-5) \pm \sqrt{(-5)^2 - 4(2)(11)}}{2(2)}$$

$$x = \frac{5 \pm \sqrt{-63}}{4}$$

$$x = \frac{5 \pm 3\sqrt{7}\,i}{4}$$

$$x = \frac{5}{4} \pm \frac{3\sqrt{7}}{4}i$$

113. $h = -16t^2 + vt$

$16t^2 - vt + h = 0$
$a = 16, b = -v, c = h$

$$t = \frac{-(-v) \pm \sqrt{(-v)^2 - 4(16)(h)}}{2(16)}$$

$$t = \frac{v \pm \sqrt{v^2 - 64h}}{32}$$

115. $16t^2 - 96t - 408 = 0$

$8(2t^2 - 12t - 51) = 0$
$a = 2, b = -12, c = -51$

$$t = \frac{-(-12) \pm \sqrt{(-12)^2 - 4(2)(-51)}}{2(2)}$$

$$t = \frac{12 \pm \sqrt{552}}{4}$$

$$t = \frac{12 \pm 2\sqrt{138}}{4}$$

$$t = \frac{6 \pm \sqrt{138}}{2}$$

$t \approx 8.87$ seconds

117. $R = x\left(40 - \frac{1}{3}x\right)$

$900 = 40x - \frac{1}{3}x^2$

$\frac{1}{3}x^2 - 40x + 900 = 0$

$x^2 - 120x + 2700 = 0$
$a = 1, b = -120, c = 2700$

$$x = \frac{-(-120) \pm \sqrt{(-120)^2 - 4(1)(2700)}}{2(1)}$$

$$x = \frac{120 \pm \sqrt{3600}}{2}$$

$$x = \frac{120 \pm 60}{2}$$

$x = 90$ or $x = 30$
30 thousand ovens

119. Let w represent the width of the doubles court.
Let $2w + 6$ represent the length of the doubles court.

$w(2w+6)=2808$
$2w^2+6w-2808=0$
$w^2+3w-1404=0$
$(w+39)(w-36)=0$
$w+39=0$ or $w-36=0$
$w=-39$ or $w=36$
The width is 36 feet.
The length is 78 feet.

121.a. $P=-x^2+122x-1965-(2x+35)$
$P=-x^2+120x-2000$

b. $x^2-120x+2000=0$
$(x-100)(x-20)=0$
$x-100=0$ or $x-20=0$
$x=100$ or $x=20$
10,000 toys

123. $260=-16t^2+144t$
$16t^2-144t+260=0$
$4t^2-36t+65=0$
$(2t-13)(2t-5)=0$
$2t-13=0$ or $2t-5=0$
$2t=13$ or $2t=5$
$t=\frac{13}{2}$ seconds or $t=\frac{5}{2}$ seconds
$t=6.5$ seconds or $t=2.5$ seconds

125. $z^2-3iz=-10$
$z^2-3iz+10=0$
$a=1, b=-3i, c=10$
$$z=\frac{-(-3i)\pm\sqrt{(-3i)^2-4(1)(10)}}{2(1)}$$
$$z=\frac{3i\pm\sqrt{9i^2-40}}{2}$$
$$z=\frac{3i\pm\sqrt{-49}}{2}$$
$$z=\frac{3i\pm 7i}{2}$$
$z=5i$ or $z=-2i$

127. $4iz^2+5z+6i=0$
$a=4i, b=5, c=6i$
$$z=\frac{-(5)\pm\sqrt{(5)^2-4(4i)(6i)}}{2(4i)}$$
$$z=\frac{-5\pm\sqrt{25-96i^2}}{8i}$$
$$z=\frac{-5\pm\sqrt{121}}{8i}$$
$$z=\frac{-5\pm 11}{8i}$$
$z=\frac{6}{8i}$ or $z=\frac{-16}{8i}$ $\left(\text{Recall}\frac{1}{i}=-i\right)$
$z=-\frac{3}{4}i$ or $z=2i$

129. $0.5z^2+(7+i)z+(6+7i)=0$
$a=0.5, b=7+i, c=6+7i$
$$z=\frac{-(7+i)\pm\sqrt{(7+i)^2-4(0.5)(6+7i)}}{2(0.5)}$$
$$z=\frac{-7-i\pm\sqrt{49+14i+i^2-12-14i}}{1}$$
$z=-7-i\pm\sqrt{36}$
$z=-7-i\pm 6$
$z=-1-i$ or $z=-13-i$

131. $2x^2+3kx+18=0$
$(3k)^2-4(2)(18)=0$
$9k^2-144=0$
$9k^2=144$
$k^2=16$
$k=\pm 4$

133. Answers will vary.

135.
$(x-3)(x^2+3x-10)+(x-2)(x^2+2x-15)=0$
$(x-3)(x-2)(x+5)+(x-2)(x-3)(x+5)=0$
$2(x-2)(x-3)(x+5)=0$
$x-2=0$ or $x-3=0$ or $x+5=0$
$x=2$ or $x=3$ or $x=-5$

137.a. $x^2-5x-36=0$
$(x-9)(x+4)=0$
$x-9=0$ or $x+4=0$
$x=9$ or $x=-4$

b. $4x^2-25=0$
$(2x+5)(2x-5)=0$
$2x+5=0$ or $2x-5=0$
$2x=-5$ or $2x=5$

$x=-\frac{5}{2}$ or $x=\frac{5}{2}$

c. $x^3+6x^2-4x-24=0$
$x^2(x+6)-4(x+6)=0$
$(x+6)(x^2-4)=0$
$(x+6)(x+2)(x-2)=0$
$x+6=0$ or $x+2=0$ or $x-2=0$
$x=-6$ or $x=-2$ or $x=2$

139. Let x represent the number of good seats sold.
Let $900-x$ represent the number of cheap seats sold.
$30x+20(900-x)=25000$
$30x+18000-20x=25000$
$10x=7000$
$x=700$
700, \$30 tickets
200, \$20 tickets

141. $\frac{6-\sqrt{-18}}{3}$
$=\frac{6-\sqrt{-1(9)(2)}}{3}$
$=\frac{6-3\sqrt{2}\,i}{3}$
$=2-\sqrt{2}\,i$

Chapter 1 Summary and Concept Review

1. a. $4x+7=5$ is linear
b. $\frac{3}{m}+2.3=7.7$ is non-linear, variable as a divisor
c. $3g(g-4)=12$ is non-linear, exponent on g is 2

2. $-2b+7=-5$
$-2b=-12$
$b=6$

3. $3(2n-6)+1=7$
$6n-18+1=7$
$6n-17=7$
$6n=24$
$n=4$

4. $4m-5=11m+2$
$-7m=7$
$m=-1$

5. $\frac{1}{2}x+\frac{2}{3}=\frac{3}{4}$
$12\left[\frac{1}{2}x+\frac{2}{3}=\frac{3}{4}\right]$
$6x+8=9$
$6x=1$
$x=\frac{1}{6}$

6. $6p-(3p+5)-9=3(p-3)$
$6p-3p-5-9=3p-9$
$3p-14=3p-9$
$-14\neq-9$
No solution.

7. $-\frac{g}{6}=3-\frac{1}{2}-\frac{5g}{12}$
$12\left[-\frac{g}{6}=3-\frac{1}{2}-\frac{5g}{12}\right]$
$-2g=36-6-5g$
$3g=30$
$g=10$

8. $V=\pi r^2h$
$\frac{V}{\pi r^2}=h$

9. $P=2L+2W$
$P-2W=2L$
$\frac{P-2W}{2}=L$

10. $ax+b=c$
$ax=c-b$
$x=\frac{c-b}{a}$

11. $2x-3y=6$
$-3y=-2x+6$
$y=\frac{2}{3}x-2$

12. Let x represent the amount of the first keg (20% sugar) used.
Let $12-x$ represent the amount of the second keg (50% sugar) used.
$0.2x+0.5(12-x)=0.4(12)$
$0.2x+6-0.5x=4.8$
$-0.3x+6=4.8$
$-0.3x=-1.2$
$x=4$

4 gallons of 20% sugar.
8 gallons of 50% sugar.

13. $3(4)+\frac{1}{2}\pi(1.5)^2$

$12+\frac{9}{8}\pi \text{ ft}^2$

14. Let t represent the time traveling.

$7t+9t=12$

$16t=12$

$t=\frac{3}{4}$ hour

15. $a \ge 35$

16. $a < 2$

17. $s \le 65$

18. $c \ge 1200$

19. $7x > 35$

$x > 5$

$(5,\infty)$

20. $-\frac{3}{5}m < 6$

$-\frac{5}{3}\left(-\frac{3}{5}m\right) < \left(-\frac{5}{3}\right)(6)$

$m > -10$

$(-10,\infty)$

21. $2(3m-2) \le 8$

$6m-4 \le 8$

$6m \le 12$

$m \le 2$

$(-\infty,2]$

22. $-1 < \frac{1}{3}x+2 \le 5$

$-3 < \frac{1}{3}x \le 3$

$-9 < x \le 9$

$(-9,9]$

23. $-4 < 2b+8$ and $3b-5 > -32$

$-12 < 2b$ and $3b > -27$

$-6 < b$ and $b > -9$

$b > -6$ and $b > -9$

$(-6,\infty)$

24. $-5(x+3) > -7$ or $x-5.2 > -2.9$

$-5x-15 > -7$ or $x > 2.3$

$-5x > 8$ or $x > 2.3$

$x < -\frac{8}{5}$ or $x > 2.3$

$\left(-\infty,-\frac{8}{5}\right)\cup(2.3,\infty)$

25. a. $\frac{7}{n-3}$

$n-3 \ne 0$

$n \ne 3$

$(-\infty,3)\cup(3,\infty)$

b. $\frac{5}{2x-3}$

$2x-3 \ne 0$

$2x \ne 3$

$x \ne \frac{3}{2}$

$\left(-\infty,\frac{3}{2}\right)\cup\left(\frac{3}{2},\infty\right)$

c. $\sqrt{x+5}$

$x+5 \ge 0$

$x \ge -5$

$[-5,\infty)$

d. $\sqrt{-3n+18}$

$-3n+18 \ge 0$

$-3n \ge -18$

$n \le 6$

$(-\infty,6]$

26. $\frac{72+95+83+79+x}{5} \ge 85$

$72+95+83+79+x \ge 425$

$329+x \ge 425$

$x \ge 96\%$

27. a. $(x+3)(x-5)(x+1)(x-4)=0$

$x+3=0$ or $x-5=0$ or $x+1=0$ or $x-4=0$

$x=-3$ or $x=5$ or $x=-1$ or $x=4$

b. $3x\left(x+\frac{5}{2}\right)(x-9)\left(x-\frac{1}{2}\right)=0$

$3x=0$ or $x+\frac{5}{2}=0$ or $x-9=0$ or $x-\frac{1}{2}=0$

$x=0$ or $x=-\frac{5}{2}$ or $x=9$ or $x=\frac{1}{2}$

28. a. $x^2+7x-18=0$

$(x+9)(x-2)=0$
$x+9=0$ or $x-2=0$
$x=-9$ or $x=2$

b. $n^2=-12n-27$

$n^2+12n+27=0$
$(n+9)(n+3)=0$
$n+9=0$ or $n+3=0$
$n=-9$ or $n=-3$

c. $2z^2-3=z$

$2z^2-z-3=0$
$(2z-3)(z+1)=0$
$2z-3=0$ or $z+1=0$
$2z=3$ or $z=-1$
$z=\frac{3}{2}$ or $z=-1$

d. $-7r^3+21r^2+28r=0$

$-7r\left(r^2-3r-4\right)=0$
$-7r(r+1)(r-4)=0$
$-7r=0$ or $r+1=0$ or $r-4=0$
$r=0$ or $r=-1$ or $r=4$

e. $-3b^3+27b=0$

$-3b\left(b^2-9\right)=0$
$-3b(b+3)(b-3)=0$
$-3b=0$ or $b+3=0$ or $b-3=0$
$b=0$ or $b=-3$ or $b=3$

f. $4a^3-6a^2-16a+24=0$

$2\left(2a^3-3a^2-8a+12\right)=0$
$2\left[a^2(2a-3)-4(2a-3)\right]=0$
$2(2a-3)\left(a^2-4\right)=0$
$2(2a-3)(a+2)(a-2)=0$
$2a-3=0$ or $a+2=0$ or $a-2=0$
$a=\frac{3}{2}$ or $a=-2$ or $a=2$

29. $\frac{3}{5x}+\frac{7}{10}=\frac{1}{4x}$

$20x\left[\frac{3}{5x}+\frac{7}{10}=\frac{1}{4x}\right]$
$12+14x=5$
$14x=-7$
$x=-\frac{1}{2}$

30. $\frac{3}{h+3}-\frac{5}{h^2+3h}=\frac{1}{h}$

$\frac{3}{h+3}-\frac{5}{h(h+3)}=\frac{1}{h}$
$h(h+3)\left[\frac{3}{h+3}-\frac{5}{h(h+3)}=\frac{1}{h}\right]$
$3h-5=h+3$
$2h=8$
$h=4$

31. $\frac{n}{n+2}-\frac{3}{n-4}=\frac{n^2+1}{n^2-2n-8}$

$\frac{n}{n+2}-\frac{3}{n-4}=\frac{n^2+1}{(n+2)(n-4)}$
$(n+2)(n-4)\left[\frac{n}{n+2}-\frac{3}{n-4}=\frac{n^2+1}{(n+2)(n-4)}\right]$
$n(n-4)-3(n+2)=n^2+1$
$n^2-4n-3n-6=n^2+1$
$-7n-6=1$
$-7n=7$
$n=-1$

32. $\frac{\sqrt{x^2+7}}{2}+3=5$

$\frac{\sqrt{x^2+7}}{2}=2$
$\sqrt{x^2+7}=4$
$x^2+7=16$
$x^2=9$
$x=\pm 3$

33. $3\sqrt{x+4}=x+4$

$\left(3\sqrt{x+4}\right)^2=(x+4)^2$
$9(x+4)=x^2+8x+16$
$9x+36=x^2+8x+16$
$0=x^2-x-20$
$0=(x-5)(x+4)$
$x-5=0$ or $x+4=0$
$x=5$ or $x=-4$

34. $\sqrt{3x+4}=2-\sqrt{x+2}$

$\left(\sqrt{3x+4}\right)^2=\left(2-\sqrt{x+2}\right)^2$
$3x+4=4-4\sqrt{x+2}+x+2$
$2x-2=-4\sqrt{x+2}$
$2(x-1)=-4\sqrt{x+2}$
$x-1=-2\sqrt{x+2}$

$(x-1)^2 = \left(-2\sqrt{x+2}\right)^2$
$x^2 - 2x + 1 = 4(x+2)$
$x^2 - 2x + 1 = 4x + 8$
$x^2 - 6x - 7 = 0$
$(x-7)(x+1) = 0$
$x - 7 = 0$ or $x + 1 = 0$
$x = 7$ or $x = -1$
$x = -1$, $x = 7$ is extraneous.

35. Let x represent the first integer.
Let $x+1$ represent the second integer.
$(x+1)^2 = 7x + 1$
$x^2 + 2x + 1 = 7x + 1$
$x^2 - 5x = 0$
$x(x-5) = 0$
$x = 0$ or $x - 5 = 0$
$x = 0$ or $x = 5$
0 and 1, or 5 and 6

36. Let w represent the width.
Let $w + 3$ represent the length.
$w(w+3) = 54$
$w^2 + 3w - 54 = 0$
$(w+9)(w-6) = 0$
$w + 9 = 0$ or $w - 6 = 0$
$w = -9$ or $w = 6$
6 inches by 9 inches
width, 6 in; length, 9 in

37. $116 = -16t^2 + 128t + 4$
$16t^2 - 128t + 112 = 0$
$t^2 - 8t + 7 = 0$
$(t-7)(t-1) = 0$
$t - 7 = 0$ or $t - 1 = 0$
$t = 7$ or $t = 1$
1 sec ond;
$h = -16(5)^2 + 128(5) + 4$
$h = 244$ feet;
$4 = -16t^2 + 128t + 4$
$16t^2 - 128t = 0$
$16t(t-8) = 0$
$16t = 0$ or $t - 8 = 0$
$t = 0$ or $t = 8$
8 seconds

38. Let x represent the number of \$2 decreases.
$(50-2x)(40+5x) = 2520$
$2000 + 250x - 80x - 10x^2 = 2520$
$0 = 10x^2 - 170x + 520$
$0 = x^2 - 17x + 52$
$0 = (x-4)(x-13)$
$x - 4 = 0$ or $x - 13 = 0$
$x = 4$ or $x = 13$
50 – 2(4)=\$42 or 50–2(13)=\$24

39. $\sqrt{-72} = \sqrt{-1(36)(2)} = 6\sqrt{2}\,i$

40. $6\sqrt{-48} = 6\sqrt{-1(16)(3)} = 24\sqrt{3}\,i$

41. $\dfrac{-10+\sqrt{-50}}{5} = \dfrac{-10+\sqrt{-1(25)(2)}}{5}$
$= \dfrac{-10+5\sqrt{2}\,i}{5} = -2 + \sqrt{2}\,i$

42. $\sqrt{3}\cdot\sqrt{-6} = \sqrt{3}\cdot\sqrt{6}\,i = \sqrt{18}\,i = 3\sqrt{2}\,i$

43. $i^{57} = \left(i^4\right)^{14} i = i$

44. $(5+2i)^2 = 25 + 20i + 4i^2 = 21 + 20i$

45. $\dfrac{5i}{1-2i}\cdot\dfrac{1+2i}{1+2i} = \dfrac{5i+10i^2}{1+2i-2i-4i^2}$
$= \dfrac{-10+5i}{5} = -2 + i$

46. $(-3+5i)-(2-2i) = -3 + 5i - 2 + 2i = -5 + 7i$

47. $(2+3i)(2-3i) = 4 - 6i + 6i - 9i^2 = 13$

48. $4i(-3+5i) = -12i + 20i^2 = -20 - 12i$

49. $x^2 - 9 = -34, x = 5i$;
$(5i)^2 - 9 = -34$
$25i^2 - 9 = -34$
$-25 - 9 = -34$
$-34 = -34$ verified;
$(-5i)^2 - 9 = -34$
$25i^2 - 9 = -34$
$-25 - 9 = -34$
$-34 = -34$ verified

50. $x^2 - 4x + 9 = 0, x = 2 + i\sqrt{5}$
$\left(2+i\sqrt{5}\right)^2 - 4\left(2+i\sqrt{5}\right) + 9 = 0$

$4+4i\sqrt{5}+5i^2-8-4i\sqrt{5}+9=0$
$0=0$ verified;
$\left(2-i\sqrt{5}\right)^2-4\left(2-i\sqrt{5}\right)+9=0$
$4-4i\sqrt{5}+5i^2-8+4i\sqrt{5}+9=0$
$0=0$ verified

51. a. $-3=2x^2$
$2x^2+3=0$
$a=2, b=0, c=3$

b. $7=-2x+11$ is not quadratic

c. $99=x^2-8x$
$x^2-8x-99=0$
$a=1, b=-8, c=-99$

d. $20=4-x^2$
$x^2+16=0$
$a=1, b=0, c=16$

52. a. $x^2-9=0$
$x^2=9$
$x=\pm 3$

b. $2(x-2)^2+1=11$
$2(x-2)^2=10$
$(x-2)^2=5$
$x-2=\pm\sqrt{5}$
$x=2\pm\sqrt{5}$

c. $3x^2+15=0$
$3x^2=-15$
$x^2=-5$
$x=\pm\sqrt{5}\,i$

d. $-2x^2+4=-46$
$-2x^2=-50$
$x^2=25$
$x=\pm 5$

53. a. $x^2+2x=15$
$x^2+2x+1=15+1$
$(x+1)^2=16$
$x+1=\pm\sqrt{16}$
$x=-1\pm 4$
$x=3$ or $x=-5$

b. $x^2+6x=16$
$x^2+6x+9=16+9$
$(x+3)^2=25$
$x+3=\pm\sqrt{25}$
$x=-3\pm 5$
$x=2$ or $x=-8$

c. $-4x+2x^2=3$
$2x^2-4x=3$
$x^2-2x=\frac{3}{2}$
$x^2-2x+1=\frac{3}{2}+1$
$(x-1)^2=\frac{5}{2}$
$x-1=\pm\sqrt{\frac{5}{2}}\,i$
$x=1\pm\frac{\sqrt{10}}{2}i$
$x\approx 2.58$ or $x\approx -0.58$

d. $3x^2-7x=-2$
$x^2-\frac{7}{3}x+\frac{49}{36}=-\frac{2}{3}+\frac{49}{36}$
$\left(x-\frac{7}{6}\right)^2=\frac{25}{36}$
$x-\frac{7}{6}=\pm\frac{5}{6}$
$x=\frac{7}{6}\pm\frac{5}{6}$
$x=2$ or $x=\frac{1}{3}$

54. a. $x^2-4x=-9$
$x^2-4x+9=0$
$a=1, b=-4, c=9$
$x=\frac{-(-4)\pm\sqrt{(-4)^2-4(1)(9)}}{2(1)}$
$x=\frac{4\pm\sqrt{-20}}{2}$
$x=\frac{4\pm 2\sqrt{5}\,i}{2}$
$x=2\pm\sqrt{5}\,i$
$x\approx 2\pm 2.24i$

b. $4x^2+7=12x$
$4x^2-12x+7=0$
$a=4, b=-12, c=7$
$x=\frac{-(-12)\pm\sqrt{(-12)^2-4(4)(7)}}{2(4)}$
$x=\frac{12\pm\sqrt{32}}{8}$

$x = \dfrac{12 \pm 4\sqrt{2}}{8}$

$x = \dfrac{3 \pm \sqrt{2}}{2}$

$x \approx 2.21$ or $x \approx 0.79$

c. $2x^2 - 6x + 5 = 0$

$a = 2, b = -6, c = 5$

$x = \dfrac{-(-6) \pm \sqrt{(-6)^2 - 4(2)(5)}}{2(2)}$

$x = \dfrac{6 \pm \sqrt{-4}}{4}$

$x = \dfrac{6 \pm 2i}{4}$

$x = \dfrac{3}{2} \pm \dfrac{1}{2}i$

55. a. $100 = -16t^2 + 96t$

$16t^2 - 96t + 100 = 0$

$4t^2 - 24t + 25 = 0$

$a = 4, b = -24, c = 25$

$t = \dfrac{-(-24) \pm \sqrt{(-24)^2 - 4(4)(25)}}{2(4)}$

$t = \dfrac{24 \pm \sqrt{176}}{8}$

$t = \dfrac{24 \pm 4\sqrt{11}}{8}$

$t = \dfrac{6 \pm \sqrt{11}}{2}$

$t \approx 4.66$ or $t \approx 1.34$

1.3 seconds

b. 4.66 seconds

c. $0 = -16t^2 + 96t$

$0 = -16t(t - 6)$

$0 = -16t$ or $t - 6 = 0$

$0 = t$ or $t = 6$

6 seconds

56. a. $120 = -16t^2 + 64t + 80$

$16t^2 - 64t + 40 = 0$

$2t^2 - 8t + 5 = 0$

$a = 2, b = -8, c = 5$

$t = \dfrac{-(-8) \pm \sqrt{(-8)^2 - 4(2)(5)}}{2(2)}$

$t = \dfrac{8 \pm \sqrt{24}}{4}$

$t = \dfrac{8 \pm 2\sqrt{6}}{4}$

$t = \dfrac{4 \pm \sqrt{6}}{2}$

$t \approx 3.22$ or $t \approx 0.78$

0.8 seconds

b. 3.2 seconds

c. $0 = -16t^2 + 64t + 80$

$0 = -16(t^2 - 4t - 5)$

$0 = (t - 5)(t + 1)$

$t - 5 = 0$ or $t + 1 = 0$

$t = 5$ or $t = -1$

5 seconds

57. Let x represent the number of \$0.25 increases.

a. $(2.50 + 0.25x)(4000 - 200x) = 11250$

$10000 - 500x + 1000x - 50x^2 = 11250$

$0 = 50x^2 - 500x + 1250$

$0 = x^2 - 10x + 25$

$0 = (x - 5)(x - 5)$

$x - 5 = 0$

$x = 5$

$2.50 + 0.25(5) = \$3.75$

b. $4000 - 200(5) = 3000$ people

58. Let x represent the time for the smaller pump.

$\dfrac{1}{x-3} + \dfrac{1}{x} = \dfrac{1}{2}$

$2x(x-3)\left[\dfrac{1}{x-3} + \dfrac{1}{x} = \dfrac{1}{2}\right]$

$2x + 2(x - 3) = x(x - 3)$

$2x + 2x - 6 = x^2 - 3x$

$0 = x^2 - 7x + 6$

$0 = (x - 6)(x - 1)$

$x - 6 = 0$ or $x - 1 = 0$

$x = 6$ or $x = 1$

It takes the smaller pump 6 hours.

Chapter 1 Mixed Review

1. a. $\dfrac{10}{\sqrt{x-8}}$

$x - 8 > 0$

$x > 8$

$x \in (8, \infty)$

b. $\dfrac{-5}{3x+4}$

$$3x+4\neq 0$$
$$3x\neq -4$$
$$x\neq -\frac{4}{3}$$
$$x\in\left(-\infty,\frac{-4}{3}\right)\cup\left(\frac{-4}{3},\infty\right)$$

3. a. x^3+10x^2+16x
$=x(x^2+10x+16)$
$=x(x+8)(x+2)$

b. $-2m^2+12m+54$
$=-2(m^2-6m-27)$
$=-2(m-9)(m+3)$

c. $18z^2-50$
$=2(9z^2-25)$
$=2(3z+5)(3z-5)$

d. $v^3+2v^2-9v-18$
$=v^2(v+2)-9(v+2)$
$=(v+2)(v^2-9)$
$=(v+2)(v+3)(v-3)$

5. $3x+4y=-12$
$4y=-3x-12$
$y=-\frac{3}{4}x-3$

7. a. $5x-(2x-3)+3x=-4(5+x)+3$
$5x-2x+3+3x=-20-4x+3$
$3+6x=-17-4x$
$10x=-20$
$x=-2$

b. $\frac{n}{5}-2=2-\frac{5}{3}-\frac{4}{15}n$
$15\left[\frac{n}{5}-2=2-\frac{5}{3}-\frac{4}{15}n\right]$
$3n-30=30-25-4n$
$3n-30=5-4n$
$7n=35$
$n=5$

9. $x^2-18x+77=0$
$(x-7)(x-11)=0$
$x-7=0$ or $x-11=0$
$x=7$ or $x=11$

11. $4x^2-5=19$
$4x^2=24$
$x^2=6$
$x=\pm\sqrt{6}$

13. $25x^2+16=40x$
$25x^2-40x+16=0$
$(5x-4)(5x-4)=0$
$5x-4=0$
$5x=4$
$x=\frac{4}{5}$

15. $2x^4-50=0$
$2\left(x^4-25\right)=0$
$2\left(x^2+5\right)\left(x^2-5\right)=0$
$x^2+5=0$ or $x^2-5=0$
$x^2=-5$ or $x^2=5$
$x=\pm\sqrt{5}\,i$ or $x=\pm\sqrt{5}$

17. a. $\sqrt{2v-3}+3=v$
$\sqrt{2v-3}=v-3$
$2v-3=v^2-6v+9$
$0=v^2-8v+12$
$0=(v-6)(v-2)$
$v-6=0$ or $v-2=0$
$v=6$ or $v=2$
$v=6$, $v=2$ is extraneous

b. $\sqrt[3]{x^2-9}+\sqrt[3]{x-11}=0$
$\sqrt[3]{x^2-9}=-\sqrt[3]{x-11}$
$x^2-9=-(x-11)$
$x^2+x-20=0$
$(x-4)(x+5)=0$
$x-4=0$ or $x+5=0$
$x=4$ or $x=-5$

19. $\frac{2(75)+2(79)+x}{5}=78$
$2(75)+2(79)+x=390$
$150+158+x=390$
$x=82$ inches
6′10″

Chapter 1 Practice Test

1. $-\frac{2}{3}x-5=7-(x+3)$
$-\frac{2}{3}x-5=7-x-3$
$-\frac{2}{3}x-5=4-x$

$\frac{1}{3}x = 9$
$x = 27$

2. $P = C + kC$
$P = C(1+k)$
$\frac{P}{1+k} = C$

3. $-5.7 + 3.1x = 14.5 - 4(x+1.5)$
$-5.7 + 3.1x = 14.5 - 4x - 6$
$-5.7 + 3.1x = 8.5 - 4x$
$7.1x = 14.2$
$x = 2$

4. $102x + 91(25) = 97(25 + x)$
$102x + 2275 = 2425 + 97x$
$5x = 150$
$x = 30$ gallons

5. $-\frac{2}{5}x + 7 > 19$
$-\frac{2}{5}x > 12$
$\left(-\frac{5}{2}\right)\left(-\frac{2}{5}x\right) < \left(-\frac{5}{2}\right)(12)$
$x < -30$

6. $-1 < 3 - x \le 8$
$-4 < -x \le 5$
$4 > x \ge -5$
$-5 \le x < 4$

7. $\frac{1}{2}x + 3 < 9$ or $\frac{2}{3}x - 1 \ge 3$
$\frac{1}{2}x < 6$ or $\frac{2}{3}x \ge 4$
$x < 12$ or $x \ge 6$
$x \in R$

8. $\frac{141 + 162 + x}{3} = 160$
$141 + 162 + x = 480$
$303 + x = 480$
$x = 177$
Jaques needs at least a 177.

9. $z^2 - 7z - 30 = 0$
$(z-10)(z+3) = 0$
$z - 10 = 0$ or $z + 3 = 0$
$z = 10$ or $z = -3$

10. $4x^2 - 25 = 0$
$(2x+5)(2x-5) = 0$
$2x + 5 = 0$ or $2x - 5 = 0$
$x = -\frac{5}{2}$ or $x = \frac{5}{2}$

11. $3x^2 - 20x = -12$
$3x^2 - 20x + 12 = 0$
$(3x-2)(x-6) = 0$
$3x - 2 = 0$ or $x - 6 = 0$
$3x = 2$ or $x = 6$
$x = \frac{2}{3}$ or $x = 6$

12. $4x^3 + 8x^2 - 9x - 18 = 0$
$4x^2(x+2) - 9(x+2) = 0$
$(x+2)\left(4x^2 - 9\right) = 0$
$(x+2)(2x+3)(2x-3) = 0$
$x + 2 = 0$ or $2x + 3 = 0$ or $2x - 3 = 0$
$x = -2$ or $2x = -3$ or $2x = 3$
$x = -2$ or $x = \frac{-3}{2}$ or $x = \frac{3}{2}$

13. $P = (120 - 2x)(3 + 0.10x)$
$P = 360 + 12x - 6x - 0.2x^2$
$P = -0.2x^2 + 6x + 360$

a. $405 = -0.2x^2 + 6x + 360$
$0.2x^2 - 6x + 45 = 0$
$2x^2 - 60x + 450 = 0$
$x^2 - 30x + 225 = 0$
$(x-15)^2 = 0$
$x = 15$
$3 + 0.10(15) = \$4.50$ per tin

b. $120 - 2(15) = 90$ tins;

14. $\frac{-8 + \sqrt{-20}}{6} = \frac{-8 + \sqrt{-4(5)}}{6} = \frac{-8 + 2\sqrt{5}\,i}{6}$
$= -\frac{4}{3} + \frac{\sqrt{5}}{3}i$

15. $i^{39} = \left(i^4\right)^9 i^3 = -i$

16. a. $\left(\frac{1}{2} + \frac{\sqrt{3}}{2}i\right) + \left(\frac{1}{2} - \frac{\sqrt{3}}{2}i\right) = 1$

b. $\left(\frac{1}{2} + \frac{\sqrt{3}}{2}i\right) - \left(\frac{1}{2} - \frac{\sqrt{3}}{2}i\right)$

$= \frac{1}{2} + \frac{\sqrt{3}}{2} i - \frac{1}{2} + \frac{\sqrt{3}}{2} i = \sqrt{3}\, i$

c. $\left(\frac{1}{2} + \frac{\sqrt{3}}{2} i\right) \cdot \left(\frac{1}{2} - \frac{\sqrt{3}}{2} i\right)$

$= \frac{1}{4} - \frac{\sqrt{3}}{4} i + \frac{\sqrt{3}}{4} i - \frac{3}{4} i^2 = 1$

17. $\frac{3i}{2-i} \cdot \frac{2+i}{2+i} = \frac{6i + 3i^2}{4 + 2i - 2i - i^2}$

$= \frac{-3+6i}{5} = -\frac{3}{5} + \frac{6}{5} i$

18. $(3i+5)(5-3i) = 15i - 9i^2 + 25 - 15i = 34$

19. $(x-1)^2 + 3 = 0$

$(x-1)^2 = -3$
$x - 1 = \pm\sqrt{3}\, i$
$x = 1 \pm \sqrt{3}\, i$

20. $x^2 - 4x + 13 = 0, x = 2 - 3i$

$(2-3i)^2 - 4(2-3i) + 13 = 0$

$4 - 12i + 9i^2 - 8 + 12i + 13 = 0$
$0 = 0$
verified

21. $2x^2 - 20x + 49 = 0$

$2x^2 - 20x = -49$

$x^2 - 10x = -\frac{49}{2}$

$x^2 - 10x + 25 = -\frac{49}{2} + 25$

$(x-5)^2 = \frac{1}{2}$

$x - 5 = \pm\sqrt{\frac{1}{2}}$

$x - 5 = \pm\sqrt{\frac{1}{2}}\sqrt{\frac{2}{2}}$

$x = 5 \pm \frac{\sqrt{2}}{2}$

22. $2x^2 - 5x + 4 = 0$

$2x^2 - 5x = -4$

$x^2 - \frac{5}{2}x = -2$

$x^2 - \frac{5}{2}x + \frac{25}{16} = -2 + \frac{25}{16}$

$\left(x - \frac{5}{4}\right)^2 = \frac{-7}{16}$

$x - \frac{5}{4} = \pm\frac{\sqrt{7}}{4} i$

$x = \frac{5}{4} \pm \frac{\sqrt{7}}{4} i$

23. $3x^2 + 2 = 6x$

$3x^2 - 6x + 2 = 0$
$a = 3, b = -6, c = 2$

$x = \frac{-(-6) \pm \sqrt{(-6)^2 - 4(3)(2)}}{2(3)}$

$x = \frac{6 \pm \sqrt{12}}{6}$

$x = \frac{6 \pm 2\sqrt{3}}{6}$

$x = \frac{3 \pm \sqrt{3}}{3} = 1 \pm \frac{\sqrt{3}}{3}$

24. $x^2 - 2x + 10 = 0$
$a = 1, b = -2, c = 10$

$x = \frac{-(-2) \pm \sqrt{(-2)^2 - 4(1)(10)}}{2(1)}$

$x = \frac{2 \pm \sqrt{-36}}{2}$

$x = \frac{2 \pm 6i}{2}$

$x = 1 \pm 3i$

25. $r = -3t^2 + 42t - 135$

a. $0 = -3t^2 + 42t - 135$
$0 = -3(t^2 - 14t + 45)$
$0 = -3(t-5)(t-9)$
$t - 5 = 0$ or $t - 9 = 0$
$t = 5$ or $t = 9$
$t = 5$ (May)

b. $t = 9$ (Sept)

c. $r = -3(7)^2 + 42(7) - 135 = 12;$

$r = -3(8)^2 + 42(8) - 135 = 9;$

July, \$3000 more.

Chapter 1 Calculator Exploration and Discovery

1. They differ by 5.

2. They differ by 0.2.

3. They differ by $\sqrt{2}$ or ≈ 1.41.

Strengthening Core Skills

1. $2x^2-5x-7=0$
$a=2, b=-5, c=-7$
$x_1=\frac{7}{2}, x_2=-1$
$\frac{7}{2}+(-1)=\frac{5}{2}=-\frac{b}{a};$
$\frac{7}{2}\cdot(-1)=\frac{-7}{2}=\frac{c}{a}$

2. $2x^2-4x-7=0$
$a=2, b=-4, c=-7$
$x_1=\frac{2+3\sqrt{2}}{2}, x_2=\frac{2-3\sqrt{2}}{2}$
$\frac{2+3\sqrt{2}}{2}+\frac{2-3\sqrt{2}}{2}=\frac{4}{2}=\frac{-b}{a};$
$\frac{2+3\sqrt{2}}{2}\cdot\frac{2-3\sqrt{2}}{2}=\frac{-14}{4}=\frac{-7}{2}=\frac{c}{a}$

3. $x^2-10x+37=0$
$a=1, b=-10, c=37$
$x_1=5+2\sqrt{3}\,i, x_2=5-2\sqrt{3}\,i$
$(5+2\sqrt{3}\,i)+(5-2\sqrt{3}\,i)=10=\frac{-b}{a};$
$(5+2\sqrt{3}\,i)(5-2\sqrt{3}\,i)=25+12=37=\frac{c}{a}$

4. $x_1=\left(\frac{-b}{2a}+\frac{\sqrt{b^2-4ac}}{2a}\right),$

$$x_2=\left(\frac{-b}{2a}-\frac{\sqrt{b^2-4ac}}{2a}\right)$$

$$x_1x_2=\left(\frac{-b}{2a}+\frac{\sqrt{b^2-4ac}}{2a}\right)\left(\frac{-b}{2a}-\frac{\sqrt{b^2-4ac}}{2a}\right)$$

$$=\frac{b^2}{4a^2}-\frac{b^2-4ac}{4a^2}$$

$$=\frac{b^2-b^2+4ac}{4a^2}=\frac{4ac}{4a^2}=\frac{c}{a} \text{ verified;}$$

$$x_1+x_2=\left(\frac{-b}{2a}+\frac{\sqrt{b^2-4ac}}{2a}\right)+\left(\frac{-b}{2a}-\frac{\sqrt{b^2-4ac}}{2a}\right)$$

$$=\frac{-2b}{2a}=\frac{-b}{a} \text{ verified}$$

2.1 Technology Highlight

Exercise 1: $Y_1 = \frac{2}{3}x + 1$; $(-1.5, 0)$, $(0, 1)$

Exercise 2: $79x - 55y = 869$

$-55y = -79x + 869$

$y = \frac{-79}{55}x - \frac{79}{5}$

x-intercept: (11, 0)

y-intercept: (0, -15.8)

2.1 Exercises

1. Lattice

3. $y = 0; x = 0$

5. $m = \frac{15}{2}$

   Answers will vary.

7. $2x + 3y = 6$

   $3y = -2x + 6$

   $y = -\frac{2}{3}x + 2$

x	y
-6	$-\frac{2}{3}(-6) + 2 = 4 + 2 = 6$
-3	$-\frac{2}{3}(-3) + 2 = 2 + 2 = 4$
0	$-\frac{2}{3}(0) + 2 = 0 + 2 = 2$
3	$-\frac{2}{3}(3) + 2 = -2 + 2 = 0$

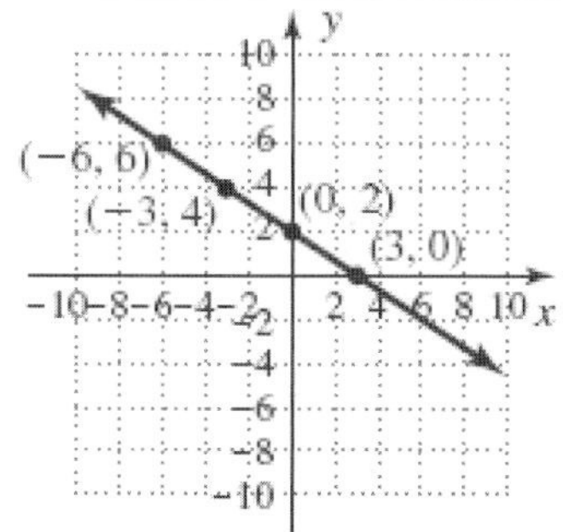

9. $y = \frac{3}{2}x + 4$

x	y
-2	$\frac{3}{2}(-2) + 4 = -3 + 4 = 1$
0	$\frac{3}{2}(0) + 4 = 0 + 4 = 4$
2	$\frac{3}{2}(2) + 4 = 3 + 4 = 7$
4	$\frac{3}{2}(4) + 4 = 6 + 4 = 10$

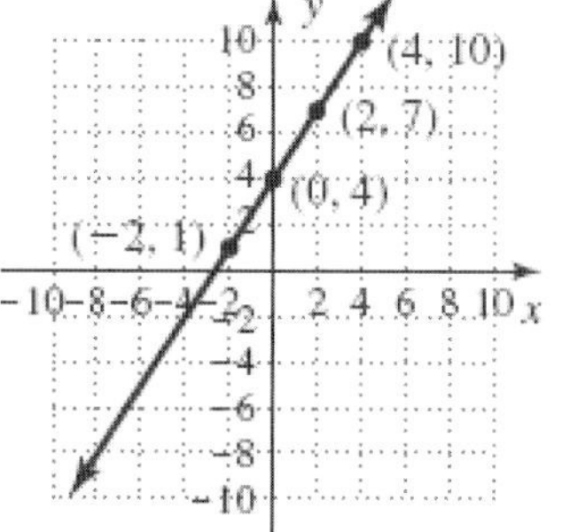

11. $y = \frac{3}{2}x + 4$

    $-0.5 = \frac{3}{2}(-3) + 4$

    $-0.5 = -\frac{9}{2} + 4$

    $-0.5 = -0.5$;

    $\frac{19}{4} = \frac{3}{2}\left(\frac{1}{2}\right) + 4$

    $\frac{19}{4} = \frac{3}{4} + 4$

    $\frac{19}{4} = \frac{19}{4}$

13. $3x + y = 6$

    x-intercept: (2, 0)

    $3x + 0 = 6$

    $3x = 6$

    $x = 2$

    y-intercept: (0, 6)

    $3(0) + y = 6$

    $y = 6$

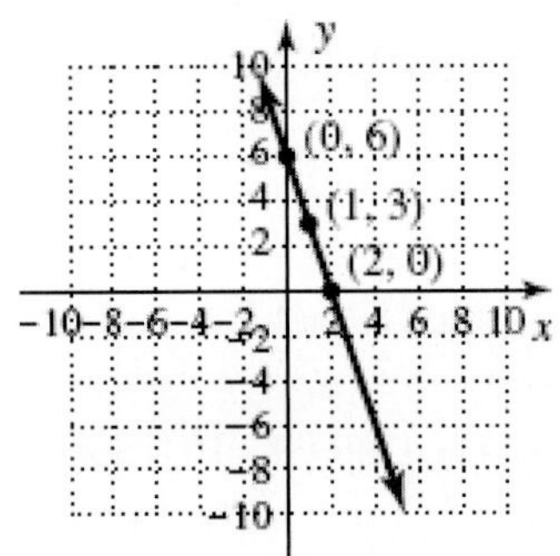

15. $5y - x = 5$

x-intercept: (-5, 0)

$$5(0) - x = 5$$
$$-x = 5$$
$$x = -5$$

y-intercept: (0, 1)

$$5y - 0 = 5$$
$$5y = 5$$
$$y = 1$$

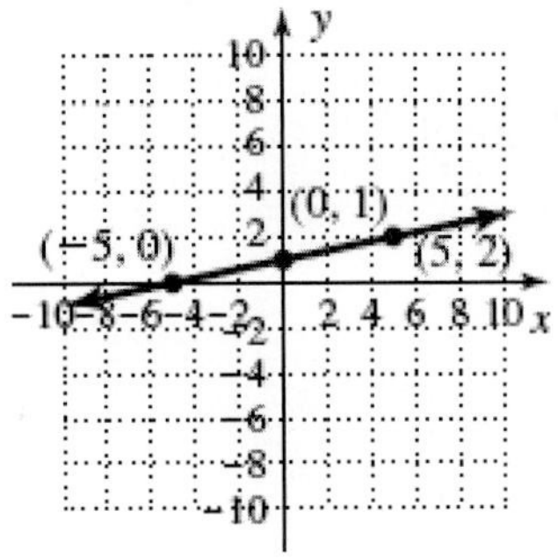

17. $-5x + 2y = 6$

x-intercept: $\left(-\frac{6}{5}, 0\right)$

$$-5x + 2(0) = 6$$
$$-5x = 6$$
$$x = -\frac{6}{5}$$

y-intercept: (0, 3)

$$-5(0) + 2y = 6$$
$$2y = 6$$
$$y = 3$$

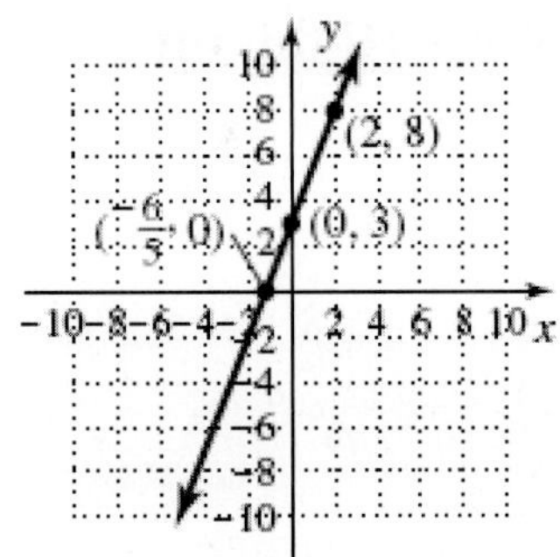

19. $2x - 5y = 4$

x-intercept: (2, 0)

$$2x - 5(0) = 4$$
$$2x = 4$$
$$x = 2$$

y-intercept: $\left(0, -\frac{4}{5}\right)$

$$2(0) - 5y = 4$$
$$-5y = 4$$
$$y = -\frac{4}{5}$$

21. $2x + 3y = -12$

x-intercept: (-6, 0)

$$2x + 3(0) = -12$$
$$2x = -12$$
$$x = -6$$

y-intercept: (0, -4)

$$2(0) + 3y = -12$$
$$3y = -12$$
$$y = -4$$

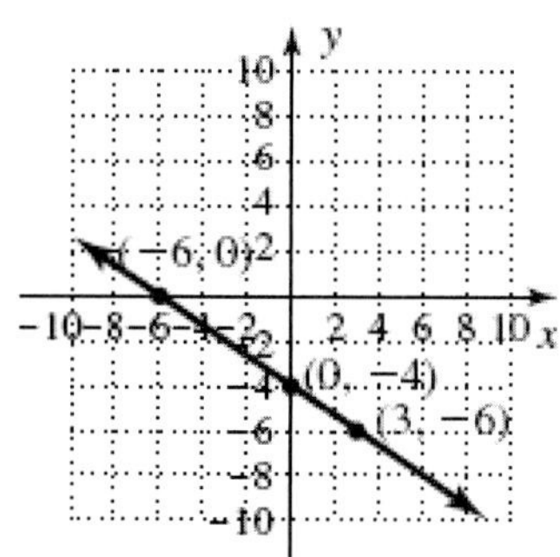

23. $y = -\frac{1}{2}x$

$y = -\frac{1}{2}(2)$

$y = -1$

(2, -1);

$y = -\frac{1}{2}x$

$y = -\frac{1}{2}(4)$

$y = -2$

(4, -2);

$y = -\frac{1}{2}x$

$y = -\frac{1}{2}(0)$

$y = 0$

(0, 0)

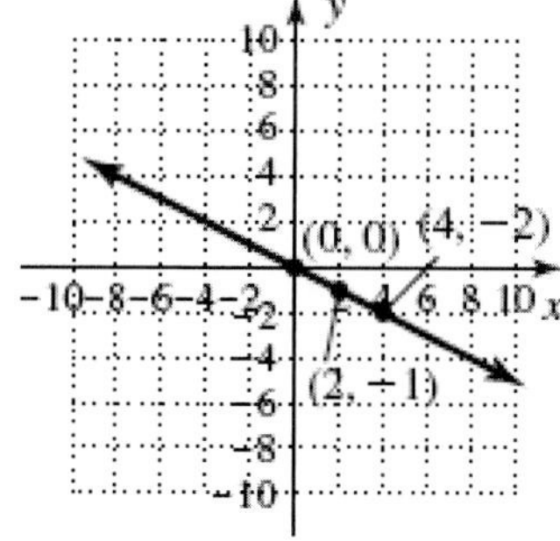

25. $y - 25 = 50x$

$y - 25 = 50(-1)$

$y - 25 = -50$

$y = -25$

(-1, -25);

$y - 25 = 50x$

$y - 25 = 50(1)$

$y - 25 = 50$

$y = 75$

(1, 75)

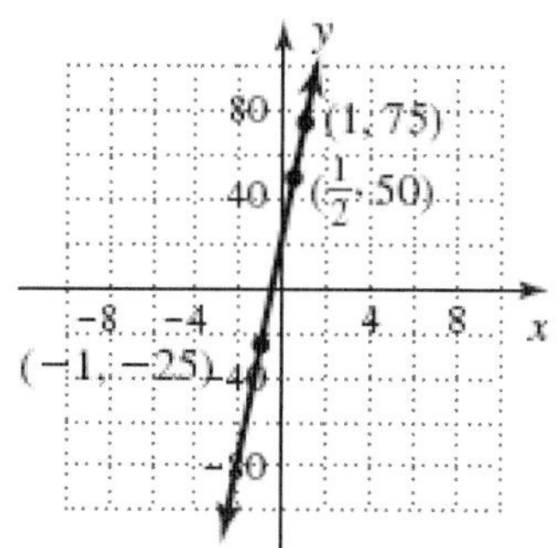

27. $y = -\frac{2}{5}x - 2$

x-intercept: (-5, 0)

$0 = -\frac{2}{5}x - 2$

$2 = -\frac{2}{5}x$

$\left(-\frac{5}{2}\right)(2) = \left(-\frac{5}{2}\right)\left(-\frac{2}{5}x\right)$

$-5 = x$

(-5, 0);

y-intercept: (0, -3)

$y = -\frac{2}{5}(0) - 2$

$y = -2$

(0, -2)

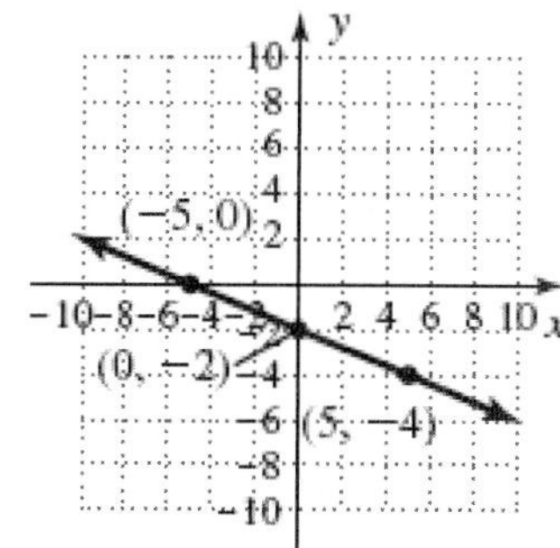

29. $2y - 3x = 0$

$2y - 3(2) = 0$

$2y - 6 = 0$

$2y = 6$

$y = 3$

(2, 3);

$2y - 3x = 0$

$2y - 3(4) = 0$

$2y - 12 = 0$

$2y = 12$

$y = 6$

(4, 6);

$$2y - 3x = 0$$
$$2y - 3(0) = 0$$
$$2y = 0$$
$$y = 0$$

(0, 0)

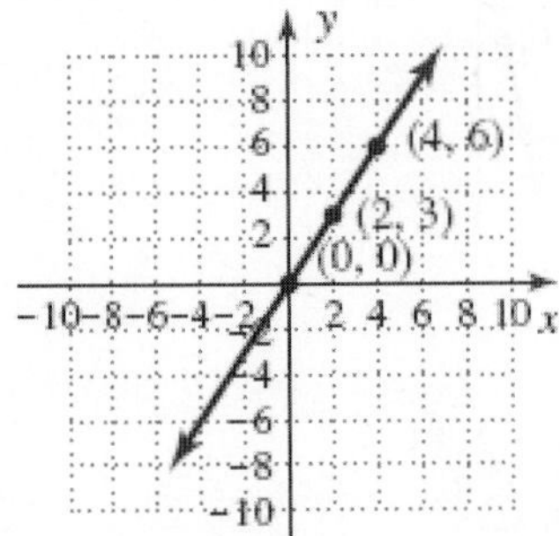

31. $3y + 4x = 12$
x-intercept: (3, 0)
$$3(0) + 4x = 12$$
$$4x = 12$$
$$x = 3$$
y-intercept: (0, 4)
$$3y + 4(0) = 12$$
$$3y = 12$$
$$y = 4$$

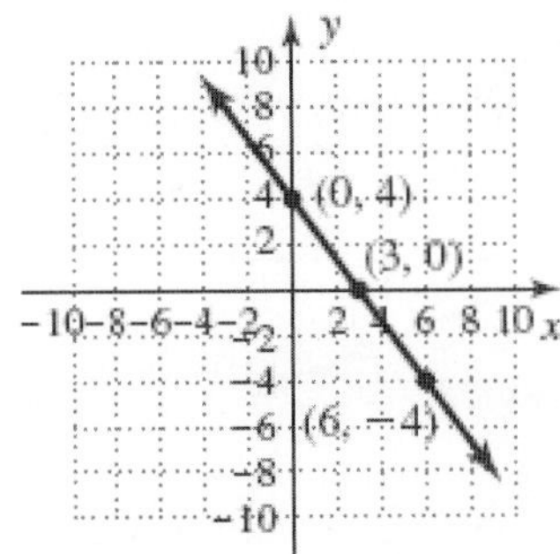

33. $x = -3$
$$x + 0y = -3$$
$$x + 0(4) = -3$$
$$x = -3$$
(-3, 4);
$$x + 0y = -3$$
$$x + 0(-4) = -3$$
$$x = -3$$
(-3, -4)

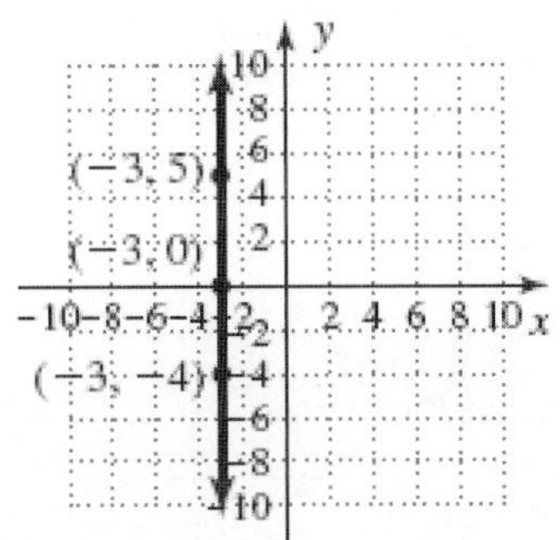

35. $x = 2$
$$x + 0y = 2$$
$$x + 0(2) = 2$$
$$x = 2$$
(2, 0)
$$x + 0y = 2$$
$$x + 0(-2) = 2$$
$$x = 2$$
(2, 0)

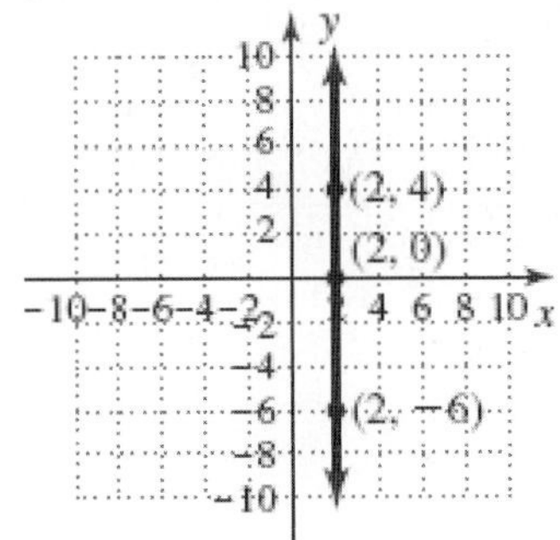

37. $L_1 : x = 2$
$L_2 : y = 4$
Point of intersection: (2, 4)

39. a. $m = \dfrac{500 - 250}{4 - 2} = \dfrac{250}{2} = 125$
Cost increased \$125,000 per 1000 square feet.
b. \$375,000

41. a. $m = \dfrac{270 - 90}{12 - 4} = \dfrac{180}{8} = 22.5$
Distance increases 22.5 miles per hour.
b. 185 miles

43. a. $m = \dfrac{165 - 142}{70 - 64} = \dfrac{23}{6}$
A person weighs 23 pounds more for each additional 6 inches in height.
b. $\dfrac{23}{6} \approx 3.8$ pounds

45. $m=\frac{6-5}{4-3}=\frac{1}{1}=1$

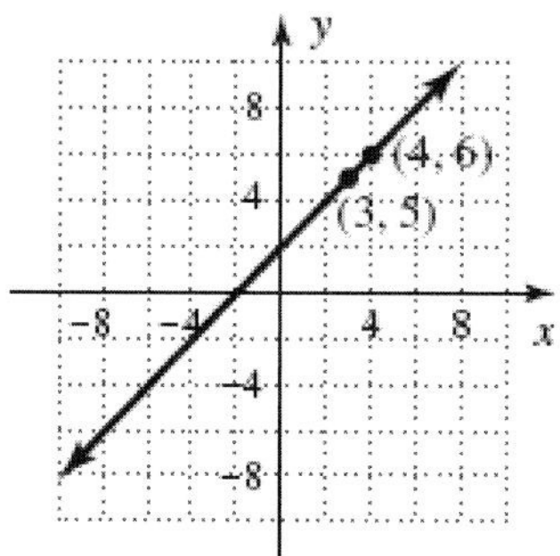

(2,4), (1,3)

47. $m=\frac{3-(-5)}{10-4}=\frac{8}{6}=\frac{4}{3}$

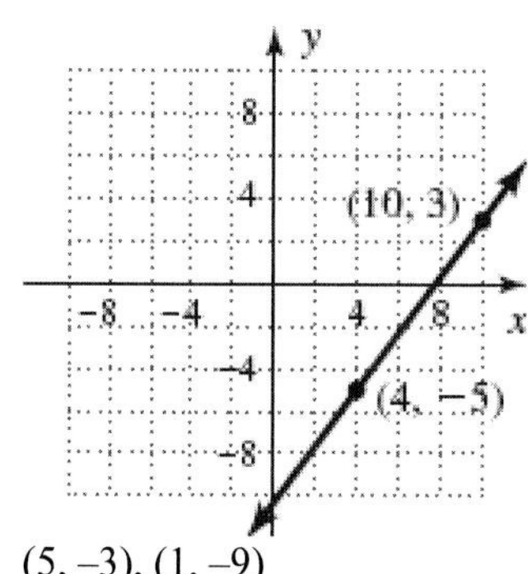

(5, –3), (1, –9)

49. $m=\frac{-8-7}{1-3}=\frac{-15}{-2}=\frac{15}{2}$

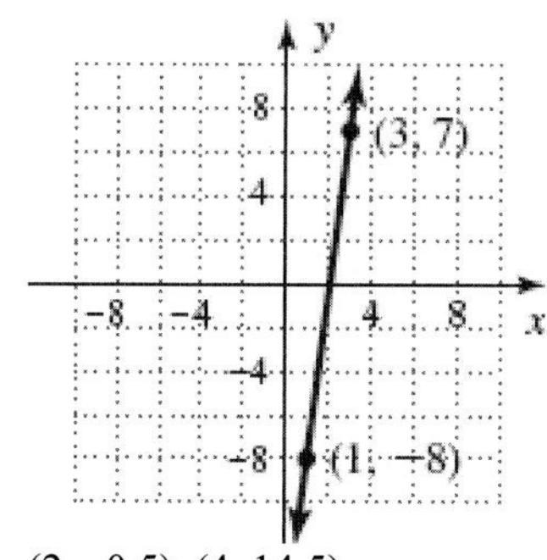

(2, –0.5), (4, 14.5)

51. $m=\frac{-3-6}{6-6}$ Undefined

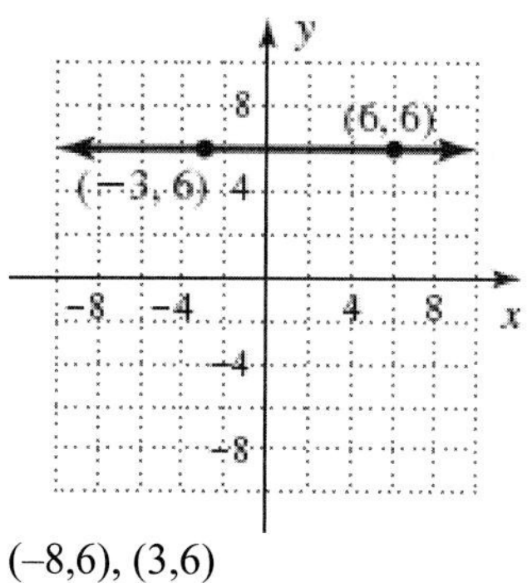

(–8,6), (3,6)

53. L_1: $m=\frac{6-0}{0-(-2)}=\frac{6}{2}=3$

L_2 : $m=\frac{5-8}{0-1}=\frac{-3}{-1}=3$

Parallel

55. L_1: $m=\frac{-4-1}{-3-0}=\frac{-5}{-3}=\frac{5}{3}$

L_2 : $m=\frac{4-0}{-4-0}=\frac{4}{-4}=-1$

Neither

57. L_1: $m=\frac{7-3}{8-6}=\frac{4}{2}=2$

L_2 : $m=\frac{2-0}{7-6}=\frac{2}{1}=2$

Parallel

59. (5, 2) (0, -3)

$m=\frac{-3-2}{0-5}=\frac{-5}{-5}=1;$

(0, -3) (4, -4)

$m=\frac{-4-(-3)}{4-0}=\frac{-4+3}{4}=\frac{-1}{4};$

(5, 2) (4, -4)

$m=\frac{-4-2}{4-5}=\frac{-6}{-1}=6$

Not a right triangle. Lines are not perpendicular. Slopes: 1; $\frac{-1}{4}$; 6

61. (-4, 3) (-7, -1)

$m=\frac{-1-3}{-7-(-4)}=\frac{-4}{-7+4}=\frac{-4}{-3}=\frac{4}{3};$

(-7, -1) (3, -2)

$m=\frac{-2-(-1)}{3-(-7)}=\frac{-2+1}{3+7}=\frac{-1}{10};$

(-4, 3) (3, -2)

$m = \frac{-2-3}{3-(-4)} = \frac{-5}{7}$

Not a right triangle. Lines are not perpendicular. Slopes: $\frac{4}{3}$; $\frac{-1}{10}$; $\frac{-5}{7}$

63. (-3, 2) (-1, 5)

$m = \frac{5-2}{-1-(-3)} = \frac{3}{-1+3} = \frac{3}{2}$;

(-3, 2) (-6, 4)

$m = \frac{4-2}{-6-(-3)} = \frac{2}{-6+3} = -\frac{2}{3}$

Right triangle because these two lines are perpendicular. Slopes: $\frac{3}{2}$; $\frac{-2}{3}$

65. $M = \left(\frac{x_1+x_2}{2}, \frac{y_1+y_2}{2}\right)$

$M = \left(\frac{1+5}{2}, \frac{8-6}{2}\right)$

$M = \left(\frac{6}{2}, \frac{2}{2}\right)$

$M = (3,1)$

67. $M = \left(\frac{x_1+x_2}{2}, \frac{y_1+y_2}{2}\right)$

$M = \left(\frac{-4.5+3.1}{2}, \frac{9.2+(-9.8)}{2}\right)$

$M = \left(\frac{-1.4}{2}, \frac{-0.6}{2}\right)$

$M = (-0.7,-0.3)$

69. $M = \left(\frac{x_1+x_2}{2}, \frac{y_1+y_2}{2}\right)$

$M = \left(\frac{\frac{1}{5}+\left(\frac{-1}{10}\right)}{2}, \frac{\frac{-2}{3}+\frac{3}{4}}{2}\right)$

$M = \left(\frac{\frac{1}{10}}{2}, \frac{\frac{1}{12}}{2}\right)$

$M = \left(\frac{1}{20}, \frac{1}{24}\right)$

71. (-5, -4) (5, 2)

$M = \left(\frac{x_1+x_2}{2}, \frac{y_1+y_2}{2}\right)$

$M = \left(\frac{-5+5}{2}, \frac{-4+2}{2}\right)$

$M = \left(\frac{0}{2}, \frac{-2}{2}\right)$

$M = (0,-1)$

73. (-4, -4) (2, 4)

$M = \left(\frac{x_1+x_2}{2}, \frac{y_1+y_2}{2}\right)$

$M = \left(\frac{-4+2}{2}, \frac{-4+4}{2}\right)$

$M = \left(\frac{-2}{2}, \frac{0}{2}\right)$

$M = (-1,0)$

The center of the circle is (-1, 0).

75. (-5, -4) (5, 2)

$d = \sqrt{(x_2-x_1)^2+(y_2-y_1)^2}$

$d = \sqrt{(5-(-5))^2+(2-(-4))^2}$

$d = \sqrt{(10)^2+(6)^2}$

$d = \sqrt{100+36}$

$d = \sqrt{136}$

$d = 2\sqrt{34}$

77. (-4, -4) (2, 4)

$d = \sqrt{(x_2-x_1)^2+(y_2-y_1)^2}$

$d = \sqrt{(4-(-4))^2+(2-(-4))^2}$

$d = \sqrt{8^2+6^2}$

$d = \sqrt{64+36}$

$d = 10$

79. $L = 0.11T + 74.2$

a. $L(20) = 0.11(20) + 74.2 = 76.4$ years

b. $77.5 = 0.11T + 74.2$
$3.3 = 0.11T$
$30 = T$
$1980 + 30 = 2010$

81. $V = 8500 - 1250y$

a. $V = 8500 - 1250(4) = \$3500$

b. $2250 = 8500 - 1250y$
$-6250 = -1250y$
$5 = y$
5 years

83. $y = 144x + 621$

a. $y = 144(12) + 621$
$y = 1728 + 621$
$y = 2349$
$2,349

b. $5000 = 144x + 621$
$4379 = 144x$
$30.4 = x$
$1980 + 31 = 2011$

85. $y = -\frac{7}{15}x + 32$

a. $y = -\frac{7}{15}(20) + 32$
$y = \frac{-28}{3} + 32$
$y = 22\frac{2}{3}$
23%

b. $20 = -\frac{7}{15}x + 32$
$-12 = -\frac{7}{15}x$
$-180 = -7x$
$25.7 = x$
$1980 + 26 = 2006$

87. Answers will vary.

89. e

91. $3x^2 - 3 + 4x + 6 = 4x^2 - 3(x+5)$
$3x^2 + 4x + 3 = 4x^2 - 3x - 15$
$-x^2 + 7x + 18 = 0$
$-(x^2 - 7x - 18) = 0$
$-(x-9)(x+2) = 0$
$x = -2$ or $x = 9$
Check:
$3(-2)^2 - 3 + 4(-2) + 6 = 4(-2)^2 - 3(-2+5)$
$3(4) - 3 - 8 + 6 = 4(4) - 3(3)$
$12 - 3 - 8 + 6 = 16 - 9$
$7 = 7;$

$3(9)^2 - 3 + 4(9) + 6 = 4(9)^2 - 3(9+5)$
$3(81) - 3 + 36 + 6 = 4(81) - 3(14)$
$243 - 3 + 36 + 6 = 324 - 42$
$282 = 282$

93. Let x represent the number of gallons of 35% brine solution.
Let y represent the total number of gallons of 45% brine solution.

$$\begin{cases} x + 12 = y \\ 0.35x + 0.55(12) = 0.45(12 + x) \end{cases}$$

$0.35x + 0.55(12) = 0.45(12 + x)$
$0.35x + 6.6 = 5.4 + 0.45x$
$-0.1x = -1.2$
$x = 12$
12 gallons

95. $\frac{1}{x-5} = \frac{-10}{x^2 - 2x - 15} + 1$
$\frac{1}{x-5} = \frac{-10}{(x-5)(x+3)} + 1$
Excluded values: $x \neq 5,\ x \neq -3$
$\left(\frac{1}{x-5} = \frac{-10}{(x-5)(x+3)} + 1\right)(x-5)(x+3)$
$x + 3 = -10 + (x-5)(x+3)$
$x + 3 = -10 + x^2 - 2x - 15$
$0 = x^2 - 3x - 28$
$0 = (x-7)(x+4)$
$x = 7$ or $x = -4$

2.2 Technology Highlight

Exercise 1:
(-8, -7): QIII
(-3, -4.5): QIII
(4, -1): QIV
(8, 1): QI
Yes

Exercise 2:
(-6, -3): QIII
(-1, -2): QIII
(3, 1): QI
(6, 8): QI
No

Exercise 3:
(-5, 7), (-2, -1), (0, 3) and (2, 7)
The points are all on the graph of Y_1.

2.2 Exercises

1. First

3. Range

5. Answers will vary.

7.

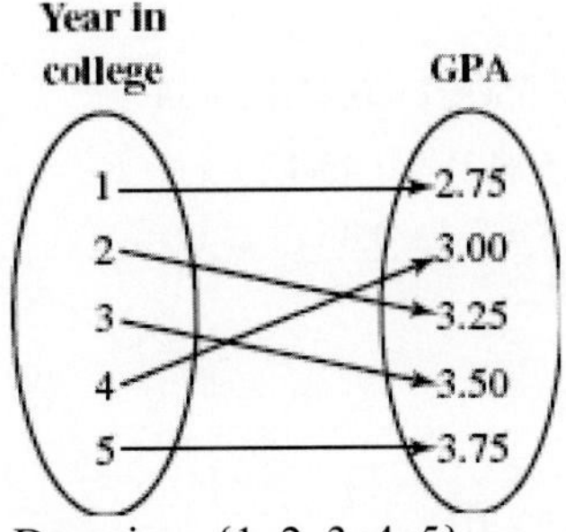

Domain = {1, 2, 3, 4, 5}
Range = {2.75, 3.00, 3.25, 3.50, 3.75}

9. D = {1, 3, 5, 7, 9}
R = {2, 4, 6, 8, 10}

11. D = {4, -1, 2, -3}
R = {0, 5, 4, 2, 3}

13. $y=-\frac{2}{3}x+1$

x	y
-6	$-\frac{2}{3}(-6)+1=4+1=5$
-3	$-\frac{2}{3}(-3)+1=2+1=3$
0	$-\frac{2}{3}(0)+1=0+1=1$
3	$-\frac{2}{3}(3)+1=-2+1=-1$
6	$-\frac{2}{3}(6)+1=-4+1=-3$
8	$-\frac{2}{3}(-8)+1=\frac{16}{3}+1=-\frac{13}{3}$

15. $x+2=|y|$

x	y
-2	0
0	2, -2
1	3, -3
3	5, -5
6	8, -8
7	9, -9

$$-2+2=|y| \quad 0=|y| \quad 0=y;$$
$$0+2=|y| \quad 2=|y| \quad \pm 2=y;$$
$$1+2=|y| \quad 3=|y| \quad \pm 3=y;$$
$$3+2=|y| \quad 5=|y| \quad \pm 5=y;$$
$$6+2=|y| \quad 8=|y| \quad \pm 8=y;$$
$$7+2=|y| \quad 9=|y| \quad \pm 9=y;$$

17. $y=x^2-1$

x	y
-3	$(-3)^2-1=9-1=8$
-2	$(-2)^2-1=4-1=3$
0	$(0)^2-1=0-1=-1$
2	$(2)^2-1=4-1=3$
3	$(3)^2-1=9-1=8$
4	$(4)^2-1=16-1=15$

19. $y=\sqrt{25-x^2}$

x	y
-4	$\sqrt{25-(-4)^2}=\sqrt{25-16}=\sqrt{9}=3$
-3	$\sqrt{25-(-3)^2}=\sqrt{25-9}=\sqrt{16}=4$
0	$\sqrt{25-(0)^2}=\sqrt{25}=5$
2	$\sqrt{25-(2)^2}=\sqrt{25-4}=\sqrt{21}$
3	$\sqrt{25-(3)^2}=\sqrt{25-9}=\sqrt{16}=4$
4	$\sqrt{25-(4)^2}=\sqrt{25-16}=\sqrt{9}=3$

21. $x-1=y^2$
$y=\pm\sqrt{x-1}$

x	y

10	$\sqrt{(10)-1}=\sqrt{9}=\pm 3$
5	$\sqrt{(5)-1}=\sqrt{4}=\pm 2$
4	$\sqrt{(4)-1}=\pm\sqrt{3}$
2	$\sqrt{(2)-1}=\sqrt{1}=\pm 1$
1.25	$\sqrt{(1.25)-1}=\sqrt{0.25}=\pm 0.5$
1	$\sqrt{(1)-1}=\sqrt{0}=0$

23. $y=\sqrt[3]{x+1}$

x	y
-9	$\sqrt[3]{(-9)+1}=\sqrt[3]{-8}=-2$
-2	$\sqrt[3]{(-2)+1}=\sqrt[3]{-1}=-1$
-1	$\sqrt[3]{(-1)+1}=\sqrt[3]{0}=0$
0	$\sqrt[3]{(0)+1}=\sqrt[3]{1}=1$
4	$\sqrt[3]{(4)+1}=\sqrt[3]{5}$
7	$\sqrt[3]{(7)+1}=\sqrt[3]{8}=2$

25. Function

27. Not a function. The Shaq is paired with two heights.

29. Not a function, 4 is paired with 2 and -5.

31. Function

33. Function

35. Not a function, -2 is paired with 3 and -4.

37. Function

39. Function

41. Not a function, 0 is paired with 4 and -4.

43. Function

45. Not a function, -3 is paired with 2 and -2.

47. Function

49. Function; $x\in[-4,-5]$ $y\in[-2,3]$

51. Function; $x\in[-4,\infty)$ $y\in[-4,\infty)$

53. Function; $x\in[-4,4]$ $y\in[-5,-1]$

55. Function; $x\in(-\infty,\infty)$ $y\in(-\infty,\infty)$

57. Not a function; $x\in[-3,5]$ $y\in[-3,3]$

59. Not a function; $x\in(-\infty,3]$ $y\in(-\infty,\infty)$

61. $y=\dfrac{3}{x-5}$

$x-5=0$

$x=5$

$x\in(-\infty,5)\cup(5,\infty)$

63. $b=\sqrt{3a+5}$

$3a+5\geq 0$

$3a\geq -5$

$a\geq -\dfrac{5}{3}$

$a\in\left[-\dfrac{5}{3},\infty\right)$

65. $y_1=\dfrac{x+2}{x^2-25}$

$x^2-25=0$

$x^2=25$

$x=\pm 5$

$x\in(-\infty,-5)\cup(-5,5)\cup(5,\infty)$

67. $u=\dfrac{v-5}{v^2-18}$

$v^2-18=0$

$v^2=18$

$v=\pm 3\sqrt{2}$

$v\in\left(-\infty,-3\sqrt{2}\right)\cup\left(-3\sqrt{2},3\sqrt{2}\right)\cup\left(3\sqrt{2},\infty\right)$

69. $y=\dfrac{17}{25}x+123$

$x\in(-\infty,\infty)$

71. $m=n^2-3n-10$

$n\in(-\infty,\infty)$

73. $y=2|x|+1$

$x \in (-\infty, \infty)$

75. $y_1 = \dfrac{x}{x^2 - 3x - 10}$

$x^2 - 3x - 10 = 0$

$(x-5)(x+2) = 0$

$x = 5$ or $x = -2$

$x \in (-\infty, -2) \cup (-2, 5) \cup (5, \infty)$

77. $y = \dfrac{\sqrt{x-2}}{2x-5}$, $x \geq 2, x \neq \dfrac{5}{2}$

$2x - 5 = 0$

$2x = 5$

$x = \dfrac{5}{2}$

$x \in \left[2, \dfrac{5}{2}\right) \cup \left(\dfrac{5}{2}, \infty\right)$

79. $f(x) = \dfrac{1}{2}x + 3$

$f(-6) = \dfrac{1}{2}(-6) + 3 = -3 + 3 = 0$;

$f\left(\dfrac{3}{2}\right) = \dfrac{1}{2}\left(\dfrac{3}{2}\right) + 3 = \dfrac{3}{4} + 3 = \dfrac{15}{4}$;

$f(2c) = \dfrac{1}{2}(2c) + 3 = c + 3$;

$f(c+2) = \dfrac{1}{2}(c+2) + 3 = \dfrac{1}{2}c + 1 + 3$

$= \dfrac{1}{2}c + 4$

81. $f(x) = 3x^2 - 4x$

$f(-6) = 3(-6)^2 - 4(-6) = 108 + 24 = 132$;

$f\left(\dfrac{3}{2}\right) = 3\left(\dfrac{3}{2}\right)^2 - 4\left(\dfrac{3}{2}\right) = 3\left(\dfrac{9}{4}\right) - 6$

$= \dfrac{27}{4} - 6 = \dfrac{3}{4}$;

$f(2c) = 3(2c)^2 - 4(2c) = 3(4c^2) - 8c$

$= 12c^2 - 8c$;

$f(c+2) = 3(c+2)^2 - 4(c+2)$

$= 3(c^2 + 4c + 4) - 4c - 8$

$= 3c^2 + 12c + 12 - 4c - 8$

$= 3c^2 + 8c + 4$

83. $h(x) = \dfrac{3}{x}$

$h(3) = \dfrac{3}{(3)} = 1$;

$h\left(-\dfrac{2}{3}\right) = \dfrac{3}{\left(-\dfrac{2}{3}\right)} = -\dfrac{9}{2}$;

$h(3a) = \dfrac{3}{3a} = \dfrac{1}{a}$;

$h(a-1) = \dfrac{3}{a-1}$

85. $h(x) = \dfrac{5|x|}{x}$

$h(3) = \dfrac{5|3|}{3} = \dfrac{5(3)}{3} = 5$;

$h\left(-\dfrac{2}{3}\right) = \dfrac{5\left|-\dfrac{2}{3}\right|}{-\dfrac{2}{3}} = \dfrac{5\left(\dfrac{2}{3}\right)}{-\dfrac{2}{3}} = -5$;

$h(3a) = \dfrac{5|3a|}{3a} = \dfrac{15|a|}{3a} = \dfrac{5|a|}{a}$;

-5 if $a < 0$; 5 if $a > 0$

$h(a-1) = \dfrac{5|a-1|}{a-1}$;

-5 if $a < 1$; 5 if $a > 1$

87. $g(r) = 2\pi r$

$g(0.4) = 2\pi(0.4) = 0.8\pi$;

$g\left(\dfrac{9}{4}\right) = 2\pi\left(\dfrac{9}{4}\right) = \dfrac{9}{2}\pi$;

$g(h) = 2\pi(h) = 2\pi h$;

$g(h+3) = 2\pi(h+3) = 2\pi h + 6\pi$

89. $g(r) = \pi r^2$

$g(0.4) = \pi(0.4)^2 = 0.16\pi$;

$g\left(\dfrac{9}{4}\right) = \pi\left(\dfrac{9}{4}\right)^2 = \dfrac{81}{16}\pi$;

$g(h) = \pi(h)^2 = \pi h^2$;

$g(h+3) = \pi(h+3)^2$

91. $p(x) = \sqrt{2x+3}$

$p(0.5)=\sqrt{2(0.5)+3}=\sqrt{1+3}=\sqrt{4}=2$;

$p\left(\frac{9}{4}\right)=\sqrt{2\left(\frac{9}{4}\right)+3}=\sqrt{\frac{9}{2}+3}=\sqrt{\frac{15}{2}}=\frac{\sqrt{30}}{2}$;

$p(a)=\sqrt{2(a)+3}=\sqrt{2a+3}$;

$p(a+3)=\sqrt{2(a+3)+3}$

$=\sqrt{2a+6+3}=\sqrt{2a+9}$

93. $p(x)=\frac{3x^2-5}{x^2}$

$p(0.5)=\frac{3(0.5)^2-5}{(0.5)^2}=\frac{3(0.25)-5}{0.25}$

$=\frac{0.75-5}{0.25}=\frac{-4.25}{0.25}=-17$

$p\left(\frac{9}{4}\right)=\frac{3\left(\frac{9}{4}\right)^2-5}{\left(\frac{9}{4}\right)^2}=\frac{3\left(\frac{81}{16}\right)-5}{\frac{81}{16}}$

$=\frac{\frac{243}{16}-5}{\frac{81}{16}}=\frac{\frac{163}{16}}{\frac{81}{16}}=\frac{163}{81}$;

$p(a)=\frac{3(a)^2-5}{(a)^2}=\frac{3a^2-5}{a^2}$;

$p(a+3)=\frac{3(a+3)^2-5}{(a+3)^2}$

$=\frac{3(a^2+6a+9)-5}{a^2+6a+9}$

$=\frac{3a^2+18a+27-5}{a^2+6a+9}$

$=\frac{3a^2+18a+22}{a^2+6a+9}$

95. a. {-1, 0, 1, 2, 3, 4, 5}
b. f(2) = 1
c. f(x) = 4; x = -1
d. {-2, -1, 0, 1, 2, 3, 4}

97. a. $x\in[-5,5]$
b. f(2) = -2
c. f(x) = 4; x = -2
d. $R\in[-3,4]$

99. a. $x\in[-3,\infty)$
b. f(2) = 2
c. $f(x)=4, x=0$
d. $y\in(-\infty,4]$

101. $W(H)=\frac{9}{2}H-151$

a. $W(75)=\frac{9}{2}(75)-151=186.5$ lbs

b. $W(72)=\frac{9}{2}(72)-151=173$ lbs
210 – 173 = 37 lbs

103.a. $N(g)=23g$
b. $g\in[0,15]$; $N\in[0,345]$

105.a. $D\in[0,\infty)$
b. $V(6.25)=(6.25)^3\approx 244$ units^3
c. $V(2x^2)=(2x^2)^3=8x^6$

107.a. $C(t)=12.50t+19.50$
b. $C(3.5)=12.50(3.5)+19.50=\$63.25$
c. $119.75=12.50t+19.50$
$100.25=12.50t$
$8\ hrs\approx t$
d. $150=12.50t+19.50$
$130.5=12.50t$
$10.44\ hrs=t$
$t\in[0,10.44]$; $C\in[0,150]$

109.a. Yes.
Each "x" is paired with exactly one "y".
b. 9 pm
c. $3\frac{1}{2}$m
d. 5 pm and 1 am

111. Answers will vary.

113.a. Son, 72.5 seconds
b. 10 meters
c. 45 seconds
d. 3

115. The y-values of the x integers between -2 and 2 become positive.

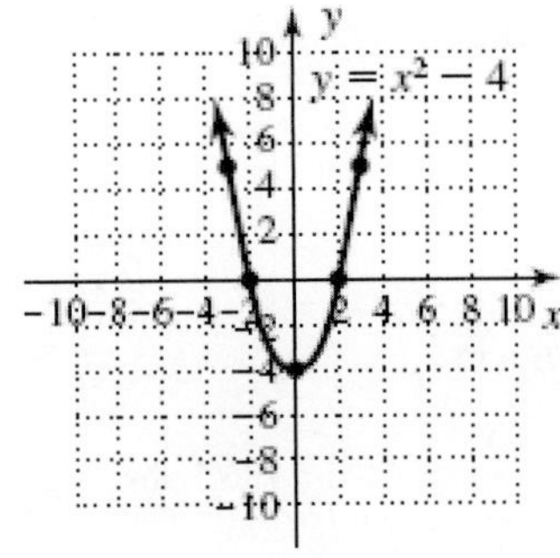

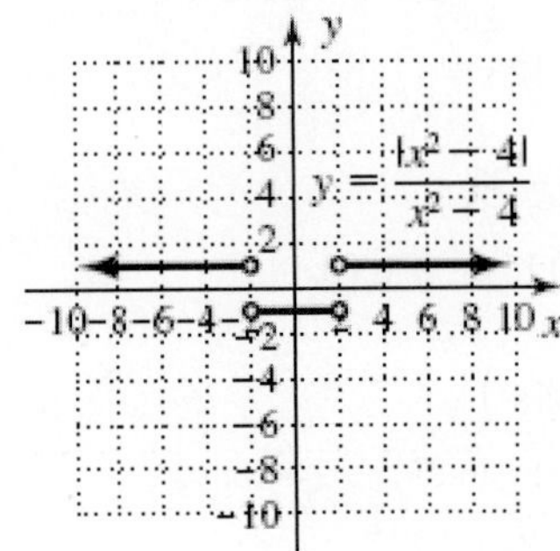

117.a. $\sqrt{24}+6\sqrt{54}-\sqrt{6}$
$=2\sqrt{6}+6\cdot 3\sqrt{6}-\sqrt{6}$
$=2\sqrt{6}+18\sqrt{6}-\sqrt{6}$
$=19\sqrt{6}$

b. $\left(2+\sqrt{3}\right)\left(2-\sqrt{3}\right)$
$=4-2\sqrt{3}+2\sqrt{3}-3$
$=1$

119.a. $x^3-3x^2-25x+75$
$=\left(x^3-3x^2\right)-(25x-75)$
$=x^2(x-3)-25(x-3)$
$=(x-3)\left(x^2-25\right)$
$=(x-3)(x-5)(x+5)$

b. $2x^2-13x-24=(2x+3)(x-8)$

c. $8x^3-125=(2x-5)\left(4x^2+10x+25\right)$

121.a. $\left(\frac{x^3y}{z^{-2}}\right)^2=\frac{x^6y^2}{z^{-4}}=x^6y^2z^4$

b. $\left(\frac{2}{3}\right)^{-3}=\left(\frac{3}{2}\right)^3=\frac{27}{8}$

2.3 Technology Highlight

Exercise 1. $y=-0.2x+68.4$
$(109.5,46.5)$ is on the line.

Exercise 2. $y=-1.5x+144$
$(32,96)$ is on the line.

Exercise 3. $y=2300x+97500$
$(4.5,\ 107{,}850)$ is on the line.

Exercise 4. $y=806x+42165$
$(4.5,\ 45{,}792)$ is on the line.

Exercise 5.In 2008, (x = 13), the median cost will be $127,400.

2.3 Exercises

1. $-\frac{7}{4}$; (0, 3)

3. 2.5

5. Answers will vary.

7. $4x+5y=10$
$5y=-4x+10$
$y=-\frac{4}{5}x+2$

x	$y=-\frac{4}{5}x+2$
-5	$y=-\frac{4}{5}(-5)+2=4+2=6$
-2	$y=-\frac{4}{5}(-2)+2=\frac{8}{5}+2=\frac{18}{5}$
0	$y=-\frac{4}{5}(0)+2=0+2=2$
1	$y=-\frac{4}{5}(1)+2=-\frac{4}{5}+2=\frac{6}{5}$
3	$y=-\frac{4}{5}(3)+2=-\frac{12}{5}+2=-\frac{2}{5}$

9. $-0.4x+0.2y=1.4$
$0.2y=0.4x+1.4$
$y=2x+7$

x	$y=2x+7$
-5	$y=2(-5)+7=-10+7=-3$
-2	$y=2(-2)+7=-4+7=3$
0	$y=2(0)+7=0+7=7$
1	$y=2(1)+7=2+7=9$
3	$y=2(3)+7=6+7=13$

11. $\frac{1}{3}x+\frac{1}{5}y=-1$

$\frac{1}{5}y=-\frac{1}{3}x-1$

$y=-\frac{5}{3}x-5$

x	$y=-\frac{5}{3}x-5$
-5	$y=-\frac{5}{3}(-5)-5=\frac{25}{3}-5=\frac{10}{3}$
-2	$y=-\frac{5}{3}(-2)-5=\frac{10}{3}-5=-\frac{5}{3}$
0	$y=-\frac{5}{3}(0)-5=0-5=-5$
1	$y=-\frac{5}{3}(1)-5=-\frac{5}{3}-5=-\frac{20}{3}$
3	$y=-\frac{5}{3}(3)-5=-5-5=-10$

13. $6x-3y=9$

$-3y=-6x+9$

$y=2x-3$

$f(x)=2x-3$

New Coefficient: 2
New Constant: -3

15. $-0.5x-0.3y=2.1$

$-0.3y=0.5x+2.1$

$y=\frac{-5}{3}x-7$

$f(x)=\frac{-5}{3}x-7$

New Coefficient: $\frac{-5}{3}$

New Constant: -7

17. $\frac{5}{6}x+\frac{1}{7}y=-\frac{4}{7}$

$\frac{1}{7}y=-\frac{5}{6}x-\frac{4}{7}$

$y=-\frac{35}{6}x-4$

$f(x)=-\frac{35}{6}x-4$

New Coefficient: $-\frac{35}{6}$

New Constant: -4

19. $y=-\frac{4}{3}x+5$

x	$y=-\frac{4}{3}x+5$
0	$y=-\frac{4}{3}(0)+5=0+5=5$
3	$y=-\frac{4}{3}(3)+5=-4+5=1$
6	$y=-\frac{4}{3}(6)+5=-8+5=-3$

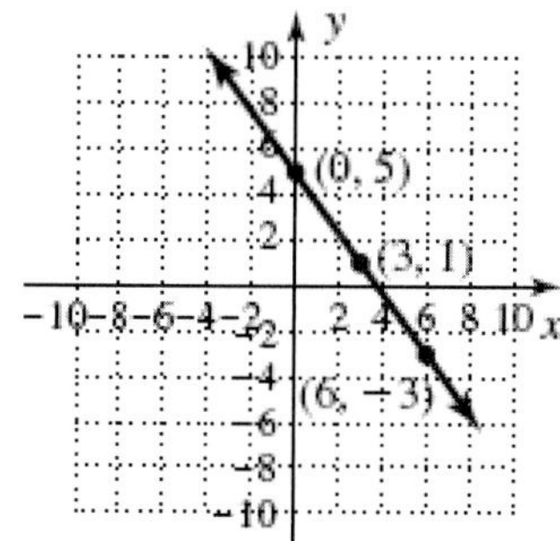

21. $y=\frac{5}{4}x+1$

x	$y=\frac{5}{4}x+1$
-4	$y=\frac{5}{4}(-4)+1=-5+1=-4$
0	$y=\frac{5}{4}(0)+1=0+1=1$
4	$y=\frac{5}{4}(4)+1=5+1=6$

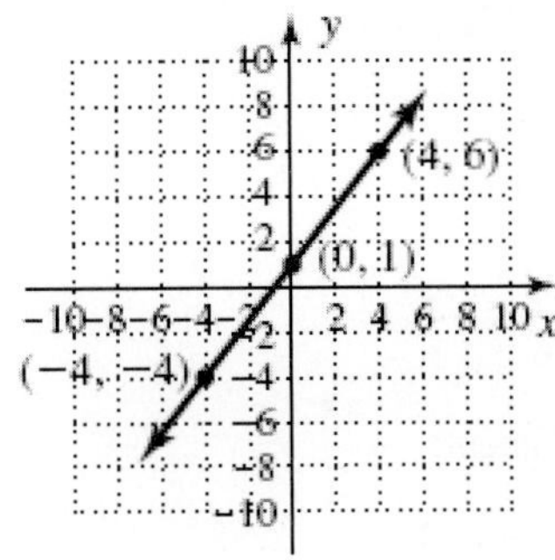

23. $y = -\frac{1}{6}x + 4$

x	$y = -\frac{1}{6}x + 4$
-6	$y = -\frac{1}{6}(-6) + 4 = 1 + 5 = 5$
0	$y = -\frac{1}{6}(0) + 4 = 0 + 4 = 4$
6	$y = -\frac{1}{6}(6) + 4 = -1 + 4 = 3$

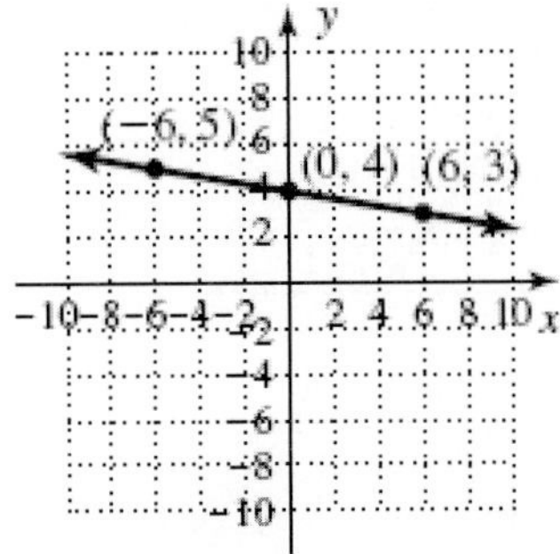

25. $3x + 4y = 12$

x-intercept: (4, 0) y-intercept: (0, 3)

$3x + 4(0) = 12 \quad 3(0) + 4y = 12$

$3x = 12 \quad 4y = 12$

$x = 4 \quad y = 3$

a. $m = \frac{0-3}{4-0} = -\frac{3}{4}$

b. $f(x) = -\frac{3}{4}x + 3$

c. The coefficient of x is the slope and the constant is the y-intercept.

27. $2x - 5y = 10$

x-intercept: (5, 0) y-intercept: (0,-2)

$2x - 5(0) = 10 \quad 2(0) - 5y = 10$

$2x = 10 \quad -5y = 10$

$x = 5 \quad y = -2$

a. $m = \frac{0-(-2)}{5-0} = \frac{2}{5}$

b. $f(x) = \frac{2}{5}x - 2$

c. The coefficient of x is the slope and the constant is the y-intercept.

29. $4x - 5y = -15$

x-intercept: $\left(-\frac{15}{4}, 0\right)$ y-intercept: (0, 3)

$4x - 5(0) = -15 \quad 4(0) - 5y = -15$

$4x = -15 \quad -5y = -15$

$x = -\frac{15}{4} \quad y = 3$

a. $m = \frac{0-3}{-\frac{15}{4} - 0} = \frac{-3}{-\frac{15}{4}} = \frac{12}{15} = \frac{4}{5}$

b. $f(x) = \frac{4}{5}x + 3$

c. The coefficient of x is the slope and the constant is the y-intercept.

31. $2x + 3y = 6$

$3y = -2x + 6$

$y = -\frac{2}{3}x + 2$

$m = -\frac{2}{3}$; y-intercept (0, 2)

33. $5x + 4y = 20$

$4y = -5x + 20$

$y = -\frac{5}{4}x + 5$

$m = -\frac{5}{4}$; y-intercept (0, 5)

35. $x = 3y$

$y = \frac{1}{3}x$

$m = \frac{1}{3}$; y-intercept (0, 0)

37. $3x + 4y - 12 = 0$

$4y = -3x + 12$

$y = -\frac{3}{4}x + 3$

$m = -\frac{3}{4}$; y-intercept (0, 3)

39. $m = \frac{2}{3}$; y-intercept (0, 1)

$y = mx + b$

$y = \frac{2}{3}x + 1$

41. $m = 3$; y-intercept (0, 3)

$y = mx + b$

$y = 3x + 3$

43. $m = 3$; y-intercept (0, 2)

$y = mx + b$

$y = 3x + 2$

45. $m = 250$; (14, 4000)

$y - y_1 = m(x - x_1)$

$y - 4000 = 250(x - 14)$

$y - 4000 = 250x - 3500$

$y = 250x + 500$

$f(x) = 250x + 500$

47. $m = \frac{75}{2}$; (24,1050)

$y - y_1 = m(x - x_1)$

$y - 1050 = \frac{75}{2}(x - 24)$

$y - 1050 = \frac{75}{2}x - 900$

$y = \frac{75}{2}x + 150$

$f(x) = \frac{75}{2}x + 150$

49. $m = 2$; (5, -3)

$y - y_1 = m(x - x_1)$

$y + 3 = 2(x - 5)$

$y + 3 = 2x - 10$

$y = 2x - 13$

$f(x) = 2x - 13$

51. $4x + 5y = 20$

$5y = -4x + 20$

$y = -\frac{4}{5}x + 4$

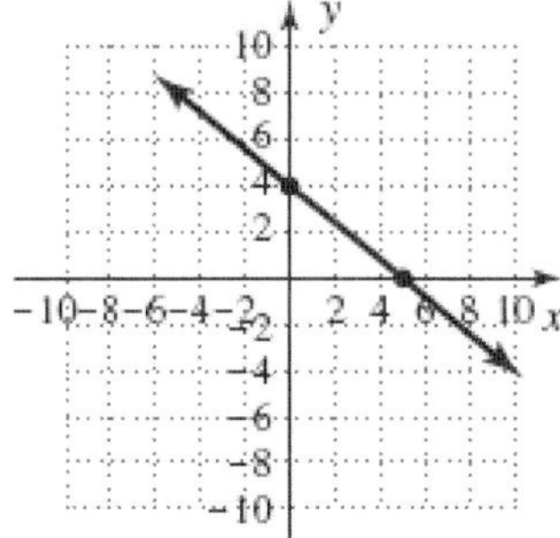

53. $5x - 3y = 15$

$-3y = -5x + 15$

$y = \frac{5}{3}x - 5$

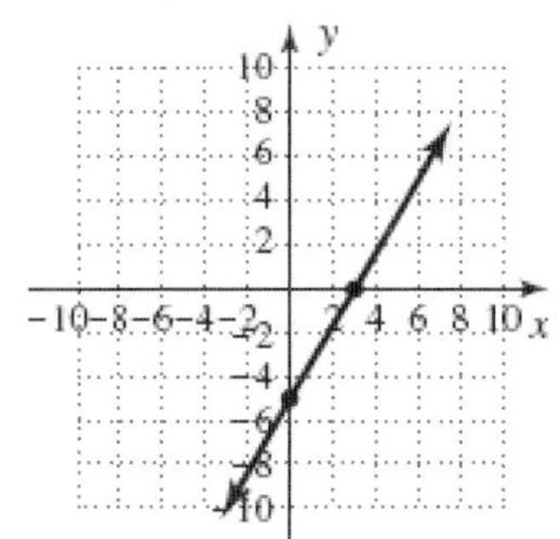

55. $y = \frac{2}{3}x + 3$

$m = \frac{2}{3}$; y-intercept (0, 3)

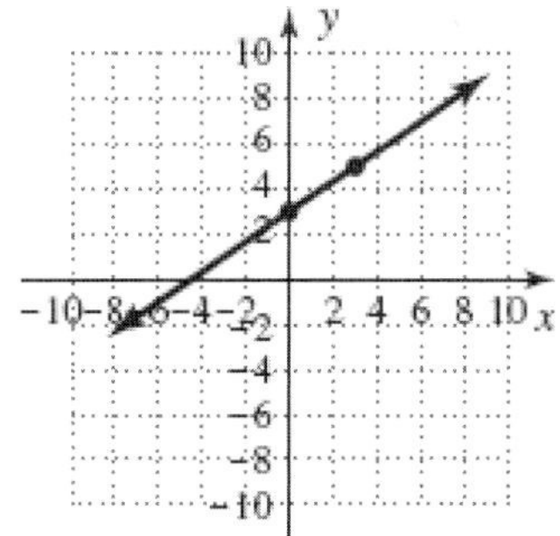

57. $y = -\frac{1}{3}x + 2$

$m = \frac{-1}{3}$; y-intercept (0, 2)

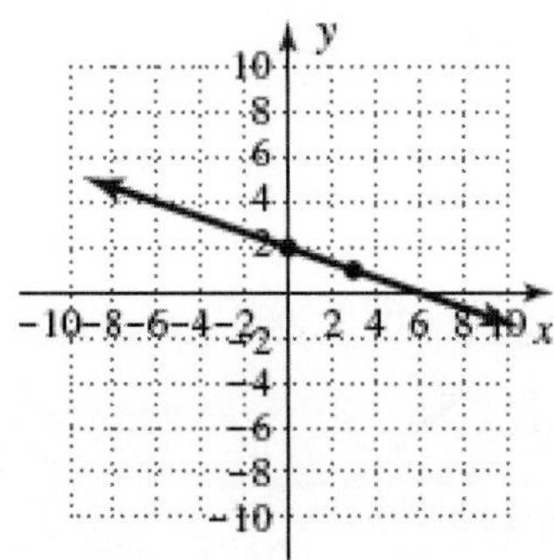

59. $y = 2x - 5$

$m = 2$; y-intercept (0, -5)

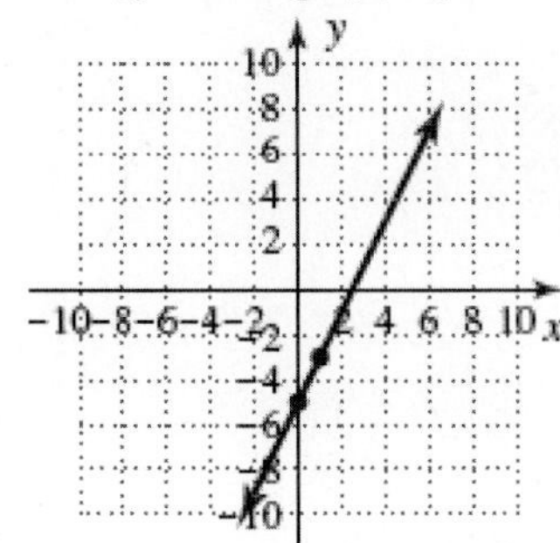

61. $f(x) = \frac{1}{2}x - 3$

$m = \frac{1}{2}$; y-intercept (0, -3)

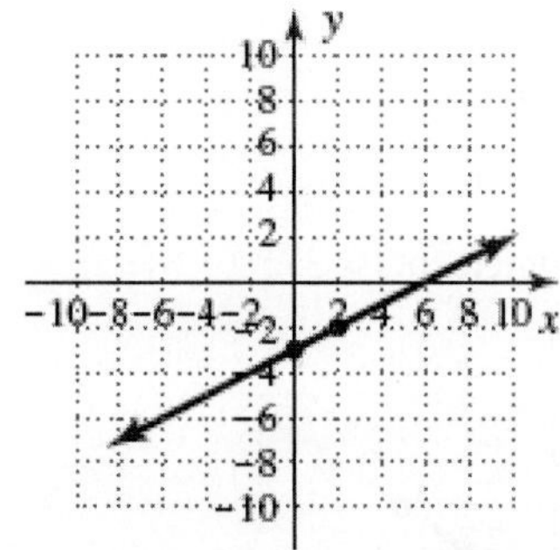

63. $2x - 5y = 10$

$-5y = -2x + 10$

$y = \frac{2}{5}x - 2$

$m = \frac{2}{5}$; (3, -2)

$y - y_1 = m(x - x_1)$

$y + 2 = \frac{2}{5}(x - 3)$

$y + 2 = \frac{2}{5}x - \frac{6}{5}$

$y = \frac{2}{5}x - \frac{16}{5}$

65. $5y - 3x = 9$

$5y = 3x + 9$

$y = \frac{3}{5}x + \frac{9}{5}$

$m = -\frac{5}{3}$; (2, 4);

$y - y_1 = m(x - x_1)$

$y - 4 \equiv -\frac{5}{3}(x - 2)$

$y - 4 = -\frac{5}{3}x + \frac{10}{3}$

$y = -\frac{5}{3}x + \frac{22}{3}$

67. $12x + 5y = 65$

$5y = -12x + 65$

$y = -\frac{12}{5}x + 13$

$m = -\frac{12}{5}$; (-2, -1)

$y - y_1 = m(x - x_1)$

$y + 1 = -\frac{12}{5}(x + 2)$

$y + 1 = -\frac{12}{5}x - \frac{24}{5}$

$y = -\frac{12}{5}x - \frac{29}{5}$

69. $4y - 5x = 8$

$4y = 5x + 8$

$y = \frac{5}{4}x + 2;$

$5y + 4x = -15$

$5y = -4x - 15$

$y = -\frac{4}{5}x - 3$

Perpendicular; slopes are opposite reciprocals.

71. $2x-5y=20$
$$-5y=-2x+20$$
$$y=\frac{2}{5}x-4;$$
$$4x-3y=18$$
$$-3y=-4x+18$$
$$y=\frac{4}{3}x-6$$
Neither

73. $-4x+6y=12$
$$6y=4x+12$$
$$y=\frac{2}{3}x+2;$$
$$2x+3y=6$$
$$3y=-2x+6$$
$$y=-\frac{2}{3}x+2$$
Neither

75. $m=2;\ P_1=(2,-5)$
$$y-y_1=m(x-x_1)$$
$$y+5=2(x-2)$$
$$y+5=2x-4$$
$$y=2x-9$$
$$f(x)=2x-9$$

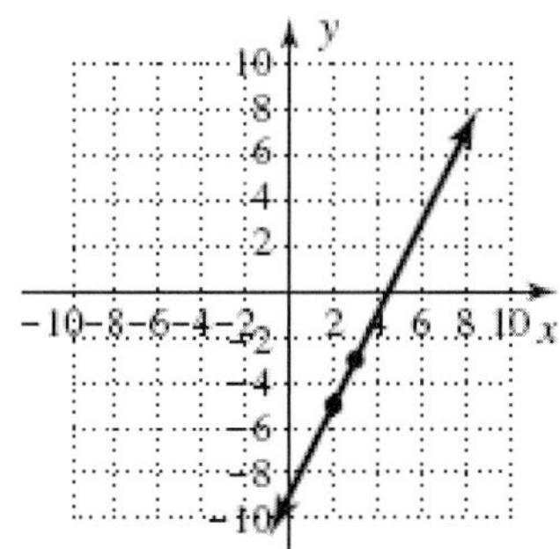

77. $m=\frac{3}{8};\ P_1=(3,-4)$
$$y-y_1=m(x-x_1)$$
$$y+4=\frac{3}{8}(x-3)$$
$$y+4=\frac{3}{8}x-\frac{9}{8}$$
$$y=\frac{3}{8}x-\frac{41}{8}$$
$$f(x)=\frac{3}{8}x-\frac{41}{8}$$

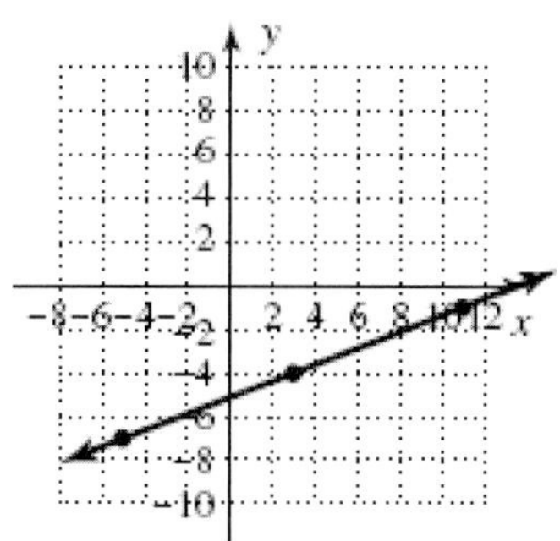

79. $m=0.5;\ P_1=(1.8,-3.1)$
$$y-y_1=m(x-x_1)$$
$$y+3.1=0.5(x-1.8)$$
$$y+3.1=0.5x-0.9$$
$$y=0.5x-4$$
$$f(x)=0.5x-4$$

81. $m=-\frac{3}{4};\ (2, -4)$

a. $m=-\frac{3}{4}$
$$y-y_1=m(x-x_1)$$
$$y+4=-\frac{3}{4}(x-2)$$
$$y+4=-\frac{3}{4}x+\frac{3}{2}$$
$$y=-\frac{3}{4}x-\frac{5}{2}$$

b. $m=\frac{4}{3}$
$$y-y_1=m(x-x_1)$$
$$y+4=\frac{4}{3}(x-2)$$
$$y+4=\frac{4}{3}x-\frac{8}{3}$$
$$y=\frac{4}{3}x-\frac{20}{3}$$

83. $m = \frac{4}{9}$; (-1, 3)

a. $m = \frac{4}{9}$

$$y - y_1 = m(x - x_1)$$
$$y - 3 = \frac{4}{9}(x+1)$$
$$y - 3 = \frac{4}{9}x + \frac{4}{9}$$
$$y = \frac{4}{9}x + \frac{31}{9}$$

b. $m = -\frac{9}{4}$

$$y - y_1 = m(x - x_1)$$
$$y - 3 = -\frac{9}{4}(x+1)$$
$$y - 3 = -\frac{9}{4}x - \frac{9}{4}$$
$$y = -\frac{9}{4}x + \frac{3}{4}$$

85. $m = -\frac{1}{2}$; (0, -2)

a. $m = -\frac{1}{2}$

$$y - y_1 = m(x - x_1)$$
$$y + 2 = -\frac{1}{2}(x - 0)$$
$$y + 2 = -\frac{1}{2}x$$
$$y = -\frac{1}{2}x - 2$$

b. $m = 2$

$$y - y_1 = m(x - x_1)$$
$$y + 2 = 2(x - 0)$$
$$y + 2 = 2x$$
$$y = 2x - 2$$

87. $m = \frac{6}{5}$; (4, 2)

$$y - y_1 = m(x - x_1)$$
$$y - 2 = \frac{6}{5}(x - 4)$$
$$y - 2 = \frac{6}{5}x - \frac{24}{5}$$
$$y = \frac{6}{5}x - \frac{14}{5}$$

For each 5000 additional sales, income rises $6000.

89. $m = -20$; (0.5, 100)

$$y - y_1 = m(x - x_1)$$
$$y - 100 = -20(x - 0.5)$$
$$y - 100 = -20x + 10$$
$$y = -20x + 110$$

For every hour of television, a students final grade falls 20%.

91. $m = \frac{35}{2}$; (0.5, 10)

$$y - y_1 = m(x - x_1)$$
$$y - 10 = \frac{35}{2}(x - 0.5)$$
$$y - 10 = \frac{35}{2}x - \frac{35}{4}$$
$$y = \frac{35}{2}x + \frac{5}{4}$$

Every 2 inches of rainfall increases the number of cattle raised per acre by 35.

93. C

95. A

97. B

99. D

101. $ax + by = c$

$$by = -ax + c$$
$$y = -\frac{a}{b}x + \frac{c}{b}$$

a. $3x + 4y = 8$

$$4y = -3x + 8$$
$$y = -\frac{3}{4}x + 2$$

$m = -\frac{3}{4}$; y-intercept (0, 2);

$m = -\frac{a}{b} = -\frac{3}{4}$;

$y - \text{int} = \frac{c}{b} = \frac{8}{4} = 2$

b. $2x + 5y = -15$

$$5y = -2x - 15$$
$$y = -\frac{2}{5}x - 3$$

$m=-\frac{2}{5}$; y-intercept (0, -3);

$m=-\frac{a}{b}=-\frac{2}{5}$;

$y-\text{int}=\frac{c}{b}=-\frac{15}{5}=-3$

c. $5x-6y=-12$

$-6y=-5x-12$

$y=\frac{5}{6}x+2$

$m=\frac{5}{6}$; y-intercept (0, 2);

$m=-\frac{5}{-6}=\frac{5}{6}$;

$y-\text{int}=\frac{c}{b}=\frac{-12}{-6}=2$

d. $3y-5x=9$

$3y=5x+9$

$y=\frac{5}{3}x+3$

$m=-\frac{-5}{3}=\frac{5}{3}$; y-intercept (0, 3);

$m=-\frac{a}{b}=-\frac{-5}{3}=\frac{5}{3}$;

$y-\text{int}=\frac{c}{b}=\frac{9}{3}=3$

103.a. As the temperature increases 5°C, the velocity of sound waves increases 3 m/s. At a temperature of 0°C, the velocity is 331 m/s.

b. $V(20)=\frac{3}{5}(20)+331=343\text{ m/s}$

c. $361=\frac{3}{5}C+331$

$30=\frac{3}{5}C$

$50=C$

50°C

105.a. $m=\frac{190-150}{2004-1998}=\frac{40}{6}=\frac{20}{3}$

$V(t)=\frac{20}{3}t+150$

b. Every three years, the coin increased in value by \$20. The initial value was \$150.

107.a. $m=\frac{51-9}{2001-1995}=\frac{42}{6}=7$

$N(t)=7t+9$

b. Every 1 year, the number of homes hooked to the internet increases by 7 million.

c. $0=7t+9$

$-9=7t$

$-\frac{9}{7}=t$

$-1.29=t$

1.29 years prior to 1995 is 1993.

109.a. $V(11)=\frac{20}{3}(11)+150=\$223.33$

b. $250<\frac{20}{3}t+150$

$100<\frac{20}{3}t$

$15<t$

The penny's value will exceed \$250 in 15 years, 2013.

c. $170=\frac{20}{3}t+150$

$20=\frac{20}{3}t$

$3=t$

Mark has owned the penny 3 years.

111.a. $N(11)=7(11)+9=86$ million

b. $100=7t+9$

$91=7t$

$13=t$

13 years

c. $115=7t+9$

$106=7t$

$15.14\approx t$

15.14 years after 1995 is 2010

113. $m=\frac{1320000-740000}{2000-1990}$

$=\frac{580000}{10}=58000$

$P(t)=58000t+740000$

Grows 58,000 every year.

$P(17)=58000(17)+740000=1726000$

115. Answers will vary.

117. Graph 1: D

Graph 2: A
Graph 3: C
Graph 4: B
Graph 5: F
Graph 6: H

119. $3x^2 - 10x = 9$

$3x^2 - 10 - 9 = 0$

$x = \dfrac{10 \pm \sqrt{(-10)^2 - 4(3)(-9)}}{2(3)}$

$x = \dfrac{10 \pm \sqrt{100 + 108}}{6}$

$x = \dfrac{10 \pm \sqrt{208}}{6}$

$x = \dfrac{10 \pm 4\sqrt{13}}{6}$

$x = \dfrac{5 \pm 2\sqrt{13}}{3}$

$x \approx 4.07$ or $x \approx -0.74$

121. $A = \pi r^2$

Larger circle: Smaller Circle

$A = \pi(10)^2$ $\quad A = \pi(8)^2$

$A = 314.16$ $\quad A = 201.06$

$314.16 - 201.06 = 113.10$ yds^2

123. $-3 < 2x + 5$ and $x - 2 < 3$

$-8 < 2x$ and $x < 5$

$-4 < x$ and $x < 5$

(-4, 5)

Chapter 2 Mid-Chapter Check

1. $4x - 3y = 12$

$-3y = -4x + 12$

$y = \dfrac{4}{3}x - 4$

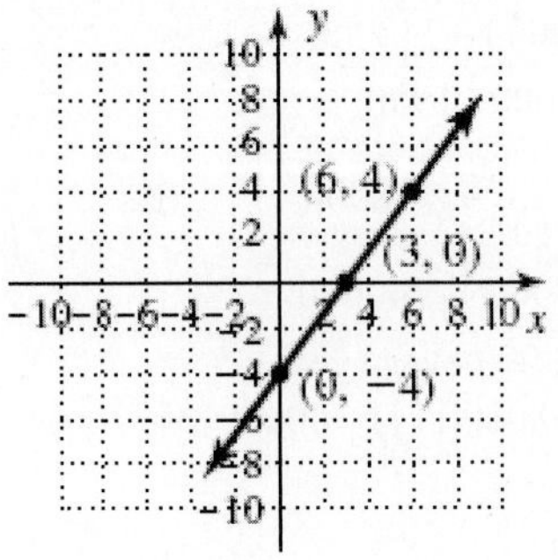

2. (–3, 8) and (4, –10)

$m = \dfrac{-10 - 8}{4 - (-3)} = \dfrac{-18}{7}$

3. Positive, loss is decreasing, profit is increasing.

$m = \dfrac{-0.5 - -2}{2003 - 2002} = \dfrac{1.5}{1} = 1.5$;

Data.com's loss decreases by 1.5 million dollars per year.

4. (1, 4) $m = -\dfrac{2}{3}$

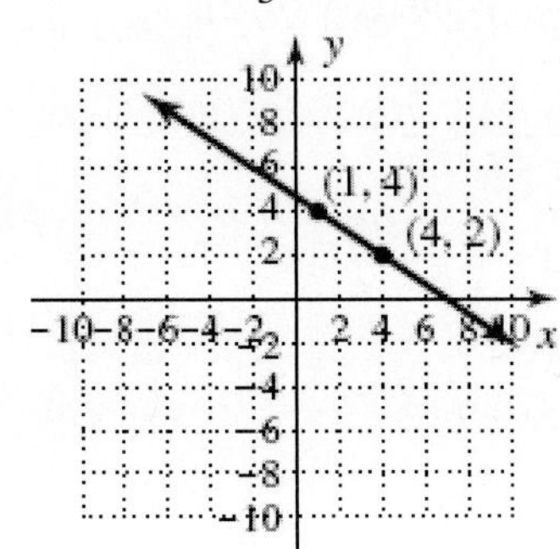

$m = \dfrac{3}{2}$ (1, 4)

$y - y_1 = m(x - x_1)$

$y - 4 = \dfrac{3}{2}(x - 1)$

$y - 4 = \dfrac{3}{2}x - \dfrac{3}{2}$

$y = \dfrac{3}{2}x + \dfrac{5}{2}$

5. $x = -3$; not a function. Input –3 is paired with more than one output.

6. $m = -\dfrac{4}{3}$; y-intercept (0, 4)

$y = -\dfrac{4}{3}x + 4$; is a function. Each input is paired with only one output.

7. a. h(2) = 0
   b. $x \in [-3,5]$
   c. $x = 3.5$ when h(x) = 4
   d. $y \in [-4,5]$

8. Rate of change from $x = 1$ and $x = 2$ is larger. It is steeper.

9. $m = \dfrac{3}{4}$; (1, 2)

$$y - y_1 = m(x - x_1)$$
$$y - 2 = \frac{3}{4}(x - 1)$$
$$y - 2 = \frac{3}{4}x - \frac{3}{4}$$
$$y = \frac{3}{4}x + \frac{5}{4}$$
$$F(p) = \frac{3}{4}p + \frac{5}{4}$$

For every 4000 pheasants, the fox population increases by 300.

$$F(20) = \frac{3}{4}(20) + \frac{5}{4} = 15 + 1.25 = 16.25$$

Fox population is 1625 when the pheasant population is 20,000.

10. a. D = {-3, -2, -1, 0, 1, 2, 3, 4}
R = {-3, -2, -1, 0, 1, 2, 3, 4}

b. $x \in [-3,4]$

$y \in [-3,4]$

c. $x \in (-\infty, \infty)$

$y \in (-\infty, \infty)$

Chapter 2 Reinforcing Basic Concepts

1. $P_1(0,5)$ $P_2(6,7)$

a. $m = \dfrac{7-5}{6-0} = \dfrac{2}{6} = \dfrac{1}{3}$; increasing

b. $y - 5 = \dfrac{1}{3}(x - 0)$

c. $y = \dfrac{1}{3}x + 5$

d. $y = \dfrac{1}{3}x + 5$

$-\dfrac{1}{3}x + y = 5$

$x - 3y = -15$

e. x-intercept: (-15, 0) y-intercept: (0, 5)

$x - 3(0) = -15$
$x = -15$

$0 - 3y = -15$
$-3y = -15$
$y = 5$

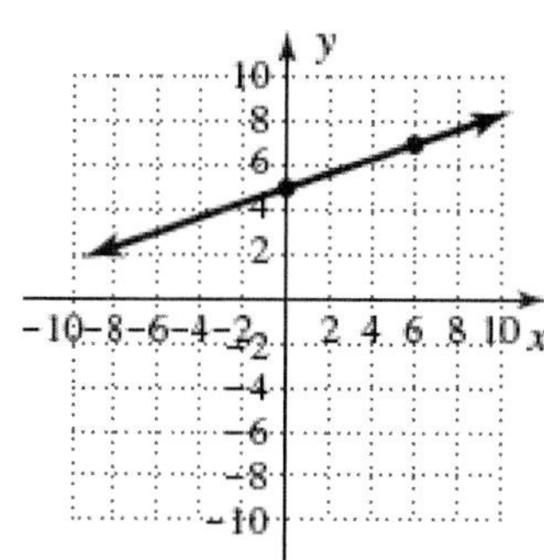

2. $P_1(3,2)$ $P_2(0,9)$

a. $m = \dfrac{9-2}{0-3} = \dfrac{7}{-3} = -\dfrac{7}{3}$; decreasing

b. $y - 9 = -\dfrac{7}{3}(x - 0)$

c. $y = -\dfrac{7}{3}x + 9$

d. $y = -\dfrac{7}{3}x + 9$

$\dfrac{7}{3}x + y = 9$

$7x + 3y = 27$

e. x-intercept: $\left(\dfrac{27}{7}, 0\right)$ y-intercept: (0, 9)

$7x + 3(0) = 27$
$x = \dfrac{27}{7}$

$7(0) + 3y = 27$
$3y = 27$
$y = 9$

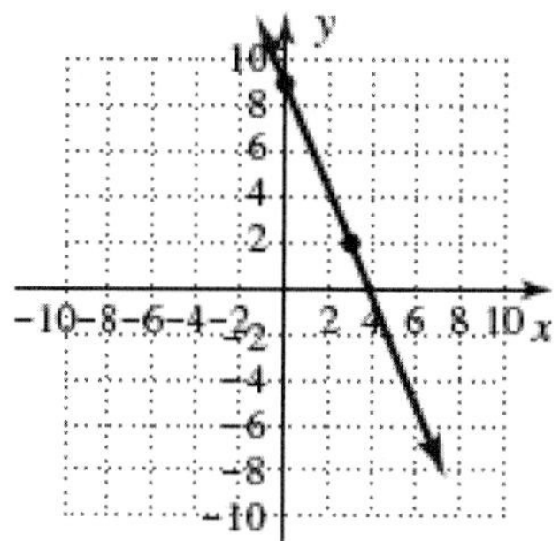

3. $P_1(3,2)$ $P_2(9,5)$

a. $m = \dfrac{5-2}{9-3} = \dfrac{3}{6} = \dfrac{1}{2}$; increasing

b. $y - 2 = \dfrac{1}{2}(x - 3)$

c. $y - 2 = \dfrac{1}{2}x - \dfrac{3}{2}$

$$y=\frac{1}{2}x+\frac{1}{2}$$

d. $y=\frac{1}{2}x+\frac{1}{2}$

$-\frac{1}{2}x+y=\frac{1}{2}$

$x-2y=-1$

e. x-intercept: (-1, 0) y-intercept: $\left(0,\frac{1}{2}\right)$

$x-2(0)=-1$
$x=-1$

$0-2y=-1$
$y=\frac{1}{2}$

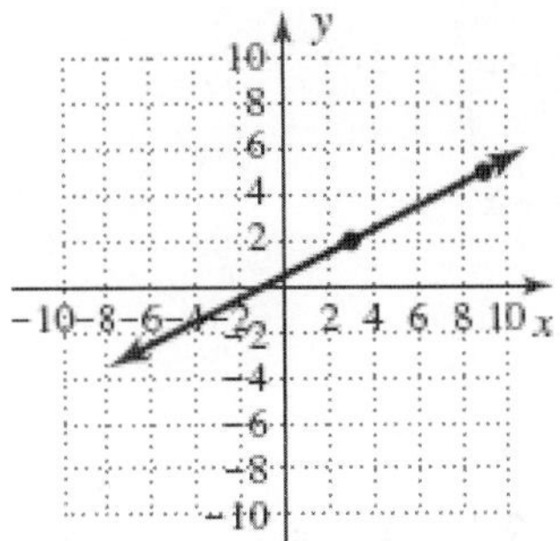

4. $P_1(-5,-4)$ $P_2(3,2)$

a. $m=\frac{2-(-4)}{3-(-5)}=\frac{6}{8}=\frac{3}{4}$; increasing

b. $y+4=\frac{3}{4}(x+5)$

c. $y+4=\frac{3}{4}x+\frac{15}{4}$

$y=\frac{3}{4}x-\frac{1}{4}$

d. $y=\frac{3}{4}x-\frac{1}{4}$

$-\frac{3}{4}x+y=-\frac{1}{4}$

$3x-4y=1$

e. x-intercept: $\left(\frac{1}{3},0\right)$

y-intercept: $\left(0,-\frac{1}{4}\right)$

$3x-4(0)=1$
$3x=1$
$x=\frac{1}{3}$

$3(0)-4y=1$
$-4y=1$
$y=-\frac{1}{4}$

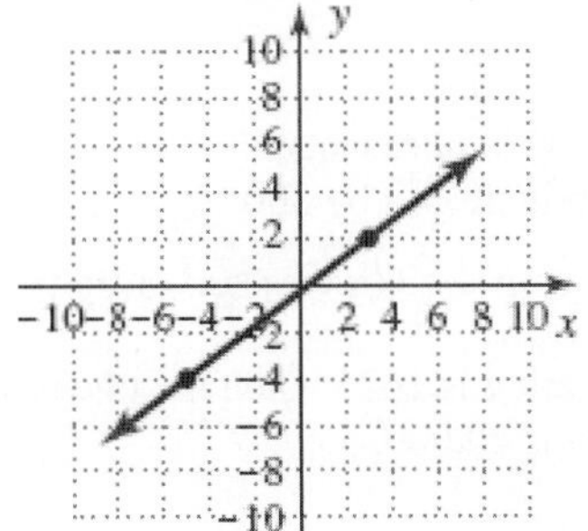

5. $P_1(-2,5)$ $P_2(6,-1)$

a. $m=\frac{-1-5}{6-(-2)}=\frac{-6}{8}=-\frac{3}{4}$; decreasing

b. $y-5=-\frac{3}{4}(x+2)$

c. $y-5=-\frac{3}{4}x-\frac{3}{2}$

$y=-\frac{3}{4}x+\frac{7}{2}$

d. $y=-\frac{3}{4}x+\frac{7}{2}$

$\frac{3}{4}x+y=\frac{7}{2}$

$3x+4y=14$

e. x-intercept: $\left(\frac{14}{3},0\right)$ y-intercept: $\left(0,\frac{7}{2}\right)$

$3x+4(0)=14$
$3x=14$
$x=\frac{14}{3}$

$3(0)+4y=14$
$4y=14$
$y=\frac{7}{2}$

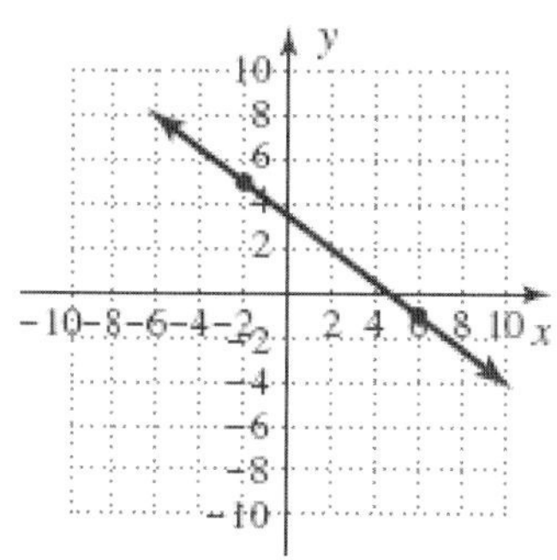

6. $P_1(2,-7)$ $P_2(-8,-2)$

a. $m=\dfrac{-2-(-7)}{-8-2}=\dfrac{5}{-10}=-\dfrac{1}{2}$; increasing

b. $y+7=-\dfrac{1}{2}(x-2)$

c. $y+7=-\dfrac{1}{2}x+1$

$y=-\dfrac{1}{2}x-6$

d. $y=-\dfrac{1}{2}x-6$

$\dfrac{1}{2}x+y=-6$

$x+2y=-12$

e. x-intercept: (-12, 0) y-intercept: (0, -6)

$x+2(0)=-12$

$x=-12$

$0+2y=-12$

$y=-6$

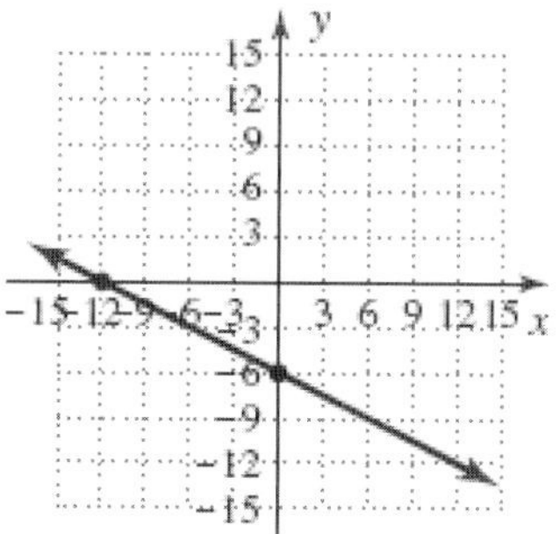

2.4 Technology Highlight

Exercise 1: [-2, 20, 2, -50, 50, 10]

Exercise 2: [-30, 10, 2, -140, 20, 10]

Exercise 3: [-20, 10, 2, -40, 80, 10]

Exercise 4: [-5, 15, 2, -20, 40, 5]

2.4 Exercises

1. Parabola

3. Point; inflection

5. Answers will vary.

7. $f(x)=x^2+2$

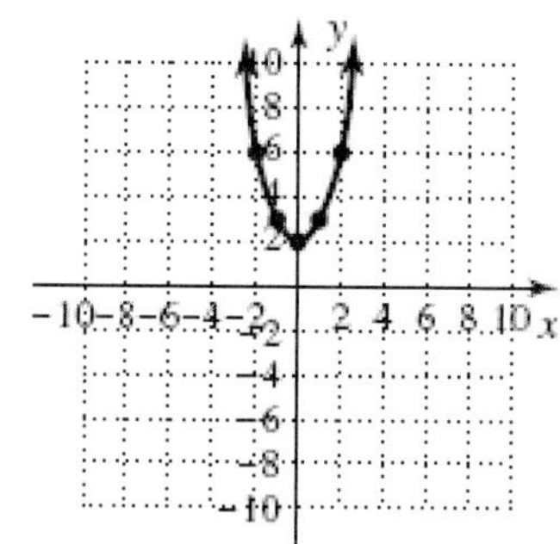

9. $p(x)=(x-1)^2$

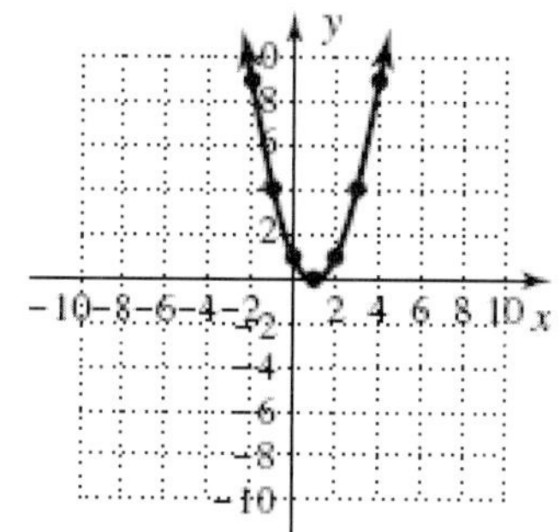

11. $f(x)=x^2+4x$

a. End Behavior: up, up

Axis of symmetry: $x=-2$

$x(x+4)=0$

$x=0;\quad x=-4$

x-intercepts: (0, 0) and (–4, 0)

$x=\dfrac{0+-4}{2}=\dfrac{-4}{2}=-2$

$y=(-2)^2+4(-2)=4-8=-4$

Vertex: (–2, –4)

y-intercept: (0, 0)

b. $D: x\in(-\infty,\infty)$

$R: y\in[-4,\infty)$

13. $p(x)=x^2-2x-3$

a. End Behavior: up, up

Axis of symmetry: $x=1$

$x^2 - 2x - 3 = 0$

$(x-3)(x+1) = 0$

$x = -3; \quad x = -1$

x-intercepts: (–1, 0) and (3, 0)

$x = \dfrac{-1+3}{2} = \dfrac{2}{2} = 1$

$p(1) = 1^2 - 2(1) - 3 = -3$

Vertex: (1, –4)

y-intercept: (0, –3)

b. $D: x \in (-\infty, \infty)$

$R: y \in [-4, \infty)$

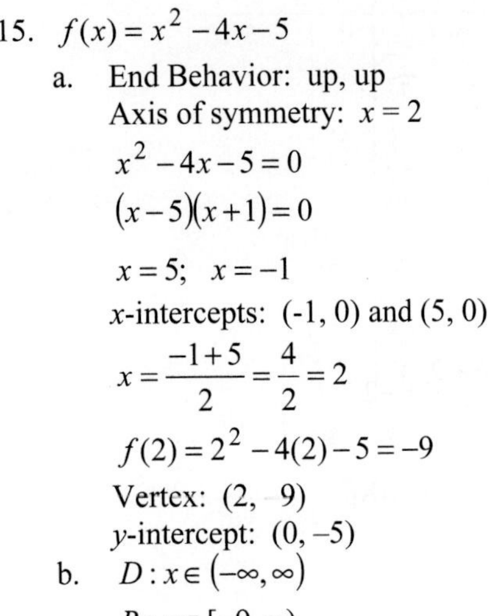

15. $f(x) = x^2 - 4x - 5$

a. End Behavior: up, up

Axis of symmetry: $x = 2$

$x^2 - 4x - 5 = 0$

$(x-5)(x+1) = 0$

$x = 5; \quad x = -1$

x-intercepts: (-1, 0) and (5, 0)

$x = \dfrac{-1+5}{2} = \dfrac{4}{2} = 2$

$f(2) = 2^2 - 4(2) - 5 = -9$

Vertex: (2, 9)

y-intercept: (0, –5)

b. $D: x \in (-\infty, \infty)$

$R: y \in [-9, \infty)$

17. $f(x) = x^2 + 5x$

Concavity: Up

$x^2 + 5x = 0$

$x(x+5) = 0$

$x = 0; \quad x = -5$

x-intercepts: (0, 0) and (–5, 0)

y-intercept: (0, 0)

$f(0) = 0 + 0 = 0$

$x = \dfrac{0+-5}{2} = -2.5$

$f(-2.5) = (-2.5)^2 + 5(-2.5) = -6.25$

Vertex: (–2.5, –6.25)

Axis of symmetry: $x = -2.5$

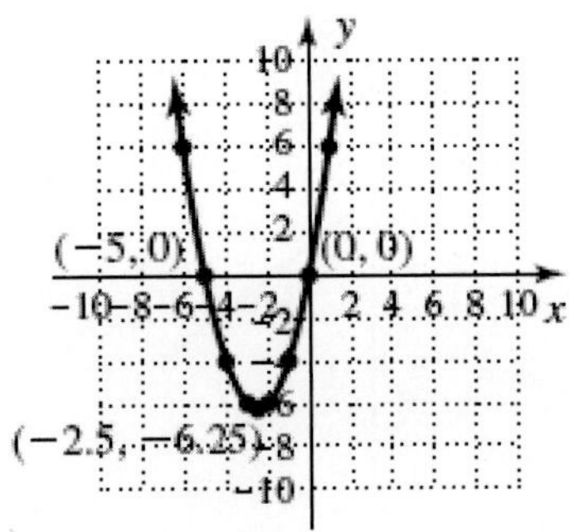

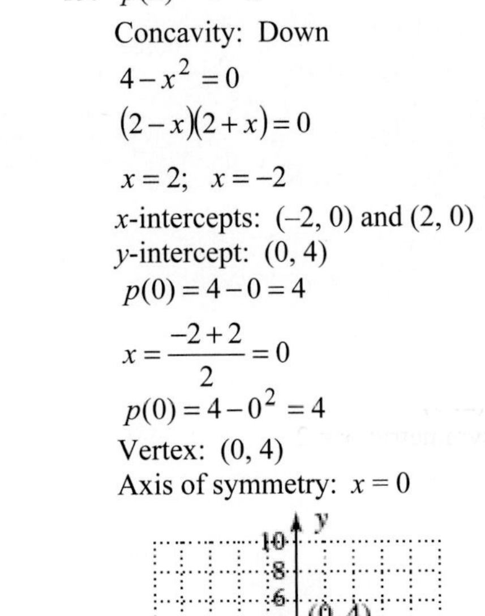

19. $p(x) = 4 - x^2$

Concavity: Down

$4 - x^2 = 0$

$(2-x)(2+x) = 0$

$x = 2; \quad x = -2$

x-intercepts: (–2, 0) and (2, 0)

y-intercept: (0, 4)

$p(0) = 4 - 0 = 4$

$x = \dfrac{-2+2}{2} = 0$

$p(0) = 4 - 0^2 = 4$

Vertex: (0, 4)

Axis of symmetry: $x = 0$

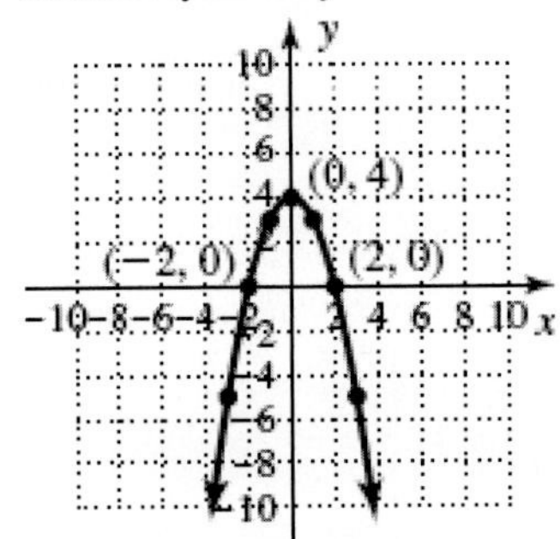

21. $r(t) = t^2 - 3t - 4$

Concavity: Up

$t^2 - 3t - 4 = 0$

$(t-4)(t+1) = 0$

$t = 4; \quad t = -1$

x-intercepts: (4, 0) and (–1, 0)

y-intercept: (0, –4)

$r(0) = 0 - 0 - 4 = -4$

$t = \dfrac{4+-1}{2} = 1.5$

$r(1.5) = (1.5)^2 - 3(1.5) - 4 = -6.25$

Vertex: (1.5, –6.25)

Axis of symmetry: $x = -2.5$

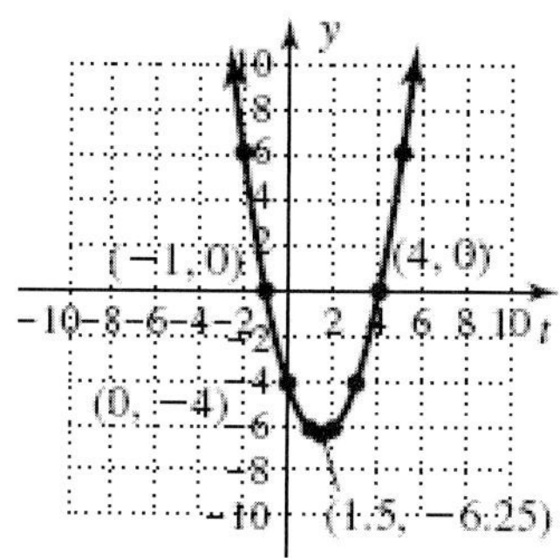

23. $f(x) = -x^2 + 4x - 4$

Concavity: Down

$-\left(x^2 - 4x + 4\right) = 0$

$-(x-2)(x-2) = 0$

$x = 2$

x-intercept: (2, 0)
y-intercept: (0, –4)

$f(0) = -0^2 + 4(0) - 4 = -4$

$x = 2$

$f(2) = -(2)^2 + 4(2) - 4 = 0$

Vertex: (2, 0)
Axis of symmetry: $x = 2$

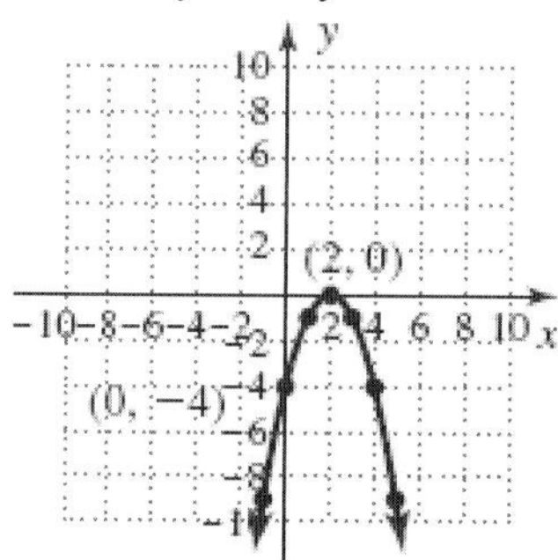

25. $y = 2x^2 + 7x - 4$

Concavity: Up

$2x^2 + 7x - 4 = 0$

$(2x-1)(x+4) = 0$

$x = \frac{1}{2};\quad x = -4$

x-intercepts: $\left(\frac{1}{2}, 0\right);\ (-4, 0)$

y-intercept: (0, –4)

$f(0) = 2(0)^2 + 7(0) - 4 = -4$

$$x = \frac{\frac{1}{2} + -4}{2} = -\frac{7}{4} = -1.75$$

$f(-1.75) = 2(-1.75)^2 + 7(-1.75) - 4 = -10.125$

Vertex: (–1.75, –10.125)

Axis of symmetry: $x = -1.75$

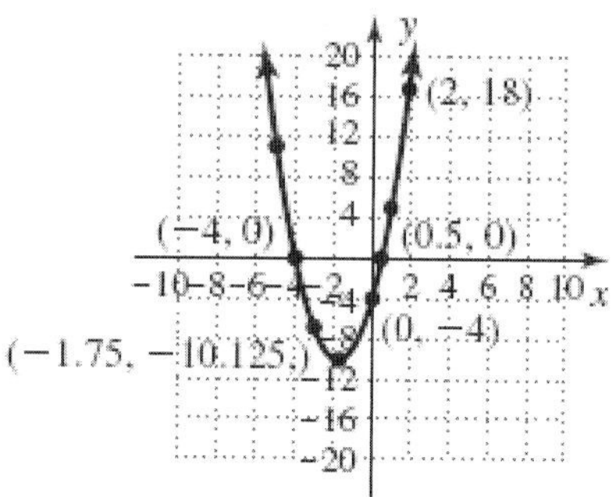

27. $p(t) = 12 - t^2 - 4t$

Concavity: Up

$-t^2 - 4t + 12 = 0$

$-\left(t^2 + 4t - 12\right) = 0$

$-(t+6)(t-2) = 0$

$t = -6;\quad t = 2$

x-intercepts: (–6, 0) and (2, 0)
y-intercept: (0, 12)

$p(0) = 12 - (0)^2 - 4(0) = 12$

$$t = \frac{-6+2}{2} = -2$$

$p(6) = 12 - (-2)^2 - 4(-2) = 16$

Vertex: (–2, 16)
Axis of symmetry: $x = -2$

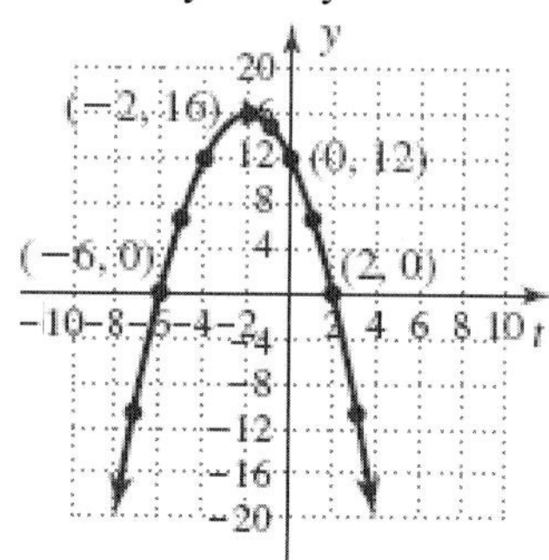

29. $f(x) = \sqrt{x} + 1$

x	$f(x) = \sqrt{x} + 1$
1	$f(1) = \sqrt{1} + 1 = 1 + 1 = 2$
4	$f(4) = \sqrt{4} + 1 = 2 + 1 = 3$
9	$f(9) = \sqrt{9} + 1 = 3 + 1 = 4$
16	$f(16) = \sqrt{16} + 1 = 4 + 1 = 5$
25	$f(25) = \sqrt{25} + 1 = 5 + 1 = 6$

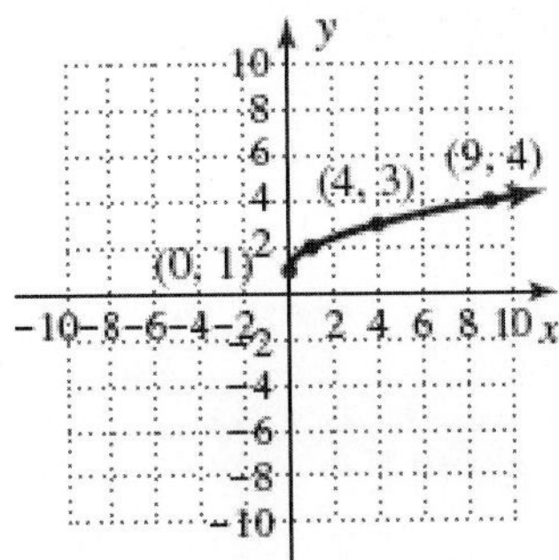

31. $p(x) = 2\sqrt{x+3}$

x	$p(x) = 2\sqrt{x+3}$
-3	$p(-3) = 2\sqrt{(-3)+3} = 2\sqrt{0} = 2(0) = 0$
-2	$p(-2) = 2\sqrt{(-2)+3} = 2\sqrt{1} = 2(1) = 2$
1	$p(1) = 2\sqrt{1+3} = 2\sqrt{4} = 2(2) = 4$
6	$p(6) = 2\sqrt{6+3} = 2\sqrt{9} = 2(3) = 6$
13	$p(13) = 2\sqrt{13+3} = 2\sqrt{16} = 2(4) = 8$
22	$p(22) = 2\sqrt{22+3} = 2\sqrt{25} = 2(5) = 10$

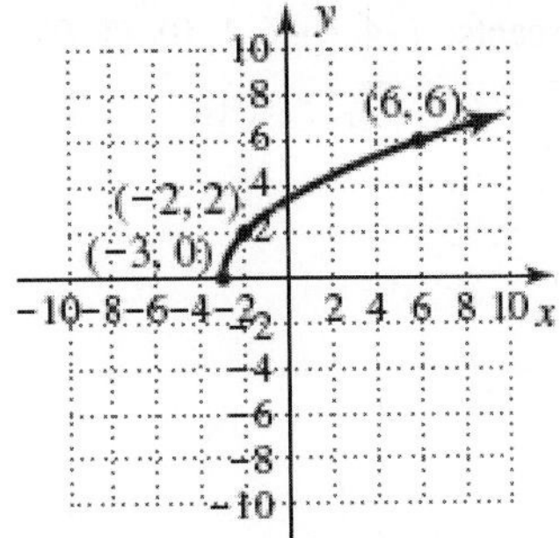

33. $f(x) = \sqrt{x+3} - 2$

Node: (-3, -2)

$x+3 \geq 0$

$x \geq -3$

End-behavior: Up to the right

x-intercept: (1, 0)

$0 = \sqrt{x+3} - 2$

$2 = \sqrt{x+3}$

$4 = x+3$

$1 = x$

y-intercept: (0, –0.268)

$y = \sqrt{0+3} - 2$

$y = \sqrt{3} - 2$

$y = 1.732 - 2$

$y = -0.268$

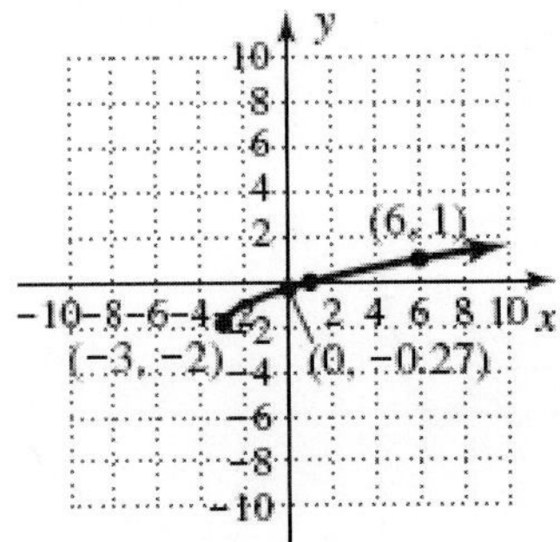

35. $r(x) = -\sqrt{x+4} - 1$

Node: (-4, -1)

$x+4 \geq 0$

$x \geq -4$

End-behavior: Down to the right

x-intercept: None

y-intercept: (0, -3)

$y = -\sqrt{0+4} - 1$

$y = -\sqrt{4} - 1$

$y = -2 - 1$

$y = -3$

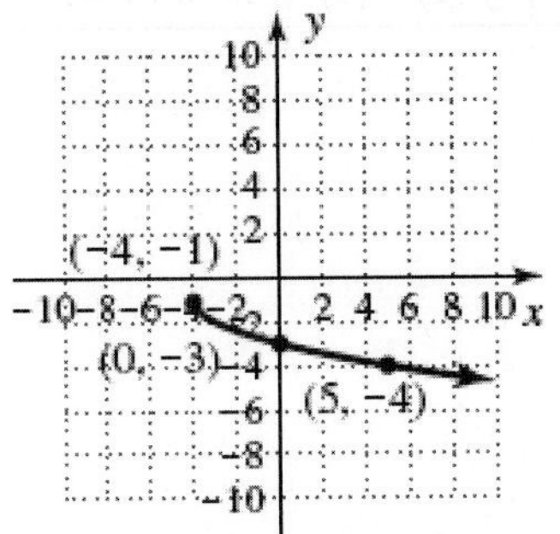

37. $p(x) = 2\sqrt{x+1} - 3$

Node: (-1, -3)

$x+1 \geq 0$

$x \geq -1$

End-behavior: Up to the right

x-intercept: (1.25, 0)

$0 = 2\sqrt{x+1} - 3$

$3 = 2\sqrt{x+1}$

$\frac{3}{2} = \sqrt{x+1}$

$\frac{9}{4} = x+1$

$\frac{5}{4} = x$

$1.25 = x$

y-intercept: (0, –1)

$y = 2\sqrt{0+1} - 3$
$y = 2\sqrt{1} - 3$
$y = 2(1) - 3$
$y = 2 - 3$
$y = -1$

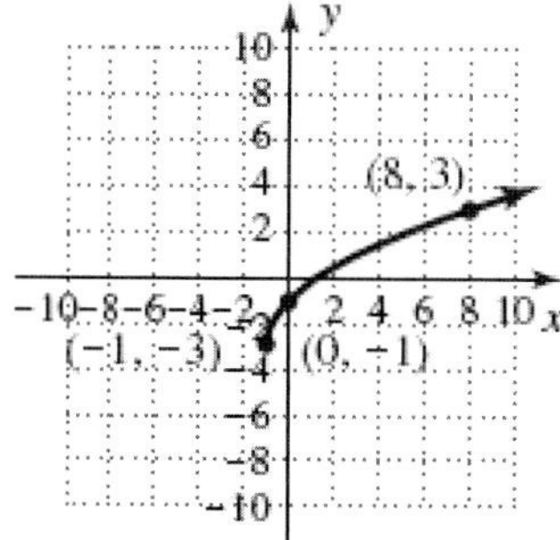

39. $f(x) = x^3 + 1$

x	$f(x) = x^3 + 1$
-2	$f(-2) = (-2)^3 + 1 = -8 + 1 = -7$
-1	$f(-1) = (-1)^3 + 1 = -1 + 1 = 0$
0	$f(0) = (0)^3 + 1 = 0 + 1 = 1$
1	$f(1) = (1)^3 + 1 = 1 + 1 = 2$
2	$f(2) = (2)^3 + 1 = 8 + 1 = 9$

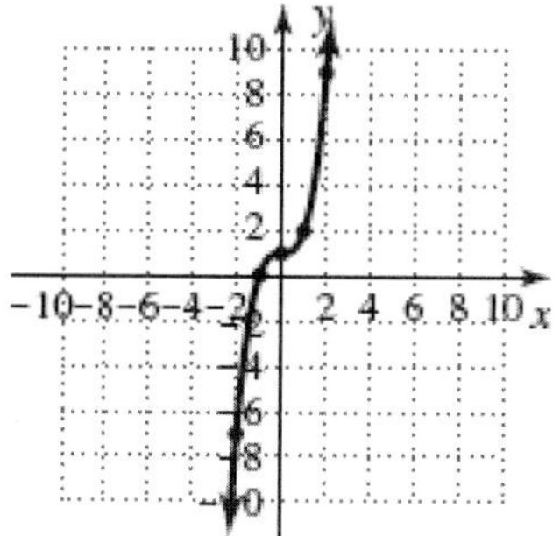

41. $p(x) = (x+2)^3$

x	$p(x) = (x+2)^3$
-4	$p(-4) = (-4+2)^3 = (-2)^3 = -8$
-3	$p(-3) = (-3+2)^3 = (-1)^3 = -1$
-2	$p(-2) = (-2+2)^3 = 0^3 = 0$
-1	$p(-1) = (-1+2)^3 = 1^3 = 1$
0	$p(0) = (0+2)^3 = 2^3 = 8$

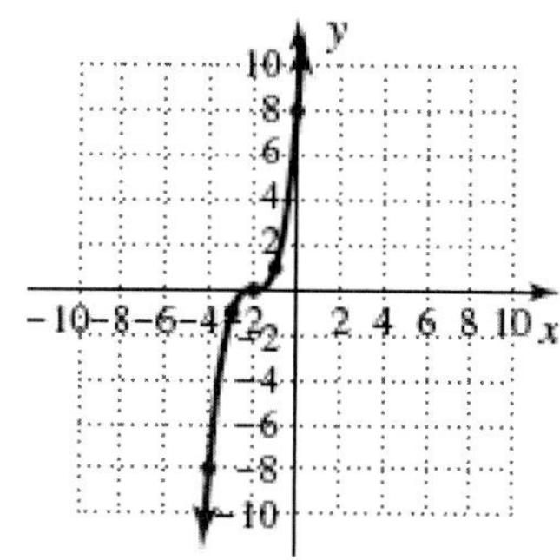

43. $f(x) = -x^3 + 3x^2 - 3x + 1$
   a. End behavior: Up on the left; down on the right
      x-intercept: (1, 0)
      y-intercept: (0, 1)
   b. $x \in (-\infty, \infty)$
      $y \in (-\infty, \infty)$
   c. Point of inflection: (1, 0)

45. $p(x) = x^3 + 4x^2 - x - 4$
   a. End behavior: Down on the left; up on the right
      x-intercepts: (–4, 0), (–1, 0), (1, 0)
      y-intercept: (0, –4)
   b. $x \in (-\infty, \infty)$
      $y \in (-\infty, \infty)$
   c. Point of inflection: $(-1.3, 2.1)$

47. $v(x) = -x^3 + 5x^2 - 2x - 8$
   a. End behavior: Up on the left; down on the right
      x-intercepts: (–1, 0), (2, 0), (4, 0)
      y-intercept: (0, –8)
   b. $x \in (-\infty, \infty)$
      $y \in (-\infty, \infty)$
   c. Point of inflection: $(1.7, -2.1)$

49. $f(x) = 4x - x^3$
   End behavior: Up on the left; down on the right
   x-intercepts: (–2, 0), (0, 0), (2, 0)

$$4x - x^3 = 0$$
$$x(4 - x^2) = 0$$
$$x(2-x)(2+x) = 0$$
$$x = 0; \quad x = 2; \quad x = -2$$

   y-intercept: (0, 0)

$$f(0) = 4(0) - (0)^3 = 0$$

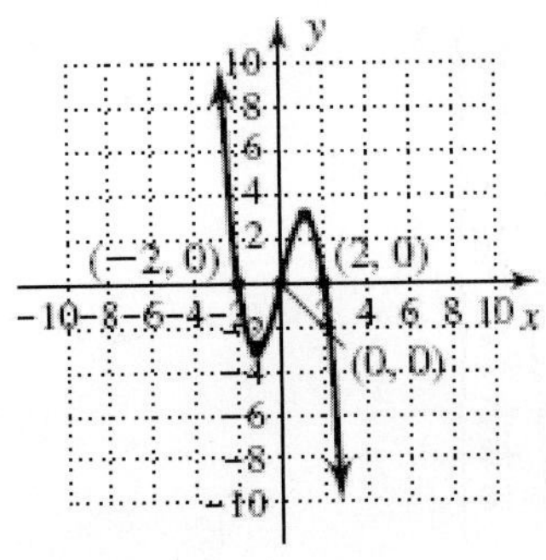

51. $v(x) = x^3 - 2x^2 - 3x$

End behavior: Down on the left; up on the right

x-intercepts: (–1, 0), (0, 0), (3, 0)

$x^3 - 2x^2 - 3x = 0$

$x(x^2 - 2x - 3) = 0$

$x(x-3)(x+1) = 0$

$x = 0;\quad x = 3;\quad x = -1$

y-intercept: (0, 0)

$v(0) = (0)^3 - 2(0)^2 - 3(0) = 0$

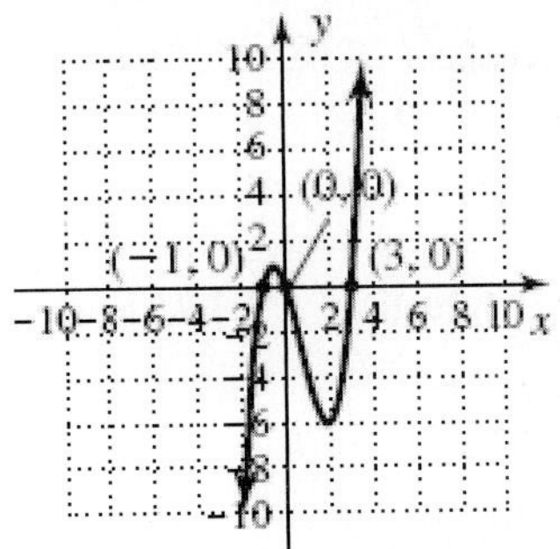

53. $r(x) = x^3 + x^2 - 4x - 4$

End behavior: Down on the left; up on the right

x-intercepts: (–2, 0), (–1, 0), (2, 0)

$x^3 + x^2 - 4x - 4 = 0$

$(x^3 + x^2) - (4x + 4) = 0$

$x^2(x+1) - 4(x+1) = 0$

$(x+1)(x^2 - 4) = 0$

$(x+1)(x-2)(x+2) = 0$

$x = -2;\quad x = -1;\quad x = 2$

y-intercept: (0, –4)

$r(0) = (0)^3 + (0)^2 - 4(0) - 4 = -4$

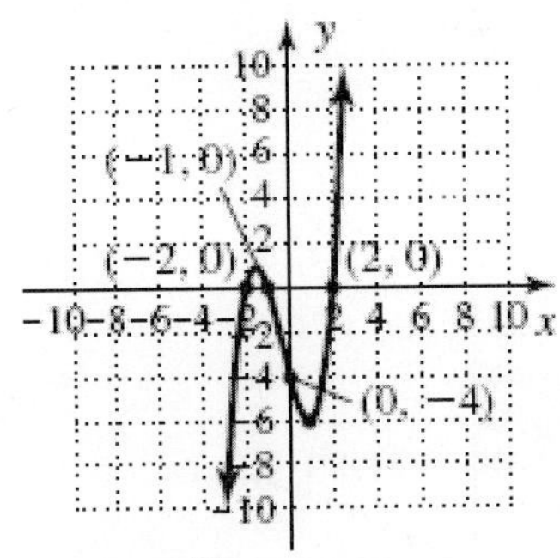

55. $f(x) = \sqrt[3]{x} + 1$

x	$f(x) = \sqrt[3]{x} + 1$
-8	$f(-8) = \sqrt[3]{-8} + 1 = -2 + 1 = -1$
-1	$f(-1) = \sqrt[3]{-1} + 1 = -1 + 1 = 0$
0	$f(0) = \sqrt[3]{0} + 1 = 0 + 1 = 1$
1	$f(1) = \sqrt[3]{1} + 1 = 1 + 1 = 2$
8	$f(8) = \sqrt[3]{8} + 1 = 2 + 1 = 3$

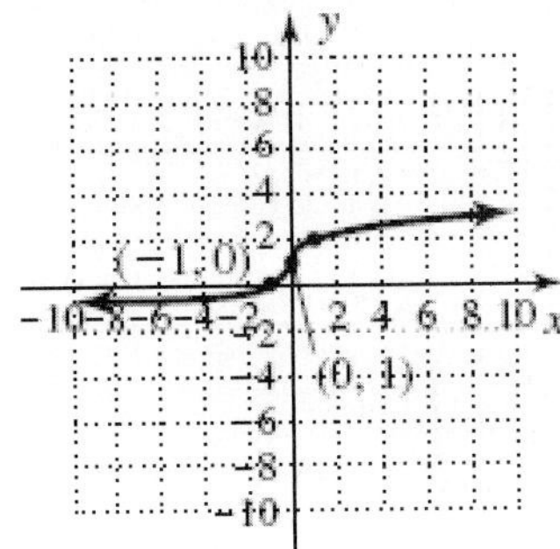

57. $p(x) = \sqrt[3]{x+2}$

x	$p(x) = \sqrt[3]{x+2}$
-10	$p(-10) = \sqrt[3]{-10+2} = \sqrt[3]{-8} = -2$
-3	$p(-3) = \sqrt[3]{-3+2} = \sqrt[3]{-1} = -1$
-2	$p(-2) = \sqrt[3]{-2+2} = \sqrt[3]{0} = 0$
-1	$p(-1) = \sqrt[3]{-1+2} = \sqrt[3]{1} = 1$
6	$p(6) = \sqrt[3]{6+2} = \sqrt[3]{8} = 2$

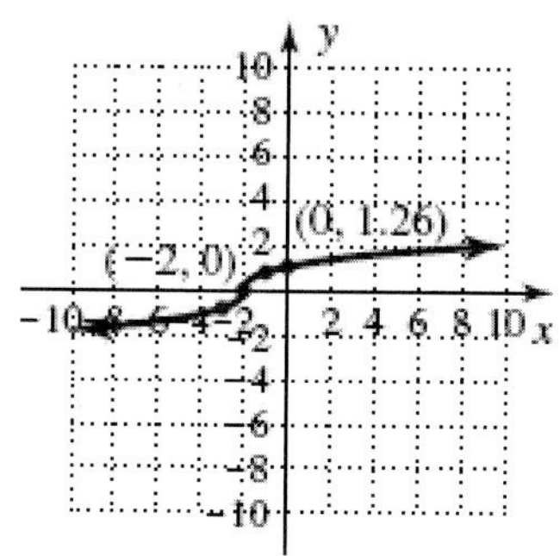

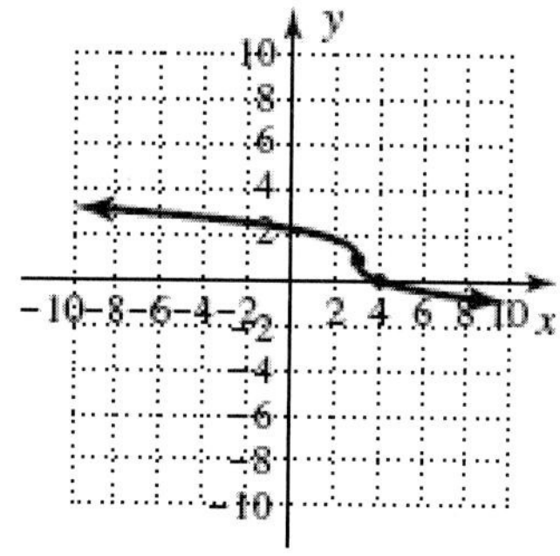

59. $f(x)=\sqrt[3]{x+1}-2$

End behavior: Increases from left to right

$$\sqrt[3]{x+1}-2=0$$
$$\sqrt[3]{x+1}=2$$
$$x+1=8$$
$$x=7$$

x-intercept: (7, 0)

$f(0)=\sqrt[3]{0+1}-2=\sqrt[3]{1}-2=1-2=-1$

y-intercept: (0, -1)

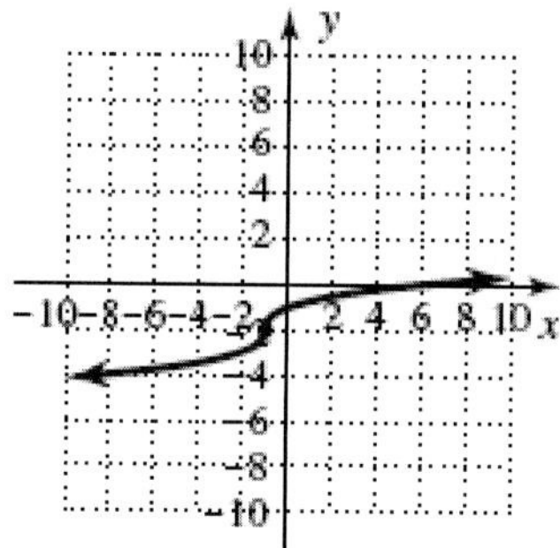

61. $r(x)=-\sqrt[3]{x-2}+1$

End behavior: Decreases from left to right

$$-\sqrt[3]{x-2}+1=0$$
$$-\sqrt[3]{x-2}=-1$$
$$\sqrt[3]{x-2}=1$$
$$x-2=1$$
$$x=3$$

x-intercept: (3, 0)

$r(0)=-\sqrt[3]{0-2}+1=-\sqrt[3]{-2}+1=2.2599$

y-intercept: (0, 2.2599)

63. $p(x)=2\sqrt[3]{x+3}-2$

End behavior: Increases from left to right

$$2\sqrt[3]{x+3}-2=0$$
$$2\sqrt[3]{x+3}=2$$
$$\sqrt[3]{x+3}=1$$
$$x+3=1$$
$$x=-2$$

x-intercept: (-2, 0)

$p(0)=2\sqrt[3]{0+3}-2=2\sqrt[3]{3}-2=0.8845$

y-intercept: (0, 0.8845)

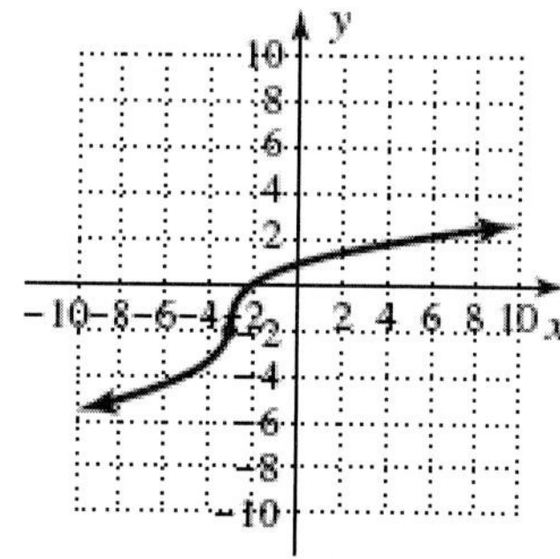

65. $f(x)=\sqrt{x+3}-1$; C

67. $p(x)=|x+1|-2$; A

69. $r(x)=-|x-1|+1$; K

71. $Y_1=\sqrt[3]{x}+1$; D

73. $f(x)=-\sqrt[3]{x+1}$; I

75. $p(x)=\frac{3}{2}x-2$; L

77. $v=\sqrt{2gs}$

a. $v=\sqrt{2(32)(25)}=\sqrt{1600}=40$ ft/sec

b. $v = \sqrt{2gs}$

$v^2 = 2gs$

$\frac{v^2}{2g} = s$

$\frac{84^2}{2(32)} = s$

$\frac{7056}{64} = s$

$110.25 ft = s$

79. $f(x) = x^3$

a. $\frac{\Delta f}{\Delta x} = \frac{f(-1)-f(-2)}{-1-(-2)} = \frac{-1-(-8)}{1} = 7$

b. $\frac{\Delta f}{\Delta x} = \frac{f(2)-f(1)}{2-1} = \frac{8-1}{1} = 7$

c. They are the same.

d. Slope of the line is the same.

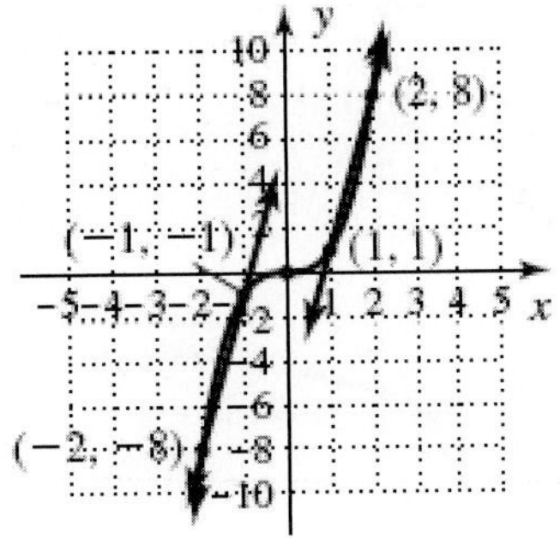

81. $h(t) = -16t^2 + 192t$

a. $h(1) = -16(1)^2 + 192(1) = 176 ft$

b. $h(2) = -16(2)^2 + 192(2) = 320 ft$

c. $\frac{\Delta h}{\Delta t} = \frac{h(2)-h(1)}{2-1} = \frac{320-176}{1}$

$= 144$ ft/sec

d. $\frac{\Delta h}{\Delta t} = \frac{h(11)-h(10)}{11-10} = \frac{176-320}{1}$

$= -144$ ft/sec

The arrow is going down.

83. $v = \sqrt{2gs}$

a. $v = \sqrt{2(32)(5)} = \sqrt{320} = 17.89$ ft/sec ;

$v = \sqrt{2(32)(10)} = \sqrt{640} = 25.30$ ft/sec

b. $v = \sqrt{2(32)(15)} = \sqrt{960} = 30.98$ ft/sec ;

$v = \sqrt{2(32)(20)} = \sqrt{1280} = 35.78$ ft/sec

c. Between $s = 5$ and $s = 10$, you are decelerating.

d. $\frac{\Delta v}{\Delta s} = \frac{v(10)-v(5)}{10-5} = \frac{25.3-17.89}{5}$

$= 1.482$ ft/sec;

$\frac{\Delta v}{\Delta s} = \frac{v(20)-v(15)}{20-15} = \frac{35.78-30.98}{5}$

$= 0.96$ ft/sec

85. $f(x) = \frac{2}{3}x - 3$

$\frac{\Delta f}{\Delta x} = \frac{f(4)-f(3)}{4-3} = \frac{-\frac{1}{3}-(-1)}{1} = \frac{2}{3}$

$\frac{\Delta f}{\Delta x} = \frac{f(9)-f(8)}{9-8} = \frac{3-\frac{7}{3}}{1} = \frac{2}{3}$

a. They are the same.

b. Slope $\left(\frac{2}{3}\right)$ is constant for a line.

c. $\frac{2}{3}$; every change of 1 in x results in $\frac{2}{3}$ unit change in y.

87. $A(L) = -L^2 + L\left(\frac{P}{2}\right)$

$0 = -L^2 + L\left(\frac{1000}{2}\right)$

$0 = -L^2 + 500L$

$0 = -L(L-500)$

$500 = L$

Total length is 500, making the length of one side

250. Leaves 500 for the width, making each width 250. A = 250(250) = 62,500ft^2.

89. $5x - 7y = 35$

x-intercept: (7, 0)	y-intercept:(0, -5)
$5x - 7(0) = 35$	$5(0) - 7y = 35$
$5x = 35$	$-7y = 35$
$x = 7$	$y = -5$

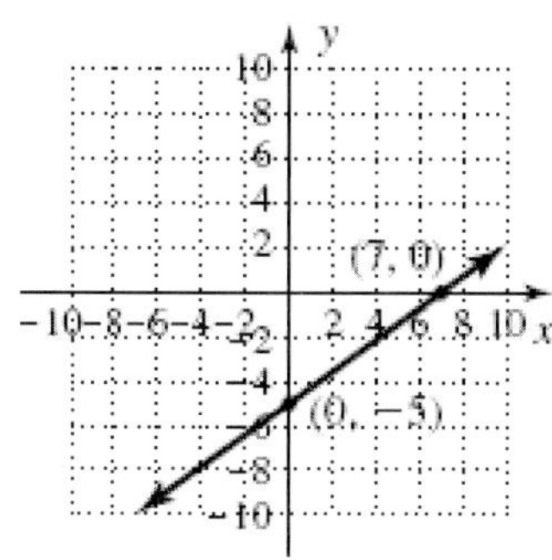

91. $-\frac{2}{3}x+7<11$

$-\frac{2}{3}x<4$

$x>4\left(-\frac{3}{2}\right)$

$x>-6$

$x\in(-6,-\infty)$

93. No; $x = 2$ is paired with $y = 2$ and $y = -2$.

2.5 Technology Highlight

Exercise 1: $y=0.2x^2+0.8x-4.2$

$f(x)\geq 0$ for $x\in(-\infty,-7]\cup[3,\infty)$

$f(x)\leq 0$ for $x\in[-7,3]$

Exercise 2: $y=-0.16x^2+0.96x+2.56$

$f(x)\geq 0$ for $x\in[-2,8]$

$f(x)\leq 0$ for $x\in(-\infty,-2]\cup[8,\infty)$

2.5 Exercises

1. Zeroes

3. All real numbers

5. $(-3,0)\cup(3,\infty)$

7. $f(x)=3x-2;\ \ f(x)>0$

$x\in\left(\frac{2}{3},\infty\right)$

$3x-2>0$

$3x>2$

$x>\frac{2}{3}$

x	$f(x)=3x-2$
0	$f(0)=3(0)-2=0-2=-2$
$\frac{1}{3}$	$f\left(\frac{1}{3}\right)=3\left(\frac{1}{3}\right)-2=1-2=-1$
$\frac{2}{3}$	$f\left(\frac{2}{3}\right)=3\left(\frac{2}{3}\right)-2=2-2=0$
1	$f(1)=3(1)-2=3-2=1$

9. $h(x)=-\frac{1}{2}x+4;\ \ h(x)\leq 0$

$x\in[8,\infty)$

$-\frac{1}{2}x+4\leq 0$

$-\frac{1}{2}x\leq -4$

$x\geq 8$

x	$h(x)=-\frac{1}{2}x+4$
7	$h(7)=-\frac{1}{2}(7)+4=-\frac{7}{2}+4=\frac{1}{2}$
8	$h(8)=-\frac{1}{2}(8)+4=-4+4=0$
9	$h(9)=-\frac{1}{2}(9)+4=-\frac{9}{2}+4=-\frac{1}{2}$
10	$h(10)=-\frac{1}{2}(10)+4=-5+4=-1$

11. $q(x)=5;\ \ q(x)<0$

No solution.

There is no value of x for which 5 is less than 0.

x	$q(x)=5$
0	$q(0)=5$
1	$q(1)=5$
2	$q(2)=5$

13. $Y_1=2x-5;\ \ Y_1<0$

$2x-5=0$

$2x=5$

$x=\frac{5}{2}$

Positive slope; $Y_1<0$ when $x<\frac{5}{2}$.

$x \in \left(-\infty, \frac{5}{2}\right)$

15. $r(x) = -\frac{3}{2}x + 2;\ \ r(x) > 0$

$-\frac{3}{2}x + 2 = 0$

$-\frac{3}{2}x = -2$

$x = -2\left(-\frac{2}{3}\right)$

$x = \frac{4}{3}$

Negative slope; $r(x) > 0$ when $x < \frac{4}{3}$.

$x \in \left(-\infty, \frac{4}{3}\right)$

17. $v(x) = -0.5x + 4;\ \ v(x) \le 0$

$-0.5x + 4 = 0$

$-0.5x = -4$

$x = 8$

Negative slope; $v(x) \le 0$ when $x \ge 8$.

$x \in [8, \infty)$

19. $f(x) = -x + 4;\ \ f(x) \le 0$

$-x + 4 = 0$

$-x = -4$

$x = 4$

Negative slope; $f(x) \le 0$ when $x \ge 4$.

$x \in [4, \infty)$

21. $f(x) = |x| - 3;\ \ f(x) \le 0$

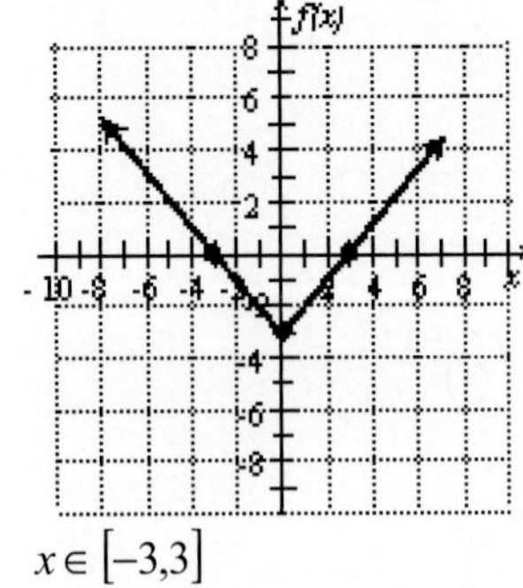

$x \in [-3, 3]$

23. $h(x) = |x - 2| + 1;\ \ h(x) < 0$

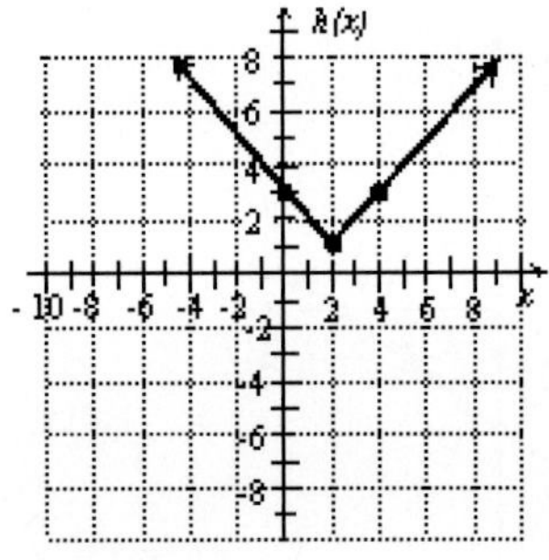

No solution

25. $q(x) = -|x| - 3;\ \ q(x) < 0$

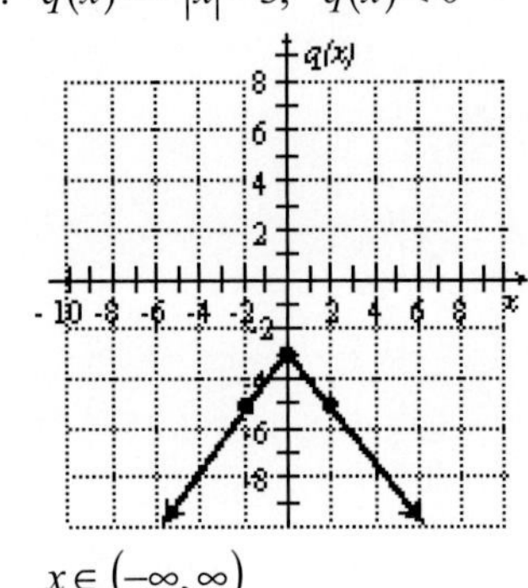

$x \in (-\infty, \infty)$

27. $f(x) > 0$

$x \in (-\infty, 3)$

29. $h(x) \ge 0$

$x \in (-\infty, -3]$

31. $g(x) < 0$

No solution

33. $f(x) > 0$

$x \in (-\infty, -2) \cup (3, \infty)$

35. $h(x) \ge 0$

$x \in [-4, 3]$

37. $g(x) < 0$

No solution

39. $f(x) > 0$

$x \in (-\infty, -4) \cup (-1, 2)$

41. $h(x) \le 0$

$x \in [-4, -3]$

43. $q(x) \ge 0$

$x \in (-\infty,-3] \cup \{2\}$

45. $f(x) = -x^2 + 4x;\ \ f(x) > 0$

$-x^2 + 4x = 0$

$-x(x-4) = 0$

$x = 0;\ \ x = 4$

Concave down

$x \in (0,4)$

47. $h(x) = x^2 + 4x - 5;\ \ h(x) \geq 0$

$x^2 + 4x - 5 = 0$

$(x+5)(x-1) = 0$

$x = -5;\ \ x = 1$

Concave up

$x \in (-\infty,-5] \cup [1,\infty)$

49. $q(x) = 2x^2 - 5x - 7;\ \ q(x) < 0$

$2x^2 - 5x - 7 = 0$

$(2x-7)(x+1) = 0$

$x = \frac{7}{2};\ \ x = -1$

Concave up

$x \in \left(-1, \frac{7}{2}\right)$

51. $s(x) = 7 - x^2;\ \ s(x) \geq 0$

$7 - x^2 = 0$

$7 = x^2$

$\pm\sqrt{7} = x$

Concave down

$x \in \left[-\sqrt{7}, \sqrt{7}\right]$

53. $Y_1 = x^2 + 3x - 6;\ \ Y_1 \geq 0$

$x^2 + 3x - 6 = 0$

$$x = \frac{-3 \pm \sqrt{3^2 - 4(1)(-6)}}{2(1)}$$

$$x = \frac{-3 \pm \sqrt{9+24}}{2}$$

$$x = \frac{-3 \pm \sqrt{33}}{2}$$

Concave up

$$x \in \left(-\infty, \frac{-3-\sqrt{33}}{2}\right] \cup \left[\frac{-3+\sqrt{33}}{2}, \infty\right)$$

55. $Y_3 = 3x^2 - 2x - 5;\ \ Y_3 \geq 0$

$3x^2 - 2x - 5 = 0$

$(3x-5)(x+1) = 0$

$x = \frac{5}{3};\ \ x = -1$

Concave up

$x \in (-\infty,-1] \cup \left[\frac{5}{3}, \infty\right)$

57. $s(x) = x^2 - 8x + 16;\ \ s(x) \geq 0$

$x^2 - 8x + 16 = 0$

$(x-4)(x-4) = 0$

$x = 4$

Concave up

$x \in (-\infty, \infty)$

59. $r(x) = 4x^2 + 12x + 9;\ \ r(x) < 0$

$4x^2 + 12x + 9 = 0$

$(2x+3)(2x+3) = 0$

$x = -\frac{3}{2}$

Concave up

No solution

61. $g(x) = -x^2 + 10x - 25;\ \ g(x) \leq 0$

$-x^2 + 10x - 25 = 0$

$-\left(x^2 - 10x + 25\right) = 0$

$-(x-5)(x-5) = 0$

$x = 5$

Concave down

$x \in (-\infty, \infty)$

63. $Y_1 = -x^2 - 2;\ \ Y_1 > 0$

$-x^2 - 2 = 0$

$-x^2 = 2$

$x^2 = -2$

$x = \sqrt{-2}$

No x-intercepts

Concave down

No solution

65. $f(x)=x^2-2x+5;\ f(x)>0$

$x^2-2x+5=0$

$x=\dfrac{2\pm\sqrt{(-2)^2-4(1)(5)}}{2(1)}$

$x=\dfrac{2\pm\sqrt{4-20}}{2}$

$x=\dfrac{2\pm\sqrt{-16}}{2}$

No x-intercepts
Concave up
$x\in(-\infty,\infty)$

67. $p(x)=2x^2-6x+9;\ p(x)\ge 0$

$2x^2-6x+9=0$

$x=\dfrac{6\pm\sqrt{(-6)^2-4(2)(9)}}{2(2)}$

$x=\dfrac{6\pm\sqrt{36-72}}{4}$

$x=\dfrac{6\pm\sqrt{-36}}{4}$

No x-intercepts
Concave up
$x\in(-\infty,\infty)$

69. $f(x)=x^2-2x-15;\ f(x)\ge 0$

$x^2-2x-15=0$

$(x-5)(x+3)=0$

$x=5;\ x=-3$

Interval $(-\infty,-3]$: positive
Interval $[-3,5]$: negative
Interval $[5,\infty)$: positive
$x\in(-\infty,-3]\cup[5,\infty)$

71. $h(x)=2x^2-7x-15;\ h(x)<0$

$2x^2-7x-15=0$

$(2x+3)(x-5)=0$

$x=-\dfrac{3}{2};\ x=5$

Concave up

Interval $\left(-\infty,-\dfrac{3}{2}\right)$: positive

Interval $\left(-\dfrac{3}{2},5\right)$: negative

Interval $(5,\infty)$: positive

$x\in\left(-\dfrac{3}{2},5\right)$

73. $Y_1=x^3-8x;\ Y_1\ge 0$

$x^3-8x=0$

$x(x^2-8)=0$

$x=0;\ x=\pm2\sqrt{2}$

Concave up
Interval $(-\infty,-2\sqrt{2}]$: negative
Interval $[-2\sqrt{2},0]$: positive
Interval $[0,2\sqrt{2}]$: negative
Interval $[2\sqrt{2},\infty)$: positive
$x\in[-2\sqrt{2},0]\cup[2\sqrt{2},\infty)$

75. $f(x)=\sqrt{r^2-x^2}$

$f(x)=\sqrt{16-x^2}$

$16-x^2=0$

$(4-x)(4+x)=0$

$x=4;\ x=-4$

$x\in[-4,4]$

77. $f(x)=\sqrt{-3x+4}$

$-3x+4\ge 0$

$-3x\ge -4$

$x\le\dfrac{4}{3}$

$x\in\left(-\infty,\dfrac{4}{3}\right]$

79. $h(x)=\sqrt{x^2-25}$

$x^2-25=0$

$x^2=25$

$x=\pm5$

$x\in(-\infty,-5)\cup(5,\infty)$

81. $q(x)=\sqrt{x^2-5x}$

$x^2-5x=0$

$x(x-5)=0$

$x=0;\quad x=5$
$x\in(-\infty,0]\cup[5,\infty)$

83. $s(x)=\sqrt{x^2-2x-15}$
$x^2-2x-15=0$
$(x-5)(x+3)=0$
$x=-3;\quad x=5$
$x\in(-\infty,-3]\cup[5,\infty)$

85. $Y_1=\sqrt{x^2-6x+9}$
$x^2-6x+9=0$
$(x-3)(x-3)=0$
$x=3$
$x\in(-\infty,\infty)$

87. $f(x)=x^3-2x^2-15x$
$x^3-2x^2-15x=0$
$x(x^2-2x-15)=0$
$x(x-5)(x+3)=0$
$x=0;\quad x=5;\quad x=-3$
$x\in[-3,0]\cup[5,\infty)$

89. The company makes a profit from April – September.
The company loses money from December – March and September – December.

91. $h(t)=5|t-3|-5$

t	$h(t)=5\|t-3\|-5$
0	$h(0)=5\|0-3\|-5=5(3)-5=10$
1	$h(1)=5\|1-3\|-5=5(2)-5=5$
2	$h(2)=5\|2-3\|-5=5(1)-5=0$
3	$h(3)=5\|3-3\|-5=5(0)-5=-5$
4	$h(4)=5\|4-3\|-5=5(1)-5=0$
5	$h(5)=5\|5-3\|-5=5(2)-5=5$
6	$h(6)=5\|6-3\|-5=5(3)-5=10$

a. 2 seconds
b. 4 – 2 = 2 seconds
c. 5 meters
d. 4 – 2 = 2 seconds

93. $x\in\{-2\}\cup[4,\infty)$
Answers will vary.

95. $f(x)=\sqrt{x^3-9x}$
$x^3-9x=0$
$x(x^2-9)=0$
$x(x-3)(x+3)=0$
$x=0;\quad x=3;\quad x=-3$
$x\in[-3,0]\cup[3,\infty)$

97. a. $(7x)^0+5x^0+2^{-1}$
$=1+5(1)+\frac{1}{2}$
$=6\frac{1}{2}$

b. $\frac{(3m^{-3}n^4)^{-2}}{3mn}$
$=\frac{3^{-2}m^6n^{-8}}{3mn}=\frac{m^5n^{-8-1}}{3^2(3)}$
$=\frac{m^5n^{-9}}{27}=\frac{m^5}{27n^9}$

99. $3x-4y=7$
$-4y=-3x+7$
$y=\frac{3}{4}x-\frac{7}{4}$
$m=\frac{3}{4}$

101. Sum:
$2+3i+2-3i=4$
Difference:
$2+3i-(2-3i)=2+3i-2+3i=6i$
Product:
$(2+3i)(2-3i)=4-9i^2=4+9=13$
Quotient:
$\frac{2+3i}{2-3i}\cdot\frac{2+3i}{2+3i}=\frac{4+12i+9i^2}{4-9i^2}$
$=\frac{4+12i-9}{4+9}=\frac{-5+12i}{13}=-\frac{5}{13}+\frac{12}{13}i$

2.6 Exercises

1. Scatterplot

3. Linear

5. Answers will vary.

7. Positive

9. Cannot be determined

11. Positive

13. a. Linear
    b. Negative

15. a. A, D, C, B
    b. a) Positive
       b) Negative
       c) Negative
       d) Positive

17. a.

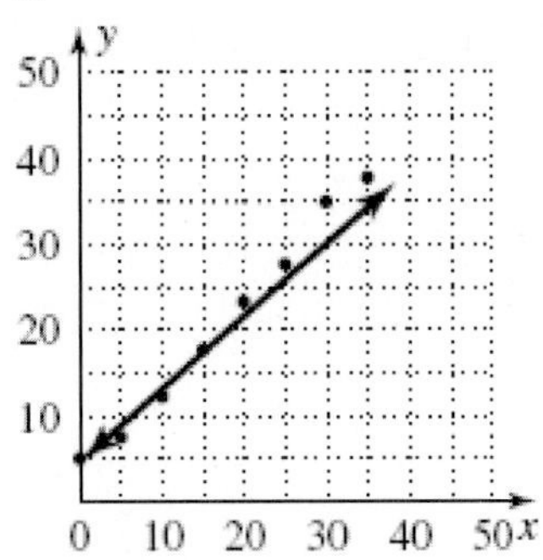

b. Positive
c. Strong

19. a.

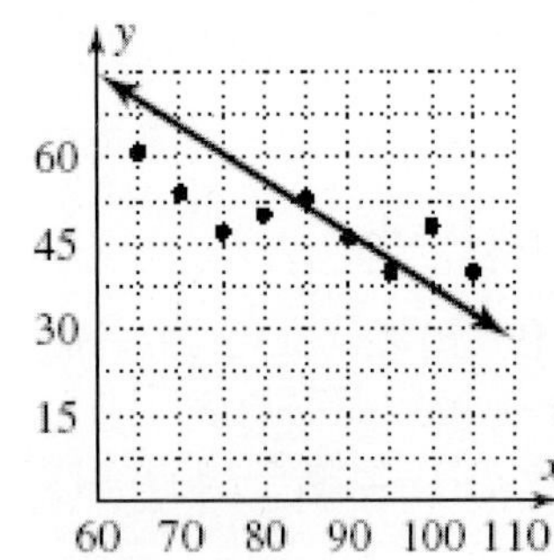

b. Negative; moderate
c. (70, 54) (80, 50)

$$m = \frac{50-54}{80-70} = -\frac{4}{10} = -0.4$$

$$y - y_1 = m(x - x_1)$$

$$y - 54 = -0.4(x - 70)$$

$$y - 54 = -0.4x + 28$$

$$y = -0.4x + 82$$

With grapher: $y = -0.4x + 82.8$

21. a.

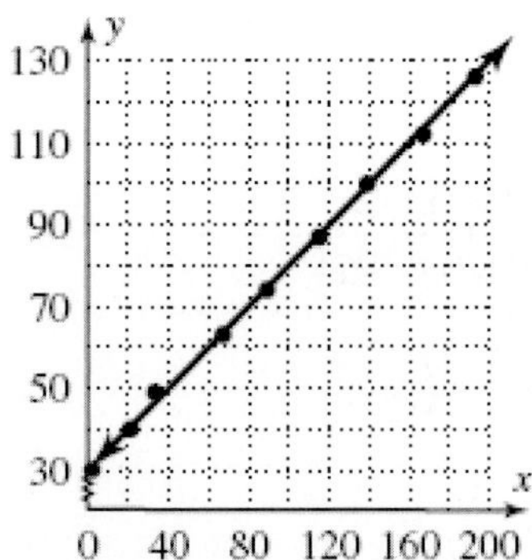

b. Positive; strong
c. (67, 63) (89, 74)

$$m = \frac{63-74}{67-89} = \frac{-11}{-22} = 0.5$$

$$y - y_1 = m(x - x_1)$$

$$y - 63 = 0.5(x - 67)$$

$$y - 63 = 0.5x - 33.5$$

$$y = 0.5x + 29.5$$

With grapher: $y = 0.5x + 30.2$

23. a.

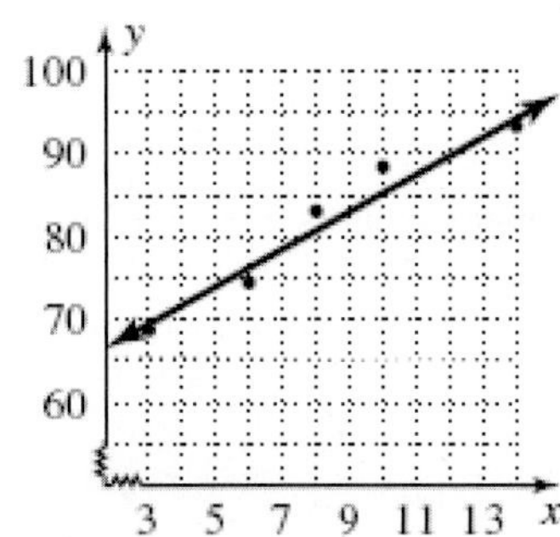

b. Positive; strong
c. (3, 74.5) (11, 93.4)

$$m = \frac{93.4-74.5}{11-3} = \frac{18.9}{8} = 2.36$$

$$y - y_1 = m(x - x_1)$$

$$y - 74.5 = 2.36(x - 3)$$

$$y - 74.5 = 2.36x - 7.08$$

$$y = 2.36x + 67.42$$

With grapher: $y = 2.4x + 69.4$

$f(2) = 2.4(2) + 69.4 = 74.2$ thousand

$f(14) = 2.4(14) + 69.4 = 103$ thousand

25. $h(t) = \frac{1}{2}gt^2 + vt$

a. $h(t) = -14.5t^2 + 90t$

b. V = 90 ft/sec

c. Venus

27. a.

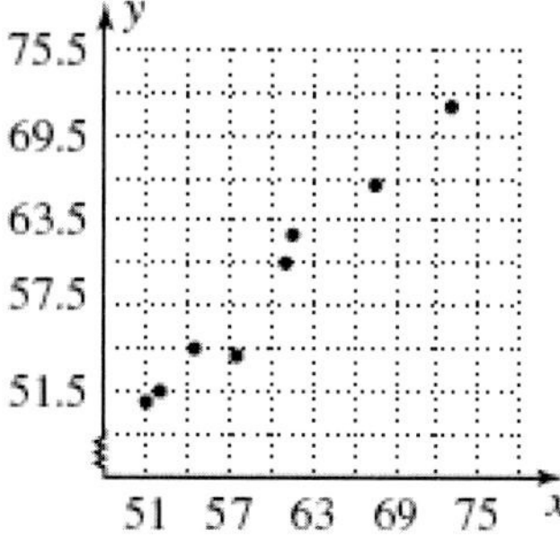

b. Linear

c. Positive

d. (73, 71.5) (51, 50.75)

$$m = \frac{71.5 - 50.75}{73 - 51} = \frac{20.75}{22} = 0.94$$

$$y - y_1 = m(x - x_1)$$

$$y - 71.5 = 0.95(x - 73)$$

$$y - 71.5 = 0.95x - 69.35$$

$$y = 0.95x + 2.15$$

With grapher: $y = 0.96x + 1.55$

$f(65) = 0.96(65) + 1.55 = 63.95in$

29. a.

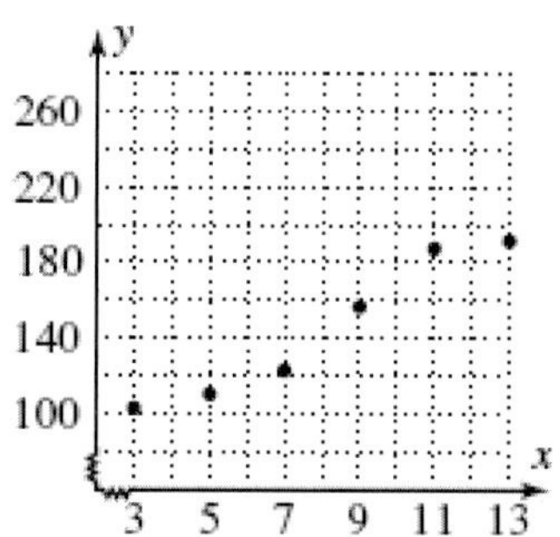

b. Linear

c. Positive

d. With grapher: $y = 9.55x + 70.42$

$f(17) = 9.55(17) + 70.42 = 232.77$

The number of patents are up because the slope is larger.

31. a.

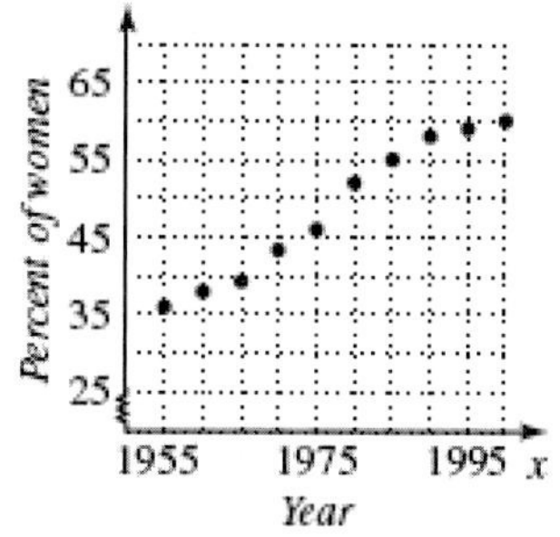

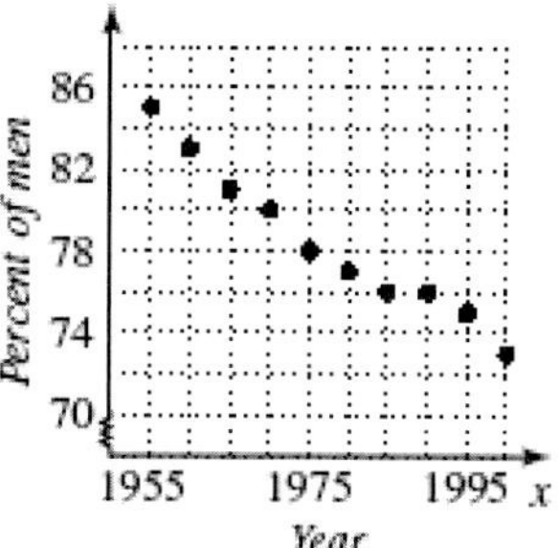

b. Women: Linear
Men: Linear

c. Women: Positive
Men: Negative

d.The percentage of females in the work force is increasing faster than the percentage of males because of the greater slope.

33. a.

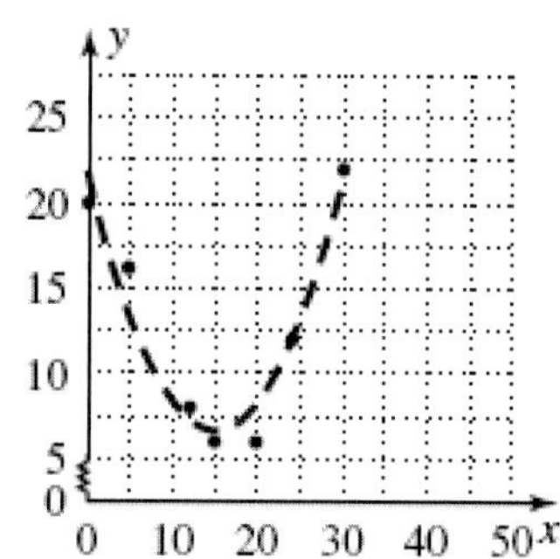

b. Strong

c. $y = 0.07x^2 - 2.02x + 21.77$

35. a.

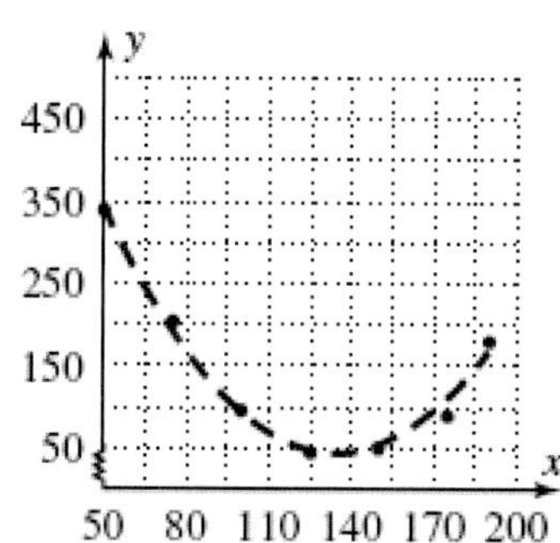

b. Strong

c. $y = 0.04x^2 - 11.32x + 807.88$

37. a. Linear

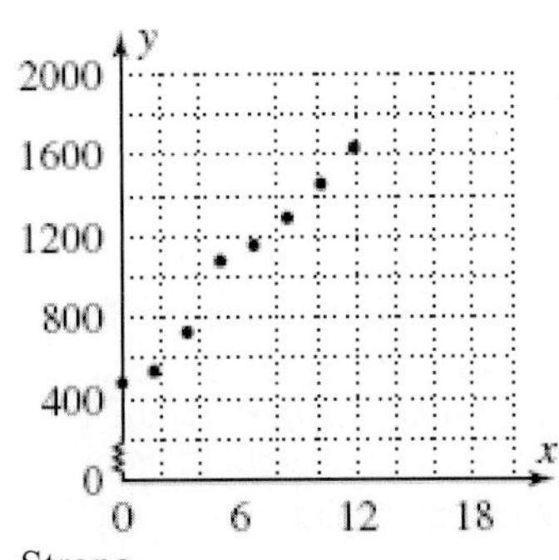

Strong

b. $y = 108.2x + 330.2$; strong

c. $f(13) = 108.2(13) + 330.2$
$= 1736.8$ billion
$f(17) = 108.2(17) + 330.2$
$= 2169.6$ billion

39. B, since there is a recognizable and fixed correspondence between the independent and dependent variables.

41. The correlation coefficient is high.
R^2, the coefficient of determination.

43. Answers will vary.

45. $P = 18 + 24 + 24 + 9\pi = 66 + 9\pi$ cm

$$A = \frac{1}{2}\pi(9)^2 + (18)(24) = 40.5\pi + 432 \text{ cm}^2$$

47. $-2(3w^2 + 5) + 4 = 7w - (w^2 + 1)$

$$-6w^2 - 10 + 4 = 7w - w^2 - 1$$

$$-6w^2 - 6 = 7w - w^2 - 1$$

$$-5w^2 - 7w - 5 = 0$$

$$w = \frac{7 \pm \sqrt{(-7)^2 - 4(-5)(-5)}}{2(-5)}$$

$$w = \frac{7 \pm \sqrt{49 - 100}}{-10}$$

$$w = \frac{7 \pm \sqrt{-51}}{-10}$$

$$w = -\frac{7}{10} \pm \frac{\sqrt{51}}{10}i$$

49. $g(x) = 2x^3 - 3x^2 + 14x - 21$

$$2x^3 - 3x^2 + 14x - 21 = 0$$

$$(2x^3 - 3x^2) + (14x - 21) = 0$$

$$x^2(2x - 3) + 7(2x - 3) = 0$$

$$(2x - 3)(x^2 + 7) = 0$$

$2x - 3 = 0$	$x^2 + 7 = 0$
$2x = 3$	$x^2 = -7$
$x = \frac{3}{2}$	$x = \pm\sqrt{7}i$

$x = \frac{3}{2}$; $x = \pm\sqrt{7}i$

Chapter 2 Summary and Review

1. a. (-4, 3) and (5, -2)

Slope triangle: $\frac{5}{9}$

(14, -7)

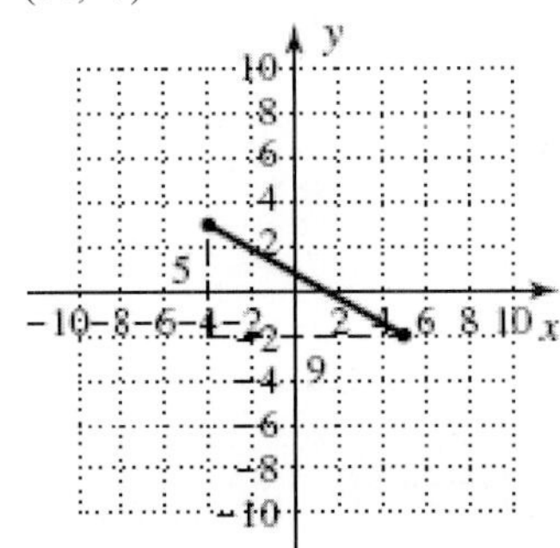

b. (3, 4) and (-6, 1)

Slope triangle: $\frac{1}{3}$

(0, 3)

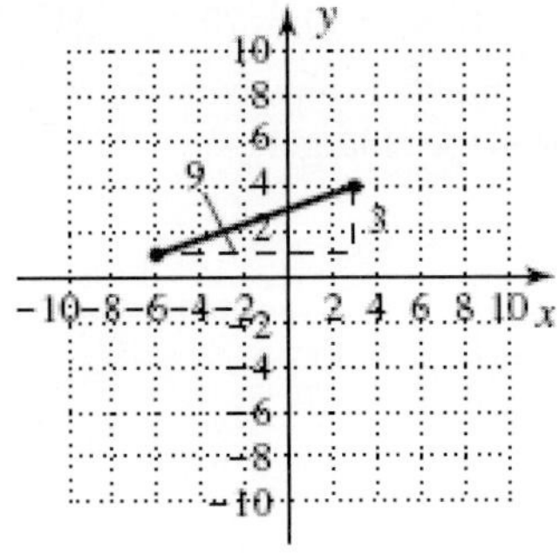

2. a. L_1: (-2, 0) and (0, 6)

$m = \dfrac{6-0}{0-(-2)} = \dfrac{6}{2} = 3$

L_2: (1, 8) and (0, 5)

$m = \dfrac{5-8}{0-1} = \dfrac{-3}{-1} = 3$

Parallel

b. L_1: (1, 10) and (-1, 7)

$m = \dfrac{7-10}{-1-1} = \dfrac{-3}{-2} = \dfrac{3}{2}$

L_2: (-2, -1) and (1, -3)

$m = \dfrac{-3-(-1)}{1-(-2)} = \dfrac{-2}{3}$

Perpendicular

3. a. $y = 3x - 2$

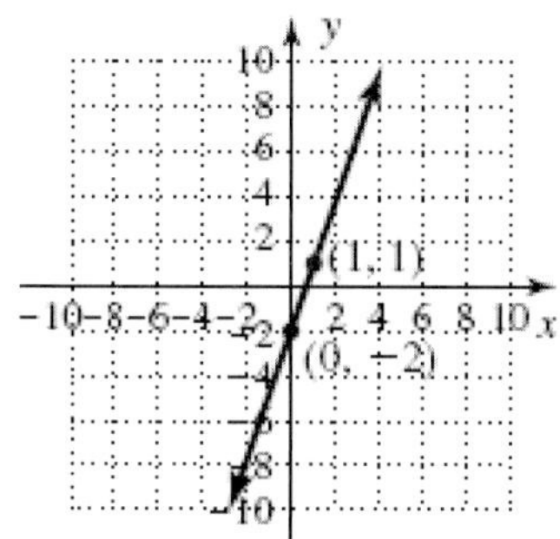

b. $y = -\dfrac{3}{2}x + 1$

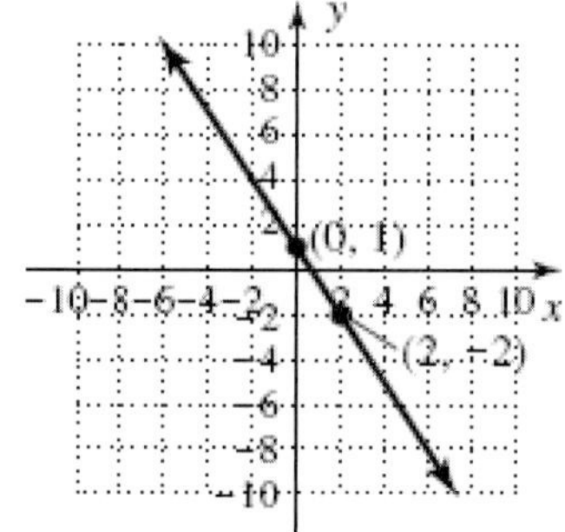

4. a. $2x + 3y = 6$

x-intercept: (3, 0) y-intercept: (0, 2)

$2x + 3(0) = 6$ $2(0) + 3y = 6$

$2x = 6$ $3y = 6$

$x = 3$ $y = 2$

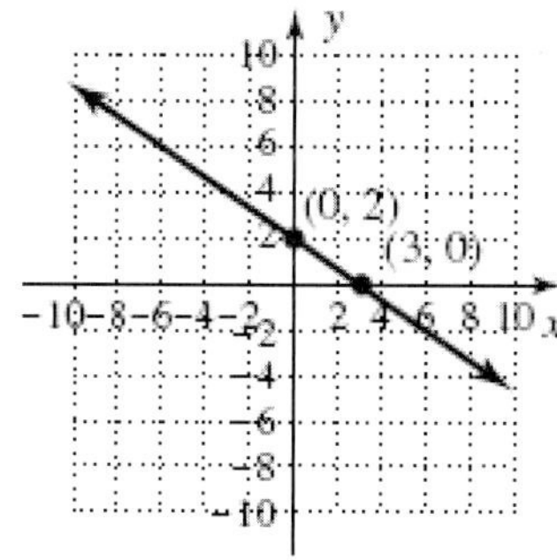

b. $y = \dfrac{4}{3}x - 2$

x-intercept: $\left(\dfrac{3}{2}, 0\right)$ y-intercept: (0,-2)

$0 = \dfrac{4}{3}x - 2$

$2 = \dfrac{4}{3}x$

$\dfrac{6}{4} = x$

$\dfrac{3}{2} = x$

$y = \dfrac{4}{3}(0) - 2$

$y = -2$

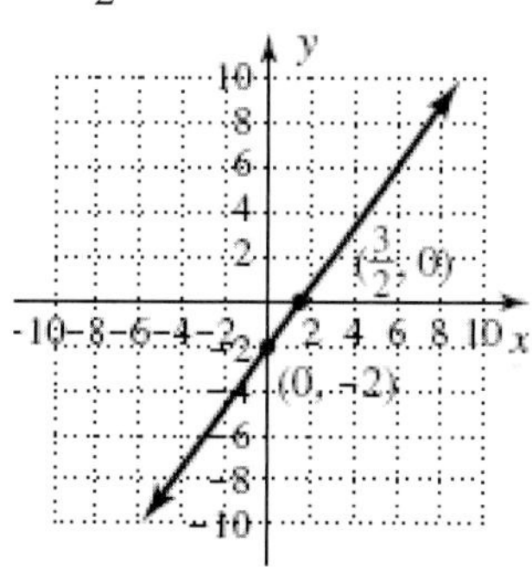

5. a. $x = 5$; vertical

b. $y = -4$; horizontal

c. $2y + x = 5$; neither

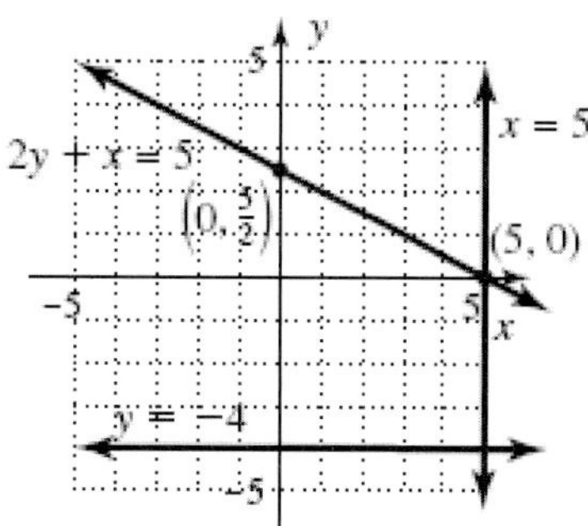

6. (-5, -4) (7, 2) (0, 16)

$m = \dfrac{16-2}{0-7} = \dfrac{14}{-7} = -2$;

$$m=\frac{2-(-4)}{7-(-5)}=\frac{6}{12}=\frac{1}{2}$$

Yes

7. $m=\frac{4}{6}=\frac{2}{3}$; y-intercept (0, 2)

   When the rodent population increases by 2000, the hawk population increases by 300.

8. (-3, -5) and (4, 5)

   Center:

   $$M=\left(\frac{-3+4}{2},\frac{-5+5}{2}\right)$$

   $$M=\left(\frac{1}{2},0\right)$$

   Diameter:

   $$d=\sqrt{(-3-4)^2+(-5-5)^2}$$

   $$d=\sqrt{(-7)^2+(-10)^2}$$

   $$d=\sqrt{49+100}$$

   $$d=\sqrt{149}$$

9. $D\in\{-7,-4,0,3,5\}$

   $R\in\{-2,0,1,3,8\}$

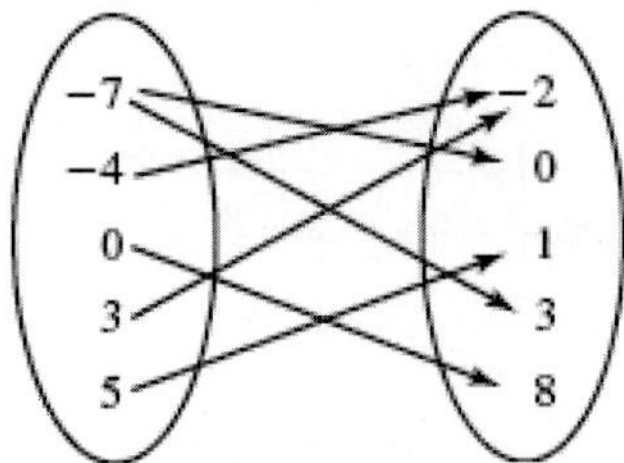

10. No, -7 is paired with 3 and 0.

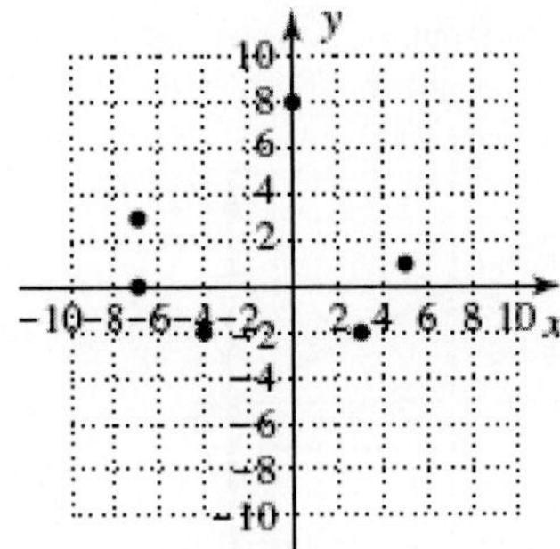

11. a. $f(x)=\sqrt{4x+5}$

   $4x+5=0$

   $4x=-5$

   $x=-\frac{5}{4}$

   $D\in\left[-\frac{5}{4},\infty\right)$

   b. $g(x)=\frac{x-4}{x^2-x-6}$

   $x^2-x-6=0$

   $(x-3)(x+2)=0$

   $x=4;\quad x=3;\quad x=-2$

   $D\in(-\infty,-2)\cup(-2,3)\cup(3,\infty)$

12. $h(x)=2x^2-3x$

   $h\left(-\frac{2}{3}\right)=2\left(-\frac{2}{3}\right)^2-3\left(-\frac{2}{3}\right)$

   $=2\left(\frac{4}{9}\right)+2=\frac{8}{9}+2=\frac{26}{9};$

   $h(3a)=2(3a)^2-3(3a)=2(9a^2)-9a$

   $=18a^2-9a;$

   $h(a-1)=2(a-1)^2-3(a-1)$

   $=2(a^2-2a+1)-3a+3$

   $=2a^2-4a+2-3a+3$

   $=2a^2-7a+5$

13. $y=\sqrt{36-x^2}$

x	$y=\sqrt{36-x^2}$
-6	$y=\sqrt{36-(-6)^2}=\sqrt{0}=0$
-4	$y=\sqrt{36-(-4)^2}=\sqrt{36-16}$ $=\sqrt{20}=4.47$
$-\sqrt{11}$	$y=\sqrt{36-(-\sqrt{11})^2}=\sqrt{36-11}$ $=\sqrt{25}=5$
0	$y=\sqrt{36-(0)^2}=\sqrt{36}=6$
$\sqrt{11}$	$y=\sqrt{36-(\sqrt{11})^2}=\sqrt{36-11}$ $=\sqrt{25}=5$
4	$y=\sqrt{36-(4)^2}=\sqrt{36-16}$ $=\sqrt{20}\approx 4.47$
6	$y=\sqrt{36-(6)^2}=\sqrt{0}=0$

$D: x\in[-6,6]; R: y\in[0,6]$

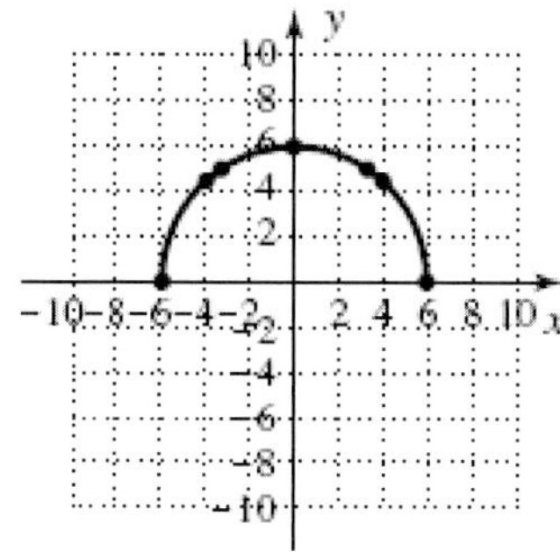

Yes, it passes the vertical line test.

14. Yes

15. I. a. $D = \{-1,0,1,2,3,4,5\}$

$R = \{-2,-1,0,1,2,3,4\}$

b. f(2) = 1

c. When f(x) = 1, x = 2

II. a. $x \in [-5,4]$

$y \in [-5,4]$

b. $f(2) = -3$

c. When $f(x) = 1$, $x = -2$

III. a. $x \in [-3, \infty)$

$y \in [-4, \infty)$

b. $f(2) = -1$

c. When $f(x) = 1$, $x = -3$ or 3

16. a. $4x + 3y - 12 = 0$

$3y = -4x + 12$

$y = -\frac{4}{3}x + 4$

$m = -\frac{4}{3}$; y-intercept (0, 4)

b. $5x - 3y = 15$

$-3y = -5x + 15$

$y = \frac{5}{3}x - 5$

$m = \frac{5}{3}$; y-intercept (0, –5)

17. a. $f(x) = -\frac{2}{3}x + 1$

$m = -\frac{2}{3}$; y-intercept (0, 1)

Slope falls

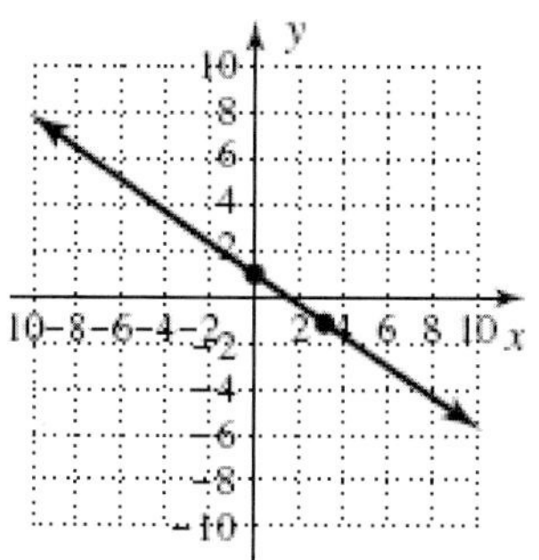

b. $h(x) = \frac{5}{2}x - 3$

$m = \frac{5}{2}$; y-intercept (0, –3)

Slope rises

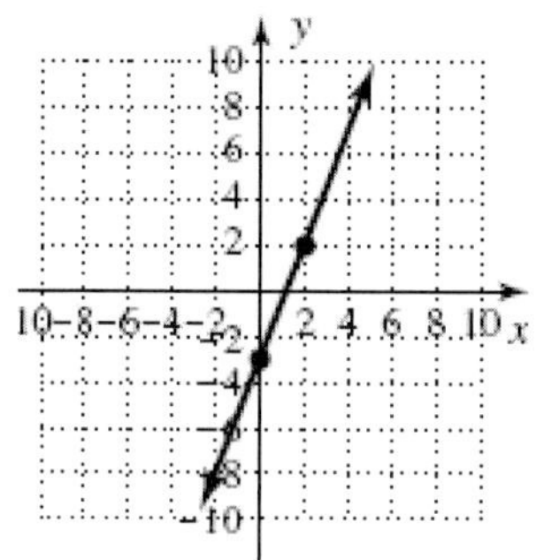

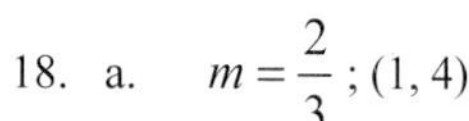

18. a. $m = \frac{2}{3}$; (1, 4)

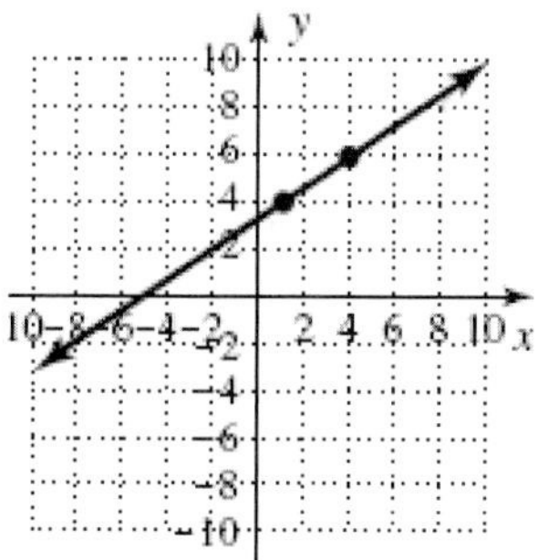

b. $m = -\frac{1}{2}$; (–2, 3)

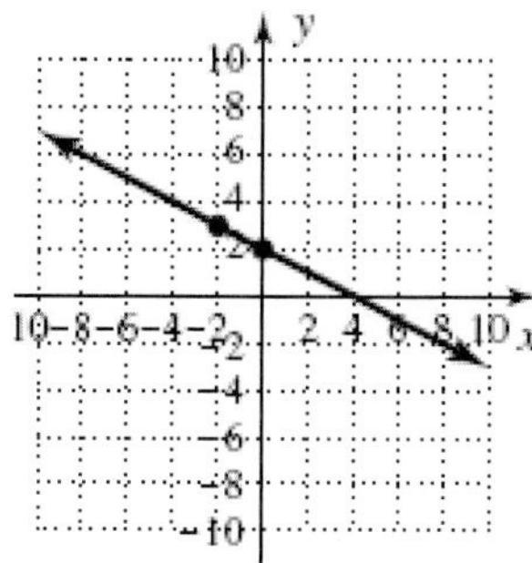

19. (–2, 5)

$x = -2; y = 5$
Point is on $y = 5$.

20. (1, 2) and (–3, 5)

$$m = \frac{5-2}{-3-1} = -\frac{3}{4}$$

$$y-2 = -\frac{3}{4}(x-1)$$

$$y-2 = -\frac{3}{4}x+\frac{3}{4}$$

$$y = -\frac{3}{4}x+\frac{11}{4}$$

21. $4x-3y=12$; (3, 4)

$$-3y = -4x+12$$

$$y = \frac{4}{3}x-4$$

$$y-4 = \frac{4}{3}(x-3)$$

$$y-4 = \frac{4}{3}x-4$$

$$y = \frac{4}{3}x$$

$$f(x) = \frac{4}{3}x$$

22. $m = \frac{2}{5}$; y-intercept (0, 2)

$$y = \frac{2}{5}x+2$$

When the rabbit population increases by 500, the wolf population increases by 200.

23. a. $m = -\frac{15}{2}$; (6, 60)

$$y-60 = -\frac{15}{2}(x-6)$$

$$y-60 = -\frac{15}{2}x+45$$

$$y = -\frac{15}{2}x+105$$

b. x-intercept: (14, 0); y-intercept: (0,105)

$$0 = -\frac{15}{2}x+105$$

$$-105 = -\frac{15}{2}x$$

$$-210 = -15x$$

$$14 = x$$

$$y = -\frac{15}{2}(0)+105$$

$$y = 105$$

c. $f(x) = -\frac{15}{2}x+105$

d. $f(20) = -\frac{15}{2}(20)+105$

$$= -150+105$$

$$= -45;$$

$$15 = -\frac{15}{2}x+105$$

$$-90 = -\frac{15}{2}x$$

$$-180 = -15x$$

$$12 = x$$

24. $f(x) = (x-2)^2$

x	$f(x) = (x-2)^2$
-1	$f(-1) = (-1-2)^2 = (-3)^2 = 9$
0	$f(0) = (0-2)^2 = (-2)^2 = 4$
1	$f(1) = (1-2)^2 = (-1)^2 = 1$
2	$f(2) = (2-2)^2 = (0^2) = 0$
3	$f(3) = (3-2)^2 = (1)^2 = 1$

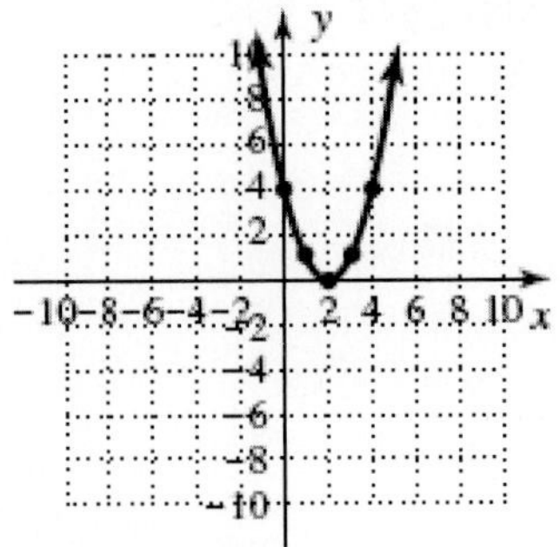

25. $g(x) = x^3 - 4x$

x	$g(x) = x^3 - 4x$
-2	$g(-2) = (-2)^3 - 4(-2) = -8+8 = 0$

-1	$g(-1)=(-1)^3-4(-1)=-1+4=3$
0	$g(0)=(0)^3-4(0)=0-0=0$
1	$g(1)=(1)^3-4(1)=1-4=-3$
2	$g(2)=(2)^3-4(2)=8-8=0$

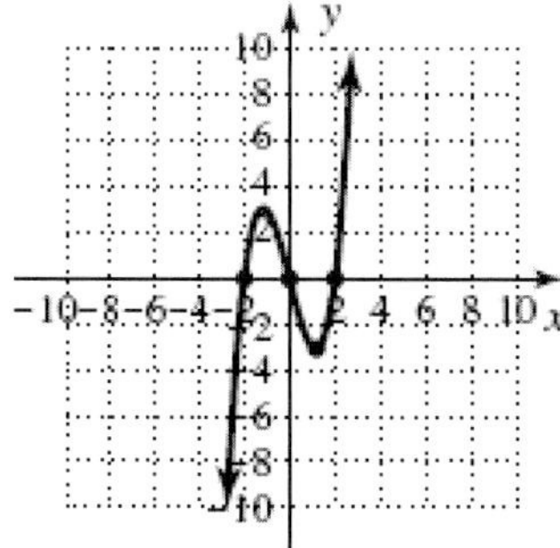

26. $p(x)=\sqrt[3]{x-1}$

x	$p(x)=\sqrt[3]{x-1}$
-5	$p(-5)=\sqrt[3]{-5-1}=\sqrt[3]{-6}\approx-1.817$
0	$p(0)=\sqrt[3]{0-1}=\sqrt[3]{-1}=-1$
1	$p(1)=\sqrt[3]{1-1}=\sqrt[3]{0}=0$
2	$p(2)=\sqrt[3]{2-1}=\sqrt[3]{1}$
5	$p(5)=\sqrt[3]{5-1}=\sqrt[3]{5}\approx1.71$

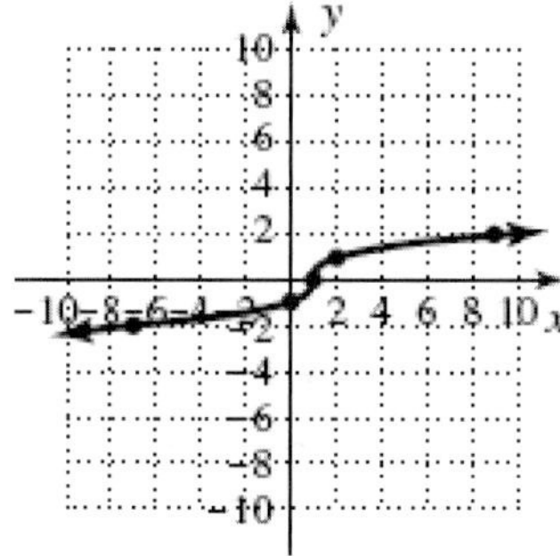

27. $q(x)=\sqrt{x}+2$

x	$q(x)=\sqrt{x}+2$
0	$q(0)=\sqrt{0}+2=2$
1	$q(1)=\sqrt{1}+2=1+2=3$
4	$q(4)=\sqrt{4}+2=2+2=4$
5	$q(9)=\sqrt{5}+2\approx4.24$

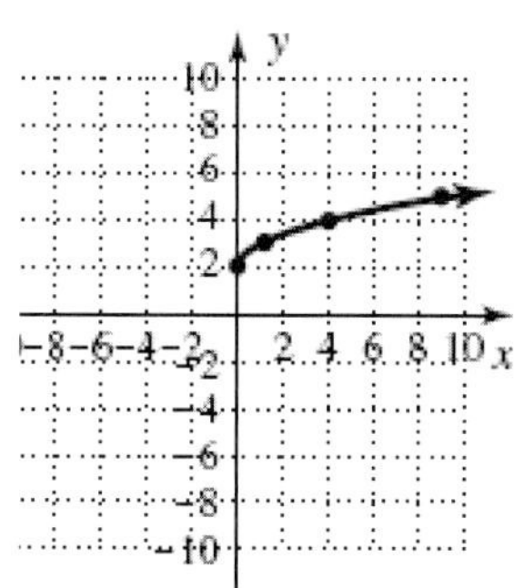

28. a. x-intercepts: (-4, 0) (1, 0)
    y-intercept: (0, -3)
    b. Up/Up
    c. Vertex: (-1.5, -4)

29. a. x-intercept: (-3, 0)
    y-intercept: (0, 2)
    b. Up on the right
    c. Node: (-4, -2)

30. a. x-intercepts: (-2, 0) (1, 0) (4, 0)
    y-intercept: (0, -1.5)
    b. Up on the left, down on the right
    c. Point of inflection: (1, 0)

31. a. x-intercepts: (-1.5, 0) (2, 0)
    y-intercept: (0, -1.5)
    b. Up/Up
    c. Vertex: $\left(\frac{1}{2},-2\right)$

32. a. x-intercept: (-1, 0)
    y-intercept: (0, -2)
    b. Up on the left, down on the right
    c. Point of inflection: (-1, -1)

33. a. x-intercept: (1, 0)
    y-intercept: (0, -2)
    b. Down on the left, up on the right
    c. NA

34. $p(x)=x^2+3x-10$

$$x^2+3x-10=0$$
$$(x+5)(x-2)=0$$

$x=-5;\quad x=2$

x-intercepts: (-5, 0), (2, 0)

y-intercept: (0, -10)

$$x=\frac{-5+2}{2}=-\frac{3}{2}=-1.5$$

$$f(-1.5)=(-1.5)^2+3(-1.5)-10=-12.25$$

Vertex: $(-1.5,-12.25)$

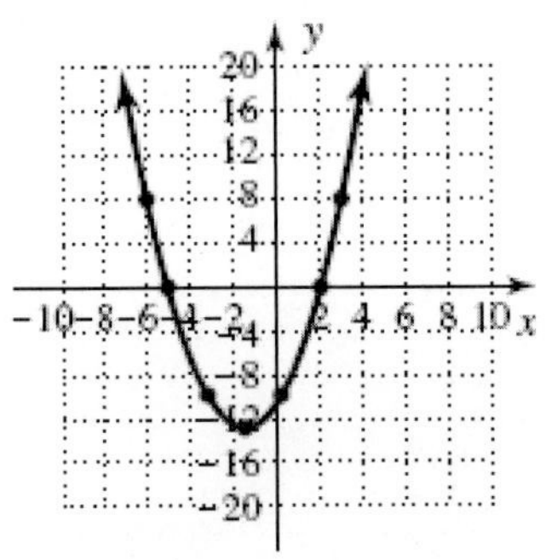

35. $q(x) = x^3 - 2x^2 - 3x$

$$x^3 - 2x^2 - 3x = 0$$
$$x(x^2 - 2x - 3) = 0$$
$$x(x-3)(x+1) = 0$$

$x = 0; \quad x = 3; \quad x = -1$

x-intercepts: (0, 0), (3, 0), (–1, 0)
y-intercept: (0, 0)
Point of inflection: $(0.67, -2.59)$

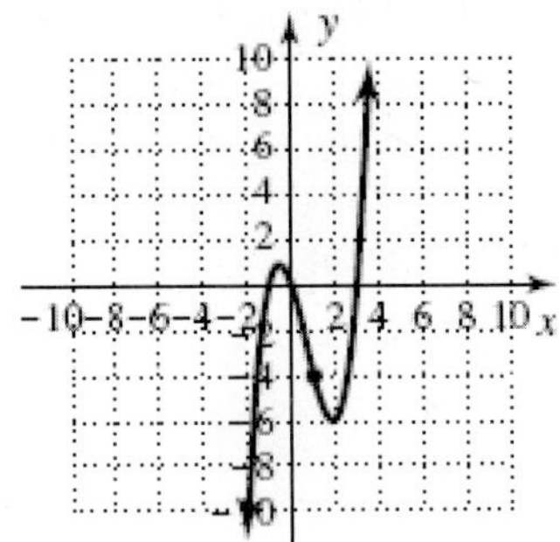

36. $f(x) = \sqrt{2x+3}$

$$2x + 3 = 0$$
$$2x = -3$$
$$x = -\frac{3}{2}$$

x-intercept: $\left(-\frac{3}{2}, 0\right)$

$y = \sqrt{2(0)+3} = \sqrt{3} = 1.7321$

y-intercept: (0, 1.7321)
Node: (–1.5,0)

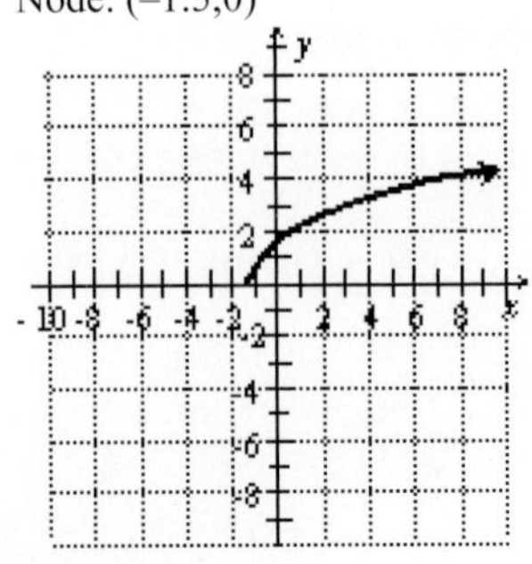

37. $g(x) = \sqrt[3]{x+1}$

$$x + 1 = 0$$
$$x = -1$$

x-intercept: (-1, 0)

$y = \sqrt[3]{0+1} = \sqrt[3]{1} = 1$

y-intercept: (0, 1)
Point of inflection: (–1, 0)

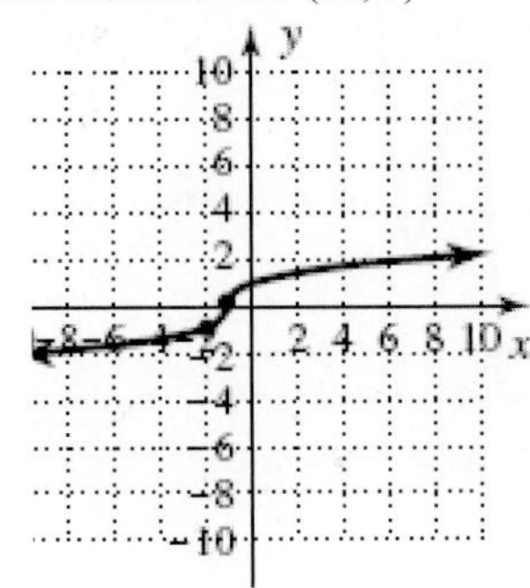

38. $P = 0.0004w^3$

a. $w = 10$ to $w = 15$

b. $\frac{\Delta P}{\Delta w} = \frac{P(10) - P(5)}{10 - 5} = \frac{0.4 - 0.05}{5}$

$= 0.07 = \frac{7}{100}$

$\frac{\Delta P}{\Delta w} = \frac{P(15) - P(10)}{15 - 10} = \frac{1.35 - 0.4}{5}$

$= 0.19 = \frac{19}{100}$

39. $f(x) > 0$

(–4, 1)

40. $g(x) > 0$

$(-\infty, -4) \cup (3, \infty)$

41. $h(x) \geq 0$

$(-\infty, -2]$

42. $f(x) \geq 0$

$(-\infty, \infty)$

43. $f(x) = 3x - 2; \quad f(x) > 0$

$$3x - 2 = 0$$
$$3x = 2$$
$$x = \frac{2}{3}$$

$\left(\frac{2}{3}, \infty\right)$

44. $g(x)=\sqrt{x-3};\quad g(x)<0$

$x-3>0$

$x>3$

No solution

45. $h(x)=x^2-5x;\quad h(x)<0$

$x^2-5x=0$

$x(x-5)=0$

$x=0,\quad x=5$

(0, 5)

46. $p(x)=x^2+4x-5;\quad p(x)\le 0$

$x^2+4x-5=0$

$(x+5)(x-1)=0$

$x=-5;\quad x=1$

[–5, 1]

47. $q(x)=-|x|+2;\quad q(x)<0$

$-|x|+2=0$

$-|x|=-2$

$|x|=2$

$x=2;\quad x=-2$

$(-\infty,-2)\cup(2,\infty)$

48. $r(x)=x^3-x;\quad r(x)<0$

$x^3-x=0$

$x(x^2-1)=0$

$x(x-1)(x+1)=0$

$x=0;\quad x=1;\quad x=-1$

$(-\infty,-1)\cup(0,1)$

49. $(-\infty,0]\cup[5,\infty)$

50. $[-1,0]\cup[1,\infty)$

51. a.

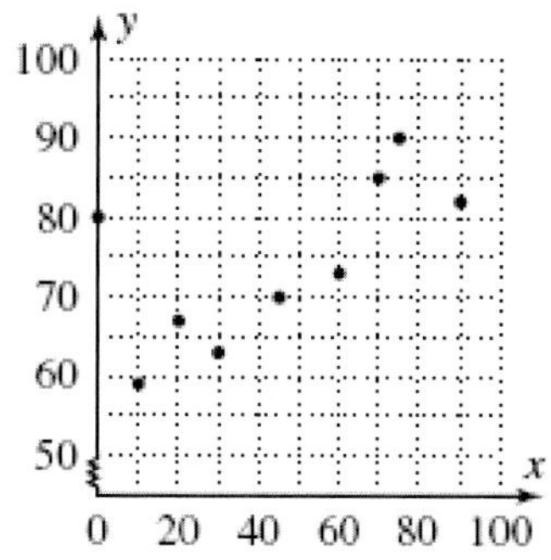

b. Linear

c. Yes

d. Positive

52. (20, 67) (70, 85)

$m=\dfrac{85-67}{70-20}=\dfrac{18}{50}=0.36$

$y-67=0.36(x-20)$

$y-67=0.36x-7.2$

$y=0.36x+59.8$

With grapher: $y=0.35x+56.10$

53. $f(120)=0.35(120)+56.10=98$

Chapter 2 Mixed Review

1. $5x+3y=-9$

$3y=-5x-9$

$y=-\dfrac{5}{3}x-3$

3. a. $f(x)=\dfrac{x+1}{x^2-25}$

$x^2-25=0$

$(x-5)(x+5)=0$

$x=5;\quad x=-5$

$x\in(-\infty,-5)\cup(-5,5)\cup(5,\infty)$

b. $g(x)=\sqrt{3x-5}$

$3x-5\ge 0$

$3x\ge 5$

$x\ge\dfrac{5}{3}$

$x\in\left[\dfrac{5}{3},\infty\right)$

5. $m=-\dfrac{3}{4}$; y-intercept (0, 1)

$y = -\frac{3}{4}x + 1$

7. (-1, -3) (3, 4)

$M = \left(\frac{-1+3}{2}, \frac{-3+4}{2}\right) = \left(1, \frac{1}{2}\right)$

$d = \sqrt{(-1-3)^2 + (-3-4)^2}$
$d = \sqrt{(-4)^2 + (-7)^2}$
$d = \sqrt{16+49}$
$d = \sqrt{65} \approx 8.06$ units

9. $f(x) = x^3 - 4x$

x	$f(x) = x^3 - 4x$
-2	$f(-2) = (-2)^3 - 4(-2) = -8 + 8 = 0$
-1	$f(-1) = (-1)^3 - 4(-1) = -1 + 4 = 3$
0	$f(0) = (0)^3 - 4(0) = 0 - 0 = 0$
1	$f(1) = (1)^3 - 4(1) = 1 - 4 = -3$
2	$f(2) = (2)^3 - 4(2) - 8 - 8 = 0$

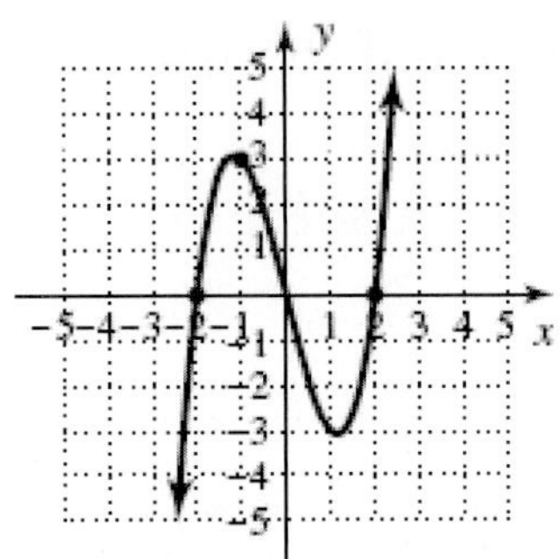

11. $h(x) = (x+3)^2$

x	$h(x) = (x+3)^2$
-6	$h(-6) = (-6+3)^2 = (-3)^2 = 9$
-3	$h(-3) = (-3+3)^2 = 0^2 = 0$
0	$h(0) = (0+3)^2 = (3)^2 = 9$

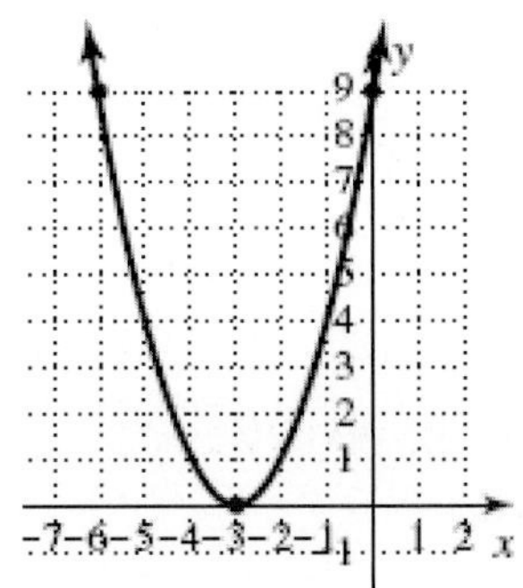

13. $2x - 5y = 10$

x-intercept: (5, 0)y-intercept: (0,-2)

$2x - 5(0) = 10$ $\quad$ $2(0) - 5y = 10$

$2x = 10$ $\quad$ $-5y = 10$

$x = 5$ $\quad$ $y = -2$

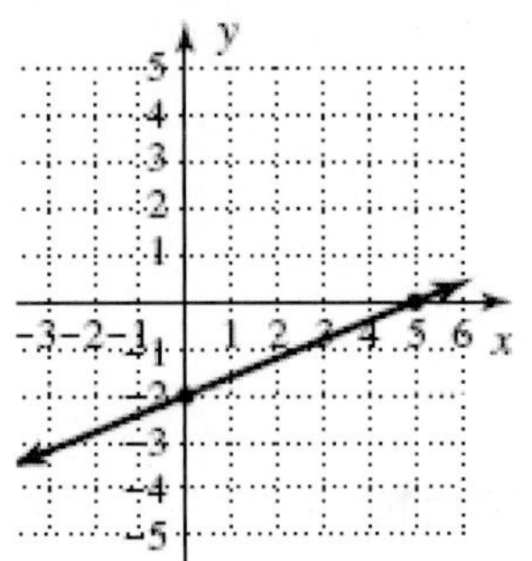

15. $f(x) = 2\sqrt{x+4} - 2;\ \ f(x) \le 0$

$x \in [-4, -3]$

17. $q(x) = \sqrt{x^2 - 9}$

$x^2 - 9 = 0$

$(x-3)(x+3) = 0$

$x = 3;\ \ x = -3$

$x \in (-\infty, -3] \cup [3, \infty)$

19. a.

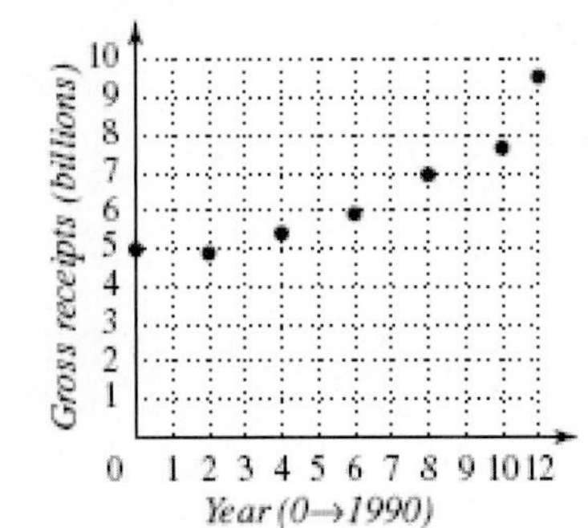

Quadratic

b. $g(t) = 0.0357t^2 - 0.0602t + 4.9795$

c. $g(15)=0.0357(15)^2-0.0602(15)+4.9795$

$g(15)\approx 12.1$ billion (2005);

$g(18)=0.0357(18)^2-0.0602(18)+4.9795$

$g(18)\approx 15.51$ billion (2008)

Chapter 2 Practice Test

1. a. $x=y^2+2y$

b. $y=\sqrt{5-2x}$

c. $|y|+1=x$

d. $y=x^2+2x$

A and C are non-functions, do not pass the vertical line test.

2. $y=-\frac{4}{3}x+2$; y-intercept: (0, 2)

$m=-\frac{4}{3}$

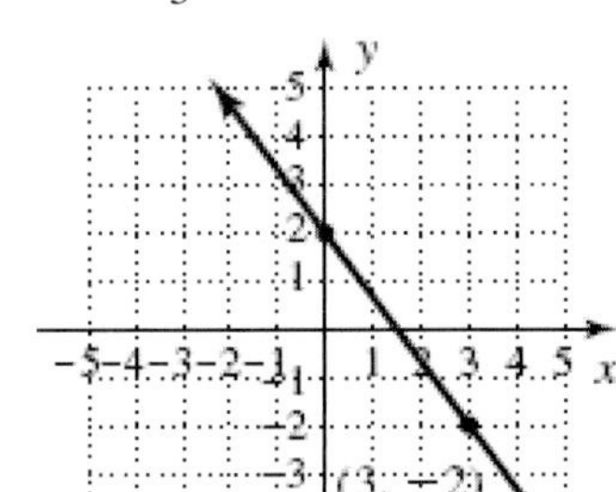

3. a. $2x+5y=-15$

$5y=-2x-15$

$y=-\frac{2}{5}x-3$

b. $y=\frac{2}{5}x+7$

Neither

4. $2x+3y=6$

$3y=-2x+6$

$y=-\frac{2}{3}x+2$

$y=-\frac{2}{3}x$

5. a. $2x+7y=31$

$2(-2)+7(5)=31$

$-4+35=31$

$31=31$

Yes

b. $2x+7y=31$

$2\left(-\frac{31}{2}\right)+7(0)=31$

$-31+0=31$

$-31\neq 31$

No

6. (2, 20) and (5, 40)

$m=\frac{40-20}{5-2}=\frac{20}{3}$;

$y-20=\frac{20}{3}(x-2)$

$y-20=\frac{20}{3}x-\frac{40}{3}$

$y=\frac{20}{3}x+\frac{20}{3}$;

$a(t)=\frac{20}{3}t+\frac{20}{3}$

$a(11)=\frac{20}{3}(11)+\frac{20}{3}=80$ mph

7. (–20, 15) and (35, –12)

a. $M=\left(\frac{-20+35}{2},\frac{15-12}{2}\right)=(7.5,1.5)$

b. $d=\sqrt{(-20-35)^2+(15+12)^2}$

$d=\sqrt{(-55)^2+(27)^2}$

$d=\sqrt{30.25+729}$

$d=\sqrt{3754}$

$d\approx 61.27$ miles

8. L_1: $x=-3$

L_2: $y=4$

9. a. $x\in\{-4,-2,0,2,4,6\}$

$y\in\{-2,-1,0,1,2,3\}$

b. $x\in[-2,6]$

$y\in[1,4]$

10. a. W(24) = 300

b. $h=30$ when $W(h)=375$

c. (20, 250) and (40, 500)

$m=\frac{500-250}{40-20}=\frac{250}{20}=\frac{25}{2}$

$W(h)=\frac{25}{2}h$

d. Wages are \$12.50 per hour.

e. $D\in[0,40]$

$R\in[0,500]$

11. $f(x)=-2x^2-3x+4$

a. $f\left(-\frac{3}{2}\right)=-2\left(-\frac{3}{2}\right)^2-3\left(-\frac{3}{2}\right)+4=4$

b. $f\left(1+\sqrt{2}\right)$
$=-2\left(1+\sqrt{2}\right)^2-3\left(1+\sqrt{2}\right)+4$
$=-2\left(1+2\sqrt{2}+2\right)-3-3\sqrt{2}+4$
$=-2-4\sqrt{2}-4-3-3\sqrt{2}+4$
$=-5-7\sqrt{2}$

c. $f(2+3i)$
$=-2(2+3i)^2-3(2+3i)+4$
$=-2\left(4+12i+9i^2\right)-6-9i+4$
$=-8-24i+18-6-9i+4$
$=8-33i$

12. $18x=3x^2+29$
$-3x^2+18x=29$
$x^2-6x=-\frac{29}{3}$
$x^2-6x+9=-\frac{29}{3}+9$
$(x-3)^2=-\frac{2}{3}$
$x-3=\pm\sqrt{\frac{2}{3}}i$
$x-3=\pm\frac{\sqrt{6}}{3}i$
$x=3\pm\frac{\sqrt{6}}{3}i$

13. $-2x^2+7x=3$
$-2x^2+7x-3=0$
$x=\frac{-7\pm\sqrt{(7)^2-4(-2)(-3)}}{2(-2)}$
$x=\frac{-7\pm\sqrt{49-24}}{-4}$
$x=\frac{-7\pm\sqrt{25}}{-4}$
$x=\frac{-7\pm5}{-4}$;
$x=\frac{-7+5}{-4}=\frac{-2}{-4}=\frac{1}{2}$;
$x=\frac{-7-5}{-4}=\frac{-12}{-4}=3$

14. $f(x)=-\frac{3}{2}x+5;\ \ f(x)>0$
$-\frac{3}{2}x+5=0$
$-\frac{3}{2}x=-5$
$x=\frac{10}{3}$
$x\in\left(-\infty,\frac{10}{3}\right)$

15. $g(x)=x^2+2x-35;\ \ f(x)\ge 0$
$x^2+2x-35=0$
$(x+7)(x-5)=0$
$x=-7;\ \ x=5$
$x\in(-\infty,-7]\cup[5,\infty)$

16. Graph I
a. Square Root
b. $x\in[-4,\infty)$
$y\in[-3,\infty)$
c. x-intercept: (–2, 0)
y-intercept: (0, 1)
d. Up on right
e. $(-2,\infty)$
f. $[-4,-2)$

Graph II
a. Cubic
b. $x\in(-\infty,\infty)$
$y\in(-\infty,\infty)$
c. x-intercept: (2, 0)
y-intercept: (0, –1)
d. Down on left, up on right
e. $(2,\infty)$
f. $(-\infty,2)$

Graph III
a. Absolute value
b. $x\in(-\infty,\infty)$
$y\in(-\infty,4]$
c. x-intercepts: (–1, 0) and (3, 0)
y-intercept: (0, 2)
d. Down/down
e. (–1, 3)
f. $(-\infty,-1)\cup(3,\infty)$

Graph IV
a. Quadratic
b. $x\in(-\infty,\infty)$
$y\in[-5.5,\infty]$
c. x-intercepts: (0, 0) and (5, 0)
y-intercept: (0, 0)

d. Up/up

e. $(-\infty,0)\cup(5,\infty)$

f. (0, 5)

17.

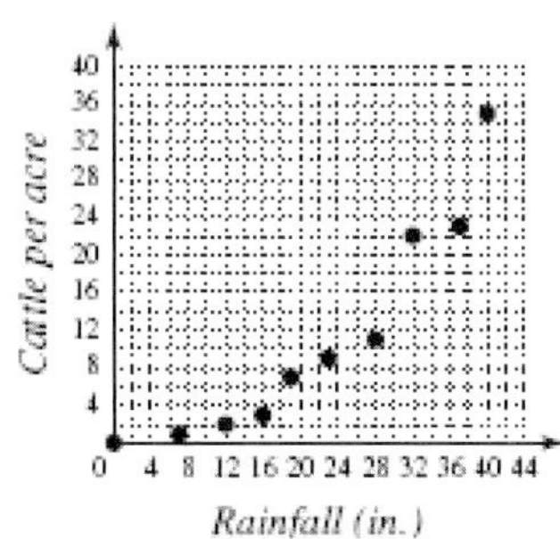

18. Yes, positive.

19. $f(x)=0.0256x^2-0.24x+0.757$

$f(50)=0.0256(50)^2-0.24(50)+0.757\approx 53$

Approximately 53 cattle per acre.

20. $S(t)=2x^2+3x$

a. No, graph is less steep.

b. $\frac{\Delta S}{\Delta t}=\frac{S(6)-S(5)}{6-5}=\frac{90-65}{1}=25$;

$\frac{\Delta S}{\Delta t}=\frac{S(7)-S(6)}{7-6}=\frac{119-90}{1}=29$

Calculator Exploration

Exercise 1:

Bounces off $x = 1$: $Y_1=(x-1)^2$

Cuts at $x = -3$: $Y_2=x+3$

Down/up: Cubic function

Contains point: (–2, 7)

$$
\begin{aligned}
y&=a(x-1)^2(x+3)\\
7&=a(-2-1)^2(-2+3)\\
-7&=a(-3)^2(1)\\
7&=9a\\
\frac{7}{9}&=a
\end{aligned}
$$

$$Y_3=\frac{7}{9}(x-1)^2(x+3)$$

Exercise 2:

Bounces off $x = -3$: $Y_1=(x+3)^2$

Cuts at $x = 4$: $Y_2=x-4$

Down/up: Cubic function

y-intercept: (0, 6)

$$
\begin{aligned}
y&=a(x+3)^2(x-4)\\
6&=a(0+3)^2(0-4)\\
6&=a(3)^2(-4)\\
6&=-36a\\
-\frac{1}{6}&=a
\end{aligned}
$$

$$Y_3=-\frac{1}{6}(x+3)^2(x-4)$$

Exercise 3: Answers will vary.

Strengthening Core Skills

Exercise 1:

x	x^2	$3x$	24	$-x^2+3x+24$
0	0	0	24	24
1	-1	3	24	26
2	-4	6	24	26
3	-9	9	24	24
4	-16	12	24	20
5	-25	15	24	14
6	-36	18	24	6
7	-49	21	24	-4
8	-64	24	24	-16

Between 6 and 7

Exercise 2:

x	x^2	-9	x^2-9
0	0	-9	-9
1	1	-9	-8
2	4	-9	-5
3	9	-9	0
4	16	-9	7
5	35	-9	16
6	46	-9	27
7	49	-9	40

x	9	x^2	$9-x^2$
0	9	0	9
1	9	-1	8
2	9	-4	5
3	9	-9	0
4	9	-16	-7
5	9	-35	-16
6	9	-46	-27
7	9	-49	-40

At 3, answers will vary, (-3, 3).

Exercise 3:

a. $p(x)=x^4-3x^3-5x^2-6x-15$

b. $q(x)=-x^5+3x^4+5x^3+2x^2+7x+9$

Answers will vary.

Cumulative Review Chapters 1-2

1. $2n - 5 = n + 3$

2. a. $2x^3 + 3x - (x-1) + x(1+x^2)$
$= 2x^3 + 3x - x + 1 + x + x^3$
$= 3x^3 + 3x + 1$

b. $(2x-3)(2x+3) = 4x^2 - 9$

3. a. $\dfrac{-15n^3m^4}{10nm^3} = \dfrac{-3n^2m}{2}$

b. $(5.1\text{x}10^{-9})\text{x}(3\text{x}10^{6})$
$= (5.1\text{x}3)\text{x}(10^{-9}\text{x}10^{6})$
$= 15.3\text{x}10^{-3}$

c. $\left(\dfrac{2ab^{-2}}{c^2}\right)^{-3} = \left(\dfrac{c^2}{2ab^{-2}}\right)^{3}$
$= \dfrac{c^6}{8a^3b^{-6}} = \dfrac{b^6c^6}{8a^3}$

d. $2x^0 + (2x)^0 + 2^{\ 1}$
$= 2(1) + 1 + \dfrac{1}{2}$
$= 3\dfrac{1}{2}$

4. a. $N \subset Z \subset W \subset Q \subset R$
False

b. $W \subset N \subset Z \subset Q \subset R$
False

c. $N \subset W \subset Z \subset Q \subset R$
True

d. $N \subset R \subset Z \subset Q \subset W$
False

5. a. $\dfrac{-2}{x^2 - 3x - 10} + \dfrac{1}{x+2}$
$= \dfrac{-2}{(x-5)(x+2)} + \dfrac{1}{x+2}$
$= \dfrac{-2}{(x-5)(x+2)} + \dfrac{1(x-5)}{(x-5)(x+2)}$
$= \dfrac{-2 + x - 5}{(x-5)(x+2)}$
$= \dfrac{x-7}{(x-5)(x+2)}$

b. $\dfrac{b^2}{4a^2} - \dfrac{c}{a} = \dfrac{b^2}{4a^2} - \dfrac{4ac}{4a^2} = \dfrac{b^2 - 4ac}{4a^2}$

6. a. $\dfrac{-10 + \sqrt{72}}{4} = \dfrac{-10 + 6\sqrt{2}}{4} = -\dfrac{5}{2} + \dfrac{3\sqrt{2}}{2}$

b. $\dfrac{1}{\sqrt{2}} = \dfrac{1}{\sqrt{2}} \cdot \dfrac{\sqrt{2}}{\sqrt{2}} = \dfrac{\sqrt{2}}{2}$

7. $-2(3-x) + 5x = 4(x+1) - 7$
$-6 + 2x + 5x = 4x + 4 - 7$
$7x - 6 = 4x - 3$
$3x = 3$
$x = 1$

8. $rt + Rt = D$
$t(r+R) = D$
$t = \dfrac{D}{r+R}$

9. $2 - x < 5$ and $3x + 2 < 8$
$-x < 3$ $\quad$ $3x < 6$
$x > -3$ $\quad$ $x < 2$
Solution set: $-3 < x < 2$

10. $x^2 - 2x + 26 = 0$
$(1+5i)^2 - 2(1+5i) + 26 = 0$
$1 + 10i + 25i^2 - 2 - 10i + 26 = 0$
$1 + 10i - 25 - 2 - 10i + 26 = 0$
$0 = 0$

11. a. $(2+5i)^2 = 4 + 20i + 25i^2$
$= 4 + 20i - 25 = -21 + 20i$

b. $\dfrac{1-2i}{1+2i} = \dfrac{1-2i}{1+2i} \cdot \dfrac{1-2i}{1-2i}$
$= \dfrac{1 - 4i + 4i^2}{1 - 4i^2} = \dfrac{1 - 4i - 4}{1+4}$
$= \dfrac{-3 - 4i}{5} = -\dfrac{3}{5} - \dfrac{4}{5}i$

12. a. $6x^2 - 7x = 20$
$6x^2 - 7x - 20 = 0$
$(3x+4)(2x-5) = 0$
$x = -\dfrac{4}{3};\ x = \dfrac{5}{2}$

b. $x^3 + 5x^2 - 15 = 3x$

$$x^3+5x^2-3x-15=0$$
$$(x^3+5x^2)-(3x+15)=0$$
$$x^2(x+5)-3(x+5)=0$$
$$(x+5)(x^2-3)=0$$
$$x=-5;\quad x=\sqrt{3};\quad x=-\sqrt{3}$$

13. $2x^2+49=-20x$
$$2x^2+20x+49=0$$
$$2x^2+20x=-49$$
$$x^2+10x=\frac{-49}{2}$$
$$x^2+10x+25=-\frac{49}{2}+25$$
$$(x+5)^2=\frac{1}{2}$$
$$x+5=\pm\sqrt{\frac{1}{2}}$$
$$x+5=\pm\frac{\sqrt{2}}{2}$$
$$x=-5\pm\frac{\sqrt{2}}{2};$$
$$x\approx-4.293$$
$$x\approx-5.707$$

14. $2x^2+20x=-51$
$$2x^2+20x+51=0$$
$$x=\frac{-20\pm\sqrt{(20)^2-4(2)(51)}}{2(2)}$$
$$x=\frac{-20\pm\sqrt{400-408}}{4}$$
$$x=\frac{-20\pm\sqrt{-8}}{4}$$
$$x=\frac{-20\pm2i\sqrt{2}}{4}$$
$$x=-5\pm\frac{i\sqrt{2}}{2}$$

15. Let w represent the width.
Let l represent the length.
$$A=lw$$
$$1457=(w+16)w$$
$$0=w^2+16w-1457$$
$$0=(w-31)(w+47)$$
$w=31$ cm; $l=47$ cm

16. $f(x)=x^3-4x;\quad f(x)\le0$
$$x^3-4x=0$$
$$x(x^2-4)=0$$
$$x(x-2)(x+2)=0$$
$$x=0;\quad x=2;\quad x=-2$$
$$x\in(-\infty,-2]\cup[0,2]$$

17. $y=\frac{1}{3}x-2$; y-intercept: $(0,-2)$; $m=\frac{1}{3}$

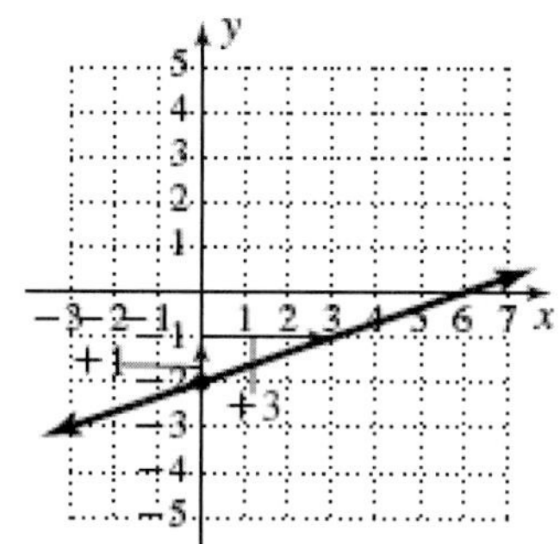

18. a.

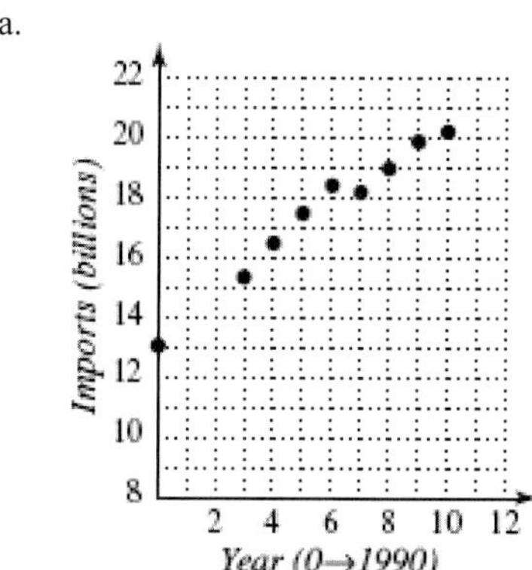

Linear

b. $I(t)=0.711t+13.470$

c. $I(15)=0.711(15)+13.470\approx24.1$
$I(18)=0.711(18)+13.470\approx26.3$

d. $22=0.711t+13.470$
$8.53=0.711t$
$11.99=t$
2002

19. $m_1=\frac{1}{2}$; $m_2=-2$; $(1, 2)$
$$y-2=-2(x-1)$$
$$y-2=-2x+2$$
$$y=-2x+4$$

20. (-4, 5), (4, -1), (0, 8)

$$d=\sqrt{(-4-4)^2+(5+1)^2}$$
$$d=\sqrt{(-8)^2+(6)^2}$$
$$d=\sqrt{100}$$
$$d=10;$$
$$d=\sqrt{(4-0)^2+(-1-8)^2}$$
$$d=\sqrt{(4)^2+(-9)^2}$$
$$d=\sqrt{97}$$
$$d\approx 9.85;$$
$$d=\sqrt{(-4-0)^2+(5-8)^2}$$
$$d=\sqrt{(-4)^2+(-3)^2}$$
$$d=\sqrt{25}$$
$$d=5;$$

P=10 + 9.85 + 5 = 24.85 units

No it is not a right triangle.

$$5^2+\left(\sqrt{97}\right)^2\neq 10^2$$

3.1 Technology Highlight

Exercise 1: $f(x)=x^2-9$ and $g(x)=x+3$

$h(x)=x-3,\ x\in(-\infty,-3)\cup(-3,\infty)$

Exercise 2: $f(x)=x^3-3x^2-4x+12$ and $g(x)=x-3$

$h(x)=x^2-4,\ x\in(-\infty,3)\cup(3,\infty)$

Exercise 3: $f(x)=x^2+x-6$ and $g(x)=\sqrt{x+2}$

$h(x)=\dfrac{x^2+x-6}{\sqrt{x+2}},\ x\in(-2,\infty)$

Exercise 4: $f(x)=|x|$ and $g(x)=\sqrt{x+5}$

$h(x)=\dfrac{|x|}{\sqrt{x+5}},\ x\in(-5,\infty)$

3.1 Exercises

1. $(f+g)(x);\ A\cap B$

3. Intersection; $g(x)$

5. Division by zero, $g(5)$ is undefined.

7. $f(x)=2x^2-x-3;\ g(x)=x^2+5x$
$$h(x)=f(x)+g(x)$$
$$=2x^2-x-3+x^2+5x$$
$$=3x^2+4x-3$$
$$x\in(-\infty,\infty)$$

9. $p(x)=2x^3+4x^2-7;\ q(x)=5x+4x^2$
$$h(x)=p(x)-q(x)$$
(a) $h(-3)=p(-3)-q(-3)$
$$=2(-3)^3+4(-3)^2-7-\left(5(-3)+4(-3)^2\right)$$
$$=2(-27)+4(9)-7-(-15+4(9))$$
$$=-54+36-7-(-15+36)$$
$$=-54+36-7-21$$
$$=-46$$
(b) $h(-3)=(p-q)(-3)$
$$=\left(2x^3+4x^2-7-5x-4x^2\right)(-3)$$
$$=\left(2x^3-5x-7\right)(-3)$$
$$=2(-3)^3-5(-3)-7$$
$$=2(-27)+15-7$$
$$=-54+15-7$$
$$=-46$$

11. $f(x)=\sqrt{4x-5};\ g(x)=8x^3+125$
$$H(x)=f(x)-g(x)=\sqrt{4x-5}-8x^3-125$$
(a) $H\left(\frac{3}{2}\right)=\sqrt{4\left(\frac{3}{2}\right)-5}-8\left(\frac{3}{2}\right)^3-125$
$$=\sqrt{6-5}-8\left(\frac{27}{8}\right)-125$$
$$=\sqrt{1}-27-125=1-27-125=-151$$
(b) $H\left(\frac{-1}{2}\right)=\sqrt{4\left(\frac{-1}{2}\right)-5}-8\left(\frac{-1}{2}\right)^3-125$
$$=\sqrt{-4-5}-8\left(\frac{-1}{8}\right)-125$$
$$=\sqrt{-9}+1-125$$
$$=\sqrt{-9}-124$$
Not defined

(c) $f(x)=\sqrt{4x-5}$
$$4x-5\ge 0$$
$$4x\ge 5$$
$$x\ge\frac{5}{4}$$
$$x\in\left[\frac{5}{4},\infty\right)$$

13. $f(x)=\sqrt{x};\ g(x)=-2$
$$h(x)=f(x)+g(x)=\sqrt{x}-2$$

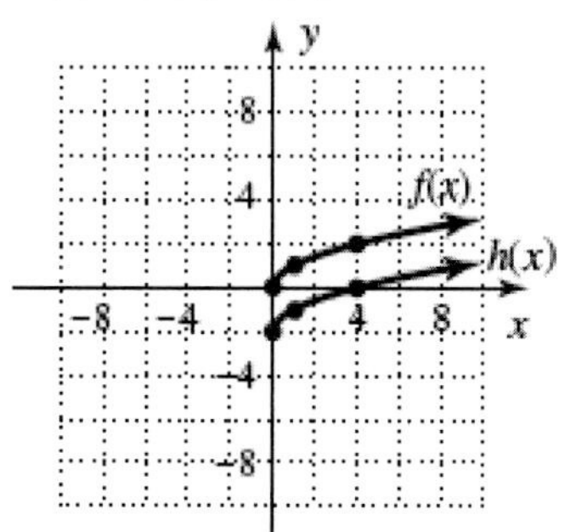

$h(x)$ shifts $f(x)$ down 2 units.

15. $f(x)=3x^2-2x-4;\ g(x)=2x+1$
$$h(x)=(f\cdot g)(x)$$

$h(x)=(3x^2-2x-4)(2x+1)$
$=6x^3+3x^2-4x^2-2x-8x-4$
$=6x^3-x^2-10x-4$
Domain: $x\in(-\infty,\infty)$

17. $p(x)=\sqrt{x+5}$; $q(x)=\sqrt{2-x}$
(a) $H(x)=(p\cdot q)(x)$
$H(x)=\sqrt{x+5}\sqrt{2-x}=\sqrt{(x+5)(2-x)}$
(b) $H(-2)=\sqrt{(-2+5)(2-(-2))}$
$=\sqrt{(3)(4)}=\sqrt{12}=2\sqrt{3}$;
$H(3)=\sqrt{(3+5)(2-3)}$
$=\sqrt{(8)(-1)}=\sqrt{-8}$ not defined
(c) Domain: $x\in[-5,2]$
$x+5\ge 0$, $x\ge -5$
$2-x\ge 0$, $-x\ge -2$, $x\le 2$

19. $f(x)=x^3-7x^2+6x$; $g(x)=x-1$
$h(x)=\left(\frac{f}{g}\right)(x)$
$h(x)=\frac{x^3-7x^2+6x}{x-1}$
$=\frac{x(x^2-7x+6)}{x-1}$
$=\frac{x(x-1)(x-6)}{x-1}$
$=x^2-6x$
Domain: $x\in(-\infty,1)\cup(1,\infty)$

21. $p(x)=2x-3$; $q(x)=\sqrt{x^2-x-6}$
(a) $H(x)=\left(\frac{p}{q}\right)(x)$
$H(x)=\frac{2x-3}{\sqrt{x^2-x-6}}$
(b) $H(-2)=\frac{2(-2)-3}{\sqrt{(-2)^2-(-2)-6}}$
$=\frac{-4-3}{\sqrt{0}}$
Not defined;

$H(5)=\frac{2(5)-3}{\sqrt{(5)^2-(5)-6}}$
$=\frac{10-3}{\sqrt{25-5-6}}=\frac{7}{\sqrt{14}}=\frac{7\sqrt{14}}{14}=\frac{\sqrt{14}}{2}$
(c) Domain: $x\in(-\infty,-2)\cup(3,\infty)$
$x^2-x-6=0$
$(x-3)(x+2)=0$
$x=3;\ x=-2$

23. $f(x)=x+1$ and $g(x)=x-5$
$h(x)=\left(\frac{f}{g}\right)(x)=\frac{x+1}{x-5}$
$D: x\in(-\infty,5)\cup(5,\infty)$

25. $f(x)=x-5$ and $g(x)=\sqrt{x-2}$
$h(x)=\left(\frac{f}{g}\right)(x)=\frac{x-5}{\sqrt{x-2}}$
$D: x\in(2,\infty)$

27. $f(x)=x^2-9$ and $g(x)=\sqrt{x+1}$
$h(x)=\left(\frac{f}{g}\right)(x)=\frac{x^2-9}{\sqrt{x+1}}$
$D: x\in(-1,\infty)$

29. $f(x)=x^2-16$ and $g(x)=x+4$
$h(x)=\left(\frac{f}{g}\right)(x)=\frac{x^2-16}{x+4}$
$=\frac{(x+4)(x-4)}{x+4}=x-4$
$D: x\in(-\infty,-4)\cup(-4,\infty)$

31. $f(x)=x^3+4x^2-2x-8$ and $g(x)=x+4$
$h(x)=\left(\frac{f}{g}\right)(x)=\frac{x^3+4x^2-2x-8}{x+4}$
$=\frac{x^2(x+4)-2(x+4)}{x+4}=\frac{(x+4)(x^2-2)}{x+4}$
$=x^2-2$
$D: x\in(-\infty,-4)\cup(-4,\infty)$

33. $f(x)=\frac{6}{x-3}$ and $g(x)=\frac{2}{x+2}$

$$h(x)=\left(\frac{f}{g}\right)(x)=\frac{\frac{6}{x-3}}{\frac{2}{x+2}}=\frac{6(x+2)}{2(x-3)}=\frac{3x+6}{x-3}$$

$D: x\in(-\infty,-2)\cup(-2,3)\cup(3,\infty)$

35. $f(x)=2x+3$ and $g(x)=x-2$

Sum:

$f(x)+g(x)=2x+3+x-2=3x+1$

Domain contains all values of x.

$D: x\in(-\infty,\infty)$

Difference:

$f(x)-g(x)=2x+3-(x-2)$

$=2x+3-x+2=x+5$

Domain contains all values of x.

$D: x\in(-\infty,\infty)$

Product:

$f(x)\cdot g(x)=(2x+3)(x-2)$

$=2x^2-4x+3x-6$

$=2x^2-x-6$

Domain contains all values of x.

$D: x\in(-\infty,\infty)$

Quotient:

$\frac{f(x)}{g(x)}=\frac{2x+3}{x-2}$

$x-2\neq 0$

$x\neq 2$

$D: x\in(-\infty,2)\cup(2,\infty)$

37. $f(x)=x^2+7$ and $g(x)=3x-2$

Sum:

$f(x)+g(x)=x^2+7+3x-2=x^2+3x+5$

Domain contains all values of x.

$D: x\in(-\infty,\infty)$

Difference:

$f(x)-g(x)=x^2+7-(3x-2)$

$=x^2+7-3x+2$

$=x^2-3x+9$

Domain contains all values of x.

$D: x\in(-\infty,\infty)$

Product:

$f(x)\cdot g(x)=(x^2+7)(3x-2)$

$=3x^3-2x^2+21x-14$

Domain contains all values of x.

$D: x\in(-\infty,\infty)$

Quotient:

$\frac{f(x)}{g(x)}=\frac{x^2+7}{3x-2}$

$3x-2\neq 0$

$3x\neq 2$

$x\neq\frac{2}{3}$

$D: x\in\left(-\infty,\frac{2}{3}\right)\cup\left(\frac{2}{3},\infty\right)$

39. $f(x)=x^2+2x-3$ and $g(x)=x-1$

Sum:

$f(x)+g(x)=x^2+2x-3+x-1$

$=x^2+3x-4$

Domain contains all values of x.

$D: x\in(-\infty,\infty)$

Difference:

$f(x)-g(x)=x^2+2x-3-(x-1)$

$=x^2+2x-3-x+1$

$=x^2+x-2$

Domain contains all values of x.

$D: x\in(-\infty,\infty)$

Product:

$f(x)\cdot g(x)=(x^2+2x-3)(x-1)$

$=x^3-x^2+2x^2-2x-3x+3$

$=x^3+x^2-5x+3$

Domain contains all values of x.

$D: x\in(-\infty,\infty)$

Quotient:

$\frac{f(x)}{g(x)}=\frac{x^2+2x-3}{x-1}$

$=\frac{(x+3)(x-1)}{x-1}=x+3$

$x-1\neq 0$

$x\neq 1$

$D: x\in(-\infty,1)\cup(1,\infty)$

41. $f(x)=3x+1$ and $g(x)=\sqrt{x-3}$

Sum:

$f(x)+g(x)=3x+1+\sqrt{x-3}$

$x-3\geq 0$

$x\geq 3$

$D: x\in[3,\infty)$

Difference:

$f(x)-g(x)=3x+1-\sqrt{x-3}$

$x-3\ge 0$

$x\ge 3$

$D: x\in [3,\infty)$

Product:

$f(x)\cdot g(x)=(3x+1)\sqrt{x-3}$

$x-3\ge 0$

$x\ge 3$

$D: x\in [3,\infty)$

Quotient:

$\frac{f(x)}{g(x)}=\frac{3x+1}{\sqrt{x-3}}$

$x-3>0$

$x>3$

$D: x\in (3,\infty)$

43. $f(x)=2x^2$ and $g(x)=\sqrt{x+1}$

Sum:

$f(x)+g(x)=2x^2+\sqrt{x+1}$

$x+1\ge 0$

$x\ge -1$

$D: x\in [-1,\infty)$

Difference:

$f(x)-g(x)=2x^2-\sqrt{x+1}$

$x+1\ge 0$

$x\ge -1$

$D: x\in [-1,\infty)$

Product:

$f(x)\cdot g(x)=2x^2\sqrt{x+1}$

$x+1\ge 0$

$x\ge -1$

$D: x\in [-1,\infty)$

Quotient:

$\frac{f(x)}{g(x)}=\frac{2x^2}{\sqrt{x+1}}$

$x+1>0$

$x>-1$

$D: x\in (-1,\infty)$

45. $f(x)=\frac{2}{x-3}$ and $g(x)=\frac{5}{x+2}$

Sum:

$f(x)+g(x)=\frac{2}{x-3}+\frac{5}{x+2}$

$=\frac{2(x+2)+5(x-3)}{(x-3)(x+2)}$

$=\frac{2x+4+5x-15}{(x-3)(x+2)}$

$=\frac{7x-11}{(x-3)(x+2)}$

$x-3\ne 0 \quad x+2\ne 0$

$x\ne 3 \quad x\ne -2$

$D: x\in (-\infty,-2)\cup(-2,3)\cup(3,\infty)$

Difference:

$f(x)-g(x)=\frac{2}{x-3}-\frac{5}{x+2}$

$=\frac{2(x+2)-5(x-3)}{(x-3)(x+2)}$

$=\frac{2x+4-5x+15}{(x-3)(x+2)}$

$=\frac{-3x+19}{(x-3)(x+2)}$

$x-3\ne 0 \quad x+2\ne 0$

$x\ne 3 \quad x\ne -2$

$D: x\in (-\infty,-2)\cup(-2,3)\cup(3,\infty)$

Product:

$f(x)\cdot g(x)=\left(\frac{2}{x-3}\right)\left(\frac{5}{x+2}\right)$

$=\frac{10}{(x-3)(x+2)}$

$=\frac{10}{x^2-x-6}$

$x-3\ne 0 \quad x+2\ne 0$

$x\ne 3 \quad x\ne -2$

$D: x\in (-\infty,-2)\cup(-2,3)\cup(3,\infty)$

Quotient:

$\frac{f(x)}{g(x)}=\frac{\frac{2}{x-3}}{\frac{5}{x+2}}=\left(\frac{2}{x-3}\right)\left(\frac{x+2}{5}\right)$

$=\frac{2(x+2)}{5(x-3)}=\frac{2x+4}{5x-15}$

$x-3\ne 0 \quad x+2\ne 0$

$x\ne 3 \quad x\ne -2$

$D: x\in (-\infty,-2)\cup(-2,3)\cup(3,\infty)$

47. $f(x)=x^2-5x-14$

$f(-2)=(-2)^2-5(-2)-14=4+10-14=0;$

$f(7)=(7)^2-5(7)-14=49-35-14=0;$

$f(a)=(a)^2-5(a)-14=a^2-5a-14;$

$f(a-2)=(a-2)^2-5(a-2)-14$
$=a^2-4a+4-5a+10-14$
$=a^2-9a$

49. $f(x)=\sqrt{x+3}$ and $g(x)=2x-5$
(a) $(f\circ g)(x)=f[g(x)]$
$=\sqrt{g(x)+3}$
$=\sqrt{(2x-5)+3}$
$=\sqrt{2x-2}$
(b) $(g\circ f)(x)=g[f(x)]$
$=2(f(x))-5$
$=2\sqrt{x+3}-5$
(c) $2x-2\geq 0$
$2x\geq 2$
$x\geq 1$
Domain of A: $x\in[1,\infty)$
$x+3\geq 0$
$x\geq -3$
Domain of B: $x\in[-3,\infty)$

51. $f(x)=\dfrac{2x}{x+3}$ and $g(x)=\dfrac{5}{x}$
(a) $(f\circ g)(x)=f[g(x)]$
$$=\frac{2(g(x))}{g(x)+3}=\frac{2\left(\frac{5}{x}\right)}{\frac{5}{x}+3}=\frac{\frac{10}{x}}{\frac{5+3x}{x}}$$
$$=\frac{10x}{x(5+3x)}=\frac{10}{5+3x}$$
(b) $(g\circ f)(x)=g[f(x)]$
$$=\frac{5}{f(x)}=\frac{5}{\frac{2x}{x+3}}=\frac{5(x+3)}{2x}=\frac{5x+15}{2x}$$
(c) $5+3x\neq 0$
$3x\neq -5$
$x\neq -\frac{5}{3}$ *and* $x\neq 0$
Domain of A: $\left\{x\middle|x\in R, x\neq -\frac{5}{3}, x\neq 0\right\}$
or $\left(-\infty,-\frac{5}{3}\right)\cup\left(-\frac{5}{3},0\right)\cup(0,\infty)$
$x+3\neq 0$
$x\neq -3$ *and* $x\neq 0$
Domain of B: $\{x|x\in R, x\neq -3, x\neq 0\}$
or $(-\infty,-3)\cup(-3,0)\cup(0,\infty)$

53. $f(x)=x^2-3x$ and $g(x)=x+2$
(a) $h(x)=(f\circ g)(x)$
$h(x)=f[g(x)]$
$h(x)=(g(x))^2-3(g(x))$
$=(x+2)^2-3(x+2)$
$=x^2+4x+4-3x-6$
$=x^2+x-2$
(b) $H(x)=(g\circ f)(x)$
$H(x)=g[f(x)]$
$H(x)=(f(x))+2$
$=x^2-3x+2$
(c) Domain contains all values of x.
Domain of A: $(-\infty,\infty)$
Domain contains all values of x.
Domain of B: $(-\infty,\infty)$

55. $f(x)=x^2+x-4$ and $g(x)=x+3$
(a) $h(x)=(f\circ g)(x)$
$h(x)=f[g(x)]$
$h(x)=(g(x))^2+g(x)-4$
$=(x+3)^2+x+3-4$
$=x^2+6x+9+x-1$
$=x^2+7x+8$
(b) $H(x)=(g\circ f)(x)$
$H(x)=g[f(x)]$
$H(x)=f(x)+3$
$=x^2+x-4+3$
$=x^2+x-1$
(c) Domain contains all values of x.
Domain of A: $(-\infty,\infty)$
Domain contains all values of x.
Domain of B: $(-\infty,\infty)$

57. $f(x)=\sqrt{x-3}$ and $g(x)=3x+4$
(a) $h(x)=(f\circ g)(x)$
$h(x)=f[g(x)]$
$h(x)=\sqrt{g(x)-3}$
$=\sqrt{3x+4-3}$
$=\sqrt{3x+1}$

(b) $H(x) = (g \circ f)(x)$
$H(x) = g[f(x)]$
$H(x) = 3(f(x)) + 4$
$= 3\sqrt{x-3} + 4$

(c) $3x + 1 \geq 0$
$3x \geq -1$
$x \geq -\frac{1\;3}{3}$

Domain of A: $\left\{x \middle| x \geq -\frac{1}{3}\right\}$

or $\left[-\frac{1}{3}, \infty\right)$;

$x - 3 \geq 0$
$x \geq 3$
Domain of B: $\{x | x \geq 3\}$
or $[3, \infty)$

59. $f(x) = |x| - 5$ and $g(x) = -3x + 1$

(a) $h(x) = (f \circ g)(x)$
$h(x) = f[g(x)]$
$h(x) = |g(x)| - 5$
$= |-3x + 1| - 5$

(b) $H(x) = (g \circ f)(x)$
$H(x) = g[f(x)]$
$H(x) = -3(f(x)) + 1$
$= -3(|x| - 5) + 1$
$= -3|x| + 15 + 1$
$= -3|x| + 16$

(c) Domain contains all values of x.
Domain of A: $(-\infty, \infty)$
Domain contains all values of x.
Domain of B: $(-\infty, \infty)$

61. $f(x) = \frac{4}{x}$ and $g(x) = \frac{1}{x-5}$

(a) $h(x) = (f \circ g)(x)$
$h(x) = f[g(x)]$
$h(x) = \frac{4}{g(x)}$
$= \frac{4}{\frac{1}{x-5}}$
$= 4(x-5)$
$= 4x - 20$

(b) $H(x) = (g \circ f)(x)$
$H(x) = g[f(x)]$
$H(x) = \frac{1}{f(x) - 5}$
$= \frac{1}{\frac{4}{x} - 5}$
$= \frac{1}{\frac{4-5x}{x}}$
$= \frac{x}{4-5x}$

(c) $x - 5 \neq 0$
$x \neq 5$
Domain of A: $\{x | x \in R, x \neq 5\}$
or $(-\infty, 5) \cup (5, \infty)$;
$4 - 5x \neq 0$
$-5x \neq -4$
$x \neq \frac{4}{5}$ *and* $x \neq 0$

Domain of B: $\left\{x \middle| x \in R, x \neq 0, x \neq \frac{4}{5}\right\}$

or $(-\infty, 0) \cup \left(0, \frac{4}{5}\right) \cup \left(\frac{4}{5}, \infty\right)$

63. $f(x) = x^2 - 8$ and $g(x) = x + 2$
$h(x) = (f \circ g)(x)$

a. $(f \circ g)(x) = f[g(x)]$
$= (g(x)^2) - 8$
$= (x+2)^2 - 8$
$= x^2 + 4x + 4 - 8$
$= x^2 + 4x - 4$;
$h(x) = x^2 + 4x - 4$
$h(5) = (5)^2 + 4(5) - 4 = 25 + 20 - 4 = 41$

b. $g(5) = 5 + 2 = 7$
$f[g(5)] = f(7)$
$= (7)^2 - 8 = 49 - 8 = 41$

65. $A = 40\pi r + 2\pi r^2$
$A = 2\pi r(20 + r)$
$A(r) = (f \cdot g)(r)$ Answers will vary.
$A(5) = 2\pi(5)(20 + 5) = 10\pi(25) = 250\pi$

67. a. $(f+g)(-4)=4$
b. $(f\cdot g)(1)=0$
c. $(f-g)(4)=2$
d. $(f+g)(0)=3$
e. $\left(\frac{f}{g}\right)(3)=\frac{2}{3}$
f. $(f\cdot g)(-2)=6$
g. $(g\cdot f)(2)=-3$
h. $(f-g)(-1)=1$
i. $(f+g)(8)=1$
j. $\left(\frac{f}{g}\right)(7)=$ undefined
k. $(f+g)(4)=8$
l. $(f\cdot g)(6)=-6$

69. a. $P(n)=R(n)-C(n)$
$P(n)=11.45n-0.1n^2$
b. $P(12)=11.45(12)-0.1(12)^2$
$=137.4-14.4=\$123$
c. $P(60)=11.45(60)-0.1(60)^2$
$=687-360=\$327$
d. At $n=115$, costs exceed revenue, $C(115)>R(115)$.

71. $f(x)=0.5x-14$; $g(x)=2x+23$
$h(x)=(f\circ g)(x)=f[g(x)]$
$h(x)=0.5(g(x))-14$
$=0.5(2x+23)-14$
$=x+11.5-14$
$=x-2.5$;
$h(13)=13-2.5=10.5$

73. $T(x)=41.6x$; $R(x)=10.9x$
(a) $T(100)=41.6(100)=4160$ baht
(b) $R(4160)=10.9(4160)=45{,}344$ Ringgit
(c) $M(x)=(R\circ T)(x)=R[T(x)]$
$M(x)=10.9(T(x))$
$=10.9(41.6x)$
$=453.44x$;
$M(100)=453.44(100)=45344$
Ringgit
Parts B and C agree.

75. $r(t)=3t$; $A=\pi r^2$
(a) $r(2)=3(2)=6$ ft
(b) $A(6)=\pi(6)^2=36\pi$ ft^2
(c) $A(t)=(A\circ r)(t)=A[r(t)]$
$A(t)=\pi(r(t))^2$
$=\pi(3t)^2$
$=9\pi t^2$;
$A(2)=9\pi(2)^2=36\pi$ ft^2
The answers do agree.

77. $h(x)=\left(\sqrt{x-2}+1\right)^3-5$
$(f\circ g)(x)=h(x)$
Answers may vary.
Ex. $f(x)=x^3-5, g(x)=\sqrt{x-2}+1$

79. $C(x)=0.0345x^4-0.8996x^3+7.5383x^2-21.7215x+40$
$L(x)=-0.0345x^4+0.8996x^3-7.5383x^2+21.7215x+10$
(a) Using the grapher, 1995 to 1996; 1999 to 2004
(b) Using the grapher, 30 seats; 1995
(c) Using the grapher, 20 seats; 1997
(d) Using the grapher, the total number in the senate (50); the number of additional seats held by the majority.

81. $f(x)=\sqrt{1-x}$ and $g(x)=\sqrt{x-2}$
Using the grapher,
$(f+g)(x)$ cannot be done because their domains do not overlap.

83. $f(x)=2x^2+3x+1$ and $g(x)=3x-5$
a. $f(2)=2(2)^2+3(2)+1=15$,
$g(15)=3(15)-5=40$;
$f(40)=2(40)^2+3(40)+1=3321$
b. $[(f+g)\circ(f\cdot g)](1)$
$f+g=2x^2+3x+1+3x-5$
$=2x^2+6x-4$
$f\cdot g=\left(2(1)^2+3(1)+1\right)(3(1)-5)$
$=(2+3+1)(3-5)$
$=6(-2)$
$=-12$

$$2(-12)^2 + 6(-12) - 4 = 212$$

85. $f(x) = \sqrt{x}$; $g(x) = \sqrt[3]{x}$; $h(x) = |x|$

(a)

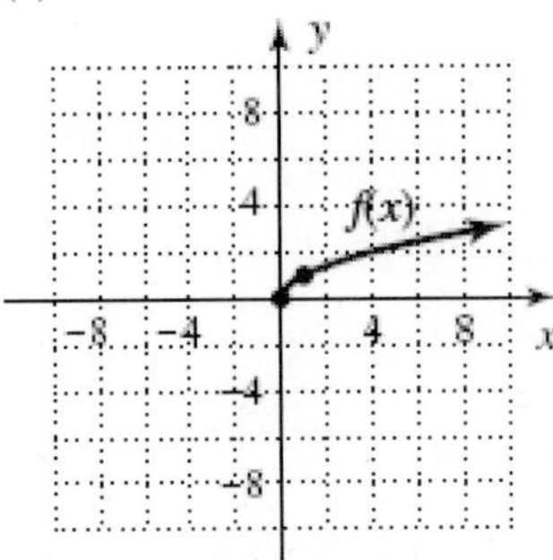

(b)

(c)

87. $f(x) = \sqrt{4 - x^2}$; $g(x) = \sqrt{x^2 - 4}$

$4 - x^2 \geq 0$

$-x^2 \geq -4$

$x^2 \leq 4$

$x \leq -2$ and $x \geq 2$

D of $f(x)$: $[-2,2]$

$x^2 - 4 \geq 0$

$x^2 \geq 4$

$x \geq 2$ and $x \leq -2$

D of $g(x)$: $(-\infty,-2) \cup (2,\infty)$

89. a. $V = \frac{1}{3}\pi r^2 h$ Volume of a cone

b. $V = \frac{4}{3}\pi r^3$ Volume of a sphere

3.2 Technology Highlight

Exercise 1: $f^{-1}(x) = \frac{x-1}{2}$

a. (-1, -1)

$f(-1) = 2(-1) + 1 = -2 + 1 = -1$

$f^{-1}(-1) = \frac{-1-1}{2} = \frac{-2}{2} = -1$

b. Use grapher with $Y_1 = \frac{x-1}{2}$,

$Y_2 = 2x + 1$

Exercise 2: $f^{-1}(x) = \sqrt{x-1}$

a. (3, 10) and (10, 3)

$f(3) = (3)^2 + 1 = 9 + 1 = 10$

$f^{-1}(10) = \sqrt{10-1} = \sqrt{9} = 3$

b. Use grapher with $Y_1 = \sqrt{x-1}$,

$Y_2 = x^2 + 1$

3.2 Exercises

1. Second; one

3. (-11, -2), (-5, 0), (1, 2), (19, 4)

5. False, Answers will vary.

7. One-to-one

9. One-to-one

11. Not a function (cannot be a one-to-one function)

13. One-to-one

15. One-to-one

17. Not one-to-one; $y = 7$ is paired with $x = -2$ and $x = 2$.

19. One-to-one

21. One-to-one

23. Not one-to-one; for $P(t) > 5$, one y corresponds to two x-values.

25. One-to-one

27. One-to-one

29. $f(x) = \{(-2, 1), (-1, 4), (0, 5), (2, 9), (5, 15)\}$
$f^{1}(x) = \{(1, -2), (4, -1), (5, 0), (9, 2), (15, 5\}$

31. $v(x) = \{(-4, 3), (-3, 2), (0, 1), (5, 0), (12, -1), (21, -2), (32, -3)\}$
$v^{-1}(x) = \{(3, -4), (2, -3), (1, 0), (0, 5), (-1, 12), (-2, 21), (-3, 32)\}$

33. $f(x) = x + 5$
$y = x + 5;$
$x = y + 5$
$x - 5 = y$
$f^{-1}(x) = x - 5$

35. $p(x) = -\frac{4}{5}x$
$y = -\frac{4}{5}x;$
$x = -\frac{4}{5}y$
$-\frac{5}{4}x = y$
$p^{-1}(x) = -\frac{5}{4}x$

37. $f(x) = 4x + 3$
Multiply by 4, add 3
Inverse:
Subtract 3, divide by 4
$f^{-1}(x) = \frac{x-3}{4}$

39. $Y_1 = \sqrt[3]{x-4}$
Subtract 4, take cube root
Inverse:
Cube x, add 4
$Y_1^{-1} = x^3 + 4$

41. $f(x) = 2x + 7$
$y = 2x + 7;$
$x = 2y + 7$
$x - 7 = 2y$
$\frac{x-7}{2} = y$
$f^{-1}(x) = \frac{x-7}{2}$
Ordered pairs will vary.

43. $f(x) = \sqrt{x-2}$
$y = \sqrt{x-2};$
$x = \sqrt{y-2}$
$x^2 = y - 2$
$x^2 + 2 = y, x \geq 0$
$f^{-1}(x) = x^2 + 2, x \geq 0$
Ordered pairs will vary.

45. $f(x) = x^2 + 3$
$y = x^2 + 3;$
$x = y^2 + 3$
$x - 3 = y^2$
$\sqrt{x-3} = y$
$f^{-1}(x) = \sqrt{x-3}, x \geq 3$
Ordered pairs will vary.

47. $f(x) = x^3 + 1$
$y = x^3 + 1;$
$x = y^3 + 1$
$x - 1 = y^3$
$\sqrt[3]{x-1} = y$
$f^{-1}(x) = \sqrt[3]{x-1}$
Ordered pairs will vary.

49. $f(x) = -2x + 5$; $g(x) = \frac{x-5}{-2}$
$(f \circ g)(x) = f[g(x)]$
$= -2(g(x)) + 5$
$= -2\left(\frac{x-5}{-2}\right) + 5$
$= x;$
$(g \circ f)(x) = g[f(x)]$
$= \frac{f(x) - 5}{-2}$
$= \frac{-2x + 5 - 5}{-2}$

$= \frac{-2x}{-2}$
$= x$

51. $f(x) = \sqrt[3]{x+5}$; $g(x) = x^3 - 5$
$(f \circ g)(x) = f[g(x)]$
$= \sqrt[3]{g(x)+5}$
$= \sqrt[3]{x^3 - 5 + 5}$
$= \sqrt[3]{x^3}$
$= x;$
$(g \circ f)(x) = g[f(x)]$
$= (f(x))^3 - 5$
$= (\sqrt[3]{x+5})^3 - 5$
$= x + 5 - 5$
$= x$

53. $f(x) = \frac{2}{3}x - 6$; $g(x) = \frac{3}{2}x + 9$
$(f \circ g)(x) = f[g(x)]$
$= \frac{2}{3}g(x) - 6$
$= \frac{2}{3}\left(\frac{3}{2}x + 9\right) - 6$
$= x + 6 - 6$
$= x;$
$(g \circ f)(x) = g[f(x)]$
$= \frac{3}{2}f(x) + 9$
$= \frac{3}{2}\left(\frac{2}{3}x - 6\right) + 9$
$= x - 9 + 9$
$= x$

55. $f(x) = x^2 - 3;\ x \geq 0$; $g(x) = \sqrt{x+3}$
$(f \circ g)(x) = f[g(x)]$
$= (g(x))^2 - 3$
$= (\sqrt{x+3})^2 - 3$
$= x + 3 - 3$
$= x;$
$(g \circ f)(x) = g[f(x)]$
$= \sqrt{f(x)+3}$
$= \sqrt{x^2 - 3 + 3}$
$= \sqrt{x^2}$
$= x$

57. $f(x) = 3x - 5$
$y = 3x - 5$
$x = 3y - 5$
$x + 5 = 3y$
$\frac{x+5}{3} = y$
$f^{-1}(x) = \frac{x+5}{3}$
$(f \circ f^{-1})(x) = f[f^{-1}(x)]$
$= 3(f^{-1}(x)) - 5$
$= 3\left(\frac{x+5}{3}\right) - 5$
$= x + 5 - 5$
$= x;$
$(f^{-1} \circ f)(x) = f^{-1}[f(x)]$
$= \frac{f(x)+5}{3}$
$= \frac{3x - 5 + 5}{3}$
$= \frac{3x}{3}$
$= x$

59. $f(x) = \frac{x-5}{2}$
$y = \frac{x-5}{2}$
$x = \frac{y-5}{2}$
$2x = y - 5$
$2x + 5 = y$
$f^{-1}(x) = 2x + 5$
$(f \circ f^{-1})(x) = f[f^{-1}(x)]$
$= \frac{f^{-1}(x) - 5}{2}$
$= \frac{2x + 5 - 5}{2}$
$= \frac{2x}{2}$
$= x;$
$(f^{-1} \circ f)(x) = f^{-1}[f(x)]$
$= 2(f(x)) + 5$

$= 2\left(\frac{x-5}{2}\right)+5$
$= x-5+5$
$= x$

61. $f(x)=\frac{1}{2}x-3$

$y=\frac{1}{2}x-3$

$x=\frac{1}{2}y-3$

$x+3=\frac{1}{2}y$

$2x+6=y$

$f^{-1}(x)=2x+6$

$(f\circ f^{-1})(x)=f[f^{-1}(x)]$

$=\frac{1}{2}(f^{-1}(x))-3$

$=\frac{1}{2}(2x+6)-3$

$=x+3-3$

$=x;$

$(f^{-1}\circ f)(x)=f^{-1}[f(x)]$

$=2(f(x))+6$

$=2\left(\frac{1}{2}x-3\right)+6$

$=x-6+6$

$=x$

63. $f(x)=x^3+3$

$y=x^3+3$

$x=y^3+3$

$x-3=y^3$

$\sqrt[3]{x-3}=y$

$f^{-1}(x)=\sqrt[3]{x-3}$

$(f\circ f^{-1})(x)=f[f^{-1}(x)]$

$=(f^{-1}(x))^3+3$

$=(\sqrt[3]{x-3})^3+3$

$=x-3+3$

$=x;$

$(f^{-1}\circ f)(x)=f^{-1}[f(x)]$

$=\sqrt[3]{f(x)-3}+3$

$=\sqrt[3]{x^3-3+3}$

$=\sqrt[3]{x^3}$

$=x$

65. $f(x)=\sqrt[3]{2x+1}$

$y=\sqrt[3]{2x+1}$

$x=\sqrt[3]{2y+1}$

$x^3=2y+1$

$x^3-1=2y$

$\frac{x^3-1}{2}=y$

$f^{-1}(x)=\frac{x^3-1}{2}$

$(f\circ f^{-1})(x)=f[f^{-1}(x)]$

$=\sqrt[3]{2(f^{-1}(x))+1}$

$=\sqrt[3]{2\left(\frac{x^3-1}{2}\right)+1}$

$=\sqrt[3]{x^3-1+1}$

$=\sqrt[3]{x^3}$

$=x;$

$(f^{-1}\circ f)(x)=f^{-1}[f(x)]$

$=\frac{(f(x))^3-1}{2}$

$=\frac{(\sqrt[3]{2x+1})^3-1}{2}$

$=\frac{2x+1-1}{2}$

$=\frac{2x}{2}$

$=x$

67. $f(x)=\frac{(x-1)^3}{8}$

$y=\frac{(x-1)^3}{8}$

$x=\frac{(y-1)^3}{8}$

$8x=(y-1)^3$

$\sqrt[3]{8x}=y-1$

$2\sqrt[3]{x}+1=y$

$f^{-1}(x)=2\sqrt[3]{x}+1$

$(f\circ f^{-1})(x)=f[f^{-1}(x)]$

$=\dfrac{(f^{-1}(x)-1)^3}{8}$

$=\dfrac{(2\sqrt[3]{x}+1-1)^3}{8}$

$=\dfrac{(2\sqrt[3]{x})^3}{8}$

$=\dfrac{8x}{8}$

$=x;$

$(f^{-1}\circ f)(x)=f^{-1}[f(x)]$

$=2\sqrt[3]{f(x)}+1$

$=2\sqrt[3]{\dfrac{(x-1)^3}{8}}+1$

$=2\left(\dfrac{x-1}{2}\right)+1$

$=x-1+1$
$=x$

69. $f(x)=\sqrt{3x+2}$

$y=\sqrt{3x+2}$

$x=\sqrt{3y+2}$

$x^2=3y+2$

$x^2-2=3y$

$\dfrac{x^2-2}{3}=y$

$f^{-1}(x)=\dfrac{x^2-2}{3}; x\geq 0;$

$(f\circ f^{-1})(x)=f(f^{-1}(x))$

$=\sqrt{3(f^{-1}(x))+2}$

$=\sqrt{3\left(\dfrac{x^2-2}{3}\right)+2}$

$=\sqrt{x^2-2+2}$

$=\sqrt{x^2}$
$=|x|$
$=x$; since x ≥ 0

$(f^{-1}\circ f)(x)=f^{-1}[f(x)]$

$=\dfrac{(f(x))^2-2}{3}$

$=\dfrac{(\sqrt{3x+2})^2-2}{3}$

$=\dfrac{3x+2-2}{3}$

$=\dfrac{3x}{3}$

$=x$

71. $p(x)=2\sqrt{x-3}$

$y=2\sqrt{x-3}$

$x=2\sqrt{y-3}$

$\dfrac{x}{2}=\sqrt{y-3}$

$\dfrac{x^2}{4}=y-3$

$\dfrac{x^2}{4}+3=y$

$p^{-1}(x)=\dfrac{x^2}{4}+3; x\geq 0;$

$(p\circ p^{-1})(x)=p(p^{-1}(x))$

$=2\sqrt{p^{-1}(x)-3}$

$=2\sqrt{\dfrac{x^2}{4}+3-3}$

$=2\sqrt{\dfrac{x^2}{4}}$

$=2\left(\dfrac{x}{2}\right)$

$=x;$

$(p^{-1}\circ p)(x)=p^{-1}[p(x)]$

$=\dfrac{(p(x))^2}{4}+3$

$=\dfrac{(2\sqrt{x-3})^2}{4}+3$

$=\dfrac{4(x-3)}{4}+3$

$=x-3+3$

$=x$

73. $v(x)=x^2+3; x\geq 0$

$y=x^2+3$

$x = y^2 + 3$
$x - 3 = y^2$
$\sqrt{x-3} = y$
$v^{-1}(x) = \sqrt{x-3}$;
$(v \circ v^{-1})(x) = v[v^{-1}(x)]$
$= (v^{-1}(x))^2 + 3$
$= (\sqrt{x-3})^2 + 3$
$= x - 3 + 3$
$= x$;
$(v^{-1} \circ v)(x) = v^{-1}[v(x)]$
$= \sqrt{v(x) - 3}$
$= \sqrt{x^2 + 3 - 3}$
$= \sqrt{x^2}$
$= |x|$
$= x$; since $x \ge 0$

75. $f(x) = 4x + 1$; $f^{-1}(x) = \dfrac{x-1}{4}$

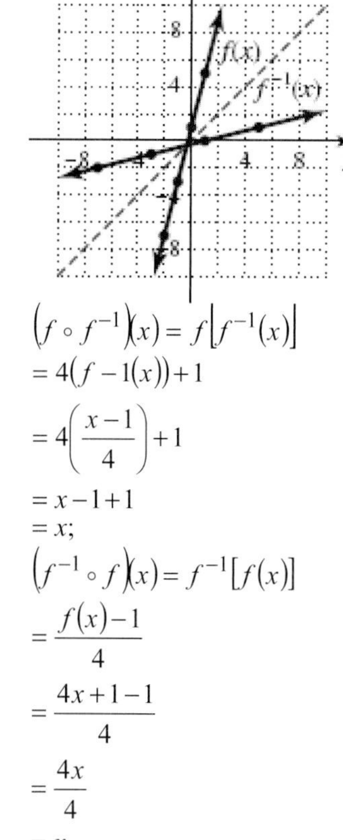

$(f \circ f^{-1})(x) = f[f^{-1}(x)]$
$= 4(f-1(x)) + 1$
$= 4\left(\dfrac{x-1}{4}\right) + 1$
$= x - 1 + 1$
$= x$;
$(f^{-1} \circ f)(x) = f^{-1}[f(x)]$
$= \dfrac{f(x) - 1}{4}$
$= \dfrac{4x + 1 - 1}{4}$
$= \dfrac{4x}{4}$
$= x$

77. $f(x) = \sqrt[3]{x+2}$; $f^{-1}(x) = x^3 - 2$

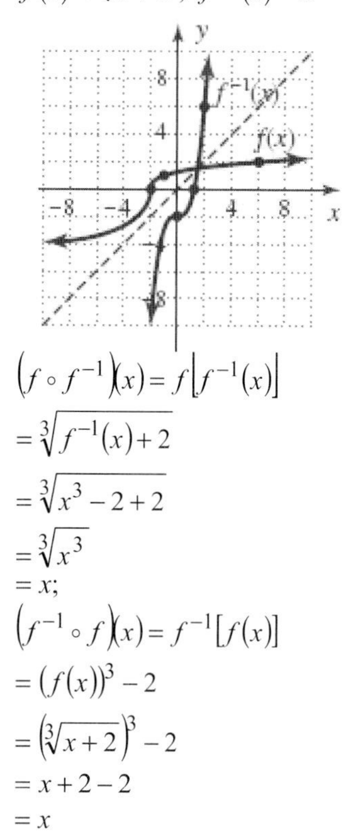

$(f \circ f^{-1})(x) = f[f^{-1}(x)]$
$= \sqrt[3]{f^{-1}(x) + 2}$
$= \sqrt[3]{x^3 - 2 + 2}$
$= \sqrt[3]{x^3}$
$= x$;
$(f^{-1} \circ f)(x) = f^{-1}[f(x)]$
$= (f(x))^3 - 2$
$= (\sqrt[3]{x+2})^3 - 2$
$= x + 2 - 2$
$= x$

79. $f(x) = 0.2x + 1$; $f^{-1}(x) = 5x - 5$

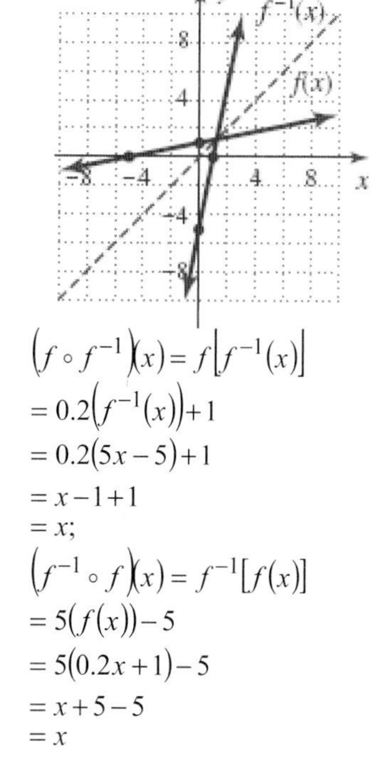

$(f \circ f^{-1})(x) = f[f^{-1}(x)]$
$= 0.2(f^{-1}(x)) + 1$
$= 0.2(5x - 5) + 1$
$= x - 1 + 1$
$= x$;
$(f^{-1} \circ f)(x) = f^{-1}[f(x)]$
$= 5(f(x)) - 5$
$= 5(0.2x + 1) - 5$
$= x + 5 - 5$
$= x$

81. $f(x) = (x+2)^2; x \ge -2$; $f^{-1}(x) = \sqrt{x} - 2$

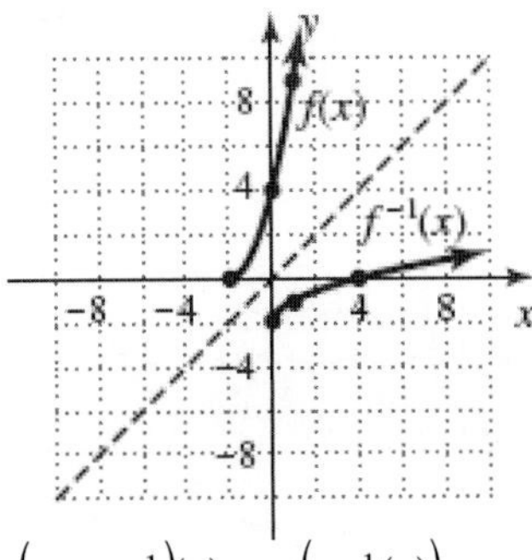

$$\left(f \circ f^{-1}\right)(x) = f\left(f^{-1}(x)\right)$$
$$= \left(f^{-1}(x) + 2\right)^2$$
$$= \left(\sqrt{x} - 2 + 2\right)^2$$
$$= \left(\sqrt{x}\right)^2$$
$$= x;$$
$$\left(f^{-1} \circ f\right)(x) = f^{-1}[f(x)]$$
$$= \sqrt{f(x)} - 2$$
$$= \sqrt{(x+2)^2} - 2$$
$$= x + 2 - 2$$
$$= x$$

83. $f(x)$ $D: [0, \infty)$ $R: [-2, \infty)$

$f^{-1}(x)$ $D: [-2, \infty)$ $R: [0, \infty)$

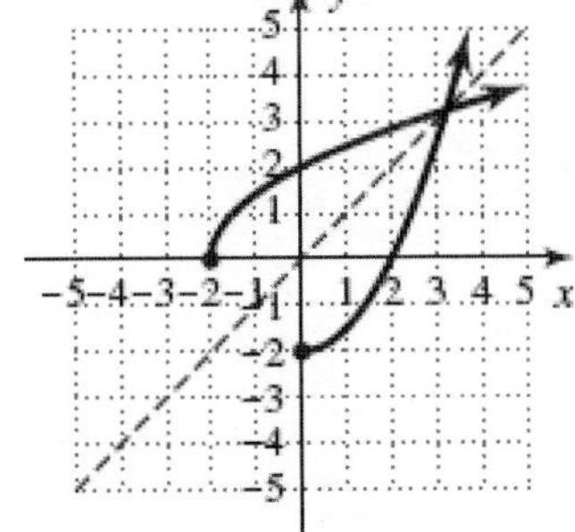

85. $f(x)$ $D: (0, \infty)$ $R: (-\infty, \infty)$

$f^{-1}(x)$ $D: (-\infty, \infty)$ $R: (0, \infty)$

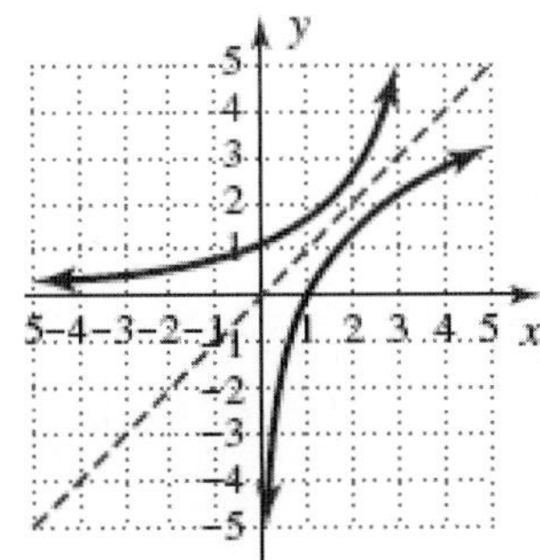

87. $f(x)$ $D: (-\infty, 4]$ $R: (-\infty, 4]$

$f^{-1}(x)$ $D: (-\infty, 4]$ $R: (-\infty, 4]$

89. $f(x) = \frac{1}{2}x - 8.5$

a. $f(80) = \frac{1}{2}(80) - 8.5 = 31.5 \text{ cm}$

b. $y = \frac{1}{2}x - 8.5$

$x = \frac{1}{2}y - 8.5$

$x + 8.5 = \frac{1}{2}y$

$2x + 17 = y$

$f^{-1}(x) = 2x + 17$

$f^{-1}(31.5) = 2(31.5) + 17 = 80$

They are the same two values, distance of projector from screen.

91. $f(x) = -\frac{7}{2}x + 59$

a. $f(35) = -\frac{7}{2}(35) + 59 = -63.5^\circ F$

b. $y = -\frac{7}{2}x + 59$

$x = -\frac{7}{2}y + 59$

$x-59=-\frac{7}{2}y$

$-\frac{2}{7}(x-59)=y$

$f^{-1}(x)=-\frac{2}{7}(x-59)$

$f^{-1}(-63.5)=-\frac{2}{7}(-63.5-59)$

$=-\frac{2}{7}(-122.5)=35$

An altitude of 35000 feet.

c. $f^{-1}(-18)=-\frac{2}{7}(-18-59)$

$=-\frac{2}{7}(-77)=22$

The approximate altitude is 22000 feet.

93. $f(x)=16x^2;x\geq 0$

a. $f(3)=16(3)^2=16(9)=144$ ft

b. $y=16x^2$

$x=16y^2$

$\frac{x}{16}=y^2$

$\frac{\sqrt{x}}{4}=y$

$f^{-1}(x)=\frac{\sqrt{x}}{4};$

$f^{-1}(144)=\frac{\sqrt{144}}{4}=\frac{12}{4}=3\text{ sec}$, the original input for $f(x)$.

c. $f^{-1}(784)=\frac{\sqrt{784}}{4}=\frac{28}{4}=7\text{ sec}$

95. $f(h)=\frac{1}{3}\pi h^3$

a. $f(30)=\frac{1}{3}\pi(30)^3=9000\pi=28260\text{ ft}^3$

b. $y=\frac{1}{3}\pi h^3$

$h=\frac{1}{3}\pi y^3$

$3h=\pi y^3$

$\frac{3h}{\pi}=y^3$

$\sqrt[3]{\frac{3h}{\pi}}=y$

$f^{-1}(h)=\sqrt[3]{\frac{3h}{\pi}};$

$f^{-1}(28260)=\sqrt[3]{\frac{3(28260)}{\pi}}=30\text{ ft}$, the original input for $f(h)$.

c. $f^{-1}(763.02)=\sqrt[3]{\frac{3(763.02)}{\pi}}=9\text{ ft}$

97. $f(x)=\pi x^3$

a. $392.5=\pi x^3$

$125=x^3$

$5=x$

5 cm

b. $y=\pi x^3$

$x=\pi y^3$

$\frac{x}{\pi}=y^3$

$\sqrt[3]{\frac{x}{\pi}}=y$

$f^{-1}(x)=\sqrt[3]{\frac{x}{\pi}};$

$f^{-1}(392.5)=\sqrt[3]{\frac{392.5}{\pi}}\approx 5$, same as original input for f(x).

c. $f^{-1}(x)$

99. $f(x)=\frac{1}{x}$

a. $y=\frac{1}{x}$

$x=\frac{1}{y}$

$xy=1$

$y=\frac{1}{x}$

$f^{-1}(x)=\frac{1}{x}$

b.

x	$f(x)=\frac{1}{x}$

-4	$f(-4)=\frac{1}{-4}$
-3	$f(-3)=\frac{1}{-3}$
-2	$f(-2)=\frac{1}{-2}$
-1	$f(-1)=\frac{1}{-1}=-1$
0	$f(0)=\frac{1}{0}$ undefined
1	$f(1)=\frac{1}{1}=1$
2	$f(2)=\frac{1}{2}$
3	$f(3)=\frac{1}{3}$
4	$f(4)=\frac{1}{4}$

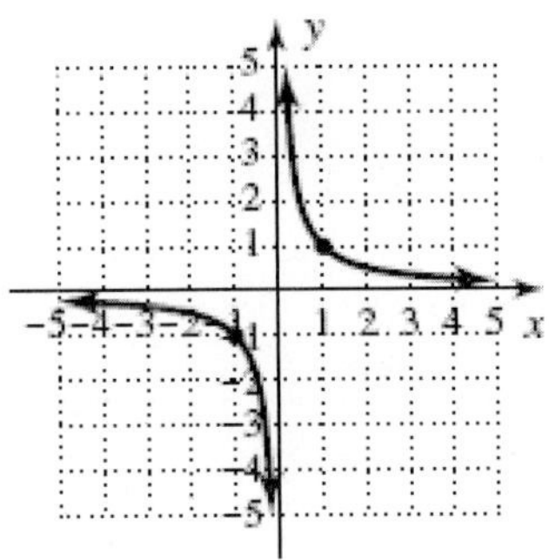

Point of intersection (1, 1) and (-1, -1); x and y coordinates are identical and on $f(x)$.

101.a. $f(1)=0$

b. $f(729)=3$

c. $f^{-1}(2)=81$

d. $f^{-1}\left(\frac{1}{2}\right)=3$

103. $y=2\sqrt{x+3}$

$$\frac{\Delta y}{\Delta x}=\frac{f(2)-f(1)}{2-1}=\frac{2\sqrt{5}-4}{1}=0.4721$$

$$\frac{\Delta y}{\Delta x}=\frac{f(5)-f(4)}{5-4}=\frac{2\sqrt{8}-2\sqrt{7}}{1}=0.3651$$

Rate of change is greater in [1, 2] due to shape of the one-wing function.

7. Function family: Square root
x-intercept: (-3, 0)

105. $f(x)=x^2-4x-41$

$$x=\frac{4\pm\sqrt{(-4)^2-4(1)(-41)}}{2(1)}$$

$$x=\frac{4\pm\sqrt{16+164}}{2}$$

$$x=\frac{4\pm\sqrt{180}}{2}$$

$$x=\frac{4\pm 6\sqrt{5}}{2}$$

$$x=2\pm 3\sqrt{5}$$

$x\approx 8.71;\ x\approx -4.71$

107.a.

Percent of U.S. Population
Year (1970 → 0)

b. Linear regression
$y=0.51x+22.51$
Strong correlation

c. $f(35)=0.51(35)+22.51=40.36\%$;
2005; 40.36%
$f(40)=0.51(40)+22.51=42.91\%$;
2010; 42.91%

3.3 Technology Highlight

Exercise 1: Shifted right 3 units; answers will vary.

Exercise 2: Shifted right 3 units; answers will vary.

3.3 Exercises

1. Stretch; compression

3. (-5, -9); up

5. Answers will vary.

y-intercept: (0, 2)
Node: (-4, -2)

End behavior: Up on right

9. Function family: Cubic
x-intercept: (-2, 0)
y-intercept: (0, -2)
Pivot point: (-1, -1)
End behavior: Up/down

11. $f(x)=\sqrt{x}$; $g(x)=\sqrt{x}+2$; $h(x)=\sqrt{x}-3$

x	$f(x)$	$g(x)$	$h(x)$
0	0	2	-3
4	2	4	-1
9	3	5	0
16	4	6	1
25	5	7	2

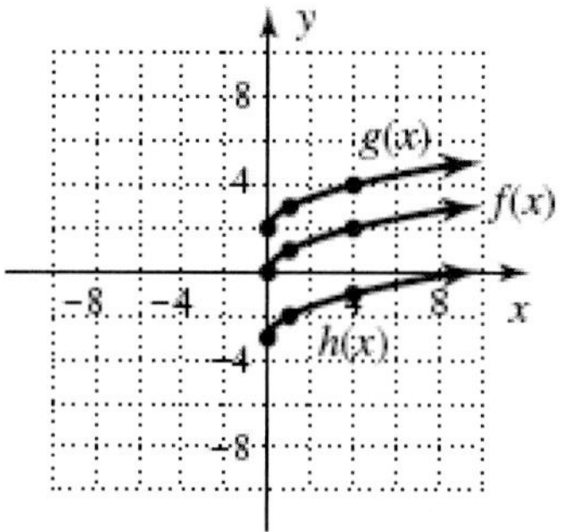

From the parent graph $f(x)=\sqrt{x}$, $g(x)$ shifts up 2 units and $h(x)$ shifts down 3 units.

13. $p(x)=|x|$; $q(x)=|x|-5$; $r(x)=|x|+2$

x	$p(x)$	$q(x)$	$r(x)$
-2	2	-3	4
-1	1	-4	3
0	0	-5	2
1	1	-4	3
2	2	-3	4

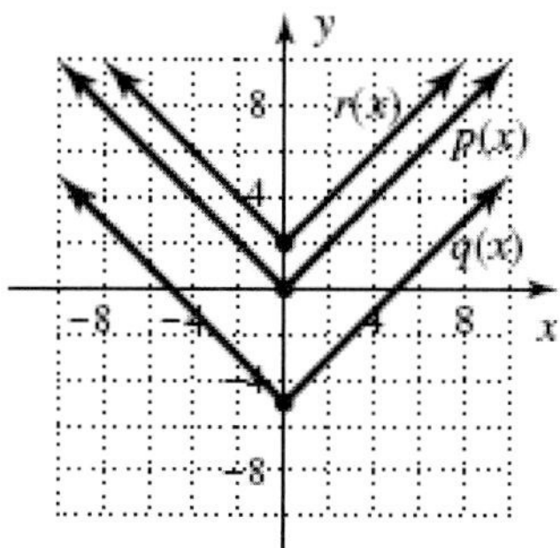

From the parent graph $p(x)=|x|$, $q(x)$ shifts down 5 units and $r(x)$ shifts up 2 units.

15. $f(x)=x^3-2$
Shifts down 2 units.

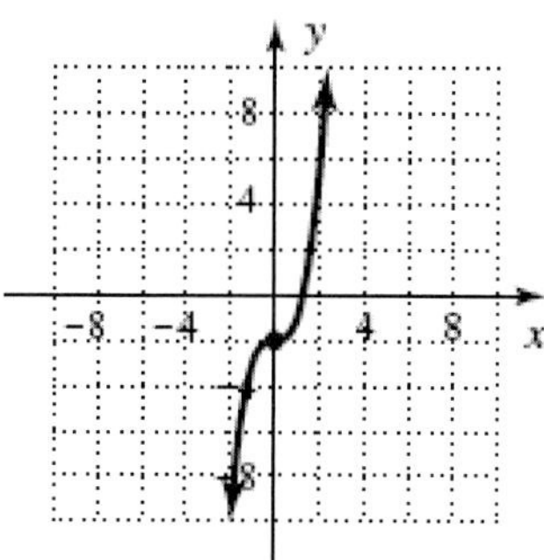

17. $h(x)=x^2+3$
Shifts up 3 units.

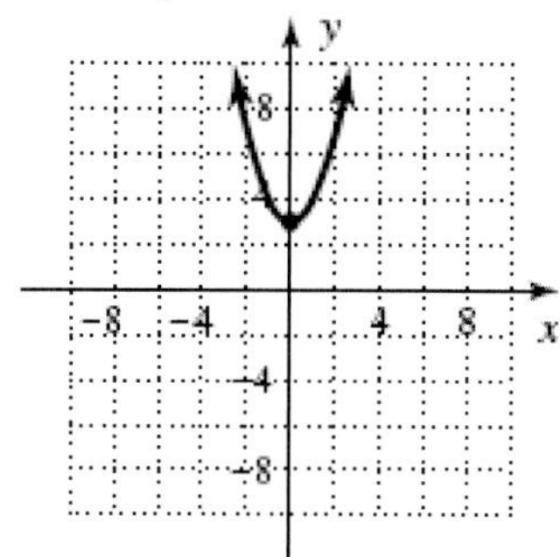

19. $p(x)=x^2$; $q(x)=(x+3)^2$

x	$p(x)=x^2$	$q(x)=(x+3)^2$
-5	25	4
-3	9	0
-1	1	4
1	1	16
3	9	36

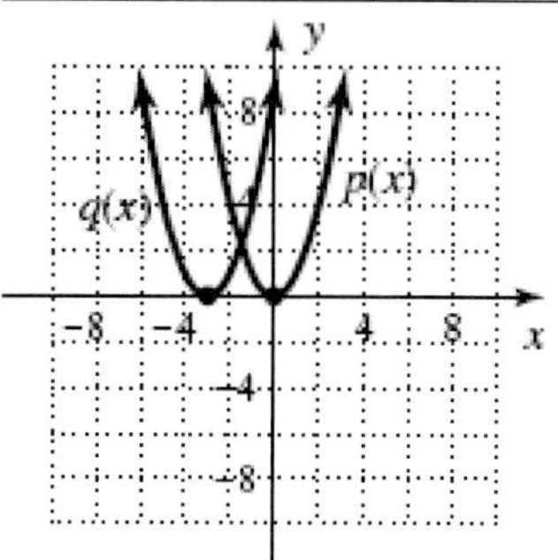

From the parent graph $p(x)=x^2$, $g(x)$ shifts left 3 units.

21. $Y_1=|x|$; $Y_2=|x-1|$

x	$Y_1=\|x\|$	$Y_2=\|x-1\|$
-2	2	3
-1	1	2
0	0	1

1	1	0
2	2	1

From the parent graph $Y_1 = |x|$, Y_2 shifts right 1 unit.

23. $p(x) = (x-3)^2$
Shifts right 3 units.

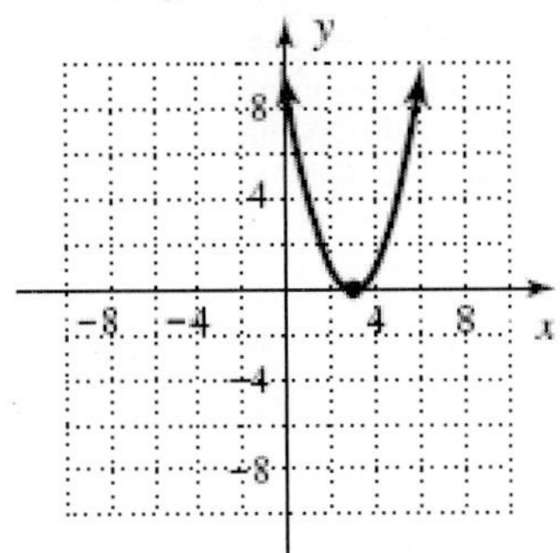

25. $h(x) = |x+3|$
Shifts left 3 units.

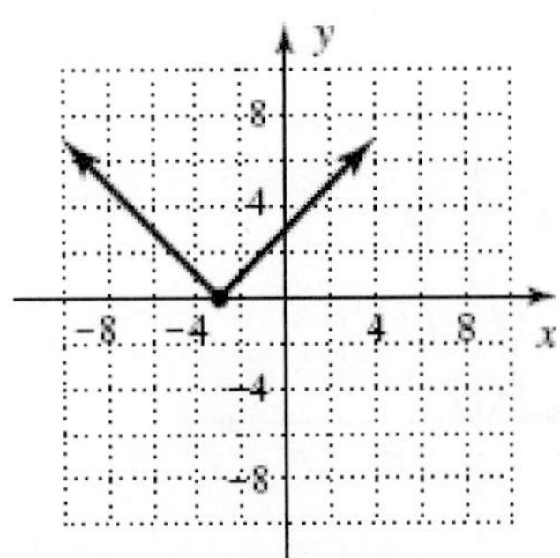

27. $g(x) = -|x|$
Reflects across the x-axis.

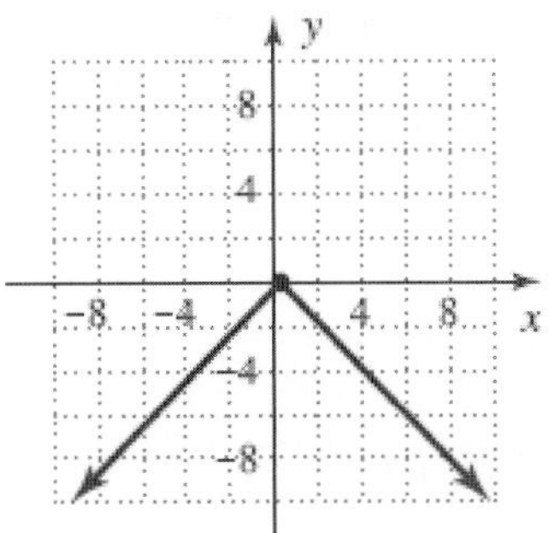

29. $f(x) = \sqrt[3]{-x}$
Reflects across the x-axis.

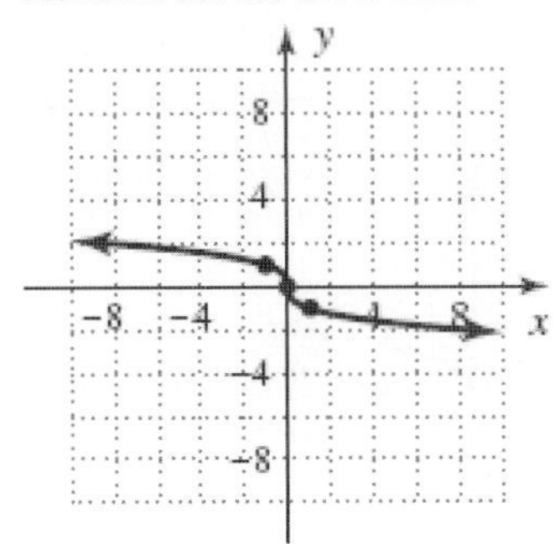

31. $p(x) = x^2$; $q(x) = 2x^2$; $r(x) = \frac{1}{2}x^2$

x	$p(x)$	$q(x)$	$r(x)$
-2	4	8	2
-1	1	2	½
0	0	0	0
1	1	2	½
2	4	8	2

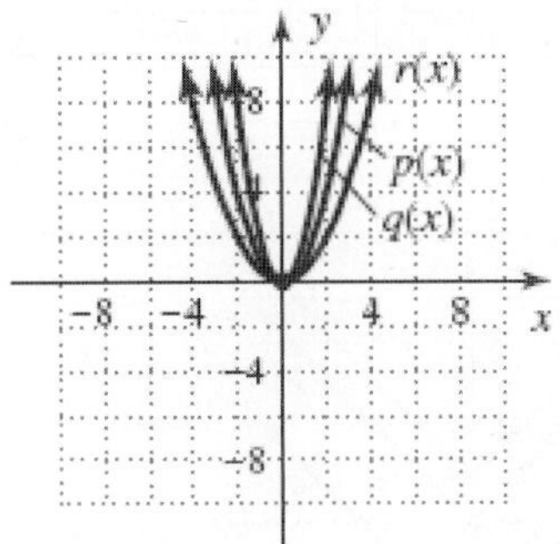

From the parent graph $p(x) = x^2$, $q(x)$ stretches upward and $r(x)$ compresses downward.

33. $Y_1 = |x|$; $Y_2 = 3|x|$; $Y_3 = \frac{1}{3}|x|$

x	Y_1	Y_2	Y_3
-2	2	6	2/3

-1	1	3	1/3
0	0	0	0
1	1	3	1/3
2	2	6	2/3

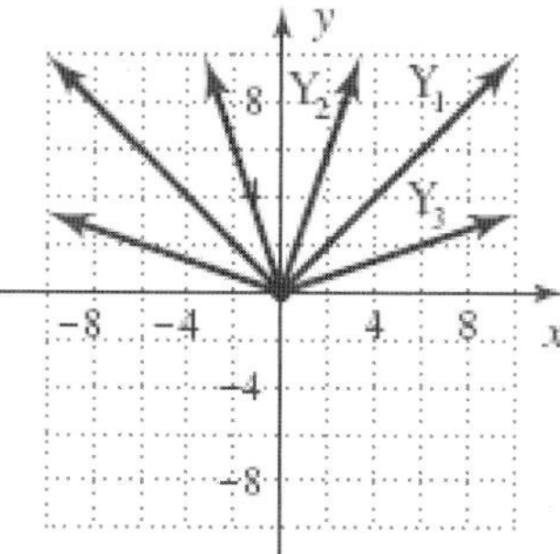

From the parent graph $Y_1 = |x|$, Y_2 stretches upward and Y_3 compresses downward.

35. $f(x) = 4\sqrt[3]{x}$
Stretches upward and downward.

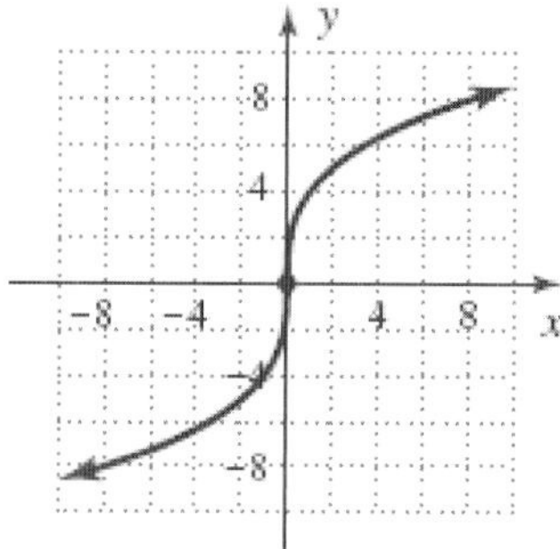

37. $p(x) = \frac{1}{3}x^3$
Compresses downward.

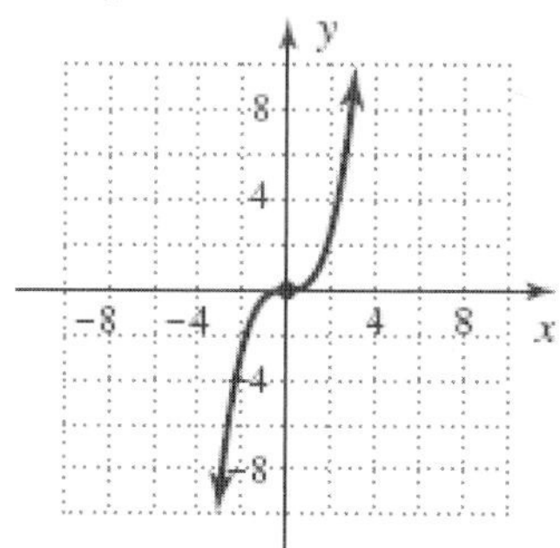

39. $f(x) = \frac{1}{2}x^3$; g

41. $f(x) = -(x-3)^2 + 2$; i

43. $f(x) = |x+4| + 1$; e

45. $f(x) = -\sqrt{x+6} - 1$; j

47. $f(x) = (x-4)^2 - 3$; l

49. $f(x) = \sqrt{x+3} - 1$; c

51. $f(x) = \sqrt{x+2} - 1$
Left 2, down 1
Node: (-2, -1)

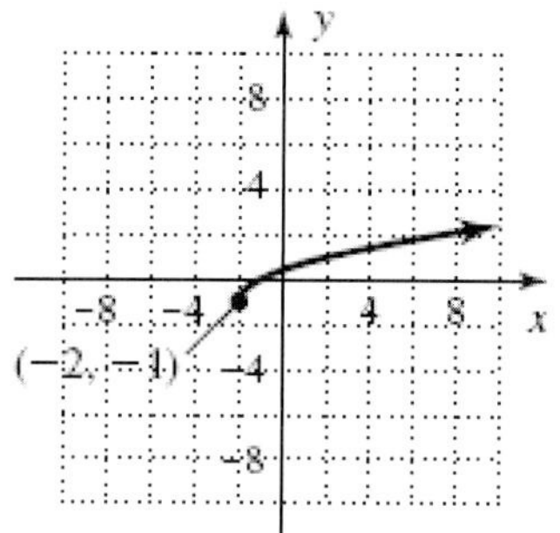

53. $h(x) = -(x+3)^2 - 2$
Reflected across x-axis, left 3, down 2
Vertex: (-3, -2)

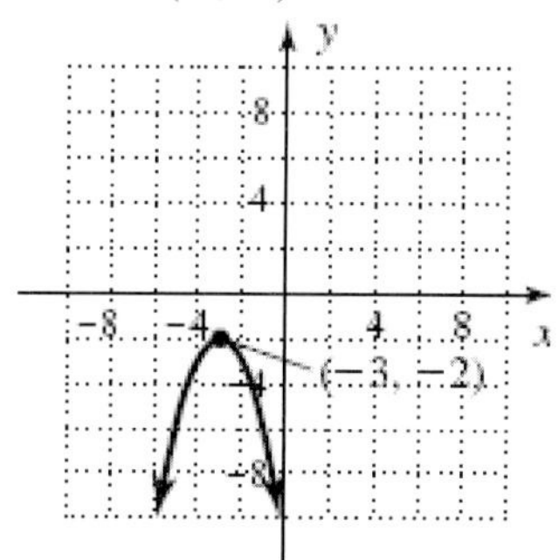

55. $p(x) = (x+3)^3 - 1$
Left 3, down 1
Pivot point: (-3, -1)

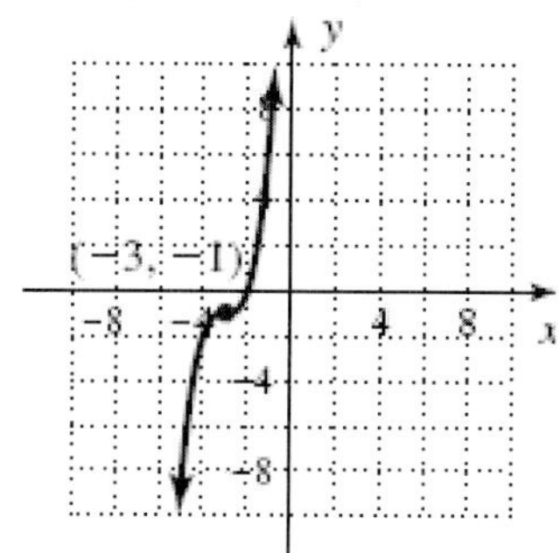

57. $Y_1 = \sqrt[3]{x+1} - 2$
Left 1, down 2
Pivot point: (-1, -2)

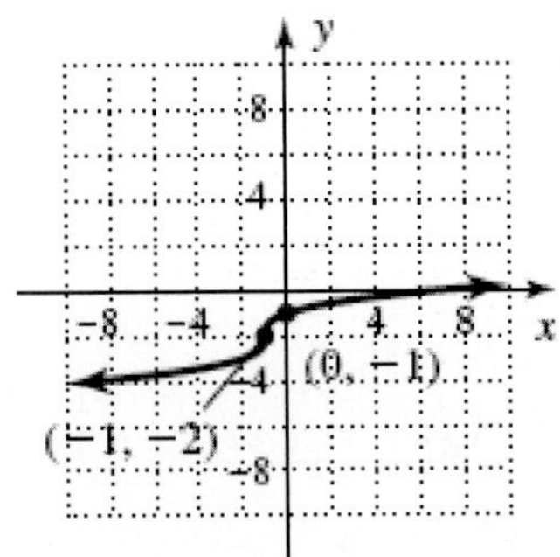

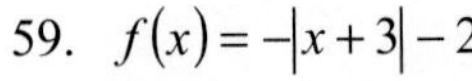

59. $f(x) = -|x+3| - 2$

Reflected across x-axis, left 3, down 2
Vertex: (-3, -2)

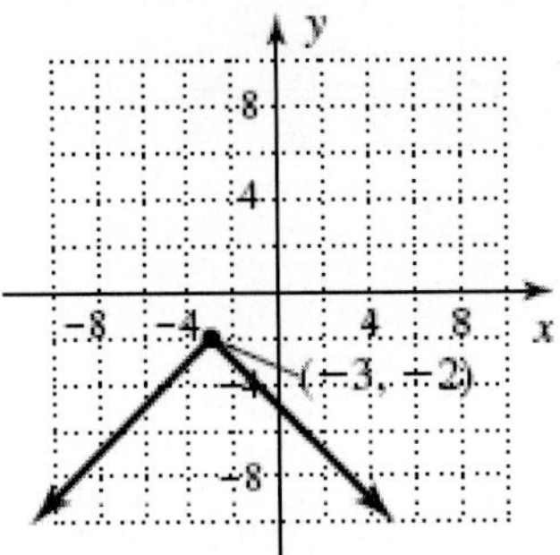

61. $h(x) = -2(x+1)^2 - 3$

Reflected across x-axis, left 1, down 3, stretched vertically
Vertex: (-1, -3)

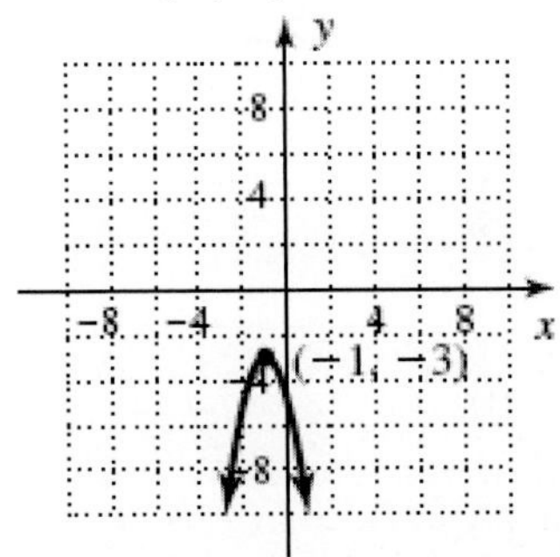

63. $p(x) = -\frac{1}{3}(x+2)^3 - 1$

Reflected across x-axis, left 2, down 1, compressed vertically
Pivot point: (-2, -1)

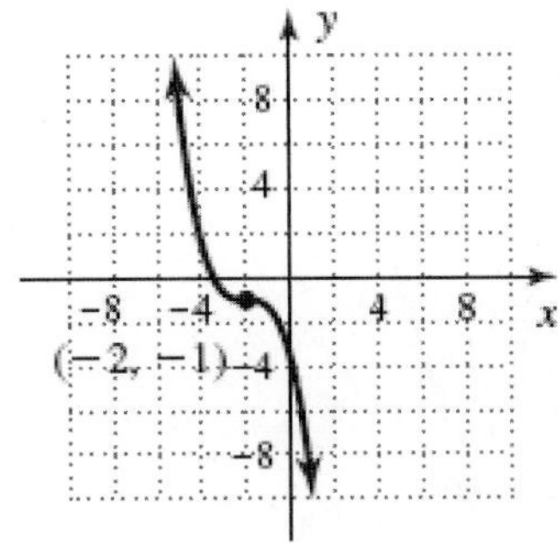

65. $Y_1 = -2\sqrt{x-4} - 3$

Reflected across x-axis, right 4, down 3, stretched vertically
Node: (4, -3)

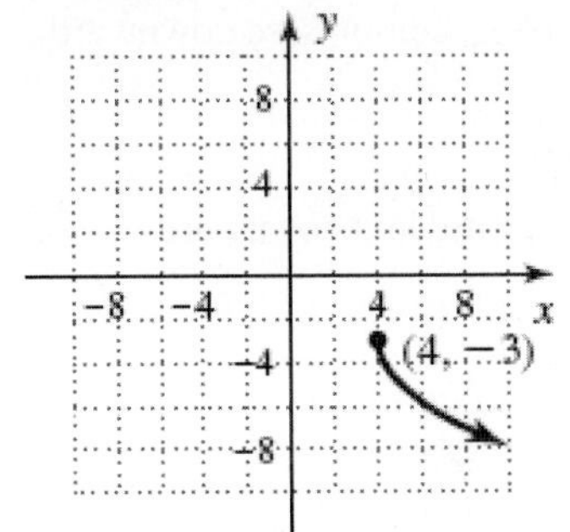

67. $h(x) = \frac{1}{5}(x-3)^2 + 1$

Right 3, up 1, compressed vertically
Vertex: (3, 1)

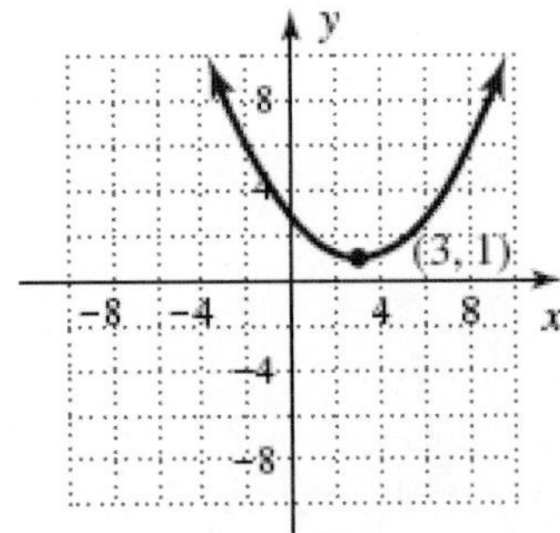

69. a. $f(x-2)$

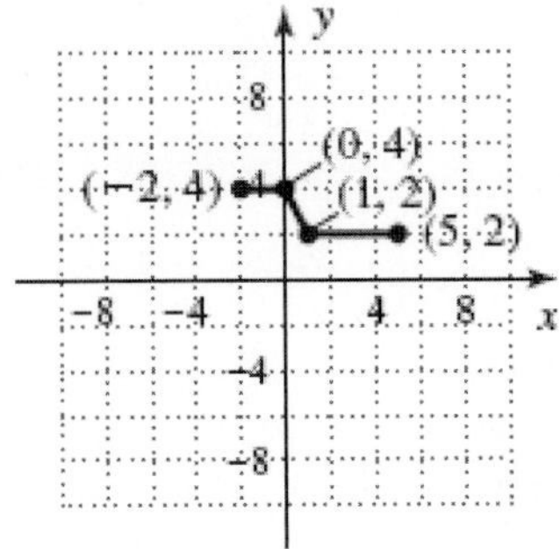

b. $-f(x) - 3$

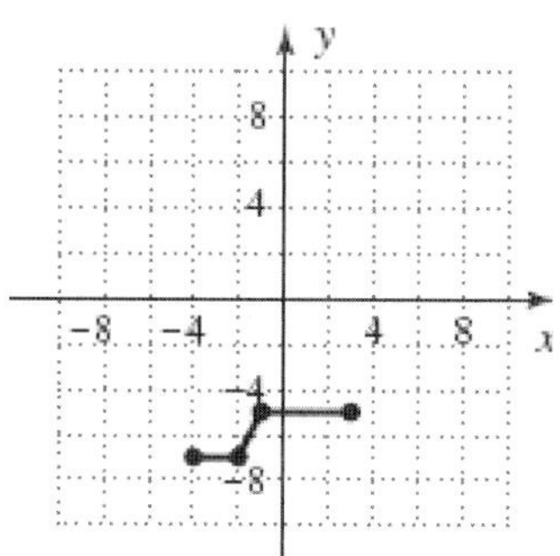

c. $\frac{1}{2}f(x+1)$

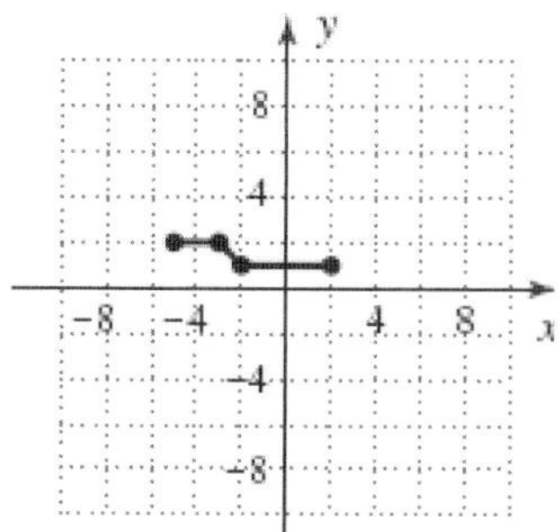

d. $f(-x)+1$

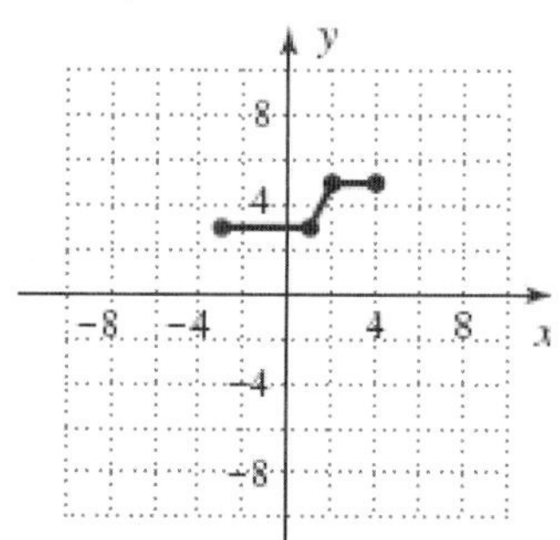

71. a. $h(x)+3$

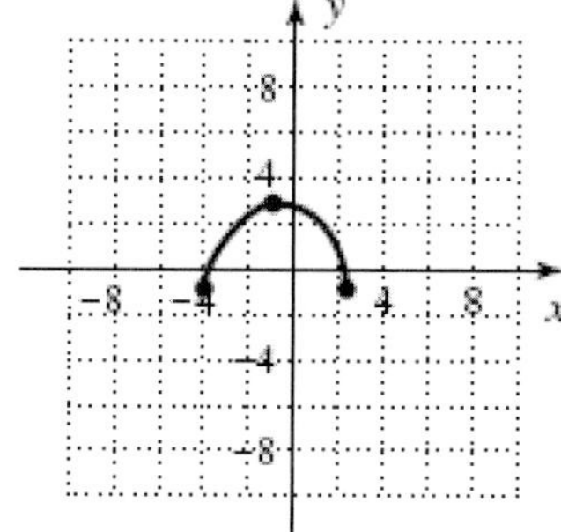

b. $-h(x-2)$

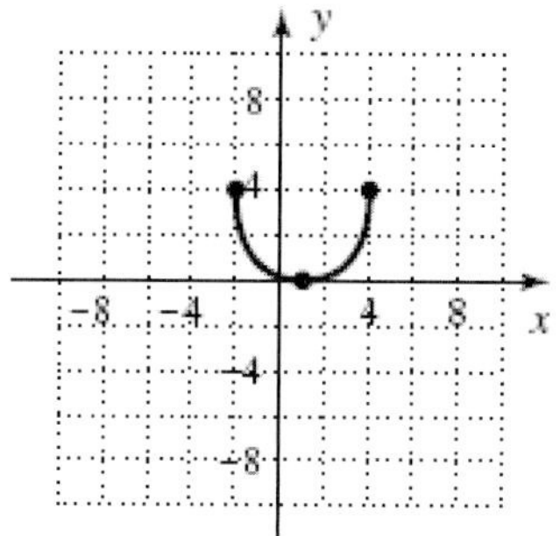

c. $h(x-2)-1$

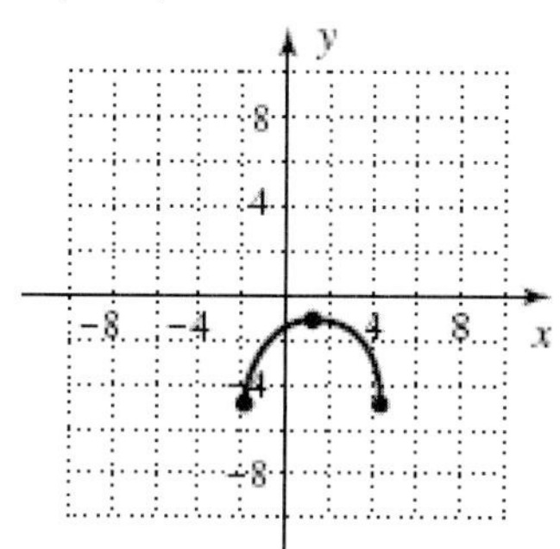

d. $\frac{1}{4}h(x)+5$

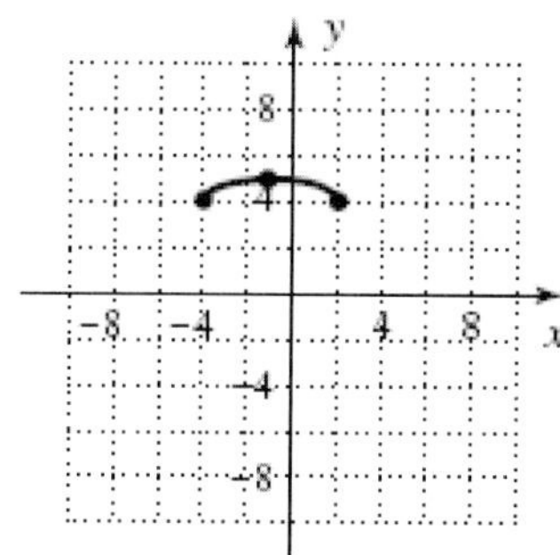

73. $f(x)=-|x-3|+6$

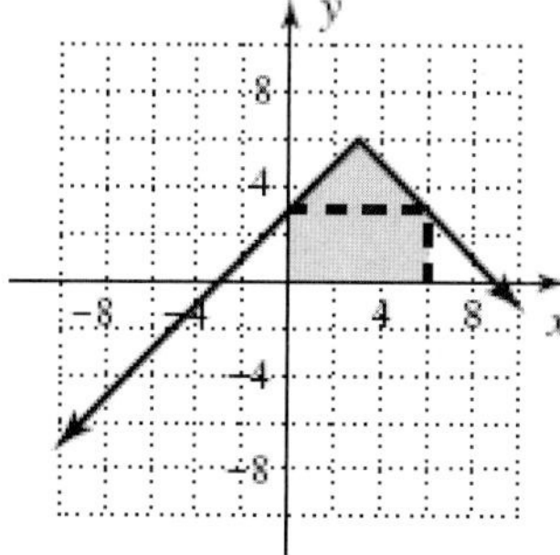

A = Area of rectangle + Area of triangle

$A = lw+\frac{1}{2}(bh)$

$A = 6(3)+\frac{1}{2}(6\cdot 3)$

$A = 27 \text{ units}^2$

75. $g(x) = -(x-2)^2 + 7$

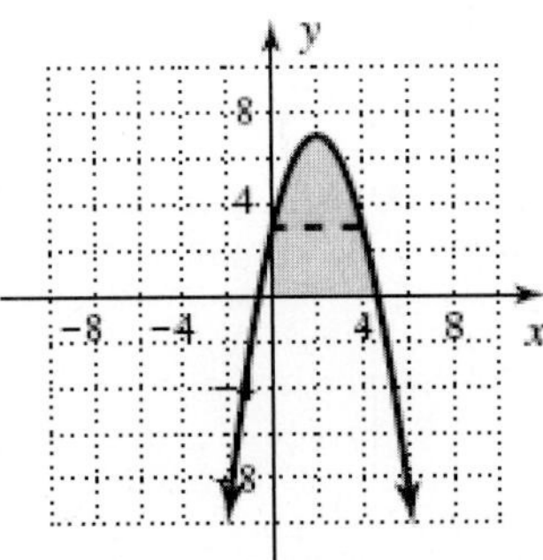

A=Area of rectangle + Area of parabolic segment

$A = lw + \frac{2}{3}ab$

$A = 4(3) + \frac{2}{3}(4)(4) = 22\frac{2}{3}$ $units^2$

77. $q(x) = \sqrt{9-(x-3)^2} + 2$

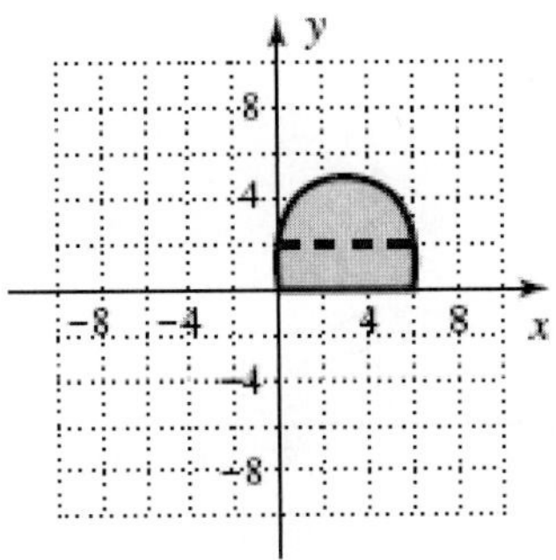

A=Area of ½ an ellipse + Area of rectangle

$A = \pi\ r_1r_2 + lw$
$A = \pi(3)(3) + 6(2)$
$A = 9\pi + 12$ units^2

79. $V = \frac{4}{3}\pi r^3$

$\frac{4}{3}\pi \approx 4.2$

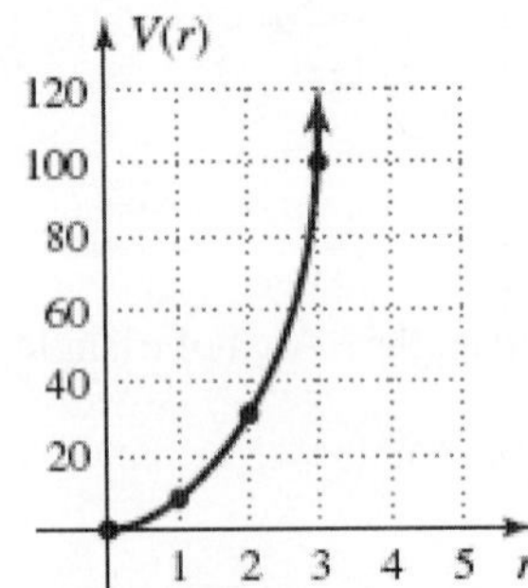

Volume estimate: 70 in^3

$V = \frac{4}{3}\pi r^3$

$V = \frac{4}{3}\pi(2.5)^3$

$V = \frac{4}{3}\pi(15.625) \approx 65.4\ \text{in}^3$

Yes

81. $T(x) = \frac{1}{4}\sqrt{x}$

The graph can be obtained from $y = \sqrt{x}$ if it is compressed vertically.

$T(81) = \frac{1}{4}\sqrt{81} = \frac{1}{4}(9) = 2.25\text{sec}$

This point is on the graph.

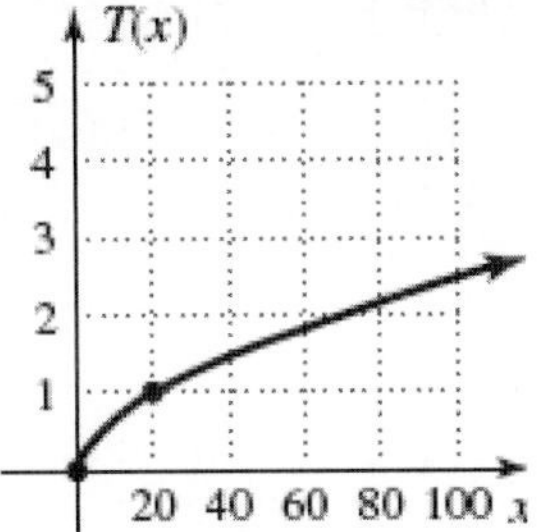

83. $P(v) = \frac{8}{125}v^3$

The graph can be obtained from $y = v^3$ if it is compressed vertically.

$P(15) = \frac{8}{125}(15)^3 = 216$ watts

This point is on the graph.

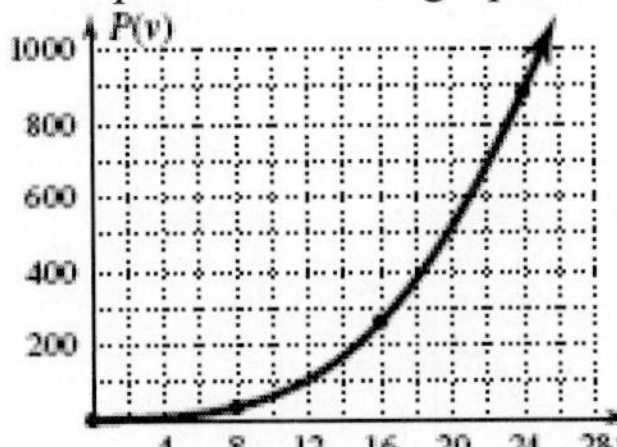

85. $v(t) = 4t$

Vertical stretch by a factor of 4.

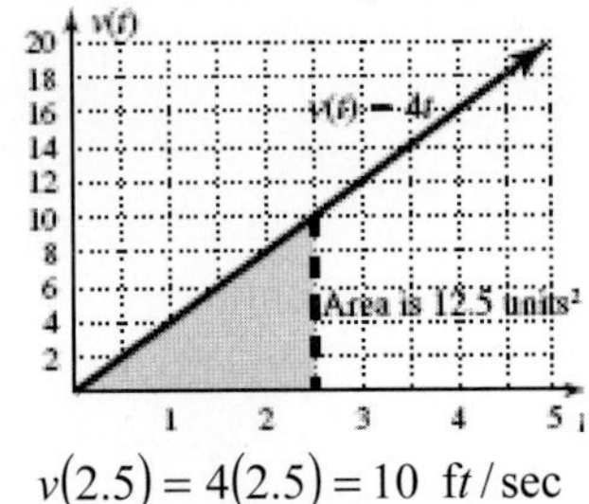

$v(2.5) = 4(2.5) = 10$ ft/sec

$$A = \frac{bh}{2}$$
$$A = \frac{2.5(10)}{2} = 12.5 \text{ ft}$$

87. $f(x) = |x|$ and $g(x) = 2\sqrt{x}$

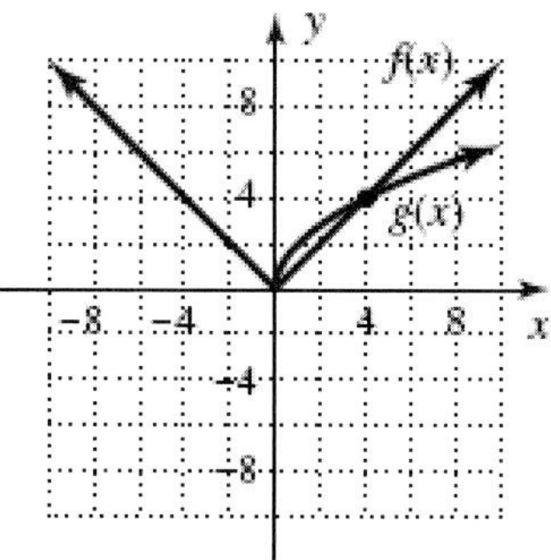

Interval: $x \in (0, 4)$
$x = 1$
$f(1) = |1| = 1$ and $g(1) = 2\sqrt{1} = 2$
$g(h) > f(h)$
Interval: $x \in (4, \infty)$
$x = 9$
$f(9) = |9| = 9$ and $g(1) = 2\sqrt{9} = 6$
$g(k) < f(k)$

89. $f(x) = 2x$
Compare $y = 2(x-3)$ with $y = 2x - 6$.
The results are identical; $2(x-3) = 2x - 6$ via the distributive property.

91. $f(x) = x^2 - 4$
$F(x) = |x^2 - 4|$

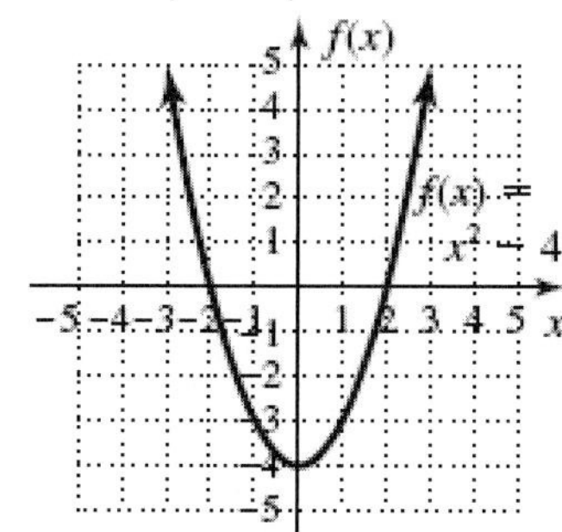

Any points in QIII and IV will reflect in x-axis and thus move to QI and II.

93. $x^3 - 8 = 0$
$(x-2)(x^2 + 2x + 4) = 0$
$$x = \frac{-2 \pm \sqrt{(2)^2 - 4(1)(4)}}{2(1)}$$
$$x = \frac{-2 \pm \sqrt{4-16}}{2}$$
$$x = \frac{-2 \pm \sqrt{-12}}{2}$$
$$x = \frac{-2 \pm 2i\sqrt{3}}{2}$$
$x = -1 \pm \sqrt{3}i$
$x = 2$

95. (-13, 9) and (7, -12)
$$d = \sqrt{(x_2 - x_1)^2 + (y_2 - y_1)^2}$$
$$d = \sqrt{(7+13)^2 + (-12-9)^2}$$
$$d = \sqrt{(20)^2 + (-21)^2}$$
$$d = \sqrt{400 + 441}$$
$$d = \sqrt{841} = 29$$
$$m = \frac{y_2 - y_1}{x_2 - x_1} = \frac{-12-9}{7+13} = \frac{-21}{20}$$

97. $\frac{2}{3}x + \frac{1}{4} = \frac{1}{2}x - \frac{7}{12}$
$$12\left(\frac{2}{3}x + \frac{1}{4} = \frac{1}{2}x - \frac{7}{12}\right)$$
$8x + 3 = 6x - 7$
$2x = -10$
$x = -5$

3.4 Technology Highlight

Exercise 1: 1.35, 6.65

Exercise 2: -0.81, 2.47

Exercise 3: -2.87, 0.87

Exercise 4: -0.80, 0.14

3.4 Exercises

1. $\frac{25}{2}$

3. Vertex

5. Answers will vary.

7. $f(x)=x^2+4x-5$

$f(x)=(x^2+4x+4)-5-4$

$f(x)=(x+2)^2-9;$

$x=\frac{-4\pm\sqrt{(4)^2-4(1)(-5)}}{2(1)}$

$x=\frac{-4\pm\sqrt{36}}{2}$

$x=\frac{-4\pm 6}{2}$

$x=1;\ \ x=-5$

Left 2, down 1

x-intercepts: (1, 0), (-5, 0)

y-intercept: (0, -5)

Vertex: (-2, -9)

$x=\frac{-b}{2a}=\frac{-4}{2(1)}=-2$

$y=(-2)^2+4(-2)-5=-9$

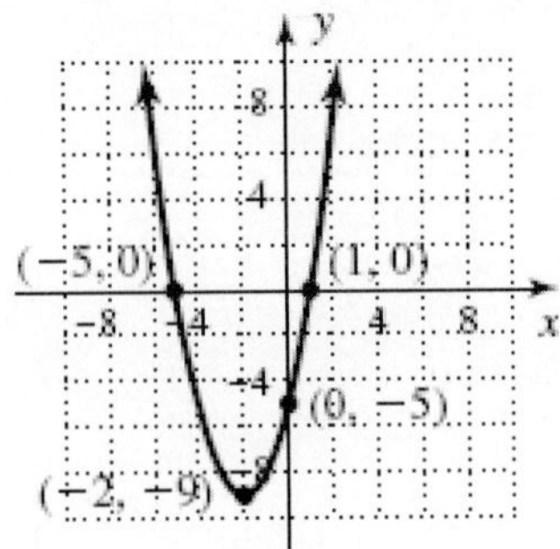

9. $h(x)=-x^2+2x+3$

$h(x)=-(x^2-2x+1)+3+1$

$h(x)=-(x-1)^2+4;$

$x=\frac{-2\pm\sqrt{(2)^2-4(-1)(3)}}{2(-1)}$

$x=\frac{-2\pm\sqrt{16}}{-2}$

$x=\frac{-2\pm 4}{-2}$

$x=-1;\ \ x=3$

Reflected in x-axis, right 1, up 4

x-intercepts: (-1, 0), (3, 0)

y-intercept: (0, 3)

Vertex: (1, 4)

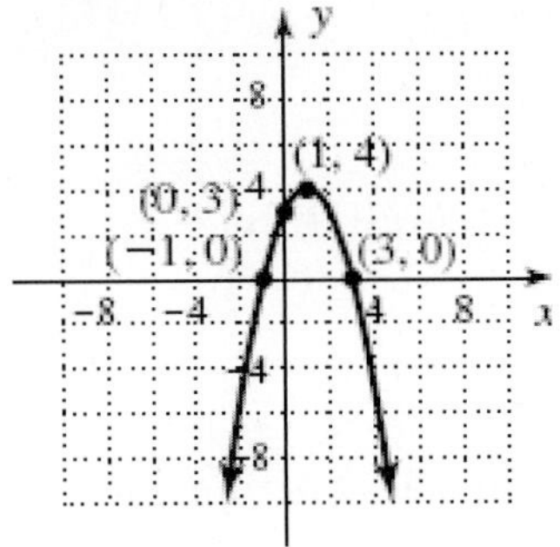

11. $p(x)=x^2-5x+2$

$p(x)=\left(x^2-5x+\frac{25}{4}\right)+2-\frac{25}{4}$

$p(x)=\left(x-\frac{5}{2}\right)^2-\frac{17}{4};$

$x=\frac{5\pm\sqrt{(-5)^2-4(1)(2)}}{2(1)}$

$x=\frac{5\pm\sqrt{17}}{2}$

$x\approx 4.56;\ \ x\approx 0.44$

Right $\frac{5}{2}$, down $-\frac{17}{4}$

x-intercepts: (4.56, 0), (0.44, 0)

y-intercept: (0, 2)

Vertex: $\left(\frac{5}{2},-\frac{17}{4}\right)$

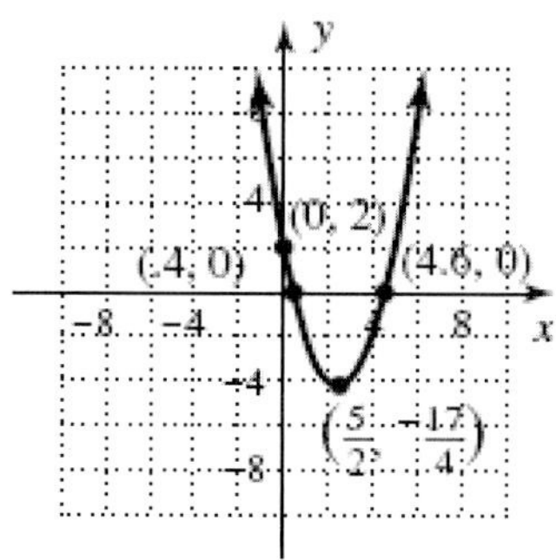

13. $Y_1 = 3x^2 + 6x - 5$

$Y_1 = 3(x^2 + 2x) - 5$

$Y_1 = 3(x^2 + 2x + 1) - 5 - 3$

$Y_1 = 3(x+1)^2 - 8;$

$$x = \frac{-6 \pm \sqrt{(6)^2 - 4(3)(-5)}}{2(3)}$$

$$x = \frac{-6 \pm \sqrt{96}}{6}$$

$x \approx 0.6; \quad x \approx -2.6$

Left 1, down 8, stretched vertically

x-intercepts: (0.6, 0), (-2.6, 0)

y-intercept: (0, -5)

Vertex: (-1, -8)

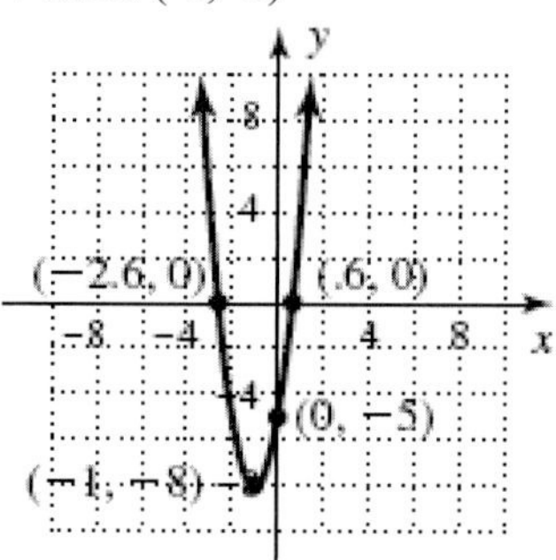

15. $f(x) = -2x^2 + 8x + 7$

$f(x) = -2(x^2 - 4x) + 7$

$f(x) = -2(x^2 - 4x + 4) + 7 + 8$

$f(x) = -2(x-2)^2 + 15;$

$$x = \frac{-8 \pm \sqrt{(8)^2 - 4(-2)(7)}}{2(-2)}$$

$$x = \frac{-8 \pm \sqrt{120}}{-4}$$

$x \approx -0.7 \quad x \approx 4.7$

Reflected in x-axis, right 2, up 15, stretched vertically

x-intercepts: (-0.7, 0), (4.7, 0)

y-intercept: (0, 7)

Vertex: (2, 15)

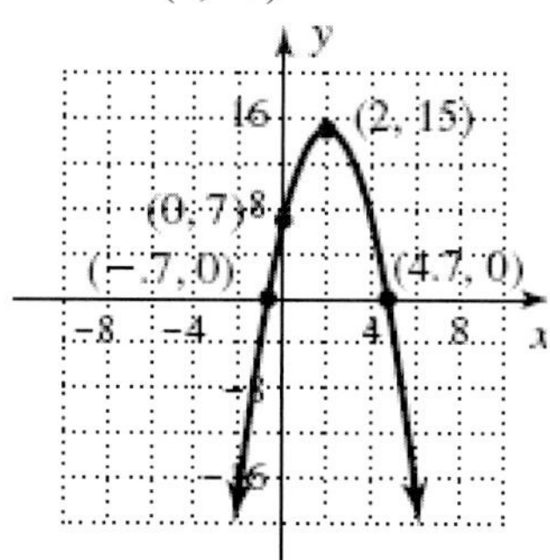

17. $h(x) = -\frac{1}{2}x^2 + 5x - 7$

$h(x) = -\frac{1}{2}(x^2 - 10x) - 7$

$h(x) = -\frac{1}{2}(x^2 - 10x + 25) - 7 + \frac{25}{2}$

$h(x) = -\frac{1}{2}(x-5)^2 + \frac{11}{2};$

$$x = \frac{-5 \pm \sqrt{(5)^2 - 4\left(-\frac{1}{2}\right)(-7)}}{2\left(-\frac{1}{2}\right)}$$

$$x = \frac{-5 \pm \sqrt{11}}{-1}$$

$$x = \frac{-5 \pm \sqrt{11}}{-1}$$

$x = 1.68; \quad x = 8.32$

Reflected in x-axis, right 5, up $\frac{11}{2}$, compressed vertically

x-intercepts: (1.68, 0), (8.32, 0)

y-intercept: (0, -7)

Vertex: $\left(5, \frac{11}{2}\right)$

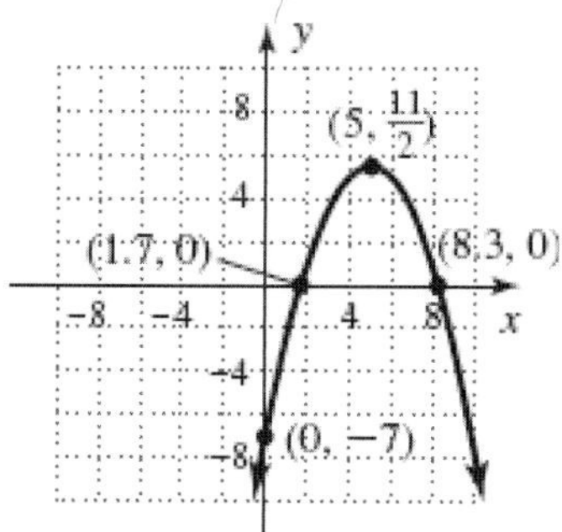

19. $p(x)=2x^2-7x+3$

$p(x)=2\left(x^2-\frac{7}{2}x\right)+3$

$p(x)=2\left(x^2-\frac{7}{2}x+\frac{49}{16}\right)+3-\frac{49}{8}$

$p(x)=2\left(x-\frac{7}{4}\right)^2-\frac{25}{8};$

$x=\frac{7\pm\sqrt{(-7)^2-4(2)(3)}}{2(2)}$

$x=\frac{7\pm\sqrt{25}}{4}$

$x=\frac{7\pm 5}{4}$

$x=3;\ x=\frac{1}{2}$

Right $\frac{7}{4}$, down $\frac{25}{8}$, stretched vertically

x-intercepts: (3, 0), $\left(\frac{1}{2},0\right)$

y-intercept: (0, 3)

Vertex: $\left(\frac{7}{4},-\frac{25}{8}\right)$

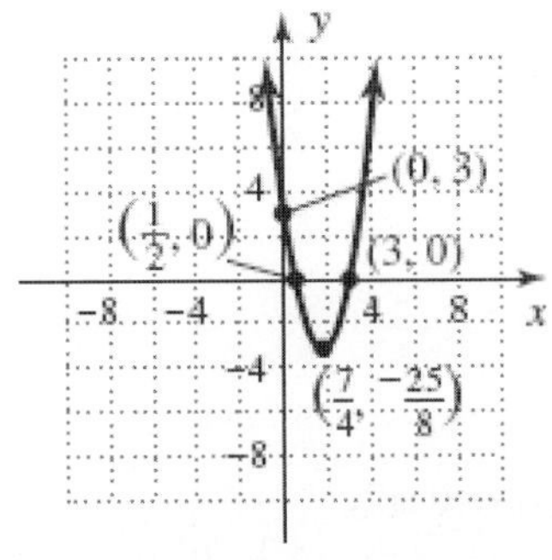

21. $f(x)=-3x^2-7x+6$

$f(x)=-3\left(x^2+\frac{7}{3}x\right)+6$

$f(x)=-3\left(x^2+\frac{7}{3}x+\frac{49}{36}\right)+6+\frac{49}{12}$

$f(x)=-3\left(x+\frac{7}{6}\right)^2+\frac{121}{12};$

$x=\frac{7\pm\sqrt{(-7)^2-4(-3)(6)}}{2(-3)}$

$x=\frac{7\pm\sqrt{121}}{-6}$

$x=\frac{7\pm 11}{-6}$

$x=-3;\ x=\frac{2}{3}$

Reflected in x-axis, left $\frac{7}{6}$, up $\frac{121}{12}$, stretched vertically

x-intercepts: (-3, 0), $\left(\frac{2}{3},0\right)$

y-intercept: (0, 6)

Vertex: $\left(-\frac{7}{6},\frac{121}{12}\right)$

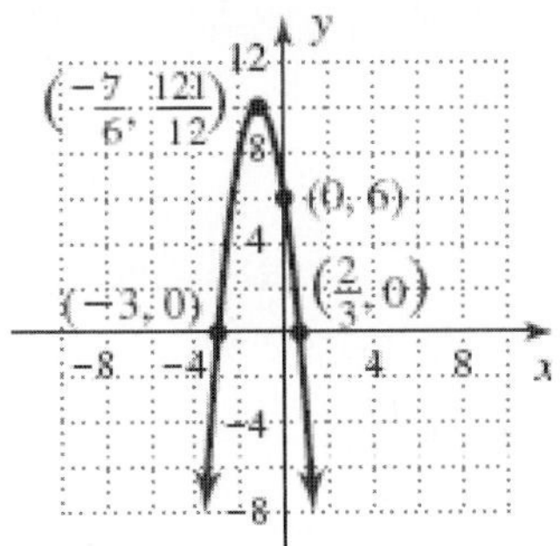

23. $Y_1=3x^2+5x-1$

$Y_1=3\left(x^2+\frac{5}{3}x\right)-1$

$Y_1=3\left(x^2+\frac{5}{3}x+\frac{25}{36}\right)-1-\frac{25}{12}$

$Y_1=3\left(x+\frac{5}{6}\right)^2-\frac{37}{12};$

$x=\frac{-5\pm\sqrt{(5)^2-4(3)(-1)}}{2(3)}$

$x=\frac{-5\pm\sqrt{37}}{6}$

$x\approx 0.18;\ x\approx -1.85$

Left $\frac{5}{6}$, down $\frac{37}{12}$, stretched vertically

x-intercepts: (0.18, 0), (-1.85, 0)

y-intercept: (0, -1)

Vertex: $\left(-\frac{5}{6},-\frac{37}{12}\right)$

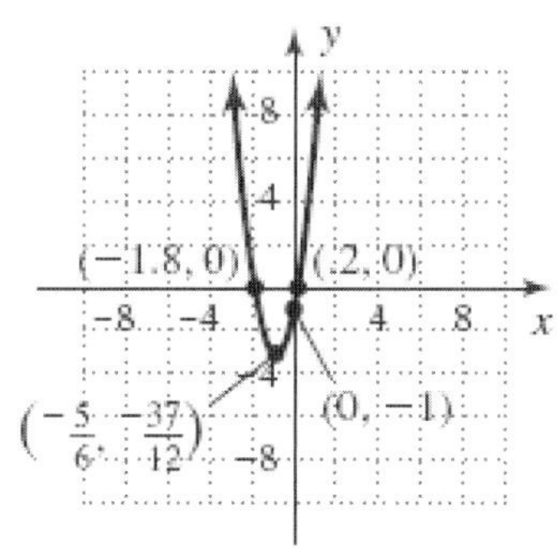

25. $y = (x+3)^2 - 5$

$$(x+3)^2 - 5 = 0$$
$$(x+3)^2 = 5$$
$$x+3 = \pm\sqrt{5}$$
$$x = -3 \pm \sqrt{5}$$

27. $y = 2(x+4)^2 - 7$

$$2(x+4)^2 - 7 = 0$$
$$2(x+4)^2 = 7$$
$$(x+4)^2 = \frac{7}{2}$$
$$x+4 = \pm\sqrt{\frac{7}{2}}$$
$$x = -4 \pm \frac{\sqrt{14}}{2}$$

29. $s(t) = 0.2(t+0.7)^2 - 0.8$

$$0.2(t+0.7)^2 - 0.8 = 0$$
$$0.2(t+0.7)^2 = 0.8$$
$$(t+0.7)^2 = 4$$
$$t+0.7 = \pm 2$$
$$t = -0.7 \pm 2$$

$t = -2.7;\ t = 1.3$

31. $f(x) = x^2 + 2x - 6$

$$f(x) = \left(x^2 + 2x\right) - 6$$
$$f(x) = \left(x^2 + 2x + 1\right) - 6 - 1$$
$$f(x) = (x+1)^2 - 7;$$
$$(x+1)^2 - 7 = 0$$
$$(x+1)^2 = 7$$
$$x+1 = \pm\sqrt{7}$$
$$x = -1 \pm \sqrt{7}$$

$x \approx 1.7;\ x \approx -3.7$

Left 1, down 7

x-intercepts: (1.7, 0), (-3.7, 0)

y-intercept: (0, -6)

Vertex: (-1, -7)

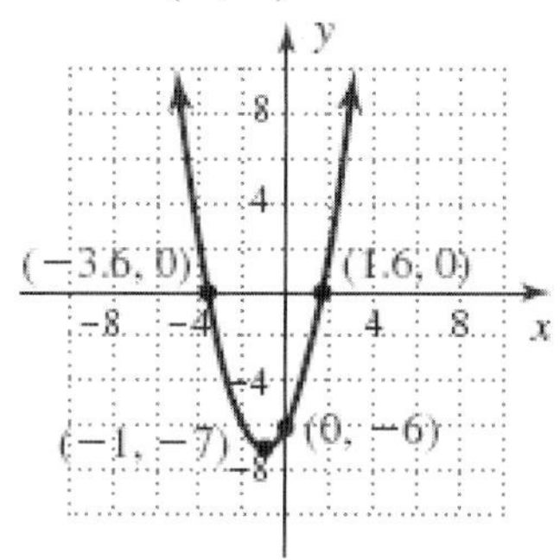

33. $h(x) = -x^2 + 4x + 2$

$$h(x) = -\left(x^2 - 4x\right) + 2$$
$$h(x) = -\left(x^2 - 4x + 4\right) + 2 + 4$$
$$h(x) = -(x-2)^2 + 6;$$
$$-(x-2)^2 + 6 = 0$$
$$-(x-2)^2 = -6$$
$$(x-2)^2 = 6$$
$$x-2 = \pm\sqrt{6}$$
$$x = 2 \pm \sqrt{6}$$

$x \approx 4.5;\ x \approx -0.5$

Reflected across x-axis, right 2, up 6,

x-intercepts: (4.5, 0), (-0.5, 0)

y-intercept: (0, 2)

Vertex: (2, 6)

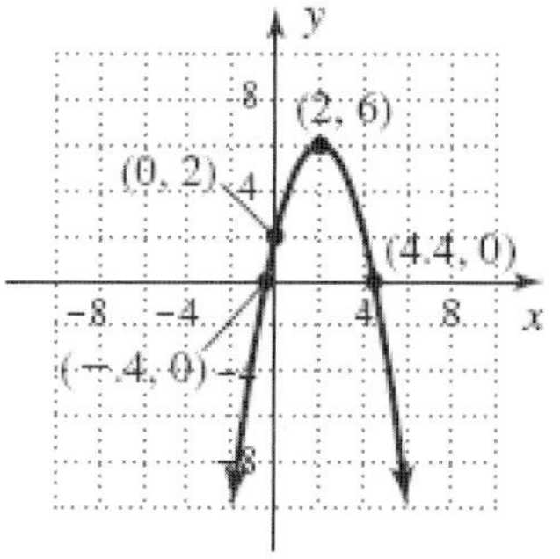

35. $Y_1 = 0.5x^2 + 3x + 7$

$Y_1 = 0.5(x^2 + 6x) + 7$
$Y_1 = 0.5(x^2 + 6x + 9) + 7 - 4.5$
$Y_1 = 0.5(x+3)^2 + 2.5;$
$0.5(x+3)^2 + 2.5 = 0$
$0.5(x+3)^2 = -2.5$
$(x+3)^2 = -5$

Left 3, up 2.5, compressed vertically
No x-intercepts
y-intercept: (0, 7)
Vertex: $\left(-3, \frac{5}{2}\right)$

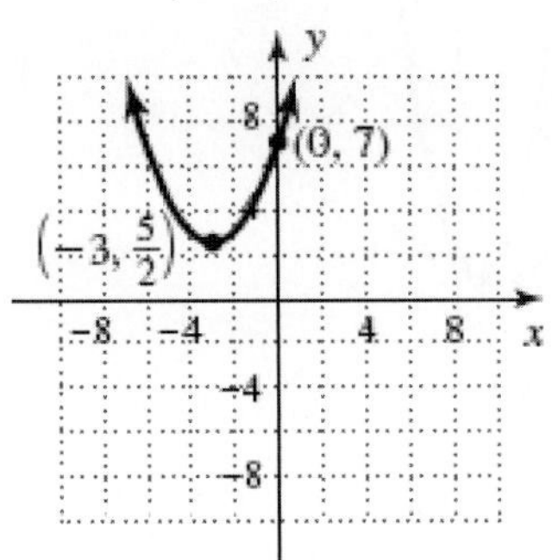

37. $Y_1 = -2x^2 + 10x - 7$
$Y_1 = -2(x^2 - 5x) - 7$
$Y_1 = -2\left(x^2 - 5x + \frac{25}{4}\right) - 7 + \frac{50}{4}$
$Y_1 = -2\left(x - \frac{5}{2}\right)^2 + \frac{11}{2};$
$-2\left(x - \frac{5}{2}\right)^2 + \frac{11}{2} = 0$
$-2\left(x - \frac{5}{2}\right)^2 = -\frac{11}{2}$
$\left(x - \frac{5}{2}\right)^2 = \frac{11}{4}$
$x - \frac{5}{2} = \pm\frac{\sqrt{11}}{2}$
$x = \frac{5}{2} \pm \frac{\sqrt{11}}{2}$

$x \approx 4.2;\ x \approx 0.8$

Reflected across x-axis, right $\frac{5}{2}$, up $\frac{11}{2}$, stretched vertically
x-intercepts: (4.2, 0), (0.8, 0)
y-intercept: (0, -7)

Vertex: $\left(\frac{5}{2}, \frac{11}{2}\right)$

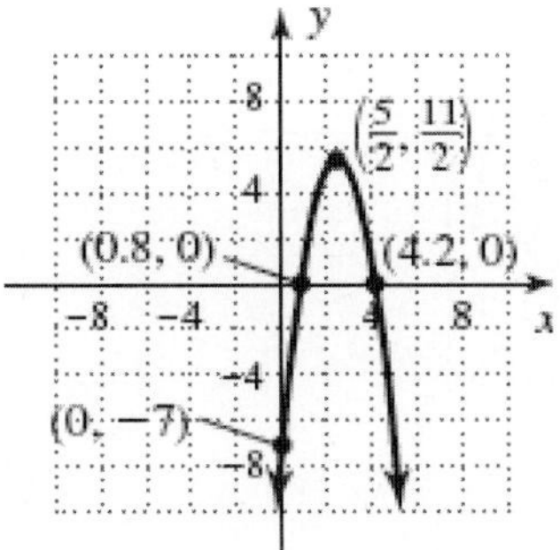

39. $f(x) = 4x^2 - 12x + 3$
$f(x) = 4(x^2 - 3x) + 3$
$f(x) = 4\left(x^2 - 3x + \frac{9}{4}\right) + 3 - 9$
$f(x) = 4\left(x - \frac{3}{2}\right)^2 - 6;$
$4\left(x - \frac{3}{2}\right)^2 - 6 = 0$
$4\left(x - \frac{3}{2}\right)^2 = 6$
$\left(x - \frac{3}{2}\right)^2 = \frac{3}{2}$
$x - \frac{3}{2} = \pm\frac{\sqrt{3}}{\sqrt{2}}$
$x = \frac{3}{2} \pm \frac{\sqrt{6}}{2}$

$x \approx 2.7;\ x \approx 0.3$

Right $\frac{3}{2}$, down 6, stretched vertically
x-intercepts: (2.7, 0), (0.3, 0)
y-intercept: (0, 3)
Vertex: $\left(\frac{3}{2}, -6\right)$

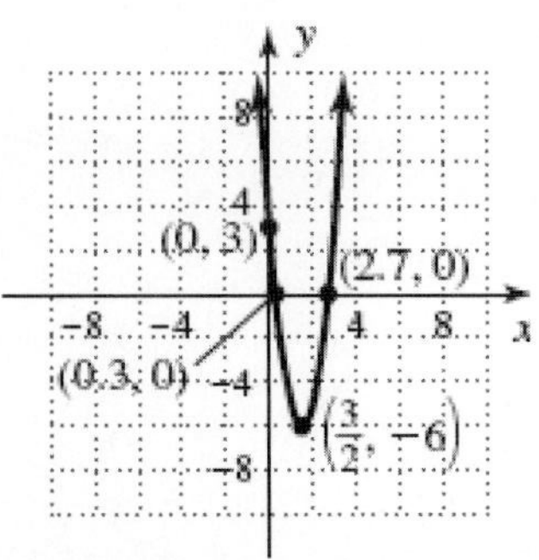

41. $p(x)=\frac{1}{2}x^2+3x-5$

$p(x)=\frac{1}{2}\left(x^2+6x\right)-5$

$p(x)=\frac{1}{2}\left(x^2+6x+9\right)-5-\frac{9}{2}$

$p(x)=\frac{1}{2}(x+3)^2-\frac{19}{2};$

$\frac{1}{2}(x+3)^2-\frac{19}{2}=0$

$\frac{1}{2}(x+3)^2=\frac{19}{2}$

$(x+3)^2=19$

$x+3=\pm\sqrt{19}$

$x=-3\pm\sqrt{19}$

$x\approx 1.4$ $x\approx -7.4$

Left 3, down $\frac{19}{2}$, compressed vertically

x-intercepts: (1.4, 0), (-7.4, 0)
y-intercept: (0, -5)

Vertex: $\left(-3,\frac{-19}{2}\right)$

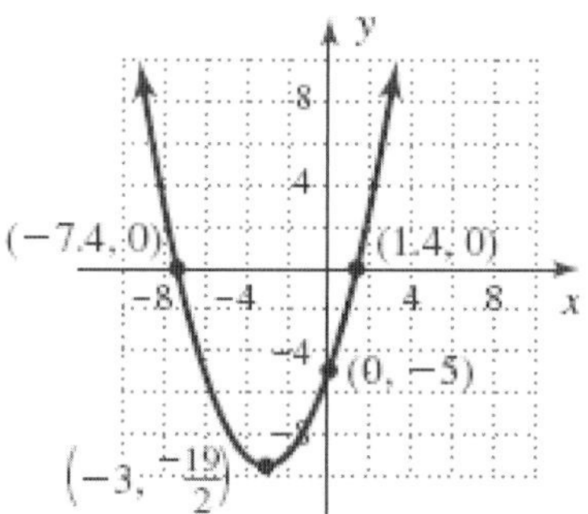

43. Vertex: (2, 0)
Point: (0, -4)

$y=a(x-h)^2+k$

$-4=a(0-2)^2+0$

$-4=4a$

$-1=a;$

$y=-(x-2)^2$

45. Node: (-3, 0)
Point: (6, 4.5)

$y=a\sqrt{x-h}+k$

$4.5=a\sqrt{6-(-3)}+0$

$4.5=3a$

$1.5=a;$

$y=1.5\sqrt{x+3}$

47. Vertex: (-4, 0)
Point: (1, 4)

$y=a|x-h|+k$

$4=a|1+4|+0$

$4=5a$

$\frac{4}{5}=a;$

$y=\frac{4}{5}|x+4|$

49. $S=\frac{1}{2}n^2+\frac{1}{2}n$

$S=\frac{1}{2}(10)^2+\frac{1}{2}(10)$

$S=\frac{1}{2}(100)+5$

$S=50+5$

$S=55$

$S=1+2+3+4+5+6+7+8+9+10=55$

51. $P(x)=-10x^2+3500x-66000$

a. (0, -66000) When no cars are produced, there is a profit loss of \$66,000.

b. $P(x)=-10x^2+3500x-66000$

$-10x^2+3500x-66000=0$

$-10\left(x^2-350+6600\right)=0$

$-10(x-20)(x-330)=0$

$x=20;\quad x=330$

(20, 0) and (330, 0)
No profit will be made if less than 20 cars or more than 330 cars are produced.

c. $x=\frac{-b}{2a}=\frac{-3500}{-20}=175$ cars

d. $P(175)=-10(175)^2+3500(175)-66000$

$P(175)=\$240{,}250$

53. $d(x)=x^2-12x$

a. $x = \frac{-b}{2a} = \frac{12}{2} = 6$ feet

b. $d(6) = (6)^2 - 12(6) = -36$

3600 feet

c. $d(4) = (4)^2 - 12(4) = -32$

3200 feet

d. $x^2 - 12x = 0$

$x(x-12) = 0$

$x = 0;\ x = 12$

12 miles

55. $P(x) = -0.5x^2 + 175x - 3300$

a. (0, -3300) If no appliances are sold, the loss will be \$3300.

b. $-0.5x^2 + 175x - 3300 = 0$

$-0.5(x^2 - 350x + 660) = 0$

$-0.5(x-20)(x-330) = 0$

$x = 20;\ x = 330$

(20, 0) and (330, 0)

If less than 20 or more than 330 appliances are made and sold, there will be no profit.

c. $0 \le x \le 200$ Most number of appliances to be produced each day is 200.

d. $x = \frac{-b}{2a} = \frac{175}{1} = 175$;

$P(175) = -0.5(175)^2 + 175(175) - 3300$
$= \$12{,}012.50$

57. $h(t) = -16t^2 + 176t$

a. $h(2) = -16(2)^2 + 176(2) = 288\,feet$

b.

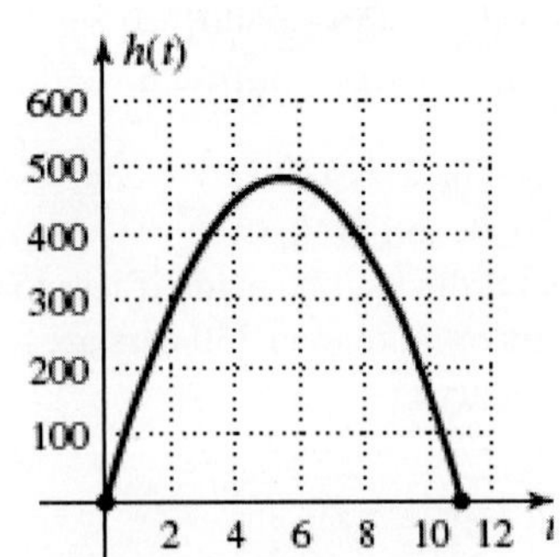

c. $x = \frac{-b}{2a} = \frac{-176}{-32} = 5.5\text{sec}$

$h(5.5) = -16(5.5)^2 + 176(5.5) = 484$ feet

d. $16t^2 - 176t = 0$

$16t(t-11) = 0$

$t = 11\text{sec}$

59. $C(x) = 2x + 35$; $R(x) = -x^2 + 122x - 365$

$P(x) = R(x) - C(x)$

$P(x) = -x^2 + 122x - 365 - 2x - 35$

$P(x) = -x^2 + 120x - 400;$

$x = \frac{-b}{2a} = \frac{-120}{-2} = 60$

6000 toys;

$P(60) = -(60)^2 + 120(60) - 400 = 3200$

\$3,200

61. Answers will vary.

63. $x = 2 \pm 3i$

$f(x) = (x-(2+3i))(x-(2-3i))$

$f(x) = (x-2-3i)(x-2+3i)$

$f(x) = ((x-2)^2 - (3i)^2)$

$f(x) = x^2 - 4x + 4 - 9i^2$

$f(x) = x^2 - 4x + 4 + 9$

$f(x) = x^2 - 4x + 13$

65. $x^{\frac{3}{4}} + 7 = 34$

$(\sqrt[4]{x})^3 = 27$

$\sqrt[4]{x} = 3$

$x = 3^4$

$x = 81$

67. $m = \frac{4}{3};\ y-\text{int}\ (0,3)$

$-4x + 3y = 9$

$3y = 4x + 9$

$y = \frac{4}{3}x + 3$

69. $f(x) = \sqrt[3]{x+3}$ and $g(x) = x^3 - 3$

$(f \circ g)(x) = f[g(x)]$

$= \sqrt[3]{g(x)+3}$

$= \sqrt[3]{x^3 - 3 + 3}$

$= \sqrt[3]{x^3}$

$= x;$

$$\begin{aligned}(g \circ f)(x) &= g[f(x)]\\ &= (f(x))^3 - 3\\ &= \left(\sqrt[3]{x+3}\right)^3 - 3\\ &= x + 3 - 3\\ &= x\end{aligned}$$

Chapter 3 Mid-Chapter Check

1. $f(x) = 3x - 5$ and $g(x) = 2x^2 + 3x$
   a. $(f+g)(3) = f(3) + g(3)$
      $f(3) = 3(3) - 5 = 4$
      $g(3) = 2(3)^2 + 3(3) = 18 + 9 = 27$
      $(f+g)(3) = 4 + 27 = 31$
   b. $(f \cdot g)(x) = f(x) \cdot g(x)$
      $= (3x-5)(2x^2+3x)$
      $= 6x^3 + 9x^2 - 10x^2 - 15x$
      $= 6x^3 - x^2 - 15x$

2. $f(x) = \frac{2}{3}x + 1$ and $g(x) = x^2 - 5x$
   a. Domain of $\left(\frac{f}{g}\right)(x)$
      $x^2 - 5x = 0$
      $x(x-5) = 0$
      $x = 0;\ x = 5$
      $x \in (-\infty, 0) \cup (0, 5) \cup (5, \infty)$
   b. $(g \circ f)(3) = g[f(3)]$
      $f(3) = \frac{2}{3}(3) + 1 = 2 + 1 = 3$
      $g(3) = (3)^2 - 5(3) = 9 - 15 = -6$

3. $h(d) = -0.0375d^2 + 1.5d$
   a. $0 = -0.0375d^2 + 1.5d$
      $0 = -0.0375d(d - 40)$
      $d = 0;\ d = 40$
      40 meters
   b. $x = \frac{-b}{2a} = \frac{-1.5}{2(-0.0375)} = 20$
      $y = -0.0375(20)^2 + 1.5(20) = 15$
      15 meters
   c. $h(10) = -0.0375(10)^2 + 1.5(10) = 11.25$
      $h(20) = -0.0375(20)^2 + 1.5(20) = 15$
      11.25 meters; 15 meters
   d. $h(37) = -0.0375(37)^2 + 1.5(37) = 14.4$
      Yes

4. a. $(f+g)(-3) = 0.5 + 2.5 = 3$
   b. $(f \cdot g)(1) = 5(-2) = -10$
   c. $(f-g)(4) = 0 - (-3) = 3$
   d. $\left(\frac{f}{g}\right)(2) = \frac{6}{-2} = -3$

5. a. $f(x) = |x+2| - 3$
      Left 2, down 3

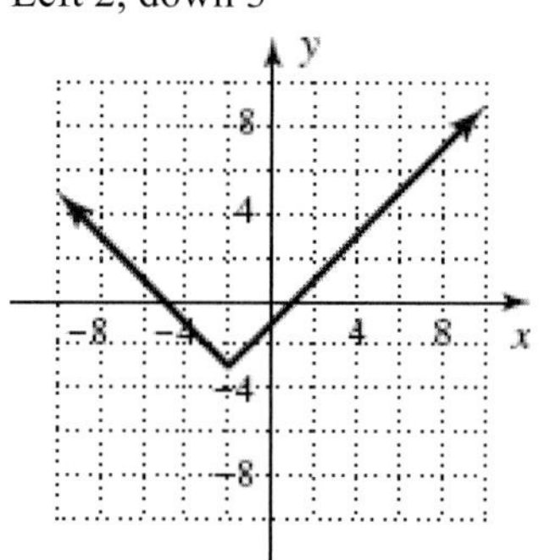

   b. $g(x) = \sqrt{x-1} + 2$
      Right 1, up 2

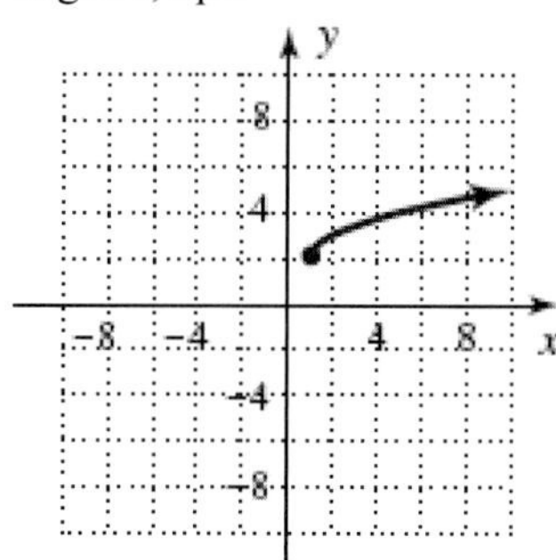

   c. $f(x) = -2(x+3)^2 - 5$
      Left 3, down 5, reflected across x-axis

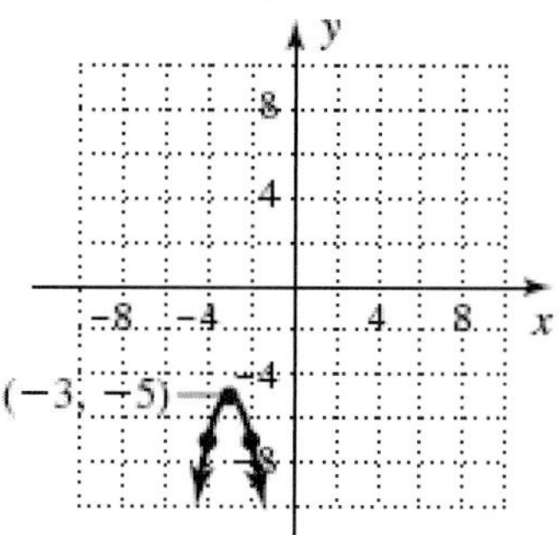

   d. $f(x) = -(x-3)^3 - 4$
      Right 3, down 4, reflected across x-axis

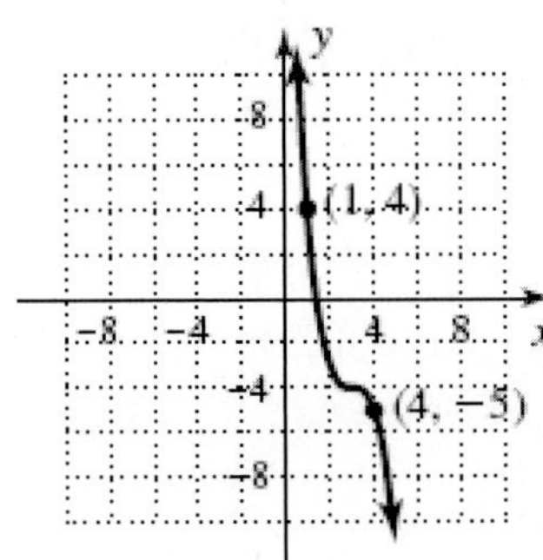

6. $f(x)=\sqrt{x-3}$

$y=\sqrt{x-3}$

$x=\sqrt{y-3}$

$x^2=y-3$

$x^2+3=y;$

$f^{-1}(x)=x^2+3$

$D_{f^{-1}(x)}: x\in[0,\infty)$

$R_{f^{-1}(x)}: x\in[3,\infty)$

$(f\circ f^{-1})(x)=f(f^{-1}(x))$

$=\sqrt{f^{-1}(x)-3}$

$=\sqrt{x^2+3-3}$

$=\sqrt{x^2}$

$=x;$

$(f^{-1}\circ f)(x)=f^{-1}[f(x)]+3$

$=(f(x))^2+3$

$=(\sqrt{x-3})^2+3$

$=x-3+3$

$=x$

7. $f(x)=-x^2+6x-7$

$f(x)=-(x^2-6x)-7$

$f(x)=-(x^2-6x+9)-7+9$

$f(x)=-(x-3)^2+2;$

Vertex: (3, 2)

$-(x-3)^2+2=0$

$-(x-3)^2=-2$

$(x-3)^2=2$

$x-3=\pm\sqrt{2}$

$x=3\pm\sqrt{2}$

$x=4.4;\ \ x=1.6$

x-intercepts: (4.4, 0) and (1.6, 0)

y-intercept: (0, -7)

Axis of symmetry: $x=3$

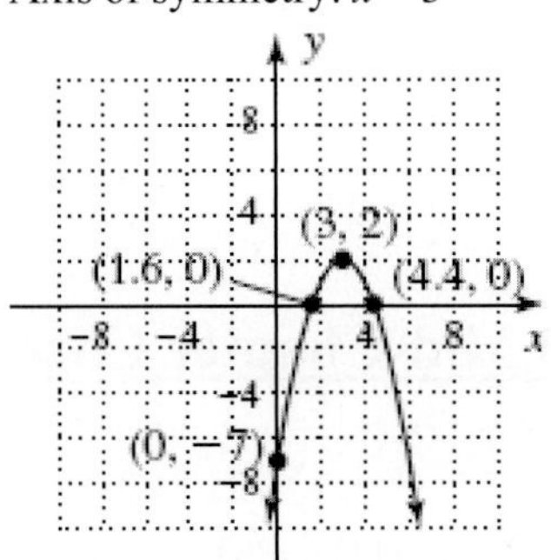

8. $g(x)=2x^2+6x-11$

$g(x)=2(x^2+3x)-11$

$g(x)=2\left(x^2+3x+\frac{9}{4}\right)-11-\frac{9}{2}$

$g(x)=2\left(x+\frac{3}{2}\right)^2-\frac{31}{2};$

Vertex: $\left(-\frac{3}{2},-\frac{31}{2}\right)$

$2\left(x+\frac{3}{2}\right)^2-\frac{31}{2}=0$

$2\left(x+\frac{3}{2}\right)^2=\frac{31}{2}$

$\left(x+\frac{3}{2}\right)^2=\frac{31}{4}$

$x+\frac{3}{2}=\pm\frac{\sqrt{31}}{2}$

$x=-\frac{3}{2}\pm\frac{\sqrt{31}}{2}$

$x=1.3;\ \ x=-4.3$

x-intercepts: (1.28, 0) and (-4.28, 0)

y-intercept: (0, -11)

Axis of symmetry: $x=-1.5$

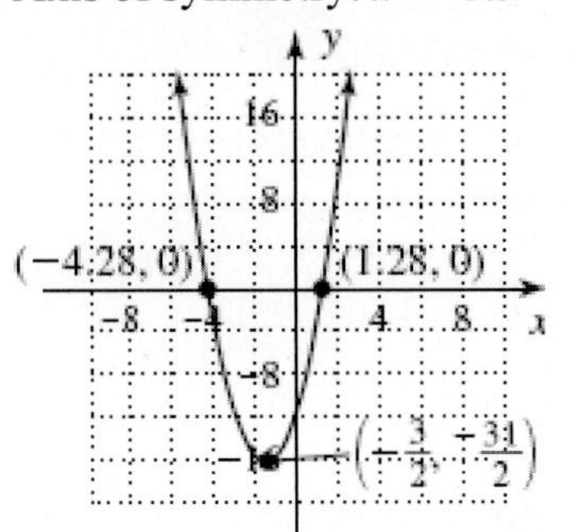

9. $h(x) = -2x^2 - 5x + 7$

$h(x) = -2\left(x^2 + \frac{5}{2}x\right) + 7$

$h(x) = -2\left(x^2 + \frac{5}{2}x + \frac{25}{16}\right) + 7 + \frac{25}{8}$

$h(x) = -2\left(x + \frac{5}{4}\right)^2 + \frac{81}{8};$

Vertex: $\left(-\frac{5}{4}, \frac{81}{8}\right)$

$-2\left(x + \frac{5}{4}\right)^2 + \frac{81}{8} = 0$

$-2\left(x + \frac{5}{4}\right)^2 = -\frac{81}{8}$

$\left(x + \frac{5}{4}\right)^2 = \frac{81}{16}$

$x + \frac{5}{4} = \pm\frac{9}{4}$

$x = -\frac{5}{4} \pm \frac{9}{4}$

$x = 1;\ \ x = -3.5$

x-intercepts: (1, 0) and (-3.5, 0)
y-intercept: (0, 7)
Axis of symmetry: $x = 1.25$

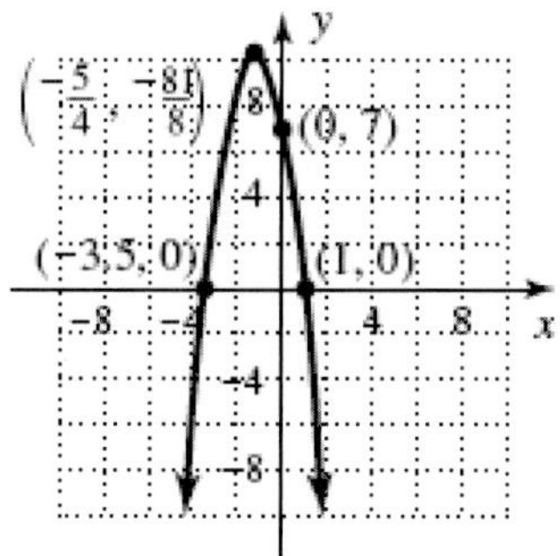

10. Answers will vary.

Chapter 3 Reinforcing Basic Concepts

Exercise 1: $f(x) = x^2 - 8x - 12$

$\frac{b}{2a} = \frac{-8}{2(1)} = -4$

$g(x) = x + 4$

$h(x) = f[g(x)] = f(x + 4)$

$= (x + 4)^2 - 8(x + 4) - 12$

$= x^2 + 8x + 16 - 8x - 32 - 12$

$= x^2 - 28$

$x^2 - 28 = 0$

$x^2 = 28$

$x = \pm 2\sqrt{7}\ ;$

$4 \pm 2\sqrt{7}$

Exercise 2: $f(x) = x^2 + 4x + 5$

$\frac{b}{2a} = \frac{4}{2(1)} = 2$

$g(x) = x - 2$

$h(x) = f[g(x)] = f(x - 2)$

$= (x - 2)^2 + 4(x - 2) + 5$

$= x^2 - 4x + 4 + 4x - 8 + 5$

$= x^2 + 1$

$x^2 + 1 = 0$

$x^2 = -1$

$x = \pm i$

$-2 \pm i$

Exercise 3: $f(x) = 2x^2 - 10x + 11$

$\frac{b}{2a} = \frac{-10}{2(2)} = -\frac{5}{2}$

$g(x) = x + \frac{5}{2}$

$h(x) = f[g(x)] = f\left(x + \frac{5}{2}\right)$

$= 2\left(x + \frac{5}{2}\right)^2 - 10\left(x + \frac{5}{2}\right) + 11$

$= 2\left(x^2 + 5x + \frac{25}{4}\right) - 10x - \frac{50}{2} + 11$

$= 2x^2 + 10x + \frac{50}{4} - 10x - \frac{50}{2} + 11$

$= 2x^2 - \frac{3}{2}$

$2x^2 - \frac{3}{2} = 0$

$2x^2 = \frac{3}{2}$

$x^2 = \frac{3}{4}$

$x = \pm\frac{\sqrt{3}}{2}$

$\frac{5}{2} \pm \frac{\sqrt{3}}{2}$

Exercise 4: $f(x) = x^3 + 3x^2 - 6x - 11$

$\frac{b}{3a} = \frac{3}{3(1)} = 1$

$g(x) = x - 1$

$h(x) = f[g(x)] = f(x-1)$

$= (x-1)^3 + 3(x-1)^2 - 6(x-1) - 11$

$= x^3 - 3x^2 + 3x - 1 + 3(x^2 - 2x + 1)$
$- 6x + 6 - 11$

$= x^3 - 3x^2 + 3x - 1 + 3x^2 - 6x + 3$
$- 6x - 5$

$= x^3 - 9x - 3$

$x^3 - 9x - 3 = 0$

Exercise 5: $f(x) = x^3 - 6x^2 + 2x + 7$

$\frac{b}{3a} = \frac{-6}{3(1)} = -2$

$g(x) = x + 2$

$h(x) = f[g(x)] = f(x+2)$

$= (x+2)^3 - 6(x+2)^2 + 2(x+2) + 7$

$= x^3 + 8x^2 + 10x + 8 - 6(x^2 + 4x + 4)$
$+ 2x + 4 + 7$

$= x^3 + 6x^2 + 12x + 8 - 6x^2 - 24x - 24$
$+ 2x + 11$

$= x^3 - 10x - 5$

$x^3 - 10x - 5 = 0$

3.5 Technology Highlight

Exercise 1: $3,420,000; cost becomes negative.

Exercise 2:The rate of change is much greater in the interval from 90 to 95.

Exercise 3: Closely

3.5 Exercises

1. $x \to -\infty,\ y \to 2^-$

3. Vertical; $y = 2$

5. In the reciprocal quadratic function, all range values are positive.

7. As $x \to +\infty,\ y \to 2^-$
   As $x \to -\infty,\ y \to 2^+$
   As $x \to -3^-,\ y \to \infty$
   As $x \to -3^+,\ y \to -\infty$
   $D: x \in (-\infty, -3) \cup (-3, \infty)$
   $R: x \in (-\infty, 2) \cup (2, \infty)$

9. As $x \to +\infty,\ y \to -2^+$
   As $x \to -\infty,\ y \to -2^+$
   As $x \to 1^-,\ y \to \infty$
   As $x \to 1^+,\ y \to \infty$
   $D: x \in (-\infty, 1) \cup (1, \infty)$
   $R: x \in (-\infty, -2) \cup (-2, \infty)$

11. Horizontal asymptote: $y = -2$
    Vertical asymptote: $x = -1$
    Parent function: $y = \frac{1}{x^2}$
    Left 1, down 2
    Equation: $y = \frac{1}{(x+1)^2} - 2$

13. Horizontal asymptote: $y = -2$
    Vertical asymptote: $x = -1$
    Parent function: $y = \frac{1}{x}$
    Left 1, down 2
    Equation: $y = \frac{1}{x+1} - 2$

15. Horizontal asymptote: $y = -5$
    Vertical asymptote: $x = -2$
    Parent function: $y = \frac{1}{x^2}$
    Left 2, down 5
    Equation: $y = \frac{1}{(x+2)^2} - 5$

17. $y \to -2^+$

19. $y \to -\infty$

21. $x \to -1,\ y \to \pm\infty$

23. $f(x) = \frac{1}{x} - 1$

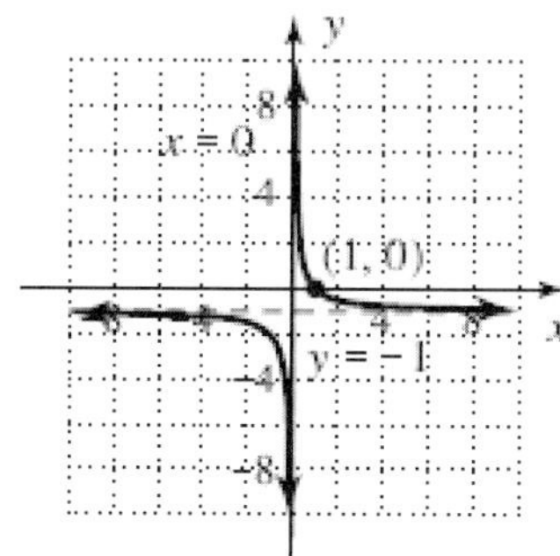

Down 1
$x \in (-\infty, 0) \cup (0, \infty)$
$y \in (-\infty, -1) \cup (-1, \infty)$

25. $h(x) = \frac{1}{x+2}$

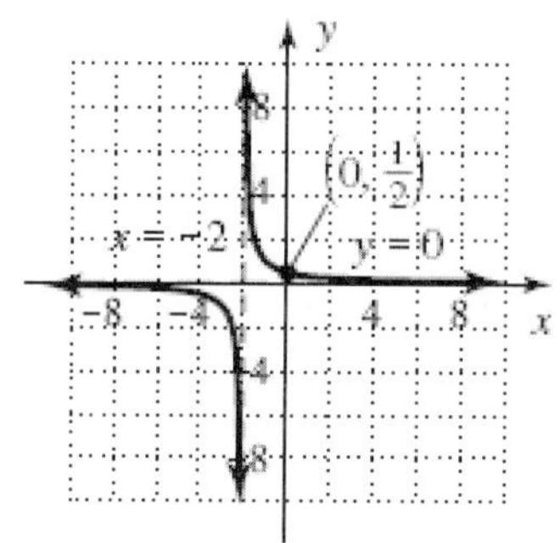

Left 2
$x \in (-\infty, -2) \cup (-2, \infty)$
$y \in (-\infty, 0) \cup (0, \infty)$

27. $g(x) = \frac{-1}{x-2}$

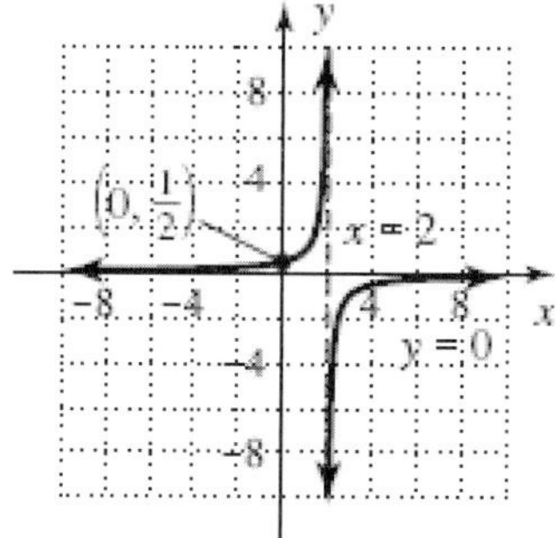

Reflected across x-axis, right 2
$x \in (-\infty, 2) \cup (2, \infty)$
$y \in (-\infty, 0) \cup (0, \infty)$

29. $f(x) = \frac{1}{x+2} - 1$

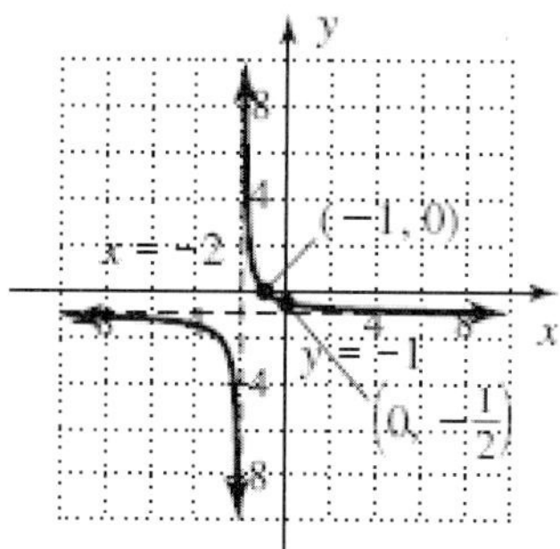

Left 2, down 1
$x \in (-\infty, -2) \cup (-2, \infty)$
$y \in (-\infty, -1) \cup (-1, \infty)$

31. $h(x) = \frac{1}{(x-1)^2}$

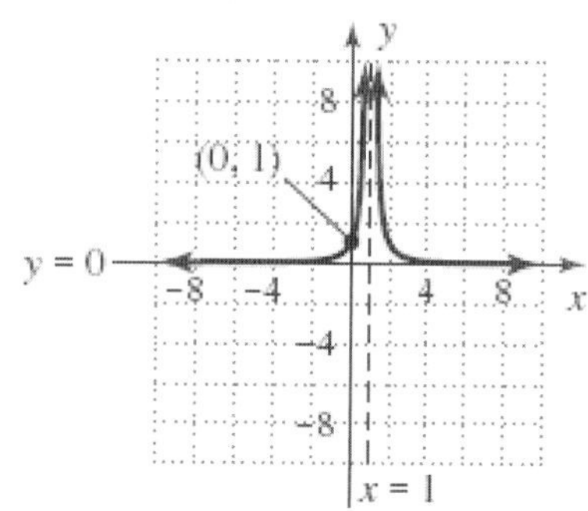

Right 1
$x \in (-\infty, 1) \cup (1, \infty)$
$y \in (0, \infty)$

33. $g(x) = \frac{-1}{(x+2)^2}$

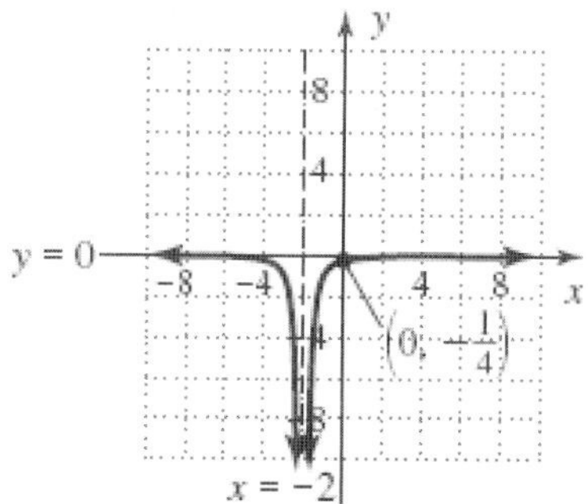

Reflected across x-axis, left 2
$x \in (-\infty, -2) \cup (-2, \infty)$
$y \in (-\infty, 0)$

35. $f(x) = \frac{1}{x^2} - 2$

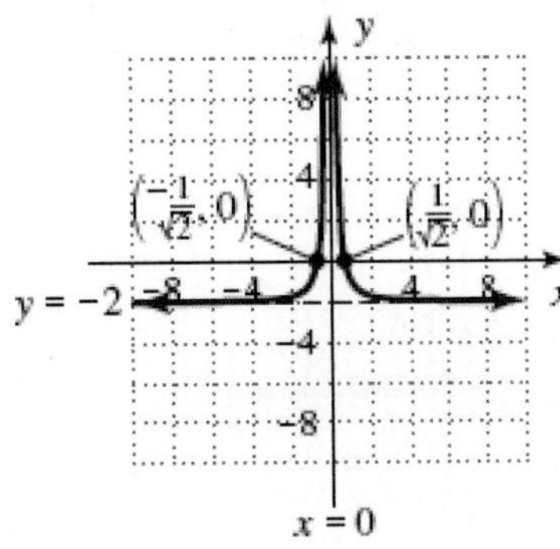

Down 2

$x \in (-\infty, 0) \cup (0, \infty)$

$y \in (-2, \infty)$

37. $h(x) = 1 + \dfrac{1}{(x+2)^2}$

Left 2, up 1

$x \in (-\infty, -2) \cup (-2, \infty)$

$y \in (1, \infty)$

39. $f(x) = 3 + \dfrac{1}{x+4}$

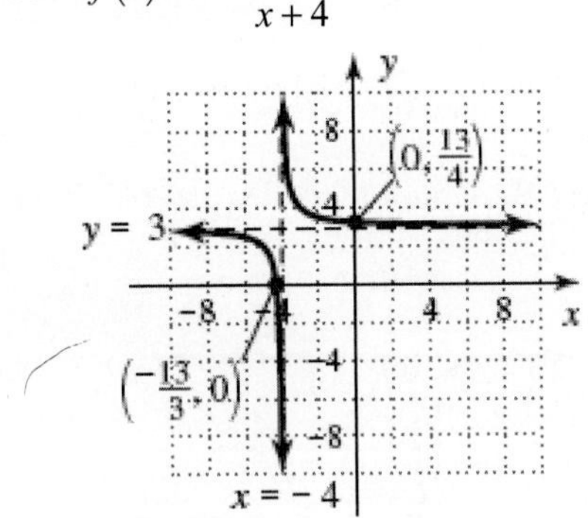

Left 4, up 3

$x \in (-\infty, -4) \cup (-4, \infty)$

$y \in (-\infty, 3) \cup (3, \infty)$

41. $f(x) = \dfrac{-1}{x-1} - 3$

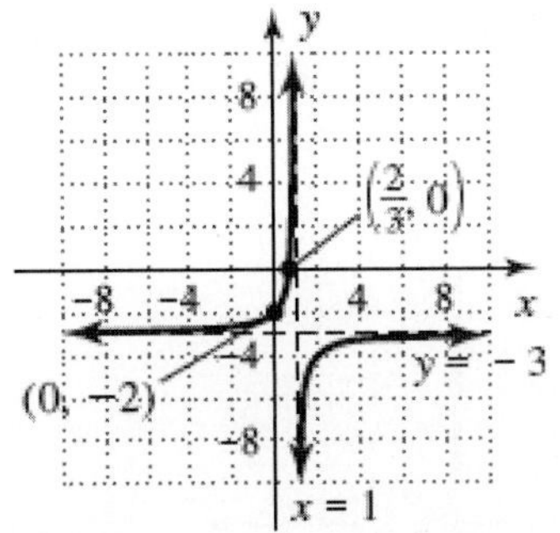

Reflected across x-axis, right 1, down 3

$x \in (-\infty, 1) \cup (1, \infty)$

$y \in (-\infty, -3) \cup (-3, \infty)$

43. $h(x) = \dfrac{-1}{(x-2)^2} + 3$

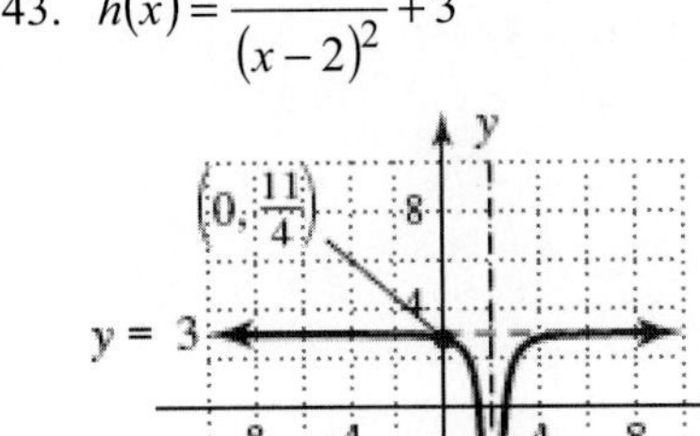

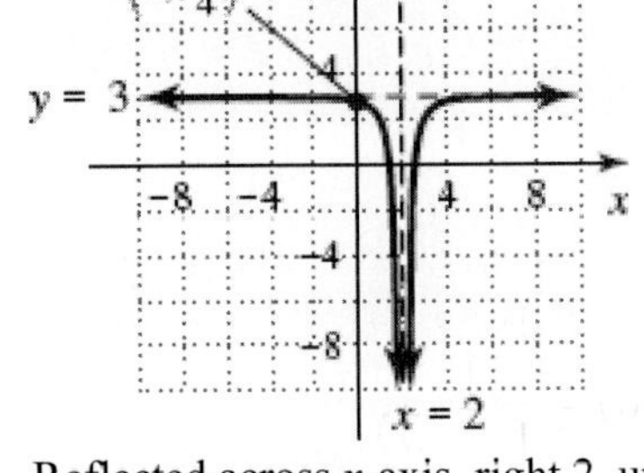

Reflected across x-axis, right 2, up 3

$x \in (-\infty, 2) \cup (2, \infty)$

$y \in (-\infty, 3)$

45. $h(x) = 3 - \dfrac{1}{(x+2)^2}$

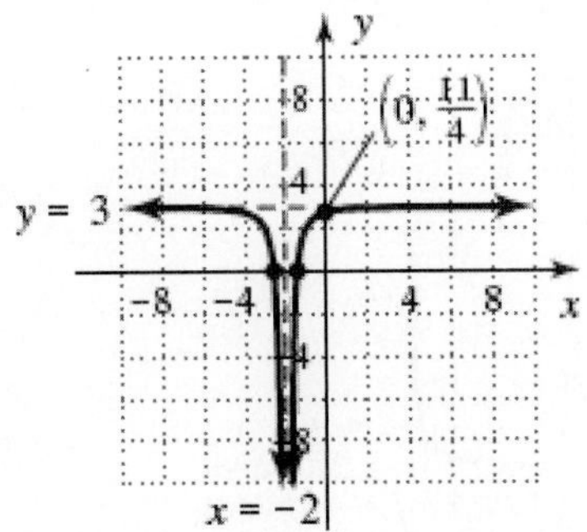

Reflected across x-axis, left 2, up 3

$x \in (-\infty, -2) \cup (-2, \infty)$

$y \in (-\infty, 3)$

47. $F = \dfrac{km_1m_2}{d^2}$

F becomes very small.

$F = \dfrac{1}{d^2}$

x	$F(d) = \dfrac{1}{d^2}$
1	$F(1) = \dfrac{1}{(1)^2} = 1$
2	$F(2) = \dfrac{1}{(2)^2} = \dfrac{1}{4}$
3	$F(3) = \dfrac{1}{(3)^2} = \dfrac{1}{9}$
4	$F(4) = \dfrac{1}{(4)^2} = \dfrac{1}{16}$

$F = \dfrac{1}{d^2}$ belongs to the family $F(x) = \dfrac{1}{x^2}$.

49. $D(p) = \dfrac{75}{p}$

a. As the number of predators increases, the deer population decreases.

$D(1) = \dfrac{75}{1} = 75$;

$D(3) = \dfrac{75}{3} = 25$;

$D(5) = \dfrac{75}{5} = 15$

b. If the number of predators becomes very large, the deer population approaches 0.

c.

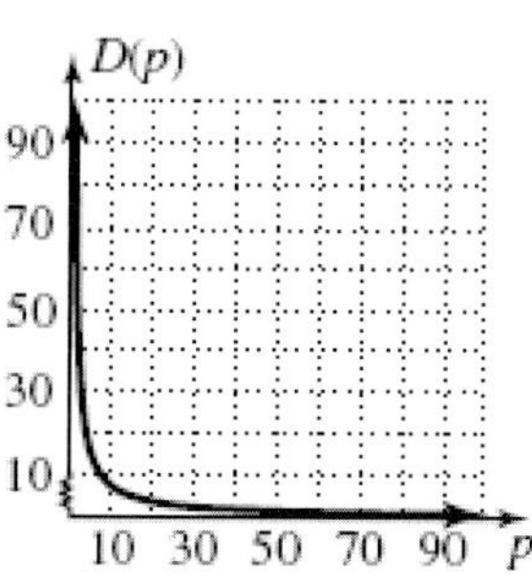

If the number of predators becomes very small, the deer population increases.

As p decreases, D becomes very large: $p \to 0,\ D \to \infty$

51. $I(d) = \dfrac{2500}{d^2}$

a. As the distance from the light bulb increases, the intensity of light decreases.

$I(5) = \dfrac{2500}{(5)^2} = 100$;

$I(10) = \dfrac{2500}{(10)^2} = 25$;

$I(15) = \dfrac{2500}{(15)^2} = 11.1$

b. If the intensity of light is increasing, the observer is moving toward the light source.

c.

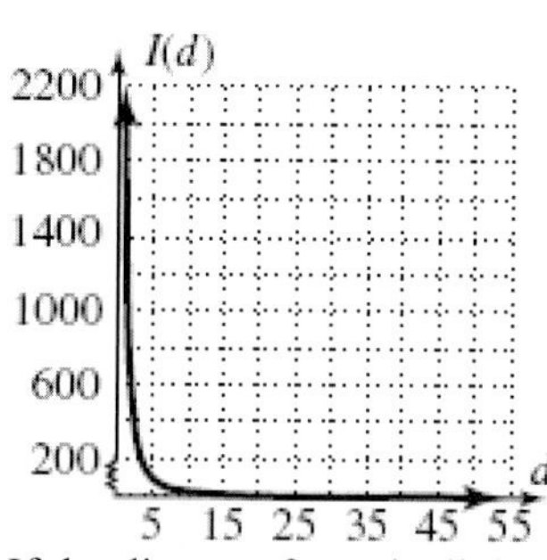

If the distance from the light bulb is very small, the intensity of light is very great.

As d decreases, Intensity becomes large: $d \to 0,\ I \to \infty$

53. $C(p) = \dfrac{-8000}{p-100} - 80$

a. $C(20) = \dfrac{-8000}{20-100} - 80 = \$20{,}000$;

$C(50) = \dfrac{-8000}{50-100} - 80 = \$80{,}000$;

$C(80) = \dfrac{-8000}{80-100} - 80 = \$320{,}000$

Cost increases dramatically

b.

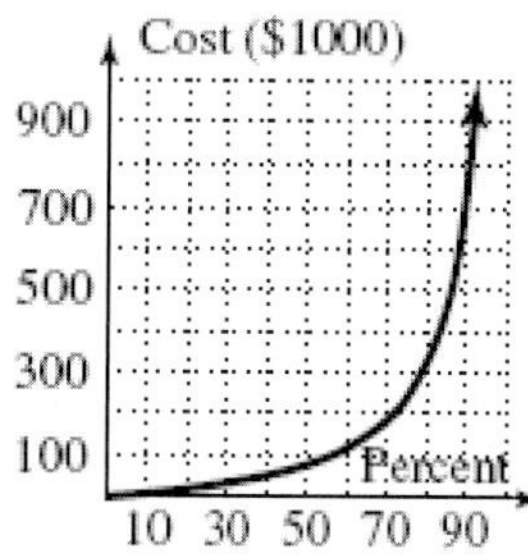

c. As $p \to 100^-,\ C \to \infty$

55. $C(h)=\dfrac{2h^2+h}{h^3+70}$

a. According to the graph, 5 hours; 0.28

b. $\dfrac{\Delta C}{\Delta h}=\dfrac{C(10)-C(8)}{10-8}=\dfrac{0.196-0.234}{2}$

$\dfrac{\Delta C}{\Delta h}=-0.019$;

$\dfrac{\Delta C}{\Delta h}=\dfrac{C(22)-C(20)}{22-20}=\dfrac{0.0924-0.102}{2}$

$\dfrac{\Delta C}{\Delta h}=-0.005$

As number of hours increases, the rate of change decreases.

c. Horizontal asymptote:

As $h\to\infty,\quad C\to 0^+$

57. $f(x)=\dfrac{1}{x}$

$y=\dfrac{1}{x}$

$x=\dfrac{1}{y}$

$xy=1$

$y=\dfrac{1}{x}$

$f^{-1}(x)=\dfrac{1}{x}$;

$(f\circ f^{-1}(x))=f\left[f^{-1}(x)\right]$

$=\dfrac{1}{f^{-1}(x)}=\dfrac{1}{\frac{1}{x}}=x$

$(f^{-1}\circ f(x))=f^{-1}[f(x)]$

$=\dfrac{1}{f(x)}=\dfrac{1}{\frac{1}{x}}=x$

Answers will vary.

59. $h(x)=\dfrac{1}{x^3}$

$h(2)=\dfrac{1}{(2)^3}=\dfrac{1}{8}$;

$h(3)=\dfrac{1}{(3)^3}=\dfrac{1}{27}$;

$h(4)=\dfrac{1}{(4)^3}=\dfrac{1}{64}$;

$h(5)=\dfrac{1}{(5)^3}=\dfrac{1}{125}$;

$A=\dfrac{1}{8}+\dfrac{1}{27}+\dfrac{1}{64}+\dfrac{1}{125}=0.186$

Pattern: Answers will vary

$H(x)=\dfrac{1}{x^4}$

$H(2)=\dfrac{1}{(2)^4}=\dfrac{1}{16}$;

$H(3)=\dfrac{1}{(3)^4}=\dfrac{1}{81}$;

$H(4)=\dfrac{1}{(4)^4}=\dfrac{1}{256}$;

$H(5)=\dfrac{1}{(5)^4}=\dfrac{1}{625}$;

$A=\dfrac{1}{16}+\dfrac{1}{81}+\dfrac{1}{256}+\dfrac{1}{625}=0.0804$

Pattern: Answers will vary

General Pattern: Answers will vary

61. $12x^2+55x-48=0$

$x=\dfrac{-55\pm\sqrt{(55)^2-4(12)(-48)}}{2(12)}$

$x=\dfrac{-55\pm\sqrt{3025+2304}}{24}$

$x=\dfrac{-55\pm\sqrt{5329}}{24}$

$x=\dfrac{3}{4};\ x=-\dfrac{16}{3}$

$(4x-3)(3x+16)=0$

63. $x^2+10x+28=0$

$(-5+i\sqrt{3})^2+10(-5+i\sqrt{3})+28=0$

$25-10i\sqrt{3}+3i^2-50+10i\sqrt{3}+28=0$

$25-3-50+28=0$

$0=0$

$-5-i\sqrt{3}$

65. $f(x)=\dfrac{1}{x-2}+3;\ g(x)=\dfrac{1}{x-3}+2$

$(f\circ g)(x)=f(g(x))$

$=\dfrac{1}{\left(\dfrac{1}{x-3}+2\right)-2}+3$

$$= \frac{1}{\frac{1}{x-3}} + 3$$
$$= x - 3 + 3$$
$$= x;$$
$$(g \circ f)(x) = g(f(x))$$
$$= \frac{1}{\left(\frac{1}{x-2} + 3\right) - 3} + 2$$
$$= \frac{1}{\frac{1}{x-2}} + 2$$
$$= x - 2 + 2$$
$$= x$$

3.6 Technology Highlight

Exercise 1:

Linear; 75; 300; 675; Fuel required increases dramatically.

Exercise 2:

$$S = \frac{kWT^2}{L}$$
$$450 = \frac{k(12)(1)^2}{10}$$
$$375 = k$$
$$S = \frac{375WT^2}{L}$$

a. $S = \frac{375(8)T^2}{8}$

$S = 375T^2$;

$S = \frac{375(8)T^2}{12}$

$S = 250T^2$;

$S = \frac{375(8)T^2}{16}$

$S = 187.5T^2$

b. $S = \frac{375(8)\left(\frac{1}{4}\right)^2}{L}$

$S = \frac{187.5}{L}$;

$S = \frac{375(8)\left(\frac{1}{2}\right)^2}{L}$

$S = \frac{750}{L}$;

$S = \frac{375(8)\left(\frac{3}{4}\right)^2}{L}$

$S = \frac{1687.5}{L}$;

3.6 Exercises

1. Constant

3. Reciprocal quadratic

5. Answers will vary.

7. $d = kr$

9. $F = ka$

11. $y = kx$

$0.6 = k(24)$

$0.025 = k$

$y = 0.025x$

x	$f(x) = 0.025x$
500	$f(500) = 0.025(500) = 12.5$
650	$16.25 = 0.025x$ $650 = x$
750	$f(750) = 0.025(750) = 18.75$

13. a. $w = kh$

$344.25 = k(37.5)$

$9.18 = k$;

$w = 9.18h$

$w = 9.18(35)$

$w = \$321.30$

b. k represents the hourly wage.

15. a. $s = kh$

$192 = k(47)$

$\frac{192}{47} = k$;

$s = \frac{192}{47}h$

b.

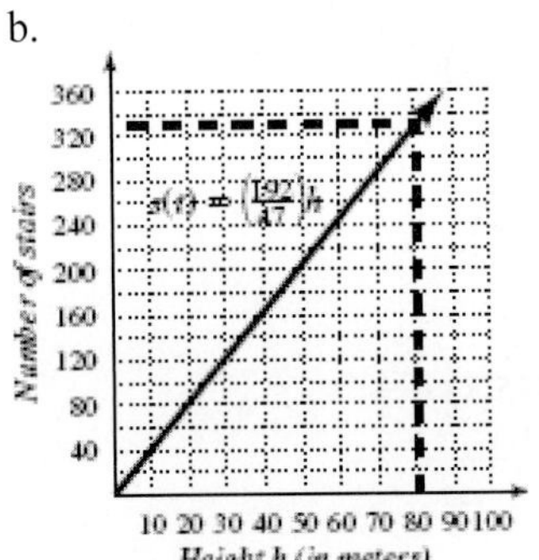

c. $s = 330$ stairs

d. $s = \frac{192}{47}(81) \approx 331$; Yes

17. $A = kS^2$

19. $P = kc^2$

21. $p = kq^2$

$280 = k(50)^2$

$\frac{280}{(50)^2} = k$

$0.112 = k;$

$p = 0.112q^2$

q	$p(q) = 0.112q^2$
45	$p(45) = 0.112(45)^2 = 226.8$
55	$338.8 = 0.112q^2$ $3025 = q^2$ $55 = q$
70	$p(70) = 0.112(70)^2 = 548.8$

23. $A = ks^2$

$3528 = k\left(14\sqrt{3}\right)^2$

$3528 = 588k$

$\frac{3528}{588} = k$

$6 = k;$

$A = 6s^2;$

$A = 6(303{,}600)^2;$

$A = 553{,}037{,}760{,}000 \text{ cm}^2$

$A = 55{,}303{,}776 \text{ m}^2$

25. a. $d = kt^2$

$169 = k(3.25)^2$

$169 = 10.5625k$

$16 = k;$

$d = 16t^2$

b.

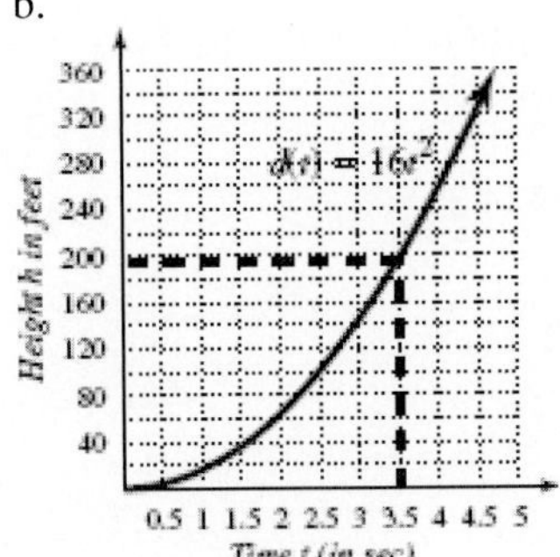

c. According to the graph, about 3.5 seconds

d. $196 = 16t^2$

$12.25 = t^2$

$3.5 \text{ sec} = t$

Yes, it was close.

e. $121 = 16t^2$

$7.5625 = t^2$

$2.75 = t$

2.75 seconds

27. $F = \frac{k}{d^2}$

29. $S = \frac{k}{L}$

31. $Y = \frac{k}{Z^2}$

$1369 = \frac{k}{3^2}$

$12321 = k;$

$Y = \frac{12321}{Z^2}$

Z	Y
37	$Y(37) = \frac{12321}{37^2} = 9$
74	$2.25 = \frac{12321}{Z^2}$ $2.25Z^2 = 12321$ $Z = 74$

111	$Y(111)=\frac{12321}{111^2}=1$

33. $W=\frac{k}{r^2}$

$75=\frac{k}{(6400)^2}$
$3072000000=k;$

$W=\frac{3072000000}{r^2}$

$W=\frac{3072000000}{(8000)^2}$
$W=48\text{ kg}$

35. $I=krt$

37. $A=kh(B+b)$

39. $V=ktr^2$

41. $C=\frac{kR}{S^2}$

$21=\frac{k(7)}{(1.5)^2}$

$47.25=7k$

$6.75=k;$

$C=\frac{6.75R}{S^2}$

R	S	C
120	6	22.5
200	12.5	8.64
350	15	10.5

$22.5=\frac{6.75(120)}{S^2}$
$22.5S^2=810$
$S=6;$

$C=\frac{6.75(200)}{(12.5)^2}=\frac{1350}{156.25}=8.64;$

$10.5=\frac{6.75R}{(15)^2}$

$2362.5=6.75R$

$350=R$

43. $E=kmv^2$

$200=k(1)(20)^2$
$0.5=k;$

$E=0.5mv^2;$

$E=0.5(1)(35)^2$

$E=612.5\text{ joules}$

45. $R(A)=\sqrt[3]{A}-1$

$f(x)=\sqrt[3]{x}$

Amount A	Rate r
1.0	0.0
1.05	0.016
1.10	0.032
1.15	0.048
1.20	0.063
1.25	0.077

Cube root family

$R(A)=\sqrt[3]{1.17}-1=0.054=5.4\%$

Interest Rate: 5.4%

47. $t=\frac{k}{v}$

$4=\frac{k}{12}$

$48=k;$

$t=\frac{48}{v};$

$t=\frac{48}{1.5}$

$t=32$ volunteers

49. $M=kE$

$16=k(96)$

$\frac{16}{96}=k$

$\frac{1}{6}=k;$

$M=\frac{1}{6}E;$

$M=\frac{1}{6}(250)$
$M=41.7\text{ kg}$

51. $D=k\sqrt{S}$

$108=k\sqrt{25}$

$21.6=k;$

$D = 21.6\sqrt{S}$;
$D = 21.6\sqrt{45}$
$D = 144.9$ ft

53. $C = kLD$

$76.50 = k(36)\left(\frac{1}{4}\right)$

$76.50 = 9k$

$8.5 = k$;

$C = 8.5LD$;

$C = 8.5(24)\left(\frac{3}{8}\right)$

$C = \$76.50$

55. $C = \frac{kp_1p_2}{d^2}$

$300 = \frac{k(300000)(420000)}{430^2}$

$55470000 = 1.26\text{x}10^{11}k$

$4.4\text{x}10^{-4} = k$;

$C = \frac{(4.4\text{x}10^{-4})p_1p_2}{d^2}$

$C = \frac{(4.4\text{x}10^{-4})(170000)(550000)}{430^2} \approx 222.5$

223 calls

57.

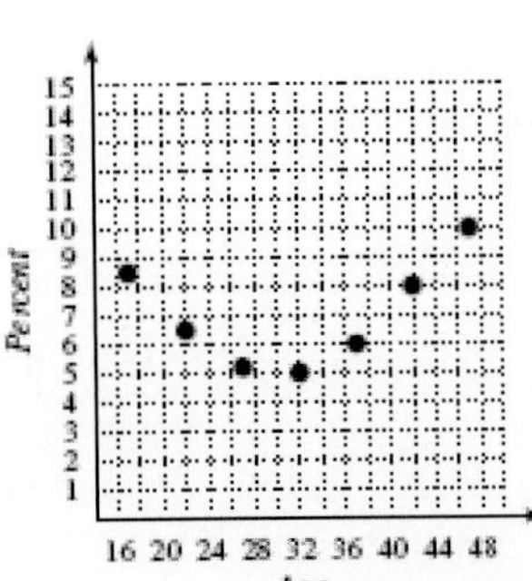

a. Scatter plot shows data is obviously nonlinear; decreasing to increasing pattern rules out a power function.

$p(t) = 0.0148t^2 - 0.9175t + 19.5601$

b.

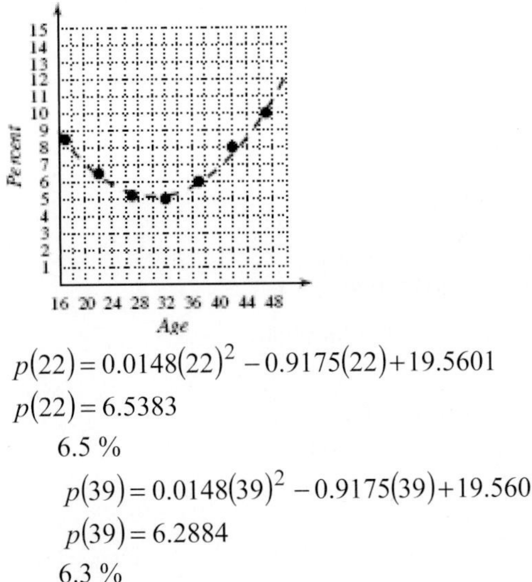

$p(22) = 0.0148(22)^2 - 0.9175(22) + 19.5601$

$p(22) = 6.5383$

6.5 %

$p(39) = 0.0148(39)^2 - 0.9175(39) + 19.5601$

$p(39) = 6.2884$

6.3 %

59. Answers will vary

61. $F = k\frac{m_1m_2}{d^2}$

$F = (6.67 \text{ x } 10^{-11})\frac{(1000)(1000)}{10^2}$
$F = (6.67 \text{ x } 10^{-11})(10000)$
$F = (6.67 \text{ x } 10^{-11})(1.0 \text{ x } 10^4)$
$F = 6.67 \text{ x } 10^{-7}$

63. $\left(\frac{2x^4}{3x^3y}\right)^{-2} = \left(\frac{2x}{3y}\right)^{-2} = \left(\frac{3y}{2x}\right)^2 = \frac{9y^2}{4x^2}$

65. $x^2 - 2x + 5 = 0$

$(1+2i)^2 - 2(1+2i) + 5 = 0$

$1 + 4i + 4i^2 - 2 - 4i + 5 = 0$

$1 - 4 - 2 + 5 = 0$

$0 = 0$

Yes

67. Let x represent the number of gallons of 25% acid solution.

$40(0.25) = 0.10(40 + x)$

$10 = 4 + 0.10x$

$6 = 0.10x$

$60 = x$

60 gallons

3.7 Technology Highlight

Exercise 1: They are approaching 4; not defined.

Exercise 2: $Y_1 = 4$ Y_2 has an error. As long as $x<2$, Y_1 will be valuated, not Y_2. Y_1 will not have an output equal to 4.

3.7 Exercises

1. Continuous

3. Smooth

5. Answers will vary.

7. a. $f(x)=\begin{cases} x^2-6x+10 & 0\le x\le 5 \\ \frac{3}{2}x-\frac{5}{2} & 5<x\le 9 \end{cases}$

   b. $y\in[1,11]$

9. a. $f(t)=\begin{cases} -t^2+6t & 0\le t\le 5 \\ 500 & t>5 \end{cases}$

   b. $y\in[0,9]$

11. $h(x)=\begin{cases} -2 & x<-2 \\ |x| & -2\le x<3 \\ 5 & x\ge 3 \end{cases}$

    $h(-5)=-2$;
    $h(-2)=|-2|=2$;
    $h(0)=|0|=0$;
    $h\left(-\frac{1}{2}\right)=\left|-\frac{1}{2}\right|=\frac{1}{2}$;
    $h(2.999)=|2.999|=2.999$;
    $h(3)=5$

13. $p(x)=\begin{cases} 5 & x<-3 \\ x^2-4 & -3\le x\le 3 \\ 2x+1 & x>3 \end{cases}$

    $p(-5)=5$;
    $p(-3)=(-3)^2-4=9-4=5$;
    $p(-2)=(-2)^2-4=4-4=0$;
    $p(0)=(0)^2-4=0-4=-4$;
    $p(3)=(3)^2-4=9-4=5$;
    $p(5)=2(5)+1=10+1=11$

15. $P(t)=\begin{cases} -0.03t^2+1.28t+1.68 & 0\le t<30 \\ 1.89t-43.5 & t\ge 30 \end{cases}$

    a. $P(5)=-0.03(5)^2+1.28(5)+1.68=7.33$
       $P(15)=-0.03(15)^2+1.28(15)+1.68=14.13$
       $P(25)=-0.03(25)^2+1.28(25)+1.68=14.93$
       $P(35)=1.89(35)-43.5=22.65$
       $P(45)=1.89(45)-43.5=41.55$
       $P(55)=1.89(55)-43.5=60.45$

    b. Each piece gives a slightly different value due to rounding of coefficients in each model. At $t=30$ we use the "first" piece: P(30)=13.08.

17. $g(x)=\begin{cases} 3x-x^2 & x\le 4 \\ 2x-12 & x>4 \end{cases}$

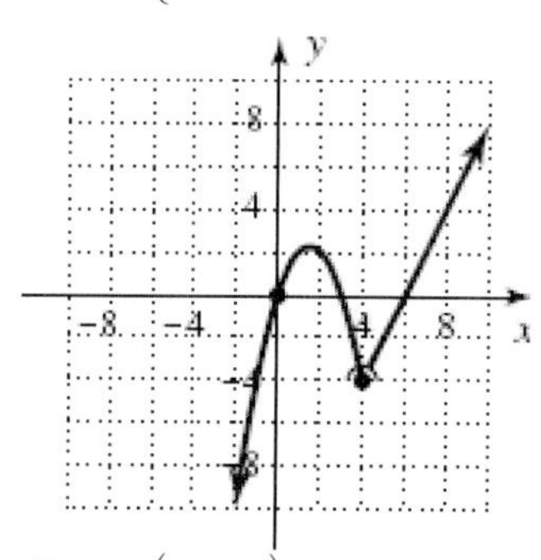

$D: x\in(-\infty,\infty)$
$R: y\in(-\infty,\infty)$

19. $p(x)=\begin{cases} -(x+5)^2 & x<-3 \\ |x|-7 & -3\le x\le 5 \\ -2 & x>5 \end{cases}$

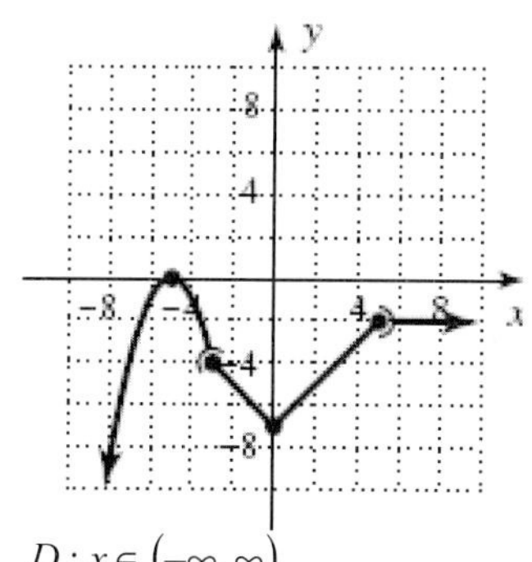

$D: x\in(-\infty,\infty)$
$R: y\in(-\infty,0]$

21. $H(x)=\begin{cases} -x+3 & x<1 \\ -|x-5|+6 & 1\le x<9 \end{cases}$

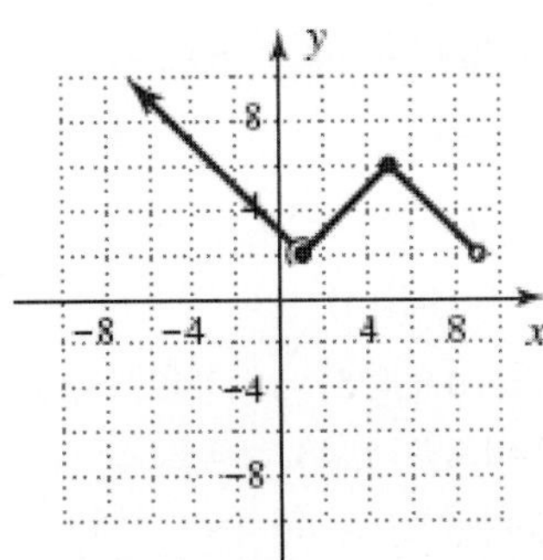

$D: x \in (-\infty, 9)$
$R: y \in [2, \infty)$

23. $f(x) = \begin{cases} -x-3 & x < -3 \\ 9-x^2 & -3 \le x < 2 \\ 4 & x \ge 2 \end{cases}$

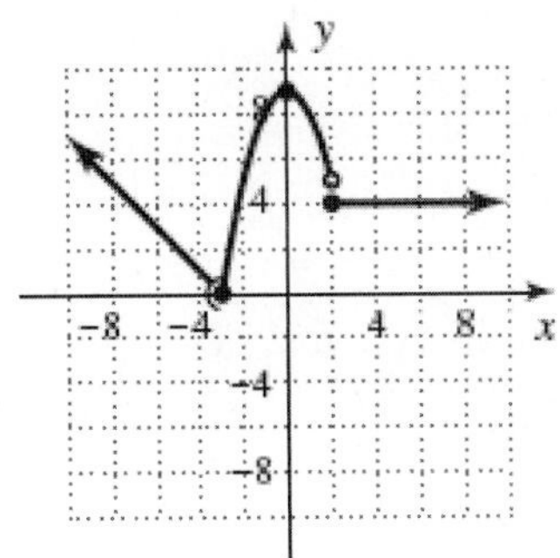

$D: x \in (-\infty, \infty)$
$R: y \in [0, \infty)$

25. $f(x) = \begin{cases} \dfrac{x^2-9}{x+3} & x \ne -3 \\ -4 & x = -3 \end{cases}$

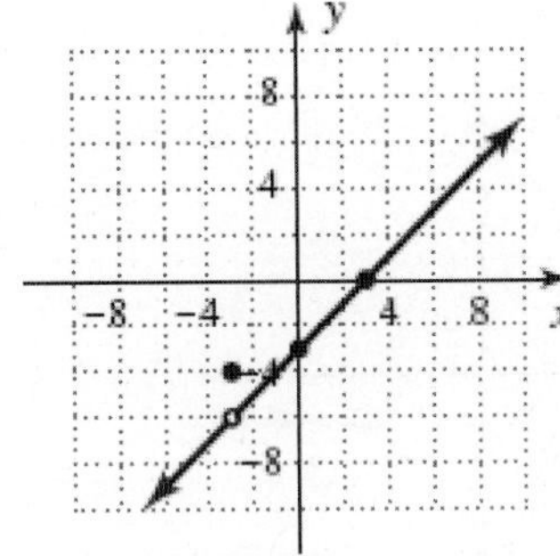

$D: x \in (-\infty, \infty)$
$R: y \in (-\infty, -6) \cup (-6, \infty)$
Discontinuity at $x = -3$
Redefine $f(x) = -6$ at $x = -3$

27. $f(x) = \begin{cases} \dfrac{x^3-1}{x-1} & x \ne 1 \\ -4 & x = 1 \end{cases}$

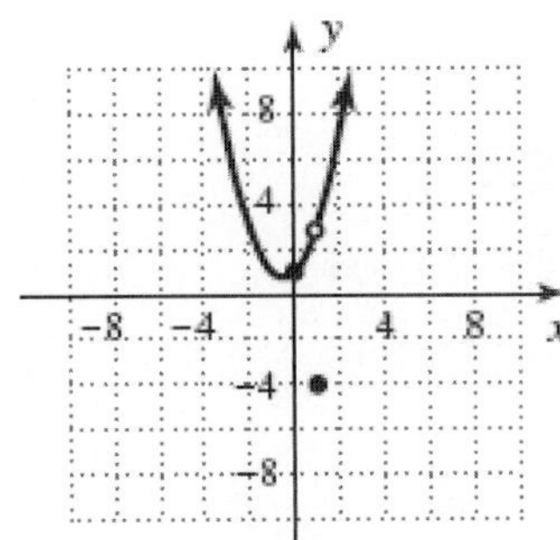

$D: x \in (-\infty, \infty)$
$R: y \in \{-4\} \cup [0.75, \infty)$
Discontinuity at $x = 1$
Redefine $f(x) = 3$ at $x = 1$

29. $|x| = \begin{cases} -x & x < 0 \\ x & x \ge 0 \end{cases}$

Graph is discontinuous at $x = 0$.

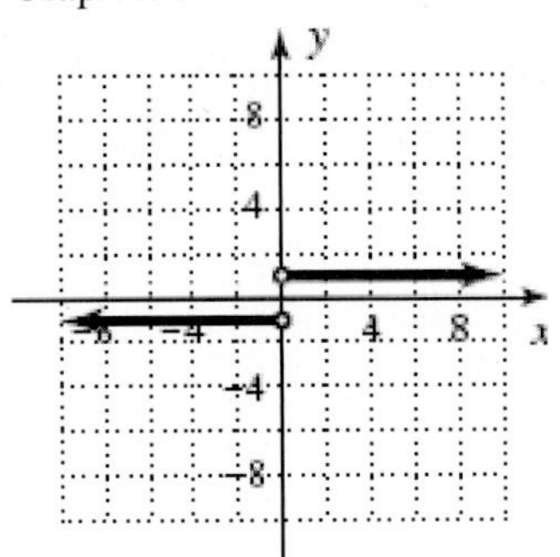

If $x < 0$, $f(x) = -1$.
If $x > 0$, $f(x) = 1$.

31. $C(p) = \begin{cases} 0.09p & 0 \le p \le 1000 \\ 0.18p - 90 & p > 1000 \end{cases}$

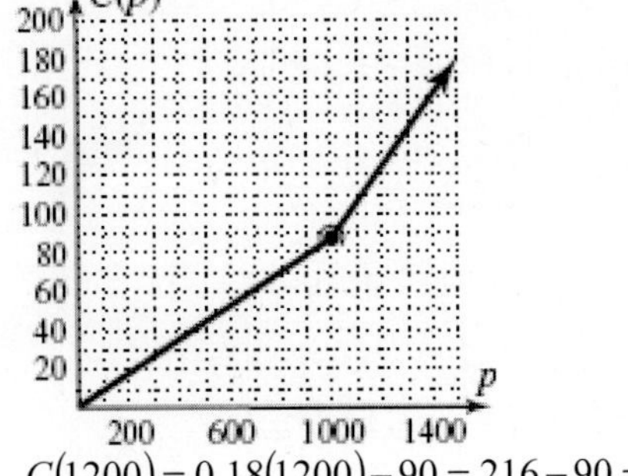

$C(1200) = 0.18(1200) - 90 = 216 - 90 = \$126$

33. $C(t) = \begin{cases} 0.75t & 0 \le t \le 25 \\ 1.5t - 18.75 & t > 25 \end{cases}$

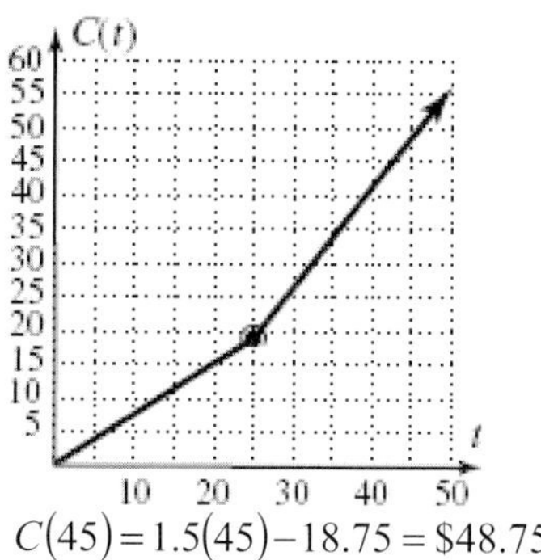

$C(45) = 1.5(45) - 18.75 = \$48.75$

35. $S(t) = \begin{cases} -1.35t^2 + 31.9t + 152 & 0 \le t \le 12 \\ 2.5t^2 - 80.6t + 950 & 12 < t \le 22 \end{cases}$

$S(25) = 2.5(25)^2 - 80.6(25) + 950$
$= 2.5(625) - 2015 + 950 = 497.5$
≈ 498 billion;

$S(28) = 2.5(28)^2 - 80.6(28) + 950$
$= 2.5(784) - 2256.8 + 950 = 653.2$
≈ 653 billion;

$S(30) = 2.5(30)^2 - 80.6(30) + 950$
$= 2.5(900) - 2418 + 950 = 782$
≈ 782 billion

37. $C(m) = \begin{cases} 3.3m & 0 \le m \le 30 \\ 3.3(30) + 7(m - 30) & m > 30 \end{cases}$

$C(m) = \begin{cases} 3.3m & 0 \le m \le 30 \\ 7m\text{-}111 & m > 30 \end{cases}$

$C(46) = 3.3(30) + 7(46 - 30) = \$2.11$

39. $C(a) = \begin{cases} 0 & a < 2 \\ 2 & 2 \le a < 13 \\ 5 & 13 \le a < 20 \\ 7 & 20 \le a < 66 \\ 5 & a \ge 66 \end{cases}$

One grandparent:
$C(1) = 5(1) = 5$;

Two adults:
$C(2) = 7(2) = 14$;

Three teenagers:
$C(3) = 5(3) = 15$;

Two children:
$C(2) = 2(2) = 4$;

One infant: 0

Total Cost: $5 + 14 + 15 + 4 + 0 = \$38$

41. $h(x) = |x - 2| - |x + 3|$

x	$h(x) = \|x - 2\| - \|x + 3\|$
-5	$h(-5) = \|-5 - 2\| - \|-5 + 3\| = 7 - 2 = 5$
-4	$h(-4) = \|-4 - 2\| - \|-4 + 3\| = 6 - 1 = 5$
-3	$h(-3) = \|-3 - 2\| - \|-3 + 3\| = 5 - 0 = 5$
-2	$h(-2) = \|-2 - 2\| - \|-2 + 3\| = 4 - 1 = 3$
-1	$h(-1) = \|-1 - 2\| - \|-1 + 3\| = 3 - 2 = 1$
0	$h(0) = \|0 - 2\| - \|0 + 3\| = 2 - 3 = -1$
1	$h(1) = \|1 - 2\| - \|1 + 3\| = 1 - 4 = -3$
2	$h(2) = \|2 - 2\| - \|2 + 3\| = 0 - 5 = -5$
3	$h(3) = \|3 - 2\| - \|3 + 3\| = 1 - 6 = -5$
4	$h(4) = \|4 - 2\| - \|4 + 3\| = 2 - 7 = -5$
5	$h(5) = \|5 - 2\| - \|5 + 3\| = 3 - 8 = -5$

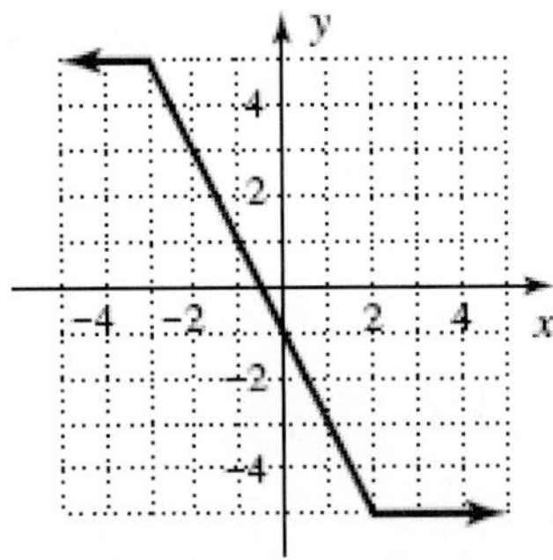

The function is continuous.

$$h(x)=\begin{cases} 5 & x \le -3 \\ -2x-1 & -3 < x < 2 \\ -5 & x \ge 2 \end{cases}$$

43. Answers will vary.

45. $f(x)=\begin{cases} x^2 & x<1 \\ 4x-3 & 1 \le x \le 3 \\ 2x+3 & x>3 \end{cases}$

47. $\dfrac{3}{x-2}+1=\dfrac{30}{x^2-4}$

$$\left(\frac{3}{x-2}+1=\frac{30}{(x-2)(x+2)}\right)(x-2)(x+2)$$

$$3(x+2)+1(x-2)(x+2)=30$$

$$3x+6+x^2-4=30$$

$$x^2+3x-28=0$$

$$(x+7)(x-4)=0$$

$$x=-7;\quad x=4$$

49. $y=2\sqrt{x+4}-1$

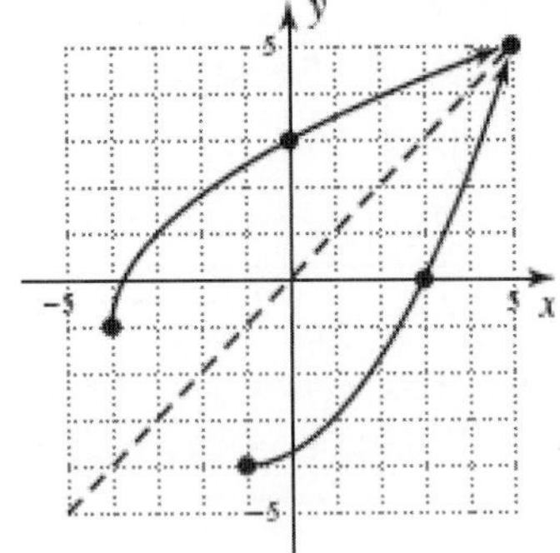

51. $f(x)=2x+1$ and $g(x)=3x^2$

a) $(f \circ g)(-2)=f[g(-2)]$

$$g(-2)=3(-2)^2=3(4)=12$$

$$f(12)=2(12)+1=24+1=25$$

b) $(f \circ g)(x)=f[g(x)]$

$$=2(g(x))+1$$

$$=2(3x^2)+1$$

$$=6x^2+1$$

3.8 Technology Highlight

Exercise 1: $y=x^2-8x+9$; (4, -7)

Exercise 2: $y=a^3-2a^2-4a+8$
(-0.67, 9.48), (2, 0)

Exercise 3: $y=x^4-5x^2-2x$
(-1.47, -3.20), (-0.20, 0.20), (1.67, -9.51)

Exercise 4: $y=x\sqrt{x+4}$; (-2.67, -3.08)

3.8 Exercises

1. Cut; linear; bounce

3. Increasing

5. Answers will vary.

7. $f(x)=\dfrac{1}{x-3}-4$
$D: x \in (-\infty,3)\cup(3,\infty)$
$R: y \in (-\infty,-4)\cup(-4,\infty)$

9. $p(x)=\begin{cases} (x+3)^3+3 & x<-3 \\ -1.5x+0.5 & -3 \le x \le 3 \\ -4 & x>3 \end{cases}$
$D: x \in (-\infty,\infty)$
$R: y \in (-\infty,5]$

11. $y=f(x)$
$D: x \in (-\infty,2]$
$R: y \in (-\infty,3]$

13.

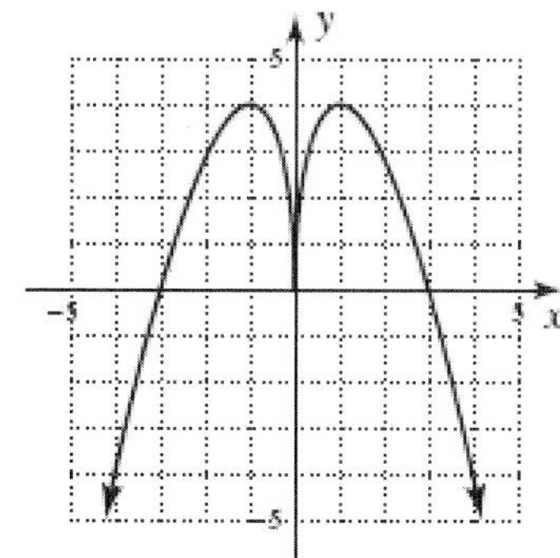

15. $f(x)=-7|x|+3x^2+5$

$f(k)=-7|k|+3(k)^2+5$
$=-7k+3k^2+5;$

$f(-k)=-7|-k|+3(-k)^2+5$
$=-7k+3k^2+5;$

$f(k)=f(-k)$
Even

17. $p(x)=2x^4-6x+1$

$p(k)=2(k)^4-6(k)+1$
$=2k^4-6k+1;$

$p(-k)=2(-k)^4-6(-k)+1$
$=2k^4+6k+1$
$p(k)\neq p(-k)$; Not even

19.

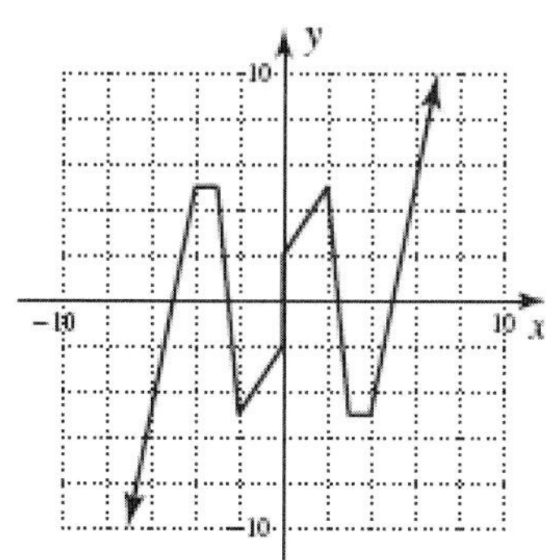

21. $f(x)=4\sqrt[3]{x}-x$

$f(k)=4\sqrt[3]{k}-(k)$
$=4\sqrt[3]{k}-(k);$

$f(-k)=4\sqrt[3]{-k}-(-k)$
$=-4\sqrt[3]{k}+(k);$
$f(k)=-f(k)$
Odd

23. $p(x)=3x^3-5x^2+1$

$p(k)=3(k)^3-5(k)^2+1$
$=3k^3-5k^2+1$
$p(-k)=3(-k)^3-5(-k)^2+1$
$=-3k^3+5k^2+1$
$p(k)\neq -p(k)$; Not Odd

25. $x\in[-1,1],\ [3,\infty)$

27. $x\in(-\infty,-1)\cup(-1,1)\cup(1,\infty)$

29. $f(x)\uparrow:(1,4)$
$f(x)\downarrow:(-2,1)\cup(4,\infty)$
Constant: $(-\infty,-2)$

31. $f(x)\uparrow:(-3,1)\cup(4,6)$
$f(x)\downarrow:(-\infty,-3),(1,4)$
Constant : None

33. $H(x)=-5|x-2|+5$

a. $D: x\in(-\infty,\infty)$
$R: y\in(-\infty,5]$
b. (1, 0), (3, 0)
c. $H(x)\geq 0: x\in[1,3]$
$H(x)\leq 0: x\in(-\infty,1]\cup[3,\infty)$
d. $H(x)\uparrow: x\in(-\infty,2)$
$H(x)\downarrow: x\in(2,\infty)$
e. Maximum: (2, 5)
f. None

35. $q(x)=-\sqrt[3]{x+1}$

a. $D: x\in(-\infty,\infty)$
$R: y\in(-\infty,\infty)$
b. (-1, 0)
c. $q(x)\geq 0: x\in(-\infty,-1]$
$q(x)\leq 0: x\in[-1,\infty)$
d. $q(x)\uparrow$: None
$q(x)\downarrow: x\in(-\infty,\infty)$
e. None
f. None

37. $y=g(x)$

a. $D: x\in(-\infty,\infty)$

$R: y \in (-\infty, \infty)$

b. (-1,0), (5, 0)

c. $g(x) \geq 0: x \in [-1, \infty)$

$g(x) \leq 0: x \in (-\infty, -1] \cup \{5\}$

d. $g(x) \uparrow: x \in (-\infty, 1) \cup (5, \infty)$

$g(x) \downarrow: x \in (1, 5)$

e. Maximum: (1,6)
Minimum: (5, 0)

f. None

39. $y = q(x)$

a. $D: x \in (-\infty, \infty)$

$R: y \in (-\infty, \infty)$

b. (-2, 0), (4, 0), (8, 0)

c. $q(x) \geq 0: x \in \{-2\} \cup [4, \infty)$

$q(x) \leq 0: x \in (-\infty, 4) \cup \{8\}$

d. $q(x) \uparrow: x \in (-\infty, -2) \cup (1, 6) \cup (8, \infty)$

$q(x) \downarrow: x \in (-2, 1) \cup (6, 8)$

e. Maximum: (-2,0), (6, 2)
Minimum: (1, -5) and (8, 0)

f. None

41. $y = Y_2$

a. $D: x \in (-\infty, \infty)$

$R: y \in (-\infty, 3]$

b. (0, 0), (2, 0)

c. $Y_2 \geq 0: x \in [0, 2]$

$Y_2 \leq 0: x \in (-\infty, 0] \cup [2, \infty)$

d. $Y_2 \uparrow: x \in (-\infty, 1)$

$Y_2 \downarrow: x \in (1, \infty)$

e. Maximum: (1, 3)

f. None

43. $Y_2 = \dfrac{-4}{(x-2)^2} + 4$

a. $D: x \in (-\infty, 2) \cup (2, \infty)$

$R: y \in (-\infty, 4)$

b. (1, 0), (3, 0)

c. $f(x) \geq 0: (-\infty, 1] \cup [3, \infty)$

$f(x) \leq 0: [1, 2) \cup (2, 3]$

d. $f(x) \uparrow: (2, \infty)$

$f(x) \downarrow: (-\infty, 2)$

e. None

f. $x = 2, y = 4$

45. $H(x) = \begin{cases} -1.5x - 6 & x \leq -2 \\ \dfrac{4}{x^2} - 1 & -2 < x < 2 \\ -4 & x \geq 3 \end{cases}$

a. $D: x \in (-\infty, 0) \cup (0, \infty)$

$R: y \in \{-4\} \cup (-3, \infty)$

b. (-4, 0)

c. $H(x) \geq 0: x \in (-\infty, -4] \cup (-2, 0) \cup (0, 2)$

$H(x) \leq 0: x \in [-4, -2] \cup [3, \infty)$

d. $H(x) \uparrow: x \in (-2, 0)$

$H(x) \downarrow: x \in (-\infty, -2) \cup (0, 2)$

Constant: $x \in (3, \infty)$

e. None

f. $x = 0$

47. $f(x) = x^2 - 4x$

$\dfrac{f(x+h) - f(x)}{h} 4$

$= \dfrac{\left[(x+h)^2 - 4(x+h)\right] - \left[x^2 - 4x\right]}{h}$

$= \dfrac{\left(x^2 + 2xh + h^2 - 4x - 4h\right) - x^2 + 4x}{h}$

$= \dfrac{2xh + h^2 - 4h}{h}$

$= \dfrac{h(2x + h - 4)}{h}$

$= 2x + h - 4$;

[0.00, 0.01]

$\dfrac{\Delta y}{\Delta x} = 2(0) + 0.01 - 4 = -3.99$;

[3.00, 3.01]

$\dfrac{\Delta y}{\Delta x} = 2(3) + 0.01 - 4 = 2.01$

Answers will vary.

49. $r(x) = \sqrt{x}$

$\dfrac{f(x+h) - f(x)}{h}$

$= \dfrac{\sqrt{x+h} - \sqrt{x}}{h} \cdot \dfrac{\sqrt{x+h} + \sqrt{x}}{\sqrt{x+h} + \sqrt{x}}$

$= \dfrac{x + h - x}{h\left(\sqrt{x+h} + \sqrt{x}\right)}$

$= \dfrac{1}{\sqrt{x+h} + \sqrt{x}}$;

[1.00, 1.01]

$\frac{\Delta y}{\Delta x} = \frac{1}{\sqrt{1+0.01}+\sqrt{1}} = 0.4987 \approx 0.5$;

[4.00, 4.01]

$\frac{\Delta y}{\Delta x} = \frac{1}{\sqrt{4+0.01}+\sqrt{4}} = 0.2498 \approx 0.25$

Answers will vary.

51. $y = \sin(x)$
   a. $R: y \in [-1,1]$
   b. (-180, 0), (0, 0), (180, 0), (360, 0)
   c. $y \uparrow: x \in (-90,90) \cup (270,360)$
      $y \downarrow: x \in (-180,-90) \cup (90,270)$
   d. Minimum: (-90, -1); (270, -1)
      Maximum: (90, 1)
   e. Odd

   $y = \cos(x)$
   a. $R: y \in [-1,1]$
   b. (-90, 0), (90, 0), (270, 0)
   c. $y \uparrow: x \in (-180,0) \cup (180,360)$
      $y \downarrow: x \in (0,180)$
   d. Minimum: (-180, -1); (180, -1)
      Maximum: (0, 1); (360, 1)
   e. Even

53. a. Increasing: $t \in (0, 1) \cup (3, 4) \cup (7, 10)$
   b. Decreasing: $t \in (4, 7)$
   c. Constant: $t \in (1, 3)$
   d. Maximum: (4, 12), (10, 16)
   e. Minimum: (7, -4)
   f. Positive: $t \in (0,6)$, (8, 10)
   g. Negative: $t \in (6, 8)$
   h. Zero: (6, 0), (8, 0)

55. $f(x) = |x| - 5$ and $g(x) = x^2 - 4$

$h(x) = (f \circ g)(x) = f[g(x)]$

$= |g(x)| - 5$

$= |x^2 - 4| - 5$

   a. $D: x \in (-\infty, \infty)$
      $R: y \in [-5, \infty)$
   b. (-3, 0), (3, 0)
   c. $h(x) \geq 0: x \in (-\infty,-3] \cup [3,\infty)$
      $h(x) \leq 0: x \in [-3,3]$
   d. $h(x) \uparrow: x \in (-2,0) \cup (2,\infty)$
      $h(x) \downarrow: x \in (-\infty,-2) \cup (0,2)$
   e. Maximum: (0, -1)
      Minimum: (-2, -5), (2, -5)
   f. None

57. a. $D: t \in [75,103]$
      $R: D \in [-300,230]$
   b. $D(t) \uparrow$ for $t \in$
      $(76,77) \cup (83,84) \cup (86,87) \cup (92,100)$
      $D(t) \downarrow$ for $t \in$
      $(75,76) \cup (77,83) \cup (84,86) \cup (89,92) \cup (100,102)$
      $D(t)$ is constant for $t \in (87,89)$
   c. Maximum: (75, -40), (77, -50),
      (84, -170), (100, 240) global maximum
      Minimum: (76, -70), (83, -210),
      (86, -220), (92, -300), (102,-140)
   d. Increase: 1996 to 1997 or 1999 to 2000
      Decrease: 2001 to 2002

59.

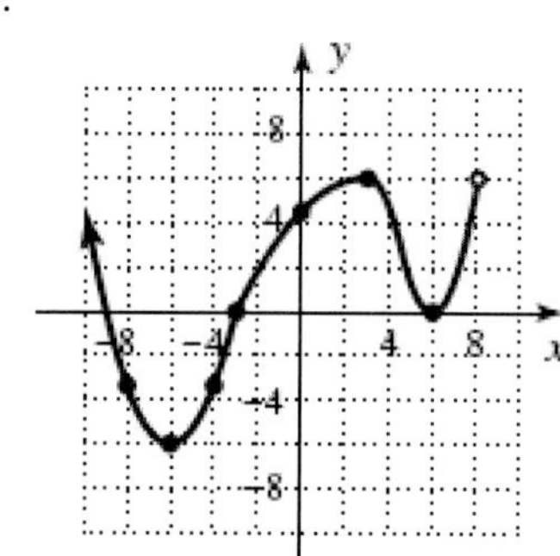

Zeroes: (-9, 0), (-3, 0), (6, 0)
Min: (-6, -6), (6, 0)
Max: (3, 6)

61. Answers will vary.

63. Answers will vary.

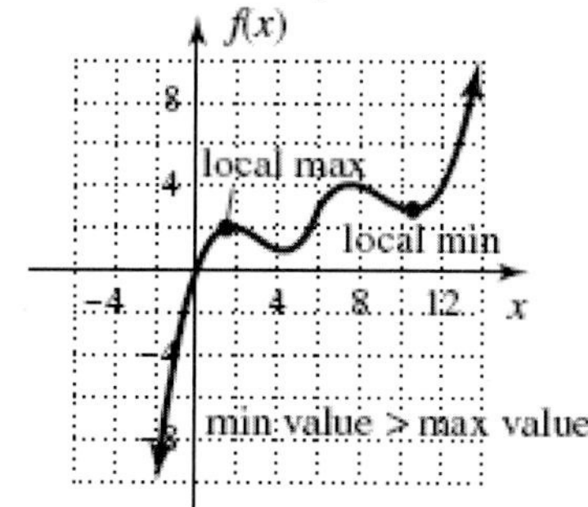

65. $x^2 - 8x - 20 = 0$
   a. $(x-10)(x+2) = 0$
      $x = 10;\ x = -2$
   b. $(x^2 - 8x) - 20 = 0$

$$\left(x^2-8x+16\right)-20-16=0$$
$$(x-4)^2-36=0$$
$$(x-4)^2=36$$
$$x-4=\pm 6$$
$$x=4\pm 6$$

$x=10;\ \ x=-2$

c. $x=\dfrac{8\pm\sqrt{(-8)^2-4(1)(-20)}}{2(1)}$

$x=\dfrac{8\pm\sqrt{64+80}}{2}$

$x=\dfrac{8\pm\sqrt{144}}{2}$

$x=\dfrac{8\pm 12}{2}$

$x=10;\ \ x=-2$

67. (1) $16^{\frac{-3}{2}}=\left(\dfrac{1}{16}\right)^{\frac{3}{2}}=\left(\sqrt{\dfrac{1}{16}}\right)^3$

$=\left(\dfrac{1}{4}\right)^3=\dfrac{1}{64}$

(2) $\left(\dfrac{27}{8}\right)^{\frac{-2}{3}}=\left(\dfrac{8}{27}\right)^{\frac{2}{3}}=\left(\sqrt[3]{\dfrac{8}{27}}\right)^2$

$=\left(\dfrac{2}{3}\right)^2=\dfrac{4}{9}$

69. General form: $y=|x|$

Node: (1, -2)

$y=|x-1|-2$

Chapter 3 Summary and Review

1. $f(x)=x^2+4x$ and $g(x)=3x-2$

$(f+g)(a)=f(a)+g(a)$
$=a^2+4a+3a-2$
$=a^2+7a-2$

2. $f(x)=x^2+4x$ and $g(x)=3x-2$

$(f\cdot g)(3)=f(3)\cdot g(3)$
$=\left((3)^2+4(3)\right)(3(3)-2)$
$=(9+12)(9-2)$
$=(21)(7)$
$=147$

3. $f(x)=x^2+4x$ and $g(x)=3x-2$

$\left(\dfrac{f}{g}\right)(x)=\dfrac{x^2+4x}{3x-2}$

$D: x\in\left(-\infty,\dfrac{2}{3}\right)\cup\left(\dfrac{2}{3},\infty\right)$

4. $p(x)=4x-3$; $q(x)=x^2+2x$; and

$r(x)=\dfrac{x+3}{4}$;

$(p\circ q)(x)=p[q(x)]$
$=4(q(x))-3$
$=4\left(x^2+2x\right)-3$
$=4x^2+8x-3$

5. $p(x)=4x-3$; $q(x)=x^2+2x$; and

$r(x)-\dfrac{x+3}{4}$;

$(q\circ p)(3)=q[p(3)]$
$p(3)=4(3)-3=12-3=9$
$q(9)=(9)^2+2(9)=81+18=99$

6. $p(x)=4x-3$; $q(x)=x^2+2x$; and

$r(x)=\dfrac{x+3}{4}$;

$(p\circ r)(x)=p[r(x)]$
$=4(r(x))-3$
$=4\left(\dfrac{x+3}{4}\right)-3$
$=x+3-3$
$=x;$

$(r\circ p)(x)=r[p(x)]$
$=\dfrac{p(x)+3}{4}$
$=\dfrac{4x-3+3}{4}$
$=\dfrac{4x}{4}$
$=x$

7. $h(x)=\sqrt{3x-2}+1$;
$f(x)=\sqrt{x}+1$;
$g(x)=3x-2$

8. $H(x)=3-\left|x^2-1\right|$;
$f(x)=3-|x|$;
$g(x)=x^2-1$

9. $s(x)=x^{\frac{2}{3}}-3x^{\frac{2}{3}}-10$
$f(x)=x^2-3x-10$
$g(x)=x^{\frac{1}{3}}$

10. $r(t)=2t+3$
$A(t)=\pi(2t+3)^2$

11. $h(x)=-|x-2|+3$; No

12. $p(x)=2x^2+7$; No

13. $s(x)=\sqrt{x-1}+5$; Yes

14. $f(x)=-3x+2$
$$y=-3x+2$$
$$x=-3y+2$$
$$x-2=-3y$$
$$\frac{x-2}{-3}=y$$
$$f^{-1}(x)=\frac{x-2}{-3}$$
$$(f\circ f^{-1})(x)=f\left[f^{-1}(x)\right]$$
$$=-3\left(f^{-1}(x)\right)+2$$
$$=-3\left(\frac{x-2}{-3}\right)+2$$
$$=x-2+2$$
$$=x;$$
$$(f^{-1}\circ f)(x)=f^{-1}[f(x)]$$
$$=\frac{f(x)-2}{-3}$$
$$=\frac{-3x+2-2}{-3}$$
$$=\frac{-3x}{-3}$$
$$=x$$

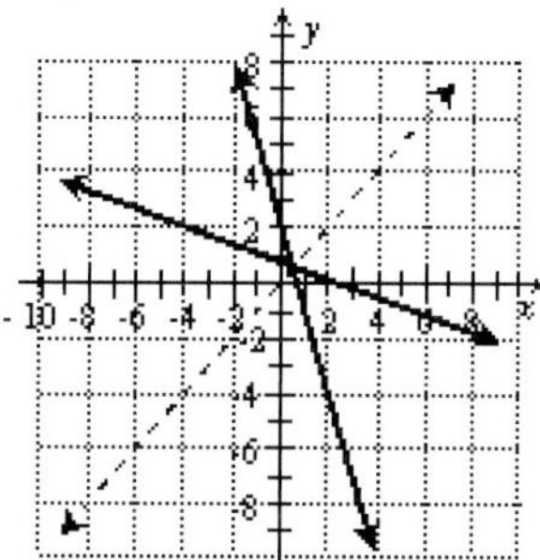

15. $f(x)=x^2-2,\quad x\geq 0$
$$y=x^2-2$$
$$x=y^2-2$$
$$x+2=y^2$$
$$\sqrt{x+2}=y$$
$$f^{-1}(x)=\sqrt{x+2}$$
$$(f\circ f^{-1})(x)=f\left[f^{-1}(x)\right]$$
$$=\left(f^{-1}(x)\right)^2-2$$
$$=\left(\sqrt{x+2}\right)^2-2$$
$$=x+2-2$$
$$=x;$$
$$(f^{-1}\circ f)(x)=f^{-1}[f(x)]$$
$$=\sqrt{f(x)+2}$$
$$=\sqrt{x^2-2+2}$$
$$=\sqrt{x^2}$$
$$=x$$

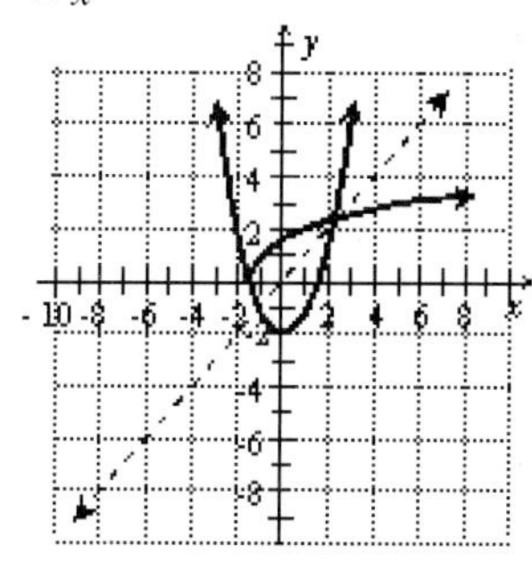

16. $f(x)=\sqrt{x-1},\quad x\ge 1$

$$y=\sqrt{x-1}$$
$$x=\sqrt{y-1}$$
$$x^2=y-1$$
$$x^2+1=y$$
$$f^{-1}(x)=x^2+1,\quad x\ge 0$$
$$(f\circ f^{-1})(x)=f[f^{-1}(x)]$$
$$=\sqrt{f^{-1}(x)-1}$$
$$=\sqrt{x^2+1-1}$$
$$=\sqrt{x^2}$$
$$=x;$$
$$(f^{-1}\circ f)(x)=f^{-1}[f(x)]$$
$$=\left(\sqrt{x-1}\right)^2+1$$
$$=x-1+1$$
$$=x$$

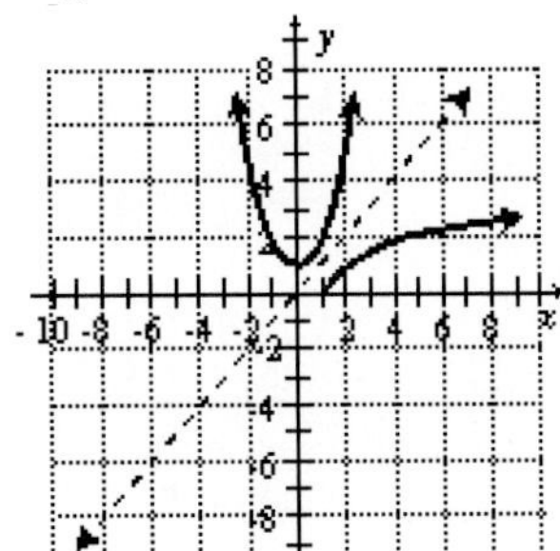

17. $f(x)$:

$\begin{cases}D: x\in[-4,\infty)\\ R: y\in[0,\infty)\end{cases}$

$f^{-1}(x)$:

$\begin{cases}D: x\in[0,\infty)\\ R: y\in[-4,\infty)\end{cases}$

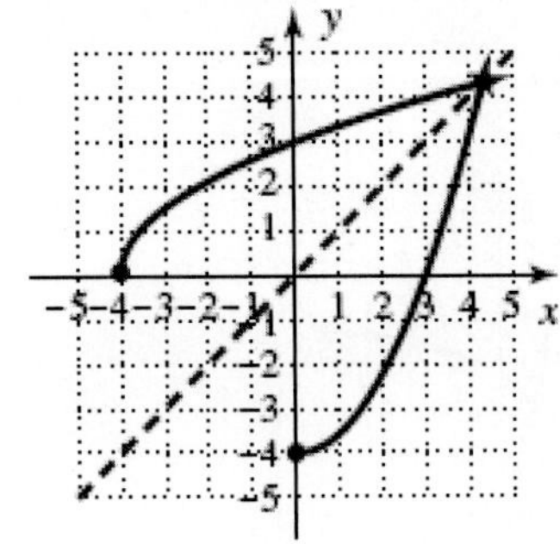

18. $f(x)$:

$\begin{cases}D: x\in(-\infty,\infty)\\ R: y\in(-\infty,\infty)\end{cases}$

$f^{-1}(x)$:

$\begin{cases}D: x\in(-\infty,\infty)\\ R: y\in(-\infty,\infty)\end{cases}$

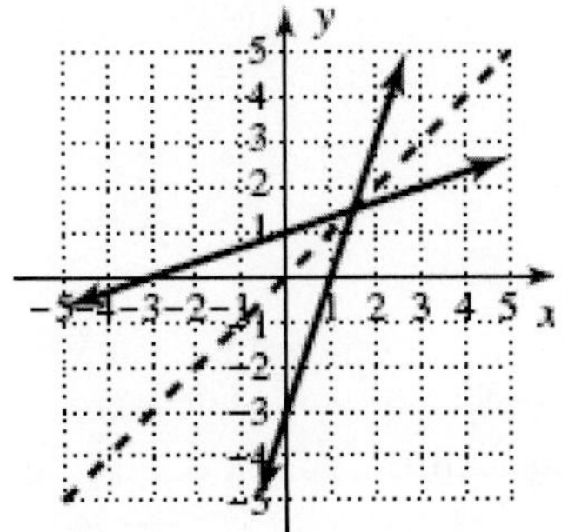

19. $f(x)$:

$\begin{cases}D: x\in(-\infty,\infty)\\ R: y\in(0,\infty)\end{cases}$

$f^{-1}(x)$:

$\begin{cases}D: x\in(0,\infty)\\ R: y\in(-\infty,\infty)\end{cases}$

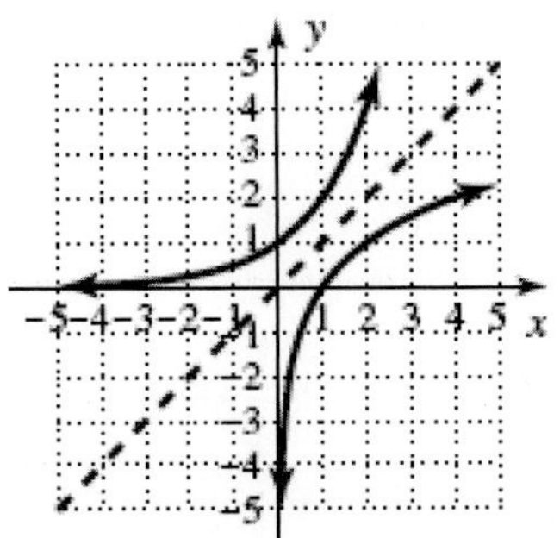

20. $f(t)=0.15t+2$

a. $f(7)=0.15(7)+2=1.05+2=\$3.05$

b. $y=0.15t+2$

$$t=0.15y+2$$
$$t-2=0.15y$$
$$\frac{t-2}{0.15}=y$$
$$f^{-1}(t)=\frac{t-2}{0.15};$$
$$f^{-1}(3.05)=\frac{3.05-2}{0.15}=\frac{1.05}{0.15}=7$$

Gives the number of days overdue if a \$3.05 fine.

c. $f^{-1}(3.80)=\frac{3.80-2}{0.15}=\frac{1.80}{0.15}=12\text{ days}$

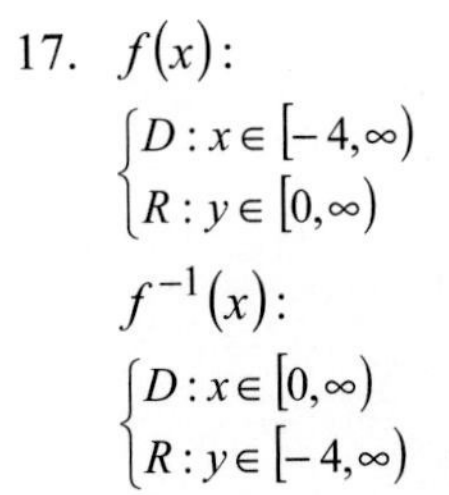

21. $f(x)=-(x+2)^2-5$; Quadratic

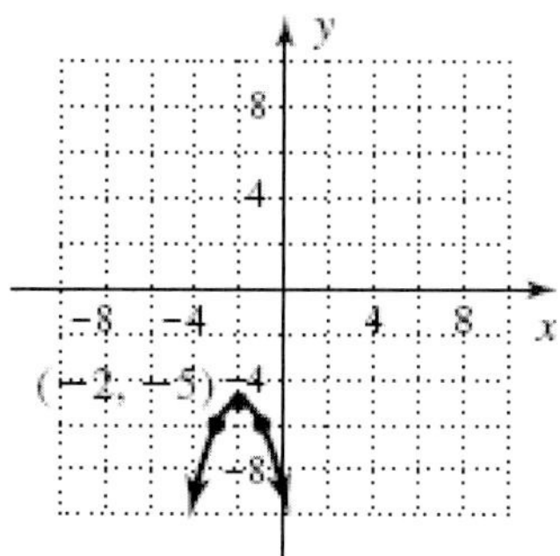

22. $f(x)=2|x+3|$; Absolute Value

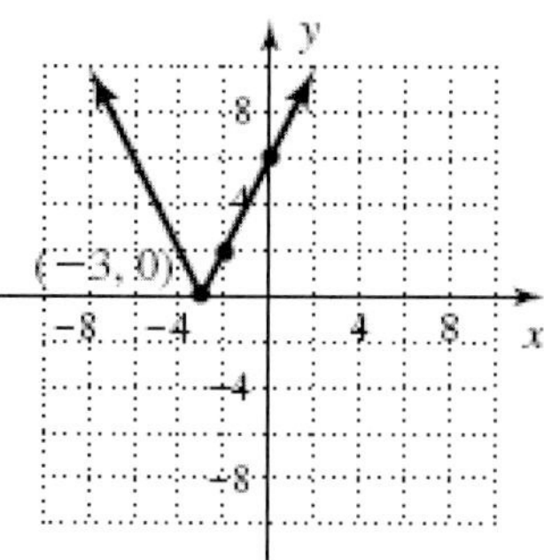

23. $f(x)=x^3-1$; Cubic

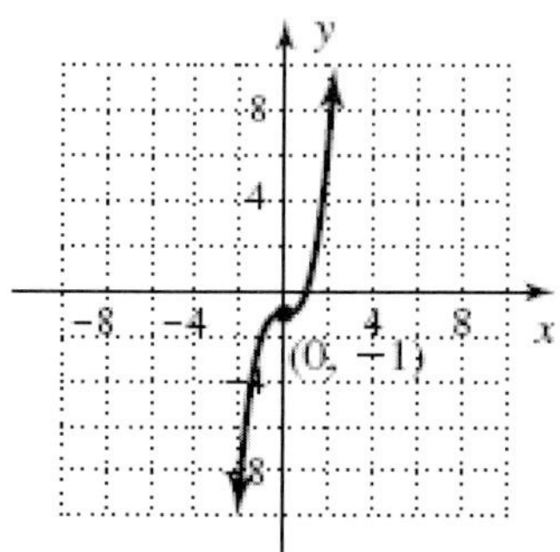

24. $f(x)=\sqrt{x-5}+2$; Square Root

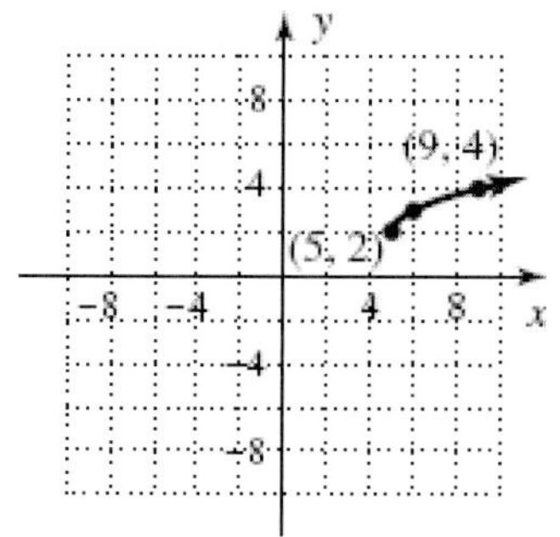

25. $f(x)=\sqrt[3]{x+2}$; Cube Root

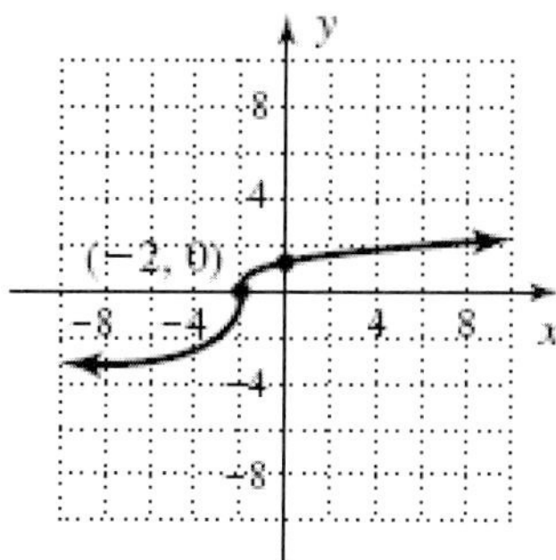

26. $f(x)=-2x+5$; Linear

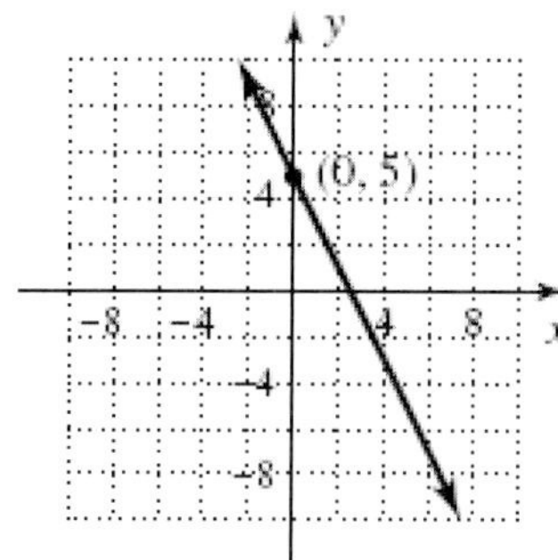

27. a. $f(x-2)$

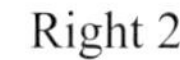

Right 2

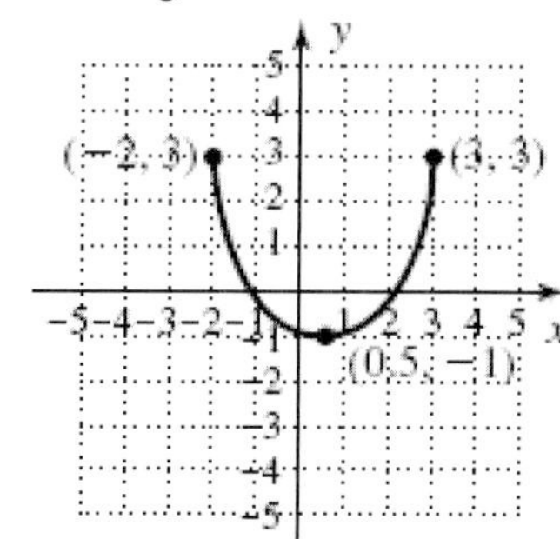

b. $-f(x)+6$

Up 6

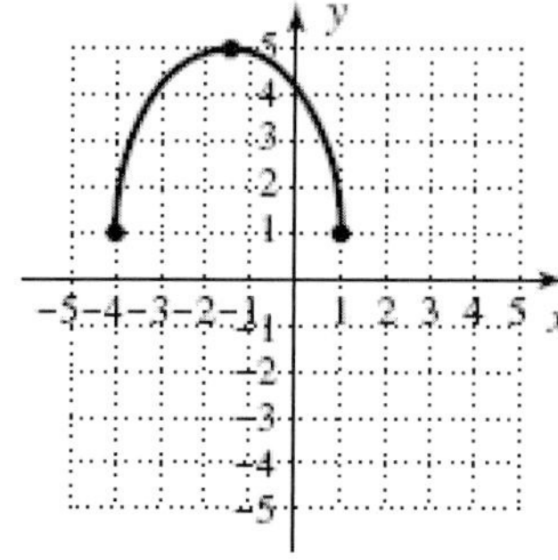

c. $\frac{1}{2}f(x)$

Compressed down

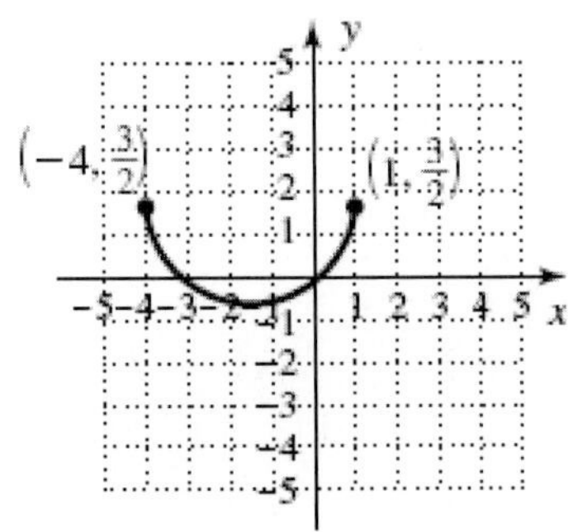

28. General form: $y = |x|$

    Node: (-1, 4)

    $q(x) = -2|x+1| + 4$

29. $f(x) = x^2 + 8x + 15$

    $0 = x^2 + 8x + 15$

    $0 = (x^2 + 8x + 16) + 15 - 16$

    $0 = (x+4)^2 - 1$

    $0 = (x+4)^2 - 1$

    $1 = (x+4)^2$

    $\pm 1 = x + 4$

    $x = -4 \pm 1$

    x-intercepts: (-5, 0) and (-3, 0)

    Vertex: (-4,-1)

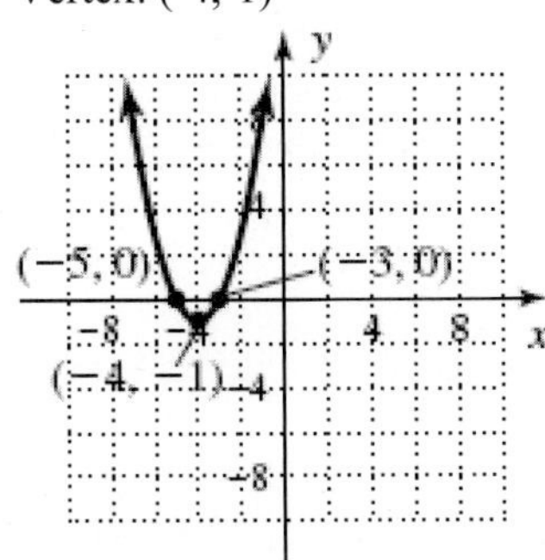

30. $f(x) = -x^2 + 4x - 5$

    $0 = -x^2 + 4x - 5$

    $0 = -(x^2 - 4x) - 5$

    $0 = -(x^2 - 4x + 4) - 5 + 4$

    $0 = -(x-2)^2 - 1$

    $0 = -(x-2)^2 - 1$

    $1 = -(x-2)^2$

    $-1 = (x-2)^2$

    x-intercepts: None; square root cannot equal a negative number.

    Vertex: (2,-1)

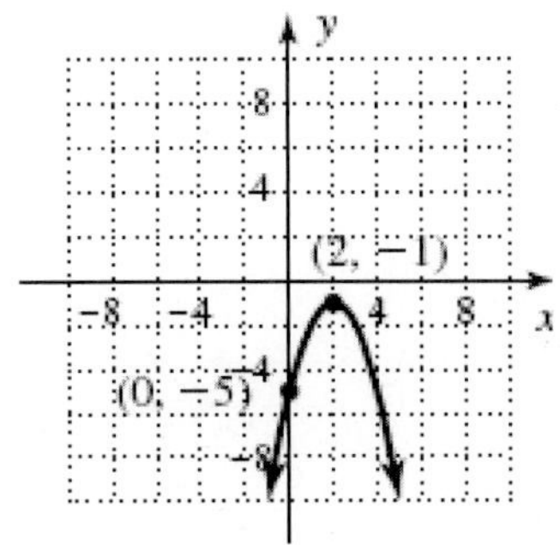

31. $f(x) = 4x^2 - 12x + 3$

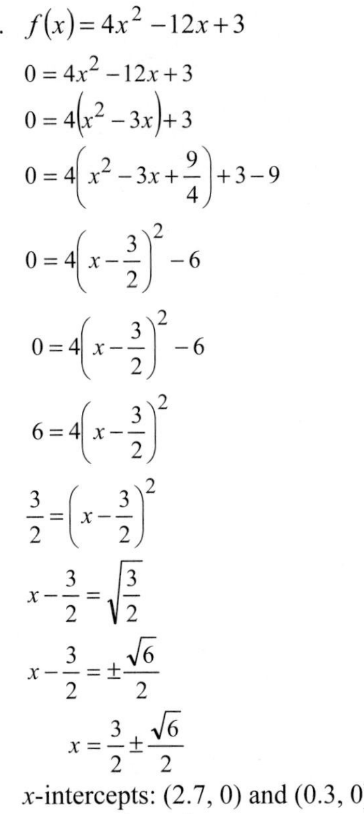

    $0 = 4x^2 - 12x + 3$

    $0 = 4(x^2 - 3x) + 3$

    $0 = 4\left(x^2 - 3x + \frac{9}{4}\right) + 3 - 9$

    $0 = 4\left(x - \frac{3}{2}\right)^2 - 6$

    $0 = 4\left(x - \frac{3}{2}\right)^2 - 6$

    $6 = 4\left(x - \frac{3}{2}\right)^2$

    $\frac{3}{2} = \left(x - \frac{3}{2}\right)^2$

    $x - \frac{3}{2} = \sqrt{\frac{3}{2}}$

    $x - \frac{3}{2} = \pm\frac{\sqrt{6}}{2}$

    $x = \frac{3}{2} \pm \frac{\sqrt{6}}{2}$

    x-intercepts: (2.7, 0) and (0.3, 0)

    Vertex: $\left(\frac{3}{2}, -6\right)$

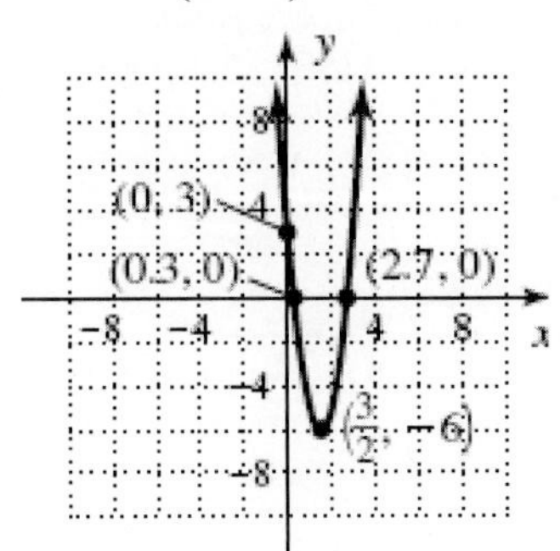

32. General form: $y = |x|$

    $y = -|x-2| + 6$

33. General form: $y = \sqrt{x}$
Node: (-4, -4)
$y = \frac{5}{2}\sqrt{x+4} - 4$

34. General form: $y = x^3$
Node: $(-3,-1)$
$y = \frac{1}{9}(x+3)^3 - 1$

35. $h(t) = -16t^2 + 96t + 2$

a. $h(0) = -16(0)^2 + 96(0) + 2 = 2$
2 feet

b. $h(2) = -16(2)^2 + 96(2) + 2$
$h(2) = -64 + 192 + 2$
$h(2) = 130$ feet

c. $130 = -16t^2 + 96t + 2$
$16t^2 - 96t + 128 = 0$
$16(t^2 - 6t + 8) = 0$
$16(t-4)(t-2) = 0$
$t = 4;\ t = 2$
4 seconds

d. $x = \frac{-b}{2a} = \frac{-96}{2(-16)} = 3$
$h(3) = -16(3)^2 + 96(3) + 2 = 146$
146 feet; 3 seconds

36. $f(x) = \frac{1}{x+2} - 1$
x-intercept: (-1, 0)
$0 = \frac{1}{x+2} - 1$
$1 = \frac{1}{x+2}$
$x + 2 = 1$
$x = -1$
y-intercept: $\left(0, -\frac{1}{2}\right)$
$y = \frac{1}{0+2} - 1$
$y = \frac{1}{2} - 1$
$y = -\frac{1}{2}$

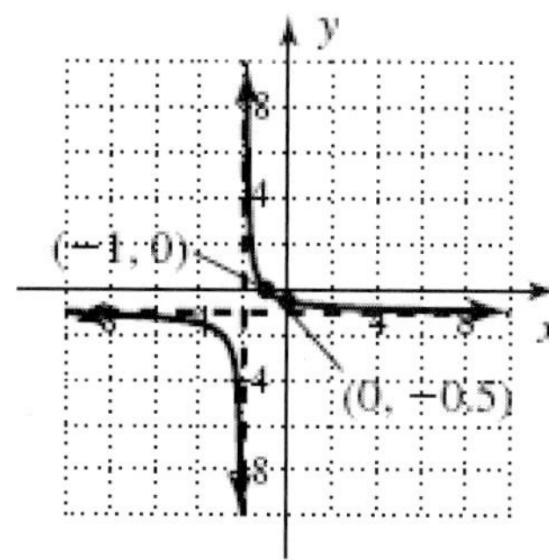

37. $h(x) = \frac{-1}{(x-2)^2} - 3$
x-intercept: None
y-intercept: $\left(0, -\frac{13}{4}\right)$
$y = \frac{-1}{(0-2)^2} - 3$
$y = \frac{-1}{4} - 3$
$y = -\frac{13}{4}$

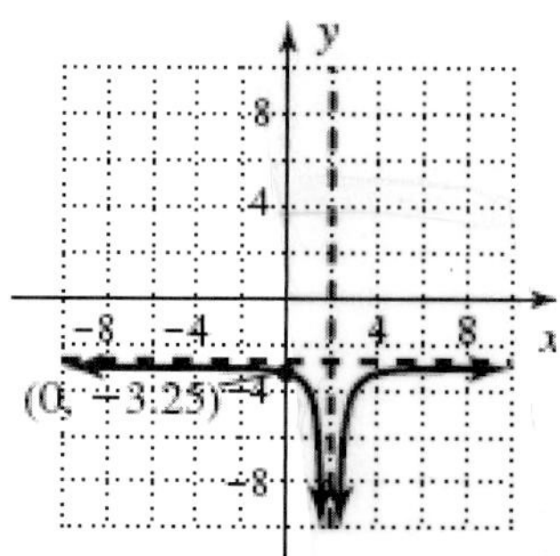

38. $C(p) = \frac{-7500}{p-100} - 75$

a. $C(30) = \frac{-7500}{30-100} - 75 = 32.143$
\$32,143
$C(50) = \frac{-7500}{50-100} - 75 = 75$
\$75,000
$C(70) = \frac{-7500}{70-100} - 75 = 175$
\$175,000
$C(90) = \frac{-7500}{90-100} - 75 = 675$
\$675,000; cost grows very fast

b.

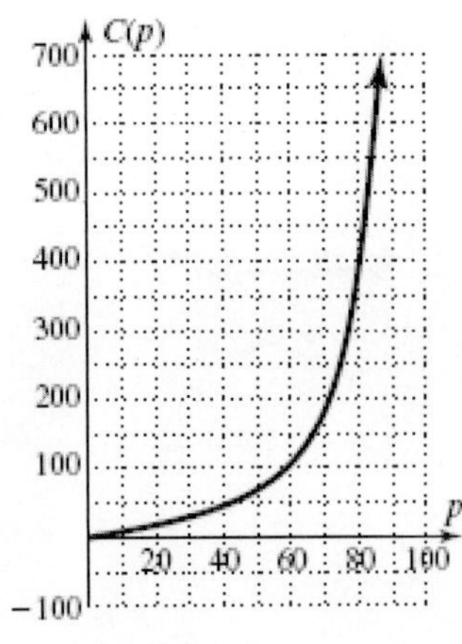

c. $p \to 100,\ C \to +\infty$

39. $y = k\sqrt[3]{x}$

$52.5 = k\sqrt[3]{27}$

$52.5 = 3k$

$17.5 = k$

$y = 17.5\sqrt[3]{x}$

x	$y = 17.5\sqrt[3]{x}$
216	$f(216) = 17.5\sqrt[3]{216} = 105$
0.343	$12.25 = 17.5\sqrt[3]{x}$ $0.7 = \sqrt[3]{x}$ $0.343 = x$
729	$f(729) = 17.5\sqrt[3]{729} = 157.5$

40. $z = \dfrac{kv}{w^2}$

$1.62 = \dfrac{k(144)}{(8)^2}$

$103.68 = 144k$

$0.72 = k$

$z = \dfrac{0.72v}{w^2}$

v	w	z
196	7	2.88
38.75	1.25	17.856
24	0.6	48

$17.856 = \dfrac{0.72v}{(1.25)^2}$

$2790 = 0.72v$

$38.75 = v;$

$48 = \dfrac{0.72(24)}{w^2}$

$48w^2 = 17.28$

$w^2 = 0.36$

$w = 0.6;$

$z = \dfrac{0.72(196)}{(7)^2} = \dfrac{141.12}{49} = 2.88$

41. $P = k\sqrt{l}$

$3 = k\sqrt{16}$

$\dfrac{3}{4} = k$

$P = \dfrac{3}{4}\sqrt{l}$

$P = \dfrac{3}{4}\sqrt{36}$

$P = \dfrac{3}{4}(6)$

$P = 4.5\text{ sec}$

42. a. $f(x) = \begin{cases} 5 & x \le -3 \\ -x+1 & -3 < x \le 3 \\ 3\sqrt{x-3}-1 & x > 3 \end{cases}$

$Y_1 = 5$; $Y_2 = -x+1$; $Y_3 = 3\sqrt{x-3}-1$

b. $R: y \in [-2, \infty)$

43. $h(x) = \begin{cases} \dfrac{x^2-2x-15}{x+3} & x \ne -3 \\ -6 & x = -3 \end{cases}$

$x \in (-\infty, \infty)$

$y \in (-\infty, -8) \cup (-8, \infty)$

Discontinuity at $x = -3$

Define $h(x) = -8$ at $x = -3$

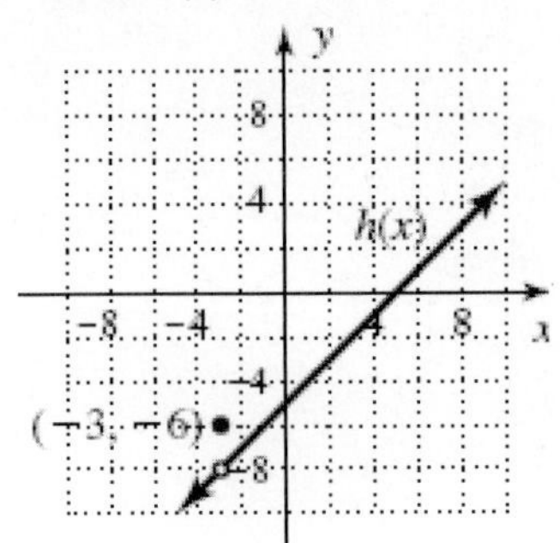

44. $p(x)=\begin{cases}-4 & x<-2\\ -|x|-2 & -2\le x<3\\ 3\sqrt{x}-9 & x\ge 3\end{cases}$

$p(-4)=-4$;

$p(-2)=-|-2|-2=-2-2=-4$;

$p(2.5)=-|2.5|-2=-2.5-2=-4.5$;

$p(2.99)=-|2.99|-2=-2.99-2=-4.99$;

$p(3)=3\sqrt{3}-9$;

$p(3.5)=3\sqrt{3.5}-9$

45. $q(x)=\begin{cases}2\sqrt{-x-3}-4 & x\le -3\\ -2|x|+2 & -3<x<3\\ 2\sqrt{x-3}-4 & x\ge 3\end{cases}$

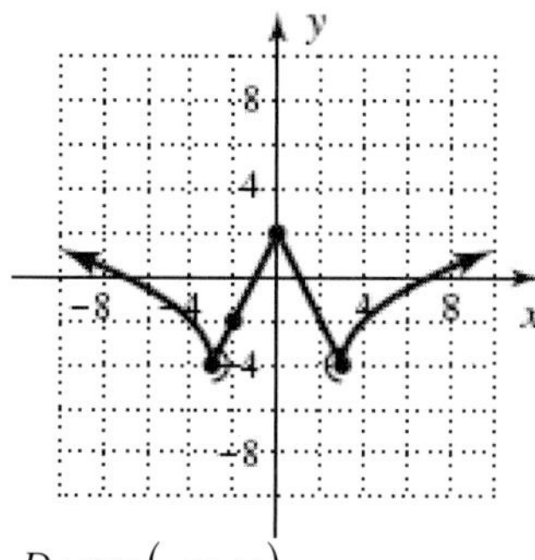

$D: x\in(-\infty,\infty)$

$R: y\in[-4,\infty)$

46. $T(x)=\begin{cases}11x^2-197.4x+1737.3 & 8\le x\le 11\\ 17x+708.67 & x>11\end{cases}$

47. $f(x)=(x-3)^2+2$

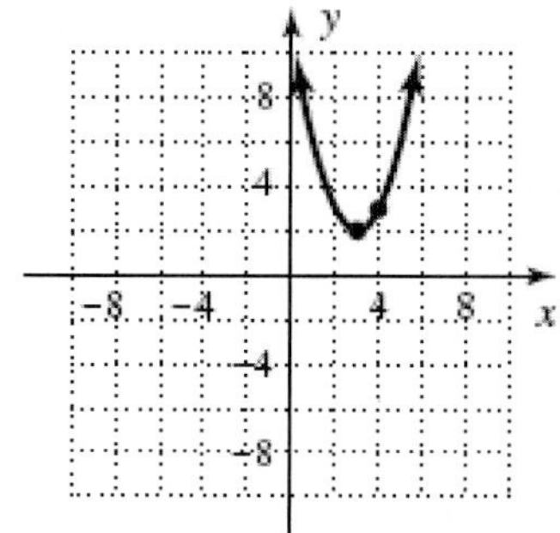

a. $D: x\in(-\infty,\infty)$

$R: y\in[2,\infty)$

b. $f(-3)=(-3-3)^2+2=(-6)^2+2=38$;

$f(-1)=(-1-3)^2+2=(-4)^2+2=18$;

$f(1)=(1-3)^2+2=(-2)^2+2=6$;

$f(2)=(2-3)^2+2=(-1)^2+2=3$;

$f(3)=(3-3)^2+2=(0)^2+2=2$

c. None

d. $f(x)<0$: None

$f(x)>0$: $x\in(-\infty,\infty)$

e. Minimum: (3, 2)

f. $f(x)\uparrow: x\in(3,\infty)$

$f(x)\downarrow: x\in(-\infty,3)$

48. $g(x)=-|x-2|+3$

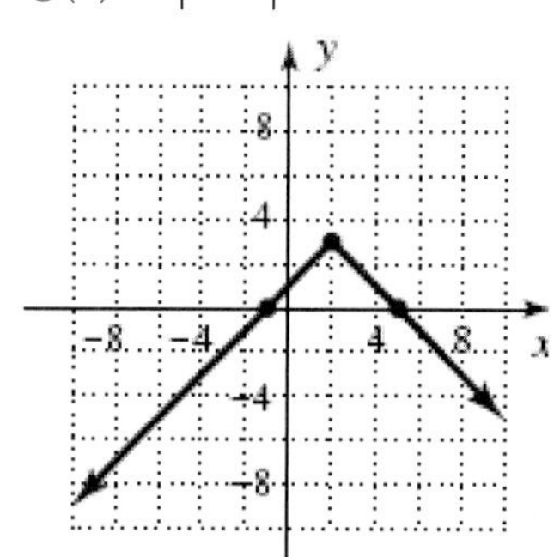

a. $D: x\in(-\infty,\infty)$

$R: y\in(-\infty,3]$

b. $f(-3)=-|-3-2|+3=-|-5|+3=-2$;

$f(-1)=-|-1-2|+3=-|-3|+3=0$;

$f(1)=-|1-2|+3=-|-1|+3=2$;

$f(2)=-|2-2|+3=-|0|+3=3$;

$f(3)=-|3-2|+3=-|1|+3=2$

c. (-1, 0), (5, 0)

d. $f(x)<0$: $x\in(-\infty,-1)\cup(5,\infty)$

$f(x)>0$: $x\in(-1,5)$

e. Maximum: (2, 3)

f. $f(x)\uparrow: x\in(-\infty,2)$

$f(x)\downarrow: x\in(2,\infty)$

49. $h(x)=\dfrac{1}{x+2}-3$

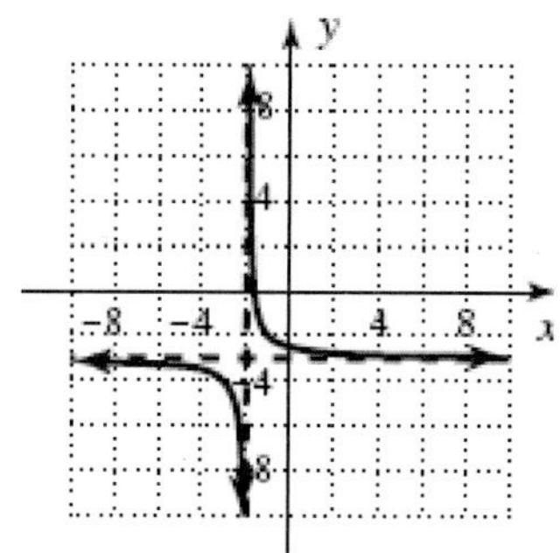

a. $D: x\in(-\infty,-2)\cup(-2,\infty)$

$R: y \in (-\infty,-3)\cup(-3,\infty)$

b. $f(-3)=\frac{1}{-3+2}-3=\frac{1}{-1}-3=-4$;

$f(-1)=\frac{1}{-1+2}-3=\frac{1}{1}-3=-2$;

$f(1)=\frac{1}{1+2}-3=\frac{1}{3}-3=-2.667$;

$f(2)=\frac{1}{2+2}-3=\frac{1}{4}-3=-2.75$;

$f(3)=\frac{1}{3+2}-3=\frac{1}{5}-3=-2.8$

c. $\left(-\frac{5}{3},0\right)$

d. $f(x)<0:\ x\in(-\infty,-2)\cup\left(-\frac{5}{3},\infty\right)$

$f(x)>0:\ x\in\left(-2,-\frac{5}{3}\right)$

e. None

f. $f(x)\uparrow$: None

$f(x)\downarrow: x\in(-\infty,-2)\cup(-2,\infty)$

50. $D: x\in(-\infty,\infty)$
$R: y\in[-5,\infty)$
$f(x)\uparrow: x\in(2,\infty)$
$f(x)\downarrow: x\in(-\infty,-2)$
$f(x)>0:\ x\in(-\infty,-1)\cup(5,\infty)$
$f(x)<0:\ x\in(-1,5)$

51. $D: x\in[-3,\infty)$
$R: y\in(-\infty,0)$
$f(x)\uparrow$: None
$f(x)\downarrow: x\in(-3,\infty)$
$f(x)>0$: None
$f(x)<0:\ x\in(-3,\infty)$

52. $D: x\in(-\infty,\infty)$
$R: y\in(-\infty,\infty)$
$f(x)\uparrow: x\in(-\infty,-3)\cup(1,\infty)$
$f(x)\downarrow: x\in(-3,1)$
$f(x)>0:\ x\in(-5,-1)\cup(4,\infty)$
$f(x)<0:\ x\in(-\infty,-5)\cup(-1,4)$

53. Zeroes: (-6, 0), (0, 0), (6, 0), (9, 0)
Minimum: (-3, -8), (7.5, -2)
Maximum: (-6, 0), (3, 4)

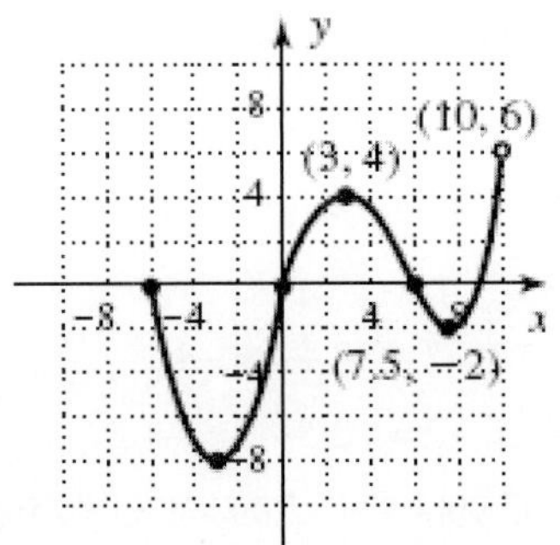

Chapter 3 Mixed Review

1. $y=\frac{k}{x^2}$

$\frac{1}{15}=\frac{k}{(9)^2}$

$\frac{1}{15}=\frac{k}{81}$

$\frac{81}{15}=k$

$5.4=k$;

$y=\frac{5.4}{x^2}$

x	$f(x)=\frac{5.4}{x^2}$
1	$f(x)=\frac{5.4}{x^2}=\frac{5.4}{1^2}=5.4$
3	$f(x)=\frac{5.4}{x^2}=\frac{5.4}{3^2}=0.6$
10	$f(x)=\frac{5.4}{x^2}=\frac{5.4}{10^2}=0.054$

3. $f(x)=\frac{1}{x-2}$ and $g(x)=x^2-2x$

$(f-g)(-1)=f(-1)-g(-1)$

$=\frac{1}{-1-2}-\left((-1)^2-2(-1)\right)$

$=\frac{1}{-3}-(3)$

$=-\frac{10}{3}$

5. $f(x)=\frac{1}{x-2}$ and $g(x)=x^2-2x$

$$\left(\frac{g}{f}\right)\left(\frac{1}{2}\right)=\frac{g\left(\frac{1}{2}\right)}{f\left(\frac{1}{2}\right)}$$

$$=\frac{\left(\frac{1}{2}\right)^2-2\left(\frac{1}{2}\right)}{\frac{1}{\frac{1}{2}-2}}=\frac{-\frac{3}{4}}{-\frac{2}{3}}=\frac{9}{8}$$

7. $f(x)=\dfrac{1}{x-2}$ and $g(x)=x^2-2x$

$(f\circ g)(x)=f[g(x)]$

$=\dfrac{1}{g(x)-2}$

$=\dfrac{1}{x^2-2x-2};$

$x=\dfrac{2\pm\sqrt{(-2)^2-4(1)(-2)}}{2(1)}$

$x=\dfrac{2\pm\sqrt{4+8}}{2}$

$x=\dfrac{2\pm\sqrt{12}}{2}$

$x=\dfrac{2\pm2\sqrt{3}}{2}$

$x=1\pm1\sqrt{3}$

$\{x|x\in R, x\neq1\pm1\sqrt{3}\}$

9. $h(x)=\sqrt[3]{x^2+5}$

$f(x)=\sqrt[3]{x};$

$g(x)=x^2+5$

11. $f(x)=5\sqrt[3]{x-8}-10$

Cube Root

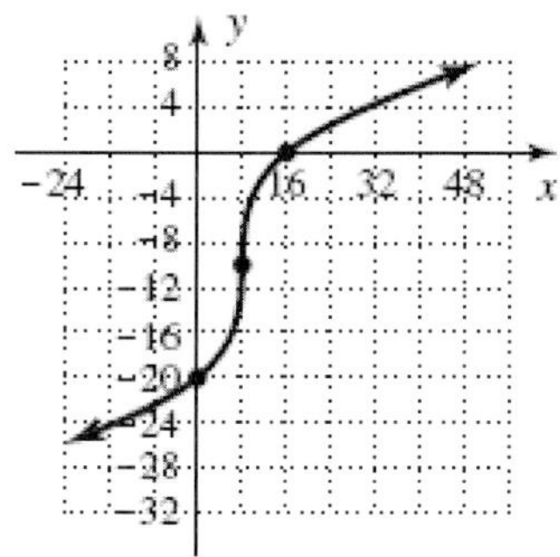

13. a. $D:x\in(-\infty,\infty)$

$R:y\in(-\infty,7]$

b. Min: (0, -2)

Max: (-3, 4) and (3, 7)

c. $f(x)\uparrow:\ x\in(-\infty,-3)\cup(0,3)$

$f(x)\downarrow:\ x\in(-3,0)\cup(3,\infty)$

d. $f(x)>0:\ x\in(-4,-1)\cup(1,5)$

$f(x)<0:\ x\in(-\infty,-4)\cup(-1,1)\cup(5,\infty)$

15. $p(x)=-x^2+10x-16$

$-(x^2-10x+16)=0$

$-(x-8)(x-2)=0$

$x=8;\quad x=2;$

x-intercepts: (8, 0), (2, 0)

$x=\dfrac{-b}{2a}=\dfrac{-10}{2(-1)}=5$

$y=-(5)^2+10(5)-16=9$

Vertex: (5, 9)

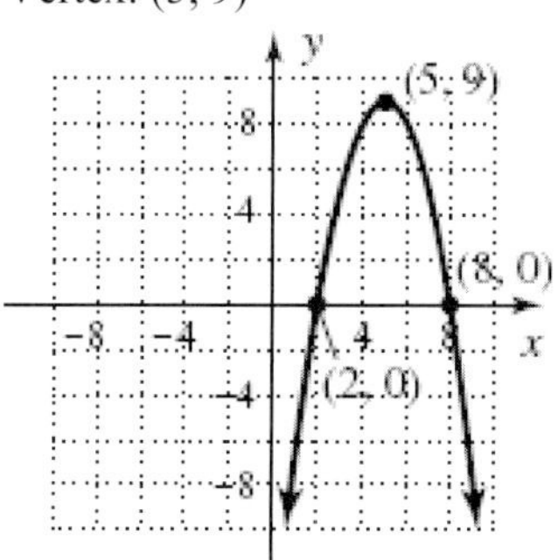

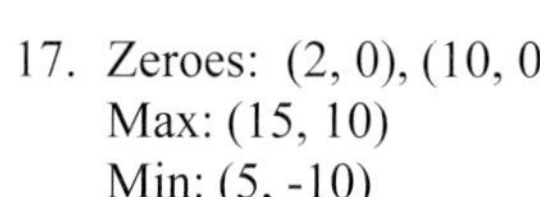

17. Zeroes: (2, 0), (10, 0)

Max: (15, 10)

Min: (5, -10)

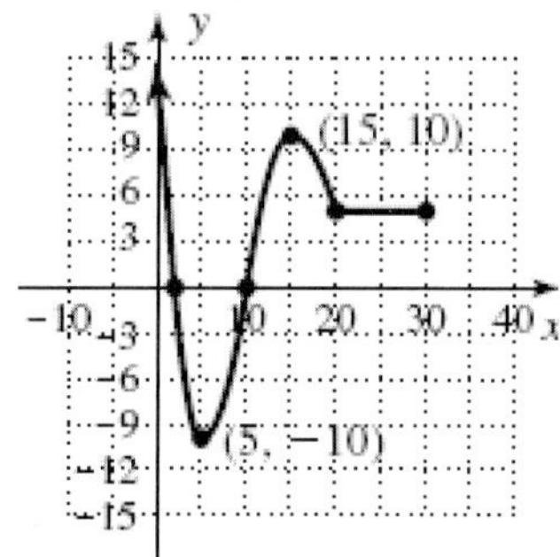

19. x-intercepts: (-1, 0), (1.5, 0)

y-intercept: (0, 3)

$f(x)=-(x+1)(x-1.5)$

$f(x)=-(2x^2-x-3)$

$f(x)=-2x^2+x+3$

Chapter 3 Practice Test

1. $f(x)=2x+1$ and $g(x)=x^2-3;\ x\ge 0$
$(f\cdot g)(3)=f(3)\cdot g(3)$
$=[2(3)+1]\cdot(3^2-3)$
$=7(6)$
$=42$

2. $f(x)=2x+1$ and $g(x)=x^2-3;\ x\ge 0$
$(g\circ f)(a)=g[f(a)]$
$f(a)=2a+1$
$g(f(a))=(f(a))^2-3$
$=(2a+1)^2-3$
$=4a^2+4a+1-3$
$=4a^2+4a-2$

3. $f(x)=2x+1$ and $g(x)=x^2-3;\ x\ge 0$
$\left(\frac{f}{g}\right)(x)=\frac{f(x)}{g(x)}$
$=\frac{2x+1}{x^2-3}$
$x^2-3=0$
$x^2=3$
$x=\pm\sqrt{3}$
$D: x\in\left[0,\sqrt{3}\right)\cup\left(\sqrt{3},\infty\right)$

4. $f(x)=2x+1$ and $g(x)=x^2-3;\ x\ge 0$
$f^{-1}(x)$:
$y=2x+1$
$x=2y+1$
$x-1=2y$
$\frac{x-1}{2}=y$
$f^{-1}(x)=\frac{x-1}{2};$
$g^{-1}(x)$:
$y=x^2-3$
$x=y^2-3$
$x+3=y^2$
$\sqrt{x+3}=y$
$g^{-1}(x)=\sqrt{x+3}$

5. $f(x)=|x-2|+3$
Right 2, up 3

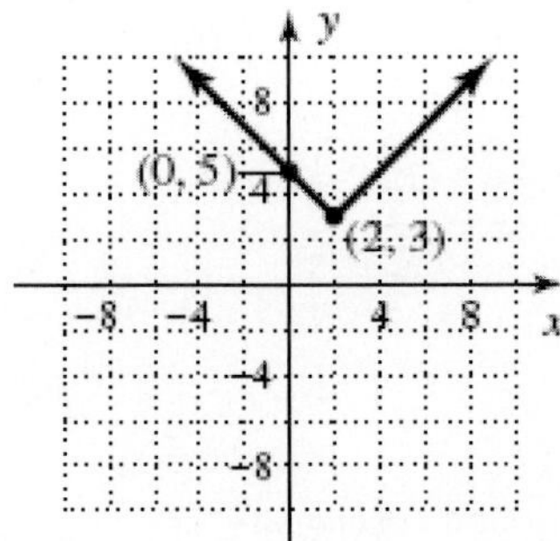

6. $g(x)=-(x+3)^2-2$
Left 3, down 2, reflected across x-axis

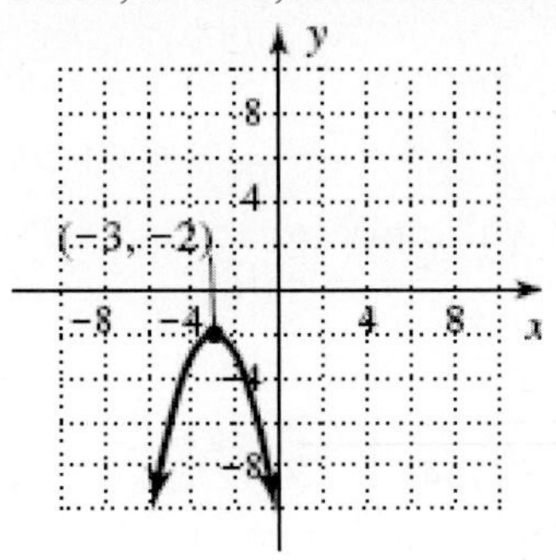

7. $f(x)=\frac{1}{x-2}+3$
Right 2, up 3

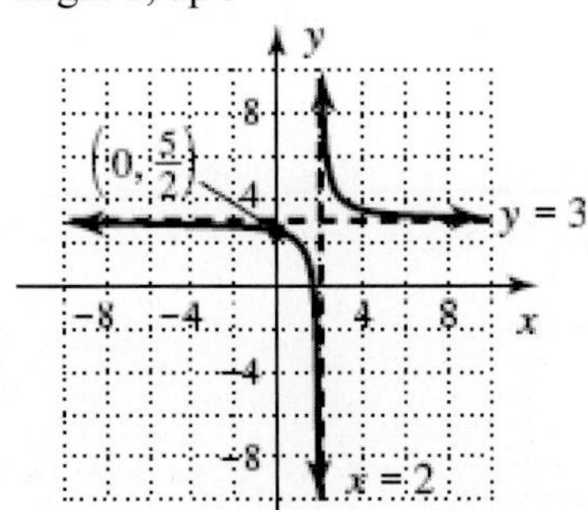

8. $g(x)=\frac{-1}{(x-3)^2}+1$
Right 3, up 1, reflected across x-axis

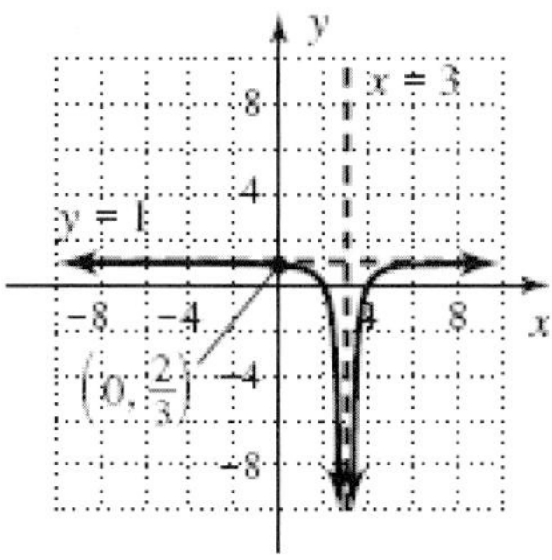

9. $f(x) = 2x^2 + 8x + 3$

$f(x) = 2(x^2 + 4x) + 3$

$f(x) = 2(x^2 + 4x + 4) + 3 - 8$

$f(x) = 2(x+2)^2 - 5$

Vertex: (-2, -5)
Axis of symmetry: x = -2
x-intercepts:

$$x = \frac{-8 \pm \sqrt{(8)^2 - 4(2)(3)}}{2(2)}$$

$$x = \frac{-8 \pm \sqrt{64 - 24}}{4}$$

$$x = \frac{-8 \pm \sqrt{40}}{4}$$

$$x = \frac{-8 \pm 2\sqrt{10}}{4}$$

$x = -0.4; \quad x = -3.6$

$(-0.4, 0), \quad (-3.6, 0)$

y-intercept: (0, 3)

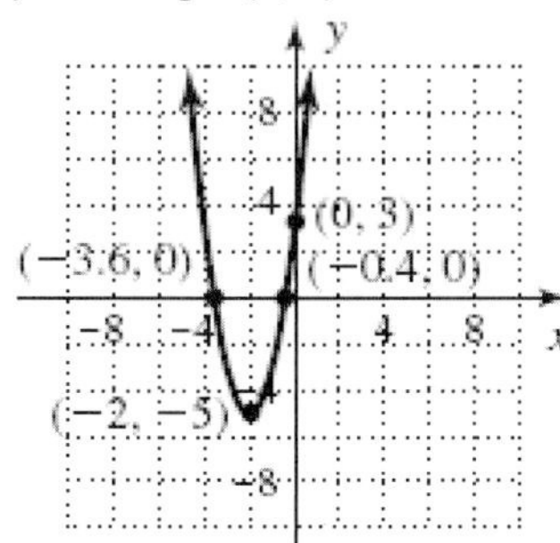

10. $f(x) = -x^2 - 4x$

$-x(x+4) = 0$

$x = 0; \quad x = -4$

x-intercepts: (0, 0), (-4, 0)
y-intercept: (0, 0)

$$x = \frac{-b}{2a} = \frac{4}{2(-1)} = -2$$

$y = -(-2)^2 - 4(-2) = 4$

Vertex: (-2, 4)
Axis of symmetry: x = -2

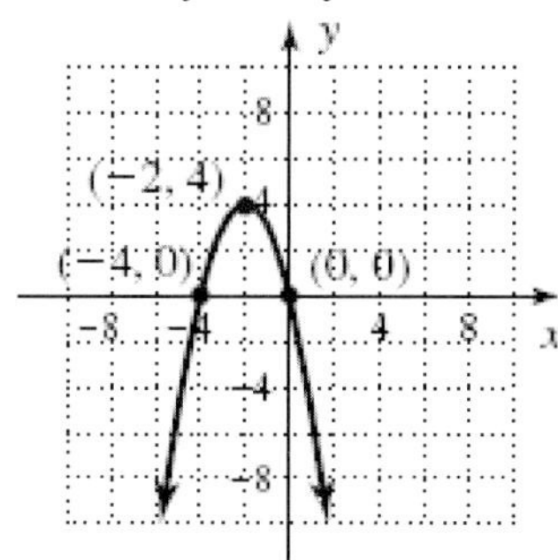

11. $f(x) = \begin{cases} 2x+3 & x < 0 \\ x^2 & 0 \le x < 2 \\ 1 & x > 2 \end{cases}$

$f(1) = (1)^2 = 1$;

$f(-3) = 2(-3) + 3 = -6 + 3 = -3$;

$f(5) = 1$

12. $h(x) = \begin{cases} 4 & x < -2 \\ 2x & -2 \le x \le 2 \\ x^2 & x > 2 \end{cases}$

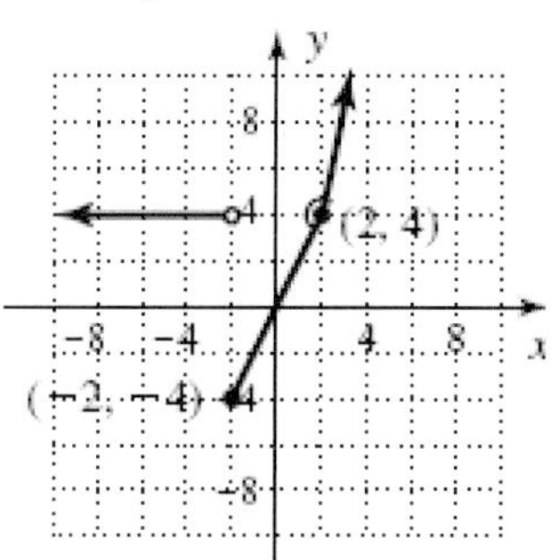

13. a. $D: x \in (-\infty, \infty)$

$R: y \in (-\infty, 4]$

b. $f(-1) = 4$

c. (-4, 0), (2, 0)

d. $f(x) > 0: \; x \in (-4, 2)$

$f(x) < 0: \; x \in (-\infty, -4) \cup (2, \infty)$

e. Min: None
Max: (-1, 4)

f. $f(x) \uparrow: x \in (-\infty, -1)$

$f(x) \downarrow: x \in (-1, \infty)$

g. $0 = a(-4+1)^2 + 4$

$$-4 = a(3)^2$$
$$\frac{-4}{9} = a;$$
$$f(x) = -\frac{4}{9}(x+1)^2 + 4$$

14. a. $D: x \in [-4, \infty)$
$R: y \in [-3, \infty)$
b. $f(-1) = 2.2$
c. (-3, 0)
d. $f(x) < 0: x \in (-4, -3)$
$f(x) > 0: x \in (-3, \infty)$
e. Min: (-4, -3)
f. $f(x)\uparrow: (-3, \infty)$
g. Parent graph: $y = \sqrt{x}$
Graph shifts left 4, down 3
$y = 3\sqrt{x+4} - 3$

15. $r(t) = \sqrt{t}$; $V(r) = \frac{4}{3}\pi r^3$
a. $(V \circ r)(t) = V[r(t)]$
$V(t) = \frac{4}{3}\pi\left(\sqrt{t}\right)^3$
b. $V(9) = \frac{4}{3}\pi\left(\sqrt{9}\right)^3$
$V(9) = \frac{4}{3}\pi(27)$
$V(9) = 36\pi \text{ in}^3$

16. a.

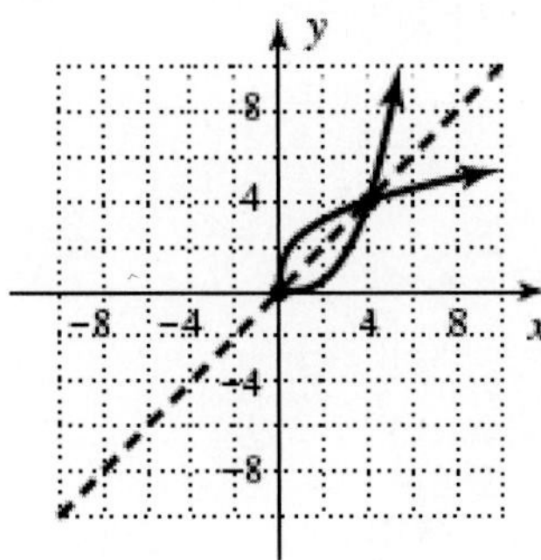

b.

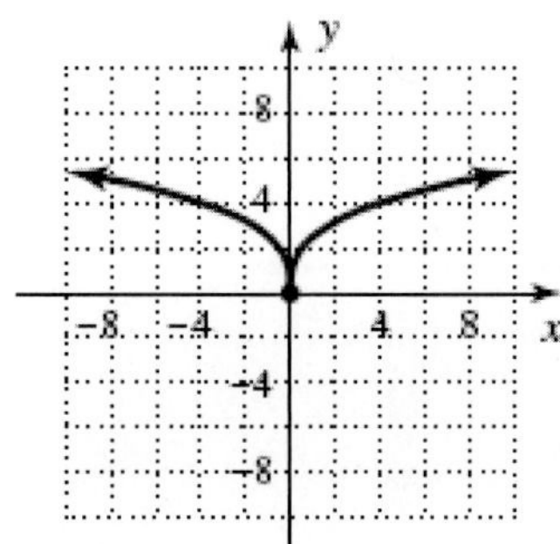

c.

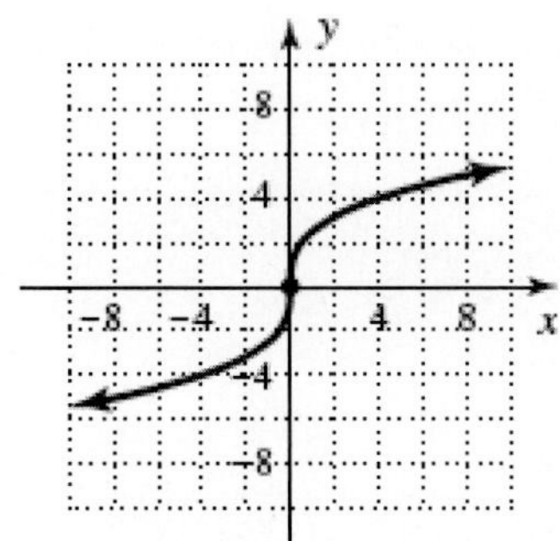

17. a. Yes, passes the vertical line test.
b. Yes, passes vertical and horizontal line tests.
c. Odd
d. (6, 4) and (-4, -2)

18. $f(x) = \frac{1}{x}$ and $g(x) = \frac{1}{x^2}$
a. $\frac{\Delta f}{\Delta x} = \frac{f(0.6) - f(0.5)}{0.6 - 0.5} = -3.333$;
$\frac{\Delta g}{\Delta x} = \frac{g(0.6) - g(0.5)}{0.6 - 0.5} = -12.222$
$g(x)$ decreases faster.
b. $\frac{\Delta f}{\Delta x} = \frac{f(1.6) - f(1.5)}{1.6 - 1.5} = -0.41666$;
$\frac{\Delta g}{\Delta x} = \frac{g(1.6) - g(1.5)}{1.6 - 1.5} = -0.53819$
$g(x)$ decreases faster

19. $h(t) = t^2 - 14t$
a. $h(4) = (4)^2 - 14(4) = -40$; 40 feet
$h(6) = (6)^2 - 14(6) = -48$; 48 feet
b. $x = \frac{14}{2(1)} = 7$
$h(7) = (7)^2 - 14(7) = -49$; 49 feet
c. 14 seconds; 7 second descent, 7 second ascent.

20. $M = \dfrac{kwh^2}{l}$

$$624 = \frac{k(3)(4)^2}{(10)}$$
$$6240 = 48k$$
$$130 = k;$$
$$M = \frac{130wh^2}{l}$$
$$M = \frac{130(3)(4)^2}{12}$$
$$M = 520 \text{ lbs}$$

Calculator Exploration and Discovery

Exercise 1:

a. Linear: $r \approx 0.99$; residuals form a quadratic pattern (not random);
Quadratic: $r \approx 0.997$; residuals appear random, we do not expect time to begin decreasing;
Power: $r \approx 0.999$; residuals appear random, context suggests a power function.

b. Using a grapher, 2.01 sec, 2.63 sec.

c. Using a grapher, 59 cm, 89 cm.

Strengthening Core Skills

Exercise 1: $f(x) = x^2 + 2x - 7$

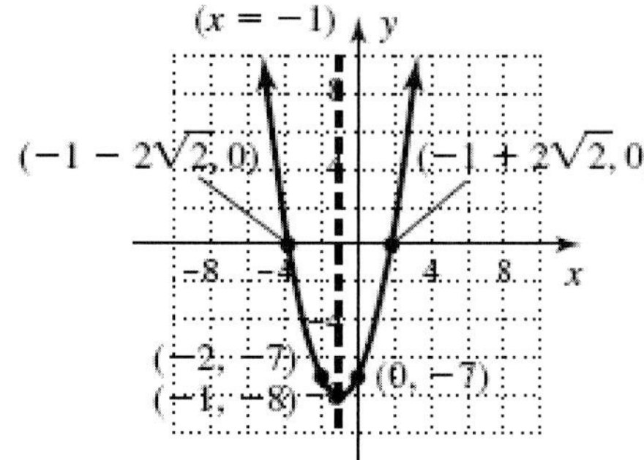

Exercise 2: $g(x) = x^2 + 5x + 9$

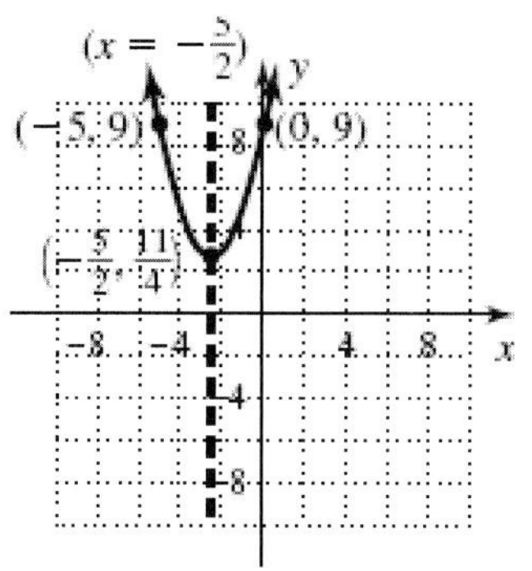

Exercise 3: $h(x) = x^2 - 6x + 11$

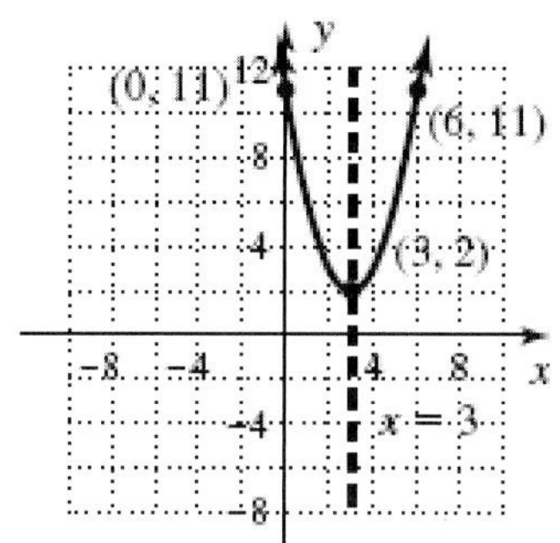

Exercise 4: $H(x) = -x^2 + 10x - 17$

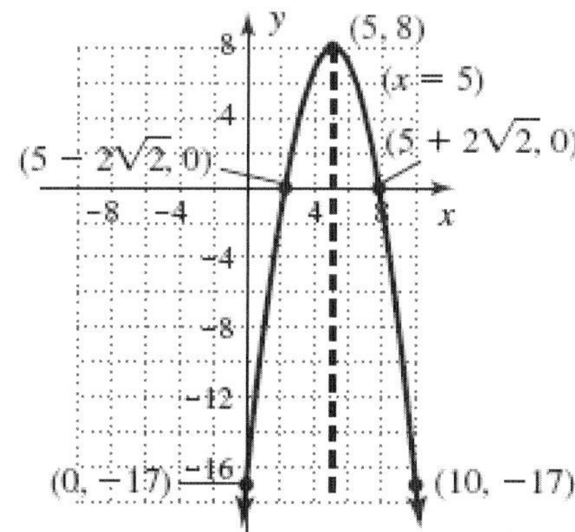

Exercise 5: $p(x) = 2x^2 + 12x + 21$

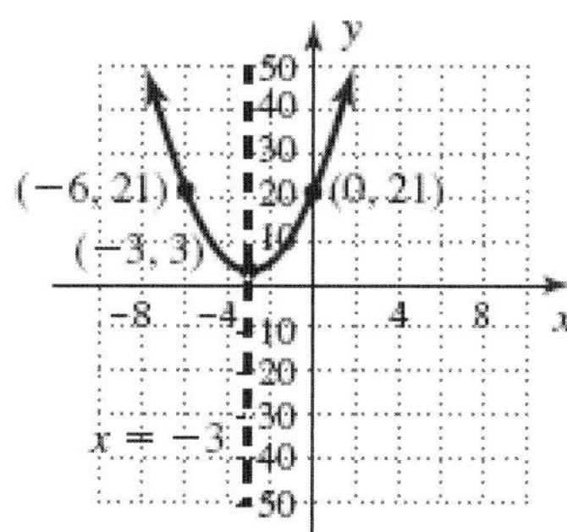

Exercise 6: $q(x) = 2x^2 - 7x + 8$

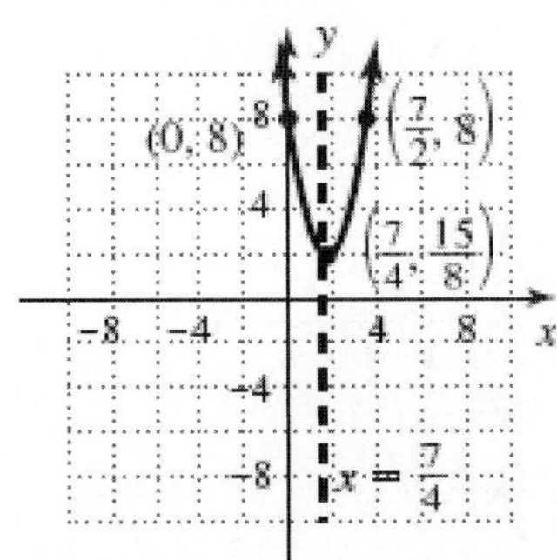

Chapters 1-3 Cumulative Review

1. $(x^3-5x^2+2x-10)\div(x-5)$

$=\dfrac{x^2(x-5)+2(x-5)}{x-5}$

$=\dfrac{(x-5)(x^2+2)}{x-5}$

$=x^2+2$

2. a. $\sqrt{18}+\sqrt{50}$

$=3\sqrt{2}+5\sqrt{2}$

$=8\sqrt{2}$

b. $\dfrac{2}{5y^2+11y+2}-\dfrac{5}{y^2-y-6}$

$=\dfrac{2}{(5y+1)(y+2)}-\dfrac{5}{(y-3)(y+2)}$

$=\dfrac{2(y-3)-5(5y+1)}{(5y+1)(y+2)(y-3)}$

$=\dfrac{2y-6-25y-5}{(5y+1)(y+2)(y-3)}$

$=\dfrac{-23y-11}{(5y+1)(y+2)(y-3)}$

3. $A=\pi r^2$

$69=\pi r^2$

$\dfrac{69}{\pi}=r^2$

$21.96=r^2$

$4.686=r;$

$C=2\pi r$

$C=2\pi(4.686)$

$C=29.45$ cm

4. $A=2\pi r^2+2\pi rh$

$2\pi r^2+2\pi rh-A=0$

$r=\dfrac{-2\pi h\pm\sqrt{(2\pi h)^2-4(2\pi)(-A)}}{2(2\pi)}$

$r=\dfrac{-2\pi h\pm\sqrt{4\pi^2h^2+4(2\pi A)}}{4\pi}$

$r=\dfrac{-2\pi h\pm\sqrt{4(\pi^2h^2+2\pi A)}}{4\pi}$

$r=\dfrac{-2\pi h\pm 2\sqrt{\pi^2h^2+2\pi A}}{4\pi}$

$r=\dfrac{-\pi h\pm\sqrt{\pi^2h^2+2\pi A}}{2\pi}$

5. $h(x)=2x^2+7x-5$

$x=\dfrac{-7\pm\sqrt{(7)^2-4(2)(-5)}}{2(2)}$

$x=\dfrac{-7\pm\sqrt{49+40}}{4}$

$x=\dfrac{-7\pm\sqrt{89}}{4}$

6. $\left(\dfrac{27}{8}\right)^{\frac{-2}{3}}=\left(\dfrac{8}{27}\right)^{\frac{2}{3}}=\left(\sqrt[3]{\dfrac{8}{27}}\right)^2=\left(\dfrac{2}{3}\right)^2=\dfrac{4}{9}$

7. a. (-4, 7) and (2, 5)

$m=\dfrac{7-5}{-4-2}=\dfrac{2}{-6}=-\dfrac{1}{3}$

b. $3x-5y=20$

$-5y=-3x+20$

$y=\dfrac{3}{5}x-4$

$m=\dfrac{3}{5}$

8. a. $f(x)=\sqrt{x-2}+3$

Right 2, up 3, reflected across x-axis

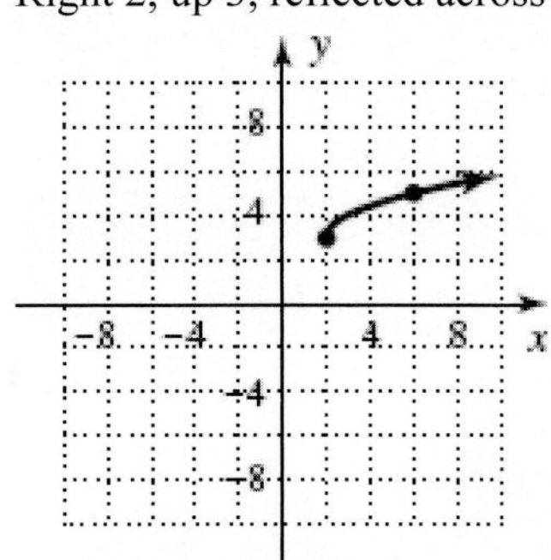

b. $f(x) = -|x+2| - 3$
Left 2, down 3

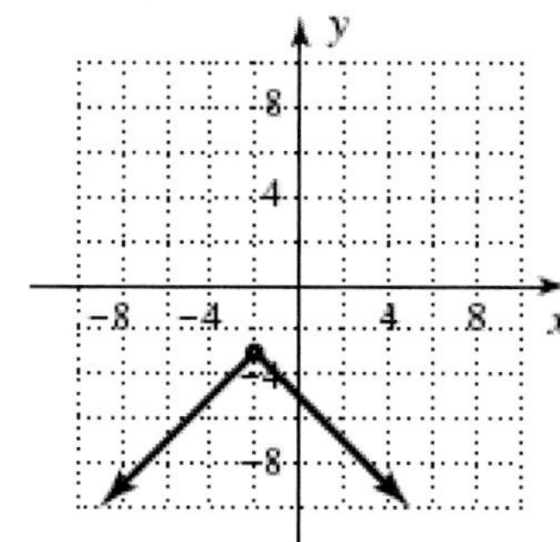

9. (-3, 2); $m = \frac{1}{2}$

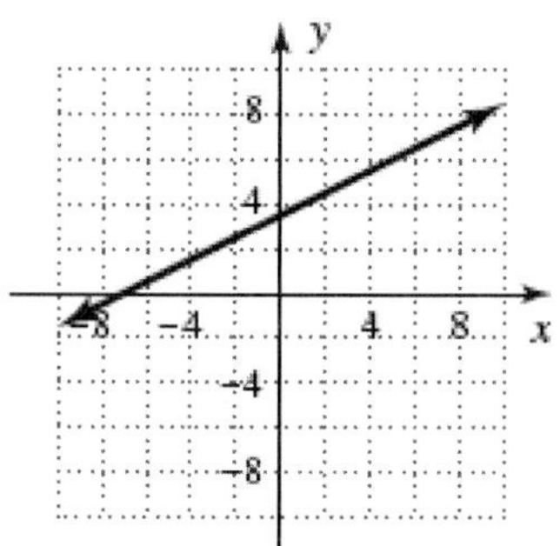

$$y - 2 = \frac{1}{2}(x+3)$$
$$y - 2 = \frac{1}{2}x + \frac{3}{2}$$
$$y = \frac{1}{2}x + \frac{7}{2}$$

10. $x^2 + 4x + 13 = 0$; $x = 2 + 3i$

$$(2+3i)^2 + 4(2+3i) + 13 = 0$$
$$4 + 12i + 9i^2 + 8 + 12i + 13 = 0$$
$$4 + 12i - 9 + 8 + 12i + 13 = 0$$
$$16 + 24i = 0$$

No

11. $g(x) = x^2 - 4x - 5$

$$g(x) = (x^2 - 4x + 4) - 5 - 4$$
$$g(x) = (x-2)^2 - 9$$

Vertex: (2, -9)

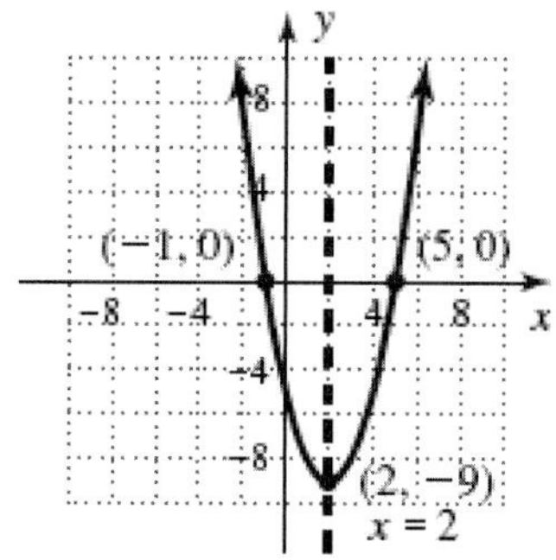

12. $x^2 - 7x + 6 \le 0$

$$(x-6)(x-1) = 0$$
$$x = 6; \quad x = 1$$

[1, 6]

13. $f(x) = 3x^2 - 6x$ and $g(x) = x - 2$

$$(f \cdot g)(x) = f(x) \cdot g(x)$$
$$= (3x^2 - 6x)(x-2)$$
$$= 3x^3 - 6x^2 - 6x^2 + 12x$$
$$= 3x^3 - 12x^2 + 12x$$
$$(f \div g)(x) = \frac{f(x)}{g(x)}$$
$$= \frac{3x^2 - 6x}{x-2}$$
$$= \frac{3x(x-2)}{x-2}$$
$$= 3x; \quad x \ne 2;$$
$$(g \circ f)(-2) = g[f(-2)];$$
$$f(-2) = 3(-2)^2 - 6(-2) = 24;$$
$$g(24) = 24 - 2 = 22$$

14. $f(x) = \frac{3}{5}x - 4$

$$y = \frac{3}{5}x - 4$$
$$x = \frac{3}{5}y - 4$$
$$x + 4 = \frac{3}{5}y$$
$$\frac{5}{3}(x+4) = y$$
$$\frac{5x+20}{3} = y$$
$$f^{-1}(x) = \frac{5x+20}{3}$$

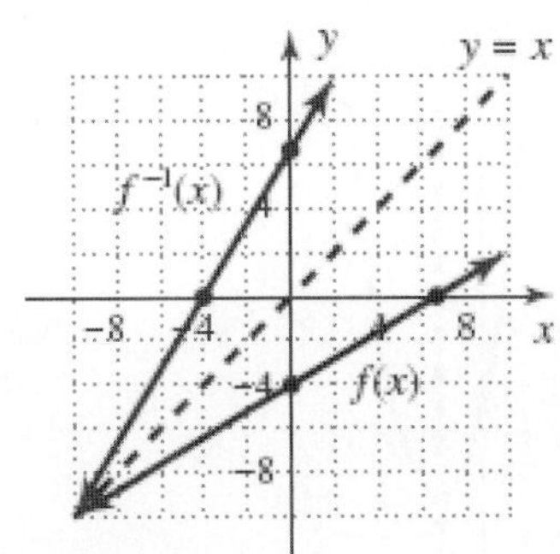

15. a. $g(x)=\frac{1}{x-2}+1$

x-intercept: (1, 0)

$$0=\frac{1}{x-2}+1$$
$$-1=\frac{1}{x-2}$$
$$-x+2=1$$
$$-x=-1$$
$$x=1;$$

y-intercept: $\left(0,\frac{1}{2}\right)$

$$y=\frac{1}{0-2}+1$$
$$y=-\frac{1}{2}+1$$
$$y=\frac{1}{2}$$

b.

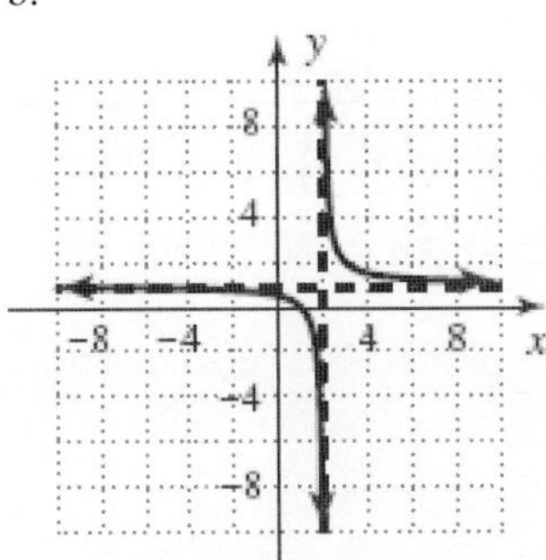

16. $f(x)=\begin{cases}x^2-4 & x<2\\ x-1 & 2\le x\le 8\end{cases}$

a. $D: x\in(-\infty,8]$
$R: y\in[-4,\infty)$

b. $f(-3)=(-3)^2-4=9-4=5$;
$f(-1)=(-1)^2-4=1-4=-3$;
$f(1)=(1)^2-4=1-4=-3$;
$f(2)=2-1=1$;
$f(3)=3-1=2$

c. (-2, 0)

d. $f(x)<0: x\in(-2,2)$
$f(x)>0: x\in(-\infty,-2),\ [2,8]$

e. Max: (8, 7)
Min: (0, -4)

f. $f(x)\uparrow: x\in\ (0,8)$
$f(x)\downarrow: x\in\ (-\infty,0)$

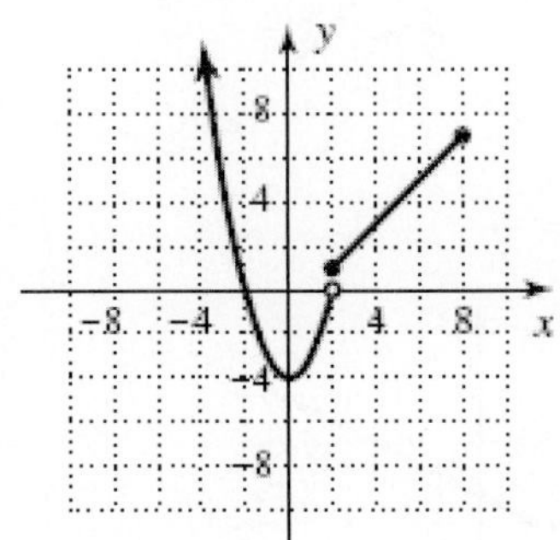

17. $f(x)=x^2$ and $g(x)=x^3$

a. $\frac{\Delta f}{\Delta x}=\frac{f(0.6)-f(0.5)}{0.6-0.5}=1.1$;
$\frac{\Delta g}{\Delta x}=\frac{g(0.6)-g(0.5)}{0.6-0.5}=0.91$;
$f(x)$ increases faster.

b. $\frac{\Delta f}{\Delta x}=\frac{f(1.6)-f(1.5)}{1.6-1.5}=3.1$;
$\frac{\Delta g}{\Delta x}=\frac{g(1.6)-g(1.5)}{1.6-1.5}=7.21$;
$g(x)$ increases faster

18. $C=kld$
$$4.98=k(3)(6)$$
$$4.98=18k$$
$$0.27666=k$$
$$C=0.27666ld$$
$$C=0.27666(2)(1.5)$$
$$C=\$0.83$$

19. $y=0.42x+0.81$
$0.42(100)+0.81=42+0.81=42.81$
≈ 43 psi

20. No; Raphael is grouped with The School of Athens and Parnassus.

4.1 Technology Highlight

1. b and c

4.1 Exercises

1. x^3+2x^2+3x-4

3. x^2+3x+6

5. $x^3+2x^2+3x-4=\left(x^2+3x+6\right)(x-1)+2$

7. $x-7$

$$\begin{array}{r} x-7 \\ x+5\overline{)x^2-2x-35} \\ -\left(x^2+5x\right) \\ \hline -7x-35 \\ -(7x-35) \\ \hline 0 \end{array}$$

9. $2r+3$

$$\begin{array}{r} 2r+3 \\ 3r-5\overline{)6r^2-r-15} \\ -\left(6r^2-10r\right) \\ \hline 9r-15 \\ -(9r-15) \\ \hline 0 \end{array}$$

11. $x^2-3x-10$

$$\begin{array}{r} x^2-3x-10 \\ x-2\overline{)x^3-5x^2-4x+20} \\ -\left(x^3-2x^2\right) \\ \hline -3x^2-4x \\ -\left(-3x^2+6x\right) \\ \hline -10x+20 \\ -(-10x+20) \\ \hline 0 \end{array}$$

13. $4n^2-2n+1$

$$\begin{array}{r} 4n^2-2n+1 \\ 2n+1\overline{)8n^3+0n^2+0n+1} \\ -\left(8n^3+4n^2\right) \\ \hline -4n^2+0n \\ -\left(-4n^2-2n\right) \\ \hline 2n+1 \\ -(2n+1) \\ \hline 0 \end{array}$$

15. $3b^2-3b-2$

$$\begin{array}{r} 3b^2-3b-2 \\ b+1\overline{)3b^3+0b^2-5b-2} \\ -\left(3b^3+3b^2\right) \\ \hline -3b^2-5b \\ -\left(-3b^2-3b\right) \\ \hline -2b-2 \\ -(-2b-2) \\ \hline 0 \end{array}$$

17. (1) $9b^2-24b+16=(3b-4)(3b-4)$;

(2) $\dfrac{9b^2-24b+16}{3b-4}=3b-4$;

$$\begin{array}{r} 3b-4 \\ 3b-4\overline{)9b^2-24b+16} \\ -\left(9b^2-12b\right) \\ \hline -12b+16 \\ -(-12b+16) \\ \hline 0 \end{array}$$

19. (1) $2n^3-n^2-19n+4$
$=\left(2n^2-7n+2\right)(n+3)-2$;

(2) $\dfrac{2n^3-n^2-19n+4}{n+3}$
$=2n^2-7n+2+\dfrac{-2}{n+3}$;

$$\begin{array}{r} 2n^2-7n+2 \\ n+3\overline{)2n^3-n^2-19n+4} \\ -\left(2n^3+6n^2\right) \\ \hline -7n^2-19n \\ -\left(-7n^2-21n\right) \\ \hline 2n+4 \\ -(2n+6) \\ \hline -2 \end{array}$$

21. (1) $g^4-15g^2+10g+24$
$=\left(g^3-4g^2+g+6\right)(g+4)$;

(2) $\dfrac{g^4-15g^2+10g+24}{g+4}$
$=g^3-4g^2+g+6$

$$
\begin{array}{r}
g^3-4g^2+g+6 \\
g+4\overline{)g^4+0g^3-15g^2+10g+24} \\
-\left(g^4+4g^3\right) \\
\hline
-4g^3-15g^2 \\
-\left(-4g^3-16g^2\right) \\
\hline
g^2+10g \\
-\left(g^2+4g\right) \\
\hline
6g+24 \\
-\left(6g+24\right) \\
\hline
0
\end{array}
$$

23. (1) $x^4-16x^2-5x-24$
$=\left(x^3-4x^2-5\right)(x+4)-4;$

(2) $\dfrac{x^4-16x^2-5x-24}{x+4}$
$=x^3-4x^2-5-\dfrac{4}{x+4}$

$$
\begin{array}{r}
x^3-4x^2\qquad-5 \\
x+4\overline{)x^4+0x^3-16x^2-5x-24} \\
-\left(x^4+4x^3\right) \\
\hline
-4x^3-16x^2 \\
-\left(-4x^3-16x^2\right) \\
\hline
-5x-24 \\
-\left(-5x-20\right) \\
\hline
-4
\end{array}
$$

25. $(x+2)$ is a factor of x^3+5x^2-x-14
verified

$$
\begin{array}{r|rrrr}
-2 & 1 & 5 & -1 & -14 \\
 & & -2 & -6 & 14 \\
\hline
 & 1 & 3 & -7 & \boxed{0}
\end{array}
$$

27. $\dfrac{x^3+12x^2+34x-7}{x+7}=x^2+5x-1$
verified

$$
\begin{array}{r|rrrr}
-7 & 1 & 12 & 34 & -7 \\
 & & -7 & -35 & 7 \\
\hline
 & 1 & 5 & -1 & \boxed{0}
\end{array}
$$

29. $x^3+3x^2-8x-13$
$=\left(x^2+2x-10\right)(x+1)-3;$
$\dfrac{x^3+3x^2-8x-13}{x+1}=x^2+2x-10+\dfrac{-3}{x+1}$

$$
\begin{array}{r|rrrr}
-1 & 1 & 3 & -8 & -13 \\
 & & -1 & -2 & 10 \\
\hline
 & 1 & 2 & -10 & \boxed{-3}
\end{array}
$$

31. $x^3-15x+12$
$=\left(x^2+3x-6\right)(x-3)-6;$
$\dfrac{x^3-15x+12}{x-3}=x^2+3x-6+\dfrac{-6}{x-3}$

$$
\begin{array}{r|rrrr}
3 & 1 & 0 & -15 & 12 \\
 & & 3 & 9 & -18 \\
\hline
 & 1 & 3 & -6 & \boxed{-6}
\end{array}
$$

33. $x^2-2x-63$
$=(x-9)(x+7)$

35. x^3+4x^2+x-6
$=(x+2)\left(x^2+2x-3\right)$

$$
\begin{array}{r|rrrr}
-2 & 1 & 4 & 1 & -6 \\
 & & -2 & -4 & 6 \\
\hline
 & 1 & 2 & -3 & \boxed{0}
\end{array}
$$

$=(x+2)(x-1)(x+3)$

37. $x^3+3x^2-16x+12$
$=(x-1)\left(x^2+4x-12\right)$

$$
\begin{array}{r|rrrr}
1 & 1 & 3 & -16 & 12 \\
 & & 1 & 4 & -12 \\
\hline
 & 1 & 4 & -12 & \boxed{0}
\end{array}
$$

$=(x-1)(x-2)(x+6)$

39. x^3-7x+6
$=(x+3)\left(x^2-3x+2\right)$

$$
\begin{array}{r|rrrr}
-3 & 1 & 0 & -7 & 6 \\
 & & -3 & 9 & -6 \\
\hline
 & 1 & -3 & 2 & \boxed{0}
\end{array}
$$

$=(x+3)(x-1)(x-2)$

41. x^3-7x^2+36
$=(x+2)\left(x^2-9x+18\right)$

$$
\begin{array}{r|rrrr}
-2 & 1 & -7 & 0 & 36 \\
 & & -2 & 18 & -36 \\
\hline
 & 1 & -9 & 18 & \boxed{0}
\end{array}
$$

$=(x+2)(x-3)(x-6)$

43. $x^3+3x^2-34x-120$
$=(x+5)\left(x^2-2x-24\right)$

$$
\begin{array}{r|rrrr}
-5 & 1 & 3 & -34 & -120 \\
 & & -5 & 10 & 120 \\
\hline
 & 1 & -2 & -24 & \boxed{0}
\end{array}
$$

$=(x+5)(x+4)(x-6)$

45. $x^4 - x^3 - 7x^2 + x + 6$

-2	1	-1	-7	1	6
		-2	6	2	-6
	1	-3	-1	3	0

$= (x+2)(x^3 - 3x^2 - x + 3)$
$= (x+2)(x^2(x-3) - 1(x-3))$
$= (x+2)(x-3)(x^2-1)$
$(x+2)(x-3)(x+1)(x-1)$

47. $A = \frac{h(B+b)}{2}$

$B = \frac{2A - hb}{h} = \frac{2A}{h} - \frac{hb}{h} = \frac{2A}{h} - b$

verified

49. $(x^2 - kx - 27)$; $(x-3)$

3	1	-k	-27
		3	9 - 3k
	1	3 - k	0

$9 - 3k - 27 = 0$
$k = -6$

51. $(5x^2 + 2x + k) \div (x+2)$

-2	5	2	k
		-10	16
	5	-8	0

$k + 16 = 0$
$k = -16$

53. $(x^3 + kx^2 - 7x + 15)$; $(x+3)$

-3	1	k	-7	15
		-3	-3k + 9	9k - 6
	1	k - 3	-3k + 2	0

$9k - 6 + 15 = 0$
$9k = -9$
$k = -1$

55. $(x^3 + 5x^2 + 2x + k) \div (x+2)$

-2	1	5	2	k
		-2	-6	8
	1	3	-4	0

$k + 8 = 0$
$k = -8$

57. $5x^2 + 228x + 756$

a. width: $x + 42$
$5x + 18$

-42	5	228	756
		-210	-756
	5	18	0

b. $\frac{\text{length}}{\text{width}} = \frac{9}{5}$

$\frac{9}{5} = \frac{5x+18}{x+42}$
$9(x+42) = 5(5x+18)$
$9x + 378 = 25x + 90$
$288 = 16x$
$18 = x$

c. 18+42 = 60 inches;
$5(18) + 18 = 108$ inches
60 x 108

59. Answers will vary.

61. $(n^3 + 3n^2 - 3kn - 10) \div (n+2)$

-2	1	3	-3k	-10
		-2	-2	6k + 4
	1	1	-3k - 2	-5

$6k + 4 - 10 = -5$
$6k = 1$
$k = \frac{1}{6}$

63. $x^3 - kx - 12$; $x - k$

k	1	0	k	-12
		k	k^2	$k^3 + k^2$
	1	k	$k^2 + k$	0

$k^3 + k^2 - 12 = 0$
$k = 2$

65. $\frac{6.48 \times 10^6}{1.62 \times 10^{-2}} = \frac{6.48}{1.62} \times \frac{10^6}{10^{-2}} = 4 \times 10^8$

67. $\frac{12}{(x-4)^2} - 3 = 0$

$(x-4)^2\left[\frac{12}{(x-4)^2} - 3 = 0\right]$

$12 - 3(x^2 - 8x + 16) = 0$
$12 - 3x^2 + 24x - 48 = 0$
$0 = 3x^2 - 24x + 36$
$0 = 3(x^2 - 8x + 12)$
$0 = 3(x-6)(x-2)$
$x - 6 = 0$ or $x - 2 = 0$
$x = 6$ or $x = 2$

69. a. $f(x)=\sqrt{x^2-3x-28}$

$x^2-3x-28\geq 0$

Critical Values: $(x-7)(x+4)=0$

$x=7$ or $x=-4$

pos neg pos

-4 7

$x\in(-\infty,-4]\cup[7,\infty)$

b. $g(x)=\sqrt{2x+3}$

$2x+3\geq 0$

$2x\geq -3$

$x\geq -\frac{3}{2}$

$x\in\left[-\frac{3}{2},\infty\right)$

4.2 Technology Highlight

1. -2, 3, $2+\sqrt{3}\,i$, $2-\sqrt{3}\,i$

4.2 Exercises

1. Linear, P(*k*), Remainder

3. a – b*i*, complex conjugate

5. Answers will vary.

7. $P(x)=x^3+2x^2-5x-6;\ x=-3$

verified

$$\begin{array}{r|rrrr} -3 & 1 & 2 & -5 & -6 \\ & & -3 & 3 & 6 \\ \hline & 1 & -1 & -2 & \boxed{0} \end{array}$$

9. $P(x)=x^3-7x+6;\ x=2$

verified

$$\begin{array}{r|rrrr} 2 & 1 & 0 & -7 & 6 \\ & & 2 & 4 & -6 \\ \hline & 1 & 2 & -3 & \boxed{0} \end{array}$$

11. $P(x)=x^3-6x^2+32;\ x=4$

verified

$$\begin{array}{r|rrrr} 4 & 1 & -6 & 0 & 32 \\ & & 4 & -8 & -32 \\ \hline & 1 & -2 & -8 & \boxed{0} \end{array}$$

13. $P(x)=x^3-6x^2+5x+12$

a. $P(-2)=-30$

$$\begin{array}{r|rrrr} -2 & 1 & -6 & 5 & 12 \\ & & -2 & 16 & -42 \\ \hline & 1 & -8 & 21 & \boxed{-30} \end{array}$$

b. $P(5)=12$

$$\begin{array}{r|rrrr} 5 & 1 & -6 & 5 & 12 \\ & & 5 & -5 & 0 \\ \hline & 1 & -1 & 0 & \boxed{12} \end{array}$$

15. $P(x)=2x^3-x^2-19x+4$

a. $P(-3)=-2$

$$\begin{array}{r|rrrr} -3 & 2 & -1 & -19 & 4 \\ & & -6 & 21 & -6 \\ \hline & 2 & -7 & 2 & \boxed{-2} \end{array}$$

b. $P(2)=-22$

$$\begin{array}{r|rrrr} 2 & 2 & -1 & -19 & 4 \\ & & 4 & 6 & -26 \\ \hline & 2 & 3 & -13 & \boxed{-22} \end{array}$$

17. $P(x)=x^4-4x^2+x+1$

a. $P(-2)=-1$

$$\begin{array}{r|rrrrr} -2 & 1 & 0 & -4 & 1 & 1 \\ & & -2 & 4 & 0 & -2 \\ \hline & 1 & -2 & 0 & 1 & \boxed{-1} \end{array}$$

b. $P(2)=3$

$$\begin{array}{r|rrrrr} 2 & 1 & 0 & -4 & 1 & 1 \\ & & 2 & 4 & 0 & 2 \\ \hline & 1 & 2 & 0 & 1 & \boxed{3} \end{array}$$

19. $P(x)=2x^3-6x^2-7x+21$

a. $P(-2)=-5$

$$\begin{array}{r|rrrr} -2 & 2 & -6 & -7 & 21 \\ & & -4 & 20 & -26 \\ \hline & 2 & -10 & 13 & \boxed{-5} \end{array}$$

b. $P(3)=0$

$$\begin{array}{r|rrrr} 3 & 2 & -6 & -7 & 21 \\ & & 6 & 0 & -21 \\ \hline & 2 & 0 & -7 & \boxed{0} \end{array}$$

21. $Px=2x^3+3x^2-9x-10$

a. $P\left(\frac{3}{2}\right)=-10$

$$\begin{array}{r|rrrr} \frac{3}{2} & 2 & 3 & -9 & -10 \\ & & 3 & 9 & 0 \\ \hline & 2 & 6 & 0 & \boxed{-10} \end{array}$$

b. $P\left(-\frac{5}{2}\right)=0$

$$\begin{array}{r|rrrr} -\frac{5}{2} & 2 & 3 & -9 & -10 \\ & & -5 & 5 & 10 \\ \hline & 2 & -2 & -4 & \underline{|0} \end{array}$$

23. $P(x)=x^3-2x^2+3x-2$

a. $P\left(\frac{1}{2}\right)=\frac{-7}{8}$

$$\begin{array}{r|rrrr} \frac{1}{2} & 1 & -2 & 3 & -2 \\ & & \frac{1}{2} & -\frac{3}{4} & \frac{9}{8} \\ \hline & 1 & -\frac{3}{2} & \frac{9}{4} & \underline{|-\frac{7}{8}} \end{array}$$

b. $P\left(\frac{1}{3}\right)=\frac{-32}{27}$

$$\begin{array}{r|rrrr} \frac{1}{3} & 1 & -2 & 3 & -2 \\ & & \frac{1}{3} & -\frac{5}{9} & \frac{22}{27} \\ \hline & 1 & -\frac{5}{3} & \frac{22}{9} & \underline{|-\frac{32}{27}} \end{array}$$

25. $x=-2,\quad x=3,\quad x=-5$; degree 3

$P(x)=(x+2)(x-3)(x+5);$

$P(x)=(x^2-x-6)(x+5)$

$P(x)=x^3+5x^2-x^2-5x-6x-30$

$P(x)=x^3+4x^2-11x-30$

27. $x=-2,\quad x=\sqrt{3},\quad x=-\sqrt{3}$; degree 3

$P(x)=(x+2)(x-\sqrt{3})(x+\sqrt{3});$

$P(x)=(x+2)(x^2+\sqrt{3}x-\sqrt{3}x-3)$

$P(x)=(x+2)(x^2-3)$

$P(x)=x^3-3x+2x^2-6$

$P(x)=x^3+2x^2-3x-6$

29. $x=-5,\quad x=2\sqrt{3},\quad x=-2\sqrt{3}$; degree 3

$P(x)=(x+5)(x-2\sqrt{3})(x+2\sqrt{3});$

$P(x)=(x+5)(x^2+2\sqrt{3}x-2\sqrt{3}x-12)$

$P(x)=(x+5)(x^2-12)$

$P(x)=x^3-12x+5x^2-60$

$P(x)=x^3+5x^2-12x-60$

31. $x=1,\quad x=-2,\quad x=\sqrt{10},\quad x=-\sqrt{10}$; degree 4

$P(x)=(x-1)(x+2)(x-\sqrt{10})(x+\sqrt{10});$

$P(x)=(x^2+x-2)(x^2+\sqrt{10}x-\sqrt{10}x-10)$

$P(x)=(x^2+x-2)(x^2-10)$

$P(x)=x^4-10x^2+x^3-10x-2x^2+20$

$P(x)=x^4+x^3-12x^2-10x+20$

33. $P(x)=x^3-4x^2+9x-36, x=3i$;

$$\begin{array}{r|rrrr} 3i & 1 & -4 & 9 & -36 \\ & & 3i & -9-12i & 36 \\ \hline & 1 & -4+3i & -12i & \underline{|0} \end{array}$$

Check:

$P(3i)=(3i)^3-4(3i)^2+9(3i)-36$

$P(3i)=27i^3-36i^2+27i-36$

$P(3i)=-27i+36+27i-36=0$

35. $P(x)=x^4+x^3+2x^2+4x-8, x=-2i$

$$\begin{array}{r|rrrrr} -2i & 1 & 1 & 2 & 4 & -8 \\ & & -2i & -4-2i & -4+4i & 8 \\ \hline & 1 & 1-2i & -2-2i & 4i & \underline{|0} \end{array}$$

Check:

$P(-2i)=(-2i)^4+(-2i)^3+2(-2i)^2$
$\quad +4(-2i)-8$

$P(-2i)=16i^4-8i^3+8i^2-8i-8$

$P(-2i)=16+8i-8-8i-8=0$

37. $P(x)=-x^3+x^2-3x-5, x=1+2i$

$$\begin{array}{r|rrrr} 1+2i & -1 & 1 & -3 & -5 \\ & & -1-2i & 4-2i & 5 \\ \hline & -1 & -2i & 1-2i & \underline{|0} \end{array}$$

Check:

$P(1+2i)=-(1+2i)^3+(1+2i)^2$
$\quad -3(1+2i)-5$

$P(1+2i)=-(1+4i+4i^2)(1+2i)$
$\quad +1+4i+4i^2-3-6i-5$

$P(1+2i)=-(-3+4i)(1+2i)-2i-11$

$P(1+2i)=-(-3-6i+4i+8i^2)-2i-11$

$P(1+2i)=3+6i-4i-8i^2-2i-11=0$

39. $x=-1,\quad x=1+3i,\quad x=1-3i$; degree 3

$P(x)=(x+1)(x-(1+3i))(x-(1-3i))$

$P(x)=(x+1)(x-1-3i)(x-1+3i)$

$P(x)=(x+1)((x-1)^2-9i^2)$

$P(x)=(x+1)(x^2-2x+1-9i^2)$

$P(x)=(x+1)(x^2-2x+10)$

$P(x) = x^3 - 2x^2 + 10x + x^2 - 2x + 10$

$P(x) = x^3 - x^2 + 8x + 10$

41. $x = \sqrt{3},\ \ x = -\sqrt{3},\ \ x = 1 + 2i,\ \ x = 1 - 2i$; degree 4

$P(x) = (x - \sqrt{3})(x + \sqrt{3})(x - (1 + 2i))(x - (1 - 2i))$

$P(x) = (x^2 - 3)(x - 1 - 2i)(x - 1 + 2i)$

$P(x) = (x^2 - 3)((x - 1)^2 - 4i^2)$

$P(x) = (x^2 - 3)(x^2 - 2x + 1 - 4i^2)$

$P(x) = (x^2 - 3)(x^2 - 2x + 5)$

$P(x) = x^4 - 2x^3 + 5x^2 - 3x^2 + 6x - 15$

$P(x) = x^4 - 2x^3 + 2x^2 + 6x - 15$

43. $x = 1,\ \ x = -3,\ \ x = 1 + \sqrt{2}i,\ \ x = 1 - \sqrt{2}i$; degree 4

$P(x) = (x - 1)(x + 3)(x - (1 + \sqrt{2}\,i))(x - (1 - \sqrt{2}\,i))$

$P(x) = (x^2 + 2x - 3)(x - 1 - \sqrt{2}\,i)(x - 1 + \sqrt{2}\,i)$

$P(x) = (x^2 + 2x - 3)((x - 1)^2 - 2i^2)$

$P(x) = (x^2 + 2x - 3)(x^2 - 2x + 1 - 2i^2)$

$P(x) = (x^2 + 2x - 3)(x^2 - 2x + 3)$

$P(x) = x^4 - 2x^3 + 3x^2 + 2x^3 - 4x^2 + 6x$

$\quad - 3x^2 + 6x - 9$

$P(x) = x^4 - 4x^2 + 12x - 9$

45. $x = 1 + \sqrt{5},\ \ x = 1 - \sqrt{5},\ \ x = 1 + i,\ \ x = 1 - i$; degree 4

$P(x) = (x - (1 + \sqrt{5}))(x - (1 - \sqrt{5}))(x - (1 + i))(x - (1 - i))$

$P(x) = (x - 1 - \sqrt{5})(x - 1 + \sqrt{5})(x - 1 - i)(x - 1 + i)$

$P(x) = ((x - 1)^2 - 5)((x - 1)^2 - i^2)$

$P(x) = (x^2 - 2x + 1 - 5)(x^2 - 2x + 1 - i^2)$

$P(x) = (x^2 - 2x - 4)(x^2 - 2x + 2)$

$P(x) = x^4 - 2x^3 + 2x^2 - 2x^3 + 4x^2 - 4x - 4x^2 + 8x - 8$

$P(x) = x^4 - 4x^3 + 2x^2 + 4x - 8$

47. $x = 4,\ \ x = -3i$

$P(x) = (x - 4)(x + 3i)(x - 3i)$

$P(x) = (x - 4)(x^2 - 9i^2)$

$P(x) = (x - 4)(x^2 + 9)$

$P(x) = x^3 + 9x - 4x^2 - 36$

$P(x) = x^3 - 4x^2 + 9x - 36$

49. $x = 5,\ \ x = 2 + 3i$

$P(x) = (x - 5)(x - (2 + 3i))(x - (2 - 3i))$

$P(x) = (x - 5)(x - 2 - 3i)(x - 2 + 3i)$

$P(x) = (x - 5)((x - 2)^2 - 9i^2)$

$P(x) = (x - 5)(x^2 - 4x + 4 + 9)$

$P(x) = (x - 5)(x^2 - 4x + 13)$

$P(x) = x^3 - 4x^2 + 13x - 5x^2 + 20x - 65$

$P(x) = x^3 - 9x^2 + 33x - 65$

51. $x = 2,\ \ x = 3 + \sqrt{2}i$

$P(x) = (x - 2)(x - (3 + \sqrt{2}\,i))(x - (3 - \sqrt{2}\,i))$

$P(x) = (x - 2)(x - 3 - \sqrt{2}\,i)(x - 3 + \sqrt{2}\,i)$

$P(x) = (x - 2)((x - 3)^2 - 2i^2)$

$P(x) = (x - 2)(x^2 - 6x + 9 + 2)$

$P(x) = (x - 2)(x^2 - 6x + 11)$

$P(x) = x^3 - 6x^2 + 11x - 2x^2 + 12x - 22$

$P(x) = x^3 - 8x^2 + 23x - 22$

53. $x = 2i,\ \ x = \sqrt{3}i$

$P(x) = (x - 2i)(x + 2i)(x - \sqrt{3}\,i)(x + \sqrt{3}\,i)$

$P(x) = (x^2 - 4i^2)(x^2 - 3i^2)$

$P(x) = (x^2 + 4)(x^2 + 3)$

$P(x) = x^4 + 3x^2 + 4x^2 + 12$

$P(x) = x^4 + 7x^2 + 12$

55. $x = 2,\ \ x = -3,\ \ x = 1 + 2i$

$P(x) = (x - 2)(x + 3)(x - (1 + 2i))(x - (1 - 2i))$

$P(x) = (x - 2)(x + 3)(x - 1 - 2i)(x - 1 + 2i)$

$P(x) = (x - 2)(x + 3)((x - 1)^2 - 4i^2)$

$P(x) = (x^2 + x - 6)(x^2 - 2x + 1 + 4)$

$P(x) = (x^2 + x - 6)(x^2 - 2x + 5)$

$P(x) = x^4 - 2x^3 + 5x^2 + x^3 - 2x^2 + 5x$

$\quad - 6x^2 + 12x - 30$

$P(x) = x^4 - x^3 - 3x^2 + 17x - 30$

57. $x = 2 + \sqrt{3},\ \ x = 1 + \sqrt{5}i$

$P(x)=\left(x-\left(2+\sqrt{3}\right)\right)\left(x-\left(2-\sqrt{3}\right)\right)\left(x-\left(1+\sqrt{5}\,i\right)\right)\left(x-\left(1-\sqrt{5}\,i\right)\right)$

$P(x)=\left(x-2-\sqrt{3}\right)\left(x-2+\sqrt{3}\right)\left(x-1-\sqrt{5}\,i\right)\left(x-1+\sqrt{5}\,i\right)$

$P(x)=\left((x-2)^2-3\right)\left((x-1)^2-5i^2\right)$

$P(x)=\left(x^2-4x+4-3\right)\left(x^2-2x+1+5\right)$

$P(x)=\left(x^2-4x+1\right)\left(x^2-2x+6\right)$

$P(x)=x^4-2x^3+6x^2-4x^3+8x^2-24x$

$+x^2-2x+6$

$P(x)=x^4-6x^3+15x^2-26x+6$

59. $P(x)=x^3-3x^2-9x+27$

$P(x)=x^2(x-3)-9(x-3)$

$P(x)=(x-3)\left(x^2-9\right)$

$P(x)=(x-3)(x+3)(x-3)$

$x=-3$; multiplicity one;

$x=3$; multiplicity two;

degree 3

61. $P(x)=x^3-6x^2+12x-8$

$P(x)=(x-2)\left(x^2-4x+4\right)$

$P(x)=(x-2)(x-2)(x-2)$

$x=2$; multiplicity three;

degree 3

63. $P(x)=\left(x^2-6x+9\right)\left(x^2-9\right)$

$P(x)=(x-3)(x-3)(x+3)(x-3)$

$x=3$; multiplicity three;

$x=-3$; multiplicity one;

degree 4

65. $P(x)=\left(x^3+4x^2-9x-36\right)\left(x^2+x-12\right)$

$P(x)=\left(x^2(x+4)-9(x+4)\right)(x+4)(x-3)$

$P(x)=(x+4)\left(x^2-9\right)(x+4)(x-3)$

$P(x)=(x+4)(x+3)(x-3)(x+4)(x-3)$

$x=-4$; multiplicity two;

$x=3$; multiplicity two;

$x=-3$; multiplicity one;

degree 5

67. $640=4x^3-84x^2+432x$

$160=x^3-21x^2+108x$

$0=x^3-21x^2+108x-160$

4	1	-21	108	-160
		4	-68	160
	1	-17	40	0

$0=(x-4)\left(x^2-17x+40\right)$

4 inch squares,

24 – 8 = 16 in

18 – 8 = 10 in

Dimensions of box:16in x 14in x 4 in

69. $x=2,\quad x=-3i$; degree 3

$P(x)=(x-2)(x+3i)(x-3i)$

$P(x)=(x-2)\left(x^2-9i^2\right)$

$P(x)=(x-2)\left(x^2+9\right)$

$P(x)=x^3+9x-2x^2-18$

$P(x)=x^3-2x^2+9x-18$

71. $x=2,\quad x=-3i$; degree 4

$P(x)=(x-2)(x-2)(x+3i)(x-3i)$

$P(x)=\left(x^2-4x+4\right)\left(x^2-9i^2\right)$

$P(x)=\left(x^2-4x+4\right)\left(x^2+9\right)$

$P(x)=x^4+9x^2-4x^3-36x+4x^2+36$

$P(x)=x^4-4x^3+13x^2-36x+36$

73. $x=1,\quad x=1+2i$; degree 3

$P(x)=(x-1)(x-(1+2i))(x-(1-2i))$

$P(x)=(x-1)(x-1-2i)(x-1+2i)$

$P(x)=(x-1)\left((x-1)^2-4i^2\right)$

$P(x)=(x-1)\left(x^2-2x+1+4\right)$

$P(x)=(x-1)\left(x^2-2x+5\right)$

$P(x)=x^3-2x^2+5x-x^2+2x-5$

$P(x)=x^3-3x^2+7x-5$

75. $x=1,\quad x=1+2i$; degree 5

$P(x)=(x-1)^3(x-(1+2i))(x-(1-2i))$

$P(x)=(x-1)^3(x-1-2i)(x-1+2i)$

77. $x=-3,\quad x=1+\sqrt{2},\quad x=5i$; degree 6

$P(x)=(x+3)^2\left(x-\left(1+\sqrt{2}\right)\right)\left(x-\left(1-\sqrt{2}\right)\right)(x-5i)(x+5i)$

$P(x)=(x+3)^2\left(x-1-\sqrt{2}\right)\left(x-1+\sqrt{2}\right)(x-5i)(x+5i)$

79. $f(x)=x^3-3x^2-13x+15$
$x=-3, x=1, x=5$
End behavior: down/up
y-intercept (0,15)

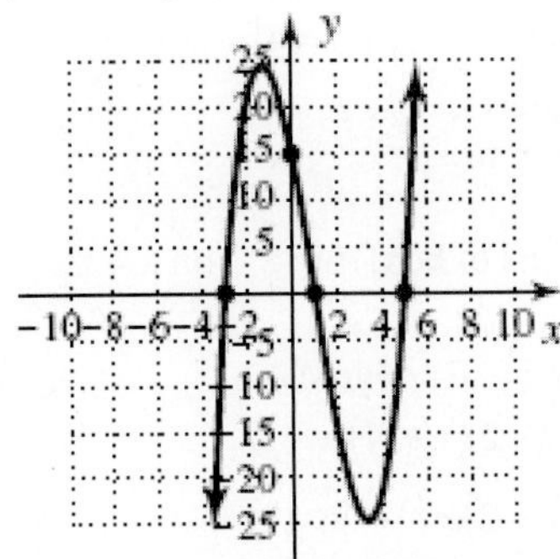

81. $f(x)=x^3-6x^2+3x+10$
$x=-1, x=2, x=5$
End behavior: down/up
y-intercept (0,10)

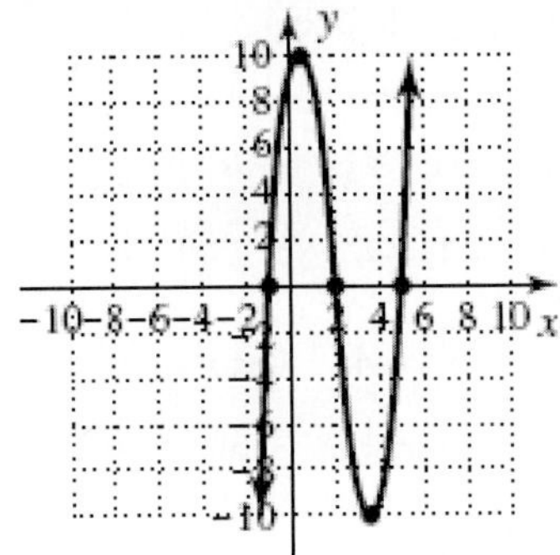

83. $f(x)=-x^3+7x-6$
$x=-3, x=1, x=2$
End behavior: up/down
y-intercept (0,-6)

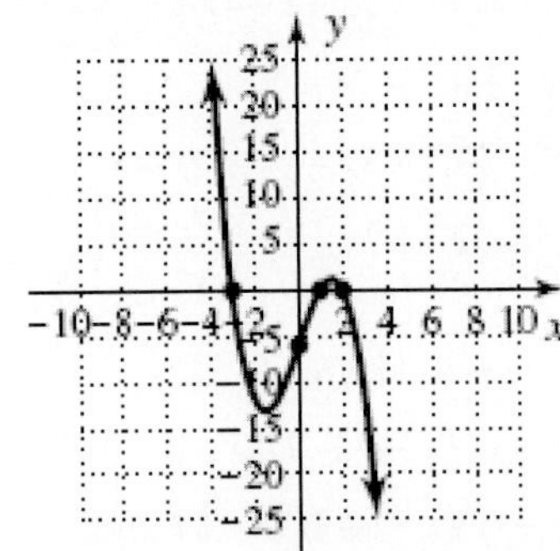

85. $P(w)=-0.1w^4+2w^3-14w^2+52w+5$

a. week 10, 22.5 thousand
$P(5)=-0.1(5)^4+2(5)^3-14(5)^2+52(5)+5=102.5$
$P(10)=-0.1(10)^4+2(10)^3-14(10)^2+52(10)+5=125$

b. one week before closing, 36 thousand
$P(1)=-0.1(1)^4+2(1)^3-14(1)^2+52(1)+5=44.9$
$P(11)=-0.1(11)^4+2(11)^3-14(11)^2+52(11)+5=80.9$

c. week 9
$P(7)=-0.1(7)^4+2(7)^3-14(7)^2+52(7)+5=128.9$
$P(8)=-0.1(8)^4+2(8)^3-14(8)^2+52(8)+5=139.4$
$P(9)=-0.1(9)^4+2(9)^3-14(9)^2+52(9)+5=140.5$
$P(10)=-0.1(10)^4+2(10)^3-14(10)^2+52(10)+5=125$

87. (a) $x=2+3i$ and $x=2-3i$
$a=2, b=\pm 3, a^2+b^2=13;$
$x^2-2(2)x+\left(2^2+(\pm 3)^2\right)$
$=x^2-4x+13$

(b) $x=1+\sqrt{2}\,i$ and $x=1-\sqrt{2}\,i$
$a=1, b=\pm\sqrt{2}, a^2+b^2=3;$
$x^2-2(1)x+\left(1^2+\left(\pm\sqrt{2}\right)^2\right)$
$=x^2-2x+3$

89. $x=2,\quad x=1-5i,\quad x=1+5i,$
$P(x)=x^3+kx+52$
$P(x)=(x+2)(x-(1-5i))(x-(1+5i))$
$P(x)=(x+2)(x-1+5i)(x-1-5i)$
$P(x)=(x+2)\left((x-1)^2-25i^2\right)$
$P(x)=(x+2)\left(x^2-2x+1+25\right)$
$P(x)=(x+2)\left(x^2-2x+26\right)$
$P(x)=x^3-2x^2+26x+2x^2-4x+52$
$P(x)=x^3+22x+52$
k=22

91. (a) $1^3+2^3+3^3=1+8+27=36$
$S_3=36$

3	1	2	1	0	0
		3	15	48	144
	1	5	16	48	144

(b) $1^3+2^3+3^3+4^3+5^3$
$=1+8+27+64+125=225$
$S_5=225$

5	1	2	1	0	0
		5	35	180	900
	1	7	36	180	900

93. $g(x)=\frac{1}{(x+3)^2}-1$

Left 3, down 1

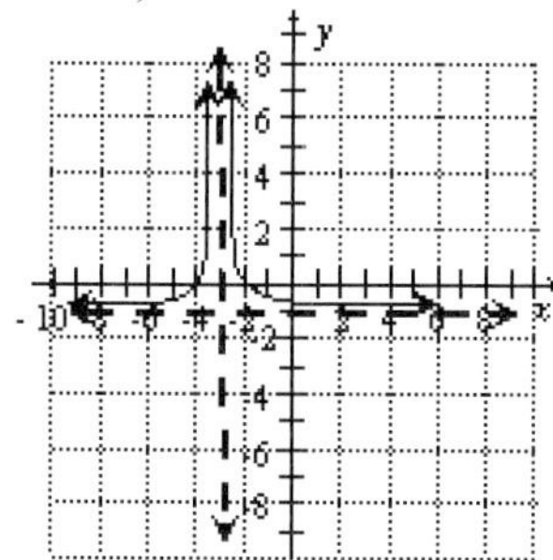

95. $2x^3-x^2-3x=0$

$x(2x^2-x-3)=0$

$x(2x-3)(x+1)=0$

$x=0$ or $2x-3=0$ or $x+1=0$

$x=0$ or $2x=3$ or $x=-1$

$x=0$ or $x=\frac{3}{2}$ or $x=-1$

97. $\frac{\Delta y}{\Delta x}=\frac{(1.1)^2-4(1.1)-((1.0)^2-4(1.0))}{1.1-1.0}\approx -1.9$

4.3 Technology Highlight p. 397

1. They give an approximate location for each zero.

2. Yes

3. Outputs change sign at each zero.

4.3 Technology Highlight

1. $2+3i$, verified

$$\begin{array}{c|cccc} 1-5i & 1 & -3+2i & 17-7i \\ & & 1-5i & -17+7i \\ \hline & 1 & -2-3i & \boxed{0} \end{array}$$

$(1-5i)(2+3i)=17-7i;$

$1-5i+2+3i=3-2i$

4.3 Exercises

1. Coefficients

3. Coefficients, sum, 0, root

5. b; 4 is not a factor of 6.

7. $C(x)=x^3+(1-4i)x^2+(-6-4i)x+24i$

$x=4i$;

$$\begin{array}{c|cccc} 4i & 1 & 1-4i & -6-4i & 24i \\ & & 4i & 4i & -24i \\ \hline & 1 & 1 & -6 & \boxed{0} \end{array}$$

$C(x)=(x-4i)(x^2+x-6)$

$C(x)=(x-4i)(x+3)(x-2)$

$x=-3, x=2$

9. $C(x)=x^3+(-2-3i)x^2+(5+6i)x-15i$;

$x=3i$;

$$\begin{array}{c|cccc} 3i & 1 & -2-3i & 5+6i & -15i \\ & & 3i & -6i & 15i \\ \hline & 1 & -2 & 5 & \boxed{0} \end{array}$$

$C(x)=(x-3i)(x^2-2x+5)$;

$x=\frac{-(-2)\pm\sqrt{(-2)^2-4(1)(5)}}{2(1)}$

$=\frac{2\pm\sqrt{-16}}{2}=1\pm 2i$;

$C(x)=(x-3i)(x-1-2i)(x-1+2i)$

$x=1\pm 2i$

11. $C(x)=x^3+(-2-6i)x^2+(4+12i)x-24i$;

$x=6i$;

$$\begin{array}{c|cccc} 6i & 1 & -2-6i & 4+12i & -24i \\ & & 6i & -12i & 24i \\ \hline & 1 & -2 & 4 & \boxed{0} \end{array}$$

$C(x)=(x-6i)(x^2-2x+4)$;

$x=\frac{-(-2)\pm\sqrt{(-2)^2-4(1)(4)}}{2(1)}$

$=\frac{2\pm\sqrt{-12}}{2}=1\pm\sqrt{3}\,i$;

$C(x)=(x-6i)(x-1-\sqrt{3}\,i)(x-1+\sqrt{3}\,i)$

$x=1\pm\sqrt{3}\,i$

13. $C(x)=x^3+(-2-i)x^2+(5+4i)x+(-6+3i)$;

$x=2-i$;

$$\begin{array}{c|cccc} 2-i & 1 & -2-i & 5+4i & -6+3i \\ & & 2-i & -2-4i & 6-3i \\ \hline & 1 & -2i & 3 & \boxed{0} \end{array}$$

$C(x)=(x-2+i)(x^2-2ix+3)$;

$$x=\frac{-(-2i)\pm\sqrt{(-2i)^2-4(1)(3)}}{2(1)}$$
$$=\frac{2i\pm\sqrt{-16}}{2}=i\pm 2i=3i \text{ or } -i;$$
$C(x)=(x-2+i)(x-3i)(x+i)$
$x=3i, x=-i$

15. $f(x)=x^3+2x^2-8x-5$
   a. $[-4,-3]$, yes
   $f(-4)$
   $=(-4)^3+2(-4)^2-8(-4)-5=-5;$
   $f(-3)$
   $=(-3)^3+2(-3)^2-8(-3)-5=10$
   b. $[-2,-1]$, no
   $f(-2)$
   $=(-2)^3+2(-2)^2-8(-2)-5=11$
   $f(-1)$
   $=(-1)^3+2(-1)^2-8(-1)-5=4$
   c. $[2,3]$, yes
   $f(2)=(2)^3+2(2)^2-8(2)-5=-5;$
   $f(3)=(3)^3+2(3)^2-8(3)-5=16$

17. $h(x)=2x^3+13x^2+3x-36$
   a. $[-5,-4]$, no
   $h(-5)$
   $=2(-5)^3+13(-5)^2+3(-5)-36=+24$
   $h(-4)$
   $=2(-4)^3+13(-4)^2+3(-4)-36=+32$
   b. $[-3,-2]$, yes
   $h(-3)$
   $=2(-3)^3+13(-3)^2+3(-3)-36=+18$
   $h(-2)$
   $=2(-2)^3+13(-2)^2+3(-2)-36=-6$
   c. $[1,2]$, yes
   $h(1)=2(1)^3+13(1)^2+3(1)-36=-18$
   $h(2)=2(2)^3+13(2)^2+3(2)-36=+38$

19. $f(x)=4x^3-19x-15$
$\frac{\{\pm1,\pm15,\pm3,\pm5\}}{\{\pm1,\pm4,\pm2\}};$
$\left\{\pm1,\pm15,\pm3,\pm5,\pm\frac{1}{4},\pm\frac{15}{4},\right.$
$\left.\pm\frac{3}{4},\pm\frac{5}{4},\pm\frac{1}{2},\pm\frac{15}{2},\pm\frac{3}{2},\pm\frac{5}{2}\right\}$

21. $h(x)=2x^3-5x^2-28x+15$
$\frac{\{\pm1,\pm15,\pm3,\pm5\}}{\{\pm1,\pm2\}};$
$\left\{\pm1,\pm15,\pm3,\pm5,\pm\frac{1}{2},\pm\frac{15}{2},\pm\frac{3}{2},\pm\frac{5}{2}\right\}$

23. $p(x)=6x^4-2x^3+5x^2-28$
$\frac{\{\pm1,\pm28,\pm2,\pm14,\pm4,\pm7\}}{\{\pm1,\pm6,\pm2,\pm3\}};$
$\left\{\pm1,\pm28,\pm2,\pm14,\pm4,\pm7,\pm\frac{1}{6},\pm\frac{14}{3},\pm\frac{1}{3},\pm\frac{7}{3},\pm\frac{2}{3},\right.$
$\left.\pm\frac{7}{6},\pm\frac{1}{2},\pm\frac{7}{2},\pm\frac{28}{3},\pm\frac{4}{3}\right\}$

25. $Y_1=32t^3-52t^2+17t+3$
$\frac{\{\pm1,\pm3\}}{\{\pm1,\pm32,\pm2,\pm16,\pm4,\pm8\}};$
$\left\{\pm1,\pm\frac{1}{32},\pm\frac{1}{2},\pm\frac{1}{16},\pm\frac{1}{4},\pm\frac{1}{8},\right.$
$\left.\pm3,\pm\frac{3}{32},\pm\frac{3}{2},\pm\frac{3}{16},\pm\frac{3}{4},\pm\frac{3}{8}\right\}$

27. $f(x)=x^3-13x+12$
Possible rational roots:
$\frac{\{\pm1,\pm12,\pm2,\pm6,\pm3,\pm4\}}{\{\pm1\}};$
$\{\pm1,\pm12,\pm2,\pm6,\pm3,\pm4\}$

```
-4| 1   0  -13   12
       -4   16  -12
   ----------------
    1  -4    3  |0
```

$f(x)=(x+4)(x^2-4x+3)$
$f(x)=(x+4)(x-1)(x-3)$
$x=-4,1,3$

29 $h(x)=x^3-19x-30$
Possible rational roots:
$\frac{\{\pm1,\pm30,\pm2,\pm15,\pm3,\pm10,\pm5,\pm6\}}{\{\pm1\}};$
$\{\pm1,\pm30,\pm2,\pm15,\pm3,\pm10,\pm5,\pm6\}$

$$\begin{array}{r|rrrr} -3 & 1 & 0 & -19 & -30 \\ & & -3 & 9 & 30 \\ \hline & 1 & -3 & -10 & \boxed{0} \end{array}$$

$h(x) = (x+3)(x^2 - 3x - 10)$

$h(x) = (x+3)(x+2)(x-5)$

$x = -3, -2, 5$

31. $p(x) = x^3 - 2x^2 - 11x + 12$

Possible rational roots:

$\dfrac{\{\pm1, \pm12, \pm2, \pm6, \pm3, \pm4\}}{\{\pm1\}};$

$\{\pm1, \pm12, \pm2, \pm6, \pm3, \pm4\}$

$$\begin{array}{r|rrrr} -3 & 1 & -2 & -11 & 12 \\ & & -3 & 15 & -12 \\ \hline & 1 & -5 & 4 & \boxed{0} \end{array}$$

$p(x) = (x+3)(x^2 - 5x + 4)$

$p(x) = (x+3)(x-1)(x-4)$

$x = -3, 1, 4$

33. $Y_1 = x^3 - 6x^2 - x + 30$

Possible rational roots:

$\dfrac{\{\pm1, \pm30, \pm2, \pm15, \pm3, \pm10, \pm5, \pm6\}}{\{\pm1\}};$

$\{\pm1, \pm30, \pm2, \pm15, \pm3, \pm10, \pm5, \pm6\}$

$$\begin{array}{r|rrrr} -2 & 1 & -6 & -1 & 30 \\ & & -2 & 16 & -30 \\ \hline & 1 & -8 & 15 & \boxed{0} \end{array}$$

$Y_1 = (x+2)(x^2 - 8x + 15)$

$Y_1 = (x+2)(x-3)(x-5)$

$x = -2, 3, 5$

35. $Y_3 = x^4 - 15x^2 + 10x + 24$

Possible rational roots:

$\dfrac{\{\pm1, \pm24, \pm2, \pm12, \pm3, \pm8, \pm4, \pm6\}}{\{\pm1\}};$

$\{\pm1, \pm24, \pm2, \pm12, \pm3, \pm8, \pm4, \pm6\}$

$$\begin{array}{r|rrrrr} -4 & 1 & 0 & -15 & 10 & 24 \\ & & -4 & 16 & -4 & -24 \\ \hline & 1 & -4 & 1 & 6 & \boxed{0} \end{array}$$

$$\begin{array}{r|rrrr} -1 & 1 & -4 & 1 & 6 \\ & & -1 & 5 & -6 \\ \hline & 1 & -5 & 6 & \boxed{0} \end{array}$$

$Y_3 = (x+4)(x+1)(x^2 - 5x + 6)$

$Y_3 = (x+4)(x+1)(x-2)(x-3)$

$x = -4, -1, 2, 3$

37. $f(x) = x^4 + 7x^3 - 7x^2 - 55x - 42$

Possible rational roots:

$\dfrac{\{\pm1, \pm42, \pm2, \pm21, \pm3, \pm14, \pm6, \pm7\}}{\{\pm1\}};$

$\{\pm1, \pm42, \pm2, \pm21, \pm3, \pm14, \pm6, \pm7\}$

$$\begin{array}{r|rrrrr} -7 & 1 & 7 & -7 & -55 & -42 \\ & & -7 & 0 & 49 & 42 \\ \hline & 1 & 0 & -7 & -6 & \boxed{0} \end{array}$$

$$\begin{array}{r|rrrr} -2 & 1 & 0 & -7 & -6 \\ & & -2 & 4 & 6 \\ \hline & 1 & -2 & -3 & \boxed{0} \end{array}$$

$f(x) = (x+7)(x+2)(x^2 - 2x - 3)$

$f(x) = (x+7)(x+2)(x+1)(x-3)$

$x = -7, -2, -1, 3$

39. $f(x) = 4x^3 - 7x + 3$

Possible rational roots: $\dfrac{\{\pm1, \pm3\}}{\{\pm1, \pm4, \pm2\}}$

$$\begin{array}{r|rrrr} 1 & 4 & 0 & -7 & 3 \\ & & 4 & 4 & -3 \\ \hline & 4 & 4 & -3 & \boxed{0} \end{array}$$

$f(x) = (x-1)(4x^2 + 4x - 3)$

$f(x) = (x-1)(2x+3)(2x-1)$

$x = \dfrac{-3}{2}, \dfrac{1}{2}, 1$

41. $h(x) = 4x^3 + 8x^2 - 3x - 9$

Possible rational roots: $\dfrac{\{\pm1, \pm9, \pm3\}}{\{\pm1, \pm4, \pm2\}}$

$\left\{\pm1, \pm9, \pm3, \pm\dfrac{1}{4}, \pm\dfrac{9}{4}, \pm\dfrac{3}{4}, \pm\dfrac{1}{2}, \pm\dfrac{9}{2}, \pm\dfrac{3}{2}\right\}$

$$\begin{array}{r|rrrr} 1 & 4 & 8 & -3 & -9 \\ & & 4 & 12 & 9 \\ \hline & 4 & 12 & 9 & \boxed{0} \end{array}$$

$h(x) = (x-1)(4x^2 + 12x + 9)$

$h(x) = (x-1)(2x+3)^2$

$x = \dfrac{-3}{2}, 1$

43. $Y_1 = 2x^3 - 3x^2 - 9x + 10$

Possible rational roots: $\dfrac{\{\pm1, \pm10, \pm2, \pm5\}}{\{\pm1, \pm2\}}$

$\left\{\pm1, \pm10, \pm2, \pm5, \pm\dfrac{1}{2}, \pm\dfrac{5}{2}\right\}$

$$\begin{array}{r|rrrr} 1 & 2 & -3 & -9 & 10 \\ & & 2 & -1 & -10 \\ \hline & 2 & -1 & -10 & \underline{|0} \end{array}$$

$Y_1 = (x-1)(2x^2 - x - 10)$

$Y_1 = (x-1)(x+2)(2x-5)$

$x = -2,\ 1, \frac{5}{2}$

45. $p(x) = 2x^4 + 3x^3 - 9x^2 - 15x - 5$

Possible rational roots: $\frac{\{\pm1, \pm5\}}{\{\pm1, \pm2\}}$

$\left\{\pm1, \pm5, \pm\frac{1}{2}, \pm\frac{5}{2}\right\}$

$$\begin{array}{r|rrrrr} -1 & 2 & 3 & -9 & -15 & -5 \\ & & -2 & -1 & 10 & 5 \\ \hline & 2 & 1 & -10 & -5 & \underline{|0} \end{array}$$

$p(x) = (x+1)(2x^3 + x^2 - 10x - 5)$

$p(x) = (x+1)(x^2(2x+1) - 5(2x+1))$

$p(x) = (x+1)(2x+1)(x^2 - 5)$

$p(x) = (x+1)(2x+1)(x-\sqrt{5})(x+\sqrt{5})$

$x = -1, \frac{-1}{2}, \sqrt{5}, -\sqrt{5}$

47. $r(x) = 3x^4 - 5x^3 + 14x^2 - 20x + 8$

Possible rational roots: $\frac{\{\pm1, \pm8, \pm2, \pm4\}}{\{\pm1, \pm3\}}$

$\left\{\pm1, \pm8, \pm2, \pm4, \pm\frac{1}{3}, \pm\frac{8}{3}, \pm\frac{2}{3}, \pm\frac{4}{3}\right\}$

$$\begin{array}{r|rrrrr} 1 & 3 & -5 & 14 & -20 & 8 \\ & & 3 & -2 & 12 & -8 \\ \hline & 3 & -2 & 12 & -8 & \underline{|0} \end{array}$$

$r(x) = (x-1)(3x^3 - 2x^2 + 12x - 8)$

$r(x) = (x-1)(x^2(3x-2) + 4(3x-2))$

$r(x) = (x-1)(3x-2)(x^2 + 4)$

$r(x) = (x-1)(3x-2)(x-2i)(x+2i)$

$x = 1, \frac{2}{3}, \pm 2i$

49. $f(x) = 2x^4 - 9x^3 + 4x^2 + 21x - 18$

Possible rational roots:

$\frac{\{\pm1, \pm18, \pm2, \pm9, \pm3, \pm6\}}{\{\pm1, \pm2\}}$

$\left\{\pm1, \pm18, \pm2, \pm9, \pm3, \pm6, \pm\frac{1}{2}, \pm\frac{9}{2}, \pm\frac{3}{2}\right\}$

$$\begin{array}{r|rrrrr} 1 & 2 & -9 & 4 & 21 & -18 \\ & & 2 & -7 & -3 & 18 \\ \hline & 2 & -7 & -3 & 18 & \underline{|0} \end{array}$$

$$\begin{array}{r|rrrr} 2 & 2 & -7 & -3 & 18 \\ & & 4 & -6 & -18 \\ \hline & 2 & -3 & -9 & \underline{|0} \end{array}$$

$f(x) = (x-1)(x-2)(2x^2 - 3x - 9)$

$f(x) = (x-1)(x-2)(x-3)(2x+3)$

$x = 1,\ 2,\ 3,\ \frac{-3}{2}$

51. $h(x) = 3x^4 + 2x^3 - 9x^2 + 4$

Possible rational roots: $\frac{\{\pm1, \pm4, \pm2\}}{\{\pm1, \pm3\}}$

$\left\{\pm1, \pm4, \pm2, \pm\frac{1}{3}, \pm\frac{4}{3}, \pm\frac{2}{3}\right\}$

$$\begin{array}{r|rrrrr} 1 & 3 & 2 & -9 & 0 & 4 \\ & & 3 & 5 & -4 & -4 \\ \hline & 3 & 5 & -4 & -4 & \underline{|0} \end{array}$$

$$\begin{array}{r|rrrr} -2 & 3 & 5 & -4 & -4 \\ & & -6 & 2 & 4 \\ \hline & 3 & -1 & -2 & \underline{|0} \end{array}$$

$h(x) = (x-1)(x+2)(3x^2 - x - 2)$

$h(x) = (x-1)(x+2)(3x+2)(x-1)$

$x = -2,\ 1,\ \frac{-2}{3}$

53. $p(x) = 2x^4 - 3x^3 - 21x^2 - 2x + 24$

Possible rational roots:

$\frac{\{\pm1, \pm24, \pm2, \pm12, \pm3, \pm8, \pm4, \pm6\}}{\{\pm1, \pm2\}}$

$\left\{\pm1, \pm24, \pm2, \pm12, \pm3, \pm8, \pm4, \pm6, \pm\frac{1}{2}, \pm\frac{3}{2}\right\}$

$$\begin{array}{r|rrrrr} 1 & 2 & -3 & -21 & -2 & 24 \\ & & 2 & -1 & -22 & -24 \\ \hline & 2 & -1 & -22 & -24 & \underline{|0} \end{array}$$

$$\begin{array}{r|rrrr} -2 & 2 & -1 & -22 & -24 \\ & & -4 & 10 & 24 \\ \hline & 2 & -5 & -12 & \underline{|0} \end{array}$$

$p(x) = (x-1)(x+2)(2x^2 - 5x - 12)$

$p(x) = (x-1)(x+2)(2x+3)(x-4)$

$x=-2,\ \frac{-3}{2},\ 1,\ 4$

55. $r(x)=3x^4-17x^3+23x^2+13x-30$

Possible rational roots: $\frac{\{\pm1,\pm30,\pm2,\pm15,\pm3,\pm10,\pm5,\pm6\}}{\{\pm1,\pm3\}}$

$\left\{\pm1,\pm30,\pm2,\pm15,\pm3,\pm10,\pm5,\pm6,\pm\frac{1}{3},\pm\frac{2}{3},\pm\frac{10}{3},\pm\frac{5}{3}\right\}$

$$\begin{array}{r|rrrrr} -1 & 3 & -17 & 23 & 13 & -30 \\ & & -3 & 20 & -43 & 30 \\ \hline & 3 & -20 & 43 & -30 & 0 \end{array}$$

$$\begin{array}{r|rrrr} 2 & 3 & -20 & 43 & -30 \\ & & 6 & -28 & 30 \\ \hline & 3 & -14 & 15 & 0 \end{array}$$

$r(x)=(x+1)(x-2)(3x^2-14x+15)$

$r(x)=(x+1)(x-2)(3x-5)(x-3)$

$x=-1,\ 2,\ 3,\ \frac{5}{3}$

57. $Y_1=x^5+6x^2-49x+42$

Possible rational roots: $\frac{\{\pm1,\pm42,\pm2,\pm21,\pm3,\pm14,\pm6,\pm7\}}{\{\pm1\}}$

$\{\pm1,\pm42,\pm2,\pm21,\pm3,\pm14,\pm6,\pm7\}$

$$\begin{array}{r|rrrrrr} 1 & 1 & 0 & 0 & 6 & -49 & 42 \\ & & 1 & 1 & 1 & 7 & -42 \\ \hline & 1 & 1 & 1 & 7 & -42 & 0 \end{array}$$

$$\begin{array}{r|rrrrr} 2 & 1 & 1 & 1 & 7 & -42 \\ & & 2 & 6 & 14 & 42 \\ \hline & 1 & 3 & 7 & 21 & 0 \end{array}$$

$Y_1=(x-1)(x-2)(x^3+3x^2+7x+21)$

$Y_1=(x-1)(x-2)(x^2(x+3)+7(x+3))$

$Y_1=(x-1)(x-2)(x+3)(x^2+7)$

$Y_1=(x-1)(x-2)(x+3)(x+\sqrt{7}\,i)(x-\sqrt{7}\,i)$

$x=1,\ 2,\ -3,\ \pm\sqrt{7}\,i$

59. $P(x)=3x^5+x^4+x^3+7x^2-24x+12$

Possible rational roots: $\frac{\{\pm1,\pm12,\pm2,\pm6,\pm3,\pm4\}}{\{\pm1,\pm3\}}$

$\left\{\pm1,\pm12,\pm2,\pm6,\pm3,\pm4,\pm\frac{1}{3},\pm\frac{2}{3},\pm\frac{4}{3}\right\}$

$$\begin{array}{r|rrrrrr} 1 & 3 & 1 & 1 & 7 & -24 & 12 \\ & & 3 & 4 & 5 & 12 & -12 \\ \hline & 3 & 4 & 5 & 12 & -12 & 0 \end{array}$$

$$\begin{array}{r|rrrrr} -2 & 3 & 4 & 5 & 12 & -12 \\ & & -6 & 4 & -18 & 12 \\ \hline & 3 & -2 & 9 & -6 & 0 \end{array}$$

$P(x)=(x-1)(x+2)(3x^3-2x^2+9x-6)$

$P(x)=(x-1)(x+2)(x^2(3x-2)+3(3x-2))$

$P(x)=(x-1)(x+2)(3x-2)(x^2+3)$

$P(x)=(x-1)(x+2)(3x-2)(x+\sqrt{3}\,i)(x-\sqrt{3}\,i)$

$x=-2,\frac{2}{3},1,\pm\sqrt{3}\,i$

61. $Y_1=x^4-5x^3+20x-16$

Possible rational roots: $\frac{\{\pm1,\pm16,\pm2,\pm8,\pm4\}}{\{\pm1\}}$

$\{\pm1,\pm16,\pm2,\pm8,\pm4\}$

$$\begin{array}{r|rrrrr} 1 & 1 & -5 & 0 & 20 & -16 \\ & & 1 & -4 & -4 & 16 \\ \hline & 1 & -4 & -4 & 16 & 0 \end{array}$$

$Y_1=(x-1)(x^3-4x^2-4x+16)$

$Y_1=(x-1)(x^2(x-4)-4(x-4))$

$Y_1=(x-1)(x-4)(x^2-4)$

$Y_1=(x-1)(x-4)(x+2)(x-2)$

$x=1,\ 2,\ 4,\ -2$

63. $r(x)=x^4+2x^3-5x^2-4x+6$

Possible rational roots: $\frac{\{\pm1,\pm6,\pm2,\pm3\}}{\{\pm1\}}$

$\{\pm1,\pm6,\pm2,\pm3\}$

$$\begin{array}{r|rrrrr} 1 & 1 & 2 & -5 & -4 & 6 \\ & & 1 & 3 & -2 & -6 \\ \hline & 1 & 3 & -2 & -6 & 0 \end{array}$$

$r(x)=(x-1)(x^3+3x^2-2x-6)$

$r(x)=(x-1)(x^2(x+3)-2(x+3))$

$r(x)=(x-1)(x+3)(x^2-2)$

$r(x)=(x+3)(x-1)(x+\sqrt{2})(x-\sqrt{2})$

$x=-3,\ 1,\pm\sqrt{2}$

65. $p(x)=2x^4-x^3+3x^2-3x-9$

Possible rational roots: $\dfrac{\{\pm 1, \pm 9, \pm 3\}}{\{\pm 1, \pm 2\}}$

$\left\{\pm 1, \pm 9, \pm 3, \pm\dfrac{1}{2}, \pm\dfrac{9}{2}, \pm\dfrac{3}{2}\right\}$

$$\begin{array}{r|rrrrr} -1 & 2 & -1 & 3 & -3 & -9 \\ & & -2 & 3 & -6 & 9 \\ \hline & 2 & -3 & 6 & -9 & \boxed{0} \end{array}$$

$p(x) = (x+1)(2x^3 - 3x^2 + 6x - 9)$

$p(x) = (x+1)(x^2(2x-3) + 3(2x-3))$

$p(x) = (x+1)(2x-3)(x^2+3)$

$p(x) = (x+1)(2x-3)(x+\sqrt{3}\,i)(x-\sqrt{3}\,i)$

$x = -1, \dfrac{3}{2}, \pm\sqrt{3}\,i$

67. $f(x) = 2x^5 - 7x^4 + 13x^3 - 23x^2 + 21x - 6$

Possible rational roots: $\dfrac{\{\pm 1, \pm 6, \pm 2, \pm 3\}}{\{\pm 1, \pm 2\}}$

$\left\{\pm 1, \pm 6, \pm 2, \pm 3, \pm\dfrac{1}{2}, \pm\dfrac{3}{2}\right\}$

$$\begin{array}{r|rrrrrr} 1 & 2 & -7 & 13 & -23 & 21 & -6 \\ & & 2 & -5 & 8 & -15 & 6 \\ \hline & 2 & -5 & 8 & -15 & 6 & \boxed{0} \end{array}$$

$$\begin{array}{r|rrrrr} 2 & 2 & -5 & 8 & -15 & 6 \\ & & 4 & -2 & 12 & -6 \\ \hline & 2 & -1 & 6 & -3 & \boxed{0} \end{array}$$

$f(x) = (x-1)(x-2)(2x^3 - x^2 + 6x - 3)$

$f(x) = (x-1)(x-2)(x^2(2x-1) + 3(2x-1))$

$f(x) = (x-1)(x-2)(2x-1)(x^2+3)$

$f(x) = (x-1)(x-2)(2x-1)(x+\sqrt{3}\,i)(x-\sqrt{3}\,i)$

$x = \dfrac{1}{2}, 1, 2, \pm\sqrt{3}\,i$

69. $f(x) = x^4 - 2x^3 + 4x - 8$

Possible rational roots: $\dfrac{\{\pm 1, \pm 8, \pm 2, \pm 4\}}{\{\pm 1\}}$

$\{\pm 1, \pm 8, \pm 2, \pm 4\}$

Zeroes of unity: none, neither 1 or -1 is a root

$(1 - 2 + 4 - 8 \neq 0), (1 + 2 - 4 - 8 \neq 0)$

\# of positive roots: 3 or 1 roots

\# of negative roots: 1 root

Bounds: roots must lie between –2 and 2.

$$\begin{array}{r|rrrrr} -2 & 1 & -2 & 0 & 4 & -8 \\ & & -2 & 8 & -16 & 24 \\ \hline & 1 & -4 & 8 & -12 & \boxed{16} \end{array}$$

$$\begin{array}{r|rrrrr} 2 & 1 & -2 & 0 & 4 & -8 \\ & & 2 & 0 & 0 & 8 \\ \hline & 1 & 0 & 0 & 4 & \boxed{0} \end{array}$$

71. $h(x) = x^5 + x^4 - 3x^3 + 5x + 2$

Possible rational roots: $\dfrac{\{\pm 1, \pm 2\}}{\{\pm 1\}}$

$\{\pm 1, \pm 2\}$

Zeroes of unity: -1 is a root

$(1 + 1 - 3 + 5 + 2 \neq 0)$, $(-1 + 1 + 3 - 5 + 2 = 0)$

\# of positive roots: 2 or 0 roots

\# of negative roots: 3 or 1 roots

Bounds: roots must lie between -3 and 2.

$$\begin{array}{r|rrrrrr} -3 & 1 & 1 & -3 & 0 & 5 & 2 \\ & & -3 & 6 & -9 & 27 & -96 \\ \hline & 1 & -2 & 3 & -9 & 32 & \boxed{-94} \end{array}$$

$$\begin{array}{r|rrrrrr} 2 & 1 & 1 & -3 & 0 & 5 & 2 \\ & & 2 & 6 & 6 & 12 & 34 \\ \hline & 1 & 3 & 3 & 6 & 17 & \boxed{36} \end{array}$$

73. $p(x) = x^5 - 3x^4 + 3x^3 - 9x^2 - 4x + 12$

Possible rational roots:

$\dfrac{\{\pm 1, \pm 12, \pm 2, \pm 6, \pm 3, \pm 4\}}{\{\pm 1\}}$

$\{\pm 1, \pm 12, \pm 2, \pm 6, \pm 3, \pm 4\}$

Zeroes of unity: $x = 1$ and $x = -1$ are roots.

$(1 - 3 + 3 - 9 - 4 + 12 = 0)$,

$(-1 - 3 - 3 - 9 + 4 + 12 = 0)$

\# of positive roots: 4, 2, or 0 roots

\# of negative roots: 1 root

Bounds: roots must lie between -1 and 4.

$$\begin{array}{r|rrrrrr} -1 & 1 & -3 & 3 & -9 & -4 & 12 \\ & & -1 & 4 & -7 & 16 & -12 \\ \hline & 1 & -4 & 7 & -16 & 12 & \boxed{0} \end{array}$$

$$\begin{array}{r|rrrrrr} 4 & 1 & -3 & 3 & -9 & -4 & 12 \\ & & 4 & 4 & 28 & 76 & 288 \\ \hline & 1 & 1 & 7 & 19 & 72 & \boxed{300} \end{array}$$

75. $r(x) = 2x^4 + 7x^2 + 11x - 20$

Possible rational roots:

$\dfrac{\{\pm 1, \pm 20, \pm 2, \pm 10, \pm 4, \pm 5\}}{\{\pm 1, \pm 2\}}$

$\left\{\pm 1, \pm 20, \pm 2, \pm 10, \pm 4, \pm 5, \pm\dfrac{1}{2}, \pm\dfrac{5}{2}\right\}$

Zeroes of unity: $x = 1$ is a root.

$(2 + 7 + 11 - 20 = 0)$, $(2 + 7 - 11 - 20 \neq 0)$

\# of positive roots: 1 root

\# of negative roots: 1 root

Bounds: roots must lie between –2 and 1.

$$\begin{array}{r|rrrrr} -2 & 2 & 0 & 7 & 11 & -20 \\ & & -4 & 8 & -30 & 38 \\ \hline & 2 & -4 & 15 & -19 & \boxed{18} \end{array}$$

$$\begin{array}{r|rrrrr} 1 & 2 & 0 & 7 & 11 & -20 \\ & & 2 & 2 & 9 & 20 \\ \hline & 2 & 2 & 9 & 20 & \boxed{0} \end{array}$$

77. $f(x)=4x^3-16x^2-9x+36$

Possible Positive roots	Possible Negative roots	Possible Complex roots	Total number of roots
2	1	0	3
0	1	2	3

Possible rational roots:
$$\frac{\{\pm1,\pm36,\pm2,\pm18,\pm3,\pm12,\pm4,\pm9,\pm6\}}{\{\pm1,\pm4,\pm2\}};$$
$$\left\{\pm1,\pm36,\pm2,\pm18,\pm3,\pm12,\pm4,\pm9,\pm6,\pm\frac{1}{4},\pm\frac{1}{2},\pm\frac{9}{2},\pm\frac{3}{4},\pm\frac{9}{4},\pm\frac{3}{2}\right\}$$

$$\begin{array}{r|rrrr} 4 & 4 & -16 & -9 & 36 \\ & & 16 & 0 & -36 \\ \hline & 4 & 0 & -9 & \boxed{0} \end{array}$$

$f(x)=(x-4)(4x^2-9)$
$f(x)=(x-4)(2x-3)(2x+3)$
$x=\frac{-3}{2},\frac{3}{2},4$

79. $h(x)=6x^3-73x^2+10x+24$

Possible Positive roots	Possible Negative roots	Possible Complex roots	Total number of roots
2	1	0	3
0	1	2	3

Possible rational roots:
$$\frac{\{\pm1,\pm24,\pm2,\pm12,\pm3,\pm8,\pm4,\pm6\}}{\{\pm1,\pm6,\pm2,\pm3\}};$$
$$\left\{\pm1,\pm24,\pm2,\pm12,\pm3,\pm8,\pm4,\pm6,\pm\frac{1}{6},\pm\frac{1}{3},\pm\frac{1}{2},\pm\frac{4}{3},\pm\frac{2}{3},\pm\frac{3}{2},\pm\frac{8}{3}\right\}$$

$$\begin{array}{r|rrrr} 12 & 6 & -73 & 10 & 24 \\ & & 72 & -12 & -24 \\ \hline & 6 & -1 & -2 & \boxed{0} \end{array}$$

$h(x)=(x-12)(6x^2-x-2)$
$h(x)=(x-12)(3x-2)(2x+1)$
$x=\frac{-1}{2},\frac{2}{3},12$

81. $p(x)=4x^4+40x^3-97x^2-10x+24$

Possible Positive roots	Possible Negative roots	Possible Complex roots	Total number of roots
2	2	0	4
0	2	2	4
2	0	2	4
0	0	4	4

Possible rational roots:
$$\frac{\{\pm1,\pm24,\pm2,\pm12,\pm3,\pm8,\pm4,\pm6\}}{\{\pm1,\pm4,\pm2\}}$$
$$\left\{\pm1,\pm24,\pm2,\pm12,\pm3,\pm8,\pm4,\pm6,\pm\frac{1}{4},\pm\frac{1}{2},\pm\frac{3}{4},\pm\frac{3}{2}\right\}$$

$$\begin{array}{r|rrrrr} 2 & 4 & 40 & -97 & -10 & 24 \\ & & 8 & 96 & -2 & -24 \\ \hline & 4 & 48 & -1 & -12 & \boxed{0} \end{array}$$

$p(x)=(x-2)(4x^3+48x^2-1x-12)$
$p(x)=(x-2)(4x^2(x+12)-1(x+12))$
$p(x)=(x-2)(x+12)(4x^2-1)$
$p(x)=(x-2)(x+12)(2x-1)(2x+1)$
$x=-12,\frac{-1}{2},\frac{1}{2},2$

83. $f(x)=9x^3+6x^2-5x-2$

Possible rational roots: $\frac{\{\pm1,\pm2\}}{\{\pm1,\pm9,\pm3\}};$
$$\left\{\pm1,\pm2,\pm\frac{1}{9},\pm\frac{2}{9},\pm\frac{1}{3},\pm\frac{2}{3}\right\}$$

$$\begin{array}{r|rrrr} -1 & 9 & 6 & -5 & -2 \\ & & -9 & 3 & 2 \\ \hline & 9 & -3 & -2 & \boxed{0} \end{array}$$

$f(x)=(x+1)(9x^2-3x-2)$
$f(x)=(x+1)(3x-2)(3x+1)$
$x=-1,-\frac{1}{3},\frac{2}{3}$

85. $q(x)=x^4-3x^3-11x^2+3x+10$

Possible rational roots: $\frac{\{\pm1,\pm10,\pm2,\pm5\}}{\{\pm1\}};$
$\{\pm1,\pm10,\pm2,\pm5\}$

$$\begin{array}{r|rrrrr} 1 & 1 & -3 & -11 & 3 & 10 \\ & & 1 & -2 & -13 & -10 \\ \hline & 1 & -2 & -13 & -10 & \boxed{0} \end{array}$$

$$\begin{array}{r|rrrr} -1 & 1 & -2 & -13 & -10 \\ & & -1 & 3 & 10 \\ \hline & 1 & -3 & -10 & \boxed{0} \end{array}$$

$q(x)=(x-1)(x+1)(x^2-3x-10)$
$q(x)=(x-1)(x+1)(x-5)(x+2)$
$x=-2,-1,1,5$

87. $r(x)=2x^3+11x^2-x-30$
Possible rational roots:
$\dfrac{\{\pm1,\pm30,\pm2,\pm15,\pm3,\pm10\}}{\{\pm1,\pm2\}}$

$\left\{\pm1,\pm30,\pm2,\pm15,\pm3,\pm10,\pm\frac{1}{2},\pm\frac{15}{2},\pm\frac{3}{2}\right\}$

$$\begin{array}{r|rrrr} -2 & 2 & 11 & -1 & -30 \\ & & -4 & -14 & 30 \\ \hline & 2 & 7 & -15 & \lfloor 0 \end{array}$$

$r(x)=(x+2)(2x^2+7x-15)$
$r(x)=(x+2)(2x-3)(x+5)$
$x=-5,-2,\frac{3}{2}$

89. $Y_1=4x^3-41x^2+94x-21$
Possible rational roots: $\dfrac{\{\pm1,\pm21,\pm3,\pm7\}}{\{\pm1,\pm4,\pm2\}}$;

$\left\{\pm1,\pm21,\pm3,\pm7,\pm\frac{1}{4},\pm\frac{21}{4},\pm\frac{3}{4},\right.$
$\left.\pm\frac{7}{4},\pm\frac{1}{2},\pm\frac{21}{2},\pm\frac{3}{2},\pm\frac{7}{2}\right\}$

$$\begin{array}{r|rrrr} 3 & 4 & -41 & 94 & -21 \\ & & 12 & -87 & 21 \\ \hline & 4 & -29 & 7 & \lfloor 0 \end{array}$$

$Y_1=(x-3)(4x^2-29x+7)$
$Y_1=(x-3)(x-7)(4x-1)$
$x=3,7,\frac{1}{4}$

91. $f(x)=3x^4+2x^3-9x^2+4$
Possible rational roots: $\dfrac{\{\pm1,\pm4,\pm2\}}{\{\pm1,\pm3\}}$;

$\left\{\pm1,\pm4,\pm2,\pm\frac{1}{3},\pm\frac{4}{3},\pm\frac{2}{3}\right\}$

$$\begin{array}{r|rrrrr} 1 & 3 & 2 & -9 & 0 & 4 \\ & & 3 & 5 & -4 & -4 \\ \hline & 3 & 5 & -4 & -4 & \lfloor 0 \end{array}$$

$$\begin{array}{r|rrrr} 1 & 3 & 5 & -4 & -4 \\ & & 3 & 8 & 4 \\ \hline & 3 & 8 & 4 & \lfloor 0 \end{array}$$

$f(x)=(x-1)(x-1)(3x^2+8x+4)$

$f(x)=(x-1)(x-1)(3x+2)(x+2)$
$x=-2,1$ multiplicity $2,\frac{-2}{3}$

93. $h(t)=32t^3-52t^2+17t+3$
Possible rational roots:
$\dfrac{\{\pm1,\pm3\}}{\{\pm1,\pm32,\pm2,\pm16,\pm4,\pm8\}}$;
$\left\{\pm1,\pm3,\pm\frac{1}{32},\pm\frac{3}{32},\pm\frac{1}{2},\pm\frac{3}{2},\pm\frac{1}{16},\right.$
$\left.\pm\frac{3}{16},\pm\frac{1}{4},\pm\frac{3}{4},\pm\frac{1}{8},\pm\frac{3}{8}\right\}$

$$\begin{array}{r|rrrr} 1 & 32 & -52 & 17 & 3 \\ & & 32 & -20 & -3 \\ \hline & 32 & -20 & -3 & \lfloor 0 \end{array}$$

$h(t)=(t-1)(32t^2-20t-3)$
$h(t)=(t-1)(8t+1)(4t-3)$
$t=1,\frac{3}{4},\frac{-1}{8}$

95. $z=a+bi:|z|=\sqrt{a^2+b^2}$

(a) $|3+4i|:=\sqrt{(3)^2+(4)^2}=5$

(b) $|-5+12i|=\sqrt{(-5)^2+(12)^2}=13$

(c) $\left|1+\sqrt{3}\,i\right|=\sqrt{(1)^2+(\sqrt{3})^2}=2$

97. $f(x)=4x^3-12x^2-24x+32$
Possible rational roots:
$\dfrac{\{\pm1,\pm32,\pm2,\pm16,\pm4,\pm8\}}{\{\pm1,\pm4,\pm2\}}$;
$\left\{\pm1,\pm32,\pm2,\pm16,\pm\frac{1}{4},\pm\frac{1}{2}\right\}$

$$\begin{array}{r|rrrr} -2 & 4 & -12 & -24 & 32 \\ & & -8 & 40 & -32 \\ \hline & 4 & -20 & 16 & \lfloor 0 \end{array}$$

$f(x)=(x+2)(4x^2-20x+16)$
$f(x)=4(x+2)(x^2-5x+4)$
$f(x)=4(x+2)(x-4)(x-1)$
$x=-2,1,4$
Yes; grapher shows maximum and minimum values occur at the zeroes of *f*.

99. $g(x)=4x^3-18x^2+2x+24$

Possible rational roots: $\frac{\{\pm1,\pm24,\pm2,\pm12,\pm3,\pm8,\pm4,\pm6\}}{\{\pm1,\pm4,\pm2\}}$;

$\left\{\pm1,\pm24,\pm2,\pm12,\pm3,\pm8,\pm4,\pm6,\pm\frac{1}{4},\pm\frac{1}{2},\pm\frac{3}{4},\pm\frac{3}{2}\right\}$

```
-1| 4  -18   2   24
        -4  22  -24
   4  -22  24  |0
```

$g(x)=(x+1)(4x^2-22x+24)$

$g(x)=2(x+1)(2x^2-11x+12)$

$g(x)=(x+1)(2x-3)(x-4)$

$x=-1,\frac{3}{2},4$

Yes; grapher shows maximum and minimum values occur at the zeroes of *f*.

101. $v=x\cdot x\cdot(x-1)=x^3-x^2$

(a) $x^3-x^2=48$

$x^3-x^2-48=0$

Possible rational roots: $\frac{\{\pm1,\pm48,\pm2,\pm24,\pm3,\pm16,\pm4,\pm12,\pm6,\pm8\}}{\{\pm1\}}$;

$\{\pm1,\pm48,\pm2,\pm24,\pm3,\pm16,\pm4,\pm12,\pm6,\pm8\}$

```
4| 1  -1    0   48
       4  -12  -48
   1   3  -12  |0
```

$x=4$

4 cm×4cm×4cm

(b) $x^3-x^2=100$

$x^3-x^2-100=0$

Possible rational roots: $\{\pm1,\pm100,\pm2,\pm50,\pm4,\pm25,\pm5,\pm20,\pm10\}$

```
5| 1  -1   0  -100
       5  20   100
   1   4  20   |0
```

$x=5$

5cm×5cm×5cm

103. V= LWH

$2w(w)(w-2)=150$

$2w^3-4w^2-150=0$

$2(w^3-2w^2-75)=0$

Possible rational roots: $\{\pm1,\pm75,\pm3,\pm25,\pm5,\pm15\}$

```
5| 2  -4   0  -150
      10  30   150
   2   6  30   |0
```

$w=5$; $2w=10$; $w-2=3$

length 10 in., width 5 in., height 3 in.

105. $f(x)=\frac{1}{4}x^4-6x^3+42x^2-72x-64$

$0=\frac{1}{4}x^4-6x^3+42x^2-72x-64$

$0=x^4-24x^3+168x^2-288x-256$

Using a grapher:

$x=4$, 8, between 12 and 13

1994, 1998, 2002; 5 years

107. $f(x)=x^5-12x^3+12x^2-13x+12$

Possible rational roots: $\{\pm1,\pm12,\pm2,\pm6,\pm3,\pm4\}$

```
1| 1  0  -12   12  -13   12
      1    1  -11    1  -12
   1  1  -11    1  -12   |0
```

```
3| 1  1  -11  1  -12
      3   12  3   12
   1  4    1  4   |0
```

$f(x)=(x-1)(x-3)(x^3+4x^2+x+4)$

$f(x)=(x-1)(x-3)(x^2(x+4)+1(x+4))$

$f(x)=(x-1)(x-3)(x+4)(x^2+1)$

$f(x)=(x-1)(x-3)(x+4)(x+i)(x-i)$

$x=-4,1,3,\pm i$

109. $w=x-\frac{b}{3}$

(a) $w^3-3w^2+6w-4=0$

$a=1,b=-3,c=6,d=-4$

$w=x-\frac{-3}{3}=x+1$;

$(x+1)^3-3(x+1)^2+6(x+1)-4=0$

$x^3+3x^2+3x+1-3x^2-6x-3+6x+6-4=0$

$x^3+3x=0$

$x\left(x^2+3\right)=0$

$x(x+\sqrt{3}\,i)(x-\sqrt{3}\,i)=0$

$x=0,\pm\sqrt{3}\,i$;

$x+1=1,1\pm\sqrt{3}\,i$

$w=1,1\pm\sqrt{3}i$

(b) $w^3-6w^2+21w-26=0$

$a=1,b=-6,c=21,d=-26$

$w = x - \frac{-6}{3} = x + 2$;

$(x+2)^3 - 6(x+2)^2 + 21(x+2) - 26 = 0$

$x^3 + 6x^2 + 12x + 8 - 6x^2$
$-24x - 24 + 21x + 42 - 26 = 0$

$x^3 + 9x = 0$

$x\left(x^2 + 9\right) = 0$

$x(x+3i)(x-3i) = 0$

$x = 0, \pm 3i$;

$x + 2 = 2, 2 \pm 3i$

$w = 2, 2 \pm 3i$

111. $f(x) = \begin{cases} 2 & x \le -1 \\ |x-1| & -1 < x < 5 \\ 4 & x \ge 5 \end{cases}$

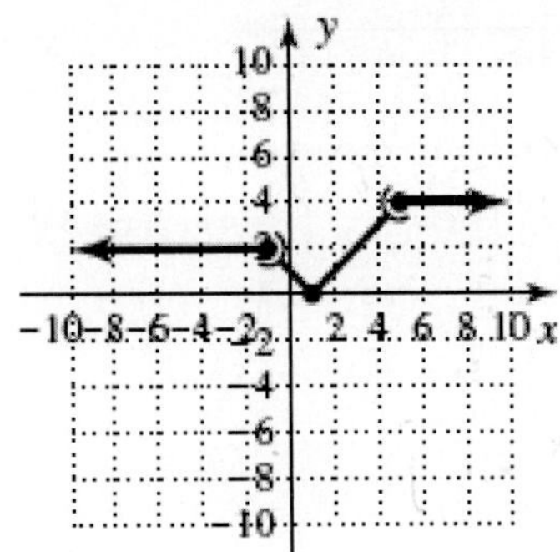

$f(-3) = 2$;

$f(2) = |2-1| = 1$;

$f(5) = 4$

113. (a) $[-4,-1] \cup [1,\infty)$
   (b) max : (-3,4)
   min : (0,-2)
   (c) $f(x)\uparrow : x \in (-\infty,-3) \cup (0,\infty)$
   $f(x)\downarrow : x \in (-3,0)$

115. $\{(-3,22),(-2,13),(-1,6),(0,1),(1,-2),(2,-3)\}$
   Inverse:
   $\{(22,-3),(13,-2),(6,-1),(1,0),(-2,1),(-3,2)\}$
   $D : \{22,13,6,1,-2,-3\}$
   $R : \{-3,-2,-1,0,1,2\}$

4.4Technology Highlight

Exercise 1:

$P(x) = (x+1)^2(x+2)^2(x-1)^3(x-2)$
Window: $x \in \{-3,3\}$, $y \in \{-12,12\}$

4.4 Exercises

1. Quartic

3. Bounce, flatter

5. Answers will vary.

7. up/down

9. down/down

11. down/up, (0, -2)

13. down/down, (0, -6)

15. up/down, (0, -6)

17. C

19. E

21. F

23. degree 6, up/up, (0, -12)

25. degree 5, up/down, (0, -24)

27. degree 6, up/up, (0, -192)

29. degree 5, up/down, (0, 2)

31. (a) even
   (b) -3, odd; -1, even; 3, odd
   (c) $f(x) = (x+3)(x+1)^2(x-3)$

33. (a) even
   (b) -3, odd; -1, odd; 2, odd; 4, odd
   (c) $f(x) = -(x+3)(x+1)(x-2)(x-4)$

35. (a) odd
   (b) -1, even; 3, odd
   (c) $f(x) = -(x+1)^2(x-3)$

37. b

39. e

41. c

43. $f(x) = (x+3)(x+1)(x-2)$
end behavior: down/up
x-intercepts: $(-3,0), (-1,0)$, and $(2,0)$;
crosses at all x-intercepts
$f(0) = (0+3)(0+1)(0-2) = -6$
y-intercept: $(0,-6)$

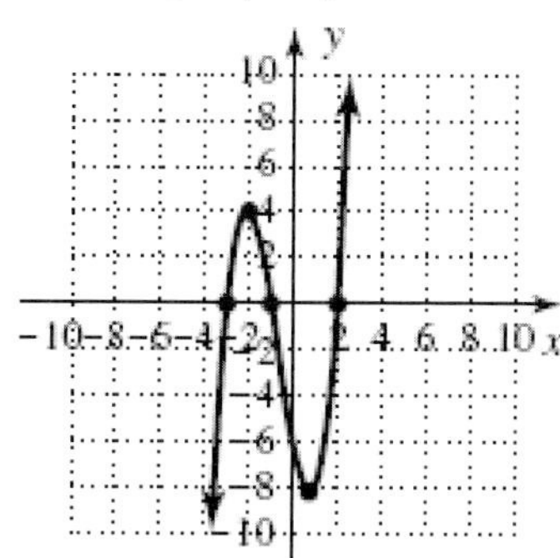

45. $p(x) = -(x+1)^2(x-3)$
end behavior: up/down
x-intercepts: $(-1,0)$ and $(3,0)$;
crosses at (3,0), bounces at (-1,0)
$p(0) = -(0+1)^2(0-3) = 3$
y-intercept: $(0,3)$

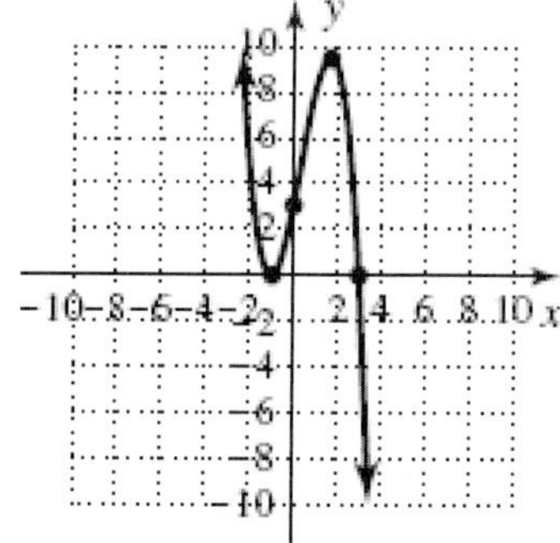

47. $Y_1 = (x+1)^2(3x-2)(x+3)$
end behavior: up/up
x-intercepts: $(-1,0), \left(\frac{2}{3},0\right)$ and $(-3,0)$;
crosses at (-3,0) and $\left(\frac{2}{3},0\right)$,
bounces at (-1,0)
$Y_1 = (0+1)^2(3(0)-2)(0+3) = -6$
y-intercept: $(0,-6)$

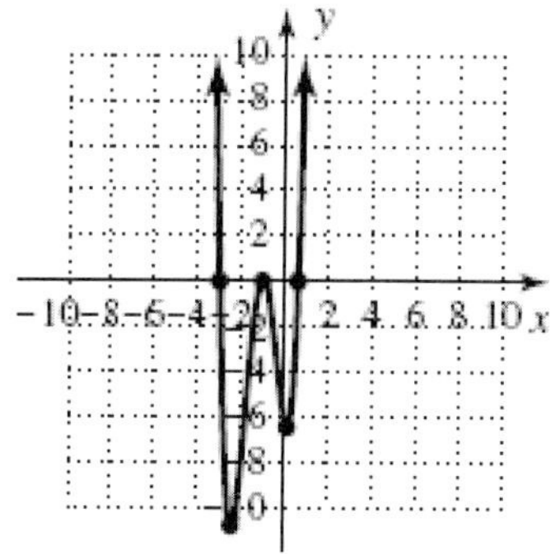

49. $r(x) = -(x+1)^2(x-2)^2(x-1)$
end behavior: up/down
x-intercepts: $(-1,0), (2,0)$ and $(1,0)$;
crosses at (1,0), bounces at (-1,0) and (2,0)
$r(0) = -(0+1)^2(0-2)^2(0-1) = 4$
y-intercept: $(0,4)$

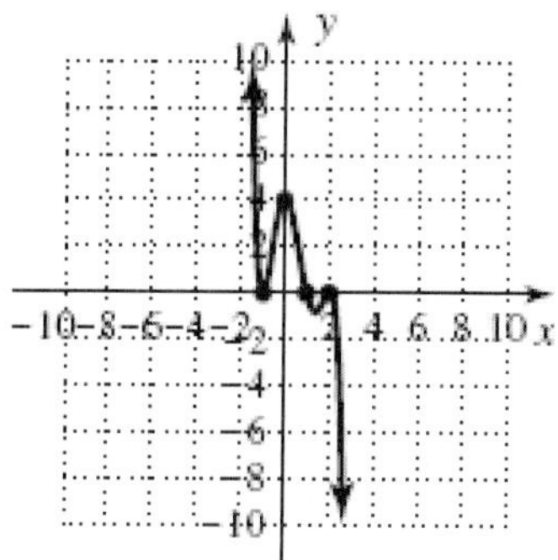

51. $f(x) = (2x+3)(x-1)^3$
end behavior: up/up
x-intercepts: $\left(-\frac{3}{2},0\right)$ and $(1,0)$;
crosses at all x-intercepts
$f(0) = (2(0)+3)(0-1)^3 = -3$
y-intercept: $(0,-3)$

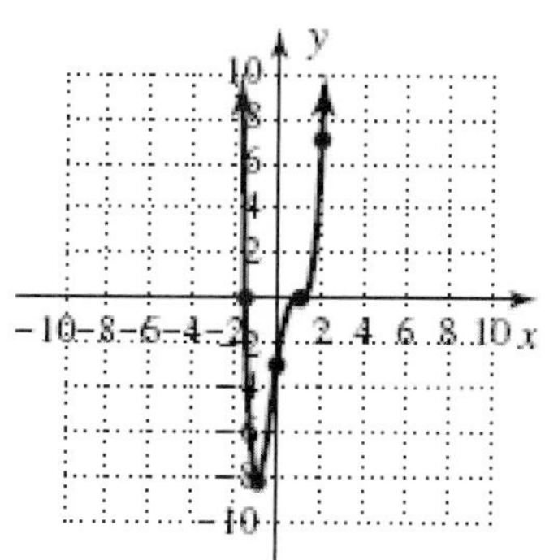

53. $h(x) = (x+1)^3(x-3)(x-2)$
end behavior: down/up

x-intercepts: $(-1,0),(3,0)$ and $(2,0)$;
crosses at all x-intercepts
$h(0)=(0+1)^3(0-3)(0-2)=6$
y-intercept: $(0,6)$

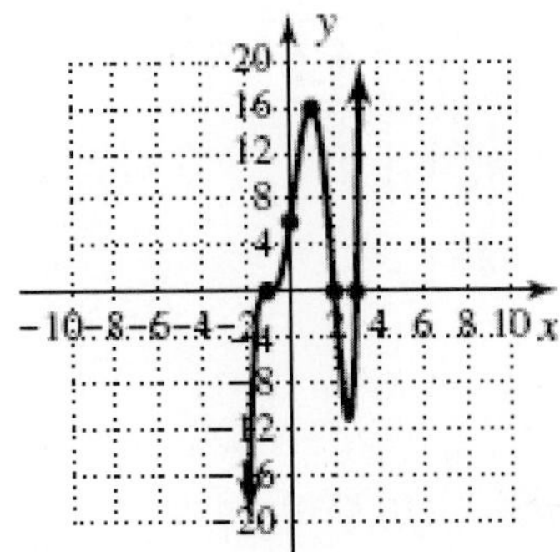

55. $Y_3=(x+1)^3(x-1)^2(x-2)$
end behavior: up/up
x-intercepts: $(-1,0),(1,0)$ and $(2,0)$;
crosses at (-1,0) and (2,0), bounces at (1,0);
$Y_3=(0+1)^3(0-1)^2(0-2)=-2$
y-intercept: $(0,-2)$

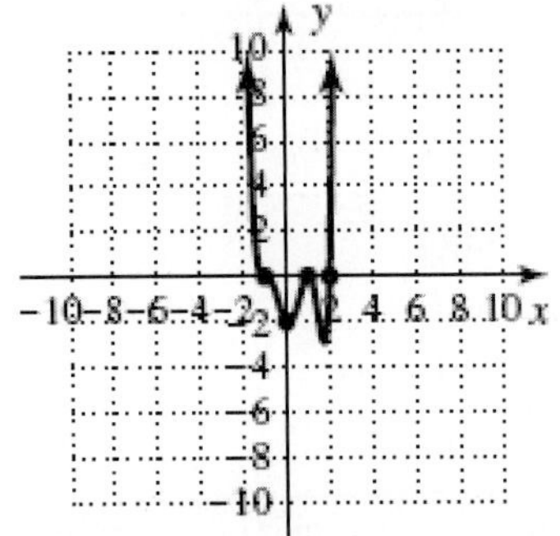

57. $P(x)=(x+3)(x+1)(x-2)(x-4)$
$P(x)=(x^2+4x+3)(x^2-6x+8)$
$P(x)=x^4-6x^3+8x^2+4x^3-24x^2+32x$
$+3x^2-18x+24$
$P(x)=x^4-2x^3-13x^2+14x+24$

59. $y=x^3+3x^2-4$
end behavior: down/up
Possible rational roots: $\{\pm1,\pm4,\pm2\}$
$y=(x+2)^2(x-1)$
x-intercepts: $(-2,0)$ and $(1,0)$
crosses at (1,0), bounces at (-2,0);
$y=0^3+3(0)^2-4=-4$
y-intercept: $(0,-4)$

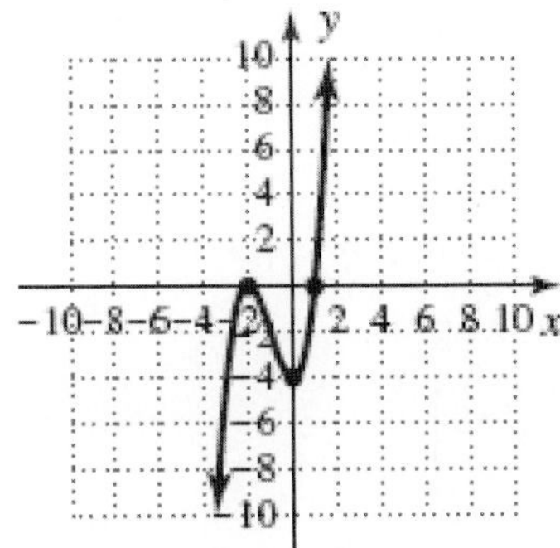

61. $f(x)=x^3-3x^2-6x+8$
end behavior: down/up
Possible rational roots: $\{\pm1,\pm8,\pm2,\pm4\}$
$f(x)=(x+2)(x-1)(x-4)$
x-intercepts: $(-2,0),(1,0)$ and $(4,0)$
crosses at all x-intercepts;
$f(0)=0^3-3(0)^2-5(0)+8=8$
y-intercept: $(0,8)$

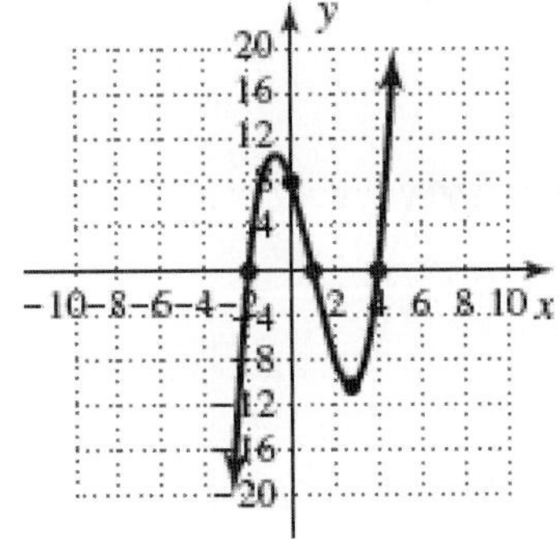

63. $h(x)=x^3+x^2-5x+3$
end behavior: down/up
Possible rational roots: $\{\pm1,\pm3\}$
$h(x)=(x+3)(x-1)^2$
x-intercepts: $(-3,0)$ and $(1,0)$
crosses at (-3,0), bounces at (1,0);
$h(0)=0^3+(0)^2-5(0)+3=3$
y-intercept: $(0,3)$

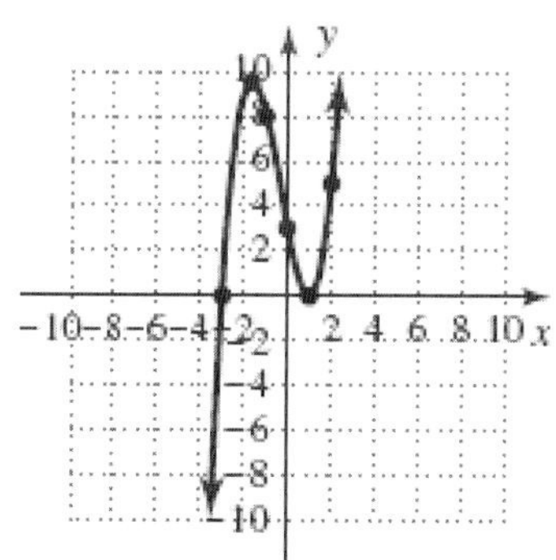

65. $p(x) = x^4 - 10x^2 + 9$
end behavior: up/up
$p(x) = (x^2 - 9)(x^2 - 1)$
$p(x) = (x+3)(x-3)(x+1)(x-1)$
x-intercepts: $(-3,0)$, $(3,0)$, $(-1,0)$ and $(1,0)$
crosses at all x-intercepts;
$p(0) = 0^4 - 10(0)^2 + 9 = 9$
y-intercept: $(0,9)$

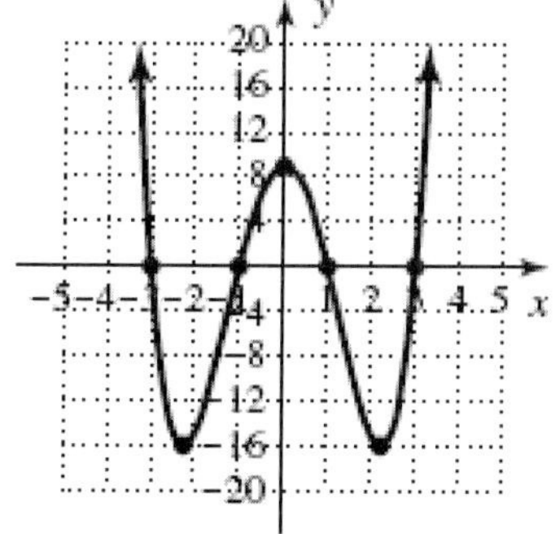

67. $r(x) = x^4 - 9x^2 - 4x + 12$
end behavior: up/up
Possible rational roots:
$\{\pm 1, \pm 12, \pm 2, \pm 6, \pm 3, \pm 4\}$
$r(x) = (x+2)^2(x-1)(x-3)$
x-intercepts: $(-2,0)$, $(1,0)$ and $(3,0)$
crosses at (1,0) and (3,0), bounces at (-2,0);
$r(0) = 0^4 - 9(0)^2 - 4(0) + 12 = 12$
y-intercept: $(0,12)$

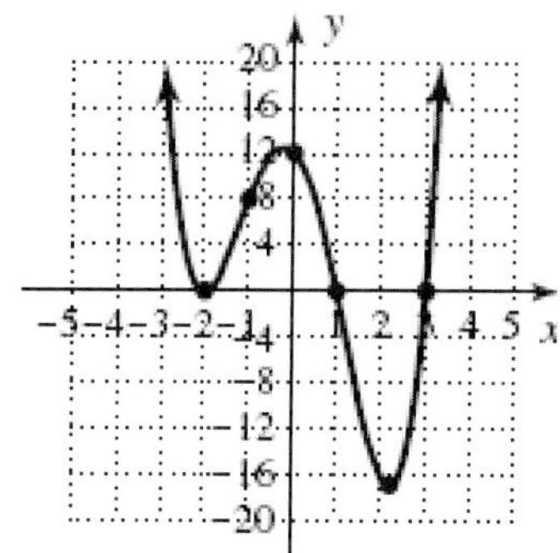

69. $Y_1 = x^4 - 6x^3 + 8x^2 + 6x - 9$
end behavior: up/up
Possible rational roots: $\{\pm 1, \pm 9, \pm 3\}$
$Y_1 = (x+1)(x-1)(x-3)^2$
x-intercepts: $(-1,0)$, $(1,0)$, and $(3,0)$
crosses at (-1,0) and (1,0), bounces at (3,0);
$Y_1 = 0^4 - 6(0)^3 + 8(0)^2 + 6(0) - 9 = -9$
y-intercept: $(0,-9)$

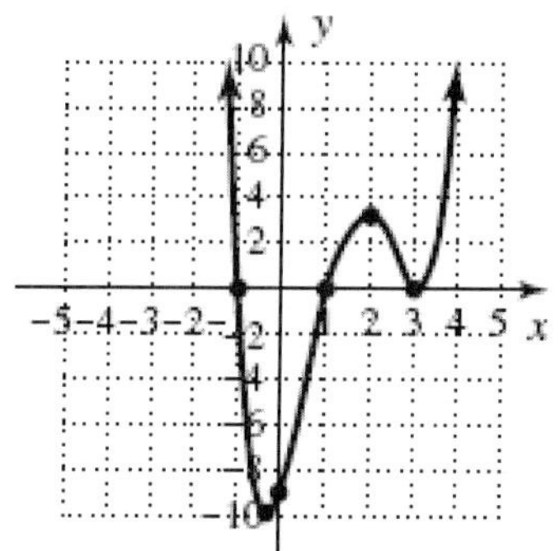

71. $Y_3 = 3x^4 + 2x^3 - 36x^2 + 24x + 32$
end behavior: up/up
Possible rational roots:
$\{\pm 1, \pm 32, \pm 2, \pm 16, \pm 4, \pm 8\}$
$Y_3 = (x+4)(3x+2)(x-2)^2$
x-intercepts: $(-4,0)$, $\left(-\frac{2}{3}, 0\right)$, and $(2,0)$
crosses at (-4,0) and $\left(-\frac{2}{3}, 0\right)$,
bounces at (2,0);
$Y_3 = 3(0)^4 + 2(0)^3 - 36(0)^2 + 24(0) + 32 = 32$
y-intercept: $(0,32)$

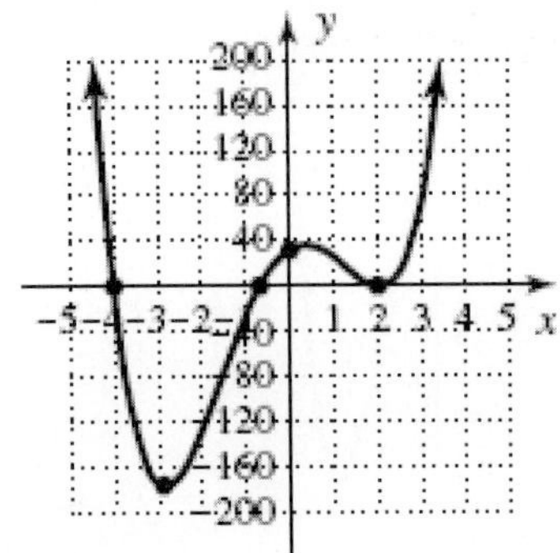

73. $F(x) = 2x^4 + 3x^3 - 9x^2$

$F(x) = x^2\left(2x^2 + 3x - 9\right)$

end behavior: up/up

Possible rational roots: $\dfrac{\{\pm1, \pm9, \pm3\}}{\{\pm1, \pm2\}}$;

$\left\{\pm1, \pm9, \pm3, \pm\frac{1}{2}, \pm\frac{9}{2}, \pm\frac{3}{2}\right\}$

$F(x) = x^2(x+3)(2x-3)$

x-intercepts: $(0,0)$, $(-3,0)$, and $\left(\frac{3}{2}, 0\right)$

crosses at (-3,0) and $\left(\frac{3}{2}, 0\right)$ bounces at (0,0);

$F(0) = 2(0)^4 + 3(0)^3 - 9(0)^2 = 0$

y-intercept: $(0,0)$

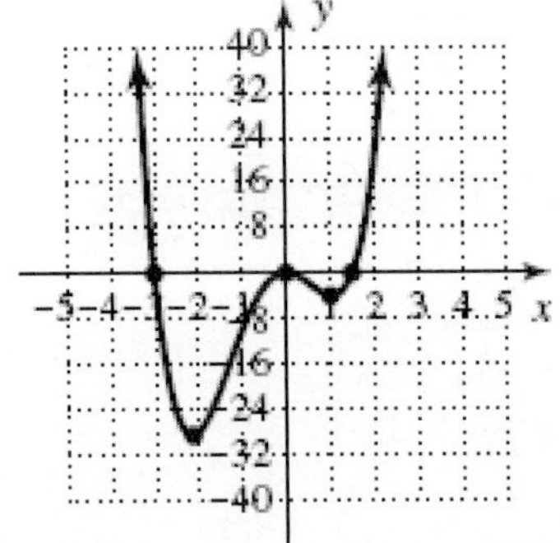

75. $f(x) = x^5 + 4x^4 - 16x^2 - 16x$

$f(x) = x\left(x^4 + 4x^3 - 16x - 16\right)$

end behavior: down/up

Possible rational roots: $\{\pm1, \pm16, \pm2, \pm8, \pm4\}$

$f(x) = x(x+2)^3(x-2)$

x-intercepts: $(0,0)$, $(-2,0)$, and $(2,0)$

crosses at all x-intercepts;

$f(0) = (0)^5 + 4(0)^4 - 16(0)^2 - 16(0) = 0$

y-intercept: $(0,0)$

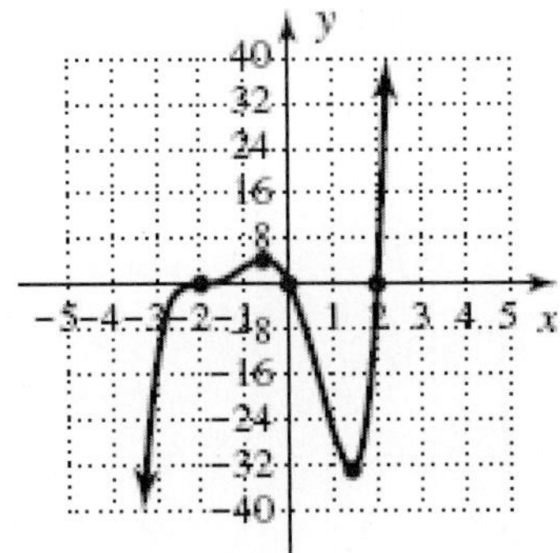

77. $h(x) = x^6 - 2x^5 - 4x^4 + 8x^3$

$h(x) = x^3\left(x^3 - 2x^2 - 4x + 8\right)$

$h(x) = x^3\left(x^2(x-2) - 4(x-2)\right)$

$h(x) = x^3(x-2)\left(x^2 - 4\right)$

$h(x) = x^3(x-2)(x+2)(x-2)$

end behavior: up/up

x-intercepts: $(0,0)$, $(-2,0)$ and $(2,0)$

crosses at (-2,0) and (0,0), bounces at (2,0);

$h(0) = (0)^6 - 2(0)^5 - 4(0)^4 + 8(0)^3 = 0$

y-intercept: $(0,0)$

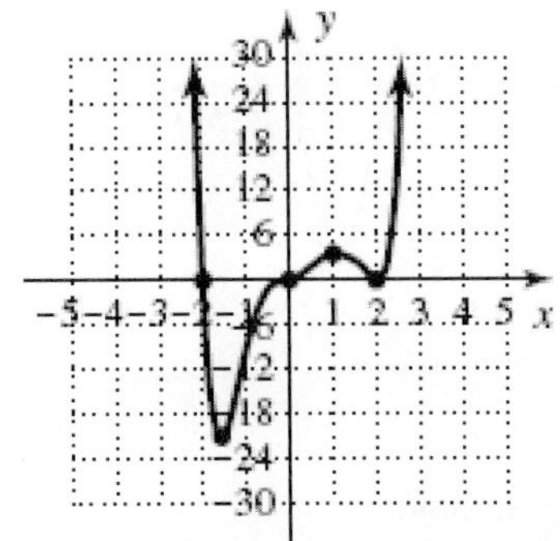

79. $h(x) = x^5 + 4x^4 - 9x - 36$

$h(x) = (x+4)\left(x - \sqrt{3}\right)\left(x + \sqrt{3}\right)\left(x^2 + 3\right)$

$h(x) = (x+4)\left(x - \sqrt{3}\right)\left(x + \sqrt{3}\right)\left(x - \sqrt{3}\,i\right)\left(x + \sqrt{3}\,i\right)$

y-intercept: (0, -36)

81. $f(x) = 2x^5 + 5x^4 - 10x^3 - 25x^2 + 12x + 30$

$f(x) = \left(x + \frac{5}{2}\right)\left(x - \sqrt{2}\right)\left(x + \sqrt{2}\right)\left(x - \sqrt{3}\right)\left(x + \sqrt{3}\right)$

$f(x) = (2x+5)\left(x - \sqrt{2}\right)\left(x + \sqrt{2}\right)\left(x - \sqrt{3}\right)\left(x + \sqrt{3}\right)$

y-intercept: (0, 30)

83. $P(x) = a(x+4)(x-1)(x-3)$;

y-intercept; (0,2)

$2 = a(0+4)(0-1)(0-3)$
$2 = 12a$
$\frac{1}{6} = a;$

$P(x) = \frac{1}{6}(x+4)(x-1)(x-3)$

85. a. 3
b. $9-4=5$
c. $P(x) = a(x-4)(x-9)$;
y-intercept: (0,8);
$8 = a(0-4)(0-9)$
$\frac{1}{4} = a$;
$P(x) = \frac{1}{4}x(x-4)(x-9)$
$P(8) = \frac{1}{4}(8)(8-4)(8-9) = -\$80{,}000$

87. End behavior precludes extended use.

89. a. $f(x) \to \infty, f(x) \to -\infty$
b. $g(x) \to \infty, g(x) \to -\infty$, $x^4 \geq 0$for all x

91. $x^5 - x^4 - x^3 + x^2 - 2x + 3 = 0$
Possible rational roots: $\{\pm1, \pm3\}$
Testing these four roots by synthetic division shows there are no rational roots.
Verified

93. $h(x) = (f \circ g)(x) = \left(\frac{1}{x}\right)^2 - 2\left(\frac{1}{x}\right)$
$= \frac{1}{x^2} - \frac{2}{x} = \frac{1-2x}{x^2};$
$D: x \in \{x | x \neq 0\}, R: y \in \{y | y \geq -1\};$
$H(x) = (g \circ f)(x) = \frac{1}{x^2 - 2x};$
$D: x \in \{x | x \neq 0, x \neq 2\}, R: y \in \{y | y \neq 0\}$

95. It is a function.

97. $SA = \pi(10)^2 + 2\pi 10(30) + \frac{1}{2}\left(4\pi(10)^2\right)$
$= 900\pi \approx 2827.4m^2$

Mid-Chapter Check

1. $x^3 + 8x^2 + 7x - 14 = (x^2 + 6x - 5)(x+2) - 4$;
$\frac{x^3 + 8x^2 + 7x - 1}{x+2} 4 = x^2 + 6x - 5 - \frac{4}{x+2};$

$$
\begin{array}{r}
x^2 + 6x - 5 \\
x+2 \overline{\smash{)}\, x^3 + 8x^2 + 7x - 14} \\
-(x^3 + 2x^2) \\
\hline
6x^2 + 7x \\
-(6x^2 + 12x) \\
\hline
-5x - 14 \\
-(-5x - 10) \\
\hline
-4
\end{array}
$$

2. $f(x) = 2x^4 - x^3 - 8x^2 + x + 6$

$$
\begin{array}{r|rrrrr}
2 & 2 & -1 & -8 & 1 & 6 \\
 & & 4 & 6 & -4 & -6 \\
\hline
 & 2 & 3 & -2 & -3 & \lfloor 0
\end{array}
$$

$$
\begin{array}{r|rrrr}
1 & 2 & 3 & -2 & -3 \\
 & & 2 & 5 & 3 \\
\hline
 & 2 & 5 & 3 & \lfloor 0
\end{array}
$$

$f(x) = (2x^2 + 5x + 3)(x-1)(x-2)$
$f(x) = (2x+3)(x+1)(x-1)(x-2)$

3. $f(-2) = 7$

$$
\begin{array}{r|rrrrr}
-2 & -3 & 0 & 7 & -8 & 11 \\
 & & 6 & -12 & 10 & -4 \\
\hline
 & -3 & 6 & -5 & 2 & \lfloor 7
\end{array}
$$

4. $f(x) = (x+2)(x-(1+i))(x-(1-i))$
$f(x) = (x+2)(x-1-i)(x-1+i)$
$f(x) = (x+2)((x-1)^2 - i^2)$
$f(x) = (x+2)(x^2 - 2x + 1 + 1)$
$f(x) = (x+2)(x^2 - 2x + 2)$
$f(x) = x^3 - 2x^2 + 2x + 2x^2 - 4x + 4$
$f(x) = x^3 - 2x + 4$

5. $g(2) = (2)^3 - 6(2) - 4 = -8;$
$g(3) = (3)^3 - 6(3) - 4 = 5;$
They have opposite signs.

6. $f(x) = x^4 + 5x^3 - 20x - 16$
Possible Rational Roots: $\{\pm1, \pm16, \pm2, \pm8, \pm4\}$;
$f(x) = (x-2)(x+1)(x+2)(x+4)$

7. $h(x)=x^4+3x^3+10x^2+6x-20$
   Possible Rational Roots:
   $\{\pm1,\pm20,\pm2,\pm10,\pm4,\pm5\}$;
   $h(x)=(x+2)(x-1)(x^2+2x+10)$
   $x=-2, x=1, x=-1\pm3i$

8. $p(x)=(x+1)^2(x-1)(x-3)$
   end behavior: up/up
   x-intercepts: (1,0),(3,0), (-1,0)
   crosses at (1,0), (3,0), bounces at (-1,0);
   $p(0)=(0+1)^2(0-1)(0-3)=3$
   y-intercept: $(0,3)$

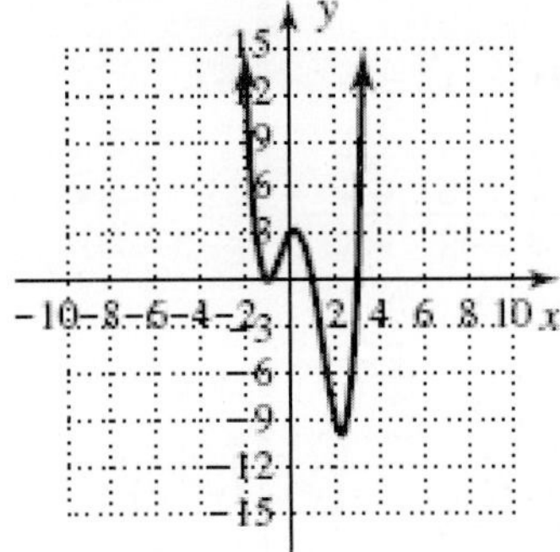

9. $q(x)=x^3+5x^2+2x-8$
   end behavior: down/up
   Possible rational roots: $\{\pm1,\pm8,\pm2,\pm4\}$
   $q(x)=(x+4)(x+2)(x-1)$
   x-intercepts: (-4,0),(-2,0), (1,0)
   crosses at all x-intercepts;
   $q(0)=0^3+5(0)^2+2(0)-8=-8$
   y-intercept: $(0,-8)$

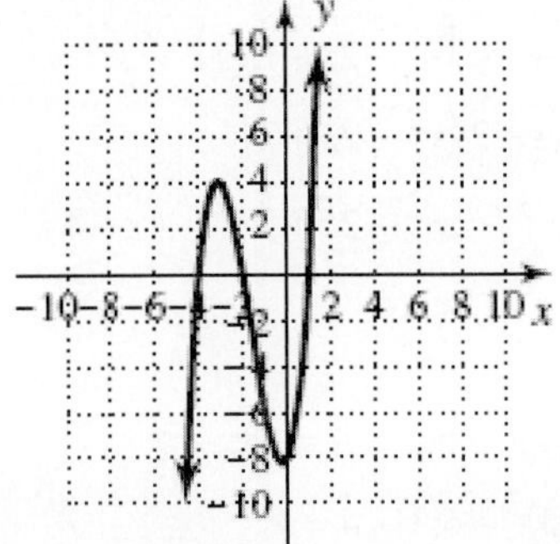

10. a. degree 4; three turning points
    b. two seconds
    c. $A(t)=a(t-1)^2(t-3)(t-5)$;
    y-intercept (0,15)

$15=a(0-1)^2(0-3)(0-5)$
$1=a$
$A(t)=(t-1)^2(t-3)(t-5)$
$A(t)=(t^2-2t+1)(t^2-8t+15)$
$A(t)=t^4-8t^3+15t^2-2t^3+16t^2$
$-30t+t^2-8t+15$
$A(t)=t^4-10t^3+32t^2-38t+15$;
$A(2)=(2)^4-10(2)^3+32(2)^2-38(2)+15=3$
Altitude is 300 ft above the hard-deck.
$A(4)=(4)^4-10(4)^3+32(4)^2-38(4)+15=-9$
Altitude is 900 ft below the hard-deck.

Reinforcing Basic Concepts

1. 1.532

2. -2.152, 1.765

4.5 Technology Highlight

1. Answers will vary.

2. Answers will vary.

4.5 Exercises

1. All real, zeroes

3. Denominator, numerator

5. Horizontal asymptote: $y=3$;
   $x=15$

7. $x-3=0$
   $x=3$
   $D:\{x|x\in R, x\neq 3\}$

9. $x^2-9=0$
   $x^2=9$
   $x=3, x=-3$
   $D:\{x|x\in R, x\neq 3, x\neq -3\}$

11. $2x^2+3x-5=0$
    $(2x+5)(x-1)=0$
    $2x+5=0$ or $x-1=0$
    $2x=-5$ or $x=1$

$x=-\frac{5}{2}$ or $x=1$

$D:\left\{x \mid x\in R, x\neq -\frac{5}{2}, x\neq 1\right\}$

13. $x^2+x+1=0, b^2-4ac<0$

no vertical asymptotes

$D:\{x \mid x\in R\}$

15. $x^2-x-6=0$

$(x-3)(x+2)=0$

$x-3=0$ or $x+2=0$

$x=3$ or $x=-2$

yes yes

17. $x^2-6x+9=0$

$(x-3)(x-3)=0$

$x-3=0$

$x=3$

no

19. $x^3+2x^2-4x-8=0$

$x^2(x+2)-4(x+2)=0$

$(x+2)(x^2-4)=0$

$(x+2)(x+2)(x-2)=0$

$x+2=0$ or $x-2=0$

$x=-2$ or $x=2$

no yes

21. $f(x)=\frac{x^2-3x}{x^2-5}$

$f(x)=\frac{x(x-3)}{x^2-5}$

x-intercepts: (0, 0) cut, (3, 0) cut;
y-intercept: (0, 0)

23. $g(x)=\frac{x^2+3x-4}{x^2-1}$

$g(x)=\frac{(x+4)(x-1)}{(x+1)(x-1)}$

x-intercept: (-4, 0) cut;
y-intercept: (0, 4)

25. $h(x)=\frac{x^3-6x^2+9x}{4-x^2}$

$h(x)=\frac{x(x^2-6x+9)}{4-x^2}$

$h(x)=\frac{x(x-3)(x-3)}{(2+x)(2-x)}$

x-intercepts: (0, 0) cut, (3, 0) bounce;
y-intercept: (0, 0)

27. $Y_1=\frac{2x-3}{x^2+1}$

(a) $HA: y=0$

(b) $2x-3=0$

$x=\frac{3}{2}$

crosses at $\left(\frac{3}{2},0\right)$

29. $r(x)=\frac{4x^2-9}{x^2-3x-18}$

(a) $HA: y=4$

(b) $4x^2-9=4(x^2-3x-18)$

$4x^2-9=4x^2-12x-72$

$12x=-63$

$12x=-63$

$x=-\frac{63}{12}=-\frac{21}{4}$

crosses at $\left(-\frac{21}{4},4\right)$

31. $p(x)=\frac{3x^2-5}{x^2-1}$

(a) $HA: y=3$

(b) $3x^2-5=3(x^2-1)$

$3x^2-5=3x^2-3$

does not cross

33. $p(x)=\frac{3x-1}{x-1}$

$$\begin{array}{r} 3 \\ x-1\overline{)3x-1} \\ -(3x-3) \\ \hline 2 \end{array}$$

$p(x)=\frac{2}{x-1}+3$

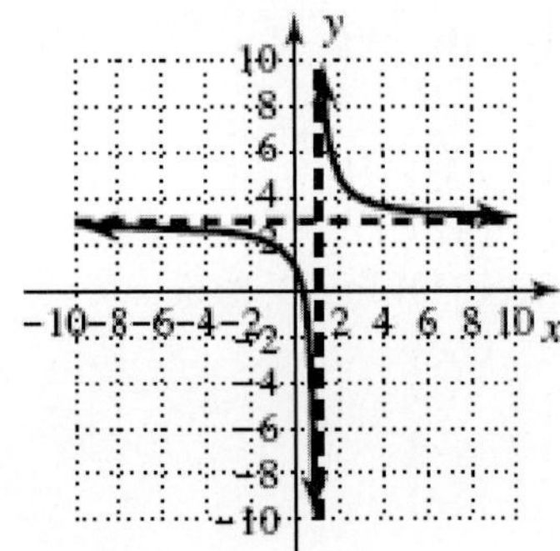

35. $q(x)=\dfrac{x+5}{x+3}$

$$\begin{array}{r} 1 \\ x+3\overline{)x+5} \\ \underline{-(x+3)} \\ 2 \end{array}$$

$q(x)=\dfrac{2}{x+3}+1$

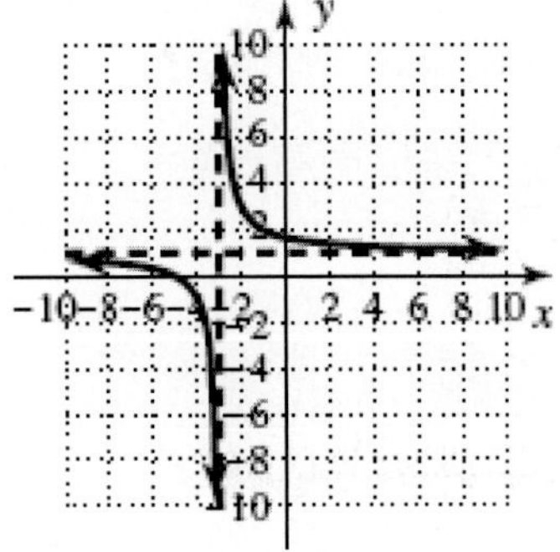

37. $h(x)=\dfrac{x^2-6x+8}{x^2-6x+9}$

$$\begin{array}{r} 1 \\ x^2-6x+9\overline{)x^2-6x+8} \\ \underline{-(x^2-6x+9)} \\ -1 \end{array}$$

$h(x)=-\dfrac{1}{(x-3)^2}+1$

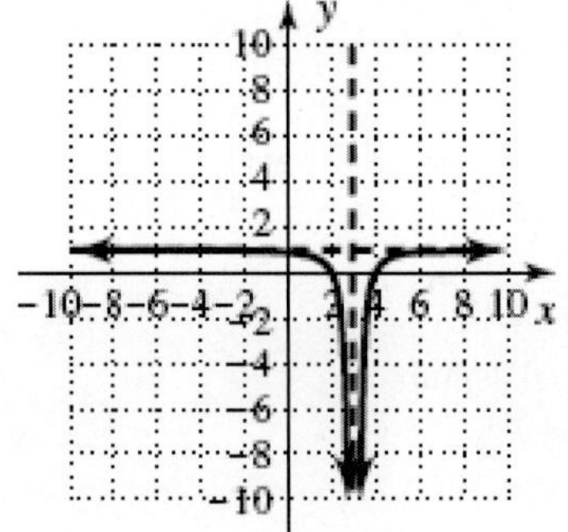

39. $f(x)=\dfrac{x+3}{x-1}$

$f(0)=\dfrac{(0)+3}{(0)-1}=-3;$

y-intercept: (0,-3);

$x-1=0$

vertical asymptote: $x=1$

x-intercept: $(-3,0)$

horizontal asymptote: $y=1$

deg num = deg den

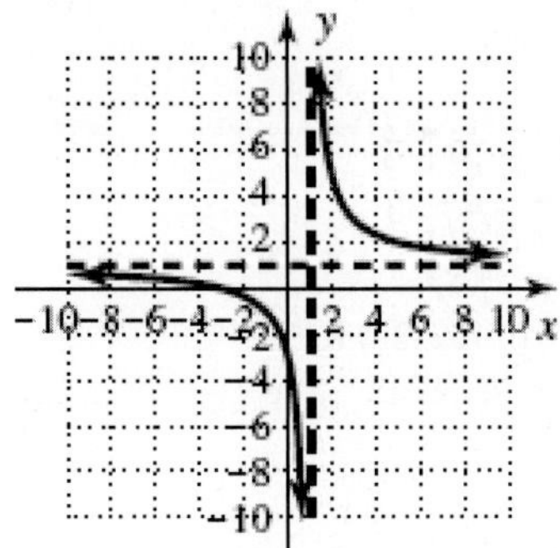

41. $F(x)=\dfrac{8x}{x^2+4}$

$F(0)=\dfrac{8(0)}{(0)^2+4}=0;$

y-intercept: (0,0);

$x^2+4\neq 0$

vertical asymptote: none

x-intercept: $(0,0)$

horizontal asymptote: $y=0$

deg num < deg den

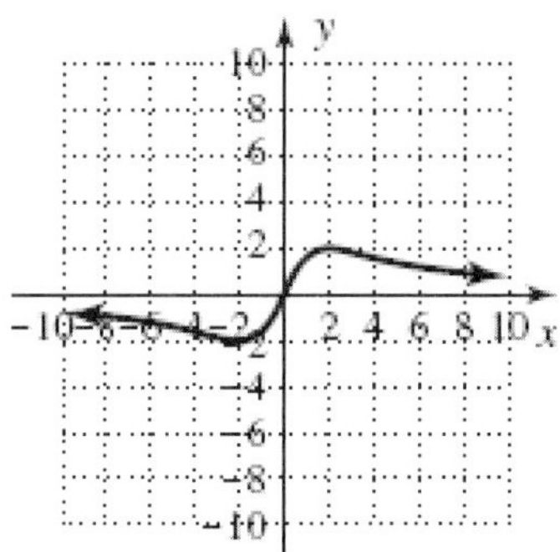

43. $p(x)=\dfrac{-2x^2}{x^2-4}$

$p(0)=\dfrac{-2(0)^2}{(0)^2-4}=0;$

y-intercept: $(0,0)$;

$x^2-4=0$

$(x+2)(x-2)=0$

vertical asymptote: $x=-2$ and $x=2$

x-intercept: $(0,0)$

horizontal asymptote: $y=-2$

deg num = deg den

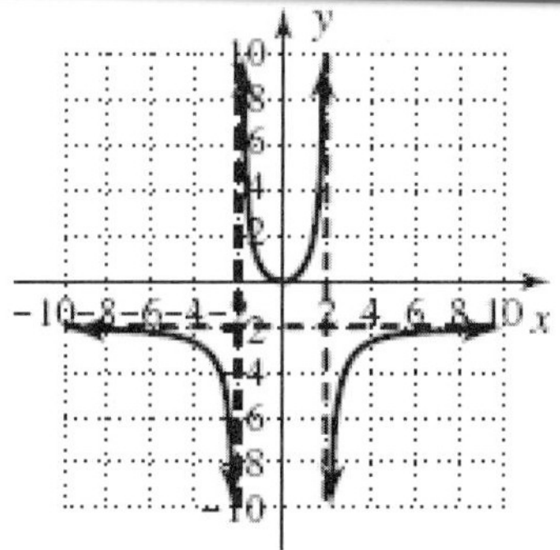

45. $q(x)=\dfrac{2x-x^2}{x^2+4x-5}$

$q(0)=\dfrac{2(0)-(0)^2}{(0)^2+4(0)-5}=0;$

y-intercept: $(0,0)$;

$x^2+4x-5=0$

$(x+5)(x-1)=0$

vertical asymptotes: $x=-5$ and $x=1$

$2x-x^2=0$

$x(2-x)=0$

$x=0$ or $x=2$

x-intercepts: $(0,0),(2,0)$

horizontal asymptote: $y=-1$

deg num = deg den

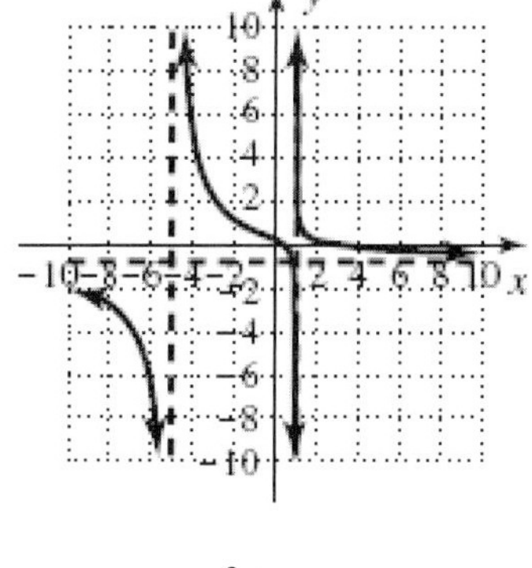

47. $h(x)=\dfrac{-3x}{x^2-6x+9}$

$h(0)=\dfrac{-3(0)}{(0)^2-6(0)+9}=0;$

y-intercept: $(0,0)$;

$x^2-6x+9=0$

$(x-3)(x-3)=0$

vertical asymptote: $x=3$

x-intercept: $(0,0)$

horizontal asymptote: $y=0$

deg num < deg den

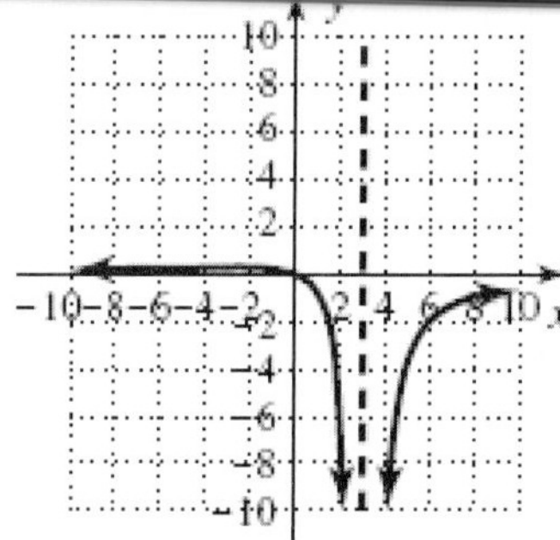

49. $y_1=\dfrac{x-1}{x^2-3x-4}$

$y_1=\dfrac{(0)-1}{(0)^2-3(0)-4}=\dfrac{1}{4};$

y-intercept: $\left(0,\dfrac{1}{4}\right)$;

$x^2-3x-4=0$

$(x-4)(x+1)=0$

vertical asymptotes: $x=4$ and $x=-1$

x-intercept: $(1,0)$

horizontal asymptote: $y=0$

deg num < deg den

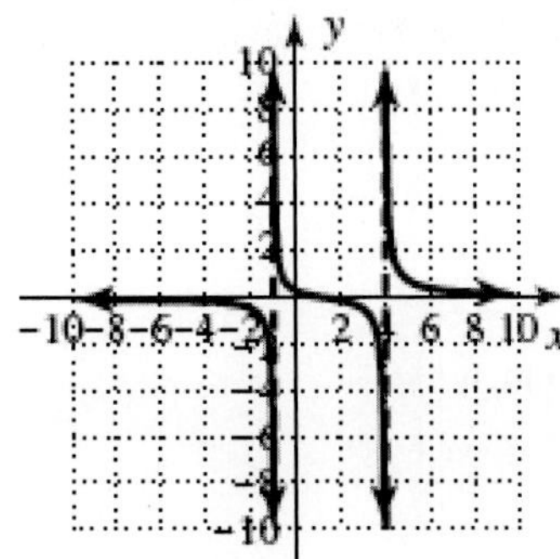

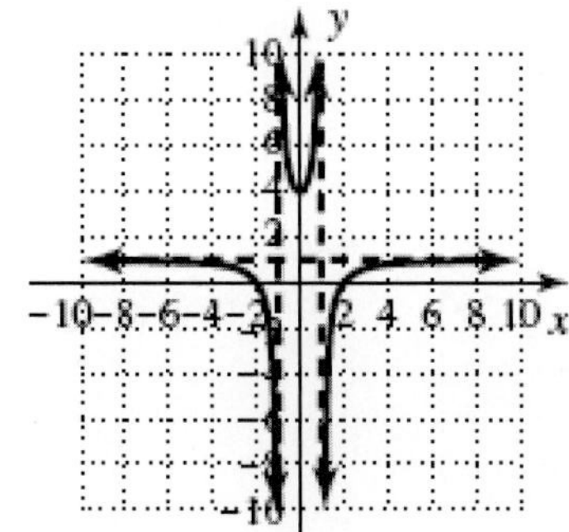

51. $s(x)=\dfrac{4x^2}{2x^2+4}$

$s(x)=\dfrac{4(0)^2}{2(0)^2+4}=0;$

y-intercept: $(0,0)$;

$2x^2+4\neq 0$

vertical asymptotes: none

x-intercept: $(0,0)$

horizontal asymptote: $y=2$

deg num = deg den

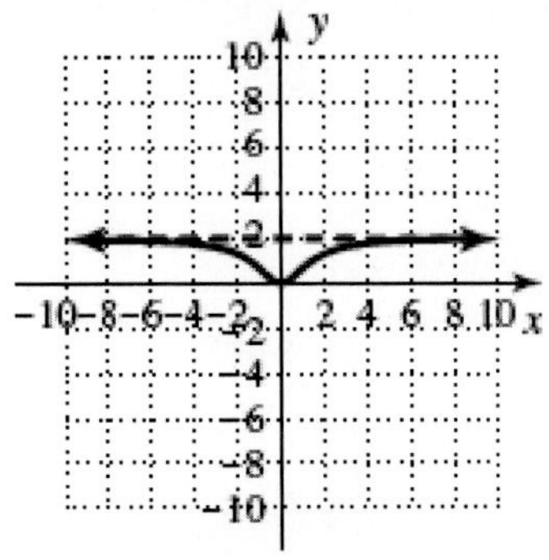

53. $Y_1=\dfrac{x^2-4}{x^2-1}$

$Y_1=\dfrac{(0)^2-4}{(0)^2-1}=4;$

y-intercept: $(0,4)$;

$Y_1=\dfrac{(x+2)(x-2)}{(x+1)(x-1)}$

vertical asymptotes: $x=-1$ and $x=1$

x-intercepts: $(-2,0)$ and $(2,0)$

horizontal asymptote: $y=1$

deg num = deg den

55. $v(t)=\dfrac{-2x}{x^3+2x^2-4x-8}$

$v(t)=\dfrac{-2(0)}{(0)^3+2(0)^2-4(0)-8}=0$

y-intercept: $(0,0)$;

$v(t)=\dfrac{-2x}{x^2(x+2)-4(x+2)}$

$v(t)=\dfrac{-2x}{(x+2)(x^2-4)}=\dfrac{-2x}{(x+2)^2(x-2)}$

vertical asymptotes: $x=-2$ and $x=2$

x-intercepts: $(0,0)$

horizontal asymptote: $y=0$

deg num < deg den

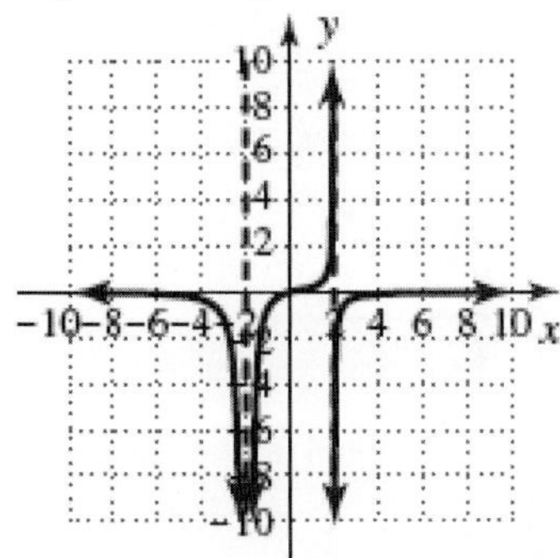

57. VA: $x=-2, x=3$

HA: $y=1$

$f(x)=\dfrac{(x-4)(x+1)}{(x+2)(x-3)}$

58. VA: $x=-3, x=3$

HA: $y=0$

$f(x)=\dfrac{5x}{(x+3)^2(x-3)}$

59. VA: $x=-3, x=3$

HA: $y=-1$

$$f(x)=\frac{x^2-4}{9-x^2}$$

61. $D(x)=\dfrac{63x}{x^2+20}$

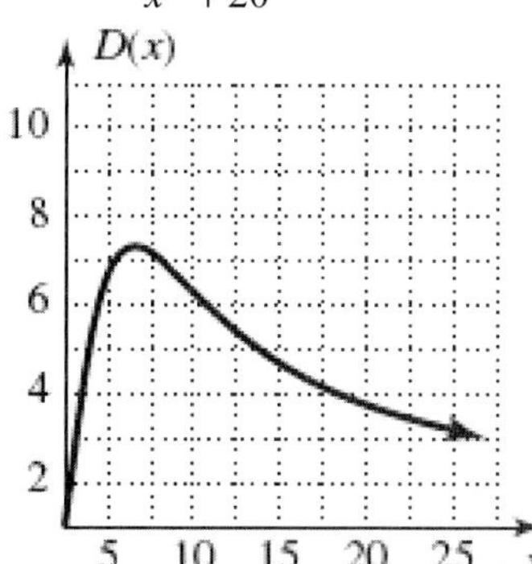

a. Population density approaches zero far from town.
b. 10 miles, 20 miles
c. 4.5 miles, 704

63. $W(t)=\dfrac{6t+40}{t}$

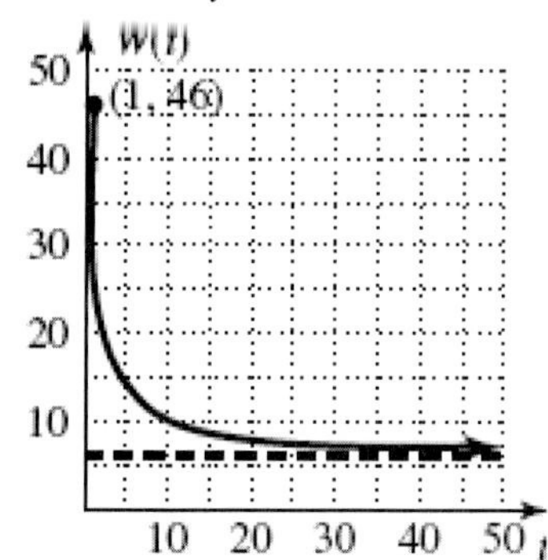

a. 2, 10
b. 10, 14
c. On the average, the number of words remembered for life, 6.

65. a. $C(x)=\dfrac{40+3x}{160+4x}$

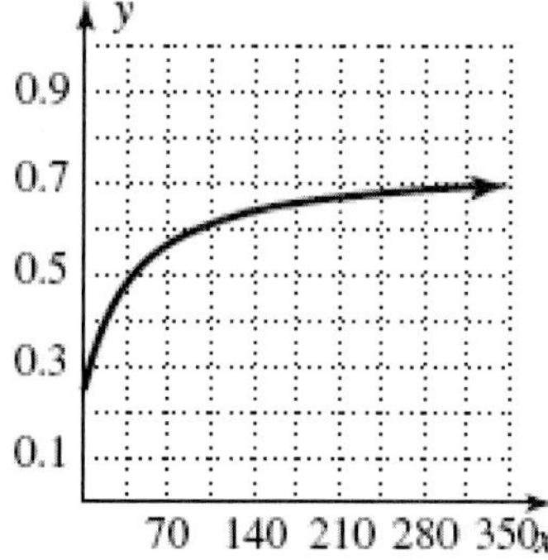

b. 35%, 62.5%, 160 gallons
c. 160 gallons; 200 gallons
d. 70%, 75%

67. $A(x)=\dfrac{125x+50000}{x};[0,5000]$

a. $C(500)=\$225$;
$C(1000)=\$175$

b. $150=\dfrac{125x+50000}{x}$
$150x=125x+50000$
$25x=50000$
$x=2000$ heaters

c. $137.50=\dfrac{125x+50000}{x}$
$137.50x=125x+50000$
$12.50x=50000$
$x=4000$ heaters

d. The horizontal asymptote at $y=125$ means the average cost approaches \$125 as monthly production gets very large. Due to the limitations on production (maximum of 5000 heaters) the average cost will never fall below $A(5000)=135$

69. $G(n)=\dfrac{336+n(95)}{4+n}$

a. $90=\dfrac{336+n(95)}{4+n}$
$90(4+n)=336+n(95)$
$360+90n=336+95n$
$24=5n$
$\dfrac{24}{5}=n$
5 tests

b. $93=\dfrac{336+n(95)}{4+n}$
$93(4+n)=336+n(95)$
$372+93n=336+95n$
$36=2n$
$18=n$

c. HA: $y=95$
$95=\dfrac{336+n(95)}{4+n}$
$95(4+n)=336+n(95)$
$380+95n=336+95n$
$380\neq 336$
The horizontal asymptote at $y=95$ means her average grade will approach

95 as the number of tests taken increases, no.

d. $93=\dfrac{336+n(100)}{4+n}$

$93(4+n)=336+n(100)$

$372+93n=336+100n$

$36=7n$

$n\approx 6$

71. a. $\dfrac{\Delta C}{\Delta x}=\dfrac{\frac{250(61)}{100-61}-\frac{250(60)}{100-60}}{61-60}=16.0;$

$\dfrac{\Delta C}{\Delta x}=\dfrac{\frac{250(71)}{100-71}-\frac{250(70)}{100-70}}{71-70}=28.7;$

$\dfrac{\Delta C}{\Delta x}=\dfrac{\frac{250(81)}{100-81}-\frac{250(80)}{100-80}}{81-80}=65.8;$

$\dfrac{\Delta C}{\Delta x}=\dfrac{\frac{250(91)}{100-91}-\frac{250(90)}{100-90}}{91-90}=277.8;$

b. 12.7, 37.1, 212.0

c. $\dfrac{\Delta C}{\Delta x}=\dfrac{\frac{350(61)}{100-61}-\frac{350(60)}{100-60}}{61-60}=22.4;$

$\dfrac{\Delta C}{\Delta x}=\dfrac{\frac{350(71)}{100-71}-\frac{350(70)}{100-70}}{71-70}=40.2;$

$\dfrac{\Delta C}{\Delta x}=\dfrac{\frac{350(81)}{100-81}-\frac{350(80)}{100-80}}{81-80}=92.1;$

$\dfrac{\Delta C}{\Delta x}=\dfrac{\frac{350(91)}{100-91}-\frac{350(90)}{100-90}}{91-90}=388.9;$

17.8, 51.9, 296.8

Answers may vary.

73. $f(x)=\dfrac{x^3-5x^2-4x+20}{x^3-5x^2-9x+45}$

$f(x)=\dfrac{x^2(x-5)-4(x-5)}{x^2(x-5)-9(x-5)}$

$f(x)=\dfrac{(x-5)(x^2-4)}{(x-5)(x^2-9)}$

$f(x)=\dfrac{(x-5)(x-2)(x+2)}{(x-5)(x-3)(x+3)};$

Vertical asymptotes at $x = -3$ and $x = 3$.

Hole at $\left(5,\dfrac{21}{16}\right);$

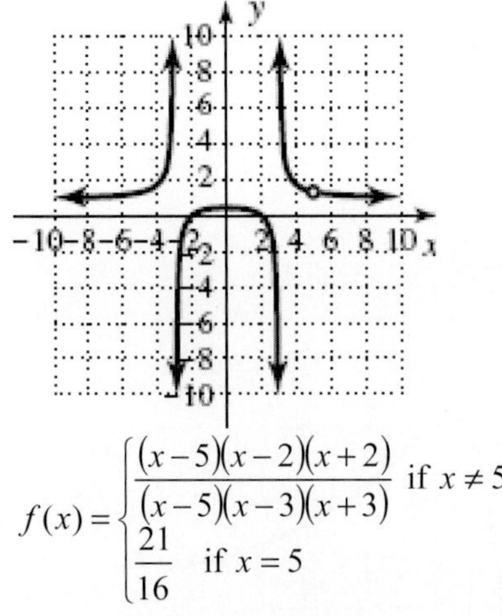

$$f(x)=\begin{cases}\dfrac{(x-5)(x-2)(x+2)}{(x-5)(x-3)(x+3)} & \text{if } x\neq 5\\ \dfrac{21}{16} & \text{if } x=5\end{cases}$$

75. $C:\text{a}+\text{b}i$ where a, b are real, $i=\sqrt{-1}$

$Q:\left\{\dfrac{a}{b}\middle| a\in Z, b\in Z, b\neq 0\right\}$

$Z:\{\ldots-2,-1,0,1,2,\ldots\}$

77. No, $f(x)$ is not one-to-one. Fails horizontal line test.

79. $f(4)=39;$

```
4| 2  -7   5   3
          8   4  36
   2   1   9 |39
```

$f\left(\dfrac{3}{2}\right)=-12;$

```
3/2| 2  -7   5    3
           3  -6  -3/2
     2  -4  -1 |3/2
```

$f(1)=1$

```
2| 2  -7   5   3
         4  -6  -2
   2  -3  -1 |1
```

4.6 Technology Highlight

1. Answers will vary.
2. Answers will vary.
3. Answers will vary.
4. Answers will vary.

4.6 Exercises

1. Non-removable

3. Two

5. Answers will vary.

7. $f(x)=\frac{(x+2)(x-2)}{x+2}$;
$x+2\neq 0$
$x\neq -2$;
$$f(x)=\begin{cases}\frac{x^2-4}{x+2}; & x\neq -2\\ -4 \quad ; & x=-2\end{cases}$$

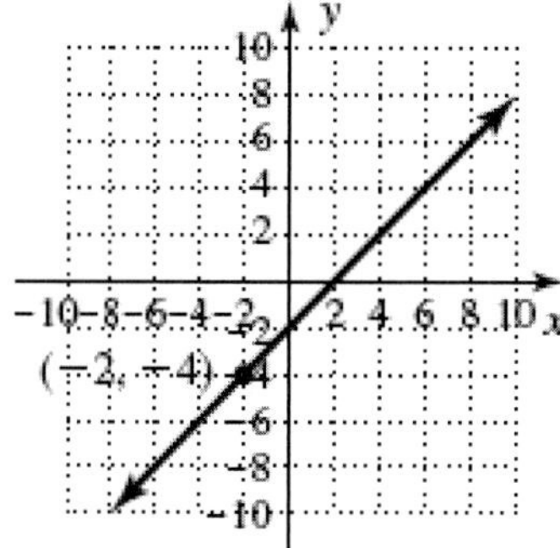

9. $g(x)=\frac{(x-3)(x+1)}{x+1}$;
$x+1\neq 0$
$x\neq -1$;
$$g(x)=\begin{cases}\frac{x^2-2x-3}{x+1}; & x\neq -1\\ -4 \quad ; & x=-1\end{cases}$$

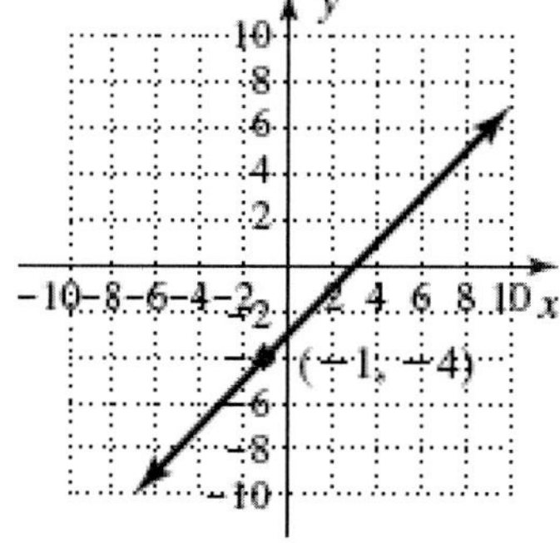

11. $h(x)=\frac{x(3-2x)}{2x-3}=\frac{-x(2x-3)}{2x-3}$
$2x-3\neq 0$
$x\neq \frac{3}{2}$;
$$h(x)=\begin{cases}\frac{3x-2x^2}{2x-3} & ; x\neq \frac{3}{2}\\ -\frac{3}{2} & ; x=\frac{3}{2}\end{cases}$$

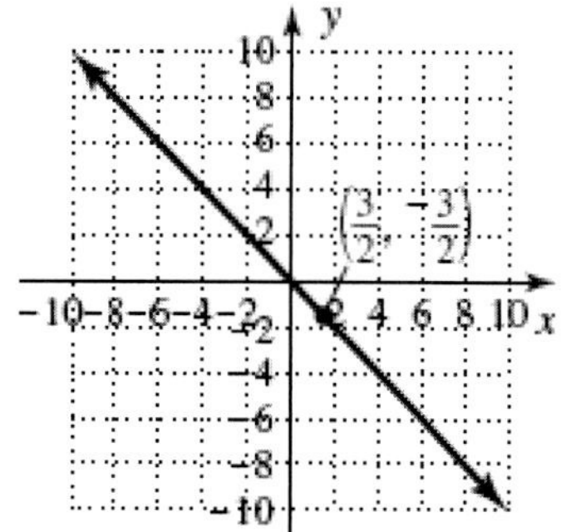

13. $p(x)=\frac{(x-2)(x^2+2x+4)}{x-2}$;
$x-2\neq 0$
$x\neq 2$;
$$p(x)=\begin{cases}\frac{x^3-8}{x-2}; & x\neq 2\\ 12 \quad ; & x=2\end{cases}$$

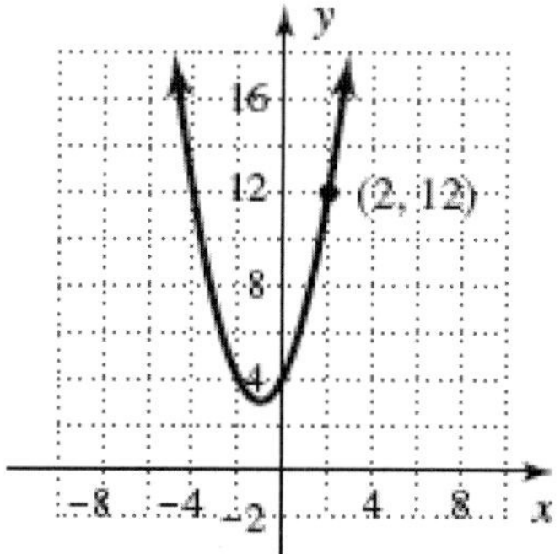

15. $q(x)=\frac{x^3-7x-6}{x+1}$;
$x+1\neq 0$
$x\neq -1$;
$$q(x)=\begin{cases}\frac{x^3-7x-6}{x+1} & x\neq -1\\ -4 & x=-1\end{cases}$$

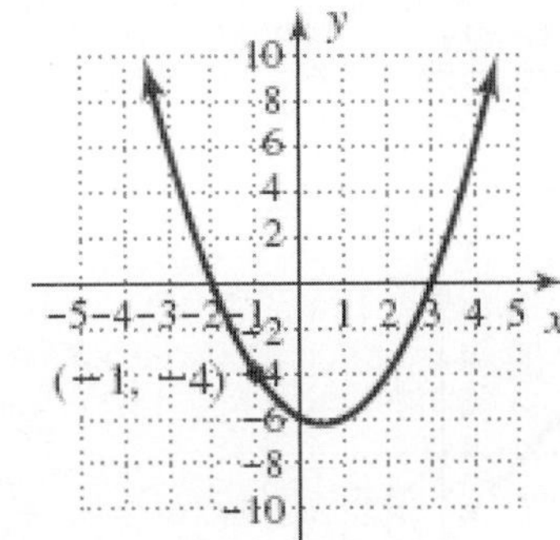

17. $r(x)=\dfrac{x^2(x+3)-(x+3)}{(x+3)(x-1)}=\dfrac{(x+3)(x^2-1)}{(x+3)(x-1)}$
$=\dfrac{(x+3)(x+1)(x-1)}{(x+3)(x-1)}$

$$r(x)=\begin{cases}\dfrac{x^3+3x^2-x-3}{x^2+2x-3} & ;\ x\neq -3, x\neq 1\\ -2 & ;\ x=-3\\ 2 & ;\ x=1\end{cases}$$

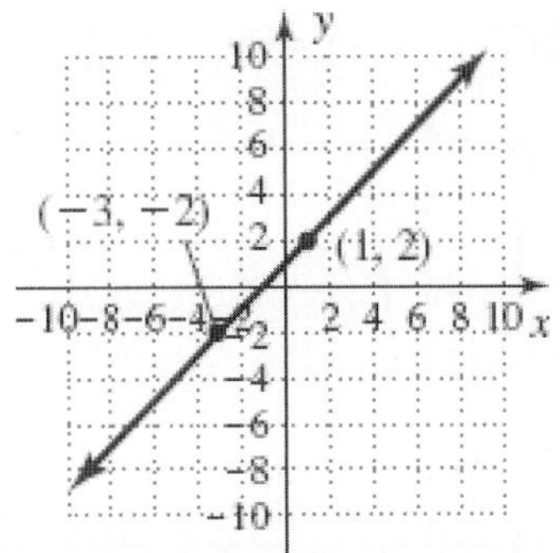

19. $Y_1=\dfrac{x^2-4}{x}$

$x^2-4=0$
$x^2=4$
$x=\pm 2$
x-intercepts: $(-2,0)$ and $(2,0)$;
y-intercept: none;
$Y_1=\dfrac{x^2}{x}-\dfrac{4}{x}=x-\dfrac{4}{x}$
$q(x)=x$
Oblique Asymptote: $y=x$
Vertical Asymptote: $x=0$

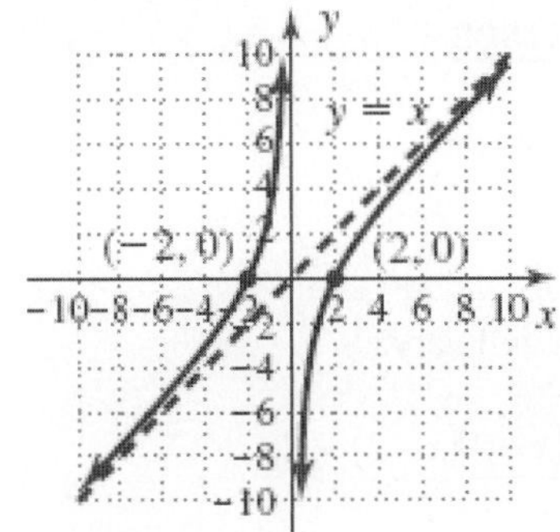

21. $v(x)=\dfrac{3-x^2}{x}$

$3-x^2=0$
$3=x^2$
$\pm\sqrt{3}=x$
x-intercepts: $(-\sqrt{3},0)$ and $(\sqrt{3},0)$
y-intercept: none;
$v(x)=\dfrac{3}{x}-\dfrac{x^2}{x}=\dfrac{3}{x}-x$
$q(x)=-x$
Oblique Asymptote: $y=-x$
Vertical Asymptote: $x=0$

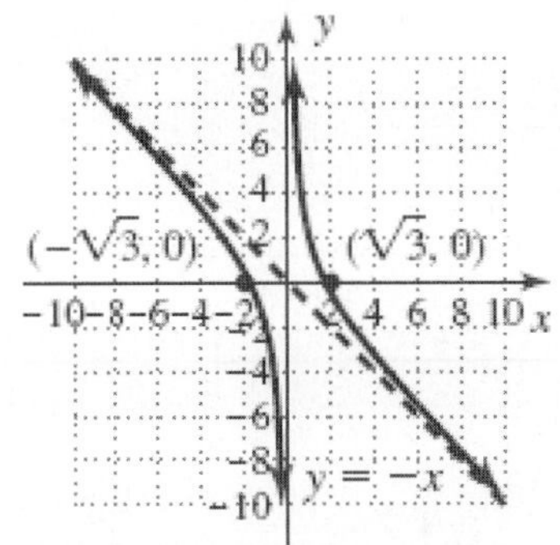

23. $w(x)=\dfrac{x^2+1}{x}$

$x^2+1\neq 0$ (complex solutions)
x-intercepts: none
y-intercept: none;
$w(x)=\dfrac{x^2}{x}+\dfrac{1}{x}=x+\dfrac{1}{x}$
$q(x)=x$
Oblique Asymptote: $y=x$
Vertical Asymptote: $x=0$

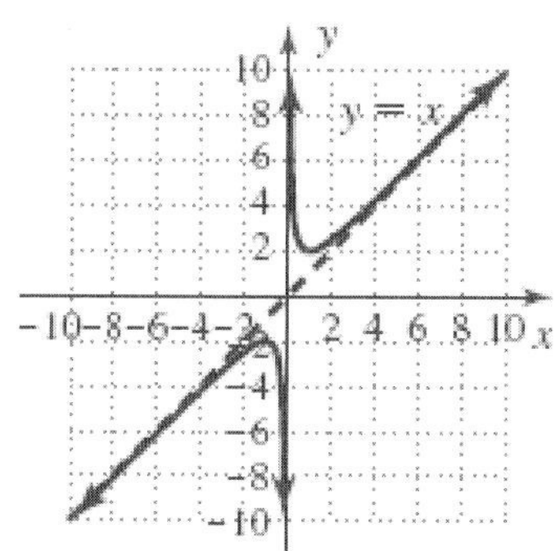

25. $h(x)=\dfrac{x^3-2x^2+3}{x^2}$

$x^3-2x^2+3=0$

Possible rational roots: $\dfrac{\{\pm1,\pm3\}}{\{\pm1\}}$;

$\pm1,\pm3$

x-intercept: (-1, 0)

y-intercept: none;

$h(x)=\dfrac{x^3}{x^2}-\dfrac{2x^2}{x^2}+\dfrac{3}{x^2}=x-2+\dfrac{3}{x^2}$

$q(x)=x-1$

Oblique Asymptote: $y=x-2$

Vertical Asymptote: $x=0$

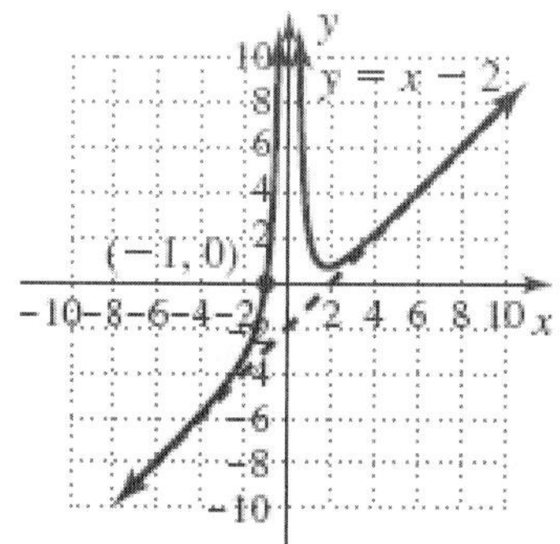

27. $Y_1=\dfrac{x^3+3x^2-4}{x^2}$

$x^3+3x^2-4=0$

Possible rational roots: $\dfrac{\{\pm1,\pm4,\pm2\}}{\{\pm1\}}$;

$\pm1,\pm4,\pm2$

x-intercept: (1, 0)

y-intercept: none;

$Y_1=\dfrac{x^3}{x^2}+\dfrac{3x^2}{x^2}-\dfrac{4}{x^2}=x+3-\dfrac{4}{x^2}$

$q(x)=x+3$

Oblique Asymptote: $y=x+3$

Vertical Asymptote: $x=0$

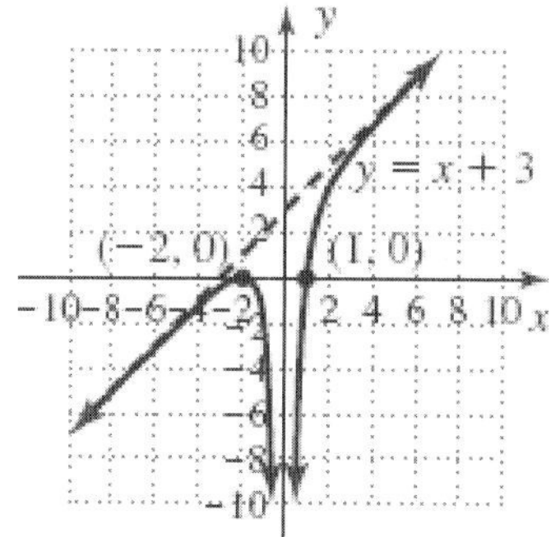

29. $f(x)=\dfrac{x^3-3x+2}{x^2}$

$x^3-3x+2=0$

Possible rational roots: $\dfrac{\{\pm1,\pm2\}}{\{\pm1\}}$;

$\pm1,\pm2$

x-intercept: (-2, 0) and (1,0)

y-intercept: none;

$f(x)=\dfrac{x^3}{x^2}-\dfrac{3x}{x^2}+\dfrac{2}{x^2}=x-\dfrac{3}{x}+\dfrac{2}{x^2}$

$q(x)=x$

Oblique Asymptote: $y=x$

Vertical Asymptote: $x=0$

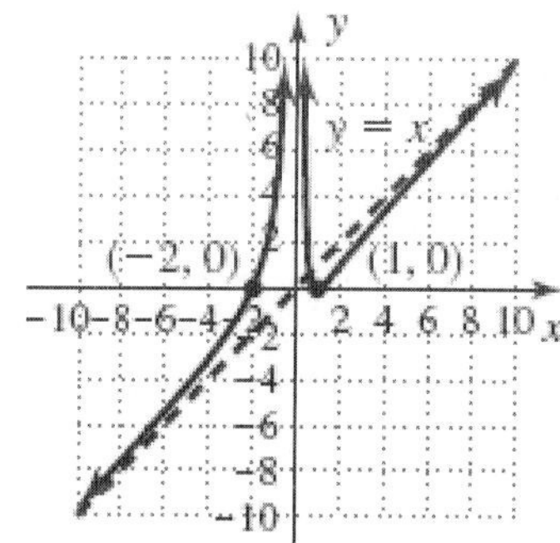

31. $Y_3=\dfrac{x^3-5x^2+4}{x^2}$

$x^3-5x^2+4=0$

Possible rational roots: $\dfrac{\{\pm1,\pm4,\pm2\}}{\{\pm1\}}$;

$\pm1,\pm4,\pm2$

$Y_3=(x-1)(x^2-4x-4)$;

$x=\dfrac{-(-4)\pm\sqrt{(-4)^2-4(1)(-4)}}{2(1)}$

$x=\dfrac{4\pm\sqrt{32}}{2}$

$x = \frac{4 \pm 4\sqrt{2}}{2}$

$x = 2 \pm 2\sqrt{2}$

x-intercepts: (1,0), $(2+\sqrt{2},0)$ and $(2-\sqrt{2},0)$

y-intercept: none;

$Y_3 = \frac{x^3}{x^2} - \frac{5x^2}{x^2} + \frac{4}{x^2} = x - 5 + \frac{4}{x^2}$

$q(x) = x - 5$

Oblique Asymptote: $y = x - 5$

Vertical Asymptote: $x = 0$

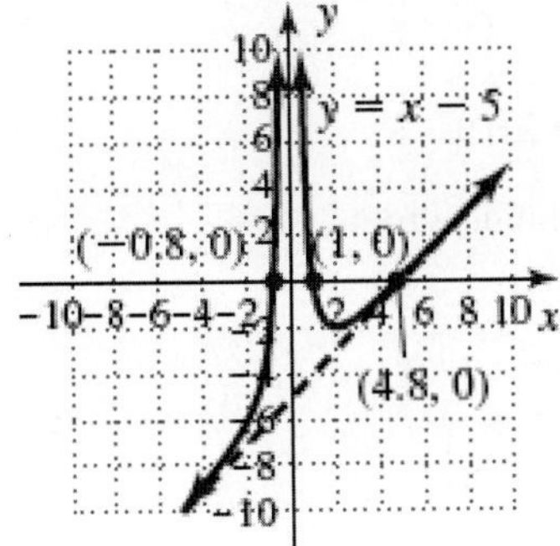

33. $r(x) = \frac{x^3 - x^2 - 4x + 4}{x^2}$

$x^3 - x^2 - 4x + 4 = 0$

$x^2(x-1) - 4(x-1) = 0$

$(x-1)(x^2-4) = 0$

$(x-1)(x+2)(x-2) = 0$

$x - 1 = 0$ or $x + 2 = 0$ or $x - 2 = 0$

$x = 1$ or $x = -2$ or $x = 2$

x-intercepts: $(-2,0)$ and $(1,0)$ and $(2,0)$

y-intercept: none;

$r(x) = \frac{x^3}{x^2} - \frac{x^2}{x^2} - \frac{4x}{x^2} + \frac{4}{x^2} = x - 1 - \frac{4}{x} + \frac{4}{x^2}$

$q(x) = x - 1$

Oblique Asymptote: $y = x - 1$

Vertical Asymptote: $x = 0$

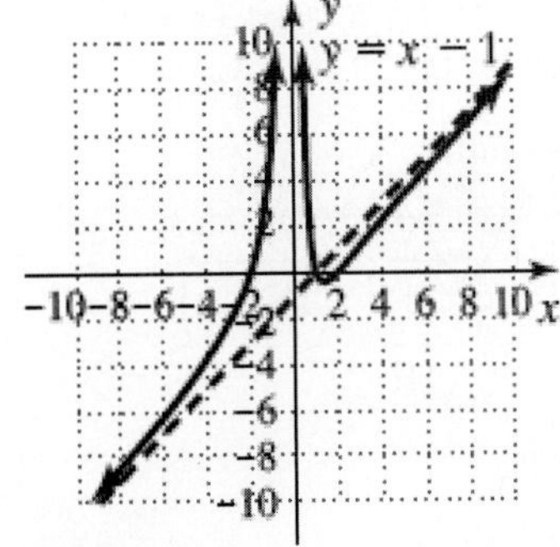

35. $g(x) = \frac{x^2 + 4x + 4}{x + 3}$

$x^2 + 4x + 4 = 0$

$(x+2)(x+2) = 0$

$x + 2 = 0$

$x = -2$

x-intercept: $(-2,0)$

y-intercept: none;

$$\begin{array}{r} x+1 \\ x+3 \overline{)\, x^2 + 4x + 4} \\ -(x^2 + 3x) \\ \hline x + 4 \\ -(x+3) \\ \hline 1 \end{array}$$

Oblique Asymptote: $y = x + 1$

$x + 3 = 0$

Vertical Asymptote: $x = -3$

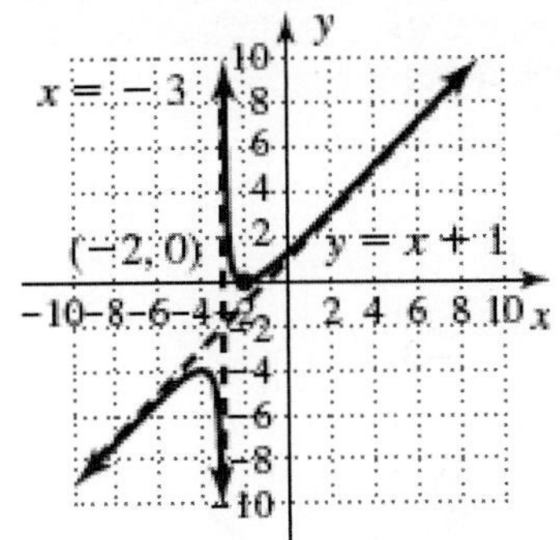

37. $f(x) = \frac{x^2 + 1}{x + 1}$

$x^2 + 1 \neq 0$ (complex solutions)

x-intercepts: none

y-intercept: (0,1);

$$\begin{array}{r} x-1 \\ x+1 \overline{)\, x^2 + 0x + 1} \\ -(x^2 + x) \\ \hline -x + 1 \\ -(-x-1) \\ \hline 2 \end{array}$$

Oblique Asymptote: $y = x - 1$

$x + 1 = 0$

Vertical Asymptote: $x = -1$

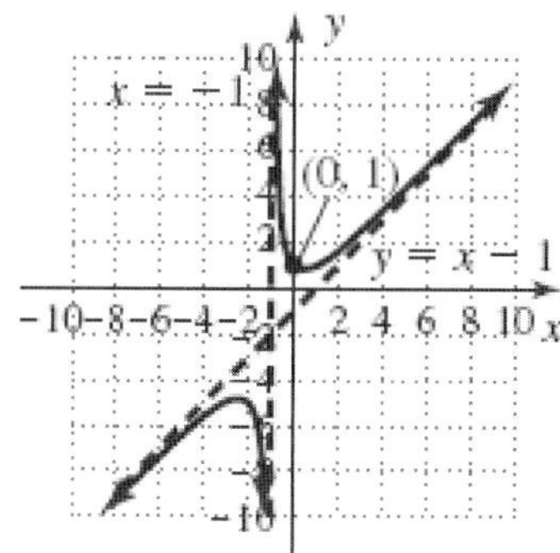

39. $Y_3 = \dfrac{x^2-4}{x+1}$

$x^2-4=0$
$x^2=4$
$x=\pm 2$
x-intercepts: $(-2,0)$ and $(2,0)$;
y-intercept: (0,-4);

$$\begin{array}{r} x-1 \\ x+1\overline{)x^2+0x-4} \\ -(x^2+x) \\ \hline -x-4 \\ -(-x-1) \\ \hline -3 \end{array}$$

Oblique Asymptote: $y=x-1$
$x+1=0$
Vertical Asymptote: $x=-1$

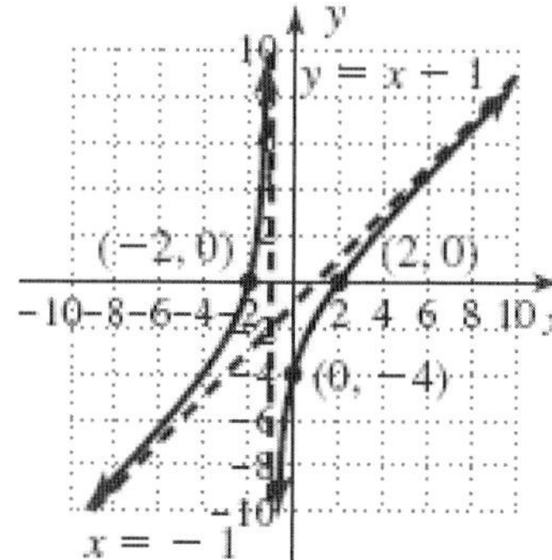

41. $v(x)=\dfrac{x^3-4x}{x^2-1}$

$x(x^2-4)=0$
$x(x+2)(x-2)=0$
$x=0$ or $x+2=0$ or $x-2=0$
$x=0$ or $x=-2$ or $x=2$
x-intercepts: $(-2,0)$, $(2,0)$ and $(0,0)$
y-intercept: (0,0);

$$\begin{array}{r} x \\ x^2-1\overline{)x^3-4x} \\ -(x^3-x) \\ \hline -3x \end{array}$$

Oblique Asymptote: $y=x$

$x^2-1=0$
$(x+1)(x-1)=0$
$x+1=0$ or $x-1=0$
$x=-1$ or $x=1$
Vertical Asymptote: $x=-1$ or $x=1$

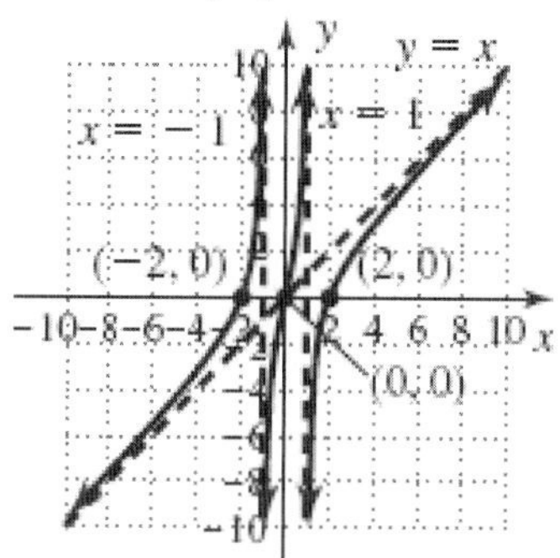

43. $w(x)=\dfrac{16x-x^3}{x^2+4}$

$x(16-x^2)=0$
$x(4+x)(4-x)=0$
$x=0$ or $4+x=0$ or $4-x=0$
$x=0$ or $x=-4$ or $x=4$
x-intercepts: $(-4,0)$, $(4,0)$ and $(0,0)$
y-intercept: (0,0);

$$\begin{array}{r} -x \\ x^2+4\overline{)-x^3+16x} \\ -(-x^3-4x) \\ \hline 20x \end{array}$$

Oblique Asymptote: $y=-x$

$x^2+4\neq 0$ (complex solutions)
Vertical Asymptote: none

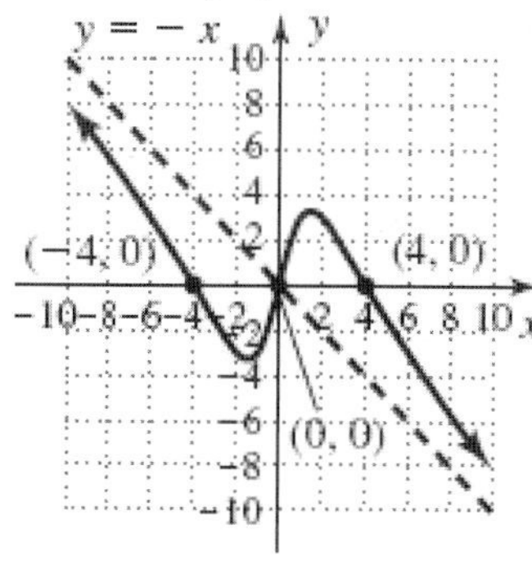

45. $W(x) = \frac{x^3 - 3x + 2}{x^2 - 9}$

$x^3 - 3x + 2 = 0$

Possible rational roots: $\frac{\pm 1, \pm 2}{\pm 1}$;

$\pm 1, \pm 2$

x-intercept: (-2, 0) and (1,0)

y-intercept: $\left(0, -\frac{2}{9}\right)$;

$$\begin{array}{r} x \\ x^2 - 9 \overline{)x^3 - 3x + 2} \\ -\left(x^3 - 9x\right) \\ \hline 6x + 2 \end{array}$$

Oblique Asymptote: $y = x$

$x^2 - 9 = 0$

$(x+3)(x-3) = 0$

$x + 3 = 0$ or $x - 3 = 0$

$x = -3$ or $x = 3$

Vertical Asymptote: $x = -3$ or $x = 3$

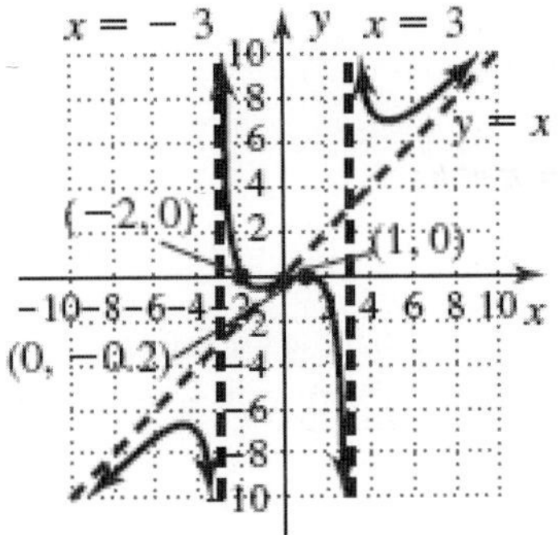

47. $p(x) = \frac{x^4 + 4}{x^2 + 1}$

$x^4 + 4 = 0$

Possible rational roots: $\frac{\pm 1, \pm 4, \pm 2}{\pm 1}$;

$\pm 1, \pm 4, \pm 2$

$x^4 + 4 \neq 0$ (complex solutions)

x-intercept: none

y-intercept: (0,4);

$$\begin{array}{r} x^2 - 1 \\ x^2 + 1 \overline{)x^4 + 0x^2 + 4} \\ -\left(x^4 + x^2\right) \\ \hline -x^2 + 4 \\ -\left(-x^2 - 1\right) \\ \hline 5 \end{array}$$

Oblique Asymptote: $y = x^2 - 1$

$x^2 + 1 \neq 0$ (complex solutions)

Vertical Asymptote: none

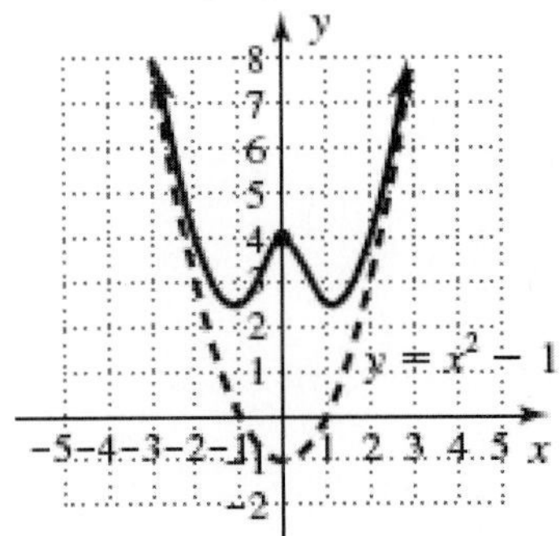

49. $q(x) = \frac{10 + 9x^2 - x^4}{x^2 + 5}$

$10 + 9x^2 - x^4 = 0$

$\left(10 - x^2\right)\left(1 + x^2\right) = 0$

$10 - x^2 = 0$ or $1 + x^2 \neq 0$

$10 = x^2$

$\pm\sqrt{10} = x$

x-intercepts: $\left(-\sqrt{10}, 0\right)$ and $\left(\sqrt{10}, 0\right)$

y-intercept: (0,2);

$$\begin{array}{r} -x^2 + 14 \\ x^2 + 5 \overline{)-x^4 + 9x^2 + 10} \\ -\left(-x^4 - 5x^2\right) \\ \hline 14x^2 + 10 \\ -\left(14x^2 + 70\right) \\ \hline -60 \end{array}$$

Oblique Asymptote: $y = -x^2 + 14$

$x^2 + 5 \neq 0$ (complex solutions)

Vertical Asymptote: none

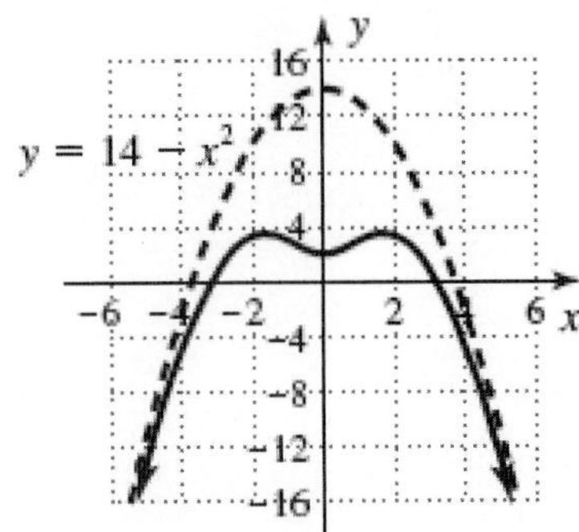

51. $f(x) = \frac{x^3}{x} + \frac{500}{x} = x^2 + \frac{500}{x}$

Oblique Asymptote: $y = x^2$

Minimum: 119.1

53. $A(a)=\frac{1}{2}\left(\frac{ka^2}{a-h}\right)$

$A(a)=\frac{1}{2}\left(\frac{6a^2}{a-5}\right)$

$A(a)=\frac{3a^2}{a-5}$

a. Oblique Asymptote: $y=3a$
$a-5=0$
Vertical Asymptote: $a=5$

b. $A(11)=\frac{3(11)^2}{11-5}=60.5$

c. (10,0)

55. a. $A(x)=\frac{4x^2+53x+250}{x}$
Vertical Asymptote: $x=0$
Oblique Asymptote: $q(x)=4x+53$

b. Cost: \$307, \$372, \$445
Avg Cost: \$307, \$186, \$148.33

c. 8, \$116.25

d. verified

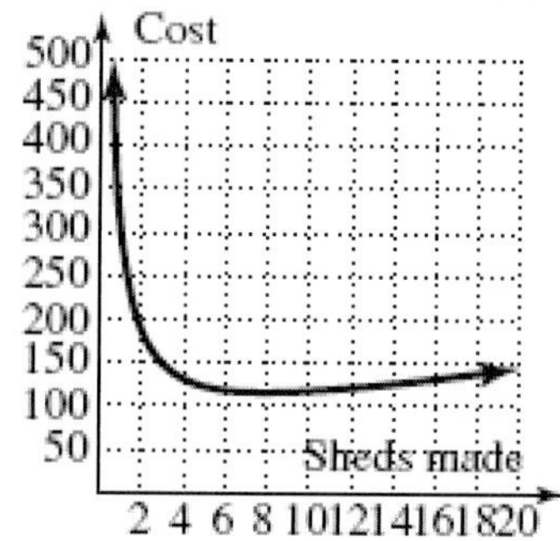

57. a. $S(x,y)=2x^2+4xy$;
$V(x,y)=x^2y$

b. $12=x^2y$

$\frac{12}{x^2}=y$;

$S(x)=2x^2+4x\left(\frac{12}{x^2}\right)$

$=2x^2+\frac{48}{x}=\frac{2x^3+48}{x}$

c. S(x) is asymptotic to $y=2x^2$

d. $x=2$ ft 3.5 in;
$y=2$ ft 3.5 in

59. a. $A(x,y)=xy$;
$R(x,y)=(x-2.5)(y-2)$

b. $60=(x-2.5)(y-2)$

$\frac{60}{x-2.5}=y-2$

$\frac{60}{x-2.5}+2=y$

$\frac{2x+55}{x-2.5}=y$;

$\text{A}(x)=x\left(\frac{60}{x-2.5}+2\right)$

$\text{A}(x)=\frac{60x}{x-2.5}+2x$

$\text{A}(x)=\frac{60x}{x-2.5}+2x\left(\frac{x-2.5}{x-2.5}\right)$

$\text{A}(x)=\frac{60x+2x^2-5x}{x-2.5}=\frac{2x^2+55x}{x-2.5}$

c. A(x) is asymptotic to $y=2x+55$

d. $x\approx 11.16$ in;
$y\approx 8.93$ in

61. a. $V=\pi r^2h$;

$\frac{V}{\pi r^2}=h$

b. $S=2\pi r^2+2\pi r\left(\frac{V}{\pi r^2}\right)=2\pi r^2+\frac{2V}{r}$

c. $S=2\pi r^2+\frac{2V}{r}=\frac{2\pi r^3+2V}{r}$

d. $\frac{1200}{\pi r^2}=h$

$r\approx 5.76$ cm, $h\approx 11.51$ cm;
$\text{S}\approx 625.13\ \text{cm}^3$

63. Answers will vary.

65. a. $m=\frac{k-0}{h-a}=\frac{k}{h-a}$;

$y-0=\frac{k}{h-a}(x-a)$

$y=\frac{k(x-a)}{h-a}$

b. x-intercept: $(a,0)$;

$y=\frac{k(0-a)}{h-a}=\frac{-ka}{h-a}$

y-intercept: $\left(0, \dfrac{-ka}{h-a}\right)$

c. $A = \dfrac{1}{2}\left(\dfrac{-ka}{h-a}\right)a = \dfrac{1}{2}\left(\dfrac{-ka^2}{h-a}\right)$

d. base $a = 2h$; height $y = 2k$; triangle is isosceles

67. $g(x) = 2x^2 - 8x + 3$

$g(x) = 2(x^2 - 4x) + 3$

$g(x) = 2(x^2 - 4x + 4) + 3 - 8$

$g(x) = 2(x-2)^2 - 5$;

$x = \dfrac{-(-8) \pm \sqrt{(-8)^2 - 4(2)(3)}}{2(2)} = \dfrac{8 \pm \sqrt{40}}{4}$

$= \dfrac{8 \pm 2\sqrt{10}}{4} = \dfrac{4 \pm \sqrt{10}}{2}$

a. $\left(-\infty, \dfrac{4-\sqrt{10}}{2}\right) \cup \left(\dfrac{4-\sqrt{10}}{2}, \infty\right)$

b. $(-\infty, 2)$

69. $\dfrac{5i}{1+2i} \cdot \dfrac{1-2i}{1-2i} = \dfrac{5i - 10i^2}{1-4i^2} = \dfrac{10+5i}{5} = 2 + i$

71. a. $b^2 - 4ac > 0, b^2 - 4ac$ a perfect square

b. $b^2 - 4ac > 0$ but not a perfect square

c. $b^2 - 4ac = 0$

d. none

e. none

f. $b^2 - 4ac < 0$

4.7 Technology Highlight

1. $P(x) < 0 : x \in (-3.1, -1.7) \cup (1.3, 2.4)$

4.7 Exercises

1. Vertical, multiplicity

3. 0

5. Answers will vary.

7. $(x+3)(x-5) < 0$

pos neg pos
-3 5

$x \in (-3, 5)$

9. $(x+1)^2(x-4) \geq 0$

neg neg pos
-1 4

$x \in [4, \infty)$

11. $(x+2)^3(x-2)^2(x-4) \leq 0$

pos neg neg pos
-2 2 4

$x \in [-2, 4]$

13. $x^2 + 4x + 1 < 0$;

$x = \dfrac{-(4) \pm \sqrt{(4)^2 - 4(1)(1)}}{2(1)} = \dfrac{-4 \pm \sqrt{12}}{2}$

$= \dfrac{-4 \pm 2\sqrt{3}}{2} = -2 \pm \sqrt{3}$;

pos neg pos
$-2-\sqrt{3}$ $-2+\sqrt{3}$

$x \in \left(-2-\sqrt{3}, -2+\sqrt{3}\right)$

15. $x^3 + x^2 - 5x + 3 \leq 0$

Possible rational roots: $\dfrac{\{\pm 1, \pm 3\}}{\{\pm 1\}}$;

$\{\pm 1, \pm 3\}$

```
1| 1   1   -5    3
       1    2   -3
   ----------------
   1   2   -3  |0
```

$(x-1)(x^2 + 2x - 3) \leq 0$

$(x-1)(x+3)(x-1) \leq 0$

$(x+3)(x-1)^2 \leq 0$

neg pos pos
-3 1

$x \in (-\infty, -3] \cup \{1\}$

17. $x^3 - 7x + 6 > 0$

Possible rational roots: $\dfrac{\{\pm 1, \pm 6, \pm 2, \pm 3\}}{\{\pm 1\}}$;

$\{\pm 1, \pm 6, \pm 2, \pm 3\}$

```
1| 1   0   -7    6
       1    1   -6
   ----------------
   1   1   -6  |0
```

$(x-1)(x^2+x-6)>0$
$(x+3)(x-1)(x-2)>0$

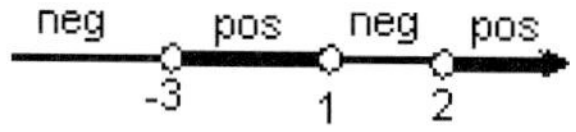

neg -3 pos 1 neg 2 pos

$x \in (-3,1) \cup (2,\infty)$

19. $x^4-10x^2>-9$
$x^4-10x^2+9>0$
$(x^2-1)(x^2-9)>0$
$(x+1)(x-1)(x+3)(x-3)>0$

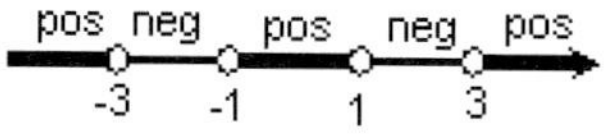

pos -3 neg -1 pos 1 neg 3 pos

$x \in (-\infty,-3) \cup (-1,1) \cup (3,\infty)$

21. $x^4-9x^2>4x-12$
$x^4-9x^2-4x+12>0$
Possible rational roots:
$\frac{\{\pm 1,\pm 12,\pm 2,\pm 6,\pm 3,\pm 4\}}{\{\pm 1\}}$;
$\{\pm 1,\pm 12,\pm 2,\pm 6,\pm 3,\pm 4\}$

1	1	0	-9	-4	12
		1	1	-8	-12
	1	1	-8	-12	0

3	1	1	-8	-12
		3	12	12
	1	4	4	0

$(x-1)(x-3)(x^2+4x+4)>0$
$(x-1)(x-3)(x+2)^2>0$

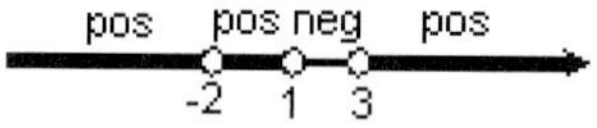

pos -2 pos 1 neg 3 pos

$x \in (-\infty,-2) \cup (-2,1) \cup (3,\infty)$

23. $x^4-6x^3 \le -8x^2-6x+9$
$x^4-6x^3+8x^2+6x-9 \le 0$
Possible rational roots: $\frac{\{\pm 1,\pm 9,\pm 3\}}{\{\pm 1\}}$;
$\{\pm 1,\pm 9,\pm 3\}$

-1	1	-6	8	6	-9
		-1	7	-15	9
	1	-7	15	-9	0

1	1	-7	15	-9
		1	-6	9
	1	-6	9	0

$(x+1)(x-1)(x^2-6x+9) \le 0$
$(x+1)(x-1)(x-3)^2 \le 0$

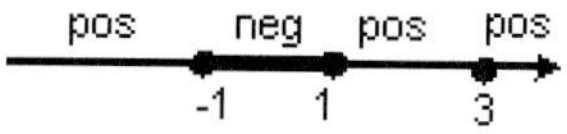

pos -1 neg 1 pos 3 pos

$x \in [-1,1] \cup \{3\}$

25. $\frac{x+3}{x-2} \le 0$

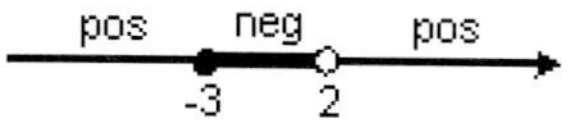

pos -3 neg 2 pos

$x \in [-3,2)$

27. $\frac{x+1}{x^2+4x+4}<0$
$\frac{x+1}{(x+2)^2}<0$

neg -2 neg -1 pos

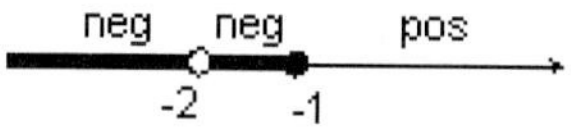

$x \in (-\infty,-2) \cup (-2,-1)$

29. $\frac{2-x}{x^2-x-6} \ge 0$
$\frac{2-x}{(x-3)(x+2)} \ge 0$

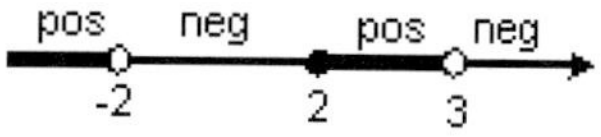

pos -2 neg 2 pos 3 neg

$x \in (-\infty,-2) \cup [2,3)$

31. $\frac{2x-x^2}{x^2+4x-5}<0$
$\frac{x(2-x)}{(x+5)(x-1)}<0$

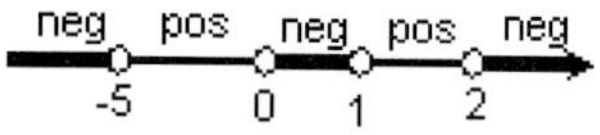

neg -5 pos 0 neg 1 pos 2 neg

$x \in (-\infty,-5) \cup (0,1) \cup (2,\infty)$

33. $\frac{x^2-4}{x^3-13x+12} \ge 0$
Possible rational roots of denominator:
$\frac{\pm 1,\pm 12,\pm 2,\pm 6,\pm 3,\pm 4}{1}$;
$\pm 1,\pm 12,\pm 2,\pm 6,\pm 3,\pm 4$

$$\begin{array}{r|rrrr} 1 & 1 & 0 & -13 & 12 \\ & & 1 & 1 & -12 \\ \hline & 1 & 1 & -12 & \boxed{0} \end{array}$$

$x^3 - 13x + 12 = (x-1)(x^2 + x - 12)$
$= (x-1)(x+4)(x-3)$;

$\dfrac{(x+2)(x-2)}{(x-1)(x+4)(x-3)} \geq 0$

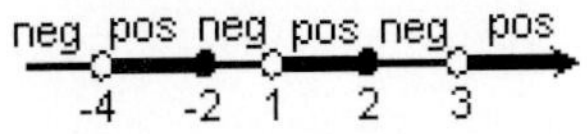

neg pos neg pos neg pos: -4, -2, 1, 2, 3

$x \in (-4,-2] \cup (1,2] \cup (3,\infty)$

35. $\dfrac{x^2+5x-14}{x^3+x^2-5x+3} > 0$

Possible rational roots of denominator:

$\dfrac{\pm 1, \pm 3}{1}$; $\pm 1, \pm 3$;

$$\begin{array}{r|rrrr} 1 & 1 & 1 & -5 & 3 \\ & & 1 & 2 & -3 \\ \hline & 1 & 2 & -3 & \boxed{0} \end{array}$$

$x^3 + x^2 - 5x + 3 = (x-1)(x^2+2x-3)$
$= (x-1)(x+3)(x-1)$;

$\dfrac{(x+7)(x-2)}{(x-1)^2(x+3)} > 0$

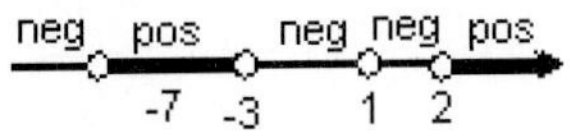

neg pos neg neg pos: -7, -3, 1, 2

$x \in (-7,-3) \cup (2,\infty)$

37. $\dfrac{2}{x-2} \leq \dfrac{1}{x}$

$\dfrac{2}{x-2} - \dfrac{1}{x} \leq 0$

$\dfrac{2x-x+2}{x(x-2)} \leq 0$

$\dfrac{x+2}{x(x-2)} \leq 0$

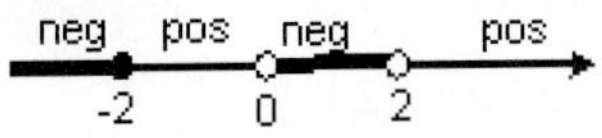

neg pos neg pos: -2, 0, 2

$x \in (-\infty,-2] \cup (0,2)$

39. $\dfrac{x-3}{x+17} > \dfrac{1}{x-1}$

$\dfrac{x-3}{x+17} - \dfrac{1}{x-1} > 0$

$\dfrac{(x-3)(x-1)-1(x+17)}{(x+17)(x-1)} > 0$

$\dfrac{x^2-4x+3-x-17}{(x+17)(x-1)} > 0$

$\dfrac{x^2-5x-14}{(x+17)(x-1)} > 0$

$\dfrac{(x-7)(x+2)}{(x+17)(x-1)} > 0$

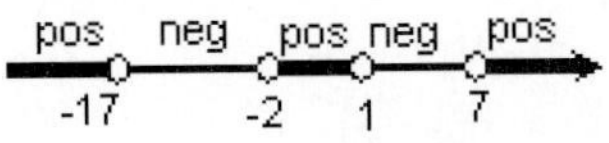

pos neg pos neg pos: -17, -2, 1, 7

$x \in (-\infty,-17) \cup (-2,1) \cup (7,\infty)$

41. $\dfrac{x+1}{x-2} \geq \dfrac{x+2}{x+3}$

$\dfrac{x+1}{x-2} - \dfrac{x+2}{x+3} \geq 0$

$\dfrac{(x+1)(x+3)-(x+2)(x-2)}{(x-2)(x+3)} \geq 0$

$\dfrac{x^2+4x+3-x^2+4}{(x-2)(x+3)} \geq 0$

$\dfrac{4x+7}{(x-2)(x+3)} \geq 0$

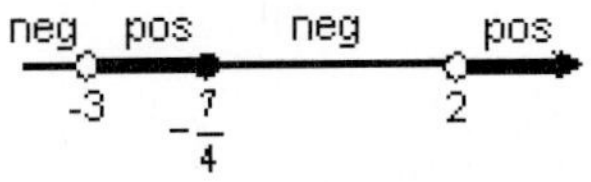

neg pos neg pos: -3, $-\frac{7}{4}$, 2

$x \in \left(-3,-\dfrac{7}{4}\right] \cup (2,\infty)$

43. $\dfrac{x+2}{x^2+9} > 0$

x^2+9 has no real roots

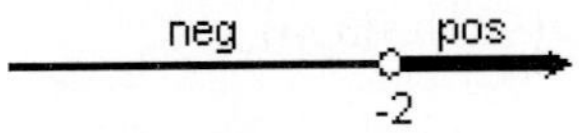

neg pos: -2

$x \in (-2,\infty)$

45. $\dfrac{x^3+1}{x^2+1} > 0$

$\dfrac{(x+1)(x^2-x+1)}{x^2+1} > 0$

x^2-x+1, x^2+1 have no real roots

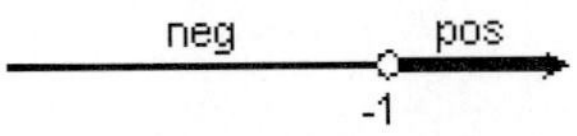

neg pos: -1

$x \in (-1,\infty)$

47. $\frac{x^4-5x^2-36}{x^2-2x+1}>0$

$\frac{(x^2-9)(x^2+4)}{(x-1)^2}>0$

$\frac{(x+3)(x-3)(x^2+4)}{(x-1)^2}>0$

x^2+4 has no real roots

pos neg neg pos
-3 1 3

$x\in(-\infty,-3)\cup(3,\infty)$

49. b

51. b

53. a. $D=-(4p^3+27(p+1)^2)$

$D=-(4p^3+27(p^2+2p+1))$

$D=-(4p^3+27p^2+54p+27)$

verified

b. $-(4p^3+27p^2+54p+27)=0$

Possible rational roots: $\frac{\{\pm1,\pm3,\pm9,\pm27\}}{\{\pm1,\pm2,\pm4\}}$

$D=-(p+3)^2\left(p+\frac{3}{4}\right)$

$p=-3, q=-3+1=-2$

$p=-\frac{3}{4}, q=-\frac{3}{4}+1=\frac{1}{4}$

c. $-(p+3)^2\left(p+\frac{3}{4}\right)>0$

$(-\infty,-3)\cup\left(-3,-\frac{3}{4}\right)$

d. Verified

55. $d(x)=k(x^3-192x+1024)$

a. $\frac{k(x^3-3(8)^2x+2(8)^3)}{k}<189$

$x^3-192x+1024<189$

$x^3-192x+835<0$

Possible rational roots:
$\pm1,\pm835,\pm5,\pm167$

$(x-5)(x^2+5x-167)<0$

$x\in(5,8]$

b. $(4)^3-192(4)+1024=320$ units

c. $\frac{k(x^3-3(8)^2x+2(8)^3)}{k}>475$

$x^3-192x+1024>475$

$x^3-192x+549>0$

Possible rational roots:

$\frac{\{\pm1,\pm3,\pm9,\pm61,\pm183,\pm549\}}{\{\pm1\}}$

$(x-3)(x^2+3x-183)>0$

$x\in[0,3)$

d. $\frac{k(x^3-3(8)^2x+2(8)^3)}{k}\le648$

$x^3-192x+1024\le648$

$x^3-192x+376\le0$

Possible rational roots:

$\frac{\{\pm1,\pm2,\pm4,\pm8,\pm47,\pm94,\pm188,\pm376\}}{\{\pm1\}}$

$(x-2)\left(x^2+2x-188\right)\le0$

2 feet

57. $f(x)=\sqrt{2x^3-x^2-16x+15}$

$2x^3-x^2-16x+15\ge0$

Possible rational roots:

$\left\{\pm1,\pm15,\pm3,\pm5,\pm\frac{1}{2},\pm\frac{15}{2},\pm\frac{3}{2},\pm\frac{5}{2}\right\}$

$(x+3)(x-1)\left(x-\frac{5}{2}\right)\ge0$

neg pos neg pos
-3 1 5/2

$x\in[-3,1]\cup\left[\frac{5}{2},\infty\right)$

59. $p(x)=\sqrt[4]{\frac{x+2}{x^2-2x-35}}$

$\frac{x+2}{(x-7)(x+5)}\ge0; x\ne7, x\ne-5$

neg pos neg pos
-5 -2 7

$x\in(-5,-2]\cup(7,\infty)$

61. a. $R=\frac{2D}{t_1+t_2}$

$$40 = \frac{2(80)}{t_1 + t_2}$$
$$1 = \frac{4}{t_1 + t_2}$$
$$1 = \frac{4}{\frac{80}{r_1} + \frac{80}{r_2}}$$
$$1 = \frac{4r_1 r_2}{80r_1 + 80r_2}$$
$$80r_1 + 80r_2 = 4r_1 r_2$$
$$20r_1 + 20r_2 = r_1 r_2$$
$$20r_2 - r_1 r_2 = -20r_1$$
$$r_2(20 - r_1) = -20r_1$$
$$r_2 = \frac{-20r_1}{20 - r_1}$$
$$r_2 = \frac{20r_1}{r_1 - 20}$$
Verified

b. Horizontal: $r_2 = 20$, as r_1 increases, r_2 decreases to maintain $R = 40$.
Vertical: $r_1 = 20$, as r_1 decreases, r_2 increases to maintain $R = 40$.

c. $$\frac{20r_1}{r_1 - 20} > r_1$$
$$\frac{20r_1}{r_1 - 20} - r_1 > 0$$
$$\frac{20r_1}{r_1 - 20} - \frac{r_1(r_1 - 20)}{r_1 - 20} > 0$$
$$\frac{20r_1 - r_1^2 + 20r_1}{r_1 - 20} > 0$$
$$\frac{40r_1 - r_1^2}{r_1 - 20} > 0$$
$$\frac{r_1(40 - r_1)}{r_1 - 20} > 0$$
Critical points: 0, 20, 40
$r_1 \in (20, 40)$

63. a. $$0.01t^2 + 0.1t + 30 < 42$$
$$0.01t^2 + 0.1t - 12 < 0$$
$$t^2 + 10t - 1200 < 0$$
$$(t + 40)(t - 30) < 0$$
$$\left(0, 30^\circ\right)$$

b. $$0.01t^2 + 0.1t + 30 > 36$$
$$0.01t^2 + 0.1t - 6 > 0$$
$$t^2 + 10t - 600 > 0$$
$$(t - 20)(t + 30) > 0$$
$$\left(20^\circ, \infty\right)$$

c. $$0.01t^2 + 0.1t + 30 > 60$$
$$0.01t^2 + 0.1t - 30 > 0$$
$$t^2 + 10t - 3000 > 0$$
$$(t + 60)(t - 50) > 0$$
$$\left(50^\circ, \infty\right)$$

65. a. $$\frac{2n^3 + 3n^2 + n}{6} \ge 30$$
$$2n^3 + 3n^2 + n \ge 180$$
$$2n^3 + 3n^2 + n - 180 \ge 0$$
Possible rational roots:
$$\left\{\pm 1, \pm 180, \pm 2, \pm 90, \pm 3, \pm 60, \pm 4, \pm 45, \pm 5, \pm 36, \pm 6, \pm 30, \pm 9, \pm 20, \pm 10, \pm 18, \pm 12, \pm 15, \pm\frac{1}{2}, \pm\frac{3}{2}, \pm\frac{45}{2}, \pm\frac{5}{2}, \pm\frac{9}{2}, \pm\frac{15}{2}\right\}$$
$$(n - 4)(2n^2 + 11n + 45) \ge 0$$
$$n \ge 4$$

b. $$\frac{2n^3 + 3n^2 + n}{6} \le 285$$
$$2n^3 + 3n^2 + n \le 1710$$
$$2n^3 + 3n^2 + n - 1710 \le 0$$
Possible rational roots:
$$\left\{\pm 1, \pm 1710, \pm 2, \pm 855, \pm 3, \pm 570, \pm 5, \pm 342, \pm 6, \pm 285, \pm 9, \pm 190, \pm 10, \pm 171, \pm 15, \pm 114, \pm 18, \pm 95, \pm 19, \pm 90, \pm 30, \pm 57, \pm 38, \pm 45, \pm\frac{1}{2}, \pm\frac{855}{2}, \pm\frac{3}{2}, \pm\frac{5}{2}, \pm\frac{285}{2}, \pm\frac{9}{2}, \pm\frac{171}{2}, \pm\frac{15}{2}, \pm\frac{95}{2}, \pm\frac{19}{2}, \pm\frac{57}{2}, \pm\frac{45}{2}\right\}$$
$$(n - 9)(2n^2 + 21n + 190) \le 0$$
$$n \le 9$$

c. $$\frac{2n^3 + 3n^2 + n}{6} \le 999$$
$$2n^3 + 3n^2 + n \le 5994$$
$$2n^3 + 3n^2 + n - 5994 \le 0$$

Possible rational roots:
$\{\pm1,\pm5994,\pm2,\pm2997,\pm3,\pm1998,\pm6,\pm999,\pm9,$
$\pm 666,\pm18,\pm333,\pm27,\pm222,\pm37,\pm162,\pm54,$
$\pm111,\pm74,\pm81,\pm\frac{1}{2},\pm\frac{3}{2},\pm\frac{999}{2},\pm\frac{9}{2},\pm\frac{333}{2},$
$\pm\frac{27}{2},\pm\frac{37}{2},\pm\frac{111}{2},\pm\frac{81}{2};\}$

$\frac{2(13)^3+3(13)^2+(13)}{6}\le 999$

$819\le 999$

$n=13$

67. a. Yes
    b. The method of this section.
    c. Lose critical values.

69. $x(x+2)(x-1)^2>0;\quad \frac{x(x+2)}{(x-1)^2}>0$

71. $x^3-3x^2-6x+8>0$

Possible rational roots: $\{\pm1,\pm8,\pm2,\pm4\}$

$(x+2)(x-1)(x-4)>0$

$f'(x)>0$ for $x\in(-2,1)\cup(4,\infty)$

73. $y=f(x+2)-3$

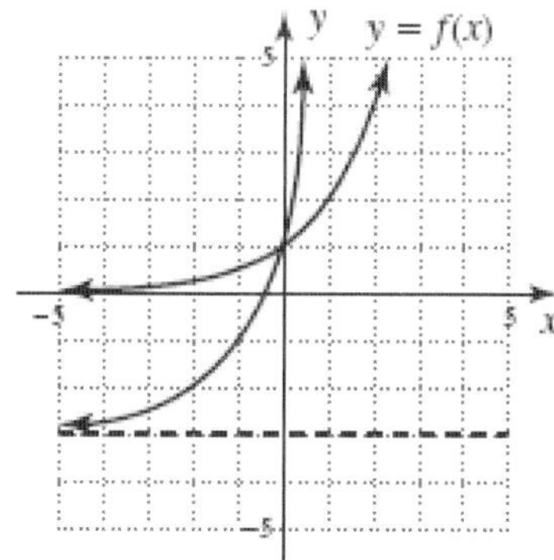

75. $y=f(x)$

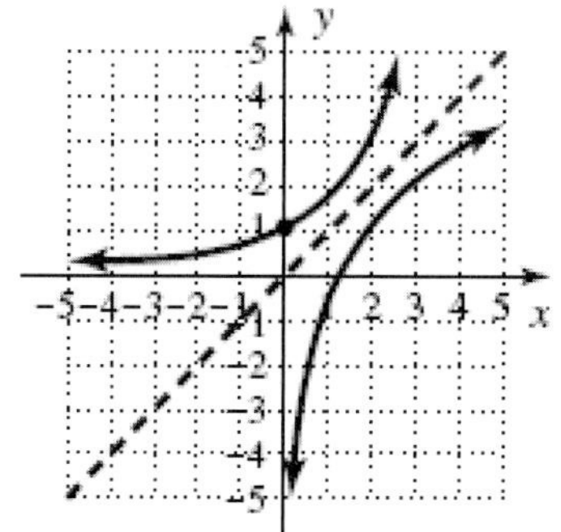

77. $\frac{1}{2}\sqrt{16-x}-\frac{x}{2}=2$

$2\left[\frac{1}{2}\sqrt{16-x}-\frac{x}{2}=2\right]$

$\sqrt{16-x}-x=4$

$\sqrt{16-x}=x+4$

$\left(\sqrt{16-x}\right)^2=(x+4)^2$

$16-x=x^2+8x+16$

$0=x^2+9x$

$0=x(x+9)$

$x=0$ or $x+9=0$

$x=0$ or $x=-9$

$x=0$, $x=-9$ is extraneous

Chapter 4 Summary and Concept Review

1. $\frac{x^3+4x^2-5x-6}{x-2}$

$$\begin{array}{r} x^2+6x+7 \\ x-2\,\overline{)\,x^3+4x^2-5x-6} \\ -(x^3-2x^2) \\ \hline 6x^2-5x \\ -(6x^2-12x) \\ \hline 7x-6 \\ -(7x-14) \\ \hline 8 \end{array}$$

$q(x)=x^2+6x+7$

$R=8$

2. $\frac{x^3+2x-4}{x^2-x}$

$$\begin{array}{r} x+1 \\ x^2-x\,\overline{)\,x^3+0x^2+2x-4} \\ -(x^3-x^2) \\ \hline x^2+2x \\ -(x^2-x) \\ \hline 3x-4 \end{array}$$

$q(x)=x+1$

$R=3x-4$

3. Since $r=0$, -7 is a root and $x+7$ is a factor.

$$\begin{array}{r|rrrrr} -7 & 2 & 13 & -6 & 9 & 14 \\ & & -14 & 7 & -7 & -14 \\ \hline & 2 & -1 & 1 & 2 & \boxed{0} \end{array}$$

4. $\frac{x^3-4x+5}{x-2}=x^2+2x+\frac{5}{x-2}$;

$x^3-4x+5=(x-2)\left(x^2+2x\right)+\frac{5}{x-2}$;

$$\begin{array}{r|l} & x^2+2x \\ x-2 & x^3+0x^2-4x+5 \\ & -\left(x^3-2x^2\right) \\ & \quad 2x^2-4x \\ & \quad -\left(2x^2-4x\right) \\ & \qquad\qquad 5 \end{array}$$

5. $P(x)=x^3+2x^2-11x-12$

Possible rational roots: $\pm1,\pm12,\pm2,\pm6,\pm3,\pm4$

$$\begin{array}{r|rrrr} -4 & 1 & 2 & -11 & -12 \\ & & -4 & 8 & 12 \\ \hline & 1 & -2 & -3 & \boxed{0} \end{array}$$

$P(x)=(x+4)\left(x^2-2x-3\right)$

$P(x)=(x+4)(x+1)(x-3)$

6. $k=16$

$$\begin{array}{r|rrrr} -4 & 1 & 3 & 0 & k \\ & & -4 & 4 & -16 \\ \hline & 1 & -1 & 4 & \boxed{0} \end{array}$$

7. $P(x)=4x^3+8x^2-3x-1$

$$\begin{array}{r|rrrr} \frac{1}{2} & 4 & 8 & -3 & -1 \\ & & 2 & 5 & 1 \\ \hline & 4 & 10 & 2 & \boxed{0} \end{array}$$

Since $r=0$, $\frac{1}{2}$ is a root and

$\left(x-\frac{1}{2}\right)$ is a factor.

8. $P(x)=x^3-2x^2+9x-18$

$$\begin{array}{r|rrrr} 3i & 1 & -2 & 9 & -18 \\ & & 3i & -9-6i & 18 \\ \hline & 1 & -2+3i & -6i & \boxed{0} \end{array}$$

Since $r=0$, $3i$ is a root and $(x-3i)$ is a factor.

9. $P(x)=x^3+9x^2+13x-10$

$$\begin{array}{r|rrrr} -7 & 1 & 9 & 13 & -10 \\ & & -7 & -14 & 7 \\ \hline & 1 & 2 & -1 & \boxed{-3} \end{array}$$

$P(-7)=-3$

10. $P(x)=(x-1)\left(x+\sqrt{5}\right)\left(x-\sqrt{5}\right)$

$P(x)=(x-1)\left(x^2-5\right)$

$P(x)=x^3-5x-x^2+5$

$P(x)=x^3-x^2-5x+5$

11. $P(x)=(x-1)^2(x+2i)(x-2i)$

$P(x)=\left(x^2-2x+1\right)\left(x^2-4i^2\right)$

$P(x)=\left(x^2-2x+1\right)\left(x^2+4\right)$

$P(x)=x^4+4x^2-2x^3-8x+x^2+4$

$P(x)=x^4-2x^3+5x^2-8x+4$

12. $C(t)=3t^3-28t^2+66t+35$

a. $C(0)=3(0)^3-28(0)^2+66(0)+35$

$C(0)=35$

350 customers;

$C(2)=3(2)^3-28(2)^2+66(2)+35$

$C(2)=79$;

$C(3)=3(3)^3-28(3)^2+66(3)+35$

$C(3)=62$;

more at 2pm, 790 – 620 = 170 more

b. $C(1)=3(1)^3-28(1)^2+66(1)+35$

$C(1)=76$;

$C(6)=3(6)^3-28(6)^2+66(6)+35$

$C(6)=71$;

busier at 1pm, $760>710$

13. $C(x)=x^3-8ix^2-19x+12i$

$$\begin{array}{r|rrrr} 3i & 1 & -8i & -19 & 12i \\ & & 3i & 15 & -12i \\ \hline & 1 & -5i & -4 & \boxed{0} \end{array}$$

$C(x)=(x-3i)\left(x^2-5ix-4\right)$

$a=1, b=-5i, c=-4$

$x=\frac{-(-5i)\pm\sqrt{(-5i)^2-4(1)(-4)}}{2(1)}$

$x=\frac{5i\pm\sqrt{25i^2+16}}{2}=\frac{5i\pm\sqrt{-9}}{2}=\frac{5i\pm3i}{2}$

$x=4i, x=i$

14. $f(x)=x^4-3x^3-8x^2+12x+6$;

$f(1)=(1)^4-3(1)^3-8(1)^2+12(1)+6=8$;

$f(2)=(2)^4-3(2)^3-8(2)^2+12(2)+6=-10$;
$f(4)=(4)^4-3(4)^3-8(4)^2+12(4)+6=-10$;
$f(5)=(5)^4-3(5)^3-8(5)^2+12(5)+6=116$;
zeroes are in $[1,2]$ and $[4,5]$

15. $f(x)=x^3-7x+2$;

```
-3| 1   0  -7   2
       -3   9  -6
    ---------------
    1  -3   2 |-4
```

Yes, -3 is a lower bound since the last row does alternate in sign.

```
-2| 1   0  -7   2
       -2   4   6
    ---------------
    1  -2  -3 | 8
```

No, -2 is not a lower bound since the last row does not alternate in sign.

16. $g(x)=x^4+3x^3-2x^2-x-30$
$g(x)$ has one variation in sign $\Rightarrow$ 1 pos root
$g(-x)$ has three variations in sign $\Rightarrow$ 3 or 1 neg root

Possible Positive roots	Possible Negative roots	Possible Complex roots	Total number of roots
1	3	0	4
1	1	2	4

A grapher shows that the second row is correct.

17. $P(x)=2x^3-3x^2-17x-12$
Possible rational roots: $\dfrac{\{\pm1,\pm12,\pm2,\pm6,\pm3,\pm4\}}{\{\pm1\}}$
$P(x)=(x-4)(2x^2+5x+3)$
$P(x)=(x-4)(x+1)(2x+3)$
$x=4,-1,-\dfrac{3}{2}$

18. $h(x)=x^4-7x^2-2x+3$
Possible rational roots: $\{\pm1,\pm3\}$
None give a zero remainder. Therefore, h has no rational roots.

19. $f(x)=-3x^5+2x^4+9x-4$
$f(0)=-3(0)^5+2(0)^4+9(0)-4=-4$
degree 5; up/down; (0,-4)

20. $g(x)=(x-1)(x+2)^2(x-2)$
$g(0)=(0-1)(0+2)^2(0-2)=8$
degree 4; up/up; (0,8)

21. $p(x)=(x+1)^3(x-2)^2$
end behavior: down/up
bounce at (2,0); cross at (-1,0)
$p(0)=(0+1)^3(0-2)^2=4$
y-intercept: $(0,4)$

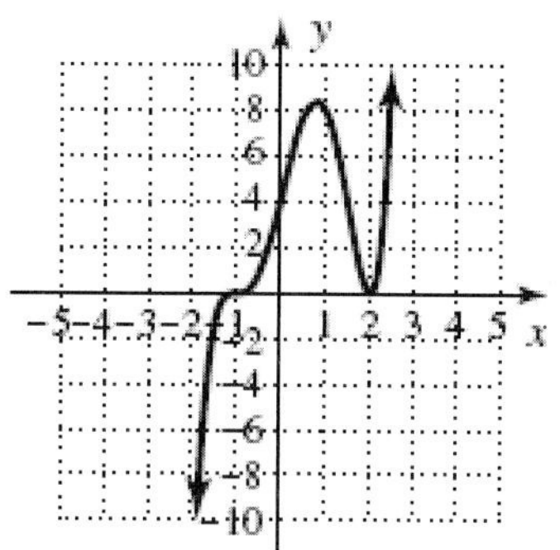

22. $q(x)=2x^3-3x^2-9x+10$
end behavior: down/up
Possible rational roots: $\{\pm1,\pm10,\pm2,\pm5\}$
$q(x)=(x+2)(2x^2-7x+5)$
$q(x)=(x+2)(2x-5)(x-1)$
cross at (-2,0), $\left(\dfrac{5}{2},0\right)$ and (1,0)
y-intercept: $(0,10)$

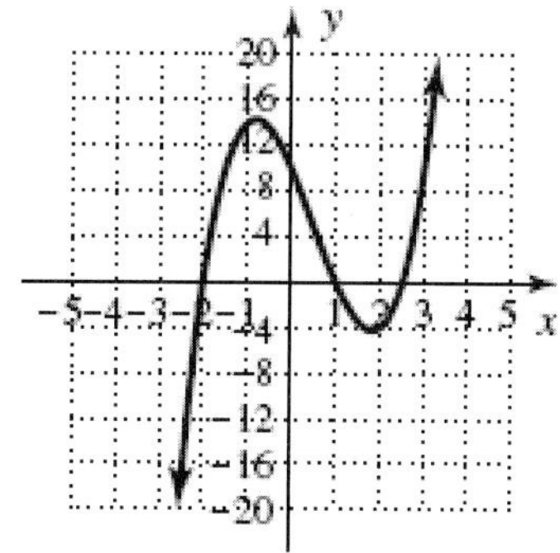

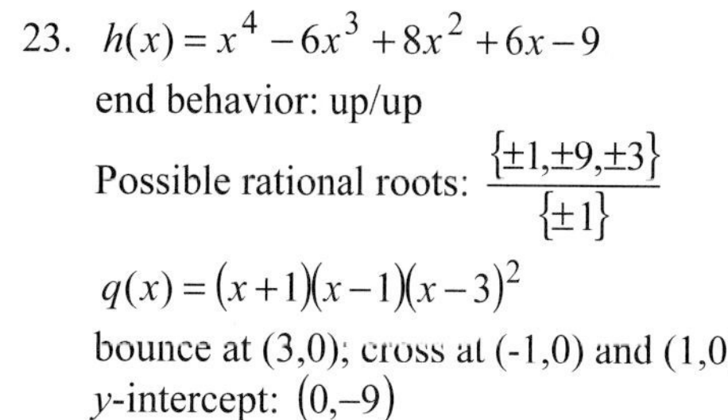

23. $h(x)=x^4-6x^3+8x^2+6x-9$
end behavior: up/up
Possible rational roots: $\dfrac{\{\pm1,\pm9,\pm3\}}{\{\pm1\}}$
$q(x)=(x+1)(x-1)(x-3)^2$
bounce at (3,0); cross at (-1,0) and (1,0)
y-intercept: $(0,-9)$

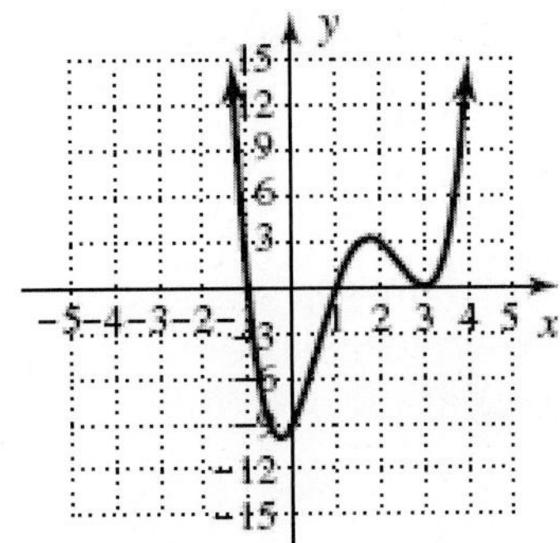

24. a. even
    b. $x = -2$, odd
       $x = -1$, even
       $x = 1$, odd
    c. degree 6: $P(x) = (x+2)(x+1)^2(x-1)^3$

25. $r(x) = \dfrac{x^2-9}{x^2-3x-4}$

    $r(x) = \dfrac{(x+3)(x-3)}{(x-4)(x+1)}$

    a. $\{x \mid x \in R, x \neq -1, 4\}$
    b. HA: $y = 1$
       (deg num = deg den)
       VA: $x = -1$, $x = 4$
    c. $r(0) = \dfrac{0^2-9}{0^2-3(0)-4} = \dfrac{9}{4}$

       y-intercept $\left(0, \dfrac{9}{4}\right)$;

       x-intercepts : (-3, 0) and (3, 0)
    d. $r(1) = \dfrac{1^2-9}{1^2-3(1)-4} = \dfrac{4}{3}$

26. $h(x) = \dfrac{3x-5}{x-2}$

```
2 | 3   -5
        6
  --------
    3 | 1
```

$h(x) = 3 + \dfrac{1}{x-2}$

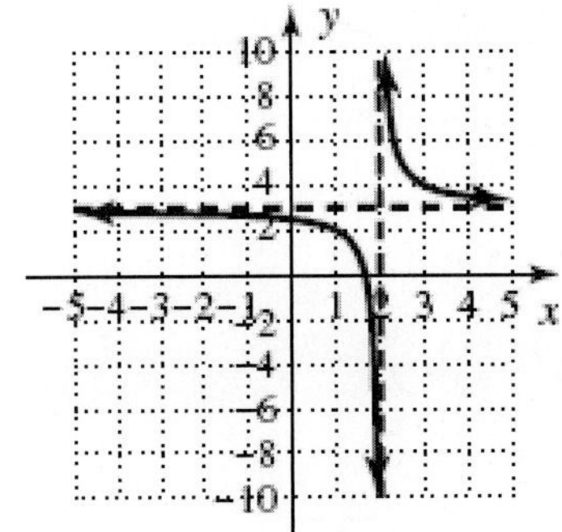

27. $r(x) = \dfrac{x^2-4x}{x^2-4}$

$r(0) = \dfrac{(0)^2-4(0)}{(0)^2-4} = 0;$

y-intercept: $(0,0)$

$r(x) = \dfrac{x(x-4)}{(x+2)(x-2)};$

vertical asymptotes: $x = -2$ and $x = 2$

x-intercepts: $(0,0)$ and $(4,0)$

horizontal asymptote: $y = 1$

(deg num = deg den)

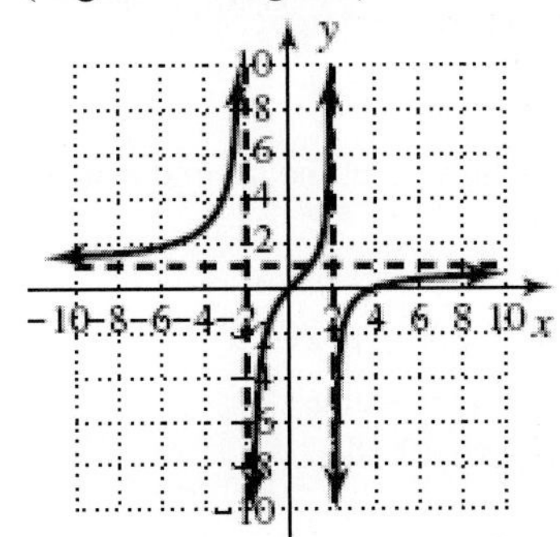

28. $t(x) = \dfrac{2x^2}{x^2-5}$

$t(0) = \dfrac{2(0)^2}{(0)^2-5} = 0;$

y-intercept: $(0,0)$

$t(x) = \dfrac{2x^2}{(x+\sqrt{5})(x-\sqrt{5})};$

vertical asymptotes: $x = -\sqrt{5}$ and $x = \sqrt{5}$

x-intercept: $(0,0)$

horizontal asymptote: $y = 2$

(deg num = deg den)

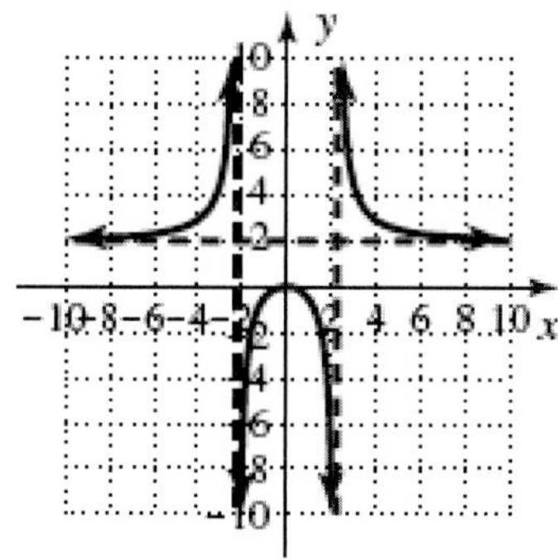

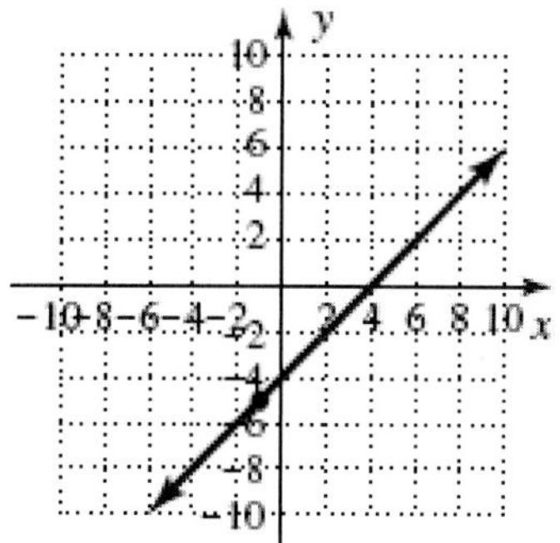

29. $r(x)=\dfrac{(x+3)(x-4)}{(x+2)(x-3)}$

$r(x)=\dfrac{x^2-x-12}{x^2-x-6};$

$r(0)=\dfrac{(0)^2-(0)-12}{(0)^2-(0)-6}=2$

30. $A(x)=\dfrac{5000+15x}{x}$

a. $y=15$; as $|x|\to+\infty$, $A(x)\to 15^+$.
As production increases, average cost decreases and approaches 15.

b. $\dfrac{5000+15x}{x}<17.50$

$\dfrac{5000+15x}{x}-17.50<0$

$\dfrac{5000+15x-17.50x}{x}<0$

$\dfrac{5000-2.5x}{x}<0;$

$5000-2.5x=0$
$x=2000;$
critical points 0, 2000
$x>2000$

31. $h(x)=\dfrac{x^2-3x-4}{x+1}=\dfrac{(x-4)(x+1)}{x+1}$

$$H(x)=\begin{cases}\dfrac{x^2-3x-4}{x+1} & x\neq -1\\ -5 & x=-1\end{cases}$$

32. $h(x)=\dfrac{x^2-2x}{x-3}$

$h(0)=\dfrac{(0)^2-2(0)}{(0)-3}=0;$

y-intercept: $(0,0)$

$h(x)=\dfrac{x(x-2)}{x-3};$

vertical asymptote: $x=3$
x-intercepts: $(0,0)$ and $(2,0)$
horizontal asymptote: none
(deg num > deg den)
oblique asymptote: $y=x+1$

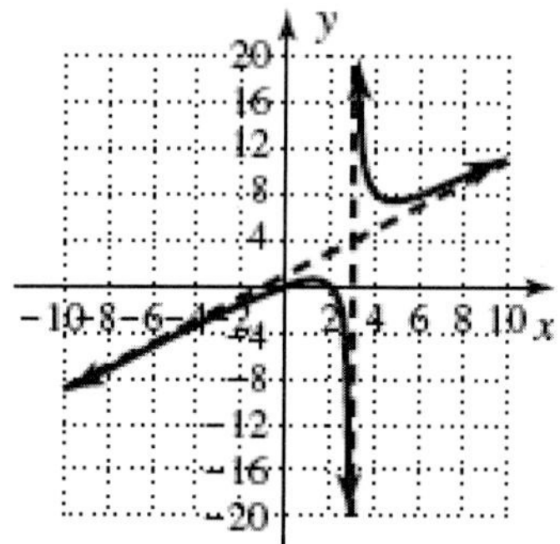

33. $t(x)=\dfrac{x^3-7x+6}{x^2}$

$t(0)=\dfrac{(0)^3-7(0)+6}{(0)^2}=\text{undefined};$

y-intercept: none
Possible rational roots for numerator:
$\dfrac{\{\pm1,\pm6,\pm2,\pm3\}}{\{\pm1\}}$

$t(x)=\dfrac{(x+3)(x-1)(x-2)}{x^2};$

vertical asymptote: $x=0$
x-intercepts: $(-3,0)$, $(1,0)$ and $(2,0)$
horizontal asymptote: none
(deg num > deg den)

oblique asymptote: $y = x$

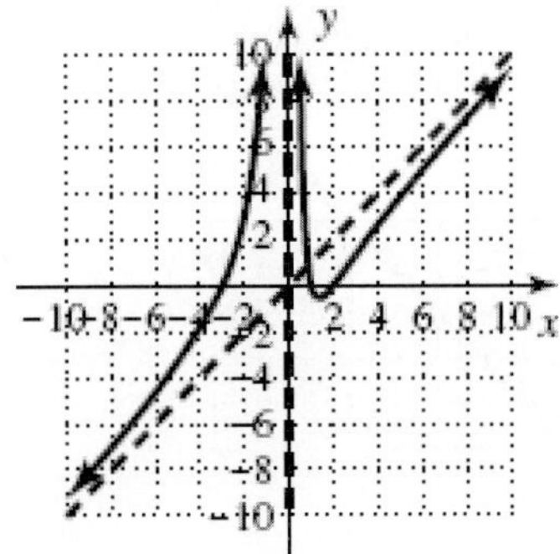

34. $x^3 + x^2 > 10x - 8$

$x^3 + x^2 - 10x + 8 > 0$

Possible rational roots: $\pm 1, \pm 8, \pm 2, \pm 4$

$(x+4)(x-1)(x-2) \geq 0$

neg pos neg pos
-4 1 2

Outputs are positive or zero for $x \in (-4,1) \cup (2, \infty)$.

35. $\dfrac{x^2 - 3x - 10}{x-2} \geq 0$

$\dfrac{(x-5)(x+2)}{x-2} \geq 0$

neg pos neg pos
-2 2 5

Outputs are positive for . $x \in [-2,2) \cup [5, \infty)$

36. $\dfrac{x}{x-2} \leq \dfrac{-1}{x}$

$\dfrac{x}{x-2} + \dfrac{1}{x} \leq 0$

$\dfrac{x^2 + x - 2}{x(x-2)} \leq 0$

$\dfrac{(x+2)(x-1)}{x(x-2)} \leq 0$

pos neg pos neg pos
-2 0 1 2

Outputs are negative or zero for $x \in [-2,0) \cup [1,2)$.

Chapter 4 Mixed Review

1. $q(x) = x^2 - 5$

$R = 8$;

$$\begin{array}{r|l} & x^2 \quad\quad -5 \\ \hline x+3 & x^3 + 3x^2 - 5x - 7 \\ & -(x^3 + 3x^2) \\ & \quad\quad -5x - 7 \\ & \quad\quad -(-5x - 15) \\ & \quad\quad\quad\quad 8 \end{array}$$

3. Possible rational roots:
$\{\pm 1, \pm 48, \pm 2, \pm 24, \pm 3, \pm 16, \pm 4, \pm 12, \pm 6, \pm 8\}$
$(x+2)(x-8)(x-3)$

(a)
$$\begin{array}{r|rrrr} -6 & 1 & -9 & 2 & 48 \\ & & -6 & -90 & 528 \\ \hline & 1 & -15 & -88 & 576 \end{array}$$

(b)
$$\begin{array}{r|rrrr} 8 & 1 & -9 & 2 & 48 \\ & & 8 & -8 & -48 \\ \hline & 1 & -1 & -6 & 0 \end{array}$$

(c)
$$\begin{array}{r|rrrr} 12 & 1 & -9 & 2 & 48 \\ & & 12 & 36 & 456 \\ \hline & 1 & 3 & 38 & 504 \end{array}$$

(d)
$$\begin{array}{r|rrrr} 4 & 1 & -9 & 2 & 48 \\ & & 4 & -20 & -72 \\ \hline & 1 & -5 & -18 & -24 \end{array}$$

(e)
$$\begin{array}{r|rrrr} -2 & 1 & -9 & 2 & 48 \\ & & -2 & 22 & -48 \\ \hline & 1 & -11 & 24 & 0 \end{array}$$

(a), (c), (d), are NOT factors.

5. $k = 6$
$$\begin{array}{r|rrrr} -2 & 1 & 4 & 7 & k \\ & & -2 & -4 & -6 \\ \hline & 1 & 2 & 3 & 0 \end{array}$$

7. $(x-3)(x+5i)(x-5i)$
$= (x-3)(x^2 - 25i^2)$
$= (x-3)(x^2 + 25)$
$= x^3 + 25x - 3x^2 - 75$
$= x^3 - 3x^2 + 25x - 75$

9. $6x^3 + x^2 - 20x - 12 = 0$
Possible rational roots:
$\{\pm 1, \pm 12, \pm 2, \pm 6, \pm 3, \pm 4, \pm \frac{1}{6}, \pm \frac{1}{3}, \pm \frac{1}{2}, \pm \frac{2}{3}, \pm \frac{3}{2}, \pm \frac{4}{3}\}$

$x = 9$ and $x = \frac{8}{3}$ CANNOT be roots.

11. $f(x) = x^3 - 13x + 12$

end behavior: down/up

Possible rational roots: $\{\pm 1, \pm 12, \pm 2, \pm 6, \pm 3, \pm 4\}$

$$\begin{array}{r|rrrr} 1 & 1 & 0 & -13 & 12 \\ & & 1 & 1 & -12 \\ \hline & 1 & 1 & -12 & \boxed{0} \end{array}$$

$f(x) = (x-1)(x^2 + x - 12)$

$f(x) = (x-1)(x+4)(x-3)$

x-intercepts: $(-4,0)$, $(1,0)$, and $(3,0)$

y-intercept: $(0,12)$

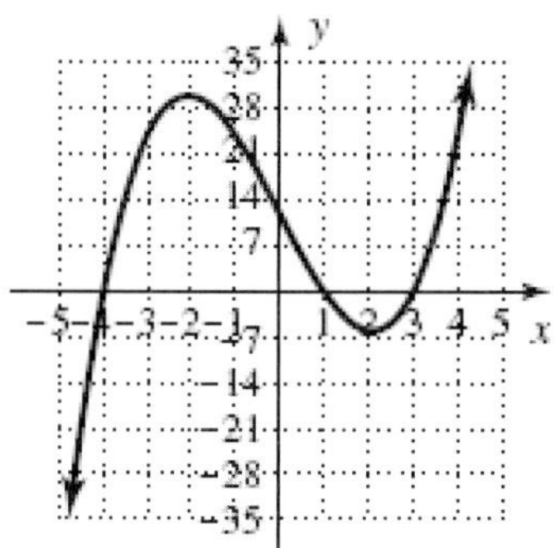

13. $h(x) = (x-1)^3(x+2)^2(x+1)$

end behavior: up/up

x-intercepts: $(1,0)$, $(-2,0)$, and $(-1,0)$

y-intercept: $(0,-4)$

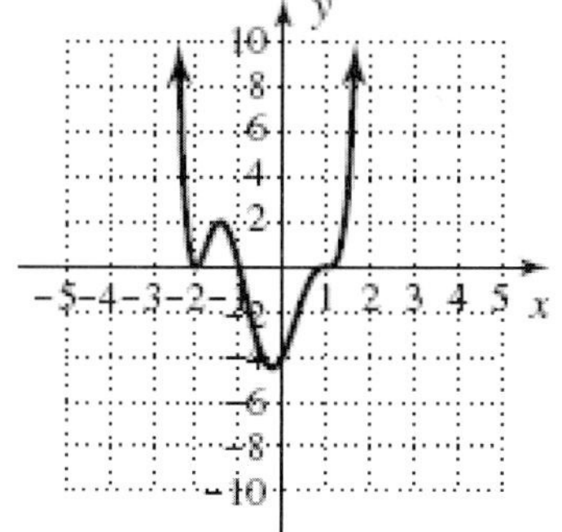

15. $q(x) = \dfrac{x^2 - 4}{x^2 - 3x - 4}$

$q(0) = \dfrac{(0)^2 - 4}{(0)^2 - 3(0) - 4} = 1;$

y-intercept: $(0,1)$

$q(x) = \dfrac{(x+2)(x-2)}{(x+1)(x-4)};$

vertical asymptotes: $x = -1$ and $x = 4$

x-intercepts: $(-2,0)$ and $(2,0)$

horizontal asymptote: $y = 1$

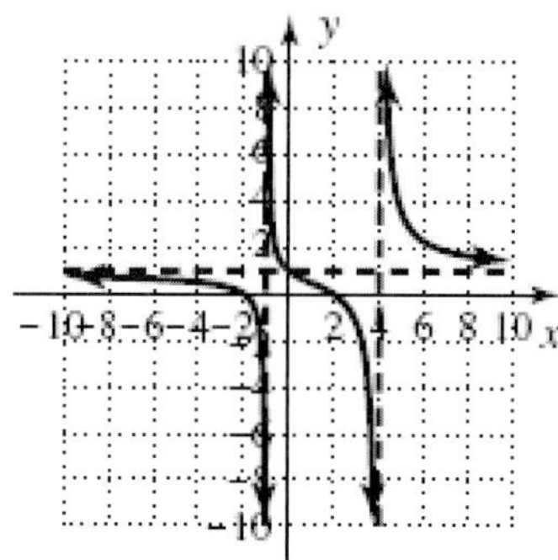

17. $y = \dfrac{x^2 - 4x}{x - 3}$

$y = \dfrac{(0)^2 - 4(0)}{(0) - 3} = 0$

y-intercept: $(0,0)$

$y = \dfrac{x(x-4)}{x-3};$

vertical asymptote: $x = 3$

x-intercepts: $(4,0)$ and $(0,0)$

horizontal asymptote: none

$$\begin{array}{r|l} & x - 1 \\ \hline x - 3 & x^2 - 4x \\ & -(x^2 - 3x) \\ \hline & -x \\ & -(-x + 3) \\ \hline & -3 \end{array}$$

oblique asymptote: $y = x - 1$

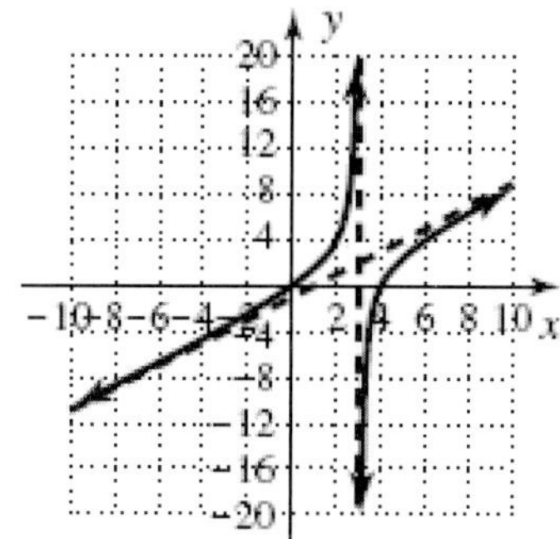

19. $\dfrac{4}{x+2} \geq \dfrac{3}{x}$

$\dfrac{4}{x+2} - \dfrac{3}{x} \geq 0$

$\dfrac{4x - 3x - 6}{x(x+2)} \geq 0$

$$\frac{x-6}{x(x+2)} \geq 0$$

neg pos neg pos
-2 0 6

$x \in (-2,0) \cup [6,\infty)$

Chapter 4 Practice Test

1. $$\frac{x^3-3x^2+5x-2}{x^2+2x+1} = x-5+\frac{14x+3}{x^2+2x+1}$$

$$\begin{array}{r} x-5 \\ x^2+2x+1 \overline{) x^3-3x^2+5x-2} \\ -(x^3+2x^2+x) \\ \hline -5x^2+4x-2 \\ -(-5x^2-10x-5) \\ \hline 14x+3 \end{array}$$

2. $$\frac{x^3+4x^2-5x-20}{x+2} = x^2+2x-9+\frac{-2}{x+2}$$

-2	1	4	-5	-20
		-2	-4	18
	1	2	-9	-2

3. $$\frac{x^4+4x^2-3x+k}{x+2}$$

$k = -35$

-2	1	0	4	-3	k
		-2	4	-16	38
	1	-2	8	-19	3

4. $(x+3);\ x^4-15x^2-10x+24$

Verified

-3	1	0	-15	-10	24
		-3	9	18	-24
	1	-3	-6	8	0

5. $f(x) = 2x^3+4x^2-5x+2$

$f(-3) = -1$

-3	2	4	-5	2
		-6	6	-3
	2	-2	1	-1

6. $P(x) = (x-2)(x-3i)(x+3i)$

$P(x) = (x-2)(x^2-9i^2)$

$P(x) = (x-2)(x^2+9)$

$P(x) = x^3+9x-2x^2-18$

$P(x) = x^3-2x^2+9x-18$

7. $Q(x) = (x^2-3x+2)(x^3-2x^2-x+2)$

$Q(x) = (x-2)(x-1)(x^2(x-2)-(x-2))$

$Q(x) = (x-2)(x-1)(x-2)(x^2-1)$

$Q(x) = (x-2)(x-1)(x-2)(x+1)(x-1)$

$Q(x) = (x-2)^2(x-1)^2(x+1)$

2 multiplicity 2

1 multiplicity 2, -1 multiplicity 1

8. $C(x) = x^3+(6-3i)x^2+(8-18i)x-24i$

3i	1	6−3i	8−18i	−24i
		3i	18i	24i
	1	6	8	0

$C(x) = (x-3i)(x^2+6x+8)$

$C(x) = (x-3i)(x+4)(x+2)$

9. $C(x) = x^4+x^3+7x^2+9x-18$

(a) Possible rational roots:

$\{\pm1, \pm18, \pm2, \pm9, \pm3, \pm6\}$

(b)

Possible Positive roots	Possible Negative roots	Possible Complex roots	Total number of roots
1	3	0	4
1	1	2	4

(c) $C(x) = (x+2)(x-1)(x^2+9)$

$C(x) = (x+2)(x-1)(x+3i)(x-3i)$

10. $f(x) = \frac{1}{2}x^3-7x^2+28x-32$

(a) $0 = \frac{1}{2}x^3-7x^2+28x-32$

$0 = x^3-14x^2+56x-64$

Possible rational roots:

$\{\pm1, \pm64, \pm2, \pm32, \pm4, \pm16, \pm8\}$

$0 = (x-2)(x-4)(x-8)$

$x = 2, x = 4, x = 8$

1992, 1994, 1998

(b) 4 years (1992-1994, 1998-2000)

(c) surplus of \$2.5 million

11. $f(x) = (x-3)(x+1)^3(x+2)^2$

end behavior: up/up

bounce at (-2,0); cross at (3,0) and (-1,0)

3 multiplicity 1, -1 multiplicity 3, -2 multiplicity 2;

$f(0)=(0-3)(0+1)^3(0+2)^2=-12$

y-intercept: $(0,-12)$

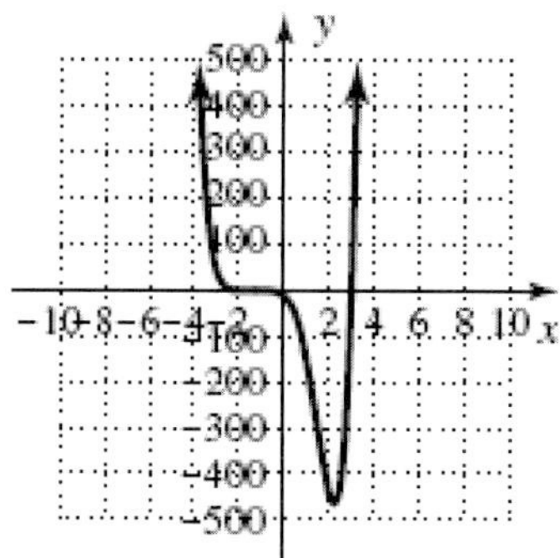

12. $g(x)=x^4-9x^2-4x+12$

end behavior: up/up

Possible rational roots: $\dfrac{\{\pm1,\pm12,\pm2,\pm6,\pm3,\pm4\}}{\{\pm1\}}$

```
-2| 1   0   -9   -4   12
       -2    4   10  -12
    ----------------------
    1  -2   -5    6  | 0

-2| 1  -2   -5    6
       -2    8   -6
    ------------------
    1  -4    3  | 0
```

$g(x)=(x+2)^2(x^2-4x+3)$

$g(x)=(x+2)^2(x-1)(x-3)$

-2 multiplicity 2, 1 multiplicity 1, 3 multiplicity 1

bounce at (2,0); cross at (1,0) and (3,0)

$g(0)=0^4-9(0)^2-4(0)+12=12$

y-intercept: $(0,12)$

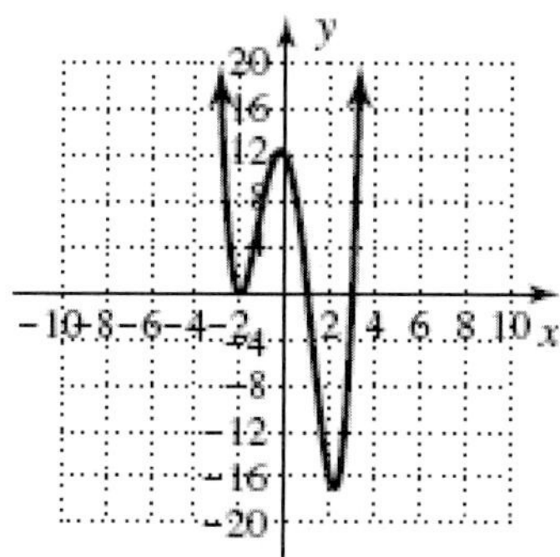

13. $h(x)=\dfrac{x-2}{x^2-3x-4}$

$h(0)=\dfrac{(0)-2}{(0)^2-3(0)-4}=\dfrac{1}{2}$;

y-intercept: $\left(0,\dfrac{1}{2}\right)$

$h(x)=\dfrac{x-2}{(x+1)(x-4)}$;

vertical asymptote: $x=-1$ and $x=4$

x-intercept: $(2,0)$

horizontal asymptote: $y=0$

deg num < deg den

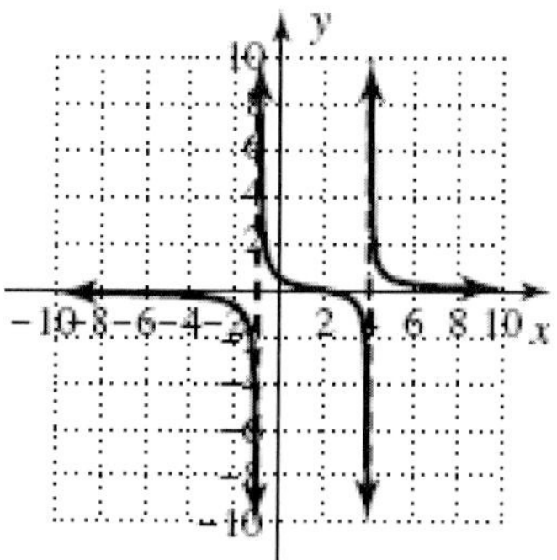

14. $C(x)=\dfrac{300x}{100-x}$

a. VA: $x=100$; removal of 100% of the contaminants

b. From 80% to 85%:

$C(85)=\dfrac{300(85)}{100-(85)}=1700$;

$1,700,000;

$C(80)=\dfrac{300(80)}{100-(80)}=1200$;

1700 – 1200 =500, $500,000;

From 90% to 95 %:

$C(95)=\dfrac{300(95)}{100-(95)}=5700$;

$5,700,000

$C(90)=\dfrac{300(90)}{100-(90)}=2700$;

5700 – 2700 =3000; $3,000,000;

It becomes cost prohibitive to remove all the contaminants.

c. $2200=\dfrac{300x}{100-x}$

$2200(100-x)=300x$

$220000-2200x=300x$

$220000=2500x$

$88=x$

$x=88\%$

15. $r(x)=\dfrac{x^3-x^2-9x+9}{x^2}$

$r(0)=\dfrac{(0)^3-(0)^2-9(0)+9}{(0)^2}=$ undefined;

y-intercept: none

$r(x)=\dfrac{x^2(x-1)-9(x-1)}{x^2}$

$r(x)=\dfrac{(x-1)(x^2-9)}{x^2}$

$r(x)=\dfrac{(x-1)(x+3)(x-3)}{x^2}$;

vertical asymptote: $x=0$

x-intercepts: $(1,0),(-3,0)$, and $(3,0)$

horizontal asymptote: none

deg num > deg den

$r(x)=\dfrac{x^3}{x^2}-\dfrac{x^2}{x^2}-\dfrac{9x}{x^2}+\dfrac{9}{x^2}$

$r(x)=x-1-\dfrac{9}{x}+\dfrac{9}{x^2}$

oblique asymptote: $y=x-1$

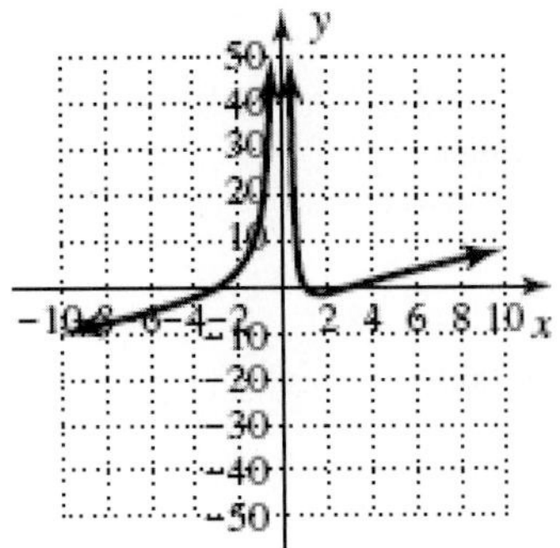

16. $R(x)=\dfrac{x^3+7x-6}{x^2-4}$

$R(0)=\dfrac{(0)^3+7(0)-6}{(0)^2-4}=\dfrac{3}{2}$;

y-intercept: $\left(0,\dfrac{3}{2}\right)$

$R(x)=\dfrac{x^3+7x-6}{(x+2)(x-2)}$;

vertical asymptotes: $x=-2$ and $x=2$

x-intercept: $\approx(0.8,0)$

horizontal asymptote: none

deg num > deg den

$$\begin{array}{r} x \\ x^2-4\overline{)x^3+7x-6} \\ -(x^3-4x) \\ \hline 11x-6 \end{array}$$

oblique asymptote: $y=x$

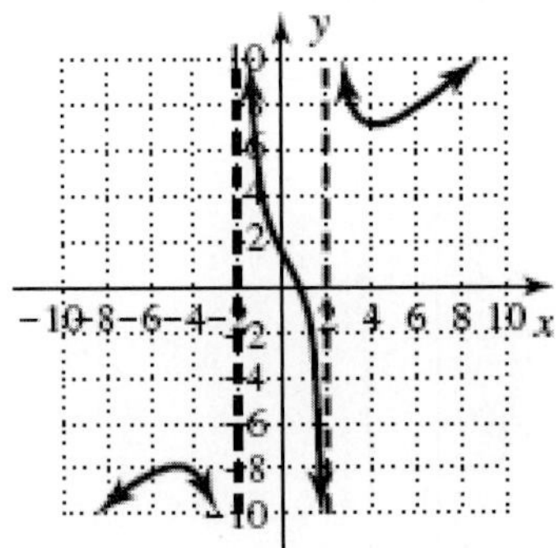

17. $\overline{C(x)}=\dfrac{2x^2+25x+128}{x}$

Using grapher: $x=8$; 800 items

Minimizes costs

18. $x^3-13x\le 12$

$x^3-13x-12\le 0$

$\{\pm1,\pm12,\pm2,\pm6,\pm3,\pm4\}$

$$\begin{array}{r|rrrr} -3 & 1 & 0 & -13 & -12 \\ & & -3 & 9 & 12 \\ \hline & 1 & -3 & -4 & 0 \end{array}$$

$(x+3)(x^2-3x-4)\le 0$

$(x+3)(x-3)(x+1)\le 0$

neg pos neg pos
-3 -1 4

$x\in(-\infty,-3]\cup[-1,4]$

19. $\dfrac{3}{x-2}<\dfrac{2}{x}$

$\dfrac{3}{x-2}-\dfrac{2}{x}<0$

$\dfrac{3x-2(x-2)}{x(x-2)}<0$

$\dfrac{3x-2x+4}{x(x-2)}<0$

$\dfrac{x+4}{x(x-2)}<0$

vertical asymptote: $x=0$, $x=2$

horizontal asymptote: $y=0$

neg pos neg pos
-4 0 2

$x\in(-\infty,-4)\cup(0,2)$

20. $C(h)=\dfrac{2h^2+5h}{h^3+55}$

a.

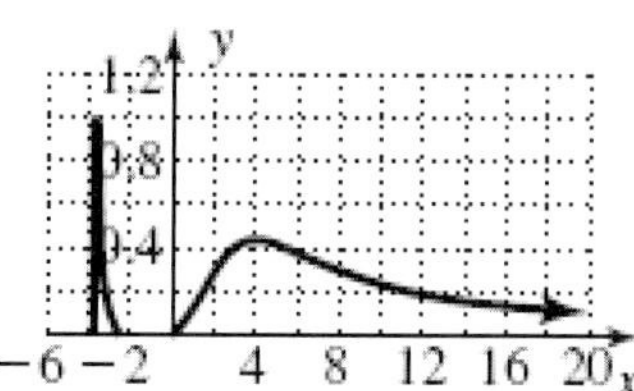

b. $h^3+55=0$
$h^3=-55$
$h=-\sqrt[3]{55}$, no

c. $C(2)=\dfrac{2(2)^2+5(2)}{(2)^3+55}\approx 0.286=28.6\%$;

$C(8)=\dfrac{2(8)^2+5(8)}{(8)^3+55}\approx 0.296=29.6\%$

d. $\dfrac{2h^2+5h}{h^3+55}<0.2$
Using grapher: $\approx$ 12 hours

e. Using grapher: 4 hours, 43.7%

f. A trace amount of the chemical will remain in the bloodstream.

Chapter 4 Calculator Exploration

1. $Y_1=(x^3-6x^2+32)(x^2+1)$
$Y_2=x^3-6x^2+32$
$Y_3=x+2$

2. $Y_1=(x+3)^2(x^3-2x^2+x-2)$
$Y_2=(x+3)^2(x-2)$
$Y_3=x-2$

3. They do not affect the solution.

4. They would not affect the answer.

Chapter 4 Strengthening Core Skills

1. $x^3-3x-18\le 0$
$\dfrac{\{\pm1,\pm18,\pm2,\pm9,\pm3,\pm6\}}{\{\pm1\}}$

$$\begin{array}{r|rrrr} 3 & 1 & 0 & -3 & -18 \\ & & 3 & 9 & 18 \\ \hline & 1 & 3 & 6 & 0 \end{array}$$

$(x-3)(x^2+3x+6)\le 0$
$x\in(-\infty,3]$

2. $\dfrac{x+1}{(x^2-4)}>0$

$\dfrac{x+1}{(x-2)(x+2)}>0$
$x\in(-2,-1)\cup(2,\infty)$

3. $x^3-13x+12<0$
$\dfrac{\{\pm1,\pm12,\pm2,\pm6,\pm3,\pm4\}}{\{\pm1\}}$

$$\begin{array}{r|rrrr} -4 & 1 & 0 & -13 & 12 \\ & & -4 & 16 & -12 \\ \hline & 1 & -4 & 3 & 0 \end{array}$$

$(x+4)(x^2-4x+3)<0$
$(x-3)(x-1)(x+4)<0$
$x\in(-\infty,-4)\cup(1,3)$

4. $x^3-3x+2\ge 0$
$(x-1)^2(x+2)\ge 0$
$x\in[-2,\infty)$

5. $x^4-x^2-12>0$
$(x^2-4)(x^2+3)>0$
$(x-2)(x+2)(x^2+3)>0$
(x^2+3) does not affect the solution set.
$x\in(-\infty,-2)\cup(2,\infty)$

6. $(x^2+5)(x^2-9)(x+2)^2(x-1)\ge 0$
$(x^2+5)(x-3)(x+3)(x+2)^2(x-1)\ge 0$
(x^2+5) does not affect the solution set and can be ignored. $(x+2)^2$ is non-negative and can be ignored.
$x\in[-3,1]\cup[3,\infty)$

Cumulative Review Chapters 1-4

1. $\dfrac{1}{R}=\dfrac{1}{R_1}+\dfrac{1}{R_2}$

$$RR_1R_2\left[\frac{1}{R}\right]=\left[\frac{1}{R_1}+\frac{1}{R_2}\right]RR_1R_2$$
$$R_1R_2 = RR_2 + RR_1$$
$$R_1R_2 = R(R_2 + R_1)$$
$$\frac{R_1R_2}{R_1 + R_2} = R$$

2. $\frac{2}{x+1}+1=\frac{5}{x^2-1}$

$$\frac{2}{x+1}+1=\frac{5}{(x+1)(x-1)}$$

$$(x+1)(x-1)\left[\frac{2}{x+1}+1\right]=\left[\frac{5}{(x+1)(x-1)}\right](x+1)(x-1)$$
$$2(x-1)+x^2-1=5$$
$$2x-2+x^2-1=5$$
$$x^2+2x-8=0$$
$$(x+4)(x-2)=0$$
$$x+4=0 \text{ or } x-2=0$$
$$x=-4 \quad \text{or } x=2$$

3. a. x^3-1
$=(x-1)(x^2+x+1)$

b. $x^3-3x^2-4x+12$
$=x^2(x-3)-4(x-3)$
$=(x-3)(x^2-4)$
$=(x-3)(x+2)(x-2)$

4. $2x^2+4x+1=0$
$a=2, b=4, c=1$

$$x=\frac{-(4)\pm\sqrt{(4)^2-4(2)(1)}}{2(2)}$$
$$x=\frac{-4\pm\sqrt{8}}{4}=\frac{-4\pm2\sqrt{2}}{4}=\frac{-2\pm\sqrt{2}}{2}$$
$$x\approx-0.29,-1.71$$

5. $x+3<5$ or $5-x<4$
$x<2$ or $-x<-1$
$x<2$ or $x>1$
$x\in(-\infty,\infty)$

6. $y=x, y=x^2, y=|x|, y=\frac{1}{x}, y=\frac{1}{x^2},$
$y=x^3, y=\sqrt{x}, y=\sqrt[3]{x}$

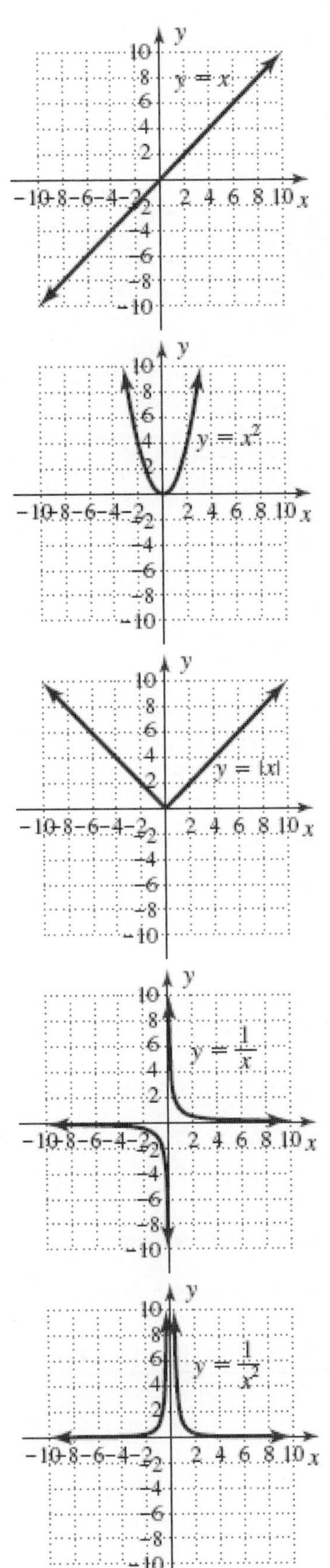

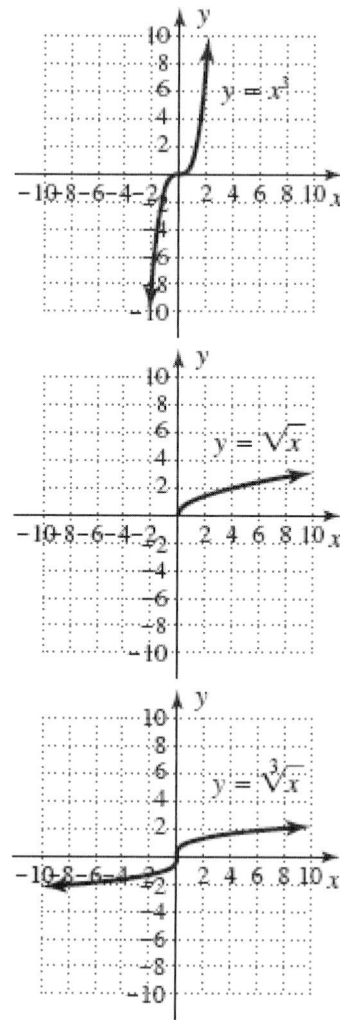

7. $(2-3i)^2-4(2-3i)+13=0$

$4-12i+9i^2-8+12i+13=0$
$4-12i-9-8+12i+13=0$
$0=0$
Verified

8. $\dfrac{x+4}{x-2}<3$

$\dfrac{x+4}{x-2}-3<0$

$\dfrac{x+4-3(x-2)}{x-2}<0$

$\dfrac{x+4-3x+6}{x-2}<0$

$\dfrac{-2x+10}{x-2}<0$

$\dfrac{-2(x-5)}{x-2}<0$

neg pos neg
2 5

$x\in(-\infty,2)\cup(5,\infty)$

9. $(1,\ 17),(61,\ 28)$

$m=\dfrac{28-17}{61-1}=\dfrac{11}{60};$

$y-17=\dfrac{11}{60}(x-1)$

$y-17=\dfrac{11}{60}x-\dfrac{11}{60}$

$y=\dfrac{11}{60}x+\dfrac{1009}{60};$

$y=\dfrac{11}{60}(121)+\dfrac{1009}{60}=39$ minutes;

Driving time increases 11 minutes every 60 days.

10. No, Herschel is paired with mathematics and astronomy.

11. $y=1.18x^2-10.99x+4.6$;

Using grapher, the profit is first earned in the 9th month.

12. $g(x)=\dfrac{-1}{(x+2)^2}+3$

Reflected in x-axis, left 2, up 3

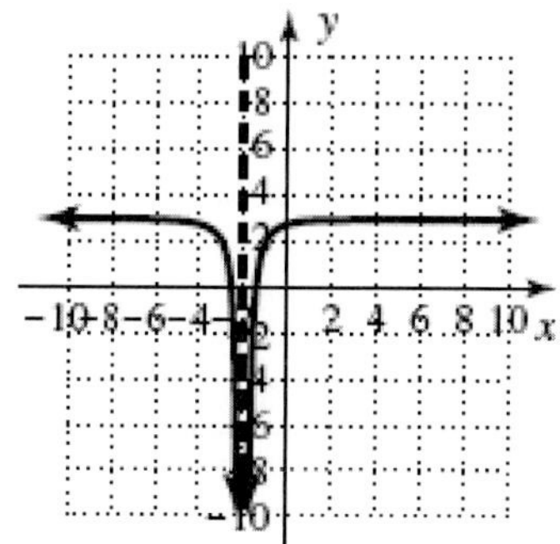

13. $f(x)=\sqrt[3]{2x-3}$;

$x=\sqrt[3]{2y-3}$

$x^3=2y-3$

$x^3+3=2y$

$\dfrac{x^3+3}{2}=y$

$f^{-1}(x)=\frac{x^3+3}{2}$

$(f\circ f^{-1})(x)=\sqrt[3]{2\left(\frac{x^3+3}{2}\right)-3}$

$=\sqrt[3]{x^3+3-3}=\sqrt[3]{x^3}=x$;

$(f^{-1}\circ f)(x)=\frac{(\sqrt[3]{2x-3})^3+3}{2}$

$=\frac{2x-3+3}{2}=\frac{2x}{2}=x$;

Verified

14. $f(x)=x^2-4x+7$

$f(x)=(x^2-4x+4)+7-4$

$f(x)=(x-2)^2+3$;

a. $x\in(-\infty,\infty)$

b. $x\in(2,\infty)$

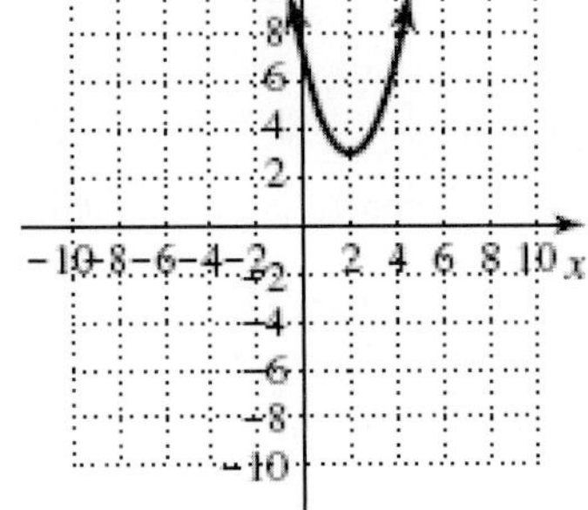

15. $F(x)=-f(x+1)+2$

Reflected in x-axis, left 1, up 2

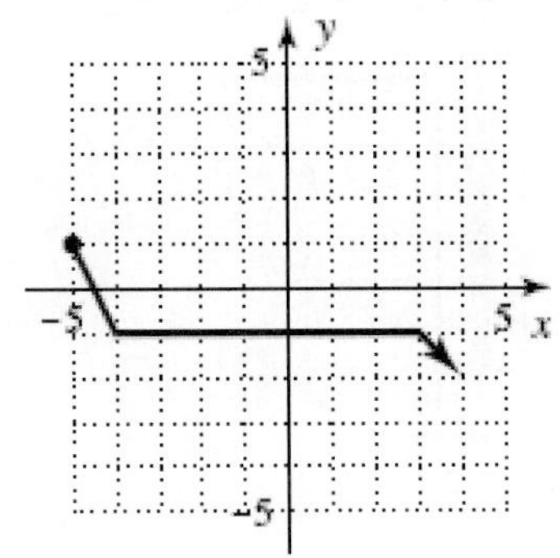

16. $f(x)=\begin{cases}-3 & x<-1\\ x & -1\le x\le 1\\ 3x & x>1\end{cases}$

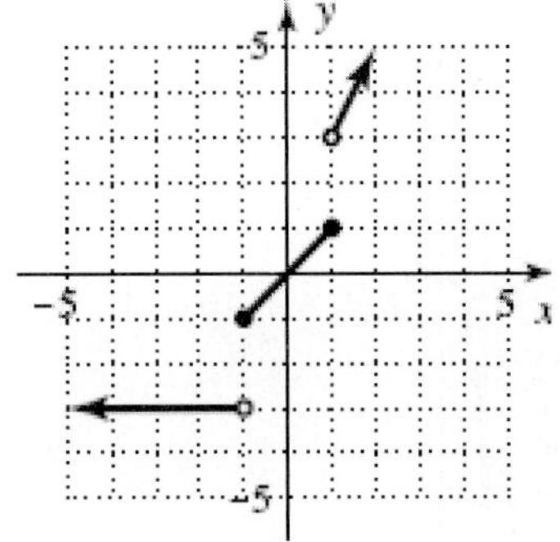

17. $Y=\frac{kX}{Z^2}$;

$10=\frac{k(32)}{(4)^2}$

$5=k$;

$Y=\frac{5X}{Z^2}$;

$1.4=\frac{5x}{(15)^2}$

$x=63$

18. $f(x)=x^4-2x^2+16x-15$

Possible rational roots: $\{\pm1,\pm15,\pm3,\pm5\}$

$$\begin{array}{r|rrrrr} 1 & 1 & 0 & -2 & 16 & -15\\ & & 1 & 1 & -1 & 15\\ \hline & 1 & 1 & -1 & 15 & 0\end{array}$$

$$\begin{array}{r|rrrr} -3 & 1 & 1 & -1 & 15\\ & & -3 & 6 & -15\\ \hline & 1 & -2 & 5 & 0\end{array}$$

$f(x)=(x-1)(x+3)(x^2-2x+5)$;

$a=1, b=-2, c=5$

$x=\frac{-(-2)\pm\sqrt{(-2)^2-4(1)(5)}}{2(1)}$

$x=\frac{2\pm\sqrt{-16}}{2}=\frac{2\pm4i}{2}=1\pm2i$;

roots: $1, -3, 1\pm2i$

19. $f(x)=x^3-3x^2-6x+8$

Possible rational roots: $\{\pm1,\pm8,\pm2,\pm4\}$

$$\begin{array}{r|rrrr} 1 & 1 & -3 & -6 & 8\\ & & 1 & -2 & -8\\ \hline & 1 & -2 & -8 & 0\end{array}$$

$f(x)=(x-1)(x^2-2x-8)$

$f(x)=(x+2)(x-1)(x-4)$

$f(x) = (x+2)(x-1)(x-4)$

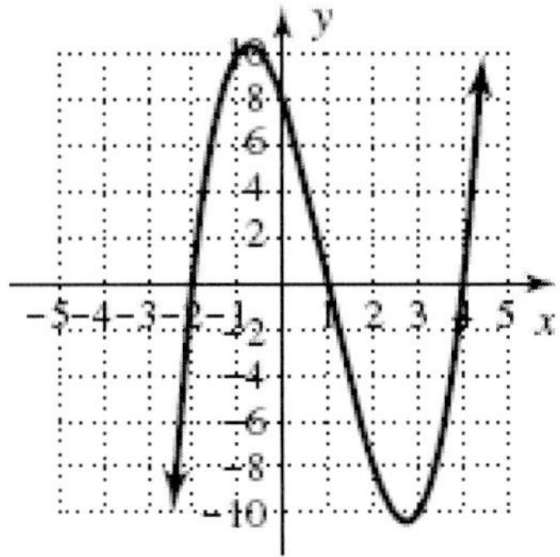

20. $h(x) = \dfrac{x-1}{x^2-4}$

$h(x) = \dfrac{x-1}{(x+2)(x-2)}$

Vertical asymptotes: $x = -2$ or $x = 2$

$x \in (-2,1] \cup (2,\infty)$

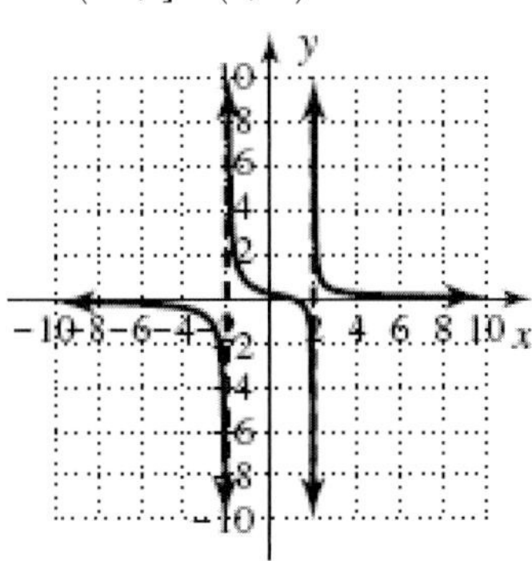

5.1 Technology Highlight

1. $3^x = 22$; $x = 2.8$

2. $2^x = 0.125$; $x = -3$

3. $5^{x-1} = 61$; $x = 3.6$

5.1 Exercises

1. b^x, b, b, x

3. a, 1

5. False; for $|b| < 1$ and $x_2 > x_1, b^{x_2} < b^{x_1}$ so the function is decreasing.

7. $P(t) = 2500 \cdot 4^t$;
$P(2) = 2500 \cdot 4^2 = 40000$;
$P\left(\frac{1}{2}\right) = 2500 \cdot 4^{\frac{1}{2}} = 5000$;
$P\left(\frac{3}{2}\right) = 2500 \cdot 4^{\frac{3}{2}} = 20000$;
$P\left(\sqrt{3}\right) = 2500 \cdot 4^{\sqrt{3}} \approx 27589.162$

9. $f(x) = 0.5 \cdot 10^x$;
$f(3) = 0.5 \cdot 10^3 = 500$;
$f\left(\frac{1}{2}\right) = 0.5 \cdot 10^{\frac{1}{2}} \approx 1.581$;
$f\left(\frac{2}{3}\right) = 0.5 \cdot 10^{\frac{2}{3}} \approx 2.321$;
$f\left(\sqrt{7}\right) = 0.5 \cdot 10^{\sqrt{7}} \approx 221.163$

11. $V(n) = 10{,}000\left(\frac{2}{3}\right)^n$;
$V(0) = 10{,}000\left(\frac{2}{3}\right)^0 = 10000$;
$V(4) = 10{,}000\left(\frac{2}{3}\right)^4 \approx 1975.309$;
$V(4.7) = 10{,}000\left(\frac{2}{3}\right)^{4.7} \approx 1487.206$;
$V(5) = 10{,}000\left(\frac{2}{3}\right)^5 \approx 1316.872$

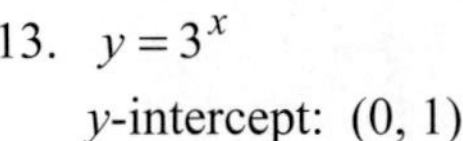

13. $y = 3^x$
y-intercept: (0, 1)

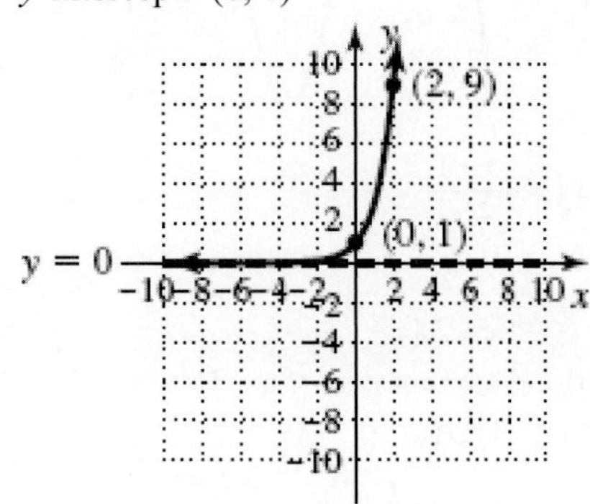

increasing

15. $y = \left(\frac{1}{3}\right)^x$
y-intercept: (0, 1)

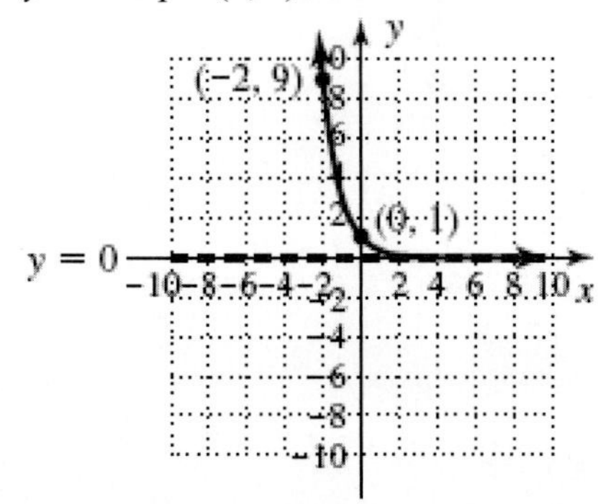

decreasing

17. $y = 3^x + 2$
up 2

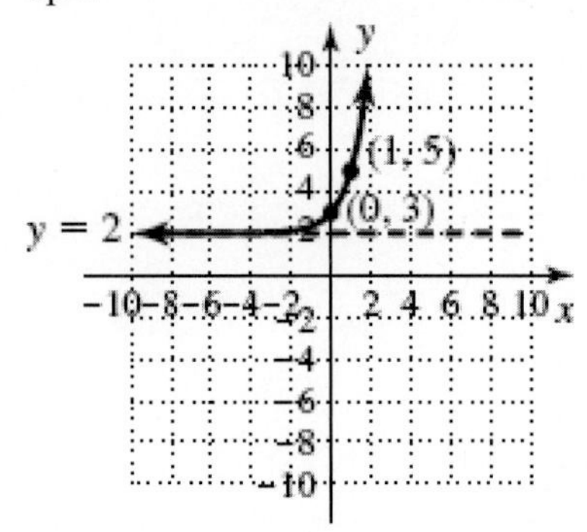

19. $y = 3^{x+3}$
left 3

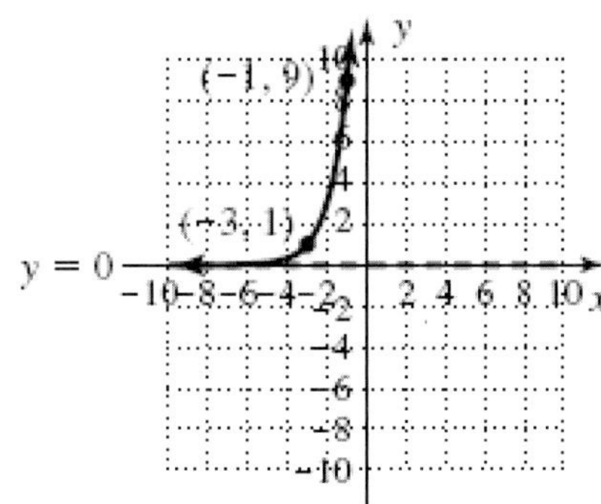

21. $y = 2^{-x}$
reflected in the y-axis

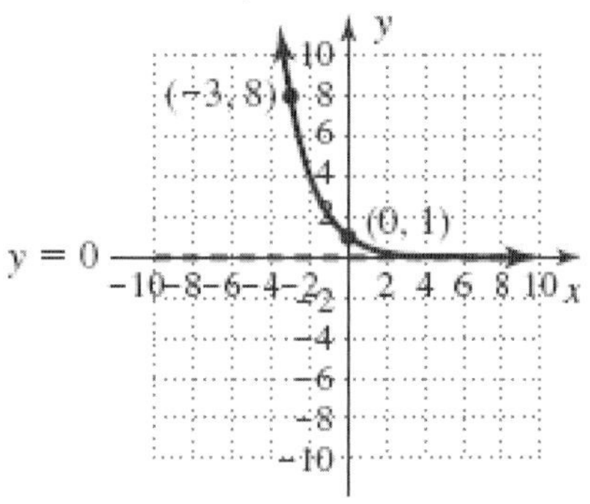

23. $y = 2^{-x} + 3$
reflected in the y-axis, up 3

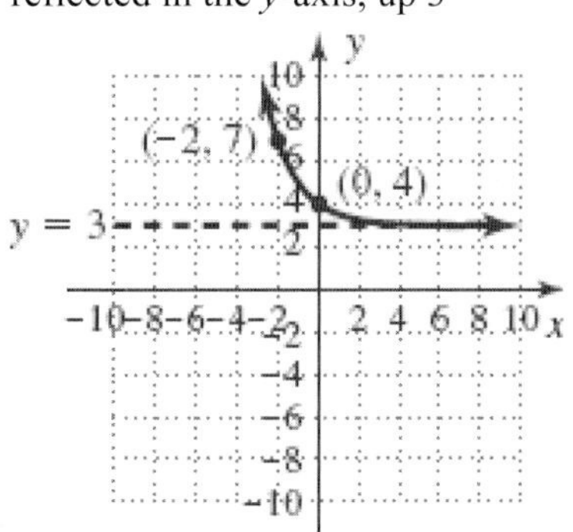

25. $y = 2^{x+1} - 3$
left 1, down 3

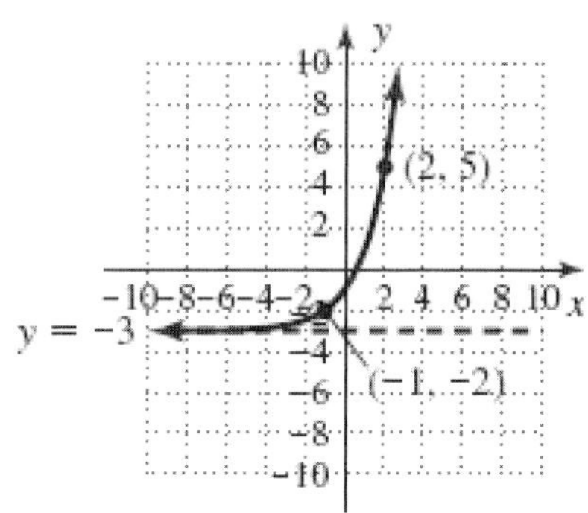

27. $y = \left(\frac{1}{3}\right)^x + 1$
up 1

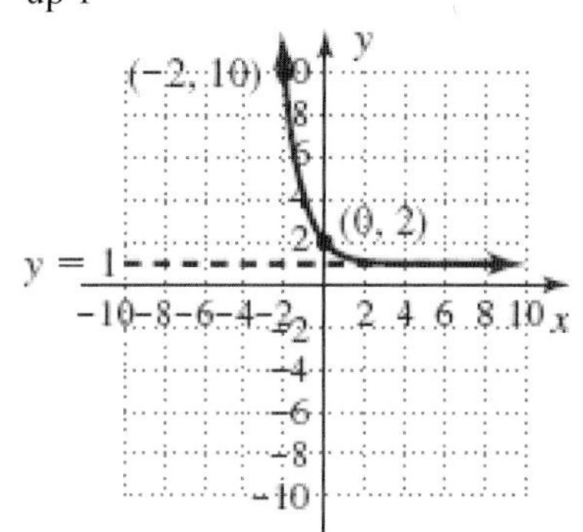

29. $y = \left(\frac{1}{3}\right)^{x-2}$
right 2

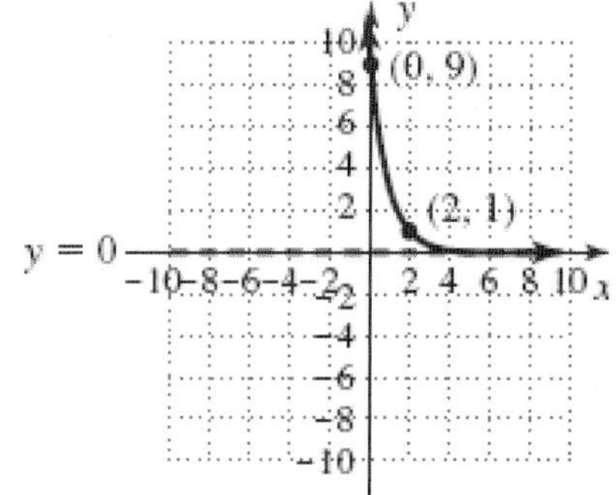

31. $y = \left(\frac{1}{3}\right)^x - 2$
right 2

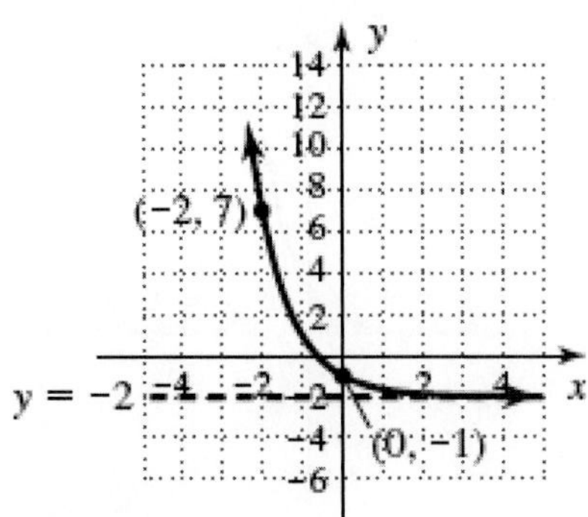

33. e; $y = 5^{-x}$

35. a; $y = 3^{-x+1}$

37. b; $y = 2^{x+1} - 2$

39. $10^x = 1000$
$10^x = 10^3$
$x = 3$

41. $25^x = 125$
$5^{2x} = 5^3$
$2x = 3$
$x = \frac{3}{2}$

43. $8^{x+2} = 32$
$2^{3(x+2)} = 2^5$
$3x + 6 = 5$
$3x = -1$
$x = -\frac{1}{3}$

45. $32^x = 16^{x+1}$
$2^{5x} = 2^{4(x+1)}$
$5x = 4x + 4$
$x = 4$

47. $\left(\frac{1}{5}\right)^x = 125$
$\left(\frac{1}{5}\right)^x = \left(\frac{1}{5}\right)^{-3}$
$x = -3$

49. $\left(\frac{1}{3}\right)^{2x} = 9^{x-6}$
$\left(\frac{1}{3}\right)^{2x} = \left(\frac{1}{3}\right)^{-2(x-6)}$
$2x = -2x + 12$
$4x = 12$
$x = 3$

51. $\left(\frac{1}{9}\right)^{x-5} = 3^{3x}$
$\left(\frac{1}{3}\right)^{2(x-5)} = \left(\frac{1}{3}\right)^{-1(3x)}$
$2x - 10 = -3x$
$5x = 10$
$x = 2$

53. $25^{3x} = 125^{x-2}$
$5^{6x} = 5^{3(x-2)}$
$6x = 3x - 6$
$3x = -6$
$x = -2$

55. $P(t) = 1000 \cdot 3^t$

(a) 12 hr $= \frac{1}{2}$ day

$P\left(\frac{1}{2}\right) = 1000 \cdot 3^{\frac{1}{2}} = 1732$;

$P(1) = 1000 \cdot 3^1 = 3000$;

$P\left(\frac{3}{2}\right) = 1000 \cdot 3^{\frac{3}{2}} = 5196$;

$P(2) = 1000 \cdot 3^2 = 9000$

(b) yes
(c) as $t \to \infty, P \to \infty$
(d)

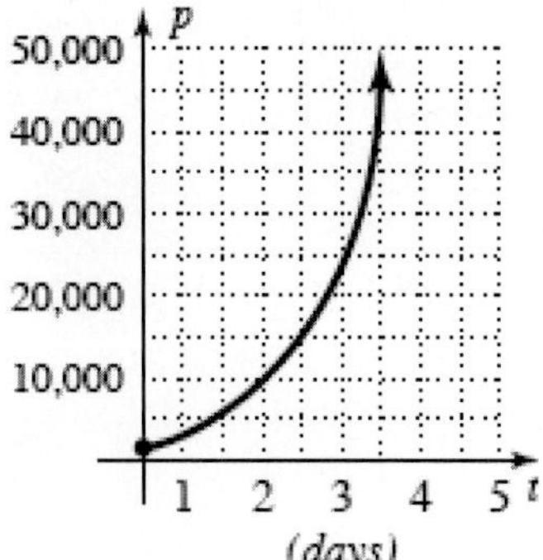

57. $A = P\left(1 + \frac{r}{n}\right)^{nt}$

(a) $A = 5000\left(1+\frac{0.06}{2}\right)^{2(16)} \approx \$12875.41$

(b) $A = 5000\left(1+\frac{0.06}{12}\right)^{12(16)} \approx \$13027.28$

$\$13027.28 - \$12875.41 = \$151.87$

59. $V(t) = V_0 \cdot \left(\frac{4}{5}\right)^t$

(a) $V(1) = 125000 \cdot \left(\frac{4}{5}\right)^1 = \$100{,}000$

(b) $64000 = 125000 \cdot \left(\frac{4}{5}\right)^t$

$\frac{64}{125} = \left(\frac{4}{5}\right)^t$

$\left(\frac{4}{5}\right)^3 = \left(\frac{4}{5}\right)^t$

$t = 3$ yrs

61. $V(t) = V_0 \cdot \left(\frac{5}{6}\right)^t$

(a) $V(5) = 216000 \cdot \left(\frac{5}{6}\right)^5 \approx \$86{,}806$

(b) $125000 = 216000 \cdot \left(\frac{5}{6}\right)^t$

$\frac{125}{216} = \left(\frac{5}{6}\right)^t$

$\left(\frac{5}{6}\right)^3 = \left(\frac{5}{6}\right)^t$

$t = 3$ yrs

63. $R(t) = R_0 \cdot 2^t$

(a) $R(4) = 2.5 \cdot 2^4 = \$40$ million

(b) $320 = 2.5 \cdot 2^t$

$128 = 2^t$

$2^7 = 2^t$

$t = 7$ yrs

65. $P(t) = P_0(1.05)^t$

$P(10) = 20000(1.05)^{10} \approx \$32577.89$

67. $Q(t) = Q_0\left(\frac{1}{2}\right)^{\frac{t}{h}}$

(a) $Q(24) = 64\left(\frac{1}{2}\right)^{\frac{24}{8}} = 8$ grams

(b) $1 = 64\left(\frac{1}{2}\right)^{\frac{t}{8}}$

$\frac{1}{64} = \left(\frac{1}{2}\right)^{\frac{t}{8}}$

$\left(\frac{1}{2}\right)^6 = \left(\frac{1}{2}\right)^{\frac{t}{8}}$

$6 = \frac{t}{8}$

$t = 48$ minutes

69. $f(20) = \left(\frac{1}{2}\right)^{20} = 9.5\text{x}10^{-7}$;

Answers will vary.

71. $10^{2x} = 25$

$\left(10^{2x}\right)^{-\frac{1}{2}} = 25^{-\frac{1}{2}}$

$10^{-x} = \frac{1}{5}$

73. $f(x) = 2x^2 - 3x$

$f(-1) = 2(-1)^2 - 3(-1) = 5$;

$f\left(\frac{1}{3}\right) = 2\left(\frac{1}{3}\right)^2 - 3\left(\frac{1}{3}\right) = -\frac{7}{9}$;

$f(a) = 2a^2 - 3a$;

$f(a+h) = 2(a+h)^2 - 3(a+h)$

$= 2\left(a^2 + 2ah + h^2\right) - 3a - 3h$

$= 2a^2 + 4ah + 2h^2 - 3a - 3h$

75. Vertex: (2, 1)

$3 = a|0-2| + 1$

$3 = 2a + 1$

$2 = 2a$

$1 = a$;

$y = 1|x-2| + 1$

77. a. $-2\sqrt{x-3}+7=21$

$-2\sqrt{x-3}=14$

$\sqrt{x-3}=-7$

no solution

b. $\frac{9}{x+3}+3=\frac{12}{x-3}$

$(x+3)(x-3)\left[\frac{9}{x+3}+3=\frac{12}{x-3}\right]$

$9(x-3)+3(x^2-9)=12(x+3)$

$9x-27+3x^2-27=12x+36$

$3x^2-3x-90=0$

$x^2-x-30=0$

$(x-6)(x+5)=0$

$x-6=0$ or $x+5=0$

$x=6$ or $x=-5$

$\{-5,6\}$

5.2 Technology Highlight

1. $-3.141592654=-\pi$

2. $3.141592654=\pi$

3. (a) $\log_{10} 10^{2.5}=2.5$

(b) $\log_{10} 10^{\pi}=\pi$

(c) $\log^{\log 10\, 2.5}=2.5$

(d) $10^{\log 10\, \pi}=\pi$

5.2 Exercises

1. $\log_b x$, b, b, greater

3. $(1,0)$; 0

5. 5; answers will vary

7. $3=\log_2 8$

$2^3=8$

9. $-1=\log_7 \frac{1}{7}$

$7^{-1}=\frac{1}{7}$

11. $0=\log_9 1$

$9^0=1$

13. $\frac{1}{3}=\log_8 2$

$8^{\frac{1}{3}}=2$

15. $1=\log_2 2$

$2^1=2$

17. $\log_7 49=2$

$7^2=49$

19. $\log_{10} 100=2$

$10^2=100$

21. $\log_{10}(0.1)=-1$

$10^{-1}=0.1$

23. $4^3=64$

$\log_4 64=3$

25. $3^{-2}=\frac{1}{9}$

$\log_3\left(\frac{1}{9}\right)=-2$

27. $9^0=1$

$\log_9 1=0$

29. $\left(\frac{1}{3}\right)^{-3}=27$

$\log_{\frac{1}{3}} 27=-3$

31. $10^3=1000$

$\log 1000=3$

33. $10^{-2}=\frac{1}{100}$

$\log\frac{1}{100}=-2$

35. $4^{\frac{3}{2}}=8$

$\log_4 8=\frac{3}{2}$

37. $4^{\frac{-3}{2}} = \frac{1}{8}$

$\log_4 \frac{1}{8} = \frac{-3}{2}$

39. $\log_{11} 121 = x$

$11^x = 121$

$11^x = 11^2$

$x = 2$

41. $\log_3 243 = x$

$3^x = 243$

$3^x = 3^5$

$x = 5$

43. $\log_7 \frac{1}{49} = x$

$7^x = \frac{1}{49}$

$7^x = 7^{-2}$

$x = -2$

45. $\log_4 4$

$=1$

47. $\log_{10} 10$

$=1$

49. $\log_4 2 = x$

$4^x = 2$

$2^{2x} = 2^1$

$2x = 1$

$x = \frac{1}{2}$

51. $\log_5 x = 2$

$5^2 = x$

$25 = x$

53. $\log_{36} x = \frac{1}{2}$

$36^{\frac{1}{2}} = x$

$6 = x$

55. $\log_x 36 = 2$

$x^2 = 36$

$x^2 = 6^2$

$x = 6$

57. $\log_x \frac{1}{4} = -2$

$x^{-2} = \frac{1}{4}$

$x^{-2} = 2^{-2}$

$x = 2$

59. $\log_{25} x = -\frac{3}{2}$

$25^{-\frac{3}{2}} = x$

$\left(\sqrt{25}\right)^{-3} = x$

$5^{-3} = x$

$\frac{1}{125} = x$

61. $\log_8 32 = x$

$8^x = 32$

$2^{3x} = 2^5$

$3x = 5$

$x = \frac{5}{3}$

63. $f(x) = \log_2 x + 3$

Shift up 3

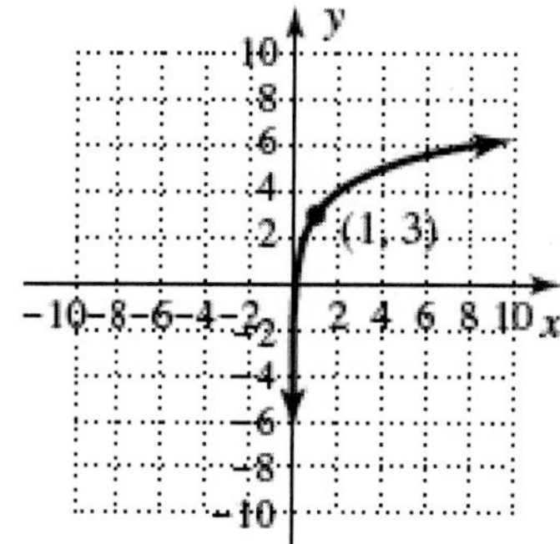

65. $h(x) = \log_2 (x-2) + 3$

Shift right 2, up 3

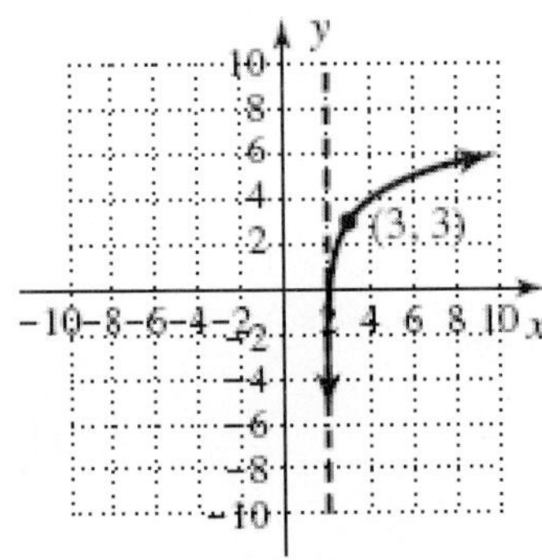

67. $q(x) = \log_3(x+1)$
Shift left 1

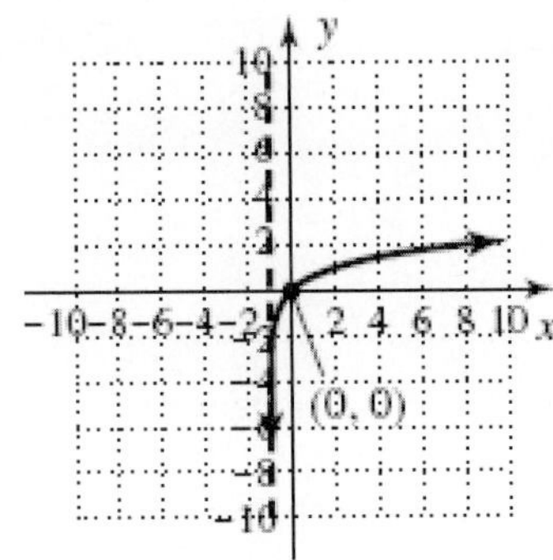

69. $Y_1 = -\log_2(x+1)$
Reflected across x–axis, shift left 1

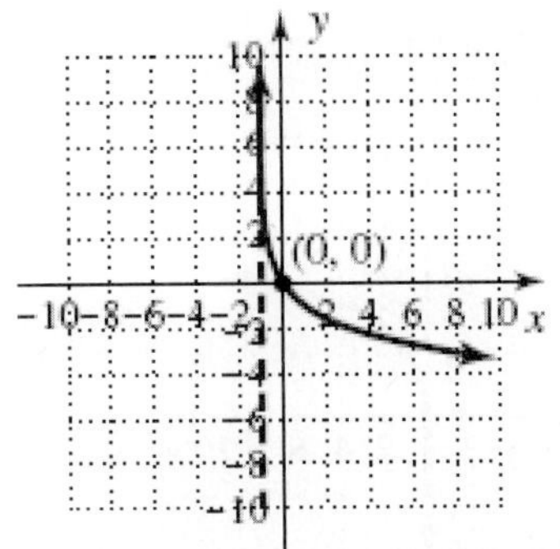

71. $y = \log_6\left(\dfrac{x+1}{x-3}\right)$

$\dfrac{x+1}{x-3} > 0, x \neq 3$

critical values: -1 and 3

pos neg pos

-1 3

$x \in (-\infty,-1) \cup (3,+\infty)$

73. $y = \log_5 \sqrt{2x-3}$

$2x - 3 > 0$

$2x > 3$

$x > \dfrac{3}{2}$

$x \in \left(\dfrac{3}{2}, \infty\right)$

75. $y = \log\left(9 - x^2\right)$

$9 - x^2 > 0$;

$(3+x)(3-x) > 0$

critical values: 3 and -3

neg pos neg

-3 3

$x \in (-3,3)$

77. $\log 175 \approx 2.2430$

Since $100<175<1000$,
log175 is bounded between 2 and 3.

79. $\log 127{,}962 \approx 5.1071$

Since $100{,}000<137{,}962<1{,}000{,}000$,
log 127,962 is bounded between 5 and 6.

80. $\log 9871 \approx 3.9944$

Since $1000<9871<10{,}000$,
log 9871 is bounded between 3 and 4.

81. $\log\dfrac{1}{5} \approx -0.6990$

Since $0.1<\dfrac{1}{5}<1$,

$\log\dfrac{1}{5}$ is bounded between -1 and 0.

83. $f(x) = -\log_{10} x$

$x = 7.94\text{x}10^{-5}$

$f\left(7.94 \text{ x } 10^{-5}\right) = -\log_{10}\left(7.94 \text{ x } 10^{-5}\right)$

pH ≈ 4.1; acid

85. $M(I) = 6 - 2.5 \cdot \log\left(\dfrac{I}{I_0}\right)$

a. $I = 27 \cdot I_0$

$M(27I_0) = 6 - 2.5 \cdot \log\left(\dfrac{27I_0}{I_0}\right)$

≈ 2.4 m

b. $I = 85 \cdot I_0$

$$M(85I_0) = 6 - 2.5 \cdot \log\left(\frac{85I_0}{I_0}\right)$$
$$\approx 1.2 \text{ m}$$

87. $M(I) = \log\left(\frac{I}{I_0}\right)$

a. $I = 50{,}000I_0$

$$M(50000I_0) = \log\left(\frac{50000I_0}{I_0}\right)$$
$$M(50000I_0) = \log(50000) \approx 4.7 \text{ m}$$

b. $I = 75{,}000I_0$

$$M(75000I_0) = \log\left(\frac{75000I_0}{I_0}\right)$$
$$M(75000I_0) = \log(75000) \approx 4.9 \text{ m}$$

89. $D(I) = 10 \cdot \log\left(\frac{I}{I_0}\right)$

a. $I = 10^{-14}$

$$D(10^{-14}) = 10 \cdot \log\left(\frac{10^{-14}}{10^{-16}}\right) = 20 \text{ dB}$$

b. $I = 10^{-4}$

$$D(10^{-4}) = 10 \cdot \log\left(\frac{10^{-4}}{10^{-16}}\right) = 120 \text{ dB}$$

91. $P(x) = 95 - 14 \cdot \log_2(x)$

a. 1 day
$P(1) = 95 - 14 \cdot \log_2(1) = 95\,\%$

b. 4 days
$P(4) = 95 - 14 \cdot \log_2(4) = 67\,\%$

c. 16 days
$P(16) = 95 - 14 \cdot \log_2(16) = 39\,\%$

93. $f(x) = \log_{10} x$

$x = 5.1\text{x}10^{-5}$

$f(5.1\text{x}10^{-5}) = \log_{10}(5.1\text{x}10^{-5})$

pH ≈ 4.3 ; acid

95. a. Threshold of audibility
0 dB

b. Lawn Mower
90 dB

c. Whisper
15 dB

d. Loud rock concert
120 dB

e. Lively party
100 dB

f. Jet engine
120 dB

Many sources give the threshold of pain as 120dB; answers may vary.

97. Answers will vary.

99. $y = \log_{\frac{1}{2}} x$

Convert to exponential form:

$$\left(\frac{1}{2}\right)^y = x$$
$$2^{-y} = x$$

Convert to exponential form:

$$-y = \log_2 x$$
$$y = -\log_2 x$$

101.a. $x < 3$ and $x > -1$

b. $x < 3$ or $x > -1$

−1 3

103.a. $x^3 - 8$
$= (x-2)(x^2 + 2x + 4)$

b. $a^2 - 49$
$= (a+7)(a-7)$

c. $n^2 - 10n + 25$
$= (n-5)(n-5)$
$= (n-5)^2$

d. $2b^2 - 7b + 6$
$= (2b-3)(b-2)$

105. $x \in (-\infty, -5)$;

$$f(x) = (x+5)(x-4)^2$$
$$f(x) = (x+5)(x^2 - 8x + 16)$$
$$f(x) = x^3 - 8x^2 + 16x + 5x^2 - 40x + 80$$
$$f(x) = x^3 - 3x^2 - 24x + 80$$

5.3 Technology Highlight

1. $y = \dfrac{\log x}{\log 2}$ if $x = 1$,

$\frac{\log 1}{\log 2} = \frac{0}{\log 2} = 0$

2. $\frac{\log 16}{\log 2} = 4$

3. $\frac{\log \sqrt{2}}{\log 2} = 0.5$

4. -2

5. $\log_2 \sqrt{3} = \frac{\log \sqrt{3}}{\log 2}$

6. $\log_2 x = \sqrt{3}$, write in exponential form

$2^{\sqrt{3}}$

5.3 Exercises

1. $\left(1 + \frac{1}{x}\right)^x, x \to \infty$

3 $\ln x$, exponent, e

5. $\log_2 9 > 3; \log_3 26 < 3$

7. $e^1 \approx 2.718282$

9. $e^2 \approx 7.389056$

11. $e^{1.5} \approx 4.481689$

13. $e^{\sqrt{2}} \approx 4.113250$

15. $f(x) = e^{x+3} - 2$

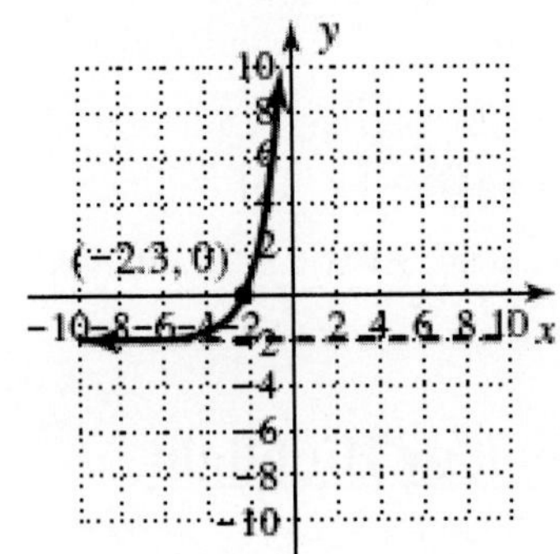

17. $r(t) = -e^t + 2$

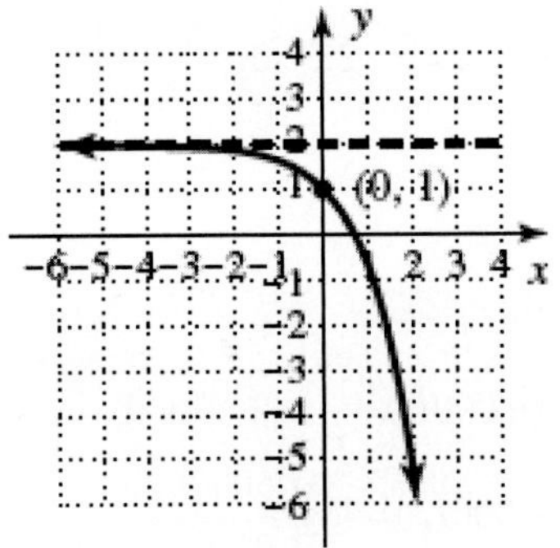

19. $p(x) = e^{-x+2} - 1$

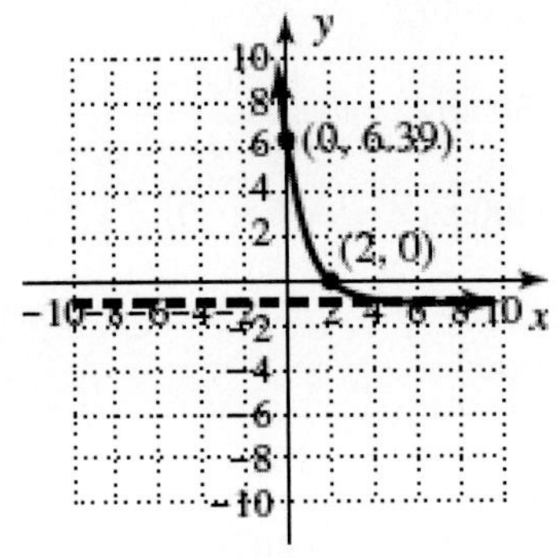

21. $\ln 50 \approx 3.912023$

23. $\ln 0.5 \approx -0.693147$

25. $\ln 225 \approx 5.416100$

27. $\ln\left(\sqrt{2}\right) \approx 0.346574$

29. $\ln(x) = 1$

$e^1 = x$

$e = x$

$x \approx 2.718$

31. $\ln(x) = -1.961$

$e^{-1.961} = x$

$x \approx 0.141$

33. $-2.4 = \ln\left(\frac{1}{x^2}\right)$

$e^{-2.4} = \frac{1}{x^2}$

$e^{-2.4}x^2 = 1$

$x^2 = \frac{1}{e^{-2.4}}$

$x^2 = e^{2.4}$

$x = \pm\sqrt{e^{2.4}}$
$x \approx \pm 3.320$

35. $\ln e^{2x} = -8.4$
$e^{-8.4} = e^{2x}$
$-8.4 = 2x$
$-4.2 = x$

36. $\ln e^{3x} = -9.6$
$e^{-9.6} = e^{3x}$
$-9.6 = 3x$
$-3.2 = x$

37. $e^{x} = 1$
By definition, $\ln x = y$ iff $e^{y} = x$
$\ln 1 = x$
$x = 0$

39. $e^{x} = 7.389$
By definition, $\ln x = y$ iff $e^{y} = x$
$x = \ln(7.389) \approx 1.99999$

41. $e^{\frac{2x}{5}} = 1.396$
By definition, $\ln x = y$ iff $e^{y} = x$
$\frac{2x}{5} = \ln(1.396)$
$x = \frac{5}{2}\ln(1.396) \approx 0.83403$

43. $e^{x} = -23.14069$
x is not a real number.

44. $e^{x} = -23.14069$
x is not a real number.

45. $\ln(2x) + \ln(x-7)$
$= \ln(2x(x-7))$
$= \ln(2x^2 - 14x)$

47. $\log(x+1) + \log(x-1)$
$= \log((x+1)(x-1))$
$= \log(x^2 - 1)$

49. $\log_3 28 - \log_3 7$
$= \log_3\left(\frac{28}{7}\right)$
$= \log_3(4)$

51. $\log x - \log(x+1)$
$= \log\left(\frac{x}{x+1}\right)$

53. $\ln(x-5) - \ln x$
$= \log\left(\frac{x-5}{x}\right)$

55. $\ln(x^2 - 4) - \ln(x+2)$
$= \ln\left(\frac{x^2 - 4}{x+2}\right)$
$= \ln\left(\frac{(x+2)(x-2)}{x+2}\right)$
$= \ln(x-2)$

57. $\log_2 7 + \log_2 6$
$= \log_2(7 \cdot 6)$
$= \log_2 42$

59. $\log_5(x^2 - 2x) + \log_5 x^{-1}$
$= \log_5(x^{-1}(x^2 - 2x))$
$= \log_5(x-2)$

61. $\log 8^{x+2} = (x+2)\log 8$

63. $\ln 5^{2x-1} = (2x-1)\ln 5$

65. $\log\sqrt{22} = \log 22^{\frac{1}{2}} = \frac{1}{2}\log 22$

67. $\log_5 81 = \log_5 3^4 = 4\log_5 3$

69. $\log(a^3 b) = 3\log a + \log b$

71. $\ln(x\sqrt[4]{y}) = \ln x + \ln y^{\frac{1}{4}}$
$= \ln x + \frac{1}{4}\ln y$

73. $\ln\left(\frac{x^2}{y}\right) = \ln x^2 - \ln y$

$= 2\ln x - \ln y$

75. $\log\left(\sqrt{\frac{x-2}{x}}\right) = \log\left(\frac{x-2}{x}\right)^{\frac{1}{2}}$

$= \frac{1}{2}\log\left(\frac{x-2}{x}\right)$

$= \frac{1}{2}[\log(x-2) - \log x]$

77. $\ln\left(\frac{7x\sqrt{3-4x}}{2(x-1)^3}\right)$

$= \ln\left(7x\sqrt{3-4x}\right) - \ln\left(2(x-1)^3\right)$

$= \ln 7x + \ln\sqrt{3-4x} - \left[\ln 2 + \ln(x-1)^3\right]$

$= \ln 7x + \ln(3-4x)^{\frac{1}{2}} - \ln 2 - \ln(x-1)^3$

$= \ln 7 + \ln x + \frac{1}{2}\ln(3-4x) - \ln 2 - 3\ln(x-1)$

79. $\log_7 60 = \frac{\ln 60}{\ln 7} = 2.104076884$

81. $\log_5 152 = \frac{\ln 152}{\ln 5} \approx 3.12151248$

83. $\log_3 1.73205 = \frac{\log 1.73205}{\log 3}$

≈ 0.499999576

85. $\log_{0.5} 0.125 = \frac{\log 0.125}{\log 0.5} = 3$

87. $f(x) = \log_3 x = \frac{\log x}{\log 3}$;

$f(5) = \frac{\log 5}{\log 3} \approx 1.4650$;

$f(15) = \frac{\log 15}{\log 3} \approx 2.4650$;

$f(45) = \frac{\log 45}{\log 3} \approx 3.4650$;

Outputs increase by 1; $f(3^3 \cdot 5) \approx 4.465$

89. $h(x) = \log_9 x = \frac{\log x}{\log 9}$;

$h(2) = \frac{\log 2}{\log 9} \approx 0.3155$;

$h(4) = \frac{\log 4}{\log 9} \approx 0.6309$;

$h(8) = \frac{\log 8}{\log 9} \approx 0.9464$;

Outputs are multiples of 0.3155;
$h(2^4) = 4(0.3155) \approx 1.2619$

91. $A(P) = -4.762\ln(0.068P)$

(a) $A(3.25) = -4.762\ln(0.068(3.25))$

≈ 7.19 m

(b) $A(7.12) = -4.762\ln(0.068(7.12))$

≈ 3.45 m

(c) $A(10.24) = -4.762\ln(0.068(10.24))$

≈ 1.72 m; Losing altitude

93. $h(T) = (30T + 80000)\cdot\ln\frac{P_0}{P}$;

(a) $h(5) = (30(5) + 80000)\cdot\ln\frac{76}{42} \approx 4833.5$ ft;

$h(2) = (30(2) + 80000)\cdot\ln\frac{76}{30} \approx 7492.1$ ft;

$h(-6) = (30(-6) + 80000)\cdot\ln\frac{76}{12} \approx 14434.4$ ft

(b)

$h(-12) = (30(-12) + 80000)\cdot\ln\frac{76}{1.7} \approx 29032.8$ ft

95. $T = \frac{1}{r}\cdot\ln\left(\frac{A}{P}\right)$

$8 = \frac{1}{0.05}\cdot\ln\left(\frac{A}{200000}\right)$

$0.4 = \ln\left(\frac{A}{200000}\right)$

By definition, $\ln x = y$ iff $e^y = x$

$e^{0.4} = \frac{A}{200000}$

$200000e^{0.4} = A$

No, \$298364.94;

$8 = \frac{1}{0.05}\cdot\ln\left(\frac{350000}{P}\right)$

$0.4 = \ln\left(\frac{350000}{P}\right)$

By definition, $\ln x = y$ iff $e^y = x$

$e^{0.4} = \frac{350000}{P}$

$\frac{350000}{e^{0.4}} = P$

$P = \$234{,}612.01$

97. $T(r) = \frac{\ln 3}{r}$

(a) $T(0.03) = \frac{\ln 3}{0.03} \approx 36.6$ years

(b) $T(0.055) = \frac{\ln 3}{0.055} \approx 20$ years

(c) $T(0.08) = \frac{\ln 3}{0.08} \approx 13.7$ years

(d) $\approx 11\%$

99. $T(p) = \frac{-\ln p}{k}$

(a) $T(0.65) = \frac{-\ln 0.65}{0.072} \approx 6$ hrs

(b) $24 = \frac{-\ln p}{0.072}$

$1.728 = -\ln p$

$e^{1.728} = e^{-\ln p}$

$e^{1.728} = p^{-1}$

$e^{1.728} = \frac{1}{p}$

$p = \frac{1}{e^{1.728}} \approx 18\%$

101. $T = -8266 \cdot \ln p$

$T = -8266 \cdot \ln(0.124) \approx 17{,}255$ yrs

103. $y = \ln x$

(a) $\frac{\Delta y}{\Delta x} = \frac{\ln 1.001 - \ln 1}{1.001 - 1} \approx 1$

(b) $\frac{\Delta y}{\Delta x} = \frac{\ln 2.001 - \ln 2}{2.001 - 2} \approx 0.5$

(c) $\frac{\Delta y}{\Delta x} = \frac{\ln 3.001 - \ln 3}{3.001 - 3} \approx 0.33$

(d) $\frac{\Delta y}{\Delta x} = \frac{\ln 4.001 - \ln 4}{4.001 - 4} \approx 0.25$

105. $y = e^x$

$\frac{\Delta y}{\Delta x} = \frac{e^{3.0001} - e^3}{3.0001 - 3} \approx 20$;

$f(3) = e^3 \approx 20$;

$\frac{\Delta y}{\Delta x} = \frac{e^{2.0001} - e^2}{2.0001 - 2} \approx 7.39$;

$f(2) = e^2 \approx 7.39$;

$\frac{\Delta y}{\Delta x} = \frac{e^{4.0001} - e^4}{4.0001 - 4} \approx 54.6$;

$f(4) = e^4 \approx 54.6$

107. (a) $D_f : x \in (0, \infty); R_f : y \in R$

$Dg : x \in [0, \infty), \mathrm{R}_g : y \in [-1, \infty]$

(b) $\ln x = \sqrt{x} - 1$

(1,0)

(c) $f(15) - g(15) = \ln 15 - \left(\sqrt{15} - 1\right)$

≈ -0.16;

$f(1500) - g(1500) = \ln 1500 - \left(\sqrt{1500} - 1\right)$

≈ -30.42;

$f(150000) - g(150000)$

$= \ln 150000 - \left(\sqrt{150000} - 1\right) \approx -374.38$

As $x \to \infty$ the difference between the 2 functions increases.

109. $\ln(p \cdot q) \neq \ln(p) \cdot \ln(q)$

$\ln(p \cdot q) = \ln p + \ln q$;

$\ln\left(\frac{p}{q}\right) \neq \frac{\ln(p)}{\ln(q)}$

$\ln\left(\frac{p}{q}\right) = \ln p - \ln q$;

$\ln(p) + \ln(q) \neq \ln(p + q)$

$\ln p + \ln q = \ln(p \cdot q)$

111. $\ln(x) = ln(10) \cdot \log(x)$

$\ln(x) = \ln(10) \cdot \frac{\ln(x)}{\ln(10)}$

$\ln(x) = \ln(x)$ verified

$\ln(e) = ln(10) \cdot \log(e) = 1$;

$\ln(10) = ln(10) \cdot \log(10) = 2.303$;

$\ln(2) = ln(10) \cdot \log(2) = 0.693$;

113. $\log_3 4 \approx 1.2619$ and $\log_3 5 \approx 1.4650$

(a) $\log_3 20 = \log_3 (4 \cdot 5) = \log_3 4 + \log_3 5$

≈ 2.7269

(b) $\log_3 \frac{4}{5} = \log_3 4 - \log_3 5$

$= \log_3 \frac{4}{5} \approx -0.2031$

(c) $\log_3 25 = \log_3 (5^2) = 2\log_3 5$

$= 2\left(\frac{\log 5}{\log 3}\right) \approx 2.9300$

115. $y = x, y = |x|, y = x^2, y = x^3, y = \sqrt{x},$

$y = \sqrt[3]{x}, y = \frac{1}{x}, y = \frac{1}{x^2}$

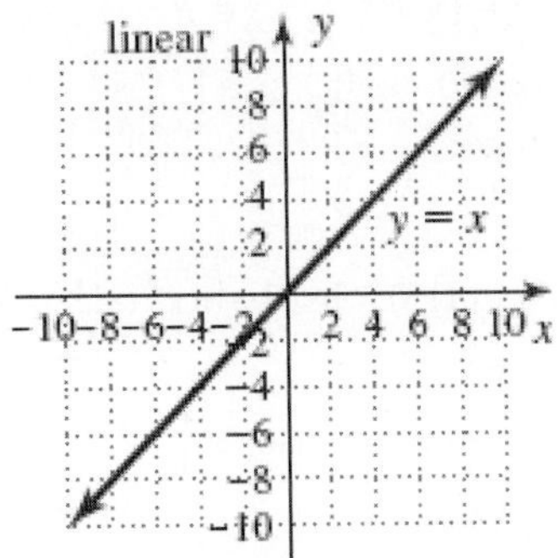

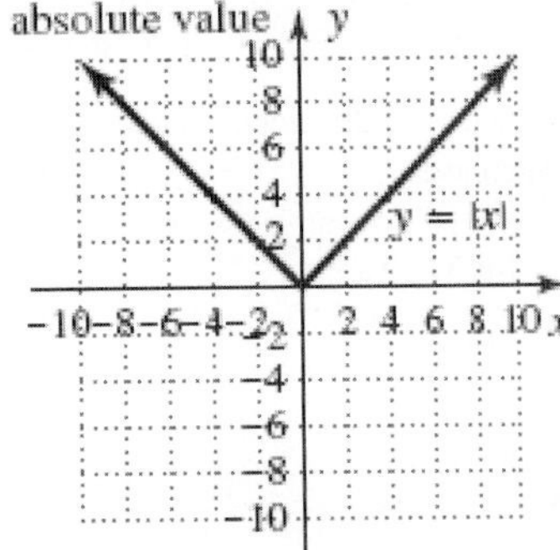

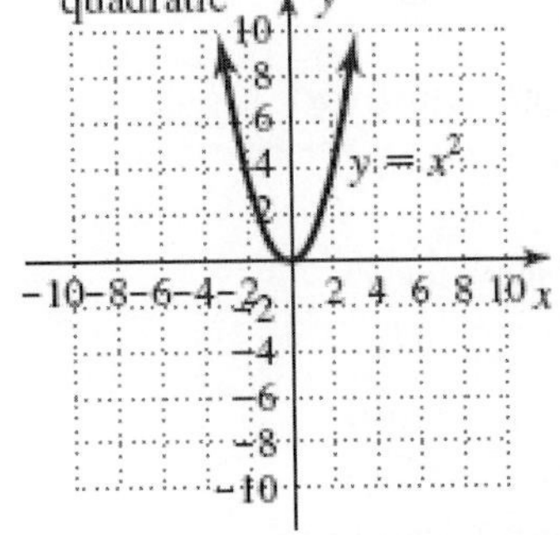

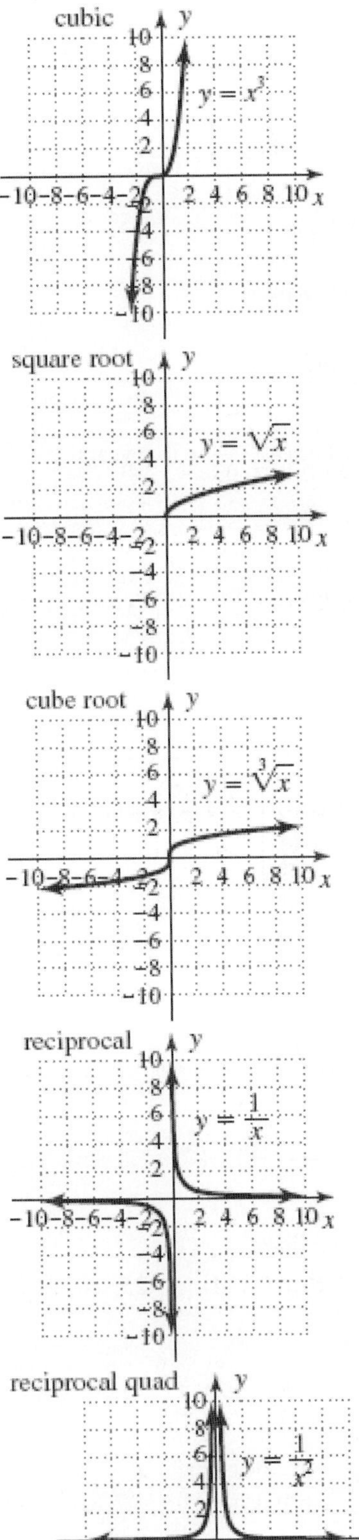

117. $r(x)=\dfrac{x^2+4x+3}{x^2+4x+4}=\dfrac{(x+3)(x+1)}{(x+2)(x+2)}$

$$\begin{array}{r} 1 \\ x^2+4x+4\overline{)x^2+4x+3} \\ -(x^2+4x+4) \\ \hline -1 \end{array}$$

$r(x)=1+\dfrac{-1}{x^2+4x+4}=1+\dfrac{-1}{(x+2)^2}$

VA: $x=-2$

HA: $y=1$

x-intercepts: $(-3,0)$ and $(-1,0)$

$r(0)=\dfrac{0^2+4(0)+3}{0^2+4(0)+4}=\dfrac{3}{4}$

y-intercept: $\left(0,\dfrac{3}{4}\right)$

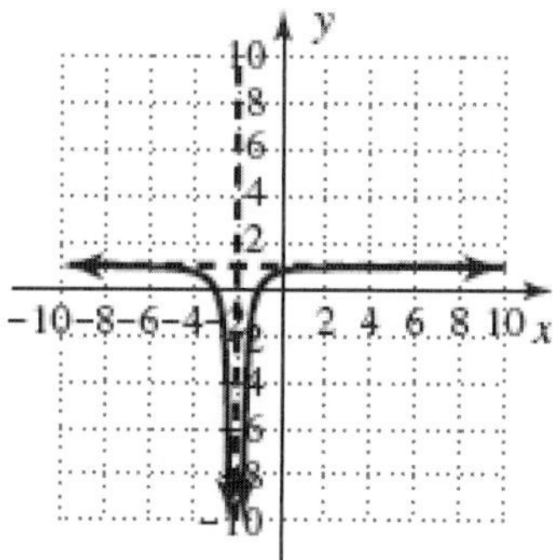

119. $y=-|x-3|+6$;

$\frac{1}{2}(12)(6)-\frac{1}{2}(3)(3)=36-4.5=31.5$ units^2

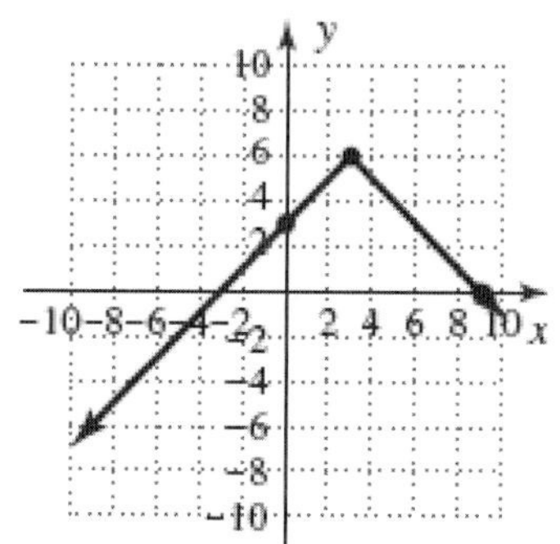

Chapter 5 Mid-Chapter Check

1. a. $27^{\frac{2}{3}}=9$

$\dfrac{2}{3}=\log_{27} 9$

b. $81^{\frac{5}{4}}=243$

$\dfrac{5}{4}=\log_{81} 243$

2. a. $\log_8 32=\dfrac{5}{3}$

$8^{\frac{5}{3}}=32$

b. $\log_{1296} 6=0.25$

$1296^{0.25}=6$

3. a. $4^{2x}=32^{x-1}$

$(2^2)^{2x}=(2^5)^{x-1}$

$2^{4x}=2^{5x-5}$

$4x=5x-5$

$x=5$

b. $\left(\dfrac{1}{3}\right)^{4b}=9^{2b-5}$

$(3^{-1})^{4b}=(3^2)^{2b-5}$

$3^{-4b}=3^{4b-10}$

$-4b=4b-10$

$-8b=-10$

$b=\dfrac{5}{4}$

4. a. $\log_{27} x=\dfrac{1}{3}$

$27^{\frac{1}{3}}=x$

$x=3$

b. $\log_b 125=3$

$b^3=125$

$b^3=5^3$

$b=5$

5. $V(t)=V_0\left(\dfrac{9}{8}\right)^t$

a. $V(3)=50000\left(\dfrac{9}{8}\right)^3=\$71191.41$

b. 6 yrs

6. a. $\log 243=2.385606274$

Since $100<175<1000$,
log243 is bounded between 2 and 3.

b. $\log 85{,}678=4.93286932$

Since $10{,}000<85678<100{,}000$,

log85678 is bounded between 4 and 5.

7. a. $\log_3 19 = \frac{\log 19}{\log 3} \approx 2.68$

Since $3^2 < 19 < 3^3$,
$\log_3 19$ is bounded between 2 and 3.

b. $\log_2 60 = \frac{\log 60}{\log 2} \approx 5.91$

Since $2^5 < 60 < 2^6$,
$\log_2 60$ is bounded between 5 and 6.

8. $R(h) = \frac{\ln 2}{h}$

a. $R(28) = \frac{\ln 2}{28} \approx 2.48\%$

b. $0.0578 = \frac{\ln 2}{h}$

$0.0578h = \ln 2$

$h = \frac{\ln 2}{0.0578} \approx 12\text{yrs}$

9. a. $\log(2x-3) + \log(2x+3)$
$\log[(2x-3)(2x+3)]$
$\log(4x^2-9)$

b. $\log(x+5) - \log(x^2-25)$

$\log\left(\frac{x+5}{x^2-25}\right)$

$\log\left(\frac{x+5}{(x+5)(x-5)}\right)$

$\log\left(\frac{1}{x-5}\right)$

10. a. $\log_7 50 = \log_7(5 \cdot 10)$
$= \log_7 5 + \log_7 10 \approx 2.0104$

b. $\log_7 2 = \log_7\left(\frac{10}{5}\right)$

$= \log_7 10 - \log_7 5 \approx 0.3562$

c. $\log_7 25 = \log_7 5^2 = 2\log_7 5 \approx 1.6542$

d. $\log_7 500 = \log_7(5 \cdot 100) = \log_7(5 \cdot 10^2)$
$= \log_7 5 + 2\log_7 10 \approx 3.1937$

Reinforcing Basic Concepts

1. Answers will vary.

2. (a) $\log x + \log(x+3) = \log[x(x+3)]$
$\log(x^2+3x)$

(b) $\ln(x+2) + \ln(x-2)$
$= \ln((x+2)(x-2)) = \ln(x^2-4)$

(c) $\log x - \log(x+3) = \log\left(\frac{x}{x+3}\right)$

3. Answers will vary.

4. (a) $\log 3^x = x\log 3$

(b) $\ln x^5 = 5\ln x$

(c) $\ln 2^{3x-1} = (3x-1)\ln 2$

5.4 Technology Highlight

1. $x = 1.1311893$, $x = -1.467671$

2. $x = -0.0506028$, $x = 1.2329626$

3. $x = 4.3556075$

4. $x = \frac{e}{2} \approx 1.3591409$, $x = \frac{1}{2e^4} \approx 0.00915782$

Let $u = \ln(2x)$; $u^2 = (\ln(2x))^2$

$u^2 + 3u - 4 = 0$
$(u+4)(u-1) = 0$
$u = -4$ or $u = 1$;
$\ln(2x) = -4$

$e^{-4} = 2x$

$\frac{e^{-4}}{2} = x$

$\frac{1}{2e^4} = x$;

$\ln(2x) = 1$

$e^1 = 2x$

$\frac{e}{2} = x$

5.4 Exercises

1. e

3. Extraneous

5. $\ln(4x+3) + \ln 2 = 3.2$
$\ln 2(4x+3) = 3.2$
$\ln(8x+6) = 3.2$

$e^{3.2} = 8x + 6$

$e^{3.2} - 6 = 8x$

$\frac{e^{3.2} - 6}{8} = x$

$x = 2.316566275$

7. $3\ln(x+4)+7=13$

$3\ln(x+4)=6$

$\ln(x+4)=2$

9. $\log(x+2)+\log x = 4$

$\log(x^2+2x)=4$

11. $2\log_2 x + \log_2(x-1)=2$

$\log_2(x^2)+\log_2(x-1)=2$

$\log_2(x^2(x-1))=2$

$\log_2(x^3-x^2)=2$

13. $4e^{x-2}+5=69$

$4e^{x-2}=64$

$e^{x-2}=16$

15. $250e^{0.05x+1}+175=512.5$

$250e^{0.05x+1}=337.5$

$e^{0.05x+1}=1.35$

17. $3^x(3^{2x-1})=81$

$3^{3x-1}=81$

19. Equating bases:

$2^x=128$

$2^x=2^7$

$x=7$

Applying a base-e logarithm:

$2^x=128$

$\ln 2^x=\ln 128$

$x\ln 2=\ln 128$

$x=\frac{\ln 128}{\ln 2}=7$

21. Equating bases:

$5^{3x}=3125$

$5^{3x}=5^5$

$3x=5$

$x=\frac{5}{3}$

Applying a base-e logarithm:

$5^{3x}=3125$

$\ln 5^{3x}=\ln 3125$

$3x\ln 5=\ln 3125$

$x=\frac{\ln 3125}{3\ln 5}=\frac{5}{3}$

23. Equating bases:

$5^{x+1}=625$

$5^{x+1}=5^4$

$x+1=4$

$x=3$

Applying a base-e logarithm:

$5^{x+1}=625$

$\ln 5^{x+1}=\ln 625$

$(x+1)\ln 5=\ln 625$

$x+1=\frac{\ln 625}{\ln 5}$

$x=\frac{\ln 625}{\ln 5}-1=3$

25. Equating bases:

$\left(\frac{1}{2}\right)^{n-1}=\frac{1}{256}$

$\left(\frac{1}{2}\right)^{n-1}=\left(\frac{1}{2}\right)^8$

$n-1=8$

$n=9$

Applying a base-e logarithm:

$\left(\frac{1}{2}\right)^{n-1}=\frac{1}{256}$

$\ln\left(\frac{1}{2}\right)^{n-1}=\ln\left(\frac{1}{256}\right)$

$(n-1)\ln\left(\frac{1}{2}\right)=\ln\left(\frac{1}{256}\right)$

$n-1=\frac{\ln\left(\frac{1}{256}\right)}{\ln\left(\frac{1}{2}\right)}$

$n=\frac{\ln\left(\frac{1}{256}\right)}{\ln\left(\frac{1}{2}\right)}+1=9$

27. Equating bases:

$$\frac{1}{625}=\left(\frac{1}{5}\right)^{n-1}$$
$$\left(\frac{1}{5}\right)^4=\left(\frac{1}{5}\right)^{n-1}$$
$$4=n-1$$
$$n=5$$

Applying a base-e logarithm:

$$\frac{1}{625}=\left(\frac{1}{5}\right)^{n-1}$$
$$\ln\left(\frac{1}{625}\right)=\ln\left(\frac{1}{5}\right)^{n-1}$$
$$\ln\left(\frac{1}{625}\right)=(n-1)\ln\left(\frac{1}{5}\right)$$
$$\frac{\ln\left(\frac{1}{625}\right)}{\ln\left(\frac{1}{5}\right)}=n-1$$
$$n=\frac{\ln\left(\frac{1}{625}\right)}{\ln\left(\frac{1}{5}\right)}+1=5$$

29. Equating bases:

$$\frac{128}{2187}=\left(\frac{2}{3}\right)^{n-1}$$
$$\left(\frac{2}{3}\right)^7=\left(\frac{2}{3}\right)^{n-1}$$
$$7=n-1$$
$$n=8$$

Applying a base-e logarithm:

$$\frac{128}{2187}=\left(\frac{2}{3}\right)^{n-1}$$
$$\ln\left(\frac{128}{2187}\right)=\ln\left(\frac{2}{3}\right)^{n-1}$$
$$\ln\left(\frac{128}{2187}\right)=(n-1)\ln\left(\frac{2}{3}\right)$$
$$\frac{\ln\left(\frac{128}{2187}\right)}{\ln\left(\frac{2}{3}\right)}=n-1$$
$$n=\frac{\ln\left(\frac{128}{2187}\right)}{\ln\left(\frac{2}{3}\right)}+1=8$$

31. Equating bases:

$$\frac{16}{625}=\left(\frac{2}{5}\right)^{n-1}$$
$$\left(\frac{2}{5}\right)^4=\left(\frac{2}{5}\right)^{n-1}$$
$$4=n-1$$
$$n=5$$

Applying a base-e logarithm:

$$\frac{16}{625}=\left(\frac{2}{5}\right)^{n-1}$$
$$\ln\left(\frac{16}{625}\right)=\ln\left(\frac{2}{5}\right)^{n-1}$$
$$\ln\left(\frac{16}{625}\right)=(n-1)\ln\left(\frac{2}{5}\right)$$
$$\frac{\ln\left(\frac{16}{625}\right)}{\ln\left(\frac{2}{5}\right)}=n-1$$
$$n=\frac{\ln\left(\frac{16}{625}\right)}{\ln\left(\frac{2}{5}\right)}+1=5$$

33. Equating bases:

$$\frac{5}{243}=5\left(\frac{1}{3}\right)^{n-1}$$
$$\frac{1}{243}=\left(\frac{1}{3}\right)^{n-1}$$
$$\left(\frac{1}{3}\right)^5=\left(\frac{1}{3}\right)^{n-1}$$
$$5=n-1$$
$$n=6$$

Applying a base-e logarithm:

$$\frac{5}{243}=5\left(\frac{1}{3}\right)^{n-1}$$
$$\frac{1}{243}=\left(\frac{1}{3}\right)^{n-1}$$

$$\ln\left(\frac{1}{243}\right) = \ln\left(\frac{1}{3}\right)^{n-1}$$

$$\ln\left(\frac{1}{243}\right) = (n-1)\ln\left(\frac{1}{3}\right)$$

$$\frac{\ln\left(\frac{1}{243}\right)}{\ln\left(\frac{1}{3}\right)} = n-1$$

$$n = \frac{\ln\left(\frac{1}{243}\right)}{\ln\left(\frac{1}{3}\right)} + 1 = 6$$

35. Equating bases:

$$\frac{56}{125} = 7\left(\frac{2}{5}\right)^{n-1}$$

$$\frac{8}{125} = \left(\frac{2}{5}\right)^{n-1}$$

$$\left(\frac{2}{5}\right)^{3} = \left(\frac{2}{5}\right)^{n-1}$$

$$3 = n-1$$

$$n = 4$$

Applying a base-e logarithm:

$$\frac{56}{125} = 7\left(\frac{2}{5}\right)^{n-1}$$

$$\frac{8}{125} = \left(\frac{2}{5}\right)^{n-1}$$

$$\ln\left(\frac{8}{125}\right) = \ln\left(\frac{2}{5}\right)^{n-1}$$

$$\ln\left(\frac{8}{125}\right) = (n-1)\ln\left(\frac{2}{5}\right)$$

$$\frac{\ln\left(\frac{8}{125}\right)}{\ln\left(\frac{2}{5}\right)} = n-1$$

$$n = \frac{\ln\left(\frac{8}{125}\right)}{\ln\left(\frac{2}{5}\right)} + 1 = 4$$

37. $10^x = 97$

$$\log 10^x = \log 97$$

$$x \log 10 = \log 97$$

$$x = \log 97$$

$$x \approx 1.9868$$

39. $879 = 10^{2x}$

$$\log 879 = \log 10^{2x}$$

$$\log 879 = 2x \log 10$$

$$x = \frac{\log 879}{2}$$

$$x \approx 1.4720$$

41. $879 = 10^{x+3}$

$$\log 879 = \log 10^{x+3}$$

$$\log 879 = (x+3)\log 10$$

$$\log 897 - 3 = x$$

$$x \approx -0.0560$$

43. $e^x = 389$

$$\ln e^x = \ln 389$$

$$x \ln e = \ln 389$$

$$x = \ln 389$$

$$x \approx 5.9636$$

45. $e^{2x} = 1389$

$$\ln e^{2x} = \ln 1389$$

$$2x \ln e = \ln 1389$$

$$x = \frac{\ln 1389}{2}$$

$$x \approx 3.6128$$

47. $e^{x+1} = 257$

$$\ln e^{x+1} = \ln 257$$

$$(x+1)\ln e = \ln 257$$

$$x = \ln 257 - 1$$

$$x \approx 4.5491$$

49. $2e^{0.25x} = 5$

$$\ln e^{0.25x} = \ln\left(\frac{5}{2}\right)$$

$$0.25x \ln e = \ln\left(\frac{5}{2}\right)$$

$$x = 4\ln\left(\frac{5}{2}\right)$$

$$x \approx 3.6652$$

51. $7^{x+2} = 231$

$$\ln 7^{x+2} = \ln 231$$
$$(x+2)\ln 7 = \ln 231$$
$$x+2 = \frac{\ln 231}{\ln 7}$$
$$x = \frac{\ln 231}{\ln 7} - 2$$
$$x \approx 0.7968$$

53. $5^{3x-2} = 128{,}965$
$$\ln 5^{3x-2} = \ln 128965$$
$$(3x-2)\ln 5 = \ln 128965$$
$$3x-2 = \frac{\ln 128695}{\ln 5}$$
$$3x = \frac{\ln 128695}{\ln 5} + 2$$
$$x = \frac{\ln 128965}{3\ln 5} + \frac{2}{3}$$
$$x \approx 3.1038$$

55. $2^{x+1} = 3^x$
$$\ln 2^{x+1} = \ln 3^x$$
$$(x+1)\ln 2 = x\ln 3$$
$$x\ln 2 + \ln 2 = x\ln 3$$
$$x\ln 2 - x\ln 3 = -\ln 2$$
$$x(\ln 2 - \ln 3) = -\ln 2$$
$$x = \frac{-\ln 2}{\ln 2 - \ln 3}$$
$$x = \frac{\ln 2}{\ln 3 - \ln 2}$$
$$x \approx 1.7095$$

57. $5^{2x+1} = 9^{x+1}$
$$\ln 5^{2x+1} = \ln 9^{x+1}$$
$$(2x+1)\ln 5 = (x+1)\ln 9$$
$$2x\ln 5 + \ln 5 = x\ln 9 + \ln 9$$
$$2x\ln 5 - x\ln 9 = \ln 9 - \ln 5$$
$$x(2\ln 5 - \ln 9) = \ln 9 - \ln 5$$
$$x = \frac{\ln 9 - \ln 5}{2\ln 5 - \ln 9}$$
$$x \approx 0.5753$$

59. $\frac{87}{1+3e^{-0.06t}} = 50$
$$87 = 50\left(1+3e^{-0.06t}\right)$$
$$87 = 50 + 150e^{-0.06t}$$
$$37 = 150e^{-0.06t}$$
$$\frac{37}{150} = e^{-0.06t}$$
$$\ln\left(\frac{37}{150}\right) = \ln e^{-0.06t}$$
$$\ln\left(\frac{37}{150}\right) = -0.06t\ln e$$
$$\ln\left(\frac{37}{150}\right) = -\frac{3}{50}t$$
$$t = \frac{-50}{3}\ln\left(\frac{37}{150}\right)$$
$$t \approx 23.3286$$

61. $160 = \frac{200}{1+59e^{-0.29t}}$
$$160\left(1+59e^{-0.29t}\right) = 200$$
$$160 + 9440e^{-0.29t} = 200$$
$$9440e^{-0.29t} = 40$$
$$e^{-0.29t} = \frac{1}{236}$$
$$\ln e^{-0.29t} = \ln\left(\frac{1}{236}\right)$$
$$-0.29t\ln e = \ln\left(\frac{1}{236}\right)$$
$$-\frac{29}{100}t = \ln\left(\frac{1}{236}\right)$$
$$t = \frac{-100}{29}\ln\left(\frac{1}{236}\right)$$
$$t = \frac{-100}{29}\ln\left(236^{-1}\right)$$
$$t = \frac{100}{29}\ln(236)$$
$$t \approx 18.8408$$

63. $\log(5x+2) = \log 2$
$$5x+2 = 2$$
$$5x = 0$$
$$x = 0$$

65. $\log_4(x+2) - \log_4 3 = \log_4(x-1)$
$$\log_4\left(\frac{x+2}{3}\right) = \log_4(x-1)$$
$$\frac{x+2}{3} = x-1$$

$x+2=3x-3$
$-2x=-5$
$x=\frac{5}{2}$

67. $\ln(8x-4)=\ln 2+\ln x$
$\ln(8x-4)=\ln(2x)$
$8x-4=2x$
$6x=4$
$x=\frac{2}{3}$

69. $\log(3x+1)=2$
Write in exponential form:
$10^2=3x+1$
$99=3x$
$x=33$

71. $\log_5(x+7)=3$
Write in exponential form:
$5^3=x+7$
$125=x+7$
$x=118$

73. $\ln(x+7)=2$
Write in exponential form:
$e^2=x+7$
$x=e^2-7$

75. $-2=\log(2x-1)$
Write in exponential form:
$10^{-2}=2x-1$
$\frac{1}{100}=2x-1$
$\frac{1}{100}+1=2x$
$\frac{101}{100}=2x$
$x=\frac{101}{200}$

77. $\log(2x)-5=-3$
Write in exponential form:
$\log(2x)=2$
$10^2=2x$
$100=2x$
$x=50$

79. $-2\cdot\ln(x+1)=-6$
Write in exponential form:
$\ln(x+1)=3$
$e^3=x+1$
$x=e^3-1$

81. $\log(2x-1)+\log 5=1$
Write in exponential form:
$\log(10x-5)=1$
$10^1=10x-5$
$15=10x$
$x=\frac{3}{2}$

83. $\log_2 9+\log_2(x+3)=3$
Write in exponential form:
$\log_2(9x+27)=3$
$2^3=9x+27$
$8=9x+27$
$-19=9x$
$x=\frac{-19}{9}$

85. $\ln(x+7)+\ln 9=2$
Write in exponential form:
$\ln(9x+63)=2$
$e^2=9x+63$
$e^2-63=9x$
$x=\frac{e^2-63}{9}$

87. $\log(x+8)+\log x=\log(x+18)$
Write in exponential form:
$\log(x^2+8x)=\log(x+18)$
$x^2+8x=x+18$
$x^2+7x-18=0$
$(x+9)(x-2)=0$
$x+9=0$ or $x-2=0$
$x=-9$ or $x=2$
$x=2$, -9 is extraneous

89. $\ln(2x+1)=3+\ln 6$
$e^{3+\ln 6}=2x+1$
$e^{3+\ln 6}-1=2x$
$\frac{e^{3+\ln 6}-1}{2}=x$

$$\frac{e^3 e^{\ln 6}-1}{2}=x$$
$$\frac{e^3(6)-1}{2}=x$$
$$\frac{6e^3-1}{2}=x$$
$$x=3e^3-\frac{1}{2}$$
$$x\approx 59.75661077$$

91. $\log(-x-1)=\log(5x)-\log x$
$\log(-x-1)=\log 5$
$-x-1=5$
$-x=6$
$x=-6$
x = - 6 is extraneous, No Solution

93. $\ln(2t+7)=\ln(3)-\ln(t+1)$
$$\ln(2t+7)=\ln\left(\frac{3}{t+1}\right)$$
$$2t+7=\frac{3}{t+1}$$
$$(t+1)(2t+7)=\left(\frac{3}{t+1}\right)(t+1)$$
$$2t^2+9t+7=3$$
$$2t^2+9t+4=0$$
$$(2t+1)(t+4)=0$$
$2t+1=0$ or $t+4=0$
$2t=-1$ or $t=-4$
$t=-\frac{1}{2}$, -4 is extraneous

95. $A=A_0\left(\frac{1}{2}\right)^{\frac{t}{h}}$
$$A=500\left(\frac{1}{2}\right)^{\frac{60}{15}}$$
31.25 grams

97. $T=T_1+(T_0-T_1)e^{-kh}$
$$45=40+(75-40)e^{-0.61h}$$
$$5=35e^{-0.61h}$$
$$\frac{1}{7}=e^{-0.61h}$$
$$\ln\left(\frac{1}{7}\right)=\ln e^{-0.61h}$$
$$\ln\left(\frac{1}{7}\right)=-0.61h$$
$$\frac{\ln\left(\frac{1}{7}\right)}{-0.61}=h$$
$h=3.19$ hrs
3 hrs 11 minutes

99. $H=(30T+8000)\ln\left(\frac{P_0}{P}\right)$
$$6194=(30(-10)+8000)\ln\left(\frac{76}{P}\right)$$
$$6194=(7700)\ln\left(\frac{76}{P}\right)$$
$$0.8044155844=\ln\left(\frac{76}{P}\right)$$
$$e^{0.8044155844}=\frac{76}{P}$$
P = 34 cmHg

101. $A=P\left(1+\frac{r}{n}\right)^{nt}$
$$6000=2500\left(1+\frac{0.08}{12}\right)^{12t}$$
$$2.4=\left(1+\frac{0.08}{12}\right)^{12t}$$
$$\ln 2.4=\ln\left(1+\frac{0.08}{12}\right)^{12t}$$
$$\ln 2.4=12t\ln\left(1+\frac{0.08}{12}\right)$$
$$\frac{\ln 2.4}{12\ln\left(1+\frac{0.08}{12}\right)}=t$$
$t\approx 10.979$
About 11 years

103. $N(A)=1500+315\cdot\ln A$
$$5000=1500+315\cdot\ln A$$
$$3500=315\cdot\ln A$$
$$\frac{100}{9}=\ln A$$
$$e^{\frac{100}{9}}=A$$
$A\approx$ \$66910

105. $V_s = V_e \cdot \ln\left(\dfrac{M_s}{M_s - M_f}\right)$

$$6 = 8 \cdot \ln\left(\frac{100}{100 - M_f}\right)$$

$$0.75 = \ln\left(\frac{100}{100 - M_f}\right)$$

$$e^{0.75} = \frac{100}{100 - M_f}$$

$$100 - M_f = \frac{100}{e^{0.75}}$$

$$-M_f = \frac{100}{e^{0.75}} - 100$$

$$M_f = \frac{-100}{e^{0.75}} + 100$$

$$M_f \approx 52.76 \text{ tons}$$

107. Answers will vary.

109. a. d
b. e
c. b
d. f
e. a
f. c

111. $3e^{2x} - 4e^x - 7 = -3$

Let $u = e^x$

$$3u^2 - 4u - 4 = 0$$

$$(3u + 2)(u - 2) = 0$$

$$3u + 2 = 0 \text{ or } u - 2 = 0$$

$$3u = -2 \quad \text{or } u = 2$$

$$u = -\frac{2}{3} \quad \text{or } u = 2$$

$$e^x = -\frac{2}{3} \quad \text{or } e^x = 2$$

$$\ln e^x = \ln\left(-\frac{2}{3}\right) \quad \text{or } \ln e^x = \ln 2$$

$$x \neq \ln\left(-\frac{2}{3}\right) \quad \text{or } x = \ln 2$$

$$x \approx 0.69319718$$

113. $\log_3 3^{2x} = \log_3 5$

$$2x \log_3 3 = \log_3 5$$

$$2x = \log_3 5$$

$$x = \frac{\log_3 5}{2}$$

115. $f(x) = 3^{x-2}; \; g(x) = \log_3 x + 2\text{f}$

$$(f \circ g)(x) = 3^{(\log_3 x + 2) - 2} = 3^{\log_3 x} = x;$$

$$(g \circ f)(x) = \log_3\left(3^{x-2}\right) + 2 =$$

$$(x - 2)\log_3 3 + 2 = x - 2 + 2 = x$$

117. $y = e^{x \ln 2} = e^{\ln 2^x} = 2^x;$

$$y = 2^x$$

$$\ln y = x \ln 2$$

$$e^{\ln y} = e^{x \ln 2}$$

$$y = e^{x \ln 2}$$

119. b

121. a. $y = \sqrt{2x + 3}$

$$2x + 3 \geq 0$$

$$2x \geq -3$$

$$x \geq \frac{-3}{2}$$

$$x \in \left[-\frac{3}{2}, \infty\right), \quad y \in [0, \infty)$$

b. $y = |x + 2| - 3$

$$x \in (-\infty, \infty), \quad y \in [-3, \infty)$$

123. $f(x) = x^3 + x + 10$

Possible rational roots: $\dfrac{\{\pm 1, \pm 10, \pm 2, \pm 5\}}{\{\pm 1\}}$

$$f(x) = (x + 2)\left(x^2 - 2x + 5\right);$$

$$a = 1, b = -2, c = 5$$

$$x = \frac{-(-2) \pm \sqrt{(-2)^2 - 4(1)(5)}}{2(1)}$$

$$x = \frac{2 \pm \sqrt{-16}}{2} = \frac{2 \pm 4i}{2} = 1 \pm 2i$$

$$x = -2, \; 1 \pm 2i$$

5.5 Technology Highlight

1. $A = P\left(1 + \dfrac{r}{n}\right)^{nt}$

Doubling time, find x when $y = 2000$

$$Y_1 = 1000\left(1+\frac{0.08}{4}\right)^{4x} \approx 8.75 \text{ yrs;}$$

$$Y_1 = 1000\left(1+\frac{0.08}{12}\right)^{12x} \approx 8.69 \text{ yrs;}$$

$$Y_1 = 1000\left(1+\frac{0.08}{365}\right)^{365x} \approx 8.665 \text{ yrs;}$$

$$Y_1 = 1000\left(1+\frac{0.08}{365\cdot 24}\right)^{365\cdot 24x} \approx 8.664 \text{ yrs}$$

8.75 yrs compounded quarterly; 8.69 yrs compounded monthly; 8.665 yrs compounded daily; 8.664 yrs compounded hourly

2. Doubling time, find x when $y = 2000$
6%; 11.64; 11.58; 11.55;

$$Y_1 = 1000\left(1+\frac{0.06}{4}\right)^{4x} \approx 11.64$$

$$Y_1 = 1000\left(1+\frac{0.06}{12}\right)^{12x} \approx 11.58$$

$$Y_1 = 1000\left(1+\frac{0.06}{365}\right)^{365x} \approx 11.55$$

8%; 8.75; 8.69; 8.67;

$$Y_1 = 1000\left(1+\frac{0.08}{4}\right)^{4x} \approx 8.75$$

$$Y_1 = 1000\left(1+\frac{0.08}{12}\right)^{12x} \approx 8.69$$

$$Y_1 = 1000\left(1+\frac{0.08}{365}\right)^{365x} \approx 8.67$$

10%; 7.02; 6.96; 6.93;

$$Y_1 = 1000\left(1+\frac{0.10}{4}\right)^{4x} \approx 7.02$$

$$Y_1 = 1000\left(1+\frac{0.10}{12}\right)^{12x} \approx 6.96$$

$$Y_1 = 1000\left(1+\frac{0.10}{365}\right)^{365x} \approx 6.93$$

12%; 5.86; 5.81; 5.78

$$Y_1 = 1000\left(1+\frac{0.12}{4}\right)^{4x} \approx 5.86$$

$$Y_1 = 1000\left(1+\frac{0.12}{12}\right)^{12x} \approx 5.81$$

$$Y_1 = 1000\left(1+\frac{0.12}{365}\right)^{365x} \approx 5.78$$

Greater interest rate has more impact.

3. No. Examples will vary.

5.5 Exercises

1. Compound

3. Q_0e^{-rt}

5. Answers will vary.

7. $I = prt$;

$9 \text{ months} = \frac{3}{4} \text{ year;}$

$229.50 = p(0.0625)(0.75)$

$\frac{229.50}{(0.0625)(0.75)} = p$

$\$4896 = p$

9. $I = prt$

$297.50 - 260 = 260r\left(\frac{3}{52}\right)$

$37.5 = 15r$

$\frac{37.5}{15} = r$

$2.50 = r$

$r = 250\%$

11. $A = p(1+rt)$

$2500 = p\left(1+0.0625\left(\frac{31}{12}\right)\right)$

$2500 = p\left(\frac{223}{192}\right)$

$p \approx \$2152.47$

13. $A = p(1+rt)$

$149925 = 120000(1+0.0475t)$

$\frac{1999}{1600} = 1+0.0475t$

$\frac{399}{1600} = 0.0475t$

$5.25 \text{ years} = t$

15. $A = p(1+r)^t$

$48428 = 38000(1+0.0625)^t$

$$\frac{12107}{9500}=(1+0.0625)^t$$
$$\ln\frac{12107}{9500}=\ln(1+0.0625)^t$$
$$\ln\frac{12107}{9500}=t\ln(1+0.0625)$$
$$\frac{\ln\frac{12107}{9500}}{\ln(1+0.0625)}=t$$
$t \approx 4$ years

17. $A=p(1+r)^t$
$$4575=1525(1+0.071)^t$$
$$3=(1+0.071)^t$$
$$\ln 3=\ln(1+0.071)^t$$
$$\ln 3=t\ln(1+0.071)$$
$$\frac{\ln 3}{\ln(1+0.071)}=t$$
$t \approx 16$ years

19. $P=\dfrac{A}{(1+r)^t}$
$$P=\frac{10000}{(1+0.0575)^5}$$
$P \approx \$7561.33$

21. $A=p\left(1+\dfrac{r}{n}\right)^{nt}$
$$129500=90000\left(1+\frac{0.07125}{52}\right)^{52t}$$
$$\frac{259}{180}=\left(1+\frac{0.07125}{52}\right)^{52t}$$
$$\ln\left(\frac{259}{180}\right)=\ln(1.001370192)^{52t}$$
$$\ln\left(\frac{259}{180}\right)=52t\ln(1.001370192)$$
$$\frac{\ln\left(\frac{259}{180}\right)}{52\ln(1.001370192)}=t$$
$t \approx 5$ years

23. $A=p\left(1+\dfrac{r}{n}\right)^{nt}$
$$10000=5000\left(1+\frac{0.0925}{365}\right)^{365t}$$
$$2=(1.000253425)^{365t}$$
$$\ln 2=\ln(1.000253425)^{365t}$$
$$\ln 2=365t\ln(1.000253425)$$
$$\frac{\ln 2}{365\ln(1.000253425)}=t$$
$t \approx 7.5$ years

25. $A=pe^{rt}$
$$2500=1750e^{0.045t}$$
$$\frac{10}{7}=e^{0.045t}$$
$$\ln\left(\frac{10}{7}\right)=\ln e^{0.045t}$$
$$\ln\left(\frac{10}{7}\right)=0.045t\ln e$$
$$\ln\left(\frac{10}{7}\right)=0.045t$$
$$\frac{\ln\left(\frac{10}{7}\right)}{0.045}=t$$
$t \approx 7.9$ years

27. $A=pe^{rt}$
$$10000=5000e^{0.0925t}$$
$$2=e^{0.0925t}$$
$$\ln 2=\ln e^{0.0925t}$$
$$\ln 2=0.0925t\ln e$$
$$\frac{\ln 2}{0.0925}=t$$
$t \approx 7.5$ years

29. $A=p+prt$

a. $A-p=prt$
$$\frac{A-p}{pr}=t$$

b. $A=p(1+rt)$
$$\frac{A}{1+rt}=p$$

31. $\frac{A}{p}=\left(1+\frac{r}{n}\right)^{nt}$

a. $\frac{A}{p}=\left(1+\frac{r}{n}\right)^{nt}$

$\sqrt[nt]{\frac{A}{p}}=1+\frac{r}{n}$

$\sqrt[nt]{\frac{A}{p}}-1=\frac{r}{n}$

$n\left(\sqrt[nt]{\frac{A}{p}}-1\right)=r$

b. $\ln\left(\frac{A}{p}\right)=\ln\left(1+\frac{r}{n}\right)^{nt}$

$\ln\left(\frac{A}{p}\right)=nt\ln\left(1+\frac{r}{n}\right)$

$\frac{\ln\left(\frac{A}{p}\right)}{n\ln\left(1+\frac{r}{n}\right)}=t$

33. $Q(t)=Q_0e^{rt}$

a. $\frac{Q(t)}{e^{rt}}=Q_0$

b. $\frac{Q(t)}{Q_0}=e^{rt}$

$\ln\left(\frac{Q(t)}{Q_0}\right)=\ln e^{rt}$

$\ln\left(\frac{Q(t)}{Q_0}\right)=rt\ln e$

$\frac{\ln\left(\frac{Q(t)}{Q_0}\right)}{r}=t$

35. $P=\frac{AR}{1-(1+R)^{-nt}}$

$P=\frac{125000\left(\frac{0.055}{12}\right)}{1-\left(1+\left(\frac{0.055}{12}\right)\right)^{-12(30)}}$

$P\approx\$709.74$

37. $I=prt$

$40=200r\left(\frac{13}{52}\right)$

$40=50r$

$0.80=r$

$r=80\%$

39. $A=p\left(1+\frac{r}{n}\right)^{nt}$

$A=10\left(1+\frac{0.10}{10}\right)^{10(10)}\approx\$27.04$, No

41. $A=p\left(1+\frac{r}{n}\right)^{nt}$

(a) $A=175000\left(1+\frac{0.0875}{2}\right)^{2(4)}$

$\approx\$246496.05$, No

(b) $r\approx 9.12\%$

43. $A=pe^{rt}$

(a) $A=12500e^{0.086(5)}=19215.72$ euros, No

(b) $20000=12500e^{r(5)}$

$\frac{8}{5}=e^{5r}$

$\ln\left(\frac{8}{5}\right)=\ln e^{5r}$

$\ln\left(\frac{8}{5}\right)=5r\ln e$

$\frac{\ln\left(\frac{8}{5}\right)}{5}=r$

$r\approx 9.4\%$

45. $A=pe^{rt}$

(a) $A=12000e^{0.055(7)}\approx 17635.37$ euros, No

(b) $20000=Pe^{0.055(7)}$

$\frac{20000}{e^{0.055(7)}}=P$

$P\approx 13{,}609$ euros

47. $Q(t)=Q_0e^{rt}$

(a) $2000=1000e^{r(12)}$

$2=e^{12r}$

$\ln 2=\ln e^{12r}$

$\ln 2 = 12r \ln e$

$\frac{\ln 2}{12} = r$

$r \approx 5.78\%$

(b) $200000 = 1000e^{(0.0578)t}$

$200 = e^{(0.0578)t}$

$\ln 200 = \ln e^{(0.0578)t}$

$\ln 200 = 0.0578t \ln e$

$\frac{\ln 200}{0.0578} = t$

$t \approx 91.67$ hours

49. $r = \frac{\ln 2}{t}$

$r = \frac{\ln 2}{8}$

$r \approx 0.087$ or $r \approx 8.7\%$;

$Q(t) = Q_0 e^{-rt}$

$0.5 = Q_0 e^{-0.087(3)}$

$\frac{0.5}{e^{-0.087(3)}} = Q_0$

$Q_0 \approx 0.65$ grams

51. $r = \frac{\ln 2}{t}$

$r = \frac{\ln 2}{432}$

$r \approx 0.0016$ or $r \approx 0.16\%$;

$Q(t) = Q_0 e^{-rt}$

$2.7 = 10e^{-0.0016t}$

$0.27 = e^{-0.0016t}$

$\ln 0.27 = \ln e^{-0.0016t}$

$\ln 0.27 = -0.0016t \ln e$

$\frac{\ln 0.27}{-0.0016} = t$

≈ 818 years

53. $A = \frac{p\left[(1+R)^{nt} - 1\right]}{R}$

$$10000 = \frac{90\left[\left(1+\frac{0.0775}{12}\right)^{12t} - 1\right]}{\frac{0.0775}{12}}$$

$$\frac{775}{12} = 90\left[\left(1+\frac{0.0775}{12}\right)^{12t} - 1\right]$$

$$\frac{155}{216} = \left(1+\frac{0.0775}{12}\right)^{12t} - 1$$

$$\frac{371}{216} = \left(1+\frac{0.0775}{12}\right)^{12t}$$

$$\ln\left(\frac{371}{216}\right) = \ln\left(1+\frac{0.0775}{12}\right)^{12t}$$

$$\ln\left(\frac{371}{216}\right) = 12t \ln\left(1+\frac{0.0775}{12}\right)$$

$$\frac{\ln\left(\frac{371}{216}\right)}{12 \ln\left(1+\frac{0.0775}{12}\right)} = t$$

≈ 7 years

55. $A = \frac{p\left[(1+R)^{nt} - 1\right]}{R}$

$$30000 = \frac{50\left[\left(1+\frac{0.062}{12}\right)^{12t} - 1\right]}{\frac{0.062}{12}}$$

$$155 = 50\left[\left(1+\frac{0.062}{12}\right)^{12t} - 1\right]$$

$$3.1 = \left(1+\frac{0.062}{12}\right)^{12t} - 1$$

$$4.1 = \left(1+\frac{0.062}{12}\right)^{12t}$$

$$\ln 4.1 = \ln\left(1+\frac{0.062}{12}\right)^{12t}$$

$$\ln(4.1) = 12t \ln\left(1+\frac{0.062}{12}\right)$$

$$\frac{\ln(4.1)}{12 \ln\left(1+\frac{0.062}{12}\right)} = t$$

≈ 22.8 years

57. $A = \frac{p\left[(1+R)^{nt} - 1\right]}{R}$

(a) $$A = \frac{250\left[\left(1+\frac{0.085}{12}\right)^{12(5)} - 1\right]}{\frac{0.085}{12}}$$

$\approx \$18610.61$, No

(b) $$22500 = \frac{p\left[\left(1+\frac{0.085}{12}\right)^{12(5)} - 1\right]}{\frac{0.085}{12}}$$

$$159.375 = p\left[\left(1+\frac{0.085}{12}\right)^{12(5)} - 1\right]$$

$$\frac{159.375}{\left[\left(1+\frac{0.085}{12}\right)^{12(5)} - 1\right]} = p$$

$p \approx \$302.25$

59. $A = pe^{rt}$

$A = 10000e^{0.062(115)} = \$12{,}488{,}769.67$

Answers will vary.

61. Answers will vary.

63. $$A = \frac{p\left[(1+R)^{nt} - 1\right]}{R}$$

$$A = \frac{50\left[\left(1+\frac{r}{12}\right)^{12(23)} - 1\right]}{\frac{r}{12}}$$

Using grapher, when A=35100,
$r \approx 7.199243\%$

65. $2000^2 + 1580^2 = x^2$

$x \approx 2548.8$ meters

67. Yes, it is a function.

$P(x) = x^4 - 4x^3 + 6x^2 - 4x - 15$

69. (a) $f(x) = x, f(x) = x^3, f(x) = \sqrt{x}, f(x) = \sqrt[3]{x}, f(x) = \frac{1}{x}$

(b) $f(x) = |x|, f(x) = x^2, f(x) = \frac{1}{x^2}$

(c) $f(x) = x, f(x) = x^3, f(x) = \sqrt{x}, f(x) = \sqrt[3]{x}$

(d) $f(x) = \frac{1}{x}, f(x) = \frac{1}{x^2}$

5.6 Technology Highlight

1. (a) 46.92 cm
   (b) 48.54 cm
   (c) 20.78 months
   Yes, very close.

5.6 Exercises

1. Data, context, situation

3. Beyond

5. 1. Clear the old data, 2. enter new data, 3. display the data, 4. calculate the regression equation, 5. display and use the regression graph and equation.

7. E

9. A

11. D

13. Linear

15. Exponential

17. Logistic

19. Exponential

21. As time increases, the amount of radioactivity decreases but it will never truly reach 0 or a negative value. Due to this and the shape, exponential with $b < 1$ and $k > 0$ is the best choice.

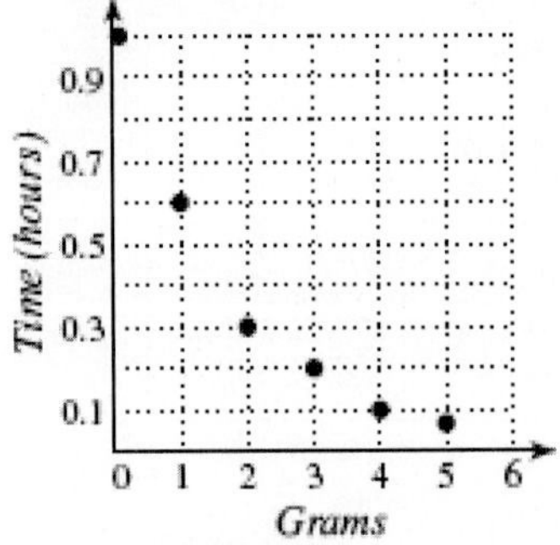

$y = 1.042(0.5626)^x$

23. $k = \ln(0.5626)$
$k \approx -0.5752$
About -57.5%

25. Sales will increase rapidly, then level off as the market is saturated with ads and advertising becomes less effective, possibly modeled by a logarithmic function.
$y \approx 120.4938 + 217.2705 \ln x$

27. $96.35 = (9.4)1.6^x$

$$\frac{96.35}{9.4} = 1.6^x$$
$$\ln\left(\frac{96.35}{9.4}\right) = \ln 1.6^x$$
$$\ln\left(\frac{96.35}{9.4}\right) = x \ln 1.6$$
$$\frac{\ln\left(\frac{96.35}{9.4}\right)}{\ln 1.6} = x$$
$$x \approx 4.95$$

29. $(-10.04)1.046^x = -396.58$

$$1.046^x = \frac{-396.58}{-10.04}$$
$$\ln 1.046^x = \ln\left(\frac{396.58}{10.04}\right)$$
$$x \ln 1.046 = \ln\left(\frac{396.58}{10.04}\right)$$
$$x = \frac{\ln\left(\frac{396.58}{10.04}\right)}{\ln 1.046}$$
$$x \approx 81.74$$

31. $4.8x^{2.5} = 468.75$

$$x^{2.5} = \frac{468.75}{4.8}$$
$$\left(x^{2.5}\right)^{\frac{2}{5}} = \left(\frac{468.75}{4.8}\right)^{\frac{2}{5}}$$
$$x = 6.25$$

33. $2.103x^{0.6} = 56.781$

$$x^{0.6} = \frac{56.781}{2.103}$$
$$\left(x^{0.6}\right)^{\frac{5}{3}} = \left(\frac{56.781}{2.103}\right)^{\frac{5}{3}}$$
$$x = 243$$

35. $498.53 + 2.3 \ln x = 2595.9$

$$2.3 \ln x = 2097.37$$
$$\ln x = 911.9$$
$$e^{\ln x} = e^{911.9}$$
$$x = e^{911.9}$$

37. $9 = 68.76 - 7.2 \ln x$

$$-59.76 = -7.2 \ln x$$
$$8.3 = \ln x$$

By definition, $\ln x = y$ iff $e^y = x$

$$e^{8.3} = x$$
$$x \approx 4023.87$$

39. $\dfrac{975}{1 + 82.3e^{-0.423x}} = 890$

$$\left(1 + 82.3e^{-0.423x}\right)\left[\frac{975}{1 + 82.3e^{-0.423x}} = 890\right]$$
$$975 = \left(1 + 82.3e^{-0.423x}\right)890$$
$$\frac{195}{178} = 1 + 82.3e^{-0.423x}$$
$$\frac{17}{178} = 82.3e^{-0.423x}$$
$$\frac{17}{14649.4} = e^{-0.423x}$$
$$\ln\left(\frac{17}{14649.4}\right) = \ln e^{-0.423x}$$
$$\ln\left(\frac{17}{14649.4}\right) = -0.423x \ln e$$
$$\frac{\ln\left(\frac{17}{14649.4}\right)}{-0.423} = x$$
$$x \approx 15.98$$

41. $5 = \dfrac{8}{1 + 9.3e^{-0.65x}}$

$$\left(1 + 9.3e^{-0.65x}\right)\left[5 = \frac{8}{1 + 9.3e^{-0.65x}}\right]$$
$$\left(1 + 9.3e^{-0.65x}\right)5 = 8$$
$$1 + 9.3e^{-0.65x} = \frac{8}{5}$$
$$9.3e^{-0.65x} = \frac{3}{5}$$

$$e^{-0.65x} = \frac{2}{31}$$

$$\ln e^{-0.65x} = \ln\left(\frac{2}{31}\right)$$

$$-0.65x \ln e = \ln\left(\frac{2}{31}\right)$$

$$x = \frac{\ln\left(\frac{2}{31}\right)}{-0.65}$$

$x \approx 4.22$

43. $N(t) = 0.325(1.057)^t$

$10 = 0.325(1.057)^t$

$$\frac{400}{13} = 1.057^t$$

$$\frac{400}{13} = 1.057^t$$

$$\ln\left(\frac{400}{13}\right) = \ln 1.057^t$$

$$\frac{\ln\left(\frac{400}{13}\right)}{\ln 1.057} = t$$

$t \approx 61.8$

1920+61.8=1981

45. $y = 53.24(1.04)^x$

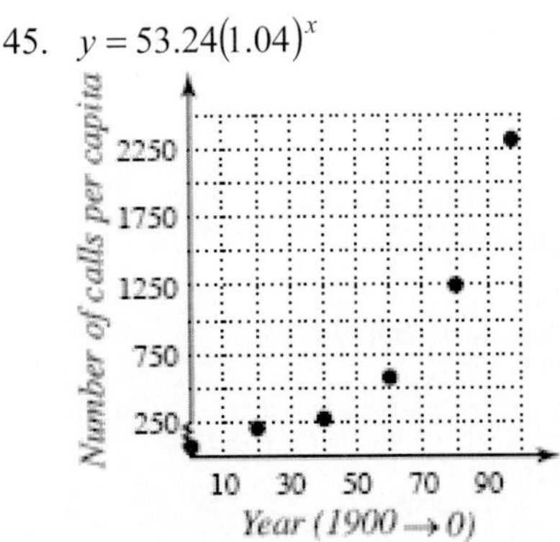

47. $y = 0.89(0.81)^x$

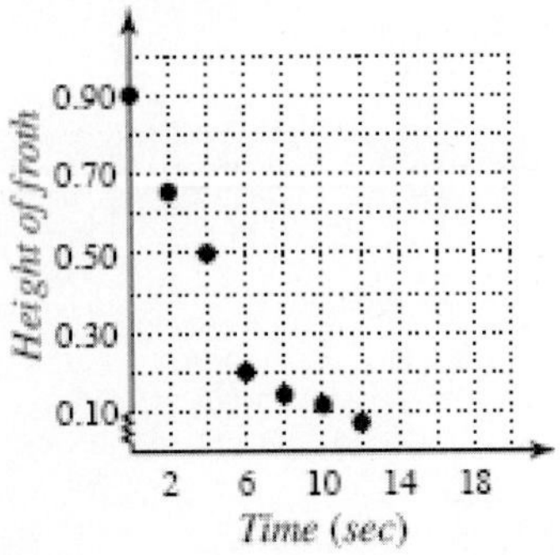

49. $y = -2635.6 + 1904.8 \ln x$

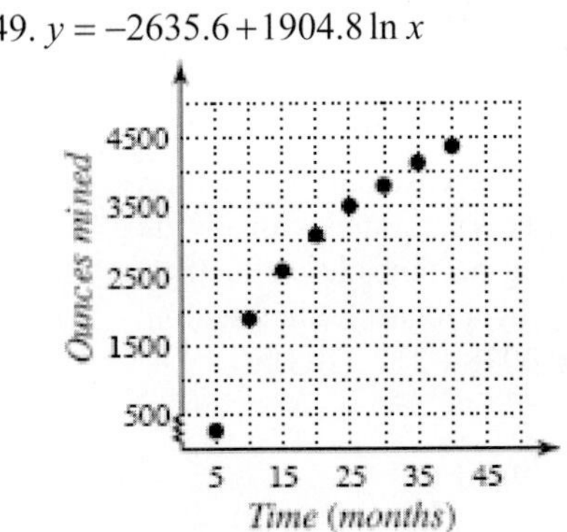

51. $y = 53.24(1.04)^x$

a. $y = 53.24(1.04)^{70} \approx 829$

$\approx$ 829 calls

b. $y = 53.24(1.04)^{105} \approx 3271$

$\approx$ 3271 calls

c. $1800 = 53.24(1.04)^x$

$$\frac{45000}{1331} = 1.04^x$$

$$\ln\left(\frac{45000}{1331}\right) = \ln 1.04^x$$

$$\ln\left(\frac{45000}{1331}\right) = x \ln 1.04$$

$$\frac{\ln\left(\frac{45000}{1331}\right)}{\ln 1.04} = x$$

$x \approx 89.8$

1900+89.2=1989.2, year 1989

53. $y = 0.89(0.81)^x$

a. $y = 0.89(0.81)^{6.5} \approx 0.23$ inches

b. $0.45 = 0.89(0.81)^x$

$$\frac{0.45}{0.89} = 0.81^x$$

$$\ln\left(\frac{0.45}{0.89}\right) = \ln 0.81^x$$

$$\ln\left(\frac{0.45}{0.89}\right) = x \ln 0.81$$

$$\frac{\ln\left(\frac{0.45}{0.89}\right)}{\ln 0.81} = x$$

$x \approx 3.2$ seconds

c. $0.02 = 0.89(0.81)^x$

$$\frac{0.02}{0.89} = 0.81^x$$
$$\ln\left(\frac{0.02}{0.89}\right) = \ln 0.81^x$$
$$\ln\left(\frac{0.02}{0.89}\right) = x \ln 0.81$$
$$\frac{\ln\left(\frac{0.02}{0.89}\right)}{\ln 0.81} = x$$
$x \approx 18$ seconds

55. $y = -2635.6 + 1904.8 \ln x$

a. $y = -2635.6 + 1904.8 \ln 18 \approx 2870$ oz

b. $4000 = -2635.6 + 1904.8 \ln x$
$$6635.6 = 1904.8 \ln x$$
$$\frac{66356}{19048} = \ln x$$
$$e^{\left(\frac{66356}{19048}\right)} = e^{\ln x}$$
$x \approx 32.6$ months

c. $y = -2635.6 + 1904.8 \ln 50 \approx 4816$ oz

57. $y = (39.86)1.16^x$

a. $y = (39.86)1.16^{18} = 576.4758263$
$\approx$ \$576,000

b. $y = (39.86)1.16^{35} = 7187.318941$
$\approx$ \$7,187,000

c. $1000 = (39.86)1.16^x$
$$25.08780733 = 1.16^x$$
$$\ln 25.08780733 = \ln 1.16^x$$
$$\ln 25.08780733 = x \ln 1.16$$
$x \approx 22$
1970+22=1992

59. $y = 19943.7 - 5231.4 \ln x$

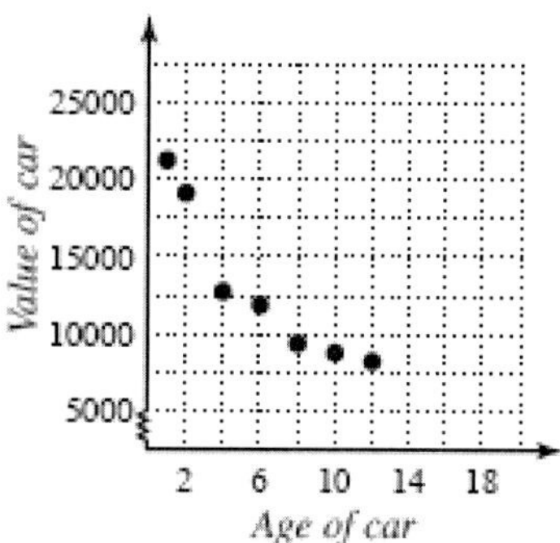

a. $y = 19943.7 - 5231.4 \ln 7.5 \approx \$9402.94$

b. $8150 = 19943.7 - 5231.4 \ln x$
$$-11793.7 = -5231.4 \ln x$$
$$\frac{117937}{52314} = \ln x$$
$$e^{\left(\frac{117937}{52314}\right)} = e^{\ln x}$$
$x \approx 9.5$ years

c. $3000 = 19943.7 - 5231.4 \ln x$
$$-16943.7 = -5231.4 \ln x$$
$$\frac{169437}{52314} = \ln x$$
$$e^{\left(\frac{169437}{52314}\right)} = e^{\ln x}$$
$x \approx 25.5$ years

61. $y = \dfrac{69.99}{1 + 4.00e^{-0.22x}}$

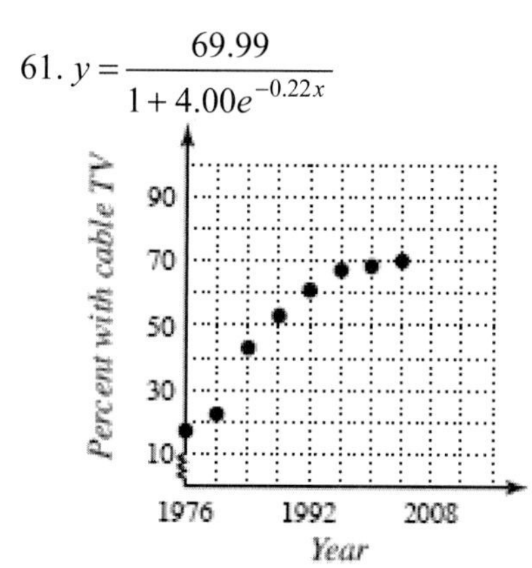

a. $y = \dfrac{69.99}{1 + 4.00e^{-0.22(14)}} \approx 59.1\%$

b. $50 = \dfrac{69.99}{1 + 4.00e^{-0.22x}}$
$$\left(1 + 4.00e^{-0.22x}\right)\left[50 = \frac{69.99}{1 + 4.00e^{-0.22x}}\right]$$
$$1 + 4.00e^{-0.22x} = \frac{69.99}{50}$$
$$4.00e^{-0.22x} = \frac{19.99}{50}$$

$$e^{-0.22x} = \frac{19.99}{200}$$

$$\ln e^{-0.22x} = \ln\left(\frac{19.99}{200}\right)$$

$$-0.22x \ln e = \ln\left(\frac{19.99}{200}\right)$$

$$x = \frac{\ln\left(\frac{19.99}{200}\right)}{-0.22}$$

$x \approx 10.5$;
1976+10=1986

c. $y = \dfrac{69.99}{1+4.00e^{-0.22(32)}} \approx 69.7\%$

63. a. Bush and Gore are tied; Gore wins.
b. Bush has his smallest lead; Bush would win.
c. Bush would always have a lead. Bush would win.
d. Exponential; outcome would be too close to call.

65. a. (2,2), (7,7)
Exponential: $Y_1 \approx 1.2117(1.2847)^x$;
Logarithmic:
$Y_2 \approx -0.7665 + 3.9912 \ln x$;
1. Symmetry about $y = x$;
2. for (a,b) on Y_1, (b,a) is on Y_2

b. $y = 1.2117(1.2847)^x$
$k = \ln 1.2117 \approx 0.2505524714$;
a. $y = 1.2117e^{0.2505524714x}$
b. $a + b\ln x = y$;
$a + b\ln y = x$
$b \ln y = x - a$
$\ln y = \dfrac{x-a}{b}$
$e^{\frac{x-a}{b}} = y$
$y = e^{\frac{x}{b}} e^{-\frac{a}{b}}$;
a = -0.7664737786,
b = 3.991178001 gives the result from a.

67. $\dfrac{\Delta f}{\Delta x} = \dfrac{e^{-1} - e^{-0.9}}{1-0.9} \approx -0.39$;
$\dfrac{\Delta g}{\Delta x} = \dfrac{-\ln 1 + \ln 0.9}{1-0.9} \approx -1.05$

69. If $x = 2, x^2 = 4 \Rightarrow (2,4)$;
If $x = 4, \sqrt{x-4} + 1 = 1 \Rightarrow (4,1)$;

$$m = \frac{4-1}{2-4} = -\frac{3}{2};$$

$$y - 4 = -\frac{3}{2}(x-2)$$

$$y - 4 = -\frac{3}{2}x + 3$$

$$y = -\frac{3}{2}x + 7$$

$$p(x) = \begin{cases} x^2 & -2 \le x < 2 \\ -\frac{3}{2}x + 7 & 2 \le x < 4 \\ \sqrt{x-4} + 1 & x \ge 4 \end{cases}$$

71. $F(x) = (x-2)^{\frac{2}{3}} + 3$

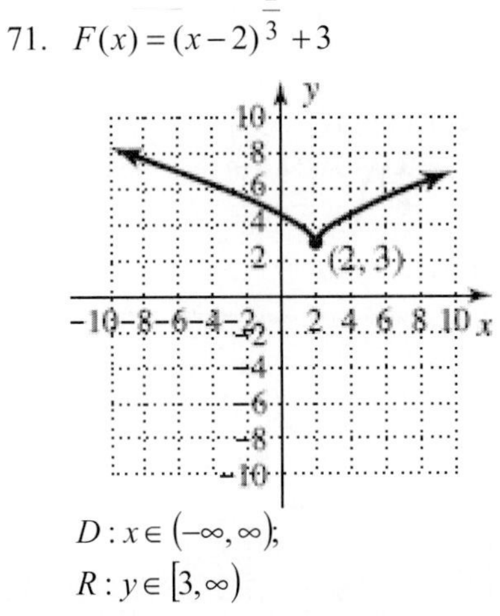

$D: x \in (-\infty, \infty)$;
$R: y \in [3, \infty)$

Chapter 5 Summary and Concept Review

1. $y = 2^x + 3$
Asymptote: $y = 3$

2. $y = 2^{-x} - 1$
Asymptote: $y = -1$

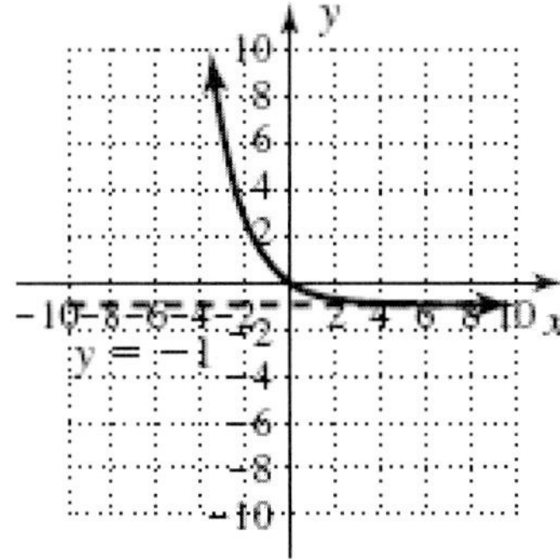

3. $y = -3^x - 2$
Asymptote: $y = -2$

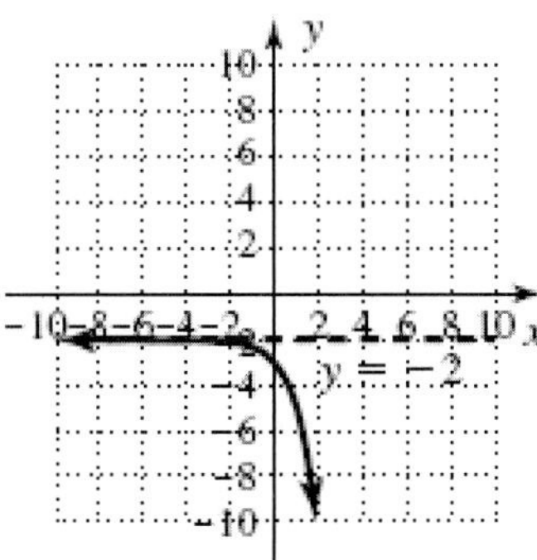

4. $3^{2x-1} = 27$
$3^{2x-1} = 3^3$
$2x - 1 = 3$
$2x = 4$
$x = 2$

5. $4^x = \frac{1}{16}$
$4^x = 4^{-2}$
$x = -2$

6. $3^x \cdot 27^{x+1} = 81$
$3^x \cdot 3^{3x+3} = 3^4$
$3^{4x+3} = 3^4$
$4x + 3 = 4$
$4x = 1$
$x = \frac{1}{4}$

7. $20000 = 142000 \cdot (0.85)^t$
$\frac{10}{71} = 0.85^t$
$\ln\left(\frac{10}{71}\right) = \ln 0.85^t$
$\ln\left(\frac{10}{71}\right) = t \ln 0.85$
$\frac{\ln\left(\frac{10}{71}\right)}{\ln 0.85} = t$
About 12.1 years

8. $\log_3 9 = 2$
$3^2 = 9$

9. $\log_5 \frac{1}{125} = -3$
$5^{-3} = \frac{1}{125}$

10. $\log_2 16 = 4$
$2^4 = 16$

11. $5^2 = 25$
$\log_5 25 = 2$

12. $2^{-3} = \frac{1}{8}$
$\log_2\left(\frac{1}{8}\right) = -3$

13. $3^4 = 81$
$\log_3 81 = 4$

14. $\log_2 32 = x$
$2^x = 32$
$2^x = 2^5$
$x = 5$

15. $\log_x 16 = 2$
$x^2 = 16$
$x^2 = 4^2$
$x = 4$

16. $\log(x - 3) = 1$
$10^1 = x - 3$
$13 = x$

17. $f(x) = \log_2 x$
Asymptote: $x = 0$

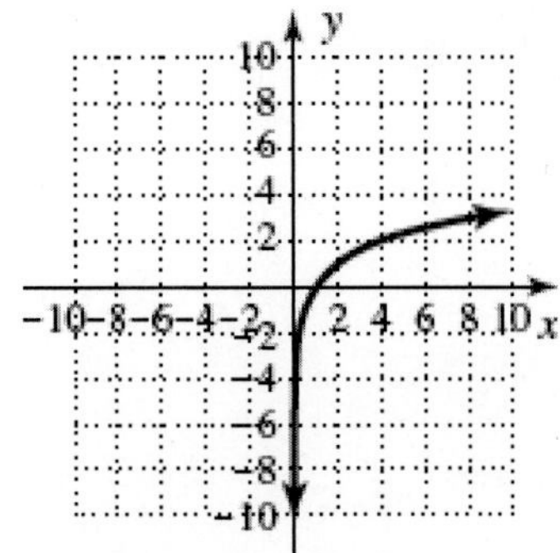

18. $f(x) = \log_2(x+3)$
Asymptote: $x = -3$

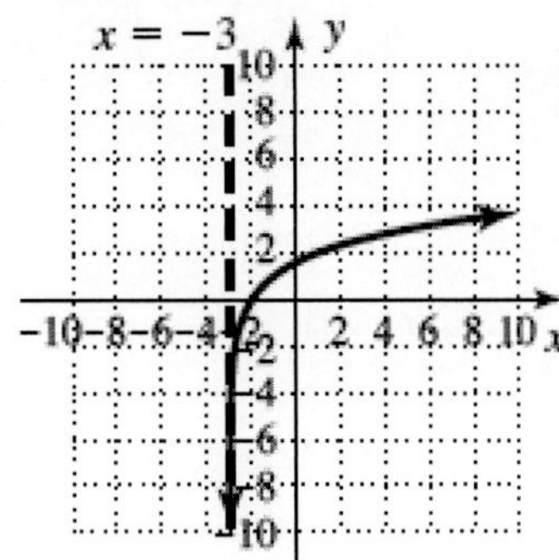

19. $f(x) = 2 + \log_2(x-1)$
Asymptote: $x = 1$

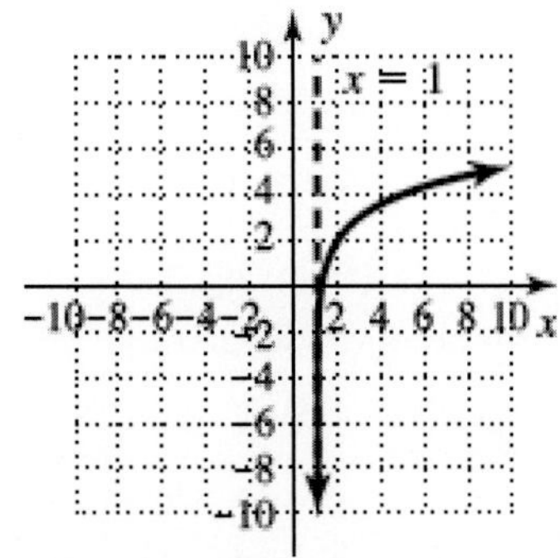

20. $M(I) = \log\frac{I}{I_0}$

a. $I = 62{,}000I_0$

$M(62000I_0) = \log\frac{62000I_0}{I_0}$

$M(62000I_0) \approx 4.79$

b. $M(I) = 7.3$

$7.3 = \log\frac{I}{I_0}$

$10^{7.3} = \frac{I}{I_0}$

$10^{7.3} I_0 = I$

21. a. $\ln x = 32$
$e^{32} = x$

b. $e^x = 9.8$
$\ln e^x = \ln 9.8$
$x \ln e = \ln 9.8$
$x = \ln 9.8$

c. $e^x = \sqrt{7}$
$\ln e^x = \ln\sqrt{7}$
$x \ln e = \ln\sqrt{7}$
$x = \ln\sqrt{7}$
$x = \frac{1}{2}\ln 7$

d. $\ln x = 2.38$
$e^{2.38} = x$

22. a. $\log_6 45$
$\frac{\log 45}{\log 6} \approx 2.125$

b. $\log_3 128$
$\frac{\log 128}{\log 3} \approx 4.417$

c. $\log_2 108$
$\frac{\log 108}{\log 2} \approx 6.755$

d. $\log_5 200$
$\frac{\log 200}{\log 5} \approx 3.292$

23. a. $\ln 7 + \ln 6$
$\ln 42$

b. $\log_9 2 + \log_9 15$
$\log_9 30$

c. $\ln(x+3) - \ln(x-1)$
$\ln\left(\frac{x+3}{x-1}\right)$

d. $\log x + \log(x+1)$
$\log(x^2 + x)$

24. a. $\log_5 9^2$
$= 2\log_5 9$

b. $\log_7 4^2$
$= 2\log_7 4$

c. $\ln 5^{2x-1}$
$=(2x-1)\ln 5$

d. $\ln 10^{3x+2}$
$=(3x+2)\ln 10$

25. a. $\log_5 \frac{2}{3}$
$=\log_5 2-\log_5 3$
$\approx 0.43-0.68$
≈ -0.25

b. $\log_5 \frac{3}{2}$
$=\log_5 3-\log_5 2$
$\approx 0.68-0.43$
≈ 0.25

c. $\log_5 12$
$=\log_5 (2^2\cdot 3)$
$=\log_5 2^2+\log_5 3$
$=2\log_5 2+\log_5 3$
$\approx 2(0.43)+0.68$
≈ 1.54

d. $\log_5 18$
$=\log_5 (2\cdot 3^2)$
$=\log_5 2+\log_5 3^2$
$=\log_5 2+2\log_5 3$
$\approx 0.43+2(0.68)$
≈ 1.79

26. a. $\ln\left(x^4\sqrt{y}\right)$
$=\ln x^4+\ln y^{\frac{1}{2}}$
$=4\ln x+\frac{1}{2}\ln y$

b. $\ln\left(\sqrt[3]{pq}\right)$
$=\ln p^{\frac{1}{3}}+\ln q$
$=\frac{1}{3}\ln p+\ln q$

c. $\log\left(\frac{\sqrt[3]{x^5y^4}}{\sqrt{x^5y^3}}\right)$
$=\log\left(\sqrt[3]{x^5y^4}\right)-\log\sqrt{x^5y^3}$
$=\log x^{\frac{5}{3}}y^{\frac{4}{3}}-\log x^{\frac{5}{2}}y^{\frac{3}{2}}$
$=\log x^{\frac{5}{3}}+\log y^{\frac{4}{3}}-\log x^{\frac{5}{2}}-\log y^{\frac{3}{2}}$
$=\frac{5}{3}\log x+\frac{4}{3}\log y-\frac{5}{2}\log x-\frac{3}{2}\log y$

d. $\log\left(\frac{4\sqrt[3]{p^5q^4}}{\sqrt{p^3q^2}}\right)$
$=\log 4\sqrt[3]{p^5q^4}-\log\sqrt{p^3q^2}$
$=\log 4p^{\frac{5}{3}}q^{\frac{4}{3}}-\log p^{\frac{3}{2}}q$
$=\log 4+\log p^{\frac{5}{3}}+\log q^{\frac{4}{3}}-(\log p+\log q)$
$=\log 4+\frac{5}{3}\log p+\frac{4}{3}\log q-\frac{3}{2}\log p-\log q$

27. $R(h)=\frac{\ln(2)}{h}$

a. $R(3.9)=\frac{\ln(2)}{3.9}$
$\approx 17.77\%$

b. $0.0289=\frac{\ln(2)}{h}$
$0.0289h=\ln 2$
$h=\frac{\ln 2}{0.0289}$
About 24 days

28. $2^x=7$
$\ln 2^x=\ln 7$
$x\ln 2=\ln 7$
$x=\frac{\ln 7}{\ln 2}$

29. $3^{x+1}=5$
$\ln 3^{x+1}=\ln 5$
$(x+1)\ln 3=\ln 5$
$x\ln 3+\ln 3=\ln 5$
$x\ln 3=\ln 5-\ln 3$
$x=\frac{\ln 5-\ln 3}{\ln 3}$
$x=\frac{\ln 5}{\ln 3}-1$

30. $4^{x-2}=3^x$
$\ln 4^{x-2}=\ln 3^x$
$(x-2)\ln 4=x\ln 3$
$x\ln 4-2\ln 4=x\ln 3$

$$x\ln 4 - x\ln 3 = 2\ln 4$$
$$x(\ln 4 - \ln 3) = 2\ln 4$$
$$x = \frac{2\ln 4}{\ln 4 - \ln 3}$$

31. $\log_5(x+1) = 2$
$$5^2 = x+1$$
$$x = 24$$

32. $\log x + \log(x-3) = 1$
$$\log x(x-3) = 1$$
$$10^1 = x(x-3)$$
$$0 = x^2 - 3x - 10$$
$$0 = (x-5)(x+2)$$
$$x = 5 \text{ or } x = -2$$
5, -2 is extraneous

33. $\log_{25}(x-2) - \log_{25}(x+3) = \frac{1}{2}$
$$\log_{25}\left(\frac{x-2}{x+3}\right) = \frac{1}{2}$$
$$25^{\frac{1}{2}} = \frac{x-2}{x+3}$$
$$5 = \frac{x-2}{x+3}$$
$$0 = \frac{x-2}{x+3} - 5$$
$$0 = \frac{x-2-5x-15}{x+3}$$
$$0 = \frac{-4x-17}{x+3}$$
$$-4x - 17 = 0$$
$$-4x = 17$$
$$x = \frac{17}{-4}$$
No solution; $x = -4.25$ is extraneous

34. $H = (30T + 8000)\ln\left(\frac{P_0}{P}\right)$
$$5657 = (30(12) + 8000)\ln\left(\frac{76}{P}\right)$$
$$5657 = (8360)\ln\left(\frac{76}{P}\right)$$
$$\frac{5657}{8360} = \ln\left(\frac{76}{P}\right)$$
$$e^{\left(\frac{5657}{8360}\right)} = e^{\ln\left(\frac{76}{P}\right)}$$
$$e^{\left(\frac{5657}{8360}\right)} = \frac{76}{P}$$
$$Pe^{\left(\frac{5657}{8360}\right)} = 76$$
$$P = \frac{76}{e^{\left(\frac{5657}{8360}\right)}}$$
$$P \approx 38.63 \text{ cmHg}$$

35. $I = prt$
$$27.75 = 600r\left(\frac{3}{12}\right)$$
$$\frac{27.75}{600} = r\left(\frac{3}{12}\right)$$
$$r = 18.5\%$$

36. $A = P\left(1 + \frac{r}{n}\right)^{nt}$
$$A = 7500\left(1 + \frac{0.078}{12}\right)^{12(6)} \approx \$11957.86$$
Almost, she needs $42.15 more.

37. $Q(t) = Q_0 e^{rt}$
$$1250 = 80e^{r5}$$
$$\frac{125}{8} = e^{5r}$$
$$\ln\left(\frac{125}{8}\right) = \ln e^{5r}$$
$$\ln\left(\frac{125}{8}\right) = 5r\ln e$$
$$\frac{\ln\left(\frac{125}{8}\right)}{5} = r$$
$$r \approx 55.0\%$$

38. $A = \frac{p\left[(1+R)^{nt} - 1\right]}{R}$

(a) $A = \frac{260\left[\left(1 + \frac{0.075}{12}\right)^{12(4)} - 1\right]}{\frac{0.075}{12}}$
$$A \approx \$14501.72,$$

No

(b) $15000 = \dfrac{p\left[\left(1+\dfrac{0.075}{12}\right)^{12(4)} - 1\right]}{\dfrac{0.075}{12}}$

$93.75 = p\left[\left(1+\dfrac{0.075}{12}\right)^{12(4)} - 1\right]$

$\dfrac{93.75}{\left[\left(1+\dfrac{0.075}{12}\right)^{12(4)} - 1\right]} = p$

$p \approx \$268.93$

39. a. Logistic: $y = \dfrac{205.85}{1+33.54e^{-0.36x}}$

Exponential: $y = 7.4(1.29)^x$

The logistic model seems to "fit" the data better.

b. Logistic; growth rate exceeds population growth

c. $y = \dfrac{205.85}{1+33.54e^{-0.36(7)}} \approx 56.1$;

$y = 7.4(1.29)^7 \approx 44.6$;

Logistic: 56.1 million,
Exponential: 44.6 million;

$y = 7.4(1.29)^{15} \approx 347.3$;

$y = 7.4(1.29)^{20} \approx 1253$

Exponential: 347.3 million for 2005,

1,253 million for 2010;

$y = \dfrac{205.85}{1+33.54e^{-0.36x}}$;

$y = \dfrac{205.85}{1+33.54e^{-0.36(15)}} \approx 179.3$;

$y = \dfrac{205.85}{1+33.54e^{-0.36(20)}} \approx 201.0$

Logistic: 178.80millino for 2005, 200.83 million for 2010.

Projections from the exponential equation are excessive.

40. Exponential

a. $y = 13.29(1.08)^x$

b. $y = 13.29(1.08)^{32} \approx 156.0$

156.0 billion dollars

c. $y = 13.29(1.08)^{45} \approx 424.2$

424.2 billion dollars

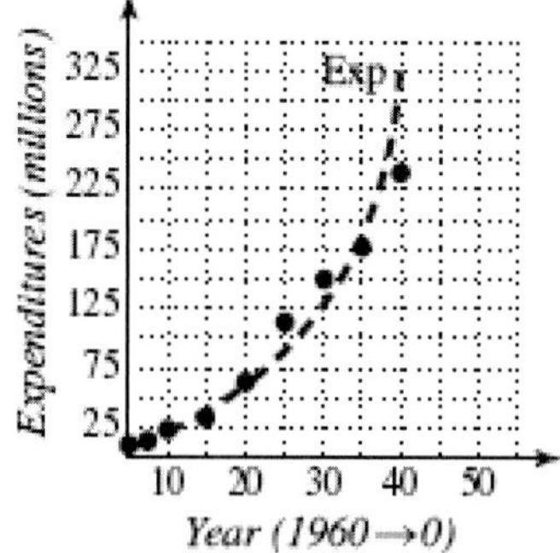

Chapter 5 Mixed Review

1. a. $\log_2 30$

$\dfrac{\log 30}{\log 2} \approx 4.9069$

b. $\log_{0.25} 8$

$\dfrac{\log 8}{\log 0.25} = -1.5$

c. $\log_8 2$

$\dfrac{\log 2}{\log 8} = \dfrac{1}{3}$

3. a. $\log_{10} 20^2$

$= 2\log_{10} 20$

$= 2[\log_{10} 10 + \log 2]$

$= 2[1 + \log 2]$

b. $\log 10^{0.05x}$

$= 0.05x \log 10$

$= 0.05x$

c. $\ln 2^{x-3}$

$= (x-3)\ln 2$

5. $y = 5 \cdot 2^{-x}$

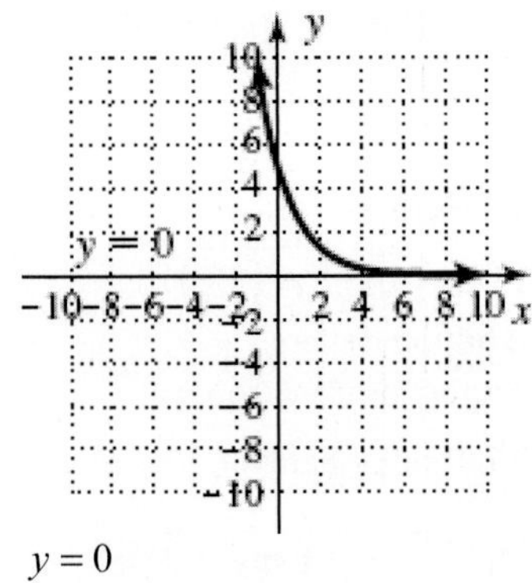

$y = 0$

7. $y = \log_2(-x) - 4$

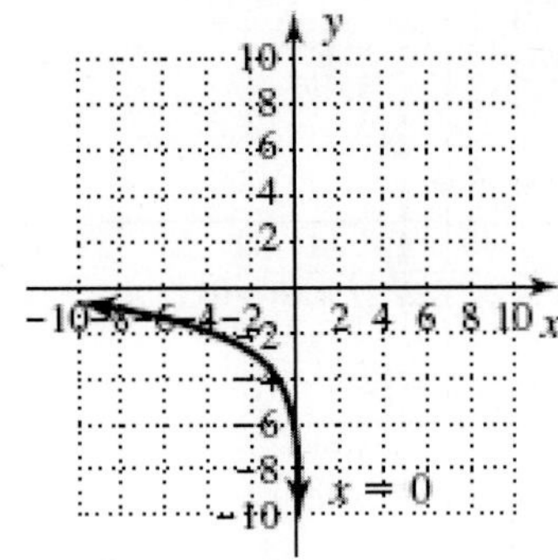

$x = 0$

9. a. $\log_5 625 = 4$

$5^4 = 625$

b. $\ln 0.15x = 0.45$

$e^{0.45} = 0.15x$

c. $\log(0.1x10^8) = 7$

$10^7 = 0.1x10^8$

11. $\log_2 128 = x$

$2^x = 128$

$2^x = 2^7$

$x = 7$

13. $10^{x-4} = 200$

$\log 10^{x-4} = \log 200$

$(x-4)\log 10 = \log 2 \cdot 10^2$

$x - 4 = \log 2 + \log 10^2$

$x - 4 = \log 2 + 2\log 10$

$x - 4 = \log 2 + 2$

$x = 6 + \log 2$

15. $\log_2(2x-5) + \log_2(x-2) = 4$

$\log_2(2x-5)(x-2) = 4$

$2^4 = (2x-5)(x-2)$

$16 = 2x^2 - 9x + 10$

$0 = 2x^2 - 9x - 6$

$a = 2, b = -9, c = -6$

$$x = \frac{-(-9) \pm \sqrt{(-9)^2 - 4(2)(-6)}}{2(2)}$$

$$x = \frac{9 \pm \sqrt{129}}{4}$$

$x = \frac{9 + \sqrt{129}}{4}$; $x = \frac{9 - \sqrt{129}}{4}$ is extraneous

17. $6.5 = \log\left(\frac{I}{2x10^{11}}\right)$

$10^{6.5} = \frac{I}{2x10^{11}}$

$10^{6.5}(2x10^{11}) = I$

$2 \cdot 10^{0.5} \cdot 10^{17} = I$

$2 \cdot \sqrt{10} \cdot 10^{17} = I$

$I = 6.3x10^{17}$

19. $r(n) = 2(0.8)^n$

$r(6) = 2(0.8)^6 = 0.524$

0.52 m;

n	$r(n) = 2(0.8^n)$
1	1.6 m
2	1.28 m
3	1.02 m
4	0.82 m
5	0.66 m
6	0.52 m

Chapter 5 Practice Test

1. $\log_3 81 = 4$

$3^4 = 81$

2. $25^{\frac{1}{2}} = 5$

$\log_{25} 5 = \frac{1}{2}$

3. $\log_b\left(\frac{\sqrt{x^5}\, y^3}{z}\right)$

$= \log_b \sqrt{x^5}\, y^3 - \log_b z$

$= \log_b x^{\frac{5}{2}} y^3 - \log_b z$

$= \log_b x^{\frac{5}{2}} + \log_b y^3 - \log_b z$

$= \frac{5}{2}\log_b x + 3\log_b y - \log_b z$

4. $\log_b m + \left(\frac{3}{2}\right)\log_b n - \frac{1}{2}\log_b p$

$= \log_b m + \log_b \sqrt{n^3} - \log_b \sqrt{p}$

$= \log_b \frac{m\sqrt{n^3}}{\sqrt{p}}$

5. $5^{x-7} = 125$
$5^{x-7} = 5^3$
$x - 7 = 3$
$x = 10$

6. $2 \cdot 4^{3x} = \frac{8^x}{16}$
$32 \cdot 4^{3x} = 8^x$
$2^5 \cdot 2^{6x} = 2^{3x}$
$2^{6x+5} = 2^{3x}$
$6x + 5 = 3x$
$3x = -5$
$x = \frac{-5}{3}$

7. $\log_a 45$
$= \log_a (3^2 \cdot 5)$
$= \log_a 3^2 + \log_a 5$
$= 2\log_a 3 + \log_a 5$
$= 2(0.48) + 1.72$
$= 2.68$

8. $\log_a 0.6$
$= \log_a \left(\frac{3}{5}\right)$
$= \log_a 3 - \log_a 5$
$= 0.48 - 1.72$
$= -1.24$

9. $g(x) = -2^{x-1} + 3$
HA: $y = 3$

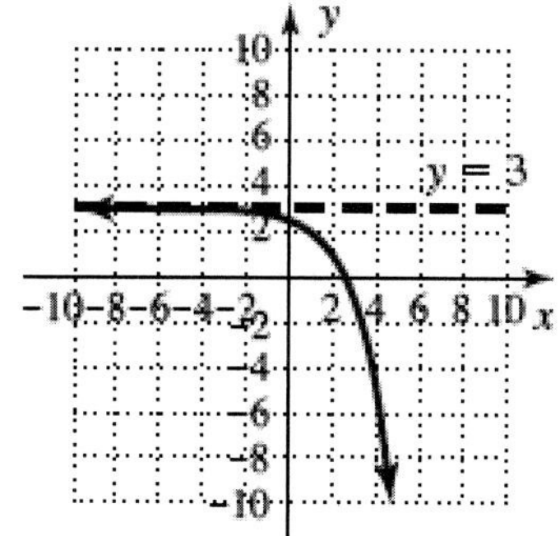

10. $h(x) = \log_2 (x - 2) + 1$
VA: $x = 2$

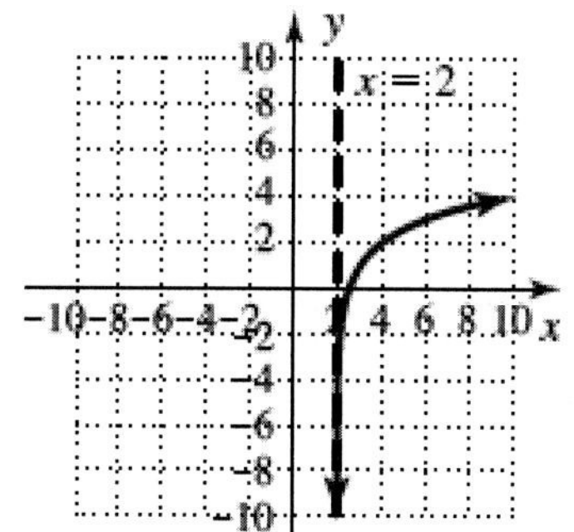

11. $\log_3 100$
$= \frac{\log 100}{\log 3}$
$= \frac{\log 10^2}{\log 3}$
$= \frac{2\log 10}{\log 3}$
$= \frac{2}{\log 3}$
≈ 4.19

12. $\log_6 0.235$
$= \frac{\log 0.235}{\log 6}$
≈ -0.81

13. $3^{x-1} = 89$
$\ln 3^{x-1} = \ln 89$
$(x-1)\ln 3 = \ln 89$
$x - 1 = \frac{\ln 89}{\ln 3}$
$x = 1 + \frac{\ln 89}{\ln 3}$

14. $\log_5 x + \log_5(x+4) = 1$

$\log_5 x(x+4) = 1$

$5^1 = x(x+4)$

$5 = x^2 + 4x$

$0 = x^2 + 4x - 5$

$0 = (x+5)(x-1)$

$x+5 = 0$ or $x-1 = 0$

$x = -5$ or $x = 1$

$x = 1$, -5 is extraneous

15. $3000 = 8000(0.82)^t$

$\frac{3}{8} = (0.82)^t$

$\ln\left(\frac{3}{8}\right) = \ln(0.82)^t$

$\ln\left(\frac{3}{8}\right) = t\ln(0.82)$

$\frac{\ln\left(\frac{3}{8}\right)}{\ln 0.82} = t$

$t \approx 5$ years

16. $2000 = 1000\left(1+\frac{0.08}{365}\right)^{365t}$

$2 = \left(1+\frac{0.08}{365}\right)^{365t}$

$\ln 2 = \ln\left(1+\frac{0.08}{365}\right)^{365t}$

$\ln 2 = 365t\ln\left(1+\frac{0.08}{365}\right)$

$\frac{\ln 2}{365\ln\left(1+\frac{0.08}{365}\right)} = t$

$t \approx 8.7$ years

17. $Q(t) = -2600 + 1900\ln x$

$3000 = -2600 + 1900\ln x$

$5600 = 1900\ln x$

$\frac{56}{19} = \ln x$

$e^{\left(\frac{56}{19}\right)} = e^{\ln x}$

$e^{\left(\frac{56}{19}\right)} = x$

$x \approx 19.1$ months

18. $A = P\left(1+\frac{r}{n}\right)^{nt}$

7% compounded semi-annually:

$A = p\left(1+\frac{0.07}{2}\right)^{2t} = p(1.035)^{2t}$

$= p(1.071225)^t$;

6.8% compounded semi-annually

$A = p\left(1+\frac{0.068}{365}\right)^{365t}$

$= p(1.000186301)^{365t} = p(1.070358529)^t$;

7% compounded semi-annually

19. $A = \frac{p\left[(1+R)^{nt} - 1\right]}{R}$

(a) $A = \frac{50\left[\left(1+\frac{0.0825}{12}\right)^{12(5)} - 1\right]}{\frac{0.0825}{12}}$

$A \approx \$3697.88$

No

(b) $4000 = \frac{p\left[\left(1+\frac{0.0825}{12}\right)^{12(5)} - 1\right]}{\frac{0.0825}{12}}$

$27.5 = p\left[\left(1+\frac{0.0825}{4}\right)^{60} - 1\right]$

$\frac{27.5}{\left[\left(1+\frac{0.0825}{4}\right)^{60} - 1\right]} = p$

$p \approx \$54.09$

20. Logistic

$y = \frac{39.1156}{1+314.6617e^{-5.9483x}}$

$15 = \frac{39.1156}{1+314.6617e^{-5.9483x}}$

$15\left(1+314.6617e^{-5.9483x}\right) = 39.1156$

$15 + 4719.9255e^{-5.9483x} = 39.1156$

$4719.9255e^{-5.9483x} = 24.1156$

$e^{-5.9483x} = \frac{24.1156}{4719.9255}$

$$\ln e^{-5.9483x} = \ln\left(\frac{24.1156}{4719.9255}\right)$$

$$-5.9483x \ln e = \ln\left(\frac{24.1156}{4719.9255}\right)$$

$$x = \frac{\ln\left(\frac{24.1156}{4719.9255}\right)}{-5.9483}$$

$x \approx 0.89$ seconds

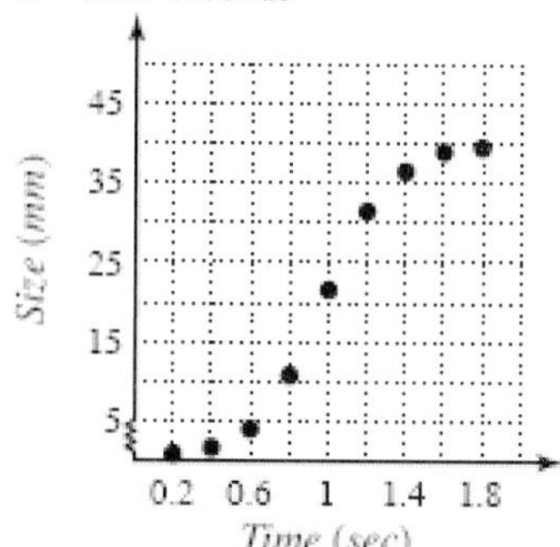

Chapter 5-Calculator Exploration and Discovery

1. $a = 25, \quad b = 0.5, \quad c = 2500$

2. $\frac{c}{1+a} = \frac{2500}{1+25} = 96.15$
   ≈ 96 ants

3. b

4. 2500 ants

5. $b = 0.6$

6. a

7. Verified

Strengthening Core Skills

1. $y = 21.303(0.842)^x$

   $5 = 21.303(0.842)^x$

   $\frac{5}{21.303} = (0.842)^x$

   $\ln\left(\frac{5}{21.303}\right) = \ln(0.842)^x$

   $\ln\left(\frac{5}{21.303}\right) = x\ln(0.842)$

   $\frac{\ln\left(\frac{5}{21.303}\right)}{\ln(0.842)} = x$

   $x \approx 8.4$

   Between eight and nine players.

Cumulative Review Chapters 1 to 5

1. $x^2 - 4x + 53 = 0$

   $a = 1, b = -4, c = 53$

   $x = \frac{-(-4) \pm \sqrt{(-4)^2 - 4(1)(53)}}{2(1)}$

   $x = \frac{4 \pm \sqrt{-196}}{2}$

   $x = \frac{4 \pm 14i}{2}$

   $x = 2 \pm 7i$

2. $6x^2 + 19x = 36$

   $6x^2 + 19x - 36 = 0$

   $a = 6, b = 19, c = -36$

   $x = \frac{-(19) \pm \sqrt{(19)^2 - 4(6)(-36)}}{2(6)}$

   $x = \frac{-19 \pm \sqrt{1225}}{12}$

   $x = \frac{-19 \pm 35}{12}$

   $x = \frac{4}{3}, x = \frac{-9}{2}$

3. $(4+5i)^2 - 8(4+5i) + 41 = 0$

   $-9 + 40i - 32 - 40i + 41 = 0$

   $0 = 0$

4. $y = 2\sqrt{x+2} - 3$

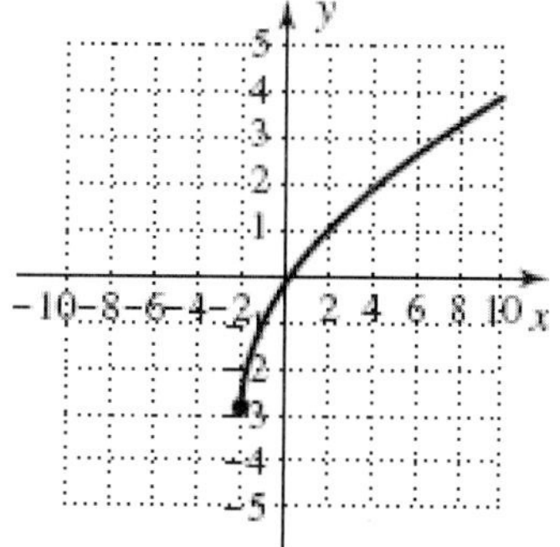

5. $f(x) = x^3 - 2$, $g(x) = \sqrt[3]{x+2}$;

$f(g(x)) = (\sqrt[3]{x+2})^3 - 2 = x + 2 + x = x$;
$g(f(x)) = \sqrt[3]{x^3 - 2 + 2} = \sqrt[3]{x^3} = x$
Since $(f \circ g)(x) = (g \circ f)(x)$, they are inverse functions.

6. $h(x) = \dfrac{\sqrt{x+3}}{x^2 + 6x + 8}$
$h(x) = \dfrac{\sqrt{x+3}}{(x+4)(x+2)}$
$x \in [-3,-2) \cup (-2, \infty)$

7. 1991 → year 1
(a) $(1,3100), (9,6740)$
$m = \dfrac{6740 - 3100}{9-1} = 455$
$y - 3100 = 455(x-1)$
$y - 3100 = 455x - 455$
$y = 455x + 2645$
$T(t) = 455t + 2645$
(b) $\dfrac{\Delta T}{\Delta t} = \dfrac{455}{1}$, triple births increase by 455 each year.
(c) In 1996, $T(6) = 455(6) + 2645 = 5375$ sets of triplets
In 2007,
$t = 17, T(17) = 455(17) + 2645 = 10{,}380$ sets of triplets

8. a. $A = \pi r^2$
b. $(\text{leg})^2 + (\text{leg})^2 = (\text{hyp})^2$
c. $P = 2L + 2W$
d. $A = \dfrac{h}{2}(B + b)$

9. $h(x) = \begin{cases} -4 & -10 \le x < -2 \\ -x^2 & -2 \le x < 3 \\ -3x - 18 & x \ge 3 \end{cases}$

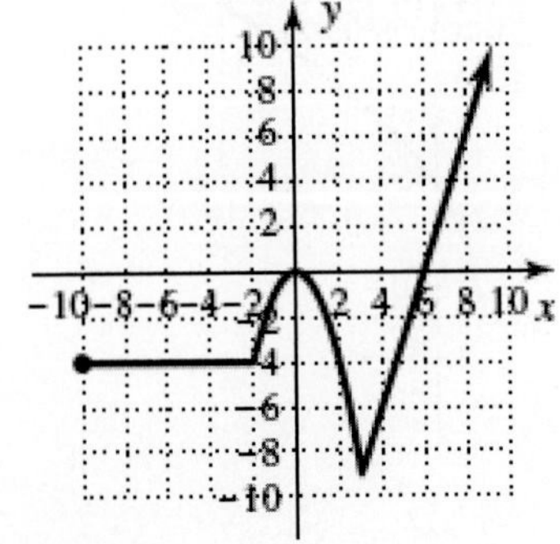

$D: x \in [-10, \infty), R: y \in [-9, \infty)$;
$h(x)\uparrow: (-2,0) \cup (3, \infty)$
$h(x)\downarrow: (0,3)$

10. $\dfrac{2x+1}{x-3} \ge 0$
$x \in \left(-\infty, -\dfrac{1}{2}\right] \cup (3, \infty)$

11. $f(x) = x^4 - 3x^3 - 12x^2 + 52x - 48$
Possible rational roots:
$\dfrac{\{\pm1, \pm48, \pm2, \pm24, \pm3, \pm16, \pm4, \pm12, \pm6, \pm8\}}{\{\pm1\}}$;
$\{\pm1, \pm48, \pm2, \pm24, \pm3, \pm16, \pm4, \pm12, \pm6, \pm8\}$

```
3 | 1  -3  -12   52  -48
  |     3    0  -36   48
  -----------------------
    1   0  -12   16 |  0

2 | 1   0  -12   16
  |     2    4  -16
  -------------------
    1   2   -8 |  0
```

$f(x) = (x-3)(x-2)(x^2 + 2x - 8)$
$f(x) = (x-3)(x-2)(x-2)(x+4)$
$x = 3, x = 2$ (multiplicity 2), $x = -4$

12. $f(c) = \dfrac{9}{5}c + 32$
$f(25) = \dfrac{9}{5}(25) + 32 = 77$
$k = 77$;
$c = \dfrac{9}{5}f + 32$
$c - 32 = \dfrac{9}{5}f$
$\dfrac{5}{9}(c - 32) = f^{-1}(c)$;
$f^{-1}(77) = \dfrac{5}{9}(77 - 32) = 25$

13. $V = \dfrac{1}{2}\pi b^2 a$
$\dfrac{2V}{\pi a} = b^2$
$\sqrt{\dfrac{2V}{\pi a}} = b$

14. $p(x) = x^3 - 4x^2 + x + 6$
end behavior: down/up

Possible rational roots: $\dfrac{\{\pm1,\pm6,\pm2,\pm3\}}{\{\pm1\}}$

$p(x)=(x+1)(x-2)(x-3)$

x-intercepts: $(-1,0),(2,0)$ and $(3,0)$

$p(0)=0^3-4(0)^2+0+6=6$

y-intercept: $(0,6)$

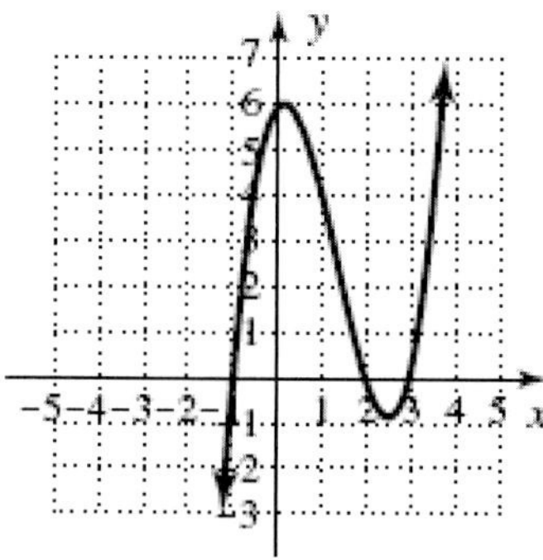

15. $r(x)\dfrac{5x^2}{x^2+4}$

$r(0)=\dfrac{5(0)^2}{(0)^2+4}=0$

y-intercept: (0,0)

Since $x^2+4\neq 0$, there are no vertical asymptotes;

x-intercept: (0,0)

horizontal asymptote: $y=5$

(deg num = deg den)

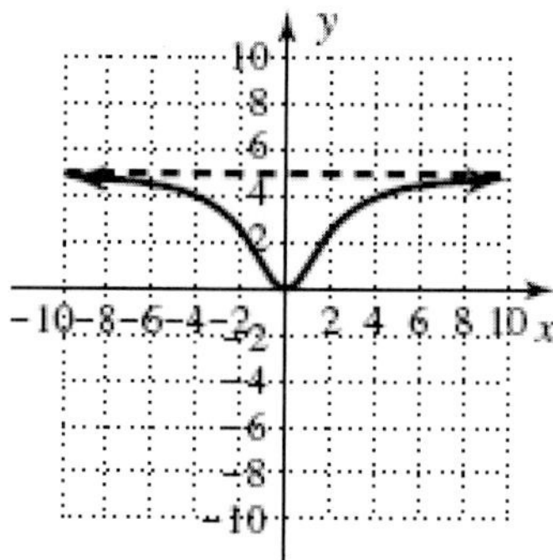

16. $10=-2e^{-0.05x}+25$

$-15=-2e^{-0.05x}$

$\dfrac{15}{2}=e^{-0.05x}$

$\ln\left(\dfrac{15}{2}\right)=\ln e^{-0.05x}$

$\ln\left(\dfrac{15}{2}\right)=-0.05x\ln e$

$\dfrac{\ln\left(\dfrac{15}{2}\right)}{-0.05}=x$

$x\approx-40.298$

17. $\ln(x+3)+\ln(x-2)=\ln 24$

$\ln(x+3)(x-2)=\ln 24$

$(x+3)(x-2)=24$

$x^2+x-6=24$

$x^2+x-30=0$

$(x+6)(x-5)=0$

$x+6=0$ or $x-5=0$

$x=-6$ or $x=5$

$x=5, x=-6$ is an extraneous root

18. $p(t)=50e^{-0.002t}$

(a) $p(183)=50e^{-0.002(183)}\approx 34.7$ watts

(b) $p(t)=\dfrac{50}{4}$

$\dfrac{50}{4}=50e^{-0.002t}$

$\dfrac{1}{4}=e^{-0.002t}$

$\ln\left(\dfrac{1}{4}\right)=\ln e^{-0.002t}$

$\ln\left(\dfrac{1}{4}\right)=-0.002t\ln e$

$\dfrac{\ln\left(\dfrac{1}{4}\right)}{-0.002}=t$

$t\approx 693$ days

Approx. 1 year 11 months

19. (a) Linear

$W=1.24L-15.83$

(b) $W=1.24(39)-15.83=32.5$ lb

$(39, 32.5)$

$W\approx 32.5$ pounds

(c) $28=1.24L-15.83$

$43.83=1.24L$

$35.3\text{ in}\approx L$

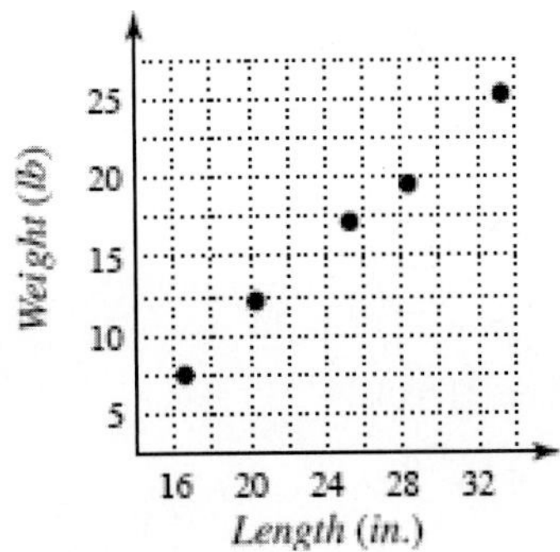

20. (a) Logarithmic

$C(a) \approx 37.9694 + 3.4229 \ln a$

(b) $C(27) = 37.9694 + 3.4229 \ln 27$

≈ 49.3 cm

(c) $50 = 37.9694 + 3.4229 \ln a$

$12.0306 = 3.4229 \ln a$

$3.514738964 = \ln a$

$e^{3.514738964} = a$

$a \approx 34$

Between 33 and 34 months

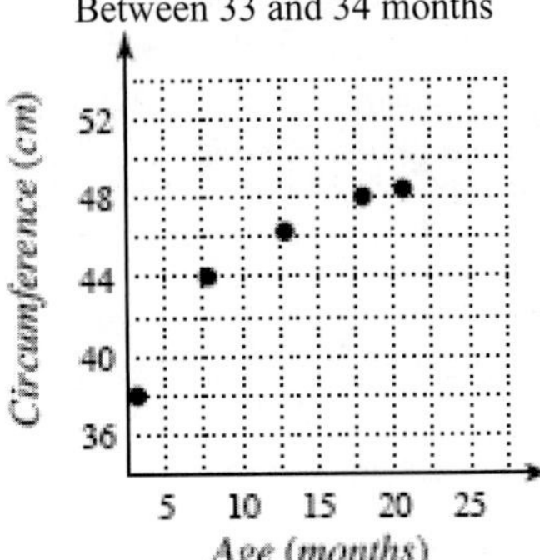

6.1 Technology Highlight

1. (-1,4)

2. $\left(\frac{1}{2},-\frac{2}{3}\right)$

6.1 Exercises

1. Inconsistent

3. Consistent; independent

5. Multiply the 1st equation by 6 and the 2nd equation by 10.

7. $\begin{cases} 7x-4y=24 \\ 4x+3y=15 \end{cases}$

$\begin{cases} y=\frac{7}{4}x-6 \\ y=-\frac{4}{3}x+5 \end{cases}$

9. A: $y=x+2$

11. C: $x+3y=-3$

13. E: $y=x+2$
$x+3y=-3$

15. $\begin{cases} 3x+y=11 \\ -5x+y=-13 \end{cases}$

(3, 2)

$$\begin{aligned} 3x+y&=11 \\ 3(3)+2&=11 \\ 9+2&=11 \\ 11&=11 \end{aligned} \qquad \begin{aligned} -5x+y&=-13 \\ -5(3)+2&=-13 \\ -15+2&=-13 \\ -13&=-13 \end{aligned}$$

Yes

17. $\begin{cases} 8x-24y=-17 \\ 12x+30y=2 \end{cases}$

$\left(-\frac{7}{8},\frac{5}{12}\right)$

$$\begin{aligned} 8x-24y&=-17 \\ 8\left(-\frac{7}{8}\right)-24\left(\frac{5}{12}\right)&=-17 \\ -7-10&=-17 \\ -17&=-17; \end{aligned}$$

$$\begin{aligned} 12x+30y&=2 \\ 12\left(-\frac{7}{8}\right)+30\left(\frac{5}{12}\right)&=2 \\ -\frac{84}{8}+\frac{150}{12}&=2 \\ 2&=2 \end{aligned}$$

Yes

19. $\begin{cases} 3x+2y=12 \\ x-y=9 \end{cases}$

$\begin{cases} y=-\frac{3}{2}x+6 \\ y=x-9 \end{cases}$

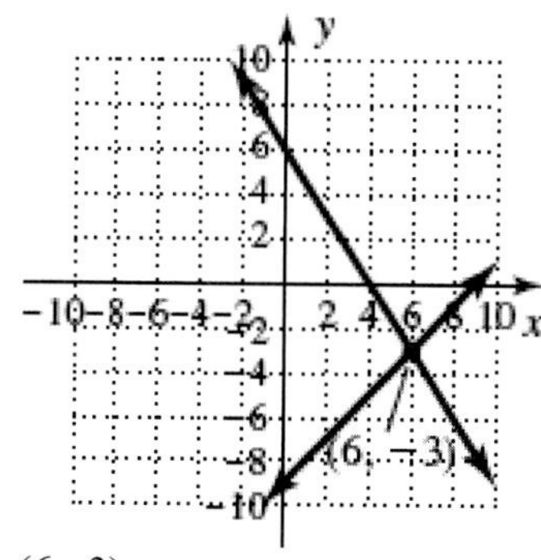

(6, -3)

21. $\begin{cases} 5x-2y=4 \\ x+3y=-15 \end{cases}$

$\begin{cases} y=\frac{5}{2}x-2 \\ y=-\frac{1}{3}x-5 \end{cases}$

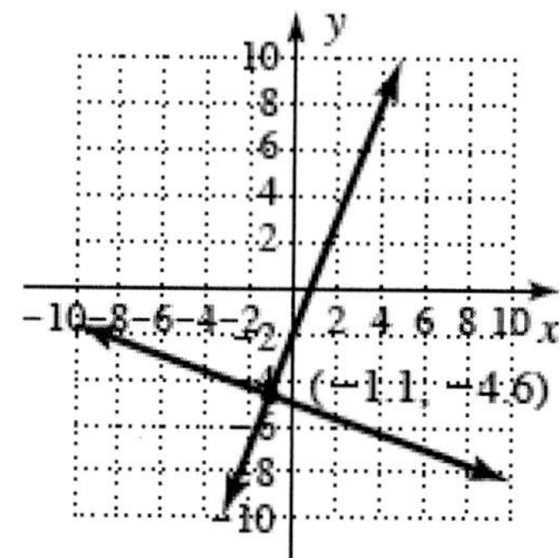

Estimate: (-1.1, -4.6)

23. $\begin{cases} x=5y-9 \\ x-2y=-6 \end{cases}$

$$\begin{aligned} x - 2y &= -6 \\ (5y - 9) - 2y &= -6 \\ 5y - 9 - 2y &= -6 \\ 3y &= 3 \\ y &= 1; \end{aligned}$$

$$\begin{aligned} x - 2y &= -6 \\ x - 2(1) &= -6 \\ x - 2 &= -6 \\ x &= -4 \end{aligned}$$

(-4, 1)

25. $\begin{cases} y = \frac{2}{3}x - 7 \\ 3x - 2y = 19 \end{cases}$

$$\begin{aligned} 3x - 2y &= 19 \\ 3x - 2\left(\frac{2}{3}x - 7\right) &= 19 \\ 3x - \frac{4}{3}x + 14 &= 19 \\ \frac{5}{3}x &= 5 \\ x &= 3; \end{aligned}$$

$$\begin{aligned} y &= \frac{2}{3}x - 7 \\ y &= \frac{2}{3}(3) - 7 \\ y &= 2 - 7 \\ y &= -5 \end{aligned}$$

(3, -5)

27. $\begin{cases} 3x - 4y = 24 \\ 5x + y = 17 \end{cases}$

Equation 2, variable y

$\begin{cases} 3x - 4y = 24 \\ y = -5x + 17 \end{cases}$

$$\begin{aligned} 3x - 4y &= 24 \\ 3x - 4(-5x + 17) &= 24 \\ 3x + 20x - 68 &= 24 \\ 23x &= 92 \\ x &= 4; \end{aligned}$$

$$\begin{aligned} 5x + y &= 17 \\ 5(4) + y &= 17 \\ 20 + y &= 17 \\ y &= -3 \end{aligned}$$

(4, -3)

29. $\begin{cases} 0.7x + 2y = 5 \\ x - 1.4y = 11.4 \end{cases}$

Equation 2, variable x

$\begin{cases} 0.7x + 2y = 5 \\ x = 1.4y + 11.4 \end{cases}$

$$\begin{aligned} 0.7x + 2y &= 5 \\ 0.7(1.4y + 11.4) + 2y &= 5 \\ 0.98y + 7.98 + 2y &= 5 \\ 2.98y &= -2.98 \\ y &= -1; \end{aligned}$$

$$\begin{aligned} x - 1.4y &= 11.4 \\ x - 1.4(-1) &= 11.4 \\ x + 1.4 &= 11.4 \\ x &= 10 \end{aligned}$$

(10, -1)

31. $\begin{cases} 5x - 6y = 2 \\ x + 2y = 6 \end{cases}$

Equation 2, variable x

$\begin{cases} 5x - 6y = 2 \\ x = -2y + 6 \end{cases}$

$$\begin{aligned} 5x - 6y &= 2 \\ 5(-2y + 6) - 6y &= 2 \\ -10y + 30 - 6y &= 2 \\ -16y &= -28 \\ y &= \frac{7}{4}; \end{aligned}$$

$$\begin{aligned} x + 2y &= 6 \\ x + 2\left(\frac{7}{4}\right) &= 6 \\ x + \frac{7}{2} &= 6 \\ x &= \frac{5}{2} \end{aligned}$$

$\left(\frac{5}{2}, \frac{7}{4}\right)$

33. $\begin{cases} 2x - 4y = 10 \\ 3x + 4y = 5 \end{cases}$

R1 + R2 = Sum

$$\begin{aligned} 2x - 4y &= 10 \\ 3x + 4y &= 5 \\ \hline 5x &= 15 \\ x &= 3; \end{aligned}$$

$$\begin{aligned} 3x + 4y &= 5 \\ 3(3) + 4y &= 5 \\ 9 + 4y &= 5 \\ 4y &= -4 \\ y &= -1 \end{aligned}$$

(3, -1)

35. $\begin{cases} 4x-3y=1 \\ 3y=-5x-19 \end{cases}$

$\begin{cases} 4x-3y=1 \\ 5x+3y=-19 \end{cases}$

R1 + R2 = Sum

$$\begin{array}{l} 4x-3y=1 \\ \underline{5x+3y=-19} \\ 9x=-18 \\ \quad x=-2; \end{array}$$

$$\begin{aligned} 3y&=-5x-19 \\ 3y&=-5(-2)-19 \\ 3y&=10-19 \\ 3y&=-9 \\ y&=-3 \end{aligned}$$

(-2, -3)

37. $\begin{cases} 2x=-3y+17 \\ 4x-5y=12 \end{cases}$

$\begin{cases} 2x+3y=17 \\ 4x-5y=12 \end{cases}$

-2R1 + R2 = Sum

$$\begin{array}{l} -4x-6y=-34 \\ \underline{\quad 4x-5y=12} \\ \quad -11y=-22 \\ \qquad y=2; \end{array}$$

$$\begin{aligned} 2x&=-3y+17 \\ 2x&=-3(2)+17 \\ 2x&=-6+17 \\ 2x&=11 \\ x&=\frac{11}{2} \end{aligned}$$

$\left(\frac{11}{2},2\right)$

39. $\begin{cases} 0.5x+0.4y=0.2 \\ 0.3y=1.3+0.2x \end{cases}$

$\begin{cases} 0.5x+0.4y=0.2 \\ -0.2x+0.3y=1.3 \end{cases}$

20R1 + 50 R2 = Sum

$$\begin{array}{l} 10x+8y=4 \\ \underline{-10x+15y=65} \\ 23y=69 \\ \quad y=3; \end{array}$$

$$\begin{aligned} 0.5x+0.4y&=0.2 \\ 0.5x+0.4(3)&=0.2 \\ 0.5x+1.2&=0.2 \\ 0.5x&=-1 \\ x&=-2 \end{aligned}$$

(-2, 3)

41. $\begin{cases} 0.32m-0.12n=-1.44 \\ -0.24m+0.08n=1.04 \end{cases}$

200 R1 + 300 R2 = Sum

$$\begin{array}{l} 64m-24n=-288 \\ \underline{-72m+24n=312} \\ -8m=24 \\ \quad m=-3; \end{array}$$

$$\begin{aligned} 0.32m-0.12n&=-1.44 \\ 0.32(-3)-0.12n&=-1.44 \\ -0.96-0.12n&=-1.44 \\ -0.12n&=-0.48 \\ n&=4 \end{aligned}$$

(-3, 4)

43. $\begin{cases} -\frac{1}{6}u+\frac{1}{4}v=4 \\ \frac{1}{2}u-\frac{2}{3}v=-11 \end{cases}$

18R1 + 6R2 = Sum

$$\begin{array}{l} -3u+4.5v=72 \\ \underline{\quad 3u-4v=-66} \\ \quad 0.5v=6 \\ \qquad v=12; \end{array}$$

$$\begin{aligned} -\frac{1}{6}u+\frac{1}{4}v&=4 \\ -\frac{1}{6}u+\frac{1}{4}(12)&=4 \\ -\frac{1}{6}u+3&=4 \\ -\frac{1}{6}u&=1 \\ u&=-6 \end{aligned}$$

(-6, 12)

45. $\begin{cases} 4x+\frac{3}{4}y=14 \\ -9x+\frac{5}{8}y=-13 \end{cases}$

9R1 + 4R2 = Sum

$$36x+\frac{27}{4}y=126$$
$$-36x+\frac{5}{2}y=-52$$
$$\frac{37}{4}y=74$$
$$y=8;$$
$$4x+\frac{3}{4}y=14$$
$$4x+\frac{3}{4}(8)=14$$
$$4x+6=14$$
$$4x=8$$
$$x=2$$
(2, 8); Consistent/independent

47. $\begin{cases}0.2y=0.3x+4\\0.6x-0.4y=-1\end{cases}$

$\begin{cases}-0.3x+0.2y=4\\0.6x-0.4y=-1\end{cases}$

2R1 + R2 = Sum
$$-0.6x+0.4y=8$$
$$0.6x-0.4y=-1$$
$$0\neq 7$$
No Solution; Inconsistent

49. $\begin{cases}6x-22=-y\\3x+\frac{1}{2}y=11\end{cases}$

$\begin{cases}6x+y=22\\3x+\frac{1}{2}y=11\end{cases}$

R1 + -2R2 = Sum
$$6x+y=22$$
$$-6x-y=-22$$
$$0=0$$
$\{(x,y)|6x-22=-y\}$
Consistent/dependent

51. $\begin{cases}-10x+35y=-5\\y=0.25x\end{cases}$

$$-10x+35y=-5$$
$$-10x+35(0.25x)=-5$$
$$-10x+8.75x=-5$$
$$-1.25x=-5$$
$$x=4;$$
$$y=0.25x$$
$$y=0.25(4)$$
$$y=1$$
(4, 1); Consistent/Independent

53. $\begin{cases}7a+b=-25\\2a-5b=14\end{cases}$

5R1 + R2 = Sum
$$35a+5b=-125$$
$$2a-5b=14$$
$$37a=-111$$
$$a=-3;$$
$$2a-5b=14$$
$$2(-3)-5b=14$$
$$-6-5b=14$$
$$-5b=20$$
$$b=-4$$
(-3, -4); Consistent/Independent

55. $\begin{cases}4a=2-3b\\6b+2a=7\end{cases}$

$\begin{cases}4a+3b=2\\2a+6b=7\end{cases}$

R1 + -2R2 = Sum
$$4a+3b=2$$
$$-4a-12b=-14$$
$$-9b=-12$$
$$b=\frac{4}{3};$$
$$4a+3b=2$$
$$4a+3\left(\frac{4}{3}\right)=2$$
$$4a+4=2$$
$$4a=-2$$
$$a=-\frac{1}{2}$$
$\left(-\frac{1}{2},\frac{4}{3}\right)$; Consistent/Independent

57. $\begin{cases}2x+4y=6\\x+12=4y\end{cases}$

$$2x+4y=6$$
$$2x+x+12=6$$
$$3x=-6$$
$$x=-2;$$

$$x+12=4y$$
$$-2+12=4y$$
$$10=4y$$
$$\frac{5}{2}=y$$
$$\left(-2,\frac{5}{2}\right)$$

59. $\begin{cases} 5x-11y=21 \\ 11y=5-8x \end{cases}$

$$5x-11y=21$$
$$5x-(5-8x)=21$$
$$5x-5+8x=21$$
$$13x=26$$
$$x=2;$$
$$11y=5-8x$$
$$11y=5-8(2)$$
$$11y=5-16$$
$$11y=-11$$
$$y=-1$$

(2, -1)

61. $\begin{cases} (R+C)T_1=D_1 \\ (R-C)T_2=D_2 \end{cases}$

$$\begin{cases} (R+C)\ 1=5 \\ (R-C)\ 3=9 \end{cases}$$
$$\begin{cases} R+C=5 \\ 3R-3C=9 \end{cases}$$

3R1 + R2 = Sum

$$3R+3C=15$$
$$3R-3C=9$$
$$6R=24$$
$$R=4;$$
$$R+C=5$$
$$4+C=5$$
$$C=1$$

The current was 1 mph.
He can row 4mph in still water.

63. a. (3, 24)

b. $\begin{cases} y=4x+12 \\ y=8x \end{cases}$

$$8x=4x+12$$
$$8x-4x=12$$
$$4x=12$$
$$x=3$$
$$y=8(3)$$
$$y=24$$

(3, 24)
These are the equations of the 2 lines. Thus, they will have the same point of intersection as the estimate point in Part A.

c. Profit = $8x-(4x+12)=8x-4x-12=4x-12$

Profit = $4(8)-12=32-12=\$20$

65. Let d represent the year the Declaration was signed.
Let c represent the year the Civil War ended.

$$\begin{cases} c+d=3641 \\ c-d=89 \end{cases}$$
$$2c=3730$$
$$c=\frac{3730}{2}$$
$$c=1865$$
$$1865+d=3641$$
$$d=3641-1865$$
$$d=1776$$

The Declaration was signed in 1776.
The Civil War ended in 1865.

67. Let L represent the length.
Let W represent the width.

$$\begin{cases} L=2W+40 \\ 2L+2W=1040 \end{cases}$$
$$2(2W+40)+2W=1040$$
$$4W+80+2W=1040$$
$$6W=1040-80$$
$$6W=960$$
$$W=\frac{960}{6}$$
$$W=160$$
$$L=2(160)+40$$
$$L=320+40$$
$$L=360$$

The football field is 160 feet wide and 360 feet long.

69. Let a represent the number of adult tickets.
Let c represent the number of children tickets.

$$\begin{cases} 9a+6.50c=30495 \\ a+c=3800 \end{cases}$$

Multiply the second equation by -9.

$$\begin{cases} 9a + 6.50c = 30495 \\ -9a - 9c = -34200 \end{cases}$$
$$-2.5c = -3705$$
$$c = \frac{-3705}{-2.5}$$
$$c = 1482$$
$$a + 1482 = 3800$$
$$a = 3800 - 1482$$
$$a = 2318$$
2318 adult tickets and 1482 children tickets were sold.

71.

	Distance	Rate	Time
Run	d	7	t
Bike	48-d	15	4-t

$$\begin{cases} d = 7t \\ 48 - d = 15(4 - t) \end{cases}$$
$$48 - 7t = 15(4 - t)$$
$$48 - 7t = 60 - 15t$$
$$-7t + 15t = 60 - 48$$
$$8t = 12$$
$$t = \frac{12}{8}$$
$$t = 1.5$$
Biked:
$$d = 15(4 - t)$$
$$d = 15(4 - 1.5)$$
$$d = 15(2.5)$$
$$d = 37.5$$
Biked for 2.5 hours, 37.5 miles.

73. Let s represent the loan made to the science major.
Let n represent the loan made to the nursing student.
$$\begin{cases} 0.07s + 0.06n = 635 \\ s + n = 10000 \end{cases}$$
$$s = 10000 - n$$
$$0.07(10000 - n) + 0.06n = 635$$
$$700 - 0.07n + 0.06n = 635$$
$$-0.01n = 635 - 700$$
$$-0.01n = -65$$
$$n = \frac{-65}{-0.01}$$
$$n = 6500$$
$$s = 10000 - 65000 = 3500$$
\$6500 was loaned to the nursing student.
\$3500 was loaned to the science major.

75. Answers will vary.

77. Tax Plan A: $0.20I$
Tax Plan B: $0.10I + 5000$
$$0.20I = 0.10I + 5000$$
$$0.20I - 0.10I = 5000$$
$$0.10I = 5000$$
$$I = \frac{5000}{0.10}$$
$$I = 50000$$
At \$50,000, both plans require the same tax.

79. $F(x) = -|x + 3| - 2$

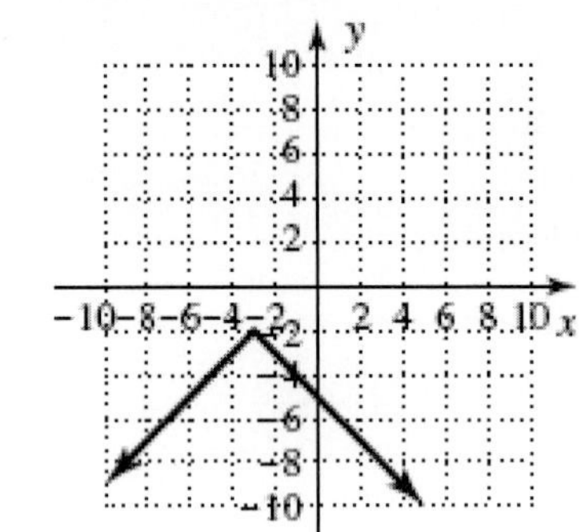

81.
$$33 = 77.5e^{-0.0052x} - 8.37$$
$$33 + 8.37 = 77.5e^{-0.0052x}$$
$$41.37 = 77.5e^{-0.0052x}$$
$$\frac{41.37}{77.5} = e^{-0.0052x}$$
$$0.533806 = e^{-0.0052x}$$
$$\ln 0.533806 = -0.0052x$$
$$\frac{\ln 0.533806}{-0.0052} = x$$
$$120.716 \approx x$$

83. Sum:
$$1 + 3i + 1 - 3i = 2$$
Difference:
$$1 + 3i - (1 - 3i)$$
$$= 1 + 3i - 1 + 3i$$
$$= 6i$$
Product:
$$(1 + 3i)(1 - 3i)$$
$$= 1 - 9i^2$$
$$= 1 + 9$$
$$= 10$$

Quotient:

$\frac{1+3i}{1-3i} \bullet \frac{1+3i}{1+3i}$

$= \frac{1+6i+9i^2}{1-9i^2}$

$= \frac{1+6i-9}{1+9}$

$= \frac{-8+6i}{10}$

$= -\frac{4}{5}+\frac{3}{5}i$

6.2 Technology Highlight

Exercise 1:

(a) Answers will vary.
(b) Answers will vary.
(c) (-4, 8, -5) and (3, -1, 2) are solutions and (6, 2,4) is not a solution.

6.2 Exercises

1. Triple

3. Equivalent systems

5. $2(2)+(-5)+z=4$
   $4+(-5)+z=4$
   $-1+z=4$
   $z=5;$
   Substitute and solve for the remaining variable.

7. $x+2y+z=9$
   Answers will vary.

9. $-x+y+2z=-6$
   Answers will vary.

11. $\begin{cases} x+y-2z=-1 \\ 4x-y+3z=3 \\ 3x+2y-z=4 \end{cases}$ (0, 3, 2)

$x+y-2z=-1$
$0+3-2(2)=-1$
$3-4=-1$
$-1=-1;$

$4x-y+3z=3$
$4(0)-3+3(2)=3$
$0-3+6=3$
$3=3;$

$3x+2y-z=4$
$3(0)+2(3)-2=4$
$0+6-2=4$
$4=4$

Yes

13. $\begin{cases} x-y-2z=-10 \\ x-z=1 \\ z=4 \end{cases}$

$x-z=1$
$x-4=1$
$x=5;$

$x-y-2z=-10$
$5-y-2(4)=-10$
$5-y-8=-10$
$-y-3=-10$
$-y=-7$
$y=7$

(5, 7, 4)

15. $\begin{cases} x+3y+2z=16 \\ -2y+3z=1 \\ 8y-13z=-7 \end{cases}$

4R2 + R3 = Sum

$-8y+12z=4$
$8y-13z=-7$

$-z=-3$
$z=3;$

$-2y+3z=1$
$-2y+3(3)=1$
$-2y+9=1$
$-2y=-8$
$y=4;$

$x+3y+2z=16$
$x+3(4)+2(3)=16$
$x+12+6=16$
$x+18=16$
$x=-2$

(-2, 4, 3)

17. $\begin{cases} 2x-y+4z=-7 \\ x+2y-5z=13 \\ y-4z=9 \end{cases}$

R1 + R3 = Sum

$2x-y+4z=-7$
$y-4z=9$

$2x = 2$

$x = 1$

Substitute $x = 1$ into R2 then new R2 + -2R3 = Sum

$1 + 2y - 5z = 13$

$2y - 5z = 12$

$2y - 5z = 12$

$-2y + 8z = -18$

$3z = -6$

$z = -2;$

$2x - y + 4z = -7$

$2(1) - y + 4(-2) = -7$

$2 - y - 8 = -7$

$-y - 6 = -7$

$-y = -1$

$y = 1$

(1, 1, -2)

19. $\begin{cases} -x + y + 2z = -10 \\ x + y - z = 7 \\ 2x + y + z = 5 \end{cases}$

R1 + R2 = Sum

$-x + y + 2z = -10$

$x + y - z = 7$

$2y + z = -3$

2R1 + R2 = Sum

$-2x + 2y + 4z = -20$

$2x + y + z = 5$

$3y + 5z = -15$

$\begin{cases} 2y + z = -3 \\ 3y + 5z = -15 \end{cases}$

-5R1 + R2 = Sum

$-10y - 5z = 15$

$3y + 5z = -15$

$-7y = 0$

$y = 0;$

$2y + z = -3$

$2(0) + z = -3$

$z = -3;$

$x + y - z = 7$

$x + 0 - (-3) = 7$

$x + 3 = 7$

$x = 4$

(4, 0, -3)

21. $\begin{cases} 3x + y - 2z = 3 \\ x - 2y + 3z = 10 \\ 4x - 8y + 5z = 5 \end{cases}$

-4R2 + R3 = Sum

$-4x + 8y - 12z = -40$

$4x - 8y + 5z = 5$

$-7z = -35$

$z = 5$

Substitute into R1 and R2

$3x + y - 2(5) = 3$

$3x + y - 10 = 3$

$3x + y = 13$

$x - 2y + 3(5) = 10$

$x - 2y + 15 = 10$

$x - 2y = -5$

$\begin{cases} 3x + y = 13 \\ x - 2y = -5 \end{cases}$

2R1 + R2 = Sum

$6x + 2y = 26$

$x - 2y = -5$

$7x = 21$

$x = 3;$

$3x + y - 2z = 3$

$3(3) + y - 2(5) = 3$

$9 + y - 10 = 3$

$y - 1 = 3$

$y = 4$

(3, 4, 5)

23. $\begin{cases} 3x - y + z = 6 \\ 2x + 2y - z = 5 \\ x - 2y + 2z = 7 \end{cases}$

R1 + R2 = Sum

$3x - y + z = 6$

$2x + 2y - z = 5$

$5x + y = 11$

2R2 + R3 = Sum

$4x + 4y - 2z = 10$

$x - 2y + 2z = 7$

$5x + 2y = 17$

$\begin{cases} 5x + y = 11 \\ 5x + 2y = 17 \end{cases}$

-R1 + R2 = Sum

$-5x - y = -11$

$5x + 2y = 17$

$y = 6;$

$$\begin{aligned}5x+y&=11\\5x+6&=11\\5x&=5\\x&=1;\end{aligned}$$

$$\begin{aligned}3x-y+z&=6\\3(1)-6+z&=6\\3-6+z&=6\\-3+z&=6\\z&=9\end{aligned}$$

(1, 6, 9)

25. $\begin{cases}3x+y+2z=3\\x-2y+3z=1\\4x-8y+12z=7\end{cases}$

2R1 + R2 = Sum

$$\begin{array}{r}6x+2y+4z=6\\x-2y+3z=1\\\hline 7x+7z=7\end{array}$$

8R1 + R3 = Sum

$$\begin{array}{r}24x+8y+16z=24\\4x-8y+12z=7\\\hline 28x+28z=31\end{array}$$

$\begin{cases}7x+7z=7\\28x+28z=31\end{cases}$

-4R1 + R2 = Sum

$$\begin{array}{r}-28x-28z=-28\\28x+28z=31\\\hline 0\neq 3\end{array}$$

No solution; inconsistent

27. $\begin{cases}4x+y+3z=3\\x-2y+3z=1\\2x-4y+6z=5\end{cases}$

-2R2 + R3 = Sum

$$\begin{array}{r}-2x+4y-6z=-2\\2x-4y+6z=5\\\hline 0\neq 3\end{array}$$

No Solution; inconsistent

29. $\begin{cases}6x-3y+7z=2\\3x-4y+z=6\end{cases}$

R1 + -2R2 = Sum

$$\begin{array}{r}6x-3y+7z=2\\-6x+8y-2z=-12\\\hline 5y+5z=-10\end{array}$$

$$\begin{aligned}5y+5z&=-10\\5y&=-5z-10\\y&=-z-2;\end{aligned}$$

$$\begin{aligned}3x-4y+z&=6\\3x-4(-z-2)+z&=6\\3x+4z+8+z&=6\\3x+5z&=-2\\3x&=-5z-2\\x&=-\frac{5}{3}z-\frac{2}{3}\end{aligned}$$

$\left(-\frac{5}{3}z-\frac{2}{3},-z-2,z\right)$

31. $\begin{cases}3x-4y+5z=5\\-x+2y-3z=-3\\3x-2y+z=1\end{cases}$

R2 + R3 = Sum

$$\begin{array}{r}-x+2y-3z=-3\\3x-2y+z=1\\\hline 2x-2z=-2\end{array}$$

$$\begin{aligned}2x-2z&=-2\\2x&=2z-2\\x&=z-1;\end{aligned}$$

$$\begin{aligned}-x+2y-3z&=-3\\-(z-1)+2y-3z&=-3\\-z+1+2y-3z&=-3\\2y-4z&=-4\\2y&=4z-4\\y&=2z-2\end{aligned}$$

$(z-1,2z-2,z)$

33. $\begin{cases}x+2y-3z=1\\3x+5y-8z=7\\x+y-2z=5\end{cases}$

R1 + - R3 = Sum

$$\begin{array}{r}x+2y-3z=1\\-x-y+2z=-5\\\hline y-z=-4\end{array}$$

$$\begin{aligned}y-z&=-4\\y&=z-4;\end{aligned}$$

$$\begin{aligned}x+y-2z&=5\\x+z-4-2z&=5\\x-z&=9\\x&=z+9\end{aligned}$$

$(z+9,z-4,z)$

35. $\begin{cases} -0.2x+1.2y-2.4z=-1 \\ 0.5x-3y+6z=2.5 \\ x-6y+12z=5 \end{cases}$

-2R2 + R3 = Sum

$$\begin{array}{r} -1x+6y-12z=-5 \\ x-6y+12z=5 \\ \hline 0=0 \end{array}$$

$\{(x,y,z)|x-6y+12z=5\}$

37. $\begin{cases} x+2y-z=1 \\ x+z=3 \\ 2x-y+z=3 \end{cases}$

R1 + 2R3 = Sum

$$\begin{array}{r} x+2y-z=1 \\ 4x-2y+2z=6 \\ \hline 5x+z=7 \end{array}$$

$\begin{cases} x+z=3 \\ 5x+z=7 \end{cases}$

-R1 + R2 = Sum

$$\begin{array}{r} -x-z=-3 \\ 5x+z=7 \\ \hline 4x=4 \end{array}$$

$x=1;$

$x+z=3$
$1+z=3$
$z=2;$

$x+2y-z=1$
$1+2y-2=1$
$2y-1=1$
$2y=2$
$y=1$

(1, 1, 2)

39. $\begin{cases} 2x-5y-4z=6 \\ x-2.5y-2z=3 \\ -3x+7.5y+6z=-9 \end{cases}$

R1 + -2R2 = Sum

$$\begin{array}{r} 2x-5y-4z=6 \\ -2x+5y+4z=-6 \\ \hline 0=0 \end{array}$$

$\left\{(x,y,z)\middle|x-\frac{5}{2}y-2z=3\right\}$

41. $\begin{cases} 4x-5y-6z=5 \\ 2x-3y+3z=0 \\ x+2y-3z=5 \end{cases}$

R1 + -2R2 = Sum

$$\begin{array}{r} 4x-5y-6z=5 \\ -4x+6y-6z=0 \\ \hline y-12z=5 \end{array}$$

R1 + -4R3 = Sum

$$\begin{array}{r} 4x-5y-6z=5 \\ -4x-8y+12z=-20 \\ \hline -13y+6z=-15 \end{array}$$

$\begin{cases} y-12z=5 \\ -13y+6z=-15 \end{cases}$

R1 + 2R2 = Sum

$$\begin{array}{r} y-12z=5 \\ -26y+12z=-30 \\ \hline -25y=-25 \end{array}$$

$y=1;$

$y-12z=5$
$1-12z=5$
$-12z=4$
$z=-\frac{1}{3};$

$x+2y-3z=5$

$x+2(1)-3\left(-\frac{1}{3}\right)=5$

$x+2+1=5$
$x+3=5$
$x=2$

$\left(2,1,-\frac{1}{3}\right)$

43. $\begin{cases} 2x+3y-5z=4 \\ x+y-2z=3 \\ x+3y-4z=-1 \end{cases}$

R1 + -2R2 = Sum

$$\begin{array}{r} 2x+3y-5z=4 \\ -2x-2y+4z=-6 \\ \hline y-z=-2 \end{array}$$

R1 + -2R3 = Sum

$$\begin{array}{r} 2x+3y-5z=4 \\ -2x-6y+8z=-2 \\ \hline -3y+3z=2 \end{array}$$

$y-z=-2;$

$y-z=-2$
$y=z-2;$

$$x+y-2z=3$$
$$x+z-2-2z=3$$
$$x-z=5$$
$$x=z+5$$
$$(z+5,z-2,z)$$

45. $\begin{cases} \frac{x}{2}+\frac{y}{3}-\frac{z}{2}=2 \\ \frac{2x}{3}-y-z=8 \\ \frac{x}{6}+2y+\frac{3z}{2}=6 \end{cases}$

3R1 + R2 = Sum

$$\frac{3}{2}x+y-\frac{3}{2}z=6$$
$$\frac{2x}{3}-y-z=8$$
$$\overline{\frac{13}{6}x-\frac{5}{2}z=14}$$

2R2 + R3 = Sum

$$\frac{4}{3}x-2y-2z=16$$
$$\frac{x}{6}+2y+\frac{3z}{2}=6$$
$$\overline{\frac{3}{2}x-\frac{1}{2}z=22}$$

$$\begin{cases} \frac{13}{6}x-\frac{5}{2}z=14 \\ \frac{3}{2}x-\frac{1}{2}z=22 \end{cases}$$

R1 + -5R2 = Sum

$$\frac{13}{6}x-\frac{5}{2}z=14$$
$$-\frac{15}{2}x+\frac{5}{2}z=-110$$
$$\overline{-\frac{16}{3}x=-96}$$
$$x=18;$$

$$\frac{3}{2}x-\frac{1}{2}z=22$$
$$\frac{3}{2}(18)-\frac{1}{2}z=22$$
$$27-\frac{1}{2}z=22$$
$$-\frac{1}{2}z=-5$$
$$z=10;$$

$$\frac{2}{3}x-y-z=8$$
$$\frac{2}{3}(18)-y-10=8$$
$$12-y-10=8$$
$$2-y=8$$
$$-y=6$$
$$y=-6$$

(18, -6, 10)

47. $\begin{cases} -A+3B+2C=11 \\ 2B+C=9 \\ B+2C=8 \end{cases}$

R2 + -2R3

$$2B+C=9$$
$$-2B-4C=-16$$
$$\overline{-3C=-7}$$
$$C=\frac{7}{3};$$

$$2B+C=9$$
$$2B+\left(\frac{7}{3}\right)=9$$
$$2B+\frac{7}{3}=9$$
$$2B=\frac{20}{3}$$
$$B=\frac{10}{3};$$

$$-A+3B+2C=11$$
$$-A+3\left(\frac{10}{3}\right)+2\left(\frac{7}{3}\right)=11$$
$$-A+10+\frac{14}{3}=11$$
$$-A+\frac{44}{3}=11$$
$$-A=-\frac{11}{3}$$
$$A=\frac{11}{3}$$

$$\left(\frac{11}{3},\frac{10}{3},\frac{7}{3}\right)$$

49. $\begin{cases} A-2B=5 \\ B+3C=7 \\ 2A-B-C=1 \end{cases}$

-2R1 + R3 = Sum

$$-2A+4B=-10$$
$$\underline{-2A-B-C=1}$$
$$3B-C=-9$$

$$\begin{cases} B+3C=7 \\ 3B-C=-9 \end{cases}$$

-3R1 + R2 = Sum

$$-3B-9C=-21$$
$$\underline{3B-C=-9}$$
$$-10C=-30$$
$$C=3;$$

$$B+3C=7$$
$$B+3(3)=7$$
$$B+9=7$$
$$B=-2;$$

$$A-2B=5$$
$$A-2(-2)=5$$
$$A+4=5$$
$$A=1$$

(1, -2, 3)

51. $\begin{cases} C=3 \\ 2A+3C=10 \\ 3B-4C=-11 \end{cases}$

$$2A+3C=10$$
$$2A+3(3)=10$$
$$2A+9=10$$
$$2A=1$$
$$A=\frac{1}{2};$$

$$3B-4C=-11$$
$$3B-4(3)=-11$$
$$3B-12=-11$$
$$3B=1$$
$$B=\frac{1}{3}$$

$$\left(\frac{1}{2},\frac{1}{3},3\right)$$

53. $\left|\dfrac{Ax+By+Cz-D}{\sqrt{A^2+B^2+C^2}}\right|$

$A=1, B=1, C=1, D=6;$

$x=3, y=4, z=5;$

$$\left|\frac{1(3)+1(4)+1(5)-6}{\sqrt{1^2+1^2+1^2}}\right|=\frac{6}{\sqrt{3}}\approx 3.464 \text{ units}$$

55. Let c represent the gestation period of a camel.

Let e represent the gestation period of an elephant.

Let r represent the gestation period of a rhinoceros.

$$\begin{cases} c+e+r=1520 \\ r=c+58 \\ 2c-162=e \end{cases}$$

$$c+e+r=1520$$
$$c+e+c+58=1520$$
$$2c+e=1462$$

$$\begin{cases} 2c-e=162 \\ 2c+e=1462 \end{cases}$$

R1 + R2 = Sum

$$2c-e=162$$
$$\underline{2c+e=1462}$$
$$4c=1624$$
$$c=406;$$

$$r=c+58$$
$$r=406+58$$
$$r=464;$$

$$2c-162=e$$
$$2(406)-162=e$$
$$812-162=e$$
$$650=e$$

Camel: 406 days

Elephant: 650 days

Rhinoceros: 464 days

57. Let x represent the amount of 20% glucose solution.

Let y represent the amount of 30% glucose solution.

Let z represent the amount of 45% glucose solution.

$$\begin{cases} x+y+z=10 \\ y=2x+1 \\ 0.20x+0.30y+0.45z=10(0.38) \end{cases}$$

$$\begin{cases} x+y+z=10 \\ y=2x+1 \\ 20x+30y+45z=380 \end{cases}$$

-45R1 + R3 = Sum

$$-45x-45y-45z=-450$$
$$\underline{20x+30y+45z=380}$$
$$-25x-15y=-70$$

$$\begin{cases} -2x+y=1 \\ -25x-15y=-70 \end{cases}$$

15R1 + R2 = Sum

$$\begin{array}{r} -30x+15y=15 \\ -25x-15y=-70 \\ \hline -55x=-55 \end{array}$$
$x=1;$

$y=2x+1$
$y=2(1)+1$
$y=2+1$
$y=3;$

$x+y+z=10$
$1+3+z=10$
$4+z=10$
$z=6$

1 liter 20% glucose solution; 3 liters 30% glucose solution; 6 liters 45% glucose solution

59. Let c represent the amount of money invested in a 4% certificate of deposit.
Let m represent the amount of money invested in a 5% money market.
Let b represent the amount of money invested in 7% Aa bonds.

$$\begin{cases} c+m+b=280000 \\ 0.04c+0.05m+0.07b=15400 \\ b=m+20000 \end{cases}$$

$$\begin{cases} c+m+b=280000 \\ 4c+5m+7b=1540000 \\ b=m+20000 \end{cases}$$

-4R1 + R2 = Sum
$$\begin{array}{r} -4c-4m-4b=-1120000 \\ 4c+5m+7b=1540000 \\ \hline m+3b=420000 \end{array}$$

$$\begin{cases} m+3b=420000 \\ -m+b=20000 \end{cases}$$

R1 + R2 = Sum
$$\begin{array}{r} m+3b=420000 \\ -m+b=20000 \\ \hline 4b=440000 \end{array}$$
$b=110000;$

$b=m+20000$
$110000=m+20000$
$90000=m;$

$c+m+b=280000$
$c+90000+110000=280000$
$c+200000=280000$
$c=80000$

$80,000 at 4%
$90,000 at 5%
$110,000 at 7%

61. Answers will vary.

63. $k=-1;\ k=\frac{1}{2}$

$$\begin{cases} x-2y-z=2 \\ x-2y+kz=5 \\ 2x-4y+4z=10 \end{cases}$$

If $k = -1$,
-R1 + R2
$$\begin{cases} -x+2y+z=-2 \\ x-2y+z=5 \end{cases}$$
$0 \neq 3;$

If $k=\frac{1}{2}$,
-2R2 + R3
$$\begin{cases} -2x+4y-z=-10 \\ 2x-4y+4z=10 \end{cases}$$
$3z=0$
$z=0;$

$$\begin{cases} x-2y=0 \\ x-2y=5 \end{cases}$$
-R1+R2
$$\begin{cases} -x+2y=0 \\ x-2y=5 \end{cases}$$
$0 \neq 3$

Inconsistent if $k=-1$ or $\frac{1}{2}$.

If $k = 2$,
$$\begin{cases} x-2y-z=2 \\ x-2y+2z=5 \end{cases}$$
-R1+R2
$$\begin{cases} -x+2y+z=-2 \\ x-2y+2z=5 \end{cases}$$
$3z=3$
$z=1;$

$$\begin{cases} x-2y+2=5 \\ 2x-4y+4=10 \end{cases}$$

$$\begin{cases} x-2y=3 \\ 2x-4y=6 \end{cases}$$
-2R1+R2
$$\begin{cases} -2x+4y=-6 \\ 2x-4y=6 \end{cases}$$
0=0

Dependent if $k = 2$.

65. b. 10 cm
$B = 15, G = 7$
implies $F = 8$;
$F = 8$ implies $I = 1$;
$I = 1$ implies $E = 9$;
$E = 9$ implies $D = 10$

67. $f(x) = 2x^3 - 3$; $f^{-1}(x) = \sqrt[3]{\frac{x+3}{2}}$

$$\begin{aligned}(f \circ f^{-1})(x) &= f(f^{-1}(x)) \\ &= 2(f^{-1}(x))^3 - 3 \\ &= 2\left(\sqrt[3]{\frac{x+3}{2}}\right)^3 - 3 \\ &= 2\left(\frac{x+3}{2}\right) - 3 \\ &= x + 3 - 3 \\ &= x;\end{aligned}$$

$$\begin{aligned}(f^{-1} \circ f)(x) &= f^{-1}(f(x)) \\ &= \sqrt[3]{\frac{f(x)+3}{2}} \\ &= \sqrt[3]{\frac{2x^3 - 3 + 3}{2}} \\ &= \sqrt[3]{\frac{2x^3}{2}} \\ &= \sqrt[3]{x^3} \\ &= x\end{aligned}$$

69. $\log(x+2) + \log(x) = \log(3)$
$\log x(x+2) = \log 3$
$x^2 + 2x - 3 = 0$
$(x+3)(x-1) = 0$
$x = -3$ or $x = 1$
$x = 1$ since $x = -3$ will not check.

71. $D: x \in (-\infty, \infty)$
$R: [-5, \infty)$
Zeroes: $x = -2.5,\ x = -1.5,\ x = 0.5,\ x = 2$
$g(x) > 0: (-\infty, -2.5) \cup (-1.5, 0.5) \cup (2, \infty)$
$g(x) < 0: (-2.5, -1.5) \cup (0.5, 2)$
Local max: (-0.5, 4)
Local min: (-2, -2), (1.5, -5)
Increasing: $x \in (-2, -0.5) \cup (1.5, \infty)$
Decreasing: $y \in (-\infty, -2) \cup (-0.5, 1.5)$

6.3 Technology Highlight:

Exercise 1:

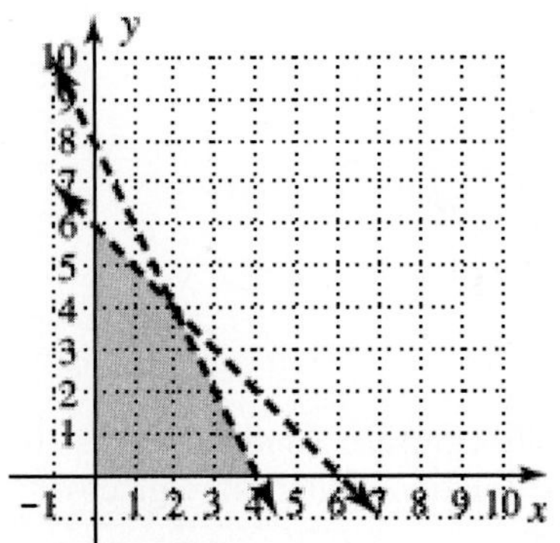

Exercise 2:

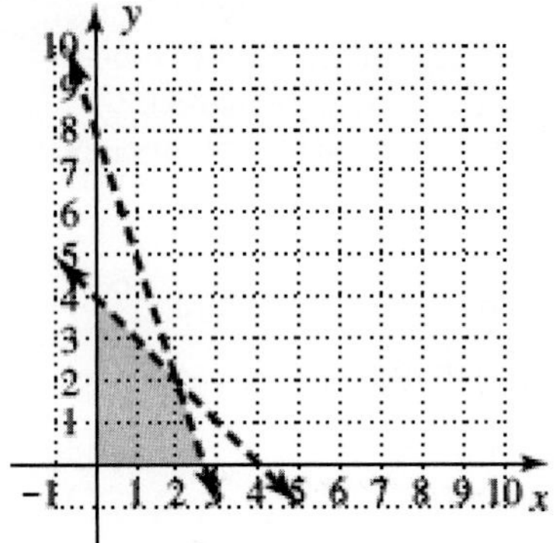

Exercise 3:

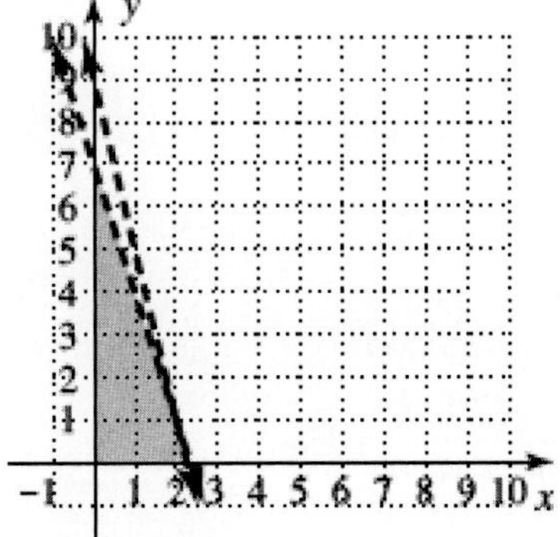

6.3 Exercises

1. Half planes

3. Solution

5. The feasible region may be bordered by three or more oblique lines, with two of them intersecting outside and away from the feasible region.

7. $2x + y > 3$

(0, 0) No

$$\begin{aligned} 2x + y &> 3 \\ 2(0) + 0 &> 3 \\ 0 + 0 &> 3 \\ 0 &> 3; \end{aligned}$$

(3, -5) No

$$\begin{aligned} 2x + y &> 3 \\ 2(3) + (-5) &> 3 \\ 6 - 5 &> 3 \\ 1 &> 3; \end{aligned}$$

(-3, -4) No

$$\begin{aligned} 2x + y &> 3 \\ 2(-3) + (-4) &> 3 \\ -6 - 4 &> 3 \\ -10 &> 3; \end{aligned}$$

(-3, 9) No

$$\begin{aligned} 2x + y &> 3 \\ 2(-3) + 9 &> 3 \\ -6 + 9 &> 3 \\ 3 &> 3 \end{aligned}$$

9. $4x - 2y \le -8$

(0, 0) No

$$\begin{aligned} 4x - 2y &\le -8 \\ 4(0) - 2(0) &\le -8 \\ 0 - 0 &\le -8 \\ 0 &\le -8; \end{aligned}$$

(-3, 5) Yes

$$\begin{aligned} 4x - 2y &\le -8 \\ 4(-3) - 2(5) &\le -8 \\ -12 - 10 &\le -8 \\ -22 &\le -8; \end{aligned}$$

(-3, -2) Yes

$$\begin{aligned} 4x - 2y &\le -8 \\ 4(-3) - 2(-2) &\le -8 \\ -12 + 4 &\le -8 \\ -8 &\le -8; \end{aligned}$$

(-1, 1) No

$$\begin{aligned} 4x - 2y &\le -8 \\ 4(-1) - 2(1) &\le -8 \\ -4 - 2 &\le -8 \\ -6 &\le -8 \end{aligned}$$

11. $x + 2y < 8$

$$\begin{aligned} 2y &< -x + 8 \\ y &< -\frac{1}{2}x + 4 \end{aligned}$$

13. $2x - 3y \ge 9$

$$\begin{aligned} -3y &\ge -2x + 9 \\ y &\le \frac{2}{3}x - 3 \end{aligned}$$

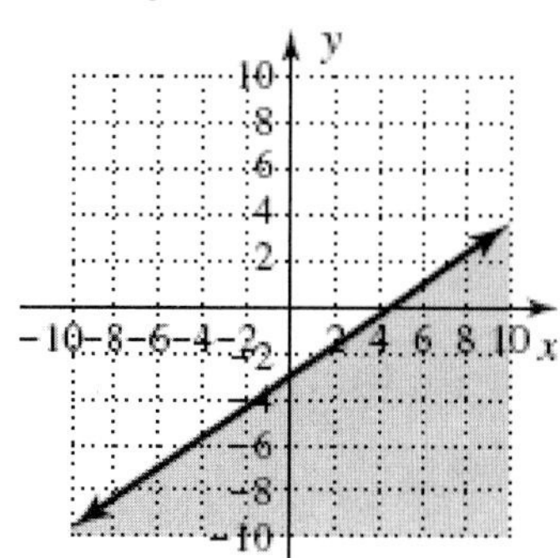

15. $\begin{cases} 5y - x \ge 10 \\ 5y + 2x \le -5 \end{cases}$

(-2, 1) No

$$\begin{aligned} 5y - x &\ge 10 \\ 5(1) - (-2) &\ge 10 \\ 5 + 2 &\ge 10 \\ 7 &\ge 10; \end{aligned}$$

(-5, -4) No

$$\begin{aligned} 5y - x &\ge 10 \\ 5(-4) - (-5) &\ge 10 \\ -20 + 5 &\ge 10 \\ -15 &\ge 10; \end{aligned}$$

(-6, 2) No

$$\begin{aligned} 5y - x &\ge 10 \\ 5(2) - (-6) &\ge 10 \\ 10 + 6 &\ge 10 \\ 16 &\ge 10; \end{aligned}$$

$$\begin{aligned} 5y + 2x &\le -5 \\ 5(2) + 2(-6) &\le -5 \\ 10 - 12 &\le -5 \\ -2 &\le -5; \end{aligned}$$

(-8, 2.2) Yes

$$\begin{aligned} 5y - x &\ge 10 \\ 5(2.2) - (-8) &\ge 10 \\ 11 + 8 &\ge 10 \\ 19 &\ge 10; \end{aligned}$$

$$5y+2x \le -5$$
$$5(2.2)+2(-8) \le -5$$
$$11-16 \le -5$$
$$-5 \le -5$$

17. $\begin{cases} x+2y \ge 1 \\ 2x-y \le -2 \end{cases}$

$\begin{cases} y \ge -\frac{1}{2}x+1 \\ y \ge 2x+2 \end{cases}$

Test Point: (-1, 2)

$$x+2y \ge 1$$
$$-1+2(2) \ge 1$$
$$-1+4 \ge 1$$
$$3 \ge 1;$$

$$2x-y \le -2$$
$$2(-1)-2 \le -2$$
$$-2-2 \le -2$$
$$-4 \le -2$$

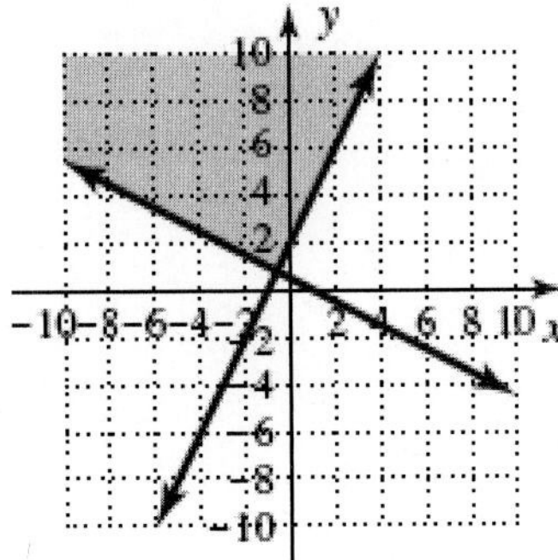

19. $\begin{cases} 3x+y > 4 \\ x > 2y \end{cases}$

$\begin{cases} y > -3x+4 \\ y < \frac{1}{2}x \end{cases}$

Test Point: (3, 0)

$$3x+y > 4$$
$$3(3)+0 > 4$$
$$9 > 4;$$

$$x > 2y$$
$$3 > 2(0)$$
$$3 > 0$$

21. $\begin{cases} 2x+y < 4 \\ 2y > 3x+6 \end{cases}$

$\begin{cases} y < -2x+4 \\ y > \frac{3}{2}x+3 \end{cases}$

Test Point: (-3, 3)

$$2x+y < 4$$
$$2(-3)+3 < 4$$
$$-6+3 < 4$$
$$-3 < 4;$$

$$2y > 3x+6$$
$$2(3) > 3(-3)+6$$
$$6 > -9+6$$
$$6 > -3$$

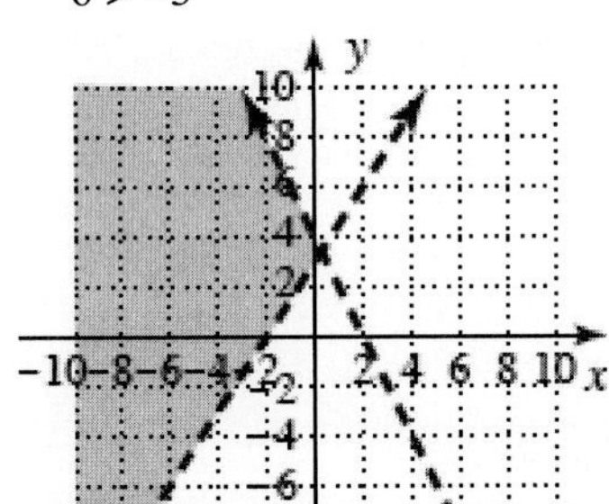

23. $\begin{cases} x > -3y-2 \\ x+3y \le 6 \end{cases}$

$\begin{cases} y > -\frac{1}{3}x-\frac{2}{3} \\ y \le -\frac{1}{3}x+2 \end{cases}$

Test Point: (0, 0)

$$x > -3y-2$$
$$0 > -3(0)-2$$
$$0 > -2;$$

$$x+3y \le 6$$
$$0+3(0) \le 6$$
$$0 \le 6$$

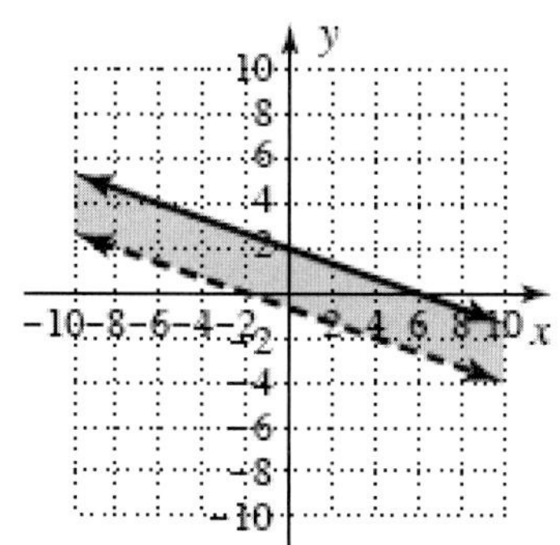

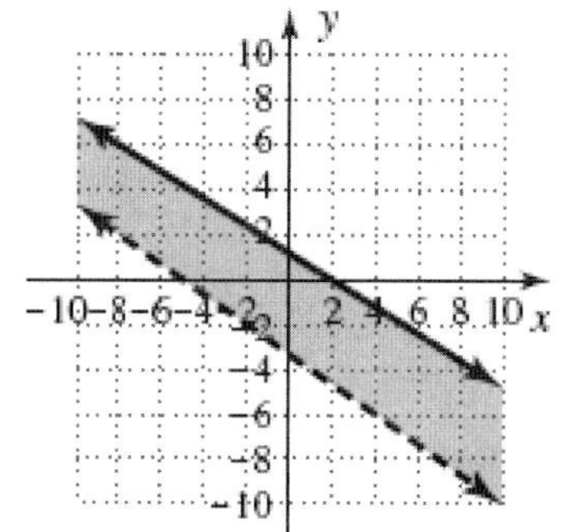

25. $\begin{cases} 5x+4y \ge 20 \\ x-1 \ge y \end{cases}$

$\begin{cases} y \ge -\dfrac{5}{4}x+5 \\ y \le x-1 \end{cases}$

Test Point: (6, 0)

$5x+4y \ge 20$
$5(6)+4(0) \ge 20$
$30 \ge 20;$

$x-1 \ge y$
$6-1 \ge 0$
$5 \ge 0$

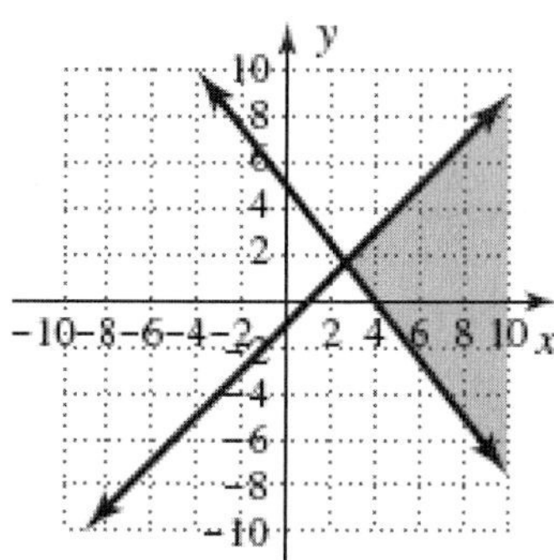

27. $\begin{cases} 0.2x > -0.3y-1 \\ 0.3x+0.5y \le 0.6 \end{cases}$

$\begin{cases} y > -\dfrac{2}{3}x-\dfrac{10}{3} \\ y \le -\dfrac{3}{5}x+\dfrac{6}{5} \end{cases}$

Test Point: (0, 0)

$0.2x > -0.3y-1$
$0.2(0) > -0.3(0)-1$
$0 > -1;$

$0.3x+0.5y \le 0.6$
$0.3(0)+0.5(0) \le 0.6$
$0 \le 0.6$

29. $\begin{cases} y \le \dfrac{3}{2}x \\ 4y \ge 6x-12 \end{cases}$

$\begin{cases} y \le \dfrac{3}{2}x \\ y \ge \dfrac{3}{2}x-3 \end{cases}$

Test Point: (1, 0)

$y \le \dfrac{3}{2}x$
$0 \le \dfrac{3}{2}(1)$
$0 \le \dfrac{3}{2};$

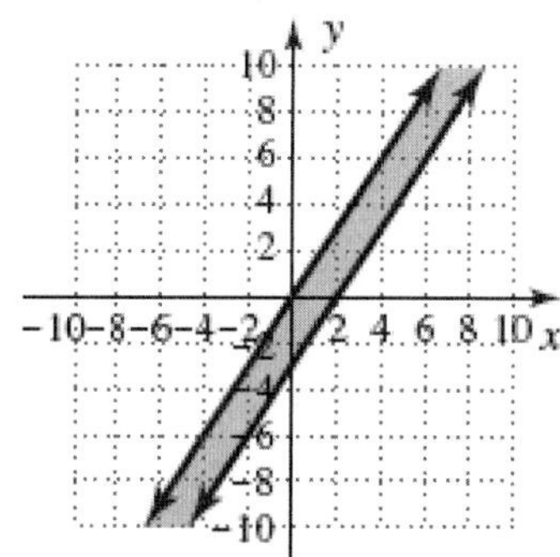

$4y \ge 6x-12$
$4(0) \ge 6(1)-12$
$0 \ge -6$

31. $\begin{cases} \dfrac{-2}{3}x+\dfrac{3}{4}y \le 1 \\ \dfrac{1}{2}x+2y \ge 3 \end{cases}$

$\begin{cases} y \le \dfrac{8}{9}x+\dfrac{4}{3} \\ y \ge -\dfrac{1}{4}x+\dfrac{3}{2} \end{cases}$

Test Point: (6, 4)

$$-\frac{2}{3}x+\frac{3}{4}y \le 1$$
$$-\frac{2}{3}(6)+\frac{3}{4}(4) \le 1$$
$$-4+3 \le 1$$
$$-1 \le 1;$$

$$\frac{1}{2}x+2y \ge 3$$

$$\frac{1}{2}(6)+2(4) \ge 3$$

$$3+8 \ge 3$$

$$11 \ge 3$$

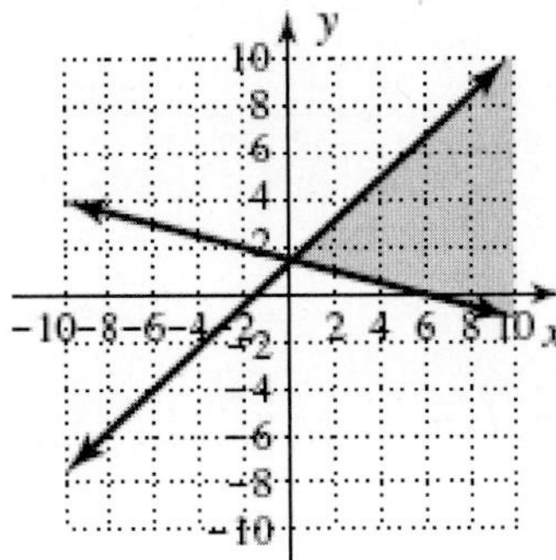

33. $\begin{cases} x-y \ge -4 \\ 2x+y \le 4 \\ x \ge 1 \end{cases}$

$\begin{cases} y \le x+4 \\ y \le -2x+4 \\ x \ge 1 \end{cases}$

Test Point: (1.5, 0.5)

$$x-y \ge -4$$
$$1.5-0.5 \ge -4$$
$$1 \ge -4;$$

$$2x+y \le 4$$
$$2(1.5)+0.5 \le 4$$
$$3+0.5 \le 4$$
$$3.5 \le 4;$$

$$x \ge 1$$

$$1.5 \ge 1$$

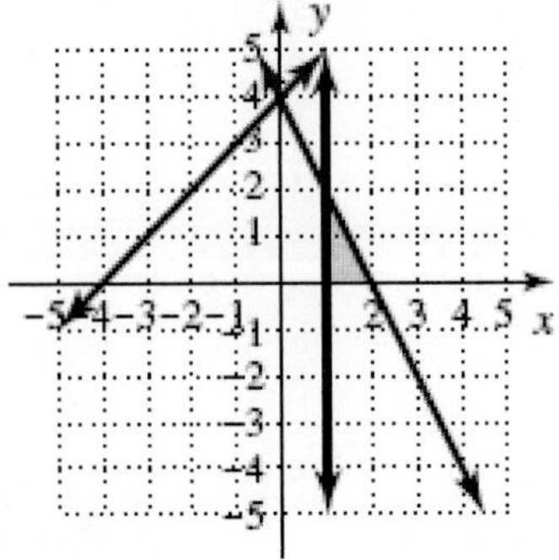

35. $\begin{cases} y \le x+3 \\ x+2y \le 4 \\ y \ge 0 \end{cases}$

$\begin{cases} y \le x+3 \\ y \le -\frac{1}{2}x+2 \\ y \ge 0 \end{cases}$

Test Point: (1, 1)

$$y \le x+3$$
$$1 \le 1+3$$
$$1 \le 4;$$

$$x+2y \le 4$$
$$1+2(1) \le 4$$
$$1+2 \le 4$$
$$3 \le 4;$$

$$y \ge 0$$

$$1 \ge 0$$

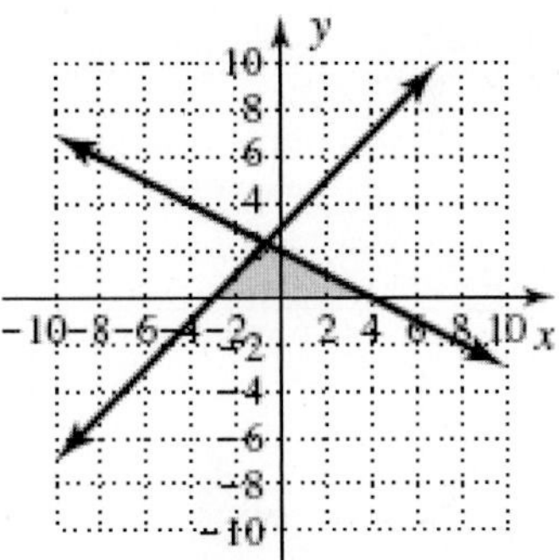

37. $\begin{cases} 2x+3y \le 18 \\ x \ge 0 \\ y \ge 0 \end{cases}$

$\begin{cases} y \le -\frac{2}{3}x+6 \\ x \ge 0 \\ y \ge 0 \end{cases}$

Test Point: (2, 2)

$$2x+3y \le 18$$
$$2(2)+3(2) \le 18$$
$$4+6 \le 18$$
$$10 \le 18;$$

$$x \ge 0$$
$$2 \ge 0;$$

$$y \ge 0$$

$$2 \ge 0$$

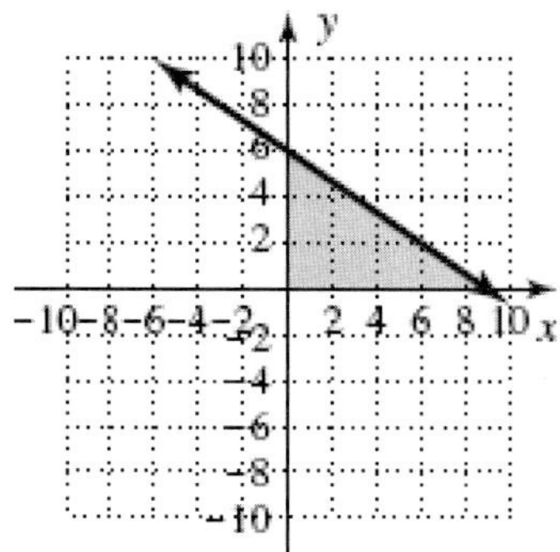

39. $\begin{cases} y - x \le 1 \\ x + y > 3 \end{cases}$

41. $\begin{cases} y - x \le 1 \\ x + y < 3 \\ y \ge 0 \end{cases}$

43.

Point	Objective Function $f(x,y)=12x+10y$	Result
(0, 0)	$f(0,0)=12(0)+10(0)$	0
(0, 8.5)	$f(0,8.5)=12(0)+10(8.5)$	85
(7, 0)	$f(7,0)=12(7)+10(0)$	84
(5, 3)	$f(5,3)=12(5)+10(3)$	90

Maximum value occurs at (5, 3).

45.

Point	Objective Function $f(x,y)=8x+15y$	Result
(0, 20)	$f(0,20)=8(0)+15(20)$	300
(35,0)	$f(35,0)=8(35)+15(0)$	280
(5, 15)	$f(5,15)=8(5)+15(15)$	265
(12,11)	$f(12,11)=8(12)+15(11)$	261

$f(x,y)=8x+15y$
(0, 20)
$f(0,20)=8(0)+15(20)=300$
(35, 0)
$f(35,0)=8(35)+15(0)=280$
(5, 15)
$f(5,15)=8(5)+15(15)=265$
(12, 11)
$f(12,11)=8(12)+15(11)=261$
Minimum value occurs at (12, 11).

47. $\begin{cases} x + 2y \le 6 \\ 3x + y \le 8 \\ x \ge 0 \\ y \ge 0 \end{cases}$

$\begin{cases} y \le -\frac{1}{2}x + 3 \\ y \le -3x + 8 \\ x \ge 0 \\ y \ge 0 \end{cases}$

Corner Point	Objective Function $f(x,y)=8x+5y$	Result
(0, 0)	$f(0,0)=8(0)+5(0)$	0
(0, 3)	$f(0,3)=8(0)+5(3)$	15
$\left(\frac{8}{3},0\right)$	$f\left(\frac{8}{3},0\right)=8\left(\frac{8}{3}\right)+5(0)$	$\frac{64}{3}$
(2, 2)	$f(2,2)=8(2)+5(2)$	26

Maximum value: (2, 2)

49. $\begin{cases} 3x + 2y \ge 18 \\ 3x + 4y \ge 24 \\ x \ge 0 \\ y \ge 0 \end{cases}$

$\begin{cases} y \ge -\frac{3}{2}x + 9 \\ y \ge -\frac{3}{4}x + 6 \\ x \ge 0 \\ y \ge 0 \end{cases}$

Corner Point	Objective Function $f(x,y)=36x+40y$	Result
(0, 9)	$f(0,9)=36(0)+40(9)$	320
(4, 3)	$f(4,3)=36(4)+40(3)$	264
(8, 0)	$f(8,0)=36(8)+40(0)$	288

Minimum value: (4, 3)

51. $\begin{cases} 20H < 200 \\ \frac{1}{2}(20)H > 50 \\ H > 0 \end{cases}$

$20H < 200$

$H < 10;$

$10H > 50$

$H > 5;$

$5 < H < 10$

53. Let J represent the amount of money given to Julius.
Let A represent the amount of money given to Anthony.
$$\begin{cases} J + A \le 50000 \\ J \ge 20000 \\ A \le 25000 \end{cases}$$

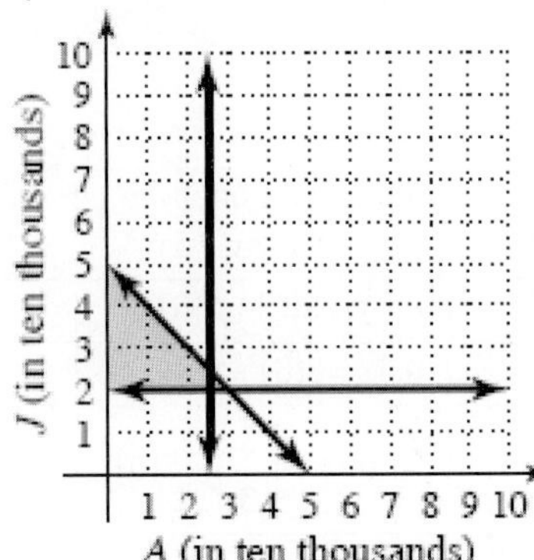

55. Let C represent the number of acres of corn.
Let S represent the number of acres of soybeans.
$$\begin{cases} C + S \le 500 \\ 3C + 2S \le 1300 \end{cases}$$
$P = 900C + 800S$
$$\begin{cases} S \le -C + 500 \\ 2S \le -3C + 1300 \end{cases}$$
$$\begin{cases} S \le -C + 500 \\ S \le \frac{-3}{2}C + 650 \end{cases}$$
Using a grapher, the pts of intersection are:
$(0,500), \left(433\frac{1}{3},0\right), (300,200)$
$P = 900(0) + 800(500) = 400{,}000;$
$P = 900\left(433\frac{1}{3}\right) + 800(0) = 390{,}000;$
$P = 900(300) + 800(200) = 430{,}000;$
300 acres of corn, 200 acres of soybeans

57. Let x represent the number of sheet metal screws.
Let y represent the number of wood screws.
$$\begin{cases} 20x + 5y \le 3(60)(60) \\ 15x + 15y \le 3(60)(60) \\ 5x + 20y \le 3(60)(60) \end{cases}$$
$R = 0.10x + 0.12y$
$$\begin{cases} 5y \le -20x + 10800 \\ 15y \le -15x + 10800 \\ 20y \le -5x + 10800 \end{cases}$$
$$\begin{cases} y \le -4x + 2160 \\ y \le -1x + 720 \\ y \le \frac{-1}{4}x + 540 \end{cases}$$
Using a grapher, the pts of intersection are:
$(0,540), (240,480), (480,240), (540,0)$
$R = 0.10(0) + 0.12(540) = 64.80;$
$R = 0.10(240) + 0.12(480) = 81.60;$
$R = 0.10(480) + 0.12(240) = 76.80;$
$R = 0.10(540) + 0.12(0) = 54;$
240 sheet metal screws; 480 wood screws

59. Let A represent the number of thousand gallons shipped from OK to CO.
Let B represent the number of thousand gallons shipped from OK to MS.
Let C represent the number of thousand gallons shipped from TX to CO.
Let D represent the number of thousand gallons shipped from TX to MS.
$\text{Cost} = 0.05A + 0.075C + 0.06B + 0.065D$
$$\begin{cases} A + C = 220 \Rightarrow C = 220 - A \\ B + D = 250 \Rightarrow D = 250 - B \end{cases}$$
Cost
$= 0.05A + 0.075(220 - A) + 0.06B + 0.065(250 - B)$
$= 0.05A + 16.5 - 0.75A + 0.06B + 16.25 - 0.065B - 0.005B + 32$
$= -0.25A - 0.005B + 32.75;$
$A + B \le 320;$
$C + D \le 240$
but $C = 220 - A$ and $D = 250 - B$
Thus,
$220 - A + 250 - B \le 240$
$-A - B \le -230$
$A + B \ge 230;$
$A \ge 0, B \ge 0, C \ge 0, D \ge 0$
$220 - A \ge 0, 250 - B \ge 0$
$A \le 220, B \le 250;$
$$\begin{cases} A + B \le 320 \\ A + B \ge 230 \\ A \le 220 \\ B \le 250 \end{cases}$$
$$\begin{cases} B \le -A + 320 \\ B \ge -A + 230 \\ A \le 220 \\ B \le 250 \end{cases}$$
Using a grapher, the pts of intersection are:
$(220,100), (220,10), (70,250), (0,250)$
$\text{Cost} = -0.25(220) - 0.005(100) + 32.75 = 26.75$
$\text{Cost} = -0.25(220) - 0.005(10) + 32.75 = 27.2;$
$\text{Cost} = -0.25(70) - 0.005(250) + 32.75 = 29.75;$

Cost $= -0.25(0) - 0.005(250) + 32.75 = 231.50$

$A = 220, B = 100,$
$C = 220 - A = 0, D = 250 - B = 150;$
220 thousand gallons from OK to CO,
100 thousand gallons from OK to MS,
0 thousand gallons from TX to CO,
150 thousand gallons from TX to MS

61. Yes, answers will vary.

63. $\begin{cases} x \geq 0 \\ y \geq 0 \\ y \leq 3 \\ x \leq 3 \end{cases}$

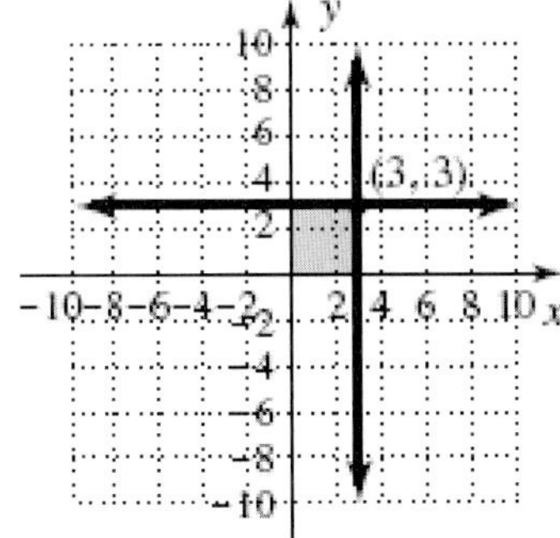

The graph is a rectangle.

Corner Point	Objective Function $f(x, y) = 4.5x + 7.2y$	Result
(0, 0)	$f(0,0) = 4.5(0) + 7.2(0)$	0
(0, 3)	$f(0,3) = 4.5(0) + 7.2(3)$	21.6
(3, 3)	$f(3,3) = 4.5(3) + 7.2(3)$	35.1
(3, 0)	$f(3,0) = 4.5(3) + 7.2(0)$	13.5

Maximum value: (3, 3)
Optimal solutions occur at vertices.

65. $\begin{cases} 2x + 5y \leq 24 \\ 3x + 4y \leq 29 \\ x + 6y \leq 26 \\ x \geq 0 \\ y \geq 0 \end{cases}$

$\begin{cases} 5y \leq -2x + 24 \\ 4y \leq -3x + 29 \\ 6y \leq -x + 26 \\ x \geq 0 \\ y \geq 0 \end{cases}$

$\begin{cases} y \leq \frac{-2}{5}x + \frac{24}{5} \\ y \leq \frac{-3}{4}x + \frac{29}{4} \\ y \leq \frac{-1}{6}x + \frac{13}{3} \end{cases}$

$f(x, y) = 22x + 15y;$

$\left(0, \frac{13}{3}\right), \left(\frac{29}{3}, 0\right), (2,4), (7,2)$

$f(x, y) = 22(0) + 15\left(\frac{13}{3}\right) = 65;$

$f(x, y) = 22\left(\frac{29}{3}\right) + 15(0) = 212.67;$

$f(x, y) = 22(2) + 15(4) = 64;$

$f(x, y) = 22(7) + 15(2) = 184;$

Maximum at $\left(\frac{29}{3}, 0\right)$

67. $\frac{x+2}{x^2 - 9} > 0$

$\frac{x+2}{(x+3)(x-3)} > 0$

Zero: (0, -2)
Restricted values: -3; 3
$(-3, -2) \cup (3, \infty)$

69. $r = \frac{kl}{d^2}$

$1500 = \frac{k(8)}{(0.004)^2}$

$0.024 = 8k$

$0.003 = k$

$r = \frac{0.003l}{d^2}$

$r = \frac{0.003(2.7)}{(0.005)^2}$

$r = \frac{0.0081}{0.000025}$

$r = 324\ \Omega$

71. $f(x) = \frac{250}{1 + 52e^{-0.75x}}$

$f(8) = \frac{250}{1 + 52e^{-0.75(8)}}$

$= \frac{250}{1 + 52e^{-6}}$

$\approx 221.46;$

$$f(10)=\frac{250}{1+52e^{-0.75(10)}}$$
$$=\frac{250}{1+52e^{-7.5}}$$
$$\approx 243.01;$$
$$f(12)=\frac{250}{1+52e^{-0.75(12)}}$$
$$=\frac{250}{1+52e^{-9}}$$
$$\approx 248.41$$

6.4 Technology Highlight:

Exercise 1: $-|x+3|+3=-1$
$$-|x+3|=-4$$
$$|x+3|=4$$
$$x+3=4 \quad \text{or} \quad x+3=-4$$
$$x=1 \qquad x=-7$$

Exercise 2: $\left|x+\frac{3}{5}\right|-\frac{7}{10}=\frac{9}{5}$
$$\left|x+\frac{3}{5}\right|=\frac{25}{10}$$
$$\left|x+\frac{3}{5}\right|=\frac{5}{2}$$
$$x+\frac{3}{5}=\frac{5}{2} \quad \text{or} \quad x+\frac{3}{5}=-\frac{5}{2}$$
$$x=\frac{19}{10} \qquad x=-\frac{31}{10}$$
$$x=1.9, x=-3.1$$

Exercise 3: $-2.5|x-2|+8.6\ge -1.4$
$$-2.5|x-2|\ge -10$$
$$|x-2|\le 4$$
$$x-2\le 4 \quad \text{or} \quad x-2\ge -4$$
$$x\le 6 \qquad x\ge -2$$
$$-2\le x\le 6$$

Exercise 4: $3.2|x-2.9|+1.6\le 4.5$
$$3.2|x-2.9|\le 2.9$$
$$|x-2.9|\le 0.90625$$
$$x-2.9\le 0.90625$$
$$x\le 3.80625$$
or
$$x-2.9\ge -0.90625$$
$$x\ge 1.99375$$
$$1.99375\le x\le 3.80625$$

6.4 Exercises

1. Reverse

3. 7; -7

5. No solution; absolute value is never less than a negative number.

7. $5|3m+7|-2.8<13.7$
$$5|3m+7|<16.5$$
$$|3m+7|<3.3$$

9. $-2|p|-3=-4$
$$-2|p|=-1$$
$$|p|=0.5$$

11. $-3\left|\frac{x}{2}+4\right|-1<-4$
$$-3\left|\frac{x}{2}+4\right|<-3$$
$$\left|\frac{x}{2}+4\right|>1$$

13. $-2|x+3|-4=-14$
$$-2|x+3|=-10$$
$$|x+3|=5$$

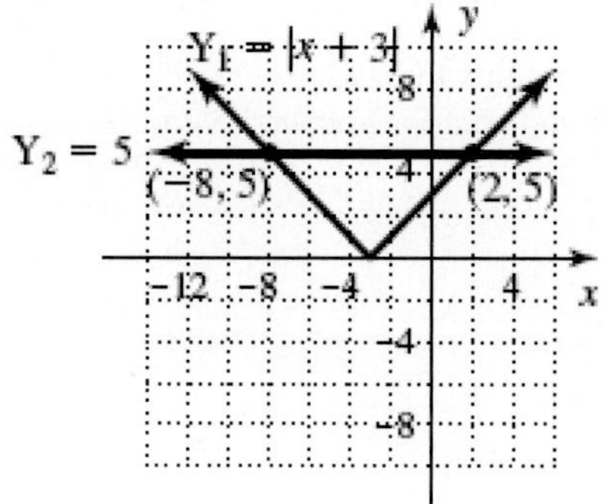

$x=2,\ x=-8$

15. $3|x+5|+6=-3$
$$3|x+5|=-9$$
$$|x+5|=-3$$

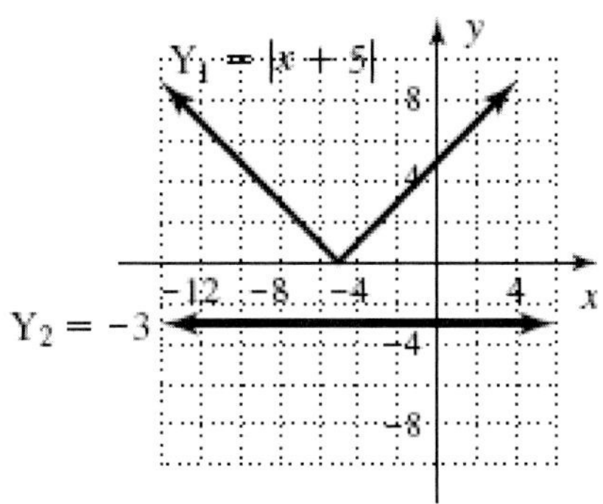

No solution

17. $-2|3x| + 5 = -19$

$-2|3x| = -24$

$|3x| = 12$

$|x| = 4$

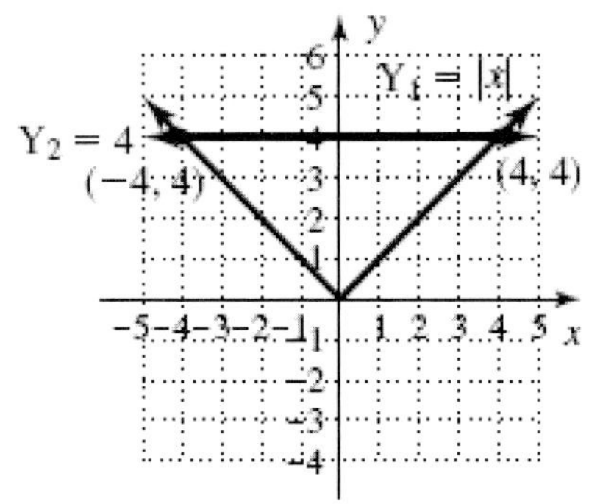

$x = 4,\ \ x = -4$

19. $|x - 2| \le 7$

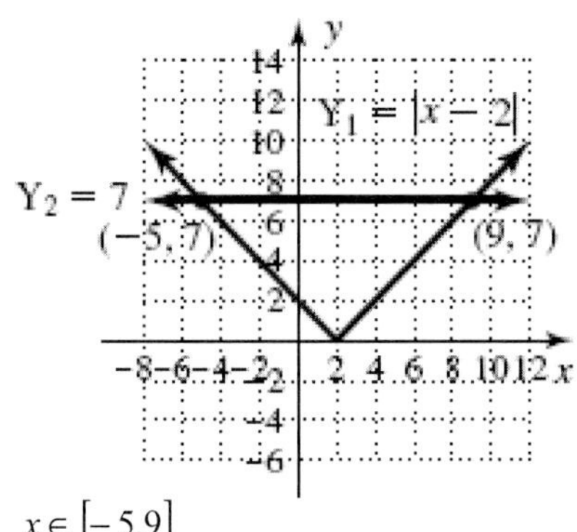

$x \in [-5, 9]$

21. $-3|x| - 2 > 4$

$-3|x| > 6$

$|x| < -2$

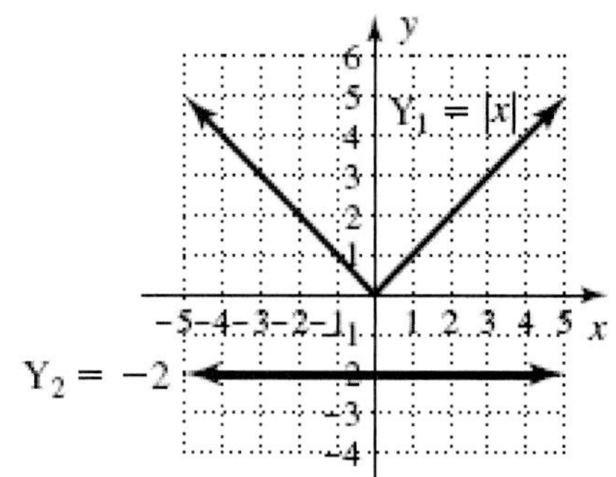

No solution

23. $\dfrac{|2x+5|}{3} + 8 < 9$

$\dfrac{|2x+5|}{3} < 1$

$|2x + 5| < 3$

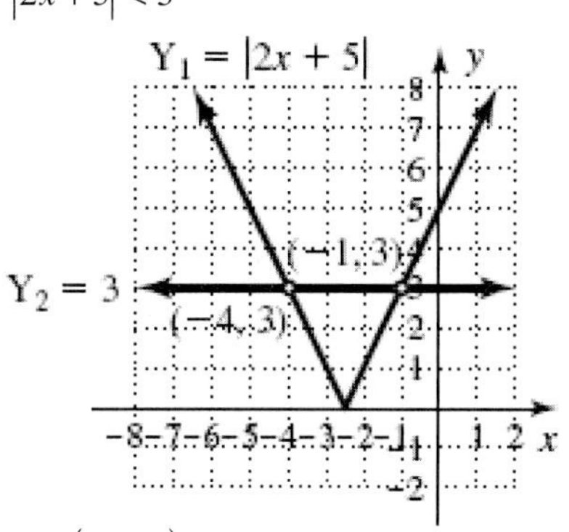

$x \in (-4, -1)$

25. $|x + 3| > 7$

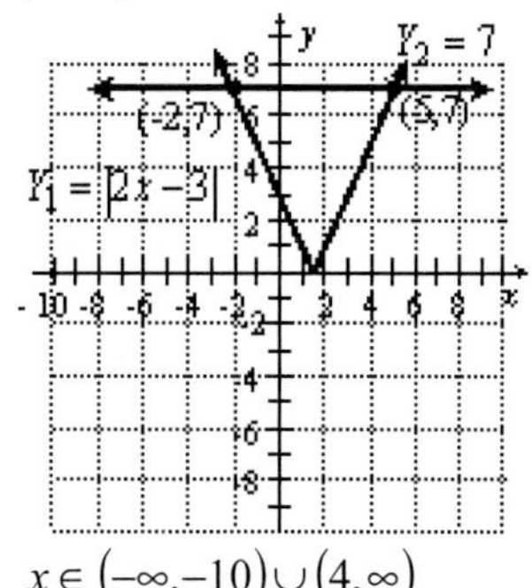

$x \in (-\infty, -10) \cup (4, \infty)$

27. $-2|x| - 5 \le -11$

$-2|x| \le -6$

$|x| \ge 3$

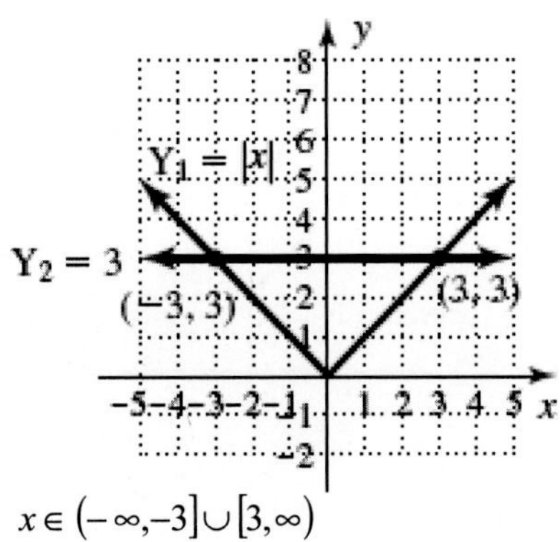

$x \in (-\infty, -3] \cup [3, \infty)$

29. $x \in \left(-\infty, -\frac{8}{3}\right] \cup \left[-\frac{2}{3}, \infty\right)$

$\begin{cases} Y_1 = |3x+5| \\ Y_2 = 3 \end{cases}$

$Y_2 \geq Y_1$

31. $|x+3| + 4 = 10$

$|x+3| = 6$

$x+3 = 6$ or $x+3 = -6$
$x = 3$ or $x = -9$

33. $|2n+7| - 1 = 9$

$|2n+7| = 10$

$2n+7 = 10$ or $2n+7 = -10$
$2n = 3$ or $2n = -17$
$n = \frac{3}{2}$ or $n = -\frac{17}{2}$

35. $-2|n+4| + 3 = -1$

$-2|n+4| = -4$

$|n+4| = 2$

$n+4 = 2$ or $n+4 = -2$
$n = -2$ or $n = -6$

37. $|x^2 - 5x - 10| = 4$

$x^2 - 5x - 10 = 4$ or $x^2 - 5x - 10 = -4$
$x^2 - 5x - 14 = 0$ or $x^2 - 5x - 6 = 0$
$(x-7)(x+2) = 0$ or $(x-6)(x+1) = 0$

$x = 7;\ x = -2;\ x = 6;\ x = -1$

39. $|x^2 + 4x - 33| = 12$

$x^2 + 4x - 33 = 12$ or $x^2 + 4x - 33 = -12$
$x^2 + 4x - 45 = 0$ or $x^2 + 4x - 21 = 0$
$(x+9)(x-5) = 0$ or $(x+7)(x-3) = 0$

$x = -9;\ x = 5;\ x = -7;\ x = 3$

41. $|3x^2 - 2x - 24| = 16$

$3x^2 - 2x - 24 = 16$ or $3x^2 - 2x - 24 = -16$
$3x^2 - 2x - 40 = 0$ or $3x^2 - 2x - 8 = 0$
$(3x+10)(x-4) = 0$ or $(3x+4)(x-2) = 0$

$x = -\frac{10}{3};\ x = 4;\ x = -\frac{4}{3};\ x = 2$

43. $1.2|n-5.8| + 6.7 = 11.5$

$1.2|n-5.8| = 4.8$

$|n-5.8| = 4$

$n - 5.8 = 4$ or $n - 5.8 = -4$
$n = 9.8$ or $n = 1.8$

45. $\left|\frac{2}{3}x + \frac{5}{6}\right| - \frac{7}{12} = \frac{11}{12}$

$\left|\frac{2}{3}x + \frac{5}{6}\right| = \frac{18}{12}$

$\left|\frac{2}{3}x + \frac{5}{6}\right| = \frac{3}{2}$

$\frac{2}{3}x + \frac{5}{6} = \frac{3}{2}$ or $\frac{2}{3}x + \frac{5}{6} = -\frac{3}{2}$
$\frac{2}{3}x = \frac{2}{3}$ or $\frac{2}{3}x = -\frac{7}{3}$
$x = 1$ or $x = -3.5$

47. $5.3|n+9.2| + 6.7 = 43.8$

$5.3|n+9.2| = 37.1$

$|n+9.2| = 7$

$n + 9.2 = 7$ or $n + 9.2 = -7$
$n = -2.2$ or $n = -16.2$

49. $5|m-2| - 7 \leq 8$

$5|m-2| \leq 15$

$|m-2| \leq 3$

$m - 2 \leq 3$ and $m - 2 \geq -3$
$m \leq 5$ and $m \geq -1$
$m \in [-1, 5]$

51. $|2n+3| - 5 < 1$

$|2n+3| < 6$

$2n+3 < 6$ and $2n+3 > -6$
$2n < 3$ and $2n > -9$
$n < \frac{3}{2}$ and $n > -\frac{9}{2}$

$n \in \left(-\frac{9}{2}, \frac{3}{2}\right)$

53. $|2-7x|+7 \le 4$

$|2-7x| \le -3$

No solution

55. $5|2x+7|+1 \ge 11$

$5|2x+7| \ge 10$

$|2x+7| \ge 2$

$2x+7 \ge 2$ or $2x+7 \le -2$

$2x \ge -5$ $\quad$ $2x \le -9$

$x \ge -\frac{5}{2}$ $\quad$ $x \le -\frac{9}{2}$

$x \in \left(-\infty, -\frac{9}{2}\right] \cup \left[-\frac{5}{2}, \infty\right)$

57. $|5p-3|+8 > 6$

$|5p-3| > -2$

$p \in (-\infty, \infty)$

59. $10|3-4x|-8 > -3$

$10|3-4x| > 5$

$|3-4x| > \frac{1}{2}$

$3-4x > \frac{1}{2}$ or $3-4x < -\frac{1}{2}$

$-4x > -\frac{5}{2}$ $\quad$ $-4x < -\frac{7}{2}$

$x < \frac{5}{8}$ $\quad$ $x > \frac{7}{8}$

$x \in \left(-\infty, \frac{5}{8}\right) \cup \left(\frac{7}{8}, \infty\right)$

61. $-3.9|3q-5|+8.7 \le -22.5$

$-3.9|3q-5| \le -31.2$

$|3q-5| \ge 8$

$3q-5 \ge 8$ or $3q-5 \le -8$

$3q \ge 13$ $\quad$ $3q \le -3$

$q \ge \frac{13}{3}$ $\quad$ $q \le -1$

$q \in (-\infty, -1] \cup \left[\frac{13}{3}, \infty\right)$

63. $|x^2-2x| \le 3$

$x^2-2x \le 3$ and $x^2-2x \ge -3$

$x^2-2x-3 \le 0$ $\quad$ $x^2-2x+3 \ge 0$

$(x+1)(x-3) \le 0$ $\quad$ No real solution

$x=-1,\ x=3$

$x \in [-1, 3]$

65. $|x^2-4| \ge 5$

$x^2-4 \ge 5$ or $x^2-4 \le -5$

$x^2-9 \ge 0$ $\quad$ $x^2 \le -1$

$(x-3)(x+3) \ge 0$ $\quad$ No real solution

$x=3,\ x=-3$

pos $\quad$ neg $\quad$ pos

-3 $\quad$ 3

$x \in (-\infty, -3] \cup [3, \infty)$

67. $|x^2-10| \ge 6$

$x^2-10 \ge 6$ or $x^2-10 \le -6$

$x^2-16 \ge 0$ $\quad$ $x^2-4 \le 0$

$(x-4)(x+4) \ge 0$ $\quad$ $(x-2)(x+2) \le 0$

$x=4,\ x=-4$ $\quad$ $x=2,\ x=-2$

pos $\quad$ neg $\quad$ pos $\quad$ neg $\quad$ pos

-4 $\quad$ -2 $\quad$ 2 $\quad$ 4

$x \in (-\infty, -4] \cup [-2, 2] \cup [4, \infty)$

69. $|d-x| \le L$

4 ft = 48 in

$|d-48| \le 3$

$d-48 \le 3$ and $d-48 \ge -3$

$d \le 51$ $\quad$ $d \ge 45$

$45 \le d \le 51$

71. $|B-2450000| \le 125000$

$B-2450000 \le 125000$

$B \le 2575000$

and

$B-2450000 \ge -125000$

$B \ge 2325000$

$[2325000, 2575000]$

73. $|s-53336| \le 11994$

$s - 53336 \le 11994$
$s \le 65330$
and
$s - 53336 \ge -11994$
$s \ge 41342$
Lowest: \$41,342
Highest: \$65,330

75. $2|d-10| < 7$
$|d-10| < 3.5$
$d - 10 < 3.5$ and $d - 10 > -3.5$
$d < 13.5$ $\quad$ $d > 6.5$
$6.5 < d < 13.5$

77. $|w-14| \le 0.1$
$w - 14 \le 0.1$ and $w - 14 \ge -0.1$
$w \le 14.1$ $\quad$ $w \ge 13.9$
$[13.9, 14.1]$; weight must be at least 13.9 oz but no more than 14.1 oz.

79. a. $|x| + x = 8$
$|x| = 8 - x$
$x = 8 - x$
$2x = 8$
$x = 4$

b. $|x-2| \le \frac{x}{2}$

$x - 2 \le \frac{x}{2}$ and $x - 2 \ge -\frac{x}{2}$
$2x - 4 \le x$ $\quad$ $2x - 4 \ge -x$
$x \le 4$ $\quad$ $3x \ge 4$
$x \ge \frac{4}{3}$

$x \in \left[\frac{4}{3}, 4\right]$

c. $x - |x| = x + |x|$
$-|x| - |x| = x - x$
$-|x| - |x| = 0$
$x = 0$

d. $|x+3| \ge 6x$
$x + 3 \ge 6x$ or $x + 3 \le -6x$
$x - 6x \ge -3$ $\quad$ $x + 6x \le -3$
$-5x \ge -3$ $\quad$ $7x \le -3$
$x \le \frac{3}{5}$ $\quad$ $x \le -\frac{3}{7}$

$x \in \left(-\infty, \frac{3}{5}\right]$

81. $f(x) = 2x + 3$, $L = 1$, $\varepsilon = 0.01$ and $|x+1| < \delta$
$|f(x) - L| < \varepsilon$
$|2x + 3 - 1| < 0.01$
$|2x + 2| < 0.01$
$2x + 2 < 0.01$ and $2x + 2 > -0.01$
$2x < -1.99$ $\quad$ $2x > -2.01$
$x < -0.995$ $\quad$ $x > -1.005$
$x \in (-1.005, -0.995)$

83. $f(x) = \begin{cases} 3 & x \le -1 \\ |x-2| & -1 < x < 4 \\ x + 2 & x \ge 4 \end{cases}$

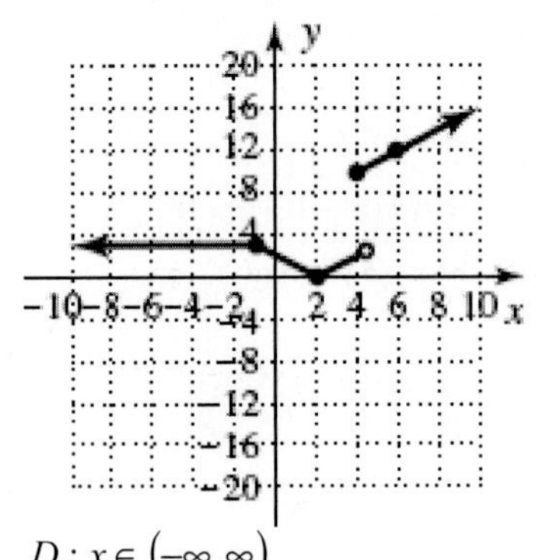

$D: x \in (-\infty, \infty)$
$R: y \in [0,3] \cup [10, \infty)$

85. $f(x) = -150e^{-0.025x+1}$
$-375 = -150e^{-0.025x+1}$
$2.5 = e^{-0.025x+1}$
$\ln 2.5 = \ln e^{-0.025x+1}$
$\ln 2.5 = -0.025x + 1$
$\ln 2.5 - 1 = -0.025x$
$\frac{\ln 2.5 - 1}{-0.025} = x$
$x \approx 3.35$

87. $f(x) = x^3 + 3x^2 - 10x$; $g(x) = x - 2$
a. $(f - g)(x) = f(x) - g(x)$

$$= x^3 + 3x^2 - 10x - (x - 2)$$
$$= x^3 + 3x^2 - 10x - x + 2$$
$$= x^3 + 3x^2 - 11x + 2$$

b. $(f \cdot g)(x) = f(x) \cdot g(x)$
$$= (x^3 + 3x^2 - 10x)(x - 2)$$
$$= x^4 - 2x^3 + 3x^3 - 6x^2 - 10x^2 + 20x$$
$$= x^4 + x^3 - 16x^2 + 20x$$

c. $\left(\frac{f}{g}\right)(x) = \frac{f(x)}{g(x)}$
$$= \frac{x^3 + 3x^2 - 10x}{x - 2}$$
$$= \frac{x(x^2 + 3x - 10)}{x - 2}$$
$$= \frac{x(x - 2)(x + 5)}{x - 2}$$
$$= x(x + 5)$$
$$= x^2 + 5x$$

d. $(f \circ g)(x) = f[g(x)]$
$$= (g(x))^3 + 3(g(x))^2 - 10(g(x))$$
$$= (x - 2)^3 + 3(x - 2)^2 - 10(x - 2)$$
$$= x^3 - 6x^2 + 12x - 8 + 3(x^2 - 4x + 4) - 10x + 20$$
$$= x^3 - 6x^2 + 12x - 8 + 3x^2 - 12x + 12 - 10x + 20$$
$$= x^3 - 3x^2 - 10x + 24$$

Chapter 6 Mid-Chapter Check

1. $\begin{cases} x - 3y = -2 \\ 2x + y = 3 \end{cases}$

$$x = 3y - 2$$
$$2x + y = 3$$
$$2(3y - 2) + y = 3$$
$$6y - 4 + y = 3$$
$$7y = 7$$
$$y = 1$$
$$x - 3y = -2$$
$$x - 3(1) = -2$$
$$x - 3 = -2$$
$$x = 1$$

(1, 1); Consistent

2. $\begin{cases} x - 3y = -4 \\ 2x + y = 13 \end{cases}$

-2R1 + R2 = Sum

$$-2x + 6y = 8$$
$$2x + y = 13$$
$$7y = 21$$
$$y = 3$$
$$x - 3y = -4$$
$$x - 3(3) = -4$$
$$x - 9 = -4$$
$$x = 5$$

(5, 3); Consistent

3. Let x represent the amount of 40% acid.
Let y represent the amount of 48% acid.

$\begin{cases} x + 10 = y \\ 0.40x + 0.64(10) = 0.48y \end{cases}$

$\begin{cases} x + 10 = y \\ 40x + 64(10) = 48y \end{cases}$

$$40x + 640 = 48(x + 10)$$
$$40x + 640 = 48x + 480$$
$$-8x = -160$$
$$x = 20$$

20 ounces

4. $\begin{cases} 5x + 2y - 4z = 22 \\ 2x - 3y + z = -1 \\ 3x - 6y + z = 2 \end{cases}$ (2, 0, -3)

$$5x + 2y - 4z = 22$$
$$5(2) + 2(0) - 4(-3) = 22$$
$$10 + 0 + 12 = 22$$
$$22 = 22;$$
$$2x - 3y + z = -1$$
$$2(2) - 3(0) + (-3) = -1$$
$$4 - 0 - 3 = -1$$
$$1 \neq -1;$$
$$3x - 6y + z = 2$$
$$3(2) - 6(0) + (-3) = 2$$
$$6 - 0 - 3 = 2$$
$$3 \neq 2$$

No

5. $\begin{cases} x + 2y - 3z = 3 \\ 2x + 4y - 6z = 6 \\ x - 2y + 5z = -1 \end{cases}$

The second equation is a multiple of the first equation.

6. $\begin{cases} x + 2y - 3z = -4 \\ 2y + z = 7 \\ 5y - 2z = 4 \end{cases}$

2R2 + R3 = Sum

$$\begin{array}{r} 4y+2z=14 \\ 5y-2z=4 \\ \hline 9y=18 \end{array}$$

$y=2;$

$$\begin{aligned} 2y+z&=7 \\ 2(2)+z&=7 \\ 4+z&=7 \\ z&=3; \end{aligned}$$

$$\begin{aligned} x+2y-3z&=-4 \\ x+2(2)-3(3)&=-4 \\ x+4-9&=-4 \\ x-5&=-4 \\ x&=1 \end{aligned}$$

(1, 2, 3)

7. $\begin{cases} 2x+3y-4z=-4 \\ x-2y+z=0 \\ -3x-2y+2z=-1 \end{cases}$

R1 + 4R2 = Sum

$$\begin{array}{r} 2x+3y-4z=-4 \\ 4x-8y+4z=0 \\ \hline 6x-5y=-4 \end{array}$$

R1 + 2R3 = Sum

$$\begin{array}{r} 2x+3y-4z=-4 \\ -6x-4y+4z=-2 \\ \hline -4x-y=-6 \end{array}$$

$\begin{cases} 6x-5y=-4 \\ -4x-y=-6 \end{cases}$

R1 + -5R2 = Sum

$$\begin{array}{r} 6x-5y=-4 \\ 20x+5y=30 \\ \hline 26x=26 \end{array}$$

$x=1;$

$$\begin{aligned} -4x-y&=-6 \\ -4(1)-y&=-6 \\ -4-y&=-6 \\ -y&=-2 \\ y&=2; \end{aligned}$$

$$\begin{aligned} x-2y+z&=0 \\ 1-2(2)+z&=0 \\ 1-4+z&=0 \\ -3+z&=0 \\ z&=3 \end{aligned}$$

(1, 2, 3)

8. $-2|x-3.5|+29\geq 34$

$$\begin{aligned} -2|x-3.5|&\geq 5 \\ |x-3.5|&\leq -2.5 \end{aligned}$$

No Solution

9. Let x represent Mozart's age.
Let y represent Morphy's age.
Let z represent Pascal's age.

$\begin{cases} x+y+z=37 \\ y=2x-3 \\ z=y+3 \end{cases}$

$\begin{cases} x+y+z=37 \\ -2x+y=-3 \\ -y+z=3 \end{cases}$

-R1 + R2 = Sum

$$\begin{array}{r} -x-y-z=-37 \\ -2x+y=-3 \\ \hline -3x-z=-40 \end{array}$$

R1 + R2 = Sum

$$\begin{array}{r} x+y+z=37 \\ -y+z=3 \\ \hline x+2z=40 \end{array}$$

$\begin{cases} -3x-z=-40 \\ x+2z=40 \end{cases}$

2R1 + R2 = Sum

$$\begin{array}{r} -6x-2z=-80 \\ x+2z=40 \\ \hline -5x=-40 \end{array}$$

$x=8;$

$$\begin{aligned} y&=2x-3 \\ y&=2(8)-3 \\ y&=16-3 \\ y&=13; \end{aligned}$$

$$\begin{aligned} z&=y+3 \\ z&=13+3 \\ z&=16 \end{aligned}$$

Mozart: 8 years
Morphy: 13 years
Pascal: 16 years

10. Let T represent the number of table candles made.
Let H represent the number of holiday candles made.

$\begin{cases} 10T+20H\leq 200 \\ 30T+40H\leq 420 \\ T\geq 0 \\ H\geq 0 \end{cases}$

$$\begin{cases} T \le -2H + 20 \\ T \le -\frac{4}{3}H + 14 \\ T \ge 0 \\ H \ge 0 \end{cases}$$

Corner Point	Objective Function $R(H,T) = 4T + 6H$	Result
(0, 0)	$R(H,T) = 4(0) + 6(0)$	0
(10, 0)	$R(H,T) = 4(0) + 6(10)$	60
(9, 2)	$R(H,T) = 4(2) + 6(9)$	62
(0,14)	$R(H,T) = 4(14) + 6(0)$	56

9 holiday, 2 table

Reinforcing Basic Concepts:

1. A. $\begin{cases} 15.3R + 35.7P = 125.97 \\ P = R + 0.10 \end{cases}$

   Regular: \$2.40 per gallon
   Premium: \$2.50 per gallon

   B. $\begin{cases} 4520x + 7980y = 1056.56 \\ y = x + 0.0345 \end{cases}$

   9.7% on \$7,980
   6.25% on \$4,520

2. A. Let L represent the length.
   Let W represent the width.
   $\begin{cases} 2L + 2W = 39 \\ L = 2W - 6 \end{cases}$
   The sheet of typing paper is 8 ½ inches wide and 11 inches long.

   B. Let J represent John's driving distance.
   Let M represent Helen's driving distance.
   $\begin{cases} J = 2M - 20 \\ J + M = 592 \end{cases}$
   John drove 388 miles and Helen drove 204 miles.

6.5 Technology Highlight

Exercise 1: (10, 12)

Exercise 2: (2, 1, -3)

6.5 Exercises

1. Square

3. 2 by 3, 1

5. Multiply R_1 by –2 and add that result to R_2.

7. $\begin{bmatrix} 1 & 0 \\ 2.1 & 1 \\ -3 & 5.8 \end{bmatrix}$

   $3 \times 2, 1$

9. $\begin{bmatrix} 1 & 0 & 4 & 2 \\ 1 & 3 & -7 & -3 \\ 5 & -1 & 2 & 9 \end{bmatrix}$

   $3 \times 4, 3$

11. $\begin{cases} x + 2y - z = 1 \\ x + z = 3 \\ 2x - y + z = 3 \end{cases}$

    $\left[\begin{array}{ccc|c} 1 & 2 & -1 & 1 \\ 1 & 0 & 1 & 3 \\ 2 & -1 & 1 & 3 \end{array}\right]$

    Diagonal entries 1, 0, 1

13. $\left[\begin{array}{cc|c} 1 & 4 & 5 \\ 0 & 1 & \frac{1}{2} \end{array}\right]$

    $\begin{cases} x + 4y = 5 \\ y = \frac{1}{2} \end{cases}$

    $$x + 4y = 5$$
    $$x + 4\left(\frac{1}{2}\right) = 5$$
    $$x + 2 = 5$$
    $$x = 3$$

    $\left(3, \frac{1}{2}\right)$

15. $\left[\begin{array}{ccc|c} 1 & 2 & -1 & 0 \\ 0 & 1 & 2 & 2 \\ 0 & 0 & 1 & 3 \end{array}\right]$

$$\begin{cases} x+2y-z=0 \\ y+2z=2 \\ z=3 \end{cases}$$

$$\begin{aligned} y+2z&=2 \\ y+2(3)&=2 \\ y+6&=2 \\ y&=-4; \\ x+2y-z&=0 \\ x+2(-4)-(3)&=0 \\ x-8-3&=0 \\ x-11&=0 \\ x&=11 \end{aligned}$$

(11, -4, 3)

17. $\left[\begin{array}{ccc|c} 1 & 3 & -4 & 29 \\ 0 & 1 & -\frac{3}{2} & \frac{21}{2} \\ 0 & 0 & 1 & 3 \end{array}\right]$

$$\begin{cases} x+3y-4z=29 \\ y-\frac{3}{2}z=\frac{21}{2} \\ z=3 \end{cases}$$

$$\begin{aligned} y-\frac{3}{2}z&=\frac{21}{2} \\ y-\frac{3}{2}(3)&=\frac{21}{2} \\ y-\frac{9}{2}&=\frac{21}{2} \\ y&=\frac{30}{2} \\ y&=15; \\ x+3y-4z&=29 \\ x+3(15)-4(3)&=29 \\ x+45-12&=29 \\ x+33&=29 \\ x&=-4 \end{aligned}$$

(-4, 15, 3)

19. $\left[\begin{array}{cc|c} \frac{1}{2} & -3 & -1 \\ -5 & 2 & 4 \end{array}\right]$ $2R1 \to R1$

$\left[\begin{array}{cc|c} 1 & -6 & -2 \\ -5 & 2 & 4 \end{array}\right]$ $5R1+R2 \to R2$

$\left[\begin{array}{cc|c} 1 & -6 & -2 \\ 0 & -28 & -6 \end{array}\right]$

21. $\left[\begin{array}{ccc|c} -2 & 1 & 0 & 4 \\ 5 & 8 & 3 & -5 \\ 1 & -3 & 3 & 2 \end{array}\right]$ $R1 \leftrightarrow R3$

$\left[\begin{array}{ccc|c} 1 & -3 & 3 & 2 \\ 5 & 8 & 3 & -5 \\ -2 & 1 & 0 & 4 \end{array}\right]$ $-5R1+R2 \to R2$

$\left[\begin{array}{ccc|c} 1 & -3 & 3 & 2 \\ 0 & 23 & -12 & -15 \\ -2 & 1 & 0 & 4 \end{array}\right]$

23. $\left[\begin{array}{ccc|c} 3 & 1 & 1 & 8 \\ 6 & -1 & -1 & 10 \\ 4 & -2 & -3 & 22 \end{array}\right]$ $-2R1+R2 \to R2$

$\left[\begin{array}{ccc|c} 3 & 1 & 1 & 8 \\ 0 & -3 & -3 & -6 \\ 4 & -2 & -3 & 34 \end{array}\right]$ $-4R1+3R3 \to R3$

$\left[\begin{array}{ccc|c} 1 & 1 & 1 & 8 \\ 0 & -3 & -3 & -6 \\ 0 & -10 & -13 & 34 \end{array}\right]$

25. $\left[\begin{array}{ccc|c} 1 & 3 & 0 & 2 \\ -2 & 4 & 1 & 1 \\ 3 & -1 & -2 & 9 \end{array}\right]$

$2R1+R2 \to R2$

$-3R1+R3 \to R3$

27. $\left[\begin{array}{ccc|c} 1 & 2 & 0 & 10 \\ 5 & 1 & 2 & 6 \\ -4 & 3 & -3 & 2 \end{array}\right]$

$-5R1+R2 \to R2$

$4R1+R3 \to R3$

29. $\begin{cases} 0.15g-0.35h=-0.5 \\ -0.12g+0.25h=0.1 \end{cases}$

$\begin{cases} 15g-35h=-50 \\ -12g+25h=10 \end{cases}$

$\left[\begin{array}{cc|c} 15 & -35 & -50 \\ -12 & 25 & 10 \end{array}\right]$ $\frac{1}{15}R1 \to R1$

$\left[\begin{array}{cc|c} 1 & -\frac{7}{3} & -\frac{10}{3} \\ -12 & 25 & 10 \end{array}\right]$ $12R1+R2 \to R2$

$$\left[\begin{array}{cc|c} 1 & -\frac{7}{3} & -\frac{10}{3} \\ 0 & -3 & -30 \end{array}\right] \quad -\frac{1}{3}R2 \to R2$$

$$\left[\begin{array}{cc|c} 1 & -\frac{7}{3} & -\frac{10}{3} \\ 0 & 1 & 10 \end{array}\right]$$

$h = 10$;

$$\begin{aligned} 0.15g - 0.35h &= -0.5 \\ 0.15g - 0.35(10) &= -0.5 \\ 0.15g - 3.5 &= -0.5 \\ 0.15g &= 3 \\ g &= 20 \end{aligned}$$

$(20, 10)$

31. $\begin{cases} x - 2y + 2z = 7 \\ 2x + 2y - z = 5 \\ 3x - y + z = 6 \end{cases}$

$$\left[\begin{array}{ccc|c} 1 & -2 & 2 & 7 \\ 2 & 2 & -1 & 5 \\ 3 & -1 & 1 & 6 \end{array}\right] \quad -2R1 + R2 \to R2$$

$$\left[\begin{array}{ccc|c} 1 & -2 & 2 & 7 \\ 0 & 6 & -5 & -9 \\ 3 & -1 & 1 & 6 \end{array}\right] \quad -3R1 + R3 \to R3$$

$$\left[\begin{array}{ccc|c} 1 & -2 & 2 & 7 \\ 0 & 6 & -5 & -9 \\ 0 & 5 & -5 & -15 \end{array}\right] \quad -\frac{5}{6}R2 + R3 \to R3$$

$$\left[\begin{array}{ccc|c} 1 & -2 & 2 & 7 \\ 0 & 6 & -5 & -9 \\ 0 & 0 & -\frac{5}{6} & -\frac{15}{2} \end{array}\right] \quad -\frac{6}{5}R3 \to R3$$

$$\left[\begin{array}{ccc|c} 1 & -2 & 2 & 7 \\ 0 & 6 & -5 & -9 \\ 0 & 0 & 1 & 9 \end{array}\right]$$

$z = 9$;

$$\begin{aligned} 6y - 5z &= -9 \\ 6y - 5(9) &= -9 \\ 6y - 45 &= -9 \\ 6y &= 36 \\ y &= 6; \end{aligned}$$

$$\begin{aligned} x - 2y + 2z &= 7 \\ x - 2(6) + 2(9) &= 7 \\ x - 12 + 18 &= 7 \\ x + 6 &= 7 \\ x &= 1 \end{aligned}$$

$(1, 6, 9)$

33. $\begin{cases} x + 2y - z = 1 \\ x + z = 3 \\ 2x - y + z = 3 \end{cases}$

$$\left[\begin{array}{ccc|c} 1 & 2 & -1 & 1 \\ 1 & 0 & 1 & 3 \\ 2 & -1 & 1 & 3 \end{array}\right] \quad -R1 + R2 \to R2$$

$$\left[\begin{array}{ccc|c} 1 & 2 & -1 & 1 \\ 0 & -2 & 2 & 2 \\ 2 & -1 & 1 & 3 \end{array}\right] \quad -2R1 + R3 \to R3$$

$$\left[\begin{array}{ccc|c} 1 & 2 & -1 & 1 \\ 0 & -2 & 2 & 2 \\ 0 & -5 & 3 & 1 \end{array}\right] \quad -\frac{5}{2}R2 + R3 \to R3$$

$$\left[\begin{array}{ccc|c} 1 & 2 & -1 & 1 \\ 0 & -2 & 2 & 2 \\ 0 & 0 & -2 & -4 \end{array}\right] \quad -\frac{1}{2}R3 \to R3$$

$$\left[\begin{array}{ccc|c} 1 & 2 & -1 & 1 \\ 0 & -2 & 2 & 2 \\ 0 & 0 & 1 & 2 \end{array}\right]$$

$z = 2$;

$$\begin{aligned} x + z &= 3 \\ x + 2 &= 3 \\ x &= 1; \end{aligned}$$

$$\begin{aligned} x + 2y - z &= 1 \\ 1 + 2y - 2 &= 1 \\ 2y - 1 &= 1 \\ 2y &= 2 \\ y &= 1 \end{aligned}$$

$(1, 1, 2)$

35. $\begin{cases} -x + y + 2z = 2 \\ x + y - z = 1 \\ 2x + y + z = 4 \end{cases}$

$$\left[\begin{array}{ccc|c} -1 & 1 & 2 & 2 \\ 1 & 1 & -1 & 1 \\ 2 & 1 & 1 & 4 \end{array}\right] \quad R1 \leftrightarrow R2$$

$$\left[\begin{array}{ccc|c} 1 & 1 & -1 & 1 \\ -1 & 1 & 2 & 2 \\ 2 & 1 & 1 & 4 \end{array}\right] \quad R1 + R2 \to R2$$

$$\left[\begin{array}{ccc|c} 1 & 1 & -1 & 1 \\ 0 & 2 & 1 & 3 \\ 2 & 1 & 1 & 4 \end{array}\right] \quad -2R1 + R3 \to R3$$

$$\left[\begin{array}{ccc|c} 1 & 1 & -1 & 1 \\ 0 & 2 & 1 & 3 \\ 0 & -1 & 3 & 2 \end{array}\right] \quad \frac{1}{2}R2 + R3 \to R3$$

$$\left[\begin{array}{ccc|c} 1 & 1 & -1 & 1 \\ 0 & 2 & 1 & 3 \\ 0 & 0 & \frac{7}{2} & \frac{7}{2} \end{array}\right] \quad \frac{2}{7}R3 \to R3$$

$$\left[\begin{array}{ccc|c} 1 & 1 & -1 & 1 \\ 0 & 2 & 1 & 3 \\ 0 & 0 & 1 & 1 \end{array}\right]$$

$z = 1$;
$2y + z = 3$
$2y + 1 = 3$
$2y = 2$
$y = 1$;
$x + y - z = 1$
$x + 1 - 1 = 1$
$x = 1$
$(1, 1, 1)$

37. $\begin{cases} 4x - 8y + 8z = 24 \\ 2x - 6y + 3z = 13 \\ 3x + 4y - z = -11 \end{cases}$

$$\left[\begin{array}{ccc|c} 4 & -8 & 8 & 24 \\ 2 & -6 & 3 & 13 \\ 3 & 4 & -1 & -11 \end{array}\right] \quad \frac{1}{4}R1 \to R1$$

$$\left[\begin{array}{ccc|c} 1 & -2 & 2 & 6 \\ 2 & -6 & 3 & 13 \\ 3 & 4 & -1 & -11 \end{array}\right] \quad -2R1 + R2 \to R2$$

$$\left[\begin{array}{ccc|c} 1 & -2 & 2 & 6 \\ 0 & -2 & -1 & 1 \\ 3 & 4 & -1 & -11 \end{array}\right] \quad -3R1 + R3 \to R3$$

$$\left[\begin{array}{ccc|c} 1 & -2 & 2 & 6 \\ 0 & -2 & -1 & 1 \\ 0 & 10 & -7 & -29 \end{array}\right] \quad 5R2 + R3 \to R3$$

$$\left[\begin{array}{ccc|c} 1 & -2 & 2 & 6 \\ 0 & -2 & -1 & 1 \\ 0 & 0 & -12 & -24 \end{array}\right] \quad -\frac{1}{12}R3 \to R3$$

$$\left[\begin{array}{ccc|c} 1 & -2 & 2 & 6 \\ 0 & -2 & -1 & 1 \\ 0 & 0 & 1 & 2 \end{array}\right]$$

$z = 2$;
$-2y - z = 1$
$-2y - 2 = 1$
$-2y = 3$
$y = -\frac{3}{2}$;
$3x + 4y - z = -11$
$3x + 4\left(-\frac{3}{2}\right) - 2 = -11$
$3x - 6 - 2 = -11$
$3x - 8 = -11$
$3x = -3$
$x = -1$
$\left(-1, \frac{-3}{2}, 2\right)$

39. $\begin{cases} y - 4z = -16 \\ x + 3y + 5z = 20 \\ 3x - 2y + 9z = 36 \end{cases}$

$$\left[\begin{array}{ccc|c} 0 & 1 & -4 & -16 \\ 1 & 3 & 5 & 20 \\ 3 & -2 & 9 & 36 \end{array}\right] \quad R1 \leftrightarrow R2$$

$$\left[\begin{array}{ccc|c} 1 & 3 & 5 & 20 \\ 0 & 1 & -4 & -16 \\ 3 & -2 & 9 & 36 \end{array}\right] \quad -3R1 + R3 \to R3$$

$$\left[\begin{array}{ccc|c} 1 & 3 & 5 & 20 \\ 0 & 1 & -4 & -16 \\ 0 & -11 & -6 & -24 \end{array}\right] \quad 11R2 + R3 \to R3$$

$$\left[\begin{array}{ccc|c} 1 & 3 & 5 & 20 \\ 0 & 1 & -4 & -16 \\ 0 & 0 & -50 & -200 \end{array}\right] \quad -\frac{1}{50}R3 \to R3$$

$$\left[\begin{array}{ccc|c} 1 & 3 & 5 & 20 \\ 0 & 1 & -4 & -16 \\ 0 & 0 & 1 & 4 \end{array}\right]$$

$z = 4$;

$$\begin{aligned} y-4z&=-16\\ y-4(4)&=-16\\ y-16&=-16\\ y&=0;\\ x+3y+5z&=20\\ x+3(0)+5(4)&=20\\ x+0+20&=20\\ x&=0 \end{aligned}$$

$(0,0,4)$

41. $\begin{cases} 3x-4y+2z=-2\\ \frac{3}{2}x-2y+z=-1\\ -6x+8y-4z=4 \end{cases}$

$$\left[\begin{array}{ccc|c} 3 & -4 & 2 & -2\\ \frac{3}{2} & -2 & 1 & -1\\ -6 & 8 & -4 & 4 \end{array}\right] \quad \frac{1}{3}R1 \to R1$$

$$\left[\begin{array}{ccc|c} 1 & -\frac{4}{3} & \frac{2}{3} & -\frac{2}{3}\\ \frac{3}{2} & -2 & 1 & -1\\ -6 & 8 & -4 & 4 \end{array}\right] \quad -\frac{3}{2}R1+R2 \to R2$$

$$\left[\begin{array}{ccc|c} 1 & -\frac{4}{3} & \frac{2}{3} & -\frac{2}{3}\\ 0 & 0 & 0 & 0\\ -6 & 8 & -4 & 4 \end{array}\right] \quad 6R1+R3 \to R3$$

$$\left[\begin{array}{ccc|c} 1 & -\frac{4}{3} & \frac{2}{3} & -\frac{2}{3}\\ 0 & 0 & 0 & 0\\ 0 & 0 & 0 & 0 \end{array}\right]$$

Dependent; $\{(x,y,z)|3x-4y+2z=-2\}$

43. $\begin{cases} 2x-y+3z=1\\ 2y+6z=2\\ x-\frac{1}{2}y+\frac{3}{2}z=5 \end{cases}$

In terms of z:

$$\left[\begin{array}{ccc|c} 2 & -1 & 3 & 1\\ 0 & 2 & 6 & 2\\ 1 & -\frac{1}{2} & \frac{3}{2} & 5 \end{array}\right] \quad R1 \leftrightarrow R3$$

$$\left[\begin{array}{ccc|c} 1 & -\frac{1}{2} & \frac{3}{2} & 5\\ 0 & 2 & 6 & 2\\ 2 & -1 & 3 & 1 \end{array}\right] \quad -2R1+R3 \to R3$$

$$\left[\begin{array}{ccc|c} 1 & -\frac{1}{2} & 3 & 5\\ 0 & 2 & 6 & 2\\ 0 & 0 & 0 & -9 \end{array}\right] \quad \frac{1}{2}R2+R1 \to R1$$

$0 \neq -9$

No solution

45. $\begin{cases} -2x+4y-3z=4\\ 5x-6y+7z=-12\\ x+2y+z=-4 \end{cases}$

In terms of z:

$$\left[\begin{array}{ccc|c} -2 & 4 & -3 & 4\\ 5 & -6 & 7 & -12\\ 1 & 2 & 1 & -4 \end{array}\right] \quad R1 \leftrightarrow R3$$

$$\left[\begin{array}{ccc|c} 1 & 2 & 1 & -4\\ 5 & -6 & 7 & -12\\ -2 & 4 & -3 & 4 \end{array}\right] \quad -5R1+R2 \leftrightarrow R2$$

$$\left[\begin{array}{ccc|c} 1 & 2 & 1 & -4\\ 0 & -16 & 2 & 8\\ -2 & 4 & -3 & 4 \end{array}\right] \quad 2R1+R3 \to R3$$

$$\left[\begin{array}{ccc|c} 1 & 2 & 1 & -4\\ 0 & -16 & 2 & 8\\ 0 & 8 & -1 & -4 \end{array}\right] \quad -\frac{1}{16}R2 \to R2$$

$$\left[\begin{array}{ccc|c} 1 & 2 & 1 & -4\\ 0 & 1 & -\frac{1}{8} & -\frac{1}{2}\\ 0 & 8 & -1 & -4 \end{array}\right] \quad -8R2+R3 \to R3$$

$$\left[\begin{array}{ccc|c} 1 & 2 & 1 & -4\\ 0 & 1 & -\frac{1}{8} & -\frac{1}{2}\\ 0 & 0 & 0 & 0 \end{array}\right] \quad -2R2+R1 \to R1$$

$$\left[\begin{array}{ccc|c} 1 & 0 & \frac{5}{4} & -3\\ 0 & 1 & -\frac{1}{8} & -\frac{1}{2}\\ 0 & 0 & 0 & 0 \end{array}\right]$$

$$x+\frac{5}{4}z=-3$$

$$x=-\frac{5}{4}z-3;$$

$$y-\frac{1}{8}z=-\frac{1}{2}$$
$$y=\frac{1}{8}z-\frac{1}{2}$$

$$\left(-\frac{5}{4}z-3,\frac{1}{8}z-\frac{1}{2},z\right)$$

47.
$$A=\pm\frac{1}{2}(x_1y_2-x_2y_1+x_2y_3-x_3y_2+x_3y_1-x_1y_3)$$
$$(6,-2),(-5,4),(-1,7)$$
$$A=\pm\frac{1}{2}(6(4)-(-5)(-2)+(-5)(7)-(-1)(4)+(-1)(-2)-6(7))$$
$$=\pm\frac{1}{2}(24-10-35+4+2-42)$$
$$=\pm\frac{1}{2}(-57)$$
$$=28.5$$

28.5 units^2

49. Let x represent the Bull's score.
Let y represent the Sun's score.

$$\begin{cases}x-y=1\\x+y=197\end{cases}$$
$$\left[\begin{array}{cc|c}1&-1&1\\1&1&197\end{array}\right]\ -R1+R2\to R2$$
$$\left[\begin{array}{cc|c}1&-1&1\\0&2&196\end{array}\right]\ \frac{1}{2}R2\to R2$$
$$\left[\begin{array}{cc|c}1&-1&1\\0&1&98\end{array}\right]$$
$$y=98;$$
$$x-y=1$$
$$x-98=1$$
$$x=99$$

98 to 99

51. Let x represent Poe's book.
Let y represent Baum's book.
Let z represent Wouk's book.

$$\begin{cases}x+y+z=100000\\x+2z=y\\z=2x\end{cases}$$

$$\begin{cases}x+y+z=100000\\x-y+2z=0\\-2x+z=0\end{cases}$$
$$\left[\begin{array}{ccc|c}1&1&1&100000\\1&-1&2&0\\-2&0&1&0\end{array}\right]\ -R1+R2\to R2$$
$$\left[\begin{array}{ccc|c}1&1&1&100000\\0&-2&1&-100000\\-2&0&1&0\end{array}\right]\ 2R1+R3\to R3$$
$$\left[\begin{array}{ccc|c}1&1&1&100000\\0&-2&1&-100000\\0&2&3&200000\end{array}\right]\ -\frac{1}{2}R2\to R2$$
$$\left[\begin{array}{ccc|c}1&1&1&100000\\0&1&-\frac{1}{2}&50000\\0&2&3&200000\end{array}\right]\ -2R2+R3\to R3$$
$$\left[\begin{array}{ccc|c}1&1&1&100000\\0&1&-\frac{1}{2}&50000\\0&0&4&100000\end{array}\right]\ \frac{1}{4}R3\to R3$$
$$\left[\begin{array}{ccc|c}1&1&1&100000\\0&1&-\frac{1}{2}&50000\\0&0&1&25000\end{array}\right]$$
$$z=25000;$$
$$y-\frac{1}{2}z=50000$$
$$y-\frac{1}{2}(25000)=50000$$
$$y-12500=50000$$
$$y=62500;$$
$$x+y+z=100000$$
$$x+62500+25000=100000$$
$$x+87500=100000$$
$$x=12500$$

Poe: \$12,500
Baum: \$62,500
Wouk: \$25,000

53. Let A represent the measure of angle A.
Let B represent the measure of angle B.
Let C represent the measure of angle C.

$$\begin{cases}A+B+C=180\\A+C=3B\\C=2B+10\end{cases}$$

$$\begin{cases} A+B+C=180 \\ -A+3B-C=0 \\ -2B+C=10 \end{cases}$$

$$\left[\begin{array}{ccc|c} 1 & 1 & 1 & 180 \\ -1 & 3 & -1 & 0 \\ 0 & -2 & 1 & 10 \end{array}\right] R1+R2 \to R2$$

$$\left[\begin{array}{ccc|c} 1 & 1 & 1 & 180 \\ 0 & 4 & 0 & 180 \\ 0 & -2 & 1 & 10 \end{array}\right] \frac{1}{2}R2+R3 \to R3$$

$$\left[\begin{array}{ccc|c} 1 & 1 & 1 & 180 \\ 0 & 4 & 0 & 180 \\ 0 & 0 & 1 & 100 \end{array}\right]$$

$C=100$;

$4B=180$

$B=45$;

$A+B+C=180$

$A+45+100=180$

$A+145=180$

$A=35$

$35^\circ, 45^\circ, 100^\circ$

55. Let x represent the amount of money invested in the 4% savings fund.
Let y represent the amount of money invested in the 7% money market.
Let z represent the amount of money invested in the 8% government bonds.

$$\begin{cases} x+y+z=2.5 \\ 0.04x+0.07y+0.08z=0.178 \\ z=2y+0.3 \end{cases}$$

$$\begin{cases} x+y+z=2.5 \\ 40x+70y+80z=178 \\ -20y+10z=3 \end{cases}$$

$$\left[\begin{array}{ccc|c} 1 & 1 & 1 & 2.5 \\ 40 & 70 & 80 & 178 \\ 0 & -20 & 10 & 3 \end{array}\right] -40R1+R2 \to R2$$

$$\left[\begin{array}{ccc|c} 1 & 1 & 1 & 2.5 \\ 0 & 30 & 40 & 78 \\ 0 & -20 & 10 & 3 \end{array}\right] \frac{1}{30}R2 \to R2$$

$$\left[\begin{array}{ccc|c} 1 & 1 & 1 & 2.5 \\ 0 & 1 & \frac{4}{3} & 2.6 \\ 0 & -20 & 10 & 3 \end{array}\right] 20R2+R3 \to R3$$

$$\left[\begin{array}{ccc|c} 1 & 1 & 1 & 2.5 \\ 0 & 1 & \frac{4}{3} & 2.6 \\ 0 & 0 & \frac{110}{3} & 55 \end{array}\right] \frac{3}{110}R3 \to R3$$

$$\left[\begin{array}{ccc|c} 1 & 1 & 1 & 2.5 \\ 0 & 1 & \frac{4}{3} & 2.6 \\ 0 & 0 & 1 & 1.5 \end{array}\right]$$

$z=1.5$;

$y+\frac{4}{3}z=2.6$

$y+\frac{4}{3}(1.5)=2.6$

$y+2=2.6$

$y=0.6$;

$x+y+z=2.5$

$x+0.6+1.5=2.5$

$x+2.1=2.5$

$x=0.4$

$0.4 million at 4%
$0.6 million at 7%
$1.5 million at 8%

57. Answers will vary

59. $\left[\begin{array}{ccc|c} 1 & a & b & 1 \\ 2b & 2a & 5 & 13 \\ 2a & 7 & 3b & -8 \end{array}\right]$ Solution set: (1, -2, 3)

$$\begin{cases} 1(1)+a(-2)+b(3)=1 \\ 2b(1)+2a(-2)+5(3)=13 \\ 2a(1)+7(-2)+3b(3)=-8 \end{cases}$$

$$\begin{cases} -2a+3b=0 \\ -4a+2b=-2 \\ 2a+9b=6 \end{cases}$$

$$\left[\begin{array}{cc|c} -2 & 3 & 0 \\ -4 & 2 & -2 \\ 2 & 9 & 6 \end{array}\right] -\frac{1}{2}R1 \to R1$$

$$\left[\begin{array}{cc|c} 1 & -\frac{3}{2} & 0 \\ -4 & 2 & -2 \\ 2 & 9 & 6 \end{array}\right] 4R1+R2 \to R2$$

$$\begin{bmatrix} 1 & -\frac{3}{2} & \vdots & 0 \\ 0 & -4 & \vdots & -2 \\ 2 & 9 & \vdots & 6 \end{bmatrix} -2R1 + R3 \to R3$$

$$\begin{bmatrix} 1 & -\frac{3}{2} & \vdots & 0 \\ 0 & -4 & \vdots & -2 \\ 0 & 12 & \vdots & 6 \end{bmatrix} 3R2 + R3 \to R3$$

$$\begin{bmatrix} 1 & -\frac{3}{2} & \vdots & 0 \\ 0 & -4 & \vdots & -2 \\ 0 & 0 & \vdots & 0 \end{bmatrix}$$

$-4b = -2$

$b = \frac{1}{2};$

$a - \frac{3}{2}b = 0$

$a - \frac{3}{2}\left(\frac{1}{2}\right) = 0$

$a - \frac{3}{4} = 0$

$a = \frac{3}{4}$

$a = \frac{3}{4}, b = \frac{1}{2}$

61. $h(x) = x^4 - x^2 - 12$

$x^4 - x^2 - 12 = 0$

$(x^2 - 4)(x^2 + 3) = 0$

$(x-2)(x+2)(x^2+3) = 0$

$x = \pm 2; x = \pm\sqrt{3}i$

63. $p(x) = x^3 - 8$

$y = x^3 - 8$

$x = y^3 - 8$

$x + 8 = y^3$

$\sqrt[3]{x+8} = y;$

$p^{-1} = \sqrt[3]{x+8};$

$(p \circ p^{-1})(x) = \left(\sqrt[3]{x+8}\right)^3 - 8$

$= x + 8 - 8 = x$

$(p^{-1} \circ p)(x) = \sqrt[3]{(x^3 - 8) + 8}$

$= \sqrt[3]{x^3} = x$

$(p \circ p^{-1})(x) = (p^{-1} \circ p)(x)$

65. $M = \dfrac{d\left[\left(1+\frac{r}{n}\right)^{nt} - 1\right]}{\frac{r}{n}}$

$M = \dfrac{250\left[\left(1+\frac{0.046}{12}\right)^{12(18)} - 1\right]}{\frac{0.046}{12}}$

$M = \dfrac{321.2791457}{\left(\frac{0.046}{12}\right)}$

$M = \$83{,}811.95$

6.6 Exercises

1. $a_{ij} = b_{ij}$

3. Scalar

5. Answers will vary.

7. $\begin{bmatrix} 1 & -3 \\ 5 & -7 \end{bmatrix}$

$2\times 2,\ a_{12} = -3,\ a_{21} = 5$

9. $\begin{bmatrix} 2 & -3 & 0.5 \\ 0 & 5 & 6 \end{bmatrix}$

$2\times 3,\ a_{12} = -3,\ a_{23} = 6,\ a_{22} = 5$

11. $\begin{bmatrix} -2 & 1 & -7 \\ 0 & 8 & 1 \\ 5 & -1 & 4 \end{bmatrix}$

$3\times 3,\ a_{12} = 1,\ a_{23} = 1,\ a_{31} = 5$

13. $\begin{bmatrix} \sqrt{1} & \sqrt{4} & \sqrt{8} \\ \sqrt{16} & \sqrt{32} & \sqrt{64} \end{bmatrix} = \begin{bmatrix} 1 & 2 & 2\sqrt{2} \\ 4 & 4\sqrt{2} & 8 \end{bmatrix}$

True.

15. $\begin{bmatrix} -2 & 3 & a \\ 2b & -5 & 4 \\ 0 & -9 & 3c \end{bmatrix} = \begin{bmatrix} c & 3 & -4 \\ 6 & -5 & -a \\ 0 & -3b & -6 \end{bmatrix}$

Conditional, $c = -2, a = -4, b = 3$

17. $A + H$

$= \begin{bmatrix} 2 & 3 \\ 5 & 8 \end{bmatrix} + \begin{bmatrix} 8 & -3 \\ -5 & 2 \end{bmatrix}$

$= \begin{bmatrix} 2+8 & 3+(-3) \\ 5+(-5) & 8+2 \end{bmatrix}$

$= \begin{bmatrix} 10 & 0 \\ 0 & 10 \end{bmatrix}$

19. $F + H$

$= \begin{bmatrix} 6 & -3 & 9 \\ 12 & 0 & -6 \end{bmatrix} + \begin{bmatrix} 8 & -3 \\ -5 & 2 \end{bmatrix}$

Not possible, different order.

21. $3H - 2A$

$= 3\begin{bmatrix} 8 & -3 \\ -5 & 2 \end{bmatrix} - 2\begin{bmatrix} 2 & 3 \\ 5 & 8 \end{bmatrix}$

$= \begin{bmatrix} 24 & -9 \\ -15 & 6 \end{bmatrix} - \begin{bmatrix} 4 & 6 \\ 10 & 16 \end{bmatrix}$

$= \begin{bmatrix} 24-4 & -9-6 \\ -15-10 & 6-16 \end{bmatrix}$

$= \begin{bmatrix} 20 & -15 \\ -25 & -10 \end{bmatrix}$

23. $\frac{1}{2}E - 3D$

$= \frac{1}{2}\begin{bmatrix} 1 & -2 & 0 \\ 0 & -1 & 2 \\ 4 & 3 & -6 \end{bmatrix} - 3\begin{bmatrix} 1 & 0 & 0 \\ 0 & 1 & 0 \\ 0 & 0 & 1 \end{bmatrix}$

$= \begin{bmatrix} \frac{1}{2} & -1 & 0 \\ 0 & -\frac{1}{2} & 1 \\ 2 & \frac{3}{2} & -3 \end{bmatrix} - \begin{bmatrix} 3 & 0 & 0 \\ 0 & 3 & 0 \\ 0 & 0 & 3 \end{bmatrix}$

$= \begin{bmatrix} \frac{1}{2}-3 & -1-0 & 0-0 \\ 0-0 & -\frac{1}{2}-3 & 1-0 \\ 2-0 & \frac{3}{2}-0 & -3-3 \end{bmatrix}$

$= \begin{bmatrix} \frac{-5}{2} & -1 & 0 \\ 0 & \frac{-7}{2} & 1 \\ 2 & \frac{3}{2} & -6 \end{bmatrix}$

25. ED

$= \begin{bmatrix} 1 & -2 & 0 \\ 0 & -1 & 2 \\ 4 & 3 & -6 \end{bmatrix}\begin{bmatrix} 1 & 0 & 0 \\ 0 & 1 & 0 \\ 0 & 0 & 1 \end{bmatrix}$

$= \begin{bmatrix} 1+0+0 & 0+(-2)+0 & 0+0+0 \\ 0+0+0 & 0+(-1)+0 & 0+0+2 \\ 4+0+0 & 0+3+0 & 0+0+(-6) \end{bmatrix}$

$= \begin{bmatrix} 1 & -2 & 0 \\ 0 & -1 & 2 \\ 4 & 3 & -6 \end{bmatrix}$

27. AH

$= \begin{bmatrix} 2 & 3 \\ 5 & 8 \end{bmatrix}\begin{bmatrix} 8 & -3 \\ -5 & 2 \end{bmatrix}$

$= \begin{bmatrix} 2(8)+3(-5) & 2(-3)+3(2) \\ 5(8)+8(-5) & 5(-3)+8(2) \end{bmatrix}$

$= \begin{bmatrix} 16-15 & -6+6 \\ 40-40 & -15+16 \end{bmatrix}$

$= \begin{bmatrix} 1 & 0 \\ 0 & 1 \end{bmatrix}$

29. FD

$= \begin{bmatrix} 6 & -3 & 9 \\ 12 & 0 & -6 \end{bmatrix}\begin{bmatrix} 1 & 0 & 0 \\ 0 & 1 & 0 \\ 0 & 0 & 1 \end{bmatrix}$

$= \begin{bmatrix} 6+0+0 & 0+(-3)+0 & 0+0+9 \\ 12+0+0 & 0+0+0 & 0+0+(-6) \end{bmatrix}$

$= \begin{bmatrix} 6 & -3 & 9 \\ 12 & 0 & -6 \end{bmatrix}$

31. HF

$= \begin{bmatrix} 8 & -3 \\ -5 & 2 \end{bmatrix}\begin{bmatrix} 6 & -3 & 9 \\ 12 & 0 & -6 \end{bmatrix}$

$= \begin{bmatrix} 8(6)+-3(12) & 8(-3)+-3(0) & 8(9)+-3(-6) \\ -5(6)+2(12) & -5(-3)+2(0) & -5(9)+2(-6) \end{bmatrix}$

$= \begin{bmatrix} 48-36 & -24+0 & 72+18 \\ -30+24 & 15+0 & -45-12 \end{bmatrix}$

$$=\begin{bmatrix} 12 & -24 & 90 \\ -6 & 15 & -57 \end{bmatrix}$$

33. H^2

$$=\begin{bmatrix} 8 & -3 \\ -5 & 2 \end{bmatrix}\begin{bmatrix} 8 & -3 \\ -5 & 2 \end{bmatrix}$$

$$=\begin{bmatrix} 8(8)+(-3)(-5) & 8(-3)+(-3)(2) \\ -5(8)+2(-5) & -5(-3)+2(2) \end{bmatrix}$$

$$=\begin{bmatrix} 64+15 & -24-6 \\ -40-10 & 15+4 \end{bmatrix}$$

$$=\begin{bmatrix} 79 & -30 \\ -50 & 19 \end{bmatrix}$$

35. FE

$$=\begin{bmatrix} 6 & -3 & 9 \\ 12 & 0 & -6 \end{bmatrix}\begin{bmatrix} 1 & -2 & 0 \\ 0 & -1 & 2 \\ 4 & 3 & -6 \end{bmatrix}$$

$$=\begin{bmatrix} 6+0+36 & -12+3+27 & 0+-6-54 \\ 12+0-24 & -24+0-18 & 0+0+36 \end{bmatrix}$$

$$=\begin{bmatrix} 42 & 18 & -60 \\ -12 & -42 & 36 \end{bmatrix}$$

37. $C+H$

$$=\begin{bmatrix} \frac{\sqrt{3}}{2} & \frac{\sqrt{3}}{3} \\ \sqrt{3} & 2\sqrt{3} \end{bmatrix}+\begin{bmatrix} -\frac{3}{19} & \frac{4}{57} \\ \frac{1}{19} & \frac{5}{57} \end{bmatrix}$$

$$=\begin{bmatrix} \frac{\sqrt{3}}{2}+\left(-\frac{3}{19}\right) & \frac{\sqrt{3}}{3}+\frac{4}{57} \\ \sqrt{3}+\frac{1}{19} & 2\sqrt{3}+\frac{5}{57} \end{bmatrix}$$

$$\approx\begin{bmatrix} 0.71 & 0.65 \\ 1.78 & 3.55 \end{bmatrix}$$

39. $E+G$

$$=\begin{bmatrix} 1 & -2 & 0 \\ 0 & -1 & 2 \\ 4 & 3 & -6 \end{bmatrix}+\begin{bmatrix} 0 & \frac{3}{4} & \frac{1}{4} \\ -\frac{1}{2} & \frac{3}{8} & \frac{1}{8} \\ -\frac{1}{4} & \frac{11}{16} & \frac{1}{16} \end{bmatrix}$$

$$=\begin{bmatrix} 1+0 & -2+\frac{3}{4} & 0+\frac{1}{4} \\ 0+\left(-\frac{1}{2}\right) & -1+\frac{3}{8} & 2+\frac{1}{8} \\ 4+\left(-\frac{1}{4}\right) & 3+\frac{11}{16} & -6+\frac{1}{16} \end{bmatrix}$$

$$\approx\begin{bmatrix} 1 & -1.25 & 0.25 \\ -0.5 & -0.63 & 2.13 \\ 3.75 & 3.69 & -5.94 \end{bmatrix}$$

41. AH

$$=\begin{bmatrix} -5 & 4 \\ 3 & 9 \end{bmatrix}\begin{bmatrix} -\frac{3}{19} & \frac{4}{57} \\ \frac{1}{19} & \frac{5}{57} \end{bmatrix}$$

$$=\begin{bmatrix} -5\left(-\frac{3}{19}\right)+4\left(\frac{1}{19}\right) & -5\left(\frac{4}{57}\right)+4\left(\frac{5}{57}\right) \\ 3\left(-\frac{3}{19}\right)+9\left(\frac{1}{19}\right) & 3\left(\frac{4}{57}\right)+9\left(\frac{5}{57}\right) \end{bmatrix}$$

$$=\begin{bmatrix} 1 & 0 \\ 0 & 1 \end{bmatrix}$$

43. EG

$$=\begin{bmatrix} 1 & -2 & 0 \\ 0 & -1 & 2 \\ 4 & 3 & -6 \end{bmatrix}\begin{bmatrix} 0 & \frac{3}{4} & \frac{1}{4} \\ -\frac{1}{2} & \frac{3}{8} & \frac{1}{8} \\ -\frac{1}{4} & \frac{11}{16} & \frac{1}{16} \end{bmatrix}$$

$$=\begin{bmatrix} 0+1+0 & \frac{3}{4}-\frac{3}{4}+0 & \frac{1}{4}-\frac{1}{4}+0 \\ 0+\frac{1}{2}-\frac{1}{2} & 0-\frac{3}{8}+\frac{11}{8} & 0-\frac{1}{8}+\frac{1}{8} \\ 0-\frac{3}{2}+\frac{3}{2} & 3+\frac{9}{8}-\frac{33}{8} & 1+\frac{3}{8}-\frac{3}{8} \end{bmatrix}$$

$$=\begin{bmatrix} 1 & 0 & 0 \\ 0 & 1 & 0 \\ 0 & 0 & 1 \end{bmatrix}$$

45. HB

$$=\begin{bmatrix} -\frac{3}{19} & \frac{4}{57} \\ \frac{1}{19} & \frac{5}{57} \end{bmatrix}\begin{bmatrix} 1 & 0 \\ 0 & 1 \end{bmatrix}$$

$$=\begin{bmatrix} \frac{-3}{19} & \frac{4}{57} \\ \frac{1}{19} & \frac{5}{57} \end{bmatrix}$$

47. DG

$$=\begin{bmatrix} 1 & 0 & 0 \\ 0 & 1 & 0 \\ 0 & 0 & 1 \end{bmatrix}\begin{bmatrix} 0 & \frac{3}{4} & \frac{1}{4} \\ -\frac{1}{2} & \frac{3}{8} & \frac{1}{8} \\ -\frac{1}{4} & \frac{11}{16} & \frac{1}{16} \end{bmatrix}$$

$$=\begin{bmatrix} 0 & \frac{3}{4} & \frac{1}{4} \\ \frac{-1}{2} & \frac{3}{8} & \frac{1}{8} \\ \frac{-1}{4} & \frac{11}{16} & \frac{1}{16} \end{bmatrix}$$

49. C^2

$$=\begin{bmatrix} \frac{\sqrt{3}}{2} & \frac{\sqrt{3}}{3} \\ \sqrt{3} & 2\sqrt{3} \end{bmatrix}\begin{bmatrix} \frac{\sqrt{3}}{2} & \frac{\sqrt{3}}{3} \\ \sqrt{3} & 2\sqrt{3} \end{bmatrix}$$

$$=\begin{bmatrix} \frac{3}{4}+1 & \frac{1}{2}+2 \\ \frac{3}{2}+6 & 1+12 \end{bmatrix}$$

$$=\begin{bmatrix} 1.75 & 2.5 \\ 7.5 & 13 \end{bmatrix}$$

51. FG

$$=\begin{bmatrix} -0.52 & 0.002 & 1.032 \\ 1.021 & -1.27 & 0.019 \end{bmatrix}\begin{bmatrix} 0 & \frac{3}{4} & \frac{1}{4} \\ -\frac{1}{2} & \frac{3}{8} & \frac{1}{8} \\ -\frac{1}{4} & \frac{11}{16} & \frac{1}{16} \end{bmatrix}$$

$$\approx\begin{bmatrix} -0.26 & 0.32 & -0.07 \\ 0.63 & 0.30 & 0.10 \end{bmatrix}$$

53. (a) $AB \neq BA$

$$\text{AB}=\begin{bmatrix} -1 & 3 & 5 \\ 2 & 7 & -1 \\ 4 & 0 & 6 \end{bmatrix}\begin{bmatrix} 0.3 & -0.4 & 1.2 \\ -2.5 & 2 & 0.9 \\ 1 & -0.5 & 0.2 \end{bmatrix}$$

$$=\begin{bmatrix} -2.8 & 3.9 & 2.5 \\ -17.9 & 13.7 & 8.6 \\ 7.2 & -4.6 & 6 \end{bmatrix}$$

$$\text{BA}=\begin{bmatrix} 0.3 & -0.4 & 1.2 \\ -2.5 & 2 & 0.9 \\ 1 & -0.5 & 0.2 \end{bmatrix}\begin{bmatrix} -1 & 3 & 5 \\ 2 & 7 & -1 \\ 4 & 0 & 6 \end{bmatrix}$$

$$=\begin{bmatrix} 3.7 & -1.9 & 9.1 \\ 10.1 & 6.5 & -9.1 \\ -1.2 & -0.5 & 6.7 \end{bmatrix}$$

$AB \neq BA$; Verified

(b) $AC \neq CA$

$$\text{AC}=\begin{bmatrix} -1 & 3 & 5 \\ 2 & 7 & -1 \\ 4 & 0 & 6 \end{bmatrix}\begin{bmatrix} 45 & -1 & 3 \\ -6 & 10 & -15 \\ 21 & -28 & 36 \end{bmatrix}$$

$$=\begin{bmatrix} 42 & -109 & 132 \\ 27 & 96 & -135 \\ 306 & -172 & 228 \end{bmatrix}$$

$$\text{CA}=\begin{bmatrix} 45 & -1 & 3 \\ -6 & 10 & -15 \\ 21 & -28 & 36 \end{bmatrix}\begin{bmatrix} -1 & 3 & 5 \\ 2 & 7 & -1 \\ 4 & 0 & 6 \end{bmatrix}$$

$$=\begin{bmatrix} -35 & 128 & 244 \\ -34 & 52 & -130 \\ 67 & -133 & 349 \end{bmatrix}$$

$AC \neq CA$; Verified

(c) $BC \neq CB$

BC

$$=\begin{bmatrix} 0.3 & -0.4 & 1.2 \\ -2.5 & 2 & 0.9 \\ 1 & -0.5 & 0.2 \end{bmatrix}\begin{bmatrix} 45 & -1 & 3 \\ -6 & 10 & -15 \\ 21 & -28 & 36 \end{bmatrix}$$

$$=\begin{bmatrix} 41.1 & -37.9 & 50.1 \\ -105.6 & -2.7 & -5.1 \\ 52.2 & -11.6 & 17.7 \end{bmatrix}$$

CB

$$=\begin{bmatrix} 45 & -1 & 3 \\ -6 & 10 & -15 \\ 21 & -28 & 36 \end{bmatrix}\begin{bmatrix} 0.3 & -0.4 & 1.2 \\ -2.5 & 2 & 0.9 \\ 1 & -0.5 & 0.2 \end{bmatrix}$$

$$=\begin{bmatrix} 19 & -21.5 & 53.7 \\ -41.8 & 29.9 & -1.2 \\ 112.3 & -82.4 & 7.2 \end{bmatrix}$$

$BC \neq CB$; Verified

55. $(B+C)A = BA + CA$

$$(B+C)A=\begin{bmatrix} 45.3 & -1.4 & 4.2 \\ -8.5 & 12 & -14.1 \\ 22 & -28.5 & 36.2 \end{bmatrix}A$$

$$=\begin{bmatrix}-31.3 & 126.1 & 253.1\\ -23.9 & 58.5 & -139.1\\ 65.8 & -133.5 & 355.7\end{bmatrix}$$

$BA+CA$

$$=\begin{bmatrix}3.7 & -1.9 & 9.1\\ 10.1 & 6.5 & -9.1\\ 1.2 & -0.5 & 6.7\end{bmatrix}+\begin{bmatrix}-35 & 128 & 244\\ -34 & 52 & -130\\ 67 & -133 & 349\end{bmatrix}$$

$$=\begin{bmatrix}-31.3 & 26.1 & 253.1\\ -23.9 & 58.5 & -139.1\\ 65.8 & -133.5 & 355.7\end{bmatrix}$$

$(B+C)A=BA+CA$; Verified

57. $\begin{bmatrix}2 & 2\\ 4.35 & 0\end{bmatrix}\cdot\begin{bmatrix}6.374\\ 4.35\end{bmatrix}$

$$=\begin{bmatrix}2(6.374)+2(4.35)\\ 4.35(6.374)+0(4.35)\end{bmatrix}$$

$$=\begin{bmatrix}21.448\\ 27.7269\end{bmatrix}$$

$P=2l+2w$

$P=2(6.374)+2(4.35)$

$P=12.748+8.7$

$P=21.448;$

$A=lw$

$A=6.374(4.35)$

$A=27.7269$

$P=21.448\text{ cm},\quad A=27.7269\text{ cm}^2$

59. a. $V\rightarrow\begin{array}{c|cc} & T & S\\ \hline S & 3820 & 1960\\ D & 2460 & 1240\\ P & 1540 & 920\end{array}$

$M\rightarrow\begin{array}{c|cc} & T & S\\ \hline S & 4220 & 2960\\ D & 2960 & 3240\\ P & 1640 & 820\end{array}$

b. $\begin{bmatrix}4220 & 2960\\ 2960 & 3240\\ 1640 & 820\end{bmatrix}-\begin{bmatrix}3820 & 1960\\ 2460 & 1240\\ 1540 & 920\end{bmatrix}$

$$=\begin{bmatrix}4220-3820 & 2960-1960\\ 2960-2460 & 3240-1240\\ 1640-1540 & 820-920\end{bmatrix}$$

$$=\begin{bmatrix}400 & 1000\\ 500 & 2000\\ 100 & -100\end{bmatrix}$$

$400+1000+500+2000+100-100$

$=3900$

3,900 more by Minsk

c. $V\rightarrow 1.04\begin{bmatrix}3820 & 1960\\ 2460 & 1240\\ 1540 & 920\end{bmatrix}$

$$=\begin{bmatrix}3972.8 & 2038.4\\ 2558.4 & 1289.6\\ 1601.6 & 956.8\end{bmatrix};$$

$M\rightarrow 1.04\begin{bmatrix}4220 & 2960\\ 2960 & 3240\\ 1640 & 820\end{bmatrix}$

$$=\begin{bmatrix}4388.8 & 3078.4\\ 3078.4 & 3369.6\\ 1705.6 & 852.8\end{bmatrix}$$

d. $\begin{bmatrix}3972.8 & 2038.4\\ 2558.4 & 1289.6\\ 1601.6 & 956.8\end{bmatrix}+\begin{bmatrix}4388.8 & 3078.4\\ 3078.4 & 3369.6\\ 1705.6 & 852.8\end{bmatrix}$

$$=\begin{bmatrix}8361.6 & 5116.8\\ 5636.8 & 4659.2\\ 3307.2 & 1809.6\end{bmatrix}$$

61. $\begin{bmatrix}1500 & 500 & 2500\end{bmatrix}\begin{bmatrix}9 & 6 & 5 & 4\\ 7 & 5 & 7 & 6\\ 2 & 3 & 5 & 2\end{bmatrix}$

$=\begin{bmatrix}22000 & 19000 & 23500 & 14000\end{bmatrix}$

Total profit for north, \$22,000.
Total profit for south, \$19,000.
Total profit for east, \$23,500.
Total profit for west, \$14,000.

63. a. $10(8)+8(1.5)+18(0.9)=\$108.20$

b. $8(7.5)+12(1.75)+20(1)=\$101.00$

c. $\begin{bmatrix}8 & 12 & 20\\ 10 & 8 & 18\end{bmatrix}\begin{bmatrix}8 & 7.5 & 10\\ 1.5 & 1.75 & 2\\ 0.9 & 1 & 0.75\end{bmatrix}$

$$=\begin{array}{r}\text{Science}\\ \text{Math}\end{array}\begin{bmatrix}100 & 101 & 119\\ 108.2 & 107 & 129.5\end{bmatrix}$$

1st row, total cost for Science from each restaurant.

2nd row, Total cost for Math from each restaurant.

65. $\begin{bmatrix} 25 & 18 & 21 \\ 22 & 19 & 18 \end{bmatrix} \begin{bmatrix} 0.6 & 0.1 & 0.3 \\ 0.5 & 0.2 & 0.3 \\ 0.4 & 0.2 & 0.4 \end{bmatrix}$

$= \begin{bmatrix} 32.4 & 10.3 & 21.3 \\ 29.9 & 9.6 & 19.5 \end{bmatrix}$

a. Approximately 10 females.
b. Approximately 19 males.
c. The number of females in the writing club.

67. $(A+B)^2 = A^2 + 2AB + B^2$
No; A & B must be square matrices.

69. 1st, 3rd rows double, 2nd row doubles +1 in a_{21}, a_{23} positions, and a_{22} stays a 1.

$\begin{bmatrix} 2^{n-1} & 0 & 2^{n-1} \\ 2^n - 1 & 1 & 2^n - 1 \\ 2^{n-1} & 0 & 2^{n-1} \end{bmatrix}$

71. $\begin{bmatrix} 2 & 1 \\ -3 & -2 \end{bmatrix} \cdot \begin{bmatrix} a & b \\ c & d \end{bmatrix} = \begin{bmatrix} 1 & 0 \\ 0 & 1 \end{bmatrix}$

$\begin{cases} 2a + c = 1 \\ -3a - 2c = 0 \end{cases}$

2R1 + R2 = Sum

$4a + 2c = 2$

$-3a - 2c = 0$

$a = 2$

$2a + c = 1$

$2(2) + c = 1$

$4 + c = 1$

$c = -3$

$\begin{cases} 2b + d = 0 \\ -3b - 2d = 1 \end{cases}$

2R1 + R2 = Sum

$4b + 2d = 0$

$-3b - 2d = 1$

$b = 1$

$2b + d = 0$

$2(1) + d = 0$

$2 + d = 0$

$d = -2$

$a = 2, b = 1, c = -3, d = -2$

73. $f(x) \geq 0$

$x \in [-3, -1] \cup [1, 3]$

75. $\dfrac{x^3 - 9x + 10}{x - 2}$

$\begin{array}{r|rrrr} 2 & 1 & 0 & -9 & 10 \\ & & 2 & 4 & -10 \\ \hline & 1 & 2 & -5 & 0 \end{array}$

$= x^2 + 2x - 5$

77. $h(x) = \dfrac{1}{x - 2} + 1$

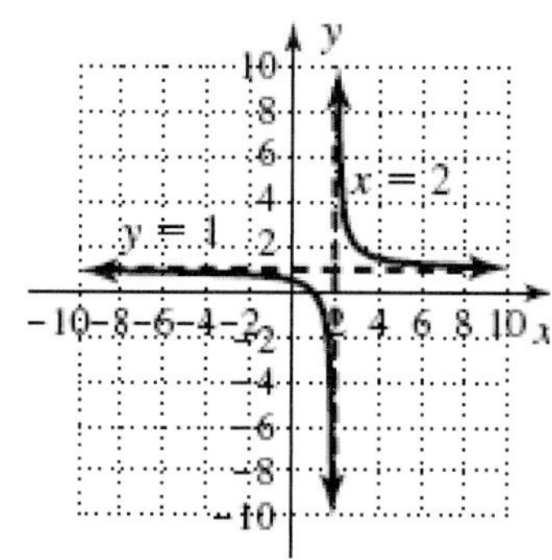

6.7 Exercises

1. Main diagonal; zeroes

3. Identity.

5. Answers will vary.

7. $A = \begin{bmatrix} 2 & 5 \\ -3 & -7 \end{bmatrix} \cdot \begin{bmatrix} a & b \\ c & d \end{bmatrix} = \begin{bmatrix} 2 & 5 \\ -3 & -7 \end{bmatrix}$

$\begin{bmatrix} 2a + 5c & 2b + 5d \\ -3a - 7c & -3b - 7d \end{bmatrix} = \begin{bmatrix} 2 & 5 \\ -3 & -7 \end{bmatrix}$

$\begin{cases} 2a + 5c = 2 \\ -3a - 7c = -3 \end{cases}$

3R1 + 2R2

$\begin{cases} 6a + 15c = 6 \\ -6a - 14c = -6 \end{cases}$

$c = 0$;

$2a + 5c = 2$

$2a + 5(0) = 2$

$2a = 2$

$a = 1$

$\begin{cases} 2b + 5d = 5 \\ -3b - 7d = -7 \end{cases}$

3R1 + 2R2

$\begin{cases} 6b + 15d = 15 \\ -6b - 14d = -14 \end{cases}$

$d = 1$;

$2b + 5d = 5$

$2b + 5(1) = 5$

$2b + 5 = 5$

$2b = 0$

$b = 0$

$$\begin{bmatrix} a & b \\ c & d \end{bmatrix} = \begin{bmatrix} 1 & 0 \\ 0 & 1 \end{bmatrix}$$

9. $A = \begin{bmatrix} 0.4 & 0.6 \\ 0.3 & 0.2 \end{bmatrix} \cdot \begin{bmatrix} a & b \\ c & d \end{bmatrix} = \begin{bmatrix} 0.4 & 0.6 \\ 0.3 & 0.2 \end{bmatrix}$

$$\begin{bmatrix} 0.4a+0.6c & 0.4b+0.6d \\ 0.3a+0.2c & 0.3b+0.2d \end{bmatrix} = \begin{bmatrix} 0.4 & 0.6 \\ 0.3 & 0.2 \end{bmatrix}$$

$$\begin{cases} 0.4a+0.6c = 0.4 \\ 0.3a+0.2c = 0.3 \end{cases}$$

30R1 + -40R2

$$\begin{cases} 12a+18c = 12 \\ -12a-8c = -12 \end{cases}$$

$$10c = 0$$
$$c = 0;$$

$$0.4a+0.6c = 0.4$$
$$0.4a+0.6(0) = 0.4$$
$$0.4a = 0.4$$
$$a = 1;$$

$$\begin{cases} 0.4b+0.6d = 0.6 \\ 0.3b+0.2d = 0.2 \end{cases}$$

30R1 + -40R2

$$\begin{cases} 12b+18d = 18 \\ -12b-8d = -8 \end{cases}$$

$$10d = 10$$
$$d = 1;$$

$$0.4b+0.6d = 0.6$$
$$0.4b+0.6(1) = 0.6$$
$$0.4b+0.6 = 0.6$$
$$0.4b = 0$$
$$b = 0;$$

$$\begin{bmatrix} a & b \\ c & d \end{bmatrix} = \begin{bmatrix} 1 & 0 \\ 0 & 1 \end{bmatrix}$$

11. $\begin{bmatrix} -3 & 8 \\ -4 & 10 \end{bmatrix} \cdot \begin{bmatrix} 1 & 0 \\ 0 & 1 \end{bmatrix}$

$$= \begin{bmatrix} -3(1)+8(0) & -3(0)+8(1) \\ -4(1)+10(0) & -4(0)+10(1) \end{bmatrix}$$

$$= \begin{bmatrix} -3 & 8 \\ -4 & 10 \end{bmatrix};$$

$$\begin{bmatrix} 1 & 0 \\ 0 & 1 \end{bmatrix} \cdot \begin{bmatrix} -3 & 8 \\ -4 & 10 \end{bmatrix}$$

$$= \begin{bmatrix} 1(-3)+0(-4) & 1(8)+0(10) \\ 0(-3)+1(-4) & 0(8)+1(10) \end{bmatrix}$$

$$= \begin{bmatrix} -3 & 8 \\ -4 & 10 \end{bmatrix}$$

$AI = IA = A$

13. $\begin{bmatrix} -4 & 1 & 6 \\ 9 & 5 & 3 \\ 0 & -2 & 1 \end{bmatrix} \cdot \begin{bmatrix} 1 & 0 & 0 \\ 0 & 1 & 0 \\ 0 & 0 & 1 \end{bmatrix}$

$$= \begin{bmatrix} -4 & 1 & 6 \\ 9 & 5 & 3 \\ 0 & -2 & 1 \end{bmatrix};$$

$$\begin{bmatrix} 1 & 0 & 0 \\ 0 & 1 & 0 \\ 0 & 0 & 1 \end{bmatrix} \cdot \begin{bmatrix} -4 & 1 & 6 \\ 9 & 5 & 3 \\ 0 & -2 & 1 \end{bmatrix}$$

$$= \begin{bmatrix} -4 & 1 & 6 \\ 9 & 5 & 3 \\ 0 & -2 & 1 \end{bmatrix}$$

$AI = IA = A$

15. $\begin{bmatrix} 5 & -4 \\ 2 & 2 \end{bmatrix}$

$$A^{-1} = \frac{1}{ad-bc}\begin{bmatrix} d & -b \\ -c & a \end{bmatrix}$$

$$A^{-1} = \frac{1}{5(2)-(-4)(2)}\begin{bmatrix} 2 & 4 \\ -2 & 5 \end{bmatrix}$$

$$A^{-1} = \frac{1}{18}\begin{bmatrix} 2 & 4 \\ -2 & 5 \end{bmatrix}$$

$$A^{-1} = \begin{bmatrix} \frac{1}{9} & \frac{2}{9} \\ \frac{-1}{9} & \frac{5}{18} \end{bmatrix}$$

17. $\begin{bmatrix} 1 & -3 \\ 4 & -10 \end{bmatrix}$

$$A^{-1} = \frac{1}{ad-bc}\begin{bmatrix} d & -b \\ -c & a \end{bmatrix}$$

$$A^{-1} = \frac{1}{1(-10)-(-3)(4)}\begin{bmatrix} -10 & 3 \\ -4 & 1 \end{bmatrix}$$

$$A^{-1} = \frac{1}{2}\begin{bmatrix} -10 & 3 \\ -4 & 1 \end{bmatrix}$$

$$A^{-1} = \begin{bmatrix} -5 & \frac{3}{2} \\ -2 & \frac{1}{2} \end{bmatrix}$$

19. $A = \begin{bmatrix} 1 & 5 \\ -2 & -9 \end{bmatrix}$ $B = \begin{bmatrix} -9 & -5 \\ 2 & 1 \end{bmatrix}$

$$AB=\begin{bmatrix}1&5\\-2&-9\end{bmatrix}\begin{bmatrix}-9&-5\\2&1\end{bmatrix}$$
$$=\begin{bmatrix}1(-9)+5(2)&1(-5)+5(1)\\-2(-9)-9(2)&-2(-5)-9(1)\end{bmatrix}$$
$$=\begin{bmatrix}1&0\\0&1\end{bmatrix};$$
$$BA=\begin{bmatrix}-9&-5\\2&1\end{bmatrix}\begin{bmatrix}1&5\\-2&-9\end{bmatrix}$$
$$=\begin{bmatrix}-9(1)-5(-2)&-9(5)-5(-9)\\2(1)+1(-2)&2(5)+1(-9)\end{bmatrix}$$
$$=\begin{bmatrix}1&0\\0&1\end{bmatrix}$$
$AB=BA=I$

21. $A=\begin{bmatrix}4&-5\\0&2\end{bmatrix}$ $B=\begin{bmatrix}\frac{1}{4}&\frac{5}{8}\\0&\frac{1}{2}\end{bmatrix}$

$$AB=\begin{bmatrix}4&-5\\0&2\end{bmatrix}\begin{bmatrix}\frac{1}{4}&\frac{5}{8}\\0&\frac{1}{2}\end{bmatrix}$$
$$=\begin{bmatrix}4\left(\frac{1}{4}\right)-5(0)&4\left(\frac{5}{8}\right)-5\left(\frac{1}{2}\right)\\0\left(\frac{1}{4}\right)+2(0)&0\left(\frac{5}{8}\right)+2\left(\frac{1}{2}\right)\end{bmatrix}$$
$$=\begin{bmatrix}1&0\\0&1\end{bmatrix};$$
$$BA=\begin{bmatrix}\frac{1}{4}&\frac{5}{8}\\0&\frac{1}{2}\end{bmatrix}\begin{bmatrix}4&-5\\0&2\end{bmatrix}$$
$$=\begin{bmatrix}\frac{1}{4}(4)+\frac{5}{8}(0)&\frac{1}{4}(-5)+\frac{5}{8}(2)\\0(4)+\frac{1}{2}(0)&0(-5)+\frac{1}{2}(2)\end{bmatrix}$$
$$=\begin{bmatrix}1&0\\0&1\end{bmatrix}$$
$AB=BA=I$

23. $A=\begin{bmatrix}-2&3&1\\5&2&4\\2&0&-1\end{bmatrix}$

$$A^{-1}=B=\begin{bmatrix}-\frac{2}{39}&\frac{1}{13}&\frac{10}{39}\\\frac{1}{3}&0&\frac{1}{3}\\-\frac{4}{39}&\frac{2}{13}&-\frac{19}{39}\end{bmatrix};$$
$$AB=\begin{bmatrix}-2&3&1\\5&2&4\\2&0&-1\end{bmatrix}\begin{bmatrix}-\frac{2}{39}&\frac{1}{13}&\frac{10}{39}\\\frac{1}{3}&0&\frac{1}{3}\\-\frac{4}{39}&\frac{2}{13}&-\frac{19}{39}\end{bmatrix}$$
$$=\begin{bmatrix}1&0&0\\0&1&0\\0&0&1\end{bmatrix}$$
$$BA=\begin{bmatrix}-\frac{2}{39}&\frac{1}{13}&\frac{10}{39}\\\frac{1}{3}&0&\frac{1}{3}\\-\frac{4}{39}&\frac{2}{13}&-\frac{19}{39}\end{bmatrix}\begin{bmatrix}-2&3&1\\5&2&4\\2&0&-1\end{bmatrix}$$
$$=\begin{bmatrix}1&0&0\\0&1&0\\0&0&1\end{bmatrix}$$
$AB=BA=I$

25. $A=\begin{bmatrix}-7&5&-3\\1&9&0\\2&-2&-5\end{bmatrix}$

$$A^{-1}=B=\begin{bmatrix}-\frac{9}{80}&\frac{31}{400}&\frac{27}{400}\\\frac{1}{80}&\frac{41}{400}&-\frac{3}{400}\\-\frac{1}{20}&-\frac{1}{100}&-\frac{17}{100}\end{bmatrix};$$
$$AB=\begin{bmatrix}-7&5&-3\\1&9&0\\2&-2&-5\end{bmatrix}\begin{bmatrix}-\frac{9}{80}&\frac{31}{400}&\frac{27}{400}\\\frac{1}{80}&\frac{41}{400}&-\frac{3}{400}\\-\frac{1}{20}&-\frac{1}{100}&-\frac{17}{100}\end{bmatrix}$$
$$=\begin{bmatrix}1&0&0\\0&1&0\\0&0&1\end{bmatrix};$$

$$BA = \begin{bmatrix} -\frac{9}{80} & \frac{31}{400} & \frac{27}{400} \\ \frac{1}{80} & \frac{41}{400} & -\frac{3}{400} \\ -\frac{1}{20} & -\frac{1}{100} & -\frac{17}{100} \end{bmatrix} \begin{bmatrix} -7 & 5 & -3 \\ 1 & 9 & 0 \\ 2 & -2 & -5 \end{bmatrix}$$

$$= \begin{bmatrix} 1 & 0 & 0 \\ 0 & 1 & 0 \\ 0 & 0 & 1 \end{bmatrix}$$

$AB = BA = I$

2627. $\begin{cases} 2x-3y=9 \\ -5x+7y=8 \end{cases}$

$$\begin{bmatrix} 2 & -3 \\ -5 & 7 \end{bmatrix} \begin{bmatrix} x \\ y \end{bmatrix} = \begin{bmatrix} 9 \\ 8 \end{bmatrix}$$

29. $\begin{cases} x + 2y - z = 1 \\ x + z = 3 \\ 2x - y + z = 3 \end{cases}$

$$\begin{bmatrix} 1 & 2 & -1 \\ 1 & 0 & 1 \\ 2 & -1 & 1 \end{bmatrix} \begin{bmatrix} x \\ y \\ z \end{bmatrix} = \begin{bmatrix} 1 \\ 3 \\ 3 \end{bmatrix}$$

31. $\begin{cases} -2w+x-4y+5=-3 \\ 2w-5x+y-3z=4 \\ -3w+x+6y+z=1 \\ w+4x-5y+z=-9 \end{cases}$

$$\begin{bmatrix} -2 & 1 & -4 & 5 \\ 2 & -5 & 1 & -3 \\ -3 & 1 & 6 & 1 \\ 1 & 4 & -5 & 1 \end{bmatrix} \begin{bmatrix} w \\ x \\ y \\ z \end{bmatrix} = \begin{bmatrix} -3 \\ 4 \\ 1 \\ -9 \end{bmatrix}$$

33. $\begin{cases} 0.05x-3.2y=-15.8 \\ 0.02x+2.4y=12.08 \end{cases}$

$$\begin{bmatrix} 0.05 & -3.2 \\ 0.02 & 2.4 \end{bmatrix} \begin{bmatrix} x \\ y \end{bmatrix} = \begin{bmatrix} -15.8 \\ 12.08 \end{bmatrix}$$

Using the grapher,

$$A^{-1} = \begin{bmatrix} \frac{300}{23} & \frac{400}{23} \\ -\frac{5}{46} & \frac{25}{92} \end{bmatrix};$$

$$A^{-1}\left(\begin{bmatrix} 0.05 & -3.2 \\ 0.02 & 2.4 \end{bmatrix} \begin{bmatrix} x \\ y \end{bmatrix}\right) = A^{-1}\begin{bmatrix} -15.8 \\ 12.08 \end{bmatrix}$$

$$\left(A^{-1}\begin{bmatrix} 0.05 & -3.2 \\ 0.02 & 2.4 \end{bmatrix}\right) \begin{bmatrix} x \\ y \end{bmatrix} = A^{-1}\begin{bmatrix} -15.8 \\ 12.08 \end{bmatrix}$$

$$\begin{bmatrix} 1 & 0 \\ 0 & 1 \end{bmatrix} \begin{bmatrix} x \\ y \end{bmatrix} = A^{-1}\begin{bmatrix} -15.8 \\ 12.08 \end{bmatrix}$$

$$\begin{bmatrix} x \\ y \end{bmatrix} = \begin{bmatrix} 4 \\ 5 \end{bmatrix}$$

$(4,5)$

35. $\begin{cases} -\frac{1}{6}u+\frac{1}{4}v=1 \\ \frac{1}{2}u-\frac{2}{3}v=-2 \end{cases}$

$$\begin{bmatrix} -\frac{1}{6} & \frac{1}{4} \\ \frac{1}{2} & -\frac{2}{3} \end{bmatrix} \begin{bmatrix} u \\ v \end{bmatrix} = \begin{bmatrix} 1 \\ -2 \end{bmatrix}$$

$$A^{-1} = \begin{bmatrix} 48 & 18 \\ 36 & 12 \end{bmatrix};$$

$$\begin{bmatrix} 48 & 18 \\ 36 & 12 \end{bmatrix}\left(\begin{bmatrix} -\frac{1}{6} & \frac{1}{4} \\ \frac{1}{2} & -\frac{2}{3} \end{bmatrix} \begin{bmatrix} u \\ v \end{bmatrix}\right) = \begin{bmatrix} 48 & 18 \\ 36 & 12 \end{bmatrix}\begin{bmatrix} 1 \\ -2 \end{bmatrix}$$

$$\left(\begin{bmatrix} 48 & 18 \\ 36 & 12 \end{bmatrix}\begin{bmatrix} -\frac{1}{6} & \frac{1}{4} \\ \frac{1}{2} & -\frac{2}{3} \end{bmatrix}\right) \begin{bmatrix} u \\ v \end{bmatrix} = \begin{bmatrix} 48 & 18 \\ 36 & 12 \end{bmatrix}\begin{bmatrix} 1 \\ -2 \end{bmatrix}$$

$$\begin{bmatrix} 1 & 0 \\ 0 & 1 \end{bmatrix} \begin{bmatrix} u \\ v \end{bmatrix} = \begin{bmatrix} 48 & 18 \\ 36 & 12 \end{bmatrix}\begin{bmatrix} 1 \\ -2 \end{bmatrix}$$

$$\begin{bmatrix} u \\ v \end{bmatrix} = \begin{bmatrix} 12 \\ 12 \end{bmatrix}$$

$(12,12)$

37. $\begin{cases} \frac{-1}{8}a+\frac{3}{5}b=\frac{5}{6} \\ \frac{5}{16}a-\frac{3}{2}b=\frac{-4}{5} \end{cases}$

$$\begin{bmatrix} -\frac{1}{8} & \frac{3}{5} \\ \frac{5}{16} & -\frac{3}{2} \end{bmatrix} \begin{bmatrix} a \\ b \end{bmatrix} = \begin{bmatrix} \frac{5}{6} \\ -\frac{4}{5} \end{bmatrix}$$

No Solution; matrix is singular

39. $\begin{cases} 0.2x-1.6y+2z=-1.9 \\ -0.4x-y+0.6z=-1 \\ 0.8x+3.2y-0.4z=0.2 \end{cases}$

$$\begin{bmatrix} 0.2 & -1.6 & 2 \\ -0.4 & -1 & 0.6 \\ 0.8 & 3.2 & -0.4 \end{bmatrix} \begin{bmatrix} x \\ y \\ z \end{bmatrix} = \begin{bmatrix} -1.9 \\ -1 \\ 0.2 \end{bmatrix}$$

Using the grapher,

$$A^{-1}=\begin{bmatrix}\frac{95}{111} & -\frac{120}{37} & -\frac{65}{111}\\ -\frac{20}{111} & \frac{35}{37} & \frac{115}{222}\\ \frac{10}{37} & \frac{40}{37} & \frac{35}{74}\end{bmatrix};$$

$$A^{-1}\left(\begin{bmatrix}0.2 & -1.6 & 2\\ -0.4 & -1 & 0.6\\ 0.8 & 3.2 & -0.4\end{bmatrix}\begin{bmatrix}x\\y\\z\end{bmatrix}\right)=A^{-1}\begin{bmatrix}-1.9\\-1\\0.2\end{bmatrix}$$

$$\left(A^{-1}\begin{bmatrix}0.2 & -1.6 & 2\\ -0.4 & -1 & 0.6\\ 0.8 & 3.2 & -0.4\end{bmatrix}\right)\begin{bmatrix}x\\y\\z\end{bmatrix}=A^{-1}\begin{bmatrix}-1.9\\-1\\0.2\end{bmatrix}$$

$$\begin{bmatrix}1 & 0 & 0\\ 0 & 1 & 0\\ 0 & 0 & 1\end{bmatrix}\begin{bmatrix}x\\y\\z\end{bmatrix}=A^{-1}\begin{bmatrix}-1.9\\-1\\0.2\end{bmatrix}$$

$$\begin{bmatrix}x\\y\\z\end{bmatrix}=\begin{bmatrix}1.5\\-0.5\\-1.5\end{bmatrix}$$

$(1.5,-0.5,-1.5)$

41. $\begin{cases}x-2y+2z=6\\ 2x-1.5y+1.8z=2.8\\ \frac{-2}{3}x+\frac{1}{2}y-\frac{3}{5}z=-\frac{11}{30}\end{cases}$

$$\begin{bmatrix}1 & -2 & 2\\ 2 & -1.5 & 1.8\\ -\frac{2}{3} & \frac{1}{2} & -\frac{3}{5}\end{bmatrix}\begin{bmatrix}x\\y\\z\end{bmatrix}=\begin{bmatrix}6\\2.8\\-\frac{11}{30}\end{bmatrix}$$

Singular; no solution

43. $\begin{cases}-2w+3x-4y+5z=-3\\ 0.2w-2.6x+y-0.4z=2.4\\ -3w+3.2x+2.8y+z=6.1\\ 1.6w+4x-5y+2.6z=-9.8\end{cases}$

$$\begin{bmatrix}-2 & 3 & -4 & 5\\ 0.2 & -2.6 & 1 & -0.4\\ -3 & 3.2 & 2.8 & 1\\ 1.6 & 4 & -5 & 2.6\end{bmatrix}\begin{bmatrix}w\\x\\y\\z\end{bmatrix}=\begin{bmatrix}-3\\2.4\\6.1\\-9.8\end{bmatrix}$$

Using the grapher,

$$A^{-1}=\begin{bmatrix}-0.35859 & 1.15741 & 0.36978 & 0.72543\\ -0.10811 & -0.13514 & 0.13514 & 0.13514\\ -0.23646 & 0.933507 & 0.44725 & 0.42633\\ -0.67737 & 1.290852 & 0.42463 & 0.55015\end{bmatrix}$$

$$A^{-1}\left(\begin{bmatrix}-2 & 3 & -4 & 5\\ 0.2 & -2.6 & 1 & -0.4\\ -3 & 3.2 & 2.8 & 1\\ 1.6 & 4 & -5 & 2.6\end{bmatrix}\begin{bmatrix}w\\x\\y\\z\end{bmatrix}\right)=A^{-1}\begin{bmatrix}-3\\2.4\\6.1\\-9.8\end{bmatrix}$$

$$\left(A^{-1}\begin{bmatrix}-2 & 3 & -4 & 5\\ 0.2 & -2.6 & 1 & -0.4\\ -3 & 3.2 & 2.8 & 1\\ 1.6 & 4 & -5 & 2.6\end{bmatrix}\right)\begin{bmatrix}w\\x\\y\\z\end{bmatrix}=A^{-1}\begin{bmatrix}-3\\2.4\\6.1\\-9.8\end{bmatrix}$$

$$\begin{bmatrix}1 & 0 & 0 & 0\\ 0 & 1 & 0 & 0\\ 0 & 0 & 1 & 0\\ 0 & 0 & 0 & 1\end{bmatrix}\begin{bmatrix}w\\x\\y\\z\end{bmatrix}=A^{-1}\begin{bmatrix}-3\\2.4\\6.1\\-9.8\end{bmatrix}$$

$$\begin{bmatrix}w\\x\\y\\z\end{bmatrix}=\begin{bmatrix}-1\\-0.5\\1.5\\0.5\end{bmatrix}$$

$(-1,-0.5,1.5,0.5)$

45. $\begin{bmatrix}4 & -7\\ 3 & -5\end{bmatrix}$

$\det A=4(-5)-3(-7)=-20+21=1$

yes

47. $\begin{bmatrix}1.2 & -0.8\\ 0.3 & -0.2\end{bmatrix}$

$\det A=1.2(-0.2)-(0.3)(-0.8)$

$=-0.24+0.24=0$

no

49. $A=\begin{bmatrix}1 & 0 & -2\\ 0 & -1 & -1\\ 2 & 1 & -4\end{bmatrix}$

$\det A$

$=1\begin{vmatrix}-1 & -1\\ 1 & -4\end{vmatrix}-0\begin{vmatrix}0 & -1\\ 2 & -4\end{vmatrix}+(-2)\begin{vmatrix}0 & -1\\ 2 & 1\end{vmatrix}$

$=1(4+1)-0-2(0+2)$

$=5-0-4$

$=1$

51. $A=\begin{bmatrix}-2 & 3 & 4\\ 0 & 6 & 2\\ 1 & -1.5 & -2\end{bmatrix}$

det A

$=0\begin{vmatrix}3 & 4\\ -1.5 & -2\end{vmatrix}+6\begin{vmatrix}-2 & 4\\ 1 & -2\end{vmatrix}-2\begin{vmatrix}-2 & 3\\ 1 & -1.5\end{vmatrix}$

$=0+6(4-4)-2(3-3)$

$=0$

Singular Matrix

53. $A=\begin{bmatrix}1 & 0 & 3 & -4\\ 2 & 5 & 0 & 1\\ 8 & 15 & 6 & -5\\ 0 & 8 & -4 & 1\end{bmatrix}$

det $A=0$

Singular

55. $\begin{cases}x-2y+2z=7\\ 2x+2y-z=5\\ 3x-y+z=6\end{cases}$

(1) $\begin{bmatrix}1 & -2 & 2\\ 2 & 2 & -1\\ 3 & -1 & 1\end{bmatrix}\begin{bmatrix}x\\ y\\ z\end{bmatrix}=\begin{bmatrix}7\\ 5\\ 6\end{bmatrix}$

(2) det A

$=1\begin{vmatrix}2 & -1\\ -1 & 1\end{vmatrix}-(-2)\begin{vmatrix}2 & -1\\ 3 & 1\end{vmatrix}+2\begin{vmatrix}2 & 2\\ 3 & -1\end{vmatrix}$

$=1(2-1)+2(2+3)+2(-2-6)$

$=1+10-16$

$=-5$

(3) $A^{-1}=\begin{bmatrix}-0.2 & 0 & 0.4\\ 1 & 1 & -1\\ 1.6 & 1 & -1.2\end{bmatrix}$

$X=A^{-1}B$

$X=\begin{bmatrix}-0.2 & 0 & 0.4\\ 1 & 1 & -1\\ 1.6 & 1 & -1.2\end{bmatrix}\begin{bmatrix}7\\ 5\\ 6\end{bmatrix}$

$X=\begin{bmatrix}1\\ 6\\ 9\end{bmatrix}$

$(1,6,9)$

57. $\begin{cases}x-3y+4z=-1\\ 4x-y+5z=7\\ 3x+2y+z=-3\end{cases}$

(1) $\begin{bmatrix}1 & -3 & 4\\ 4 & -1 & 5\\ 3 & 2 & 1\end{bmatrix}\begin{bmatrix}x\\ y\\ z\end{bmatrix}=\begin{bmatrix}-1\\ 7\\ -3\end{bmatrix}$

(2) det A

$=1\begin{vmatrix}-1 & 5\\ 2 & 1\end{vmatrix}-(-3)\begin{vmatrix}4 & 5\\ 3 & 1\end{vmatrix}+4\begin{vmatrix}4 & -1\\ 3 & 2\end{vmatrix}$

$=1(-1-10)+3(4-15)+4(8+3)$

$=-11-33+44$

$=0$

Singular

59. $A=\begin{bmatrix}3 & -5\\ 2 & 1\end{bmatrix}$

$A^{-1}=\dfrac{1}{ad-bc}\begin{bmatrix}d & -b\\ -c & a\end{bmatrix}$

$A^{-1}=\dfrac{1}{3(1)-(-5)(2)}\begin{bmatrix}1 & 5\\ -2 & 3\end{bmatrix}$

$A^{-1}=\dfrac{1}{13}\begin{bmatrix}1 & 5\\ -2 & 3\end{bmatrix}$

$A^{-1}=\begin{bmatrix}\frac{1}{13} & \frac{5}{13}\\ \frac{-2}{13} & \frac{3}{13}\end{bmatrix}$;

$AA^{-1}=\begin{bmatrix}3 & -5\\ 2 & 1\end{bmatrix}\begin{bmatrix}\frac{1}{13} & \frac{5}{13}\\ \frac{-2}{13} & \frac{3}{13}\end{bmatrix}=\begin{bmatrix}1 & 0\\ 0 & 1\end{bmatrix}$

$A^{-1}A=\begin{bmatrix}\frac{1}{13} & \frac{5}{13}\\ \frac{-2}{13} & \frac{3}{13}\end{bmatrix}\begin{bmatrix}3 & -5\\ 2 & 1\end{bmatrix}=\begin{bmatrix}1 & 0\\ 0 & 1\end{bmatrix}$

$AA^{-1}=A^{-1}A=I$

61. $C=\begin{bmatrix}0.3 & -0.4\\ -0.6 & 0.8\end{bmatrix}$

$C^{-1}=\dfrac{1}{ad-bc}\begin{bmatrix}d & -b\\ -c & a\end{bmatrix}$

$C^{-1}=\dfrac{1}{0.3(0.8)-(-0.4)(-0.6)}\begin{bmatrix}0.8 & 0.4\\ 0.6 & 0.3\end{bmatrix}$

$C^{-1}=\dfrac{1}{0}\begin{bmatrix}0.8 & 0.4\\ 0.6 & 0.3\end{bmatrix}$

Singular

63. Let B represent the number of Behemoth slushies sold.

Let G represent the number of Gargantuan slushies sold.
Let M represent the number of Mammoth slushies sold.
Let J represent the number of Jumbo slushies sold.

$$\begin{cases} 2.59B+2.29G+1.99M+1.59J=402.29 \\ 60B+48G+36M+24J=7884 \\ B+G+M+J=191 \\ B=J+1 \end{cases}$$

$$\begin{cases} 2.59B+2.29G+1.99M+1.59J=402.29 \\ 60B+48G+36M+24J=7884 \\ B+G+M+J=191 \\ B-J=1 \end{cases}$$

$$\begin{bmatrix} 2.59 & 2.29 & 1.99 & 1.59 \\ 60 & 48 & 36 & 24 \\ 1 & 1 & 1 & 1 \\ 1 & 0 & 0 & -1 \end{bmatrix}\begin{bmatrix} B \\ G \\ M \\ J \end{bmatrix}=\begin{bmatrix} 402.29 \\ 7884 \\ 191 \\ 1 \end{bmatrix}$$

$$A^{-1}=\begin{bmatrix} -10 & \frac{1}{4} & \frac{109}{10} & 1 \\ 10 & -\frac{1}{6} & -\frac{139}{10} & -2 \\ 10 & -\frac{1}{3} & -\frac{69}{10} & 1 \\ -10 & \frac{1}{4} & \frac{109}{10} & 0 \end{bmatrix};$$

$$X=A^{-1}B$$

$$X=\begin{bmatrix} -10 & \frac{1}{4} & \frac{109}{10} & 1 \\ 10 & -\frac{1}{6} & -\frac{139}{10} & -2 \\ 10 & -\frac{1}{3} & -\frac{69}{10} & 1 \\ -10 & \frac{1}{4} & \frac{109}{10} & 0 \end{bmatrix}\begin{bmatrix} 402.29 \\ 7884 \\ 191 \\ 1 \end{bmatrix}$$

$$X=\begin{bmatrix} 31 \\ 52 \\ 78 \\ 30 \end{bmatrix}$$

31 Behemoth
52 Gargantuan
78 Mammoth
30 Jumbo

65. Let J represent the playing time of *Jumpin' Jack Flash.*
Let T represent the playing time of *Tumbling Dice.*
Let W represent the playing time of *Wild Horses.*
Let Y represent the playing time of *You Can't Always Get What You Want.*

$$\begin{cases} J+T+W+Y=20.75 \\ J+T=Y \\ W=J+2 \\ Y=2T \end{cases}$$

$$\begin{cases} J+T+W+Y=20.75 \\ J+T-Y=0 \\ -J+W=2 \\ -2T+Y=0 \end{cases}$$

$$\begin{bmatrix} 1 & 1 & 1 & 1 \\ 1 & 1 & 0 & -1 \\ -1 & 0 & 1 & 0 \\ 0 & -2 & 0 & 1 \end{bmatrix}\begin{bmatrix} J \\ T \\ W \\ Y \end{bmatrix}=\begin{bmatrix} 20.75 \\ 0 \\ 2 \\ 0 \end{bmatrix}$$

$$A^{-1}=\begin{bmatrix} 0.2 & 0.6 & -0.2 & 0.4 \\ 0.2 & -0.4 & -0.2 & -0.6 \\ 0.2 & 0.6 & 0.8 & 0.4 \\ 0.4 & -0.8 & -0.4 & -0.2 \end{bmatrix};$$

$$X=A^{-1}B$$

$$X=\begin{bmatrix} 0.2 & 0.6 & -0.2 & 0.4 \\ 0.2 & -0.4 & -0.2 & -0.6 \\ 0.2 & 0.6 & 0.8 & 0.4 \\ 0.4 & -0.8 & -0.4 & -0.2 \end{bmatrix}\begin{bmatrix} 20.75 \\ 0 \\ 2 \\ 0 \end{bmatrix}$$

$$X=\begin{bmatrix} 3.75 \\ 3.75 \\ 5.75 \\ 7.5 \end{bmatrix}$$

Jumping' Jack Flash: 3.75 min
Tumbling Dice: 3.75 min
Wild Horses: 5.75 min
You Can't Always Get What You Want: 7.5 min

67. Let A represent the number of Clock A manufactured.
Let B represent the number of Clock B manufactured.
Let C represent the number of Clock C manufactured.
Let D represent the number of Clock D manufactured.

$$\begin{cases} 2.2A+2.5B+2.75C+3D=262 \\ 1.2A+1.4B+1.8C+2D=160 \\ 0.2A+0.25B+0.3C+0.5D=29 \\ 0.5A+0.55B+0.75C+D=68 \end{cases}$$

$$\begin{bmatrix} 2.2 & 2.5 & 2.75 & 3 \\ 1.2 & 1.4 & 1.8 & 2 \\ 0.2 & 0.25 & 0.3 & 0.5 \\ 0.5 & 0.55 & 0.75 & 1 \end{bmatrix}\begin{bmatrix} A \\ B \\ C \\ D \end{bmatrix}=\begin{bmatrix} 262 \\ 160 \\ 29 \\ 68 \end{bmatrix}$$

$$A^{-1}=\begin{bmatrix} \frac{30}{11} & -\frac{205}{22} & -\frac{210}{11} & 20 \\ 0 & 5 & 20 & -20 \\ -\frac{20}{11} & \frac{50}{11} & -\frac{80}{11} & 0 \\ 0 & -\frac{3}{2} & 4 & 2 \end{bmatrix};$$

$X=A^{-1}B$

$$X=\begin{bmatrix} \frac{30}{11} & -\frac{205}{22} & -\frac{210}{11} & 20 \\ 0 & 5 & 20 & -20 \\ -\frac{20}{11} & \frac{50}{11} & -\frac{80}{11} & 0 \\ 0 & -\frac{3}{2} & 4 & 2 \end{bmatrix}\begin{bmatrix} 262 \\ 160 \\ 29 \\ 68 \end{bmatrix}$$

$$X=\begin{bmatrix} 30 \\ 20 \\ 40 \\ 12 \end{bmatrix}$$

30 of clock A
20 of clock B
40 of clock C
12 of clock D

69. $ax^3+bx^2+cx+d=0$
(-4, -6), (-1, 0), (1, -16) and (3, 8)

$a(-4)^3+b(-4)^2+c(-4)+d=-6$
$-64a+16b-4c+d=-6;$

$a(-1)^3+b(-1)^2+c(-1)+d=0$
$-a+b-c+d=0;$

$a(1)^3+b(1)^2+c(1)+d=-16$
$a+b+c+d=-16;$

$a(3)^3+b(3)^2+c(3)+d=8$
$27a+9b+3c+d=8;$

$$\begin{cases} -64a+16b-4c+d=-6 \\ -a+b-c+d=0 \\ a+b+c+d=-16 \\ 27a+9b+3c+d=8 \end{cases}$$

$$\begin{bmatrix} -64 & 16 & -4 & 1 \\ -1 & 1 & -1 & 1 \\ 1 & 1 & 1 & 1 \\ 27 & 9 & 3 & 1 \end{bmatrix}\begin{bmatrix} a \\ b \\ c \\ d \end{bmatrix}=\begin{bmatrix} -6 \\ 0 \\ -16 \\ 8 \end{bmatrix}$$

$$A^{-1}=\begin{bmatrix} -\frac{1}{105} & \frac{1}{24} & -\frac{1}{20} & \frac{1}{56} \\ \frac{1}{35} & 0 & -\frac{1}{10} & \frac{1}{14} \\ \frac{1}{105} & -\frac{13}{24} & \frac{11}{20} & -\frac{1}{56} \\ -\frac{1}{35} & \frac{1}{2} & \frac{3}{5} & -\frac{1}{14} \end{bmatrix};$$

$X=A^{-1}B$

$$X=\begin{bmatrix} -\frac{1}{105} & \frac{1}{24} & -\frac{1}{20} & \frac{1}{56} \\ \frac{1}{35} & 0 & -\frac{1}{10} & \frac{1}{14} \\ \frac{1}{105} & -\frac{13}{24} & \frac{11}{20} & -\frac{1}{56} \\ -\frac{1}{35} & \frac{1}{2} & \frac{3}{5} & -\frac{1}{14} \end{bmatrix}\begin{bmatrix} -6 \\ 0 \\ -16 \\ 8 \end{bmatrix}$$

$$X=\begin{bmatrix} 1 \\ 2 \\ -9 \\ -10 \end{bmatrix}$$

$y=x^3+2x^2-9x-10$

71. Let x represent the number of ounces of food for Food I.
Let y represent the number of ounces of food for Food II.
Let z represent the number of ounces of food for Food III.

$$\begin{cases} 2x+4y+3z=20 \\ 4x+2y+5z=30 \\ 5x+6y+7z=44 \end{cases}$$

$$\begin{bmatrix} 2 & 4 & 3 \\ 4 & 2 & 5 \\ 5 & 6 & 7 \end{bmatrix}\begin{bmatrix} x \\ y \\ z \end{bmatrix}=\begin{bmatrix} 20 \\ 30 \\ 44 \end{bmatrix}$$

$$A^{-1}=\begin{bmatrix} 8 & 5 & -7 \\ 1.5 & 0.5 & -1 \\ -7 & -4 & 6 \end{bmatrix};$$

$X=A^{-1}B$

$$X=\begin{bmatrix} 8 & 5 & -7 \\ 1.5 & 0.5 & -1 \\ -7 & -4 & 6 \end{bmatrix}\begin{bmatrix} 20 \\ 30 \\ 44 \end{bmatrix}$$

$$X=\begin{bmatrix}2\\1\\4\end{bmatrix}$$

2 oz food I
1 oz food II
4 oz food III

73. $A=\begin{bmatrix}2 & 3\\-5 & -4\end{bmatrix}, \text{B}=\begin{bmatrix}4\\9\end{bmatrix}, \text{C}=\begin{bmatrix}12\\-4\end{bmatrix}, \text{X}=\begin{bmatrix}x\\y\end{bmatrix}$

Answers will vary

75. $A=\begin{bmatrix}1 & -1 & 0\\1 & 2 & 1\\2 & 3 & 2\end{bmatrix}$

$$\begin{bmatrix}1 & -1 & 0\\1 & 2 & 1\\2 & 3 & 2\end{bmatrix}\begin{bmatrix}a & b & c\\d & e & f\\g & h & i\end{bmatrix}=\begin{bmatrix}1 & 0 & 0\\0 & 1 & 0\\0 & 0 & 1\end{bmatrix}$$

$$\begin{bmatrix}a-d & b-e & c-f\\a+2d+g & b+2e+h & c+2f+i\\2a+3d+2g & 2b+3e+2h & 2c+3f+2i\end{bmatrix}=\begin{bmatrix}1 & 0 & 0\\0 & 1 & 0\\0 & 0 & 1\end{bmatrix}$$

$$\begin{cases}a-d=1\\a+2d+g=0\\2a+3d+2g=0\end{cases}$$

-2R2 + R3 $\rightarrow$ R3

$-d=0$

$d=0;$

$a-d=1$

$a-0=1$

$a=1;$

$a+2d+g=0$

$1+2(0)+g=0$

$g=-1;$

$$\begin{cases}b-e=0\\b+2e+h=1\\2b+3e+2h=0\end{cases}$$

-2R2 + R3 $\rightarrow$ R3

$-e=-2$

$e=2;$

$b-e=0$

$b-2=0$

$b=2;$

$b+2e+h=1$

$2+2(2)+h=1$

$2+4+h=1$

$6+h=1$

$h=-5;$

$$\begin{cases}c-f=0\\c+2f+i=0\\2c+3f+2i=1\end{cases}$$

-2R2 + R3 $\rightarrow$ R3

$-f=1$

$f=-1;$

$c-f=0$

$c-(-1)=0$

$c+1=0$

$c=-1;$

$c+2f+i=0$

$-1+2(-1)+i=0$

$-1-2+i=0$

$-3+i=0$

$i=3;$

$$A^{-1}=\begin{bmatrix}1 & 2 & -1\\0 & 2 & -1\\-1 & -5 & 3\end{bmatrix}$$

$$A^{-1}A=\begin{bmatrix}1 & 2 & -1\\0 & 2 & -1\\-1 & -5 & 3\end{bmatrix}\begin{bmatrix}1 & -1 & 0\\1 & 2 & 1\\2 & 3 & 2\end{bmatrix}=\begin{bmatrix}1 & 0 & 0\\0 & 1 & 0\\0 & 0 & 1\end{bmatrix}$$

$$AA^{-1}=\begin{bmatrix}1 & -1 & 0\\1 & 2 & 1\\2 & 3 & 2\end{bmatrix}\begin{bmatrix}1 & 2 & -1\\0 & 2 & -1\\-1 & -5 & 3\end{bmatrix}=\begin{bmatrix}1 & 0 & 0\\0 & 1 & 0\\0 & 0 & 1\end{bmatrix}$$

$A^{-1}A=AA^{-1}=I$

77. Let x represent the smaller fraction.
Let y represent the middle fraction.
Let z represent the larger fraction.

$$\begin{cases}x+y=z\\z-x=y\\4x=y+z\end{cases}$$

$$\begin{cases}x+y-z=0\\-x-y+z=0\\4x-y-z=0\end{cases}$$

$$\begin{bmatrix} 1 & 1 & -1 \\ -1 & -1 & 1 \\ 4 & -1 & -1 \end{bmatrix}\begin{bmatrix} x \\ y \\ z \end{bmatrix} = \begin{bmatrix} 0 \\ 0 \\ 0 \end{bmatrix}$$

Singular matrix. Answers will vary.

79. $f(x) = x^3 + 2 \qquad g(x) = \sqrt[3]{x-2}$

$(f \circ g)(x) = f(g(x))$
$= (g(x))^3 + 2$
$= \left(\sqrt[3]{x-2}\right)^3 + 2$
$= x - 2 + 2$
$= x;$
$(g \circ f)(x) = g(f(x))$
$= \sqrt[3]{f(x) - 2}$
$= \sqrt[3]{x^3 + 2 - 2}$
$= \sqrt[3]{x^3}$
$= x$

They are inverses.

81. $x^3 - 7x^2 = -36$

$x^3 - 7x^2 + 36 = 0$

Possible roots:

$$\frac{\{\pm1, \pm36, \pm2, \pm18, \pm3, \pm12, \pm4, \pm9, \pm6\}}{\{\pm1\}}$$

$$\begin{array}{r|rrrr} 6 & 1 & -7 & 0 & 36 \\ & & 6 & -6 & -36 \\ \hline & 1 & -1 & -6 & \boxed{0} \end{array}$$

$(x-6)(x^2 - x - 6) = 0$
$(x-6)(x-3)(x+2) = 0$
$x = 6, x = 3, x = -2$

83. $y = \log_2(x-2)$

Graph A; shifts right 2

$y = \log_2(x) - 2$

Graph B; shifts down 2

6.8 Exercises

1. $a_{11}a_{22} - a_{21}a_{12}$

3. Constant

5. Answers will vary.

7. $\begin{cases} 2x + 5y = 7 \\ -3x + 4y = 1 \end{cases}$

$D = \begin{vmatrix} 2 & 5 \\ -3 & 4 \end{vmatrix}$; $D_x = \begin{vmatrix} 7 & 5 \\ 1 & 4 \end{vmatrix}$; $D_y = \begin{vmatrix} 2 & 7 \\ -3 & 1 \end{vmatrix}$

9. $\begin{cases} 4x + y = -11 \\ 3x - 5y = -60 \end{cases}$

$D = \begin{vmatrix} 4 & 1 \\ 3 & -5 \end{vmatrix} = -20 - 3 = -23;$

$D_x = \begin{vmatrix} -11 & 1 \\ -60 & -5 \end{vmatrix} = 55 + 60 = 115;$

$D_y = \begin{vmatrix} 4 & -11 \\ 3 & -60 \end{vmatrix} = -240 + 33 = -207;$

$x = \frac{D_x}{D} = \frac{115}{-23} = -5;$

$y = \frac{D_y}{D} = \frac{-207}{-23} = 9$

$(-5, 9)$

11. $\begin{cases} \frac{x}{8} + \frac{y}{4} = 1 \\ \frac{y}{5} = \frac{x}{2} + 6 \end{cases}$

$\begin{cases} \frac{x}{8} + \frac{y}{4} = 1 \\ -\frac{x}{2} + \frac{y}{5} = 6 \end{cases}$

$D = \begin{vmatrix} \frac{1}{8} & \frac{1}{4} \\ -\frac{1}{2} & \frac{1}{5} \end{vmatrix} = \frac{1}{40} + \frac{1}{8} = \frac{3}{20};$

$D_x = \begin{vmatrix} 1 & \frac{1}{4} \\ 6 & \frac{1}{5} \end{vmatrix} = \frac{1}{5} - \frac{3}{2} = -\frac{13}{10};$

$D_y = \begin{vmatrix} \frac{1}{8} & 1 \\ -\frac{1}{2} & 6 \end{vmatrix} = \frac{3}{4} + \frac{1}{2} = \frac{5}{4};$

$x = \frac{D_x}{D} = \frac{-\frac{13}{10}}{\frac{3}{20}} = -\frac{260}{30} = -\frac{26}{3};$

$y = \frac{D_y}{D} = \frac{\frac{5}{4}}{\frac{3}{20}} = \frac{100}{12} = \frac{25}{3}$

$\left(\frac{-26}{3}, \frac{25}{3}\right)$

13. $\begin{cases} 0.6x - 0.3y = 8 \\ 0.8x - 0.4y = -3 \end{cases}$

$$D = \begin{vmatrix} 0.6 & -0.3 \\ 0.8 & -0.4 \end{vmatrix} = -0.24 + 0.24 = 0$$

No Solution; determinant cannot be zero.

15. a. $\begin{cases} 4x - y + 2z = -5 \\ -3x + 2y - z = 8 \\ x - 5y + 3z = -3 \end{cases}$

$$D = \begin{vmatrix} 4 & -1 & 2 \\ -3 & 2 & -1 \\ 1 & -5 & 3 \end{vmatrix}$$

$$D = 4\begin{vmatrix} 2 & -1 \\ -5 & 3 \end{vmatrix} + 1\begin{vmatrix} -3 & -1 \\ 1 & 3 \end{vmatrix} + 2\begin{vmatrix} -3 & 2 \\ 1 & -5 \end{vmatrix}$$

$D = 4(1) + 1(-8) + 2(13) = 22$; solutions possible

$$D_x = \begin{vmatrix} -5 & -1 & 2 \\ 8 & 2 & -1 \\ -3 & -5 & 3 \end{vmatrix};$$

$$D_y = \begin{vmatrix} 4 & -5 & 2 \\ -3 & 8 & -1 \\ 1 & -3 & 3 \end{vmatrix};$$

$$D_z = \begin{vmatrix} 4 & -1 & -5 \\ -3 & 2 & 8 \\ 1 & -5 & -3 \end{vmatrix}$$

b. $\begin{cases} 4x - y + 2z = -5 \\ -3x + 2y - z = 8 \\ x + y + z = -3 \end{cases}$

$$D = \begin{vmatrix} 4 & -1 & 2 \\ -3 & 2 & -1 \\ 1 & 1 & 1 \end{vmatrix}$$

$$D = 4\begin{vmatrix} 2 & -1 \\ 1 & 1 \end{vmatrix} + 1\begin{vmatrix} -3 & -1 \\ 1 & 1 \end{vmatrix} + 2\begin{vmatrix} -3 & 2 \\ 1 & 1 \end{vmatrix}$$

$D = 4(3) + 1(-2) + 2(-5) = 0$

$D = 0$; Cramer's Rule cannot be used.

17. $\begin{cases} x + 2y + 5z = 10 \\ 3x + 4y - z = 10 \\ x - y - z = -2 \end{cases}$

$$D = \begin{vmatrix} 1 & 2 & 5 \\ 3 & 4 & -1 \\ 1 & -1 & -1 \end{vmatrix} = -36;$$

$$D_x = \begin{vmatrix} 10 & 2 & 5 \\ 10 & 4 & -1 \\ -2 & -1 & -1 \end{vmatrix} = -36;$$

$$D_y = \begin{vmatrix} 1 & 10 & 5 \\ 3 & 10 & -1 \\ 1 & -2 & -1 \end{vmatrix} = -72;$$

$$D_z = \begin{vmatrix} 1 & 2 & 10 \\ 3 & 4 & 10 \\ 1 & -1 & -2 \end{vmatrix} = -36;$$

$$x = \frac{D_x}{D} = \frac{-36}{-36} = 1;$$

$$y = \frac{D_y}{D} = \frac{-72}{-36} = 2;$$

$$z = \frac{D_z}{D} = \frac{-36}{-36} = 1$$

$(1, 2, 1)$

19. $\begin{cases} y + 2z = 1 \\ 4x - 5y + 8z = -8 \\ 8x - 9z = 9 \end{cases}$

$$D = \begin{vmatrix} 0 & 1 & 2 \\ 4 & -5 & 8 \\ 8 & 0 & -9 \end{vmatrix} = 180;$$

$$D_x = \begin{vmatrix} 1 & 1 & 2 \\ -8 & -5 & 8 \\ 9 & 0 & -9 \end{vmatrix} = 135;$$

$$D_y = \begin{vmatrix} 0 & 1 & 2 \\ 4 & -8 & 8 \\ 8 & 9 & -9 \end{vmatrix} = 300;$$

$$D_z = \begin{vmatrix} 0 & 1 & 1 \\ 4 & -5 & -8 \\ 8 & 0 & 9 \end{vmatrix} = -60;$$

$$x = \frac{D_x}{D} = \frac{135}{180} = \frac{3}{4};$$

$$y = \frac{D_y}{D} = \frac{300}{180} = \frac{5}{3};$$

$$z = \frac{D_z}{D} = \frac{-60}{180} = -\frac{1}{3}$$

$$\left(\frac{3}{4},\frac{5}{3},-\frac{1}{3}\right)$$

21. $$\begin{cases} w+2x-3y=-8 \\ x-3y+5z=-22 \\ 4w-5x=5 \\ -y+3z=-11 \end{cases}$$

$$D=\begin{vmatrix} 1 & 2 & -3 & 0 \\ 0 & 1 & -3 & 5 \\ 4 & -5 & 0 & 0 \\ 0 & 0 & -1 & 3 \end{vmatrix}=-16;$$

$$D_w=\begin{vmatrix} -8 & 2 & -3 & 0 \\ -22 & 1 & -3 & 5 \\ 5 & -5 & 0 & 0 \\ -11 & 0 & -1 & 3 \end{vmatrix}=0;$$

$$D_x=\begin{vmatrix} 1 & -8 & -3 & 0 \\ 0 & -22 & -3 & 5 \\ 4 & 5 & 0 & 0 \\ 0 & -11 & -1 & 3 \end{vmatrix}=16;$$

$$D_y=\begin{vmatrix} 1 & 2 & -8 & 0 \\ 0 & 1 & -22 & 5 \\ 4 & -5 & 5 & 0 \\ 0 & 0 & -11 & 3 \end{vmatrix}=-32;$$

$$D_z=\begin{vmatrix} 1 & 2 & -3 & -8 \\ 0 & 1 & -3 & -22 \\ 4 & -5 & 0 & 5 \\ 0 & 0 & -1 & -11 \end{vmatrix}=48;$$

$$w=\frac{D_w}{D}=\frac{0}{-16}=0;$$

$$x=\frac{D_x}{D}=\frac{16}{-16}=-1;$$

$$y=\frac{D_y}{D}=\frac{-32}{-16}=2;$$

$$z=\frac{D_z}{D}=\frac{48}{-16}=-3$$

$(0,-1,2,-3)$

23. $$A=\begin{vmatrix} L & r^2 \\ -\frac{\pi}{2} & W \end{vmatrix}$$

$$A=\begin{vmatrix} 20 & 8^2 \\ -\frac{\pi}{2} & 16 \end{vmatrix}=320-64\left(-\frac{\pi}{2}\right)$$

$$=320+32\pi\approx 420.5\text{ in}^2$$

25. $(2,1),(3,7),(5,3)$

$$A=\frac{\left|\begin{vmatrix} x_1 & y_1 & 1 \\ x_2 & y_2 & 1 \\ x_3 & y_3 & 1 \end{vmatrix}\right|}{2}$$

$$A=\frac{\left|\begin{vmatrix} 2 & 1 & 1 \\ 3 & 7 & 1 \\ 5 & 3 & 1 \end{vmatrix}\right|}{2}=\left|\frac{-16}{2}\right|=8\text{ cm}^2$$

27. $(-4,2),(-6,-1),(3,-1,),(5,2)$

$$A=\frac{\left|\begin{vmatrix} x_1 & y_1 & 1 \\ x_2 & y_2 & 1 \\ x_3 & y_3 & 1 \end{vmatrix}\right|}{2}$$

For triangle use (-4, 2), (-6, -1) and (5, 2).

$$A=\frac{\left|\begin{vmatrix} -4 & 2 & 1 \\ -6 & -1 & 1 \\ 5 & 2 & 1 \end{vmatrix}\right|}{2}=\left|\frac{27}{2}\right|=\frac{27}{2}$$

$$2\left(\frac{27}{2}\right)=2\text{ ft}^2$$

29. $h=6\text{m}$; vertices $(3,5),(-4,2),(-1,6)$

$$A=\frac{\left|\begin{vmatrix} 3 & 5 & 1 \\ -4 & 2 & 1 \\ -1 & 6 & 1 \end{vmatrix}\right|}{2}=\left|\frac{-19}{2}\right|=9.5$$

$$V=\frac{1}{3}Bh$$

$$V=\frac{1}{3}(9.5)(6)$$

$$V=19\text{ m}^3$$

31. $(1,5),(-2,-1),(4,11)$

$$|A|=\begin{vmatrix} 1 & 5 & 1 \\ -2 & -1 & 1 \\ 4 & 11 & 1 \end{vmatrix}=0$$

Yes

33. $(-2.5,5.2),(1.2,-5.6),(2.2,-8.5)$

$$|A|=\begin{vmatrix} -2.5 & 5.2 & 1 \\ 1.2 & -5.6 & 1 \\ 2.2 & -8.5 & 1 \end{vmatrix}=0.07$$

No

35. $2x-3y=7$; (2, -1), (-1.3, -3.2), (-3.1, -4.4)

$(2,-1)$ Yes

$2(2)-3(-1)=7$

$4+3=7$

$7=7;$

$(-1.3,-3.2)$ Yes

$2(-1.3)-3(-3.2)=7$

$-2.6+9.6=7$

$7=7;$

$(-3.1,-4.4)$ Yes

$$|A|=\begin{vmatrix} 2 & -1 & 1 \\ -1.3 & -3.2 & 1 \\ -3.1 & -4.4 & 1 \end{vmatrix}=0$$

37. $\dfrac{3x+2}{(x+3)(x-2)}$

$=\dfrac{A}{x+3}+\dfrac{B}{x-2}$

39. $\dfrac{3x^2-2x+5}{(x-1)(x+2)(x-3)}$

$=\dfrac{A}{x-1}+\dfrac{B}{x+2}+\dfrac{C}{x-3}$

41. $\dfrac{x^2+5}{x(x-3)(x+1)}$

$=\dfrac{A}{x}+\dfrac{B}{x-3}+\dfrac{C}{x+1}$

43. $\dfrac{x^2+x-1}{x^2(x+2)}$

$=\dfrac{A}{x}+\dfrac{B}{x^2}+\dfrac{C}{x+2}$

45. $\dfrac{x^3+3x-2}{(x+1)(x^2+2)^2}$

$=\dfrac{A}{x+1}+\dfrac{Bx+C}{(x^2+2)}+\dfrac{Dx+E}{(x^2+2)^2}$

47. $\dfrac{4-x}{x^2+x}=\dfrac{4-x}{x(x+1)}=\dfrac{A}{x}+\dfrac{B}{x+1}$

$4-x=A(x+1)+Bx$

$4-x=Ax+A+Bx$

$4-x=(A+B)x+A$

$$\begin{cases} A+B=-1 \\ A=4 \end{cases}$$

$$\begin{bmatrix} 1 & 1 \\ 1 & 0 \end{bmatrix}\begin{bmatrix} A \\ B \end{bmatrix}=\begin{bmatrix} -1 \\ 4 \end{bmatrix}$$

$$A^{-1}=\begin{bmatrix} 0 & 1 \\ 1 & -1 \end{bmatrix}$$

$X=A^{-1}B$

$$X=\begin{bmatrix} 4 \\ -5 \end{bmatrix}$$

$\dfrac{4}{x}-\dfrac{5}{x+1}$

49. $\dfrac{2x-27}{2x^2+x-15}=\dfrac{2x-27}{(2x-5)(x+3)}$

$=\dfrac{A}{2x-5}+\dfrac{B}{x+3}$

$2x-27=A(x+3)+B(2x-5)$

$2x-27=Ax+3A+B2x-5B$

$2x-27=(A+2B)x+3A-5B$

$$\begin{cases} A+2B=2 \\ 3A-5B=-27 \end{cases}$$

$$\begin{bmatrix} 1 & 2 \\ 3 & -5 \end{bmatrix}\begin{bmatrix} A \\ B \end{bmatrix}=\begin{bmatrix} 2 \\ -27 \end{bmatrix}$$

$$A^{-1}=\begin{bmatrix} \frac{5}{11} & \frac{2}{11} \\ \frac{3}{11} & \frac{-1}{11} \end{bmatrix}$$

$X=A^{-1}B$

$$X=\begin{bmatrix} -4 \\ 3 \end{bmatrix}$$

$\dfrac{-4}{2x-5}+\dfrac{3}{x+3}$

51. $\dfrac{8x^2-3x-7}{x^3-x}=\dfrac{8x^2-3x-7}{x(x^2-1)}$

$=\dfrac{8x^2-3x-7}{x(x-1)(x+1)}=\dfrac{A}{x}+\dfrac{B}{x+1}+\dfrac{C}{x-1}$

$8x^2-3x-7=A(x+1)(x-1)+Bx(x-1)+Cx(x+1)$

$8x^2-3x-7=Ax^2-A+Bx^2-Bx+Cx^2+Cx$

$8x^2-3x-7=x^2(A+B+C)+x(-B+C)-A$

$$\begin{cases} A+B+C=8 \\ -B+C=-3 \\ -A=-7 \end{cases}$$

$$\begin{bmatrix} 1 & 1 & 1 \\ 0 & -1 & 1 \\ -1 & 0 & 0 \end{bmatrix}\begin{bmatrix} A \\ B \\ C \end{bmatrix}=\begin{bmatrix} 8 \\ -3 \\ -7 \end{bmatrix}$$

$$A^{-1}=\begin{bmatrix} 0 & 0 & -1 \\ \frac{1}{2} & \frac{-1}{2} & \frac{1}{2} \\ \frac{1}{2} & \frac{1}{2} & \frac{1}{2} \end{bmatrix}$$

$$X=A^{-1}B$$

$$X=\begin{bmatrix} 7 \\ 2 \\ -1 \end{bmatrix}$$

$$\frac{7}{x}+\frac{2}{x+1}-\frac{1}{x-1}$$

53. $\dfrac{3x^2+7x-1}{x^3+2x^2+x}=\dfrac{3x^2+7x-1}{x(x^2+2x+1)}$

$$=\frac{3x^2+7x-1}{x(x+1)^2}=\frac{A}{x}+\frac{B}{x+1}+\frac{C}{(x+1)^2}$$

$$3x^2+7x-1=A(x+1)^2+Bx(x+1)+Cx$$

$$3x^2+7x-1=Ax^2+2Ax+A+Bx^2+Bx+Cx$$

$$3x^2+7x-1=x^2(A+B)+x(2A+B+C)+A$$

$$\begin{cases} A+B=3 \\ 2A+B+C=7 \\ A=-1 \end{cases}$$

$$\begin{bmatrix} 1 & 1 & 0 \\ 2 & 1 & 1 \\ 1 & 0 & 0 \end{bmatrix}\begin{bmatrix} A \\ B \\ C \end{bmatrix}=\begin{bmatrix} 3 \\ 7 \\ -1 \end{bmatrix}$$

$$A^{-1}=\begin{bmatrix} 0 & 0 & 1 \\ 1 & 0 & -1 \\ -1 & 1 & -1 \end{bmatrix}$$

$$X=A^{-1}B$$

$$X=\begin{bmatrix} -1 \\ 4 \\ 5 \end{bmatrix}$$

$$\frac{-1}{x}+\frac{4}{x+1}+\frac{5}{(x+1)^2}$$

55. $\dfrac{3x^2+10x+4}{8-x^3}=\dfrac{3x^2+10x+4}{(2-x)(4+2x+x^2)}$

$$=\frac{A}{2-x}+\frac{Bx+C}{4+2x+x^2}$$

$$3x^2+10x+4=A(4+2x+x^2)+(Bx+C)(2-x)$$

$$3x^2+10x+4$$
$$=4A+2Ax+Ax^2+2Bx-Bx^2+2C-Cx$$

$$3x^2+10x+4$$
$$=x^2(A-B)+x(2A+2B-C)+4A+2C$$

$$\begin{cases} A-B=3 \\ 2A+2B-C=10 \\ 4A+2C=4 \end{cases}$$

$$\begin{bmatrix} 1 & -1 & 0 \\ 2 & 2 & -1 \\ 4 & 0 & 2 \end{bmatrix}\begin{bmatrix} A \\ B \\ C \end{bmatrix}=\begin{bmatrix} 3 \\ 10 \\ 4 \end{bmatrix}$$

$$A^{-1}=\begin{bmatrix} \frac{1}{3} & \frac{1}{6} & \frac{1}{12} \\ \frac{-2}{3} & \frac{1}{6} & \frac{1}{12} \\ \frac{-2}{3} & \frac{-1}{3} & \frac{1}{3} \end{bmatrix}$$

$$X=A^{-1}B$$

$$X=\begin{bmatrix} 3 \\ 0 \\ -4 \end{bmatrix}$$

$$\frac{3}{2-x}-\frac{4}{4+2x+x^2}$$

57. $\dfrac{x^4-x^2-2x+1}{x^5-2x^3+x}=\dfrac{x^4-x^2-2x+1}{x(x^4-2x^2+1)}$

$$=\frac{x^4-x^2-2x+1}{x(x^2-1)^2}=\frac{A}{x}+\frac{Bx+C}{x^2-1}+\frac{Dx+E}{(x^2-1)^2}$$

$$x^4-x^2-2x+1$$
$$=A(x^2-1)^2+x(Bx+C)(x^2-1)+x(Dx+E)$$

$$x^4-x^2-2x+1$$
$$=Ax^4-2Ax^2+A+Bx^4-Bx^2$$
$$+Cx^3-Cx+Dx^2+Ex$$

$$x^4-x^2-2x+1$$
$$=x^4(A+B)+Cx^3+x^2(-2A-B+D)$$
$$+x(-C+E)+A$$

$$\begin{cases} A+B=1 \\ C=0 \\ -2A-B+D=-1 \\ -C+E=-2 \\ A=1 \end{cases}$$

$$\begin{bmatrix} 1 & 1 & 0 & 0 & 0 \\ 0 & 0 & 1 & 0 & 0 \\ -2 & -1 & 0 & 1 & 0 \\ 0 & 0 & -1 & 0 & 1 \\ 1 & 0 & 0 & 0 & 0 \end{bmatrix}\begin{bmatrix} A \\ B \\ C \\ D \\ E \end{bmatrix}=\begin{bmatrix} 1 \\ 0 \\ -1 \\ -2 \\ 1 \end{bmatrix}$$

$$A^{-1}=\begin{bmatrix} 0 & 0 & 0 & 0 & 1 \\ 1 & 0 & 0 & 0 & -1 \\ 0 & 1 & 0 & 0 & 0 \\ 1 & 0 & 1 & 0 & 1 \\ 0 & 1 & 0 & 1 & 0 \end{bmatrix}$$

$$X=A^{-1}B$$

$$X=\begin{bmatrix} 1 \\ 0 \\ 0 \\ 1 \\ -2 \end{bmatrix}$$

$$\frac{1}{x}+\frac{x-2}{\left(x^2-1\right)^2}$$

59. Let x represent the rate of the \$15000 investment.
Let y represent the rate of the \$25000 investment.

$$\begin{cases} 15000x+25000y=2900 \\ 25000x+15000y=2700 \end{cases}$$

$$D=\begin{vmatrix} 15000 & 25000 \\ 25000 & 15000 \end{vmatrix}=-400000000;$$

$$D_x=\begin{vmatrix} 2900 & 25000 \\ 2700 & 15000 \end{vmatrix}=-24000000;$$

$$D_y=\begin{vmatrix} 15000 & 2900 \\ 25000 & 2700 \end{vmatrix}=-32000000;$$

$$x=\frac{D_x}{D}=\frac{-24000000}{-400000000}=0.06;$$

$$y=\frac{D_y}{D}=\frac{-32000000}{-400000000}=0.08$$

\$15000 invested at 6%
\$25000 invested at 8%

61. Model:

$$f(x)=0.0257x^4-0.715x^3+6.7852x^2-25.1022x+35.1111$$

May: 5.9 x 1000; Revenue ≈ \$5,917.60

July: 8.3 x 1000; Revenue ≈ \$8,331.20

November: (4.6048>4.0847); more revenue is earned in November than April.

63. $\begin{cases} x+3y+5z=6 \\ 2x-4y+6z=14 \\ 9x-6y+3z=3 \end{cases}$

(1) -2R1 + R2

$$\begin{cases} -2x-6y-10z=-12 \\ 2x-4y+6z=14 \end{cases}$$
$$-10y-4z=2$$

-9R1 + R3.

$$\begin{cases} -9x-27y-45z=-54 \\ 9x-6y+3z=3 \end{cases}$$
$$-33y-42z=-51$$

$$\begin{cases} -10y-4z=2 \\ -33y-42z=-51 \end{cases}$$

-21R1 + 2R2

$$\begin{cases} 210y+84z=-42 \\ -66y-84z=-102 \end{cases}$$
$$144y=-144$$
$$y=-1;$$

$$210y+84z=-42$$
$$210(-1)+84z=-42$$
$$-210+84z=-42$$
$$84z=168$$
$$z=2;$$

$$x+3y+5z=6$$
$$x+3(-1)+5(2)=6$$
$$x-3+10=6$$
$$x=-1$$

(-1, -1, 2)

(2) $\left[\begin{array}{ccc|c} 1 & 3 & 5 & 6 \\ 2 & -4 & 6 & 14 \\ 9 & -6 & 3 & 3 \end{array}\right]$ -2R1 + R2 → R2

$\left[\begin{array}{ccc|c} 1 & 3 & 5 & 6 \\ 0 & -10 & -4 & 2 \\ 9 & -6 & 3 & 3 \end{array}\right]$ -9R1 + R3 → R3

$\left[\begin{array}{ccc|c} 1 & 3 & 5 & 6 \\ 0 & -10 & -4 & 2 \\ 0 & -33 & -42 & -51 \end{array}\right]$ $-\frac{1}{10}R2 \rightarrow R2$

$$\left[\begin{array}{ccc|c} 1 & 3 & 5 & 6 \\ 0 & 1 & \frac{2}{5} & -\frac{1}{5} \\ 0 & -33 & -42 & -51 \end{array}\right] 33R2 + R3 \rightarrow R3$$

$$\left[\begin{array}{ccc|c} 1 & 3 & 5 & 6 \\ 0 & 1 & \frac{2}{5} & -\frac{1}{5} \\ 0 & 0 & -\frac{144}{5} & -\frac{288}{5} \end{array}\right] -\frac{5}{144}R3 \rightarrow R3$$

$$\left[\begin{array}{ccc|c} 1 & 3 & 5 & 6 \\ 0 & 1 & \frac{2}{5} & -\frac{1}{5} \\ 0 & 0 & 1 & 2 \end{array}\right]$$

$z = 2$;

$$y + \frac{2}{5}z = -\frac{1}{5}$$
$$y + \frac{2}{5}(2) = -\frac{1}{5}$$
$$y + \frac{4}{5} = -\frac{1}{5}$$
$$y = -1;$$
$$x + 3y + 5z = 6$$
$$x + 3(-1) + 5(2) = 6$$
$$x - 3 + 10 = 6$$
$$x + 7 = 6$$
$$x = -1$$

(-1, -1, 2)

(3) $\begin{cases} x + 3y + 5z = 6 \\ 2x - 4y + 6z = 14 \\ 9x - 6y + 3z = 3 \end{cases}$

$$D = \begin{bmatrix} 1 & 3 & 5 \\ 2 & -4 & 6 \\ 9 & -6 & 3 \end{bmatrix} = 288;$$

$$D_x = \begin{bmatrix} 6 & 3 & 5 \\ 14 & -4 & 6 \\ 3 & -6 & 3 \end{bmatrix} = -288;$$

$$D_y = \begin{bmatrix} 1 & 6 & 5 \\ 2 & 14 & 6 \\ 9 & 3 & 3 \end{bmatrix} = -288;$$

$$D_z = \begin{bmatrix} 1 & 3 & 6 \\ 2 & -4 & 14 \\ 9 & -6 & 3 \end{bmatrix} = 576;$$

$$x = \frac{D_x}{D} = \frac{-288}{288} = -1;$$

$$y = \frac{D_y}{D} = \frac{-288}{288} = -1;$$

$$z = \frac{D_z}{D} = \frac{576}{288} = 2$$

(-1, -1, 2)

(4) $\begin{bmatrix} 1 & 3 & 5 \\ 2 & -4 & 6 \\ 9 & -6 & 3 \end{bmatrix}\begin{bmatrix} x \\ y \\ z \end{bmatrix} = \begin{bmatrix} 6 \\ 14 \\ 3 \end{bmatrix}$

$$A^{-1} = \begin{bmatrix} \frac{1}{12} & \frac{-13}{96} & \frac{19}{144} \\ \frac{1}{6} & \frac{-7}{48} & \frac{1}{72} \\ \frac{1}{12} & \frac{11}{96} & \frac{-5}{144} \end{bmatrix}$$

$$X = A^{-1}B$$

$$X = \begin{bmatrix} -1 \\ -1 \\ 2 \end{bmatrix}$$

(-1, -1, 2)

Answers will vary.

65. $(-5,-5), (5,-5), (8,6), (-8,6), (0,12.5)$

First triangle vertices:
(-8, 6), (0, 12.5), (8, 6)

$$A = \frac{\left|\begin{vmatrix} -8 & 6 & 1 \\ 0 & 12.5 & 1 \\ 8 & 6 & 1 \end{vmatrix}\right|}{2} = \left|\frac{-104}{2}\right| = 52$$

Second triangle vertices:
(-8, 6), (8, 6) (-5, -5)

$$A = \frac{\left|\begin{vmatrix} -8 & 6 & 1 \\ 8 & 6 & 1 \\ -5 & -5 & 1 \end{vmatrix}\right|}{2} = \left|\frac{-176}{2}\right| = 88$$

Area of third triangle is same as second.
Third triangle vertices:
(8, 6), (-5, -5), (5, -5)

$$A = \frac{\left|\begin{vmatrix} 8 & 6 & 1 \\ -5 & -5 & 1 \\ 5 & -5 & 1 \end{vmatrix}\right|}{2} = \left|\frac{-110}{2}\right| = 55$$

Total Area: 52 + 88 + 55 = 195 u^2

67. $2 - \sqrt{3}i$ and $2 + \sqrt{3}i$

Sum: $2 - \sqrt{3}i + 2 + \sqrt{3}i = 4$

Difference:

$2-\sqrt{3}i-\left(2+\sqrt{3}i\right)$
$=2-\sqrt{3}i-2-\sqrt{3}i$
$=-2\sqrt{3}i$

Product:

$\left(2-\sqrt{3}i\right)\left(2+\sqrt{3}i\right)$
$=4-3i^2=4+3=7$

Quotient:

$$\frac{2-\sqrt{3}i}{2+\sqrt{3}i}\cdot\frac{2-\sqrt{3}i}{2-\sqrt{3}i}$$
$$=\frac{4-4\sqrt{3}i+3i^2}{4-3i^2}$$
$$=\frac{4-4\sqrt{3}i-3}{4+3}$$
$$=\frac{1-4\sqrt{3}i}{7}=\frac{1}{7}-\frac{4\sqrt{3}}{7}i$$

69. $3^{2x-1}=9^{2-x}$

$$\ln 3^{2x-1}=\ln 9^{2-x}$$
$$(2x-1)\ln 3=(2-x)\ln 9$$
$$2x\ln 3-\ln 3=2\ln 9-x\ln 9$$
$$2x\ln 3+x\ln 9=2\ln 9+\ln 3$$
$$x(2\ln 3+\ln 9)=2\ln 9+\ln 3$$
$$x=\frac{2\ln 9+\ln 3}{2\ln 3+\ln 9}$$
$$x=1.25;$$

$3^{2x-1}=9^{2-x}$
$3^{2x-1}=\left(3^2\right)^{2-x}$
$3^{2x-1}=3^{4-2x}$
$2x-1=4-2x$
$4x=5$
$x=1.25$

71. $g(x)=-|x+1|+3$

Left graph, graph shifts left 1
Vertex: (-1, 3)
Opens down

Chapter 6 Summary and Review

1. $\begin{cases}3x-2y=4\\-x+3y=8\end{cases}$

$\begin{cases}y=\frac{3}{2}x-2\\y=\frac{1}{3}x+\frac{8}{3}\end{cases}$

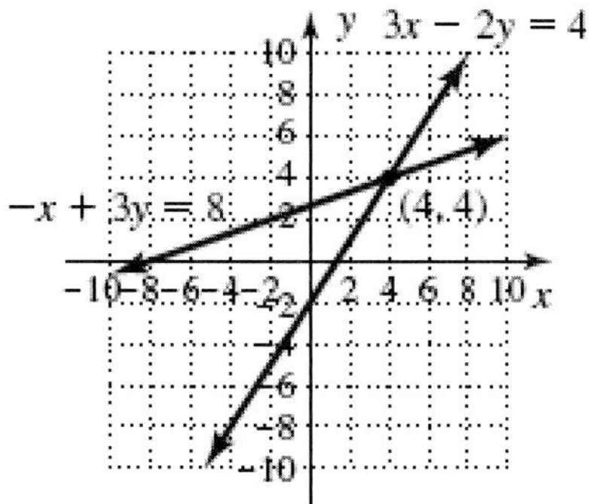

2. $\begin{cases}0.2x+0.5y=-1.4\\x-0.3y=1.4\end{cases}$

$\begin{cases}y=-\frac{2}{5}x-\frac{14}{5}\\y=\frac{10}{3}x-\frac{14}{3}\end{cases}$

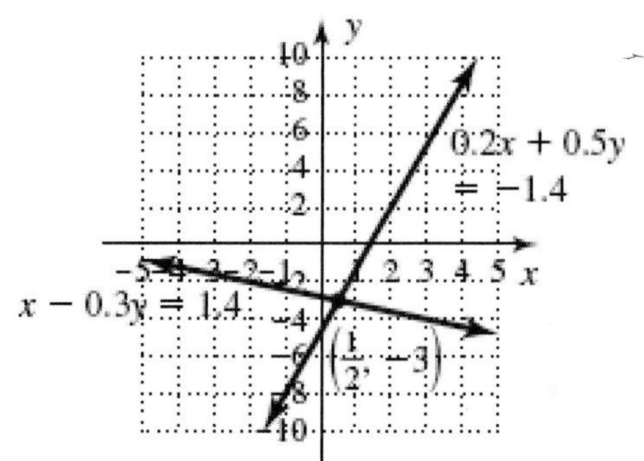

3. $\begin{cases}2x+y=2\\x-2y=4\end{cases}$

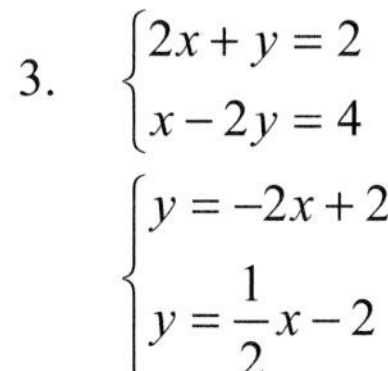
$\begin{cases}y=-2x+2\\y=\frac{1}{2}x-2\end{cases}$

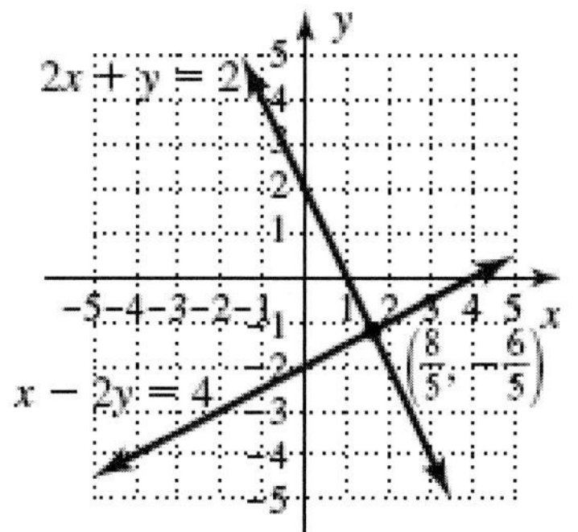

4. $\begin{cases} y = 5 - x \\ 2x + 2y = 13 \end{cases}$

$$2x + 2y = 13$$
$$2x + 2(5 - x) = 13$$
$$2x + 10 - 2x = 13$$
$$10 \neq 13$$

No solution; inconsistent

5. $\begin{cases} x + y = 4 \\ 0.4x + 0.3y = 1.7 \end{cases}$

$$x = 4 - y$$
$$0.4x + 0.3y = 1.7$$
$$0.4(4 - y) + 0.3y = 1.7$$
$$1.6 - 0.4y + 0.3y = 1.7$$
$$-0.1y = 0.1$$
$$y = -1;$$
$$x + y = 4$$
$$x + (-1) = 4$$
$$x - 1 = 4$$
$$x = 5$$

(5, -1); consistent

6. $\begin{cases} x - 2y = 3 \\ x - 4y = -1 \end{cases}$

$$x = 3 + 2y$$
$$x - 4y = -1$$
$$3 + 2y - 4y = -1$$
$$3 - 2y = -1$$
$$-2y = -4$$
$$y = 2;$$
$$x - 2y = 3$$
$$x - 2(2) = 3$$
$$x - 4 = 3$$
$$x = 7$$

(7, 2); consistent

7. $\begin{cases} 2x - 4y = 10 \\ 3x + 4y = 5 \end{cases}$

R1 + R2 = Sum

$$2x - 4y = 10$$
$$\underline{3x + 4y = 5}$$
$$5x = 15$$
$$x = 3;$$
$$2x - 4y = 10$$
$$2(3) - 4y = 10$$
$$6 - 4y = 10$$
$$-4y = 4$$
$$y = -1$$

(3, -1); consistent

8. $\begin{cases} -x + 5y = 8 \\ x + 2y = 6 \end{cases}$

R1 + R2 = Sum

$$-x + 5y = 8$$
$$\underline{x + 2y = 6}$$
$$7y = 14$$
$$y = 2;$$
$$x + 2y = 6$$
$$x + 2(2) = 6$$
$$x + 4 = 6$$
$$x = 2$$

(2, 2); consistent

9. $\begin{cases} 2x = 3y + 6 \\ 2.4x + 3.6y = 6 \end{cases}$

$\begin{cases} 2x - 3y = 6 \\ 24x + 36y = 60 \end{cases}$

12R1 + R2 = Sum

$$24x - 36y = 72$$
$$\underline{24x + 36y = 60}$$
$$48x = 132$$
$$x = \frac{11}{4};$$
$$2x = 3y + 6$$
$$2\left(\frac{11}{4}\right) = 3y + 6$$
$$\frac{11}{2} = 3y + 6$$
$$-\frac{1}{2} = 3y$$
$$-\frac{1}{6} = y$$

$\left(\frac{11}{4}, -\frac{1}{6}\right)$; consistent

10. Let h represent the height of the John Hancock building.
Let s represent the height of the Sears Tower.

$\begin{cases} h + s = 2577 \\ s = h + 323 \end{cases}$

$$h + s = 2577$$
$$h + h + 323 = 2577$$
$$2h = 2254$$
$$h = 1127$$

$$\begin{aligned} h+s &= 2577 \\ 1127+s &= 2577 \\ s &= 1450 \end{aligned}$$

Sears Tower: 1450 feet
Hancock Building: 1127 feet

11. $\begin{cases} x+y-2z=-1 \\ 4x-y+3z=3 \\ 3x+2y-z=4 \end{cases}$

R1 + R2 = Sum

$$\begin{array}{r} x+y-2z=-1 \\ 4x-y+3z=3 \\ \hline 5x+z=2 \end{array}$$

-2R1 + R3 = Sum

$$\begin{array}{r} -2x-2y+4z=2 \\ 3x+2y-z=4 \\ \hline x+3z=6 \end{array}$$

$\begin{cases} 5x+z=2 \\ x+3z=6 \end{cases}$

-3R1 + R2 = Sum

$$\begin{array}{r} -15x-3z=-6 \\ x+3z=6 \\ \hline -14x=0 \\ x=0; \end{array}$$

$$\begin{aligned} x+3z &= 6 \\ 0+3z &= 6 \\ 3z &= 6 \\ z &= 2; \end{aligned}$$

$$\begin{aligned} x+y-2z &= -1 \\ 0+y-2(2) &= -1 \\ y-4 &= -1 \\ y &= 3 \end{aligned}$$

(0, 3, 2)

12. $\begin{cases} -x+y+2z=2 \\ x+y-z=1 \\ 2x+y+z=4 \end{cases}$

R1 + R2 = Sum

$$\begin{array}{r} -x+y+2z=2 \\ x+y-z=1 \\ \hline 2y+z=3 \end{array}$$

2R1 + R3 = Sum

$$\begin{array}{r} -2x+2y+4z=4 \\ 2x+y+z=4 \\ \hline 3y+5z=8 \end{array}$$

$\begin{cases} 2y+z=3 \\ 3y+5z=8 \end{cases}$

-5R1 + R2 = Sum

$$\begin{array}{r} -10y-5z=-15 \\ 3y+5z=8 \\ \hline -7y=-7 \\ y=1; \end{array}$$

$$\begin{aligned} 2y+z &= 3 \\ 2(1)+z &= 3 \\ 2+z &= 3 \\ z &= 1; \end{aligned}$$

$$\begin{aligned} x+y-z &= 1 \\ x+1-1 &= 1 \\ x &= 1 \end{aligned}$$

(1, 1, 1)

13. $\begin{cases} 3x+y+2z=3 \\ x-2y+3z=1 \\ 4x-8y+12z=7 \end{cases}$

-4R2 + R3 = Sum

$$\begin{array}{r} -4x+8y-12z=-4 \\ 4x-8y+12z=7 \\ \hline 0 \neq 3 \end{array}$$

No solution; inconsistent

14. Let n represent the number of nickels.
Let d represent the number of dimes.
Let q represent the number of quarters.

$\begin{cases} n+d+q=217 \\ n=q+12 \\ 0.10d=0.05n+4.90 \end{cases}$

$\begin{cases} n+d+q=217 \\ n-q=12 \\ -5n+10d=490 \end{cases}$

-10R1 + R3 = Sum

$$\begin{array}{r} -10n-10d-10q=-2170 \\ -5n+10d=490 \\ \hline -15n-10q=-1680 \end{array}$$

$\begin{cases} n-q=12 \\ -15n-10q=-1680 \end{cases}$

15R1 + R2 = Sum

$$\begin{array}{r} 15n-15q=180 \\ -15n-10q=-1680 \\ \hline -25q=-1500 \\ q=60; \end{array}$$

$$\begin{aligned} n-q &= 12 \\ n-60 &= 12 \\ n &= 72; \end{aligned}$$

$$n + d + q = 217$$
$$72 + d + 60 = 217$$
$$d + 132 = 217$$
$$d = 85$$

72 nickels, 85 dimes, 60 quarters

15. Let c represent the amount invested in the certificate of deposit.
Let m represent the amount invested in the money market certificate.
Let b represent the amount invested in Aa bonds.

$$\begin{cases} c + m + b = 280 \\ 0.04c + 0.05m + 0.07b = 15.4 \\ b = m + 20 \end{cases}$$

$$\begin{cases} c + m + b = 280 \\ 4c + 5m + 7b = 1540 \\ b - m = 20 \end{cases}$$

-4R1 + R2 = Sum

$$-4c - 4m - 4b = -1120$$
$$\underline{4c + 5m + 7b = 1540}$$
$$m + 3b = 420$$

$$\begin{cases} b - m = 20 \\ 3b + m = 420 \end{cases}$$

R1 + R2 = Sum

$$b - m = 20$$
$$\underline{3b + m = 420}$$
$$4b = 440$$
$$b = 110;$$

$$b = m + 20$$
$$110 = m + 20$$
$$90 = m;$$

$$c + m + b = 280$$
$$c + 90 + 110 = 280$$
$$c + 200 = 280$$
$$c = 80$$

\$80,000 at 4%
\$90,000 at 5%
\$110,000 at 7%

16. $\begin{cases} -x - y > -2 \\ -x + y < -4 \end{cases}$

$$\begin{cases} y < -x + 2 \\ y < x - 4 \end{cases}$$

Test point: (2, -5)

$$-x - y > -2$$
$$-2 - (-5) > -2$$
$$-2 + 5 > -2$$
$$3 > -2;$$

$$-x + y < -4$$
$$-2 + (-5) < -4$$
$$-7 < -4$$

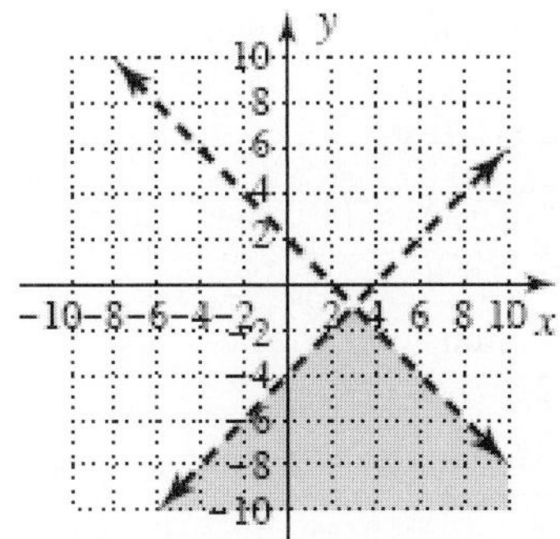

17. $\begin{cases} x - 4y \le 5 \\ -x + 2y \le 0 \end{cases}$

$$\begin{cases} y \ge \frac{1}{4}x - \frac{5}{4} \\ y \le \frac{1}{2}x \end{cases}$$

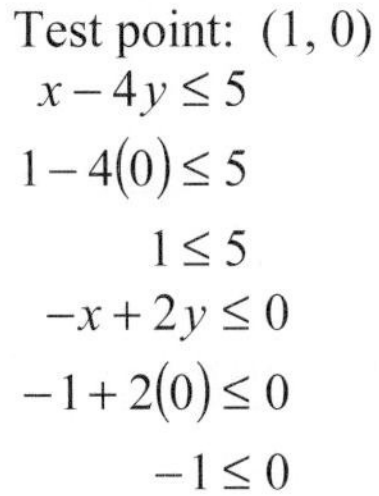

Test point: (1, 0)

$$x - 4y \le 5$$
$$1 - 4(0) \le 5$$
$$1 \le 5$$
$$-x + 2y \le 0$$
$$-1 + 2(0) \le 0$$
$$-1 \le 0$$

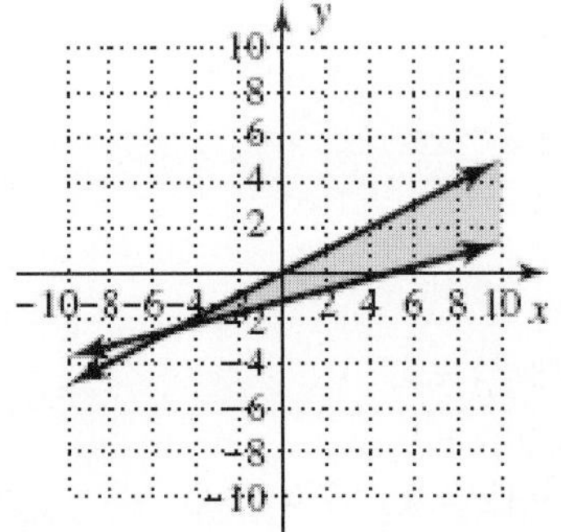

18. $\begin{cases} x + 2y \ge 1 \\ 2x - y \le -2 \end{cases}$

$$\begin{cases} y \ge -\frac{1}{2}x + \frac{1}{2} \\ y \ge 2x + 2 \end{cases}$$

Test point: (0, 4)

$$x + 2y \ge 1$$
$$0 + 2(4) \ge 1$$
$$8 \ge 1;$$

$$2x - y \le -2$$
$$2(0) - 4 \le -2$$
$$-4 \le -2$$

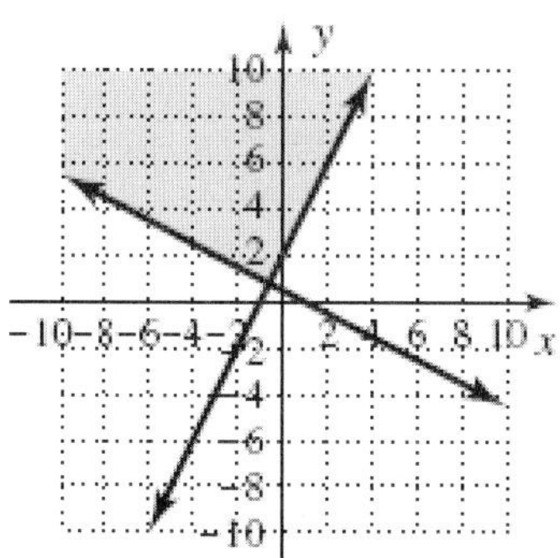

19. $\begin{cases} x + y \le 7 \\ 2x + y \le 10 \\ 2x + 3y \le 18 \\ x \ge 0 \\ y \ge 0 \end{cases}$

$\begin{cases} y \le -x + 7 \\ y \le -2x + 10 \\ y \le -\dfrac{2}{3}x + 6 \\ x \ge 0 \\ y \ge 0 \end{cases}$

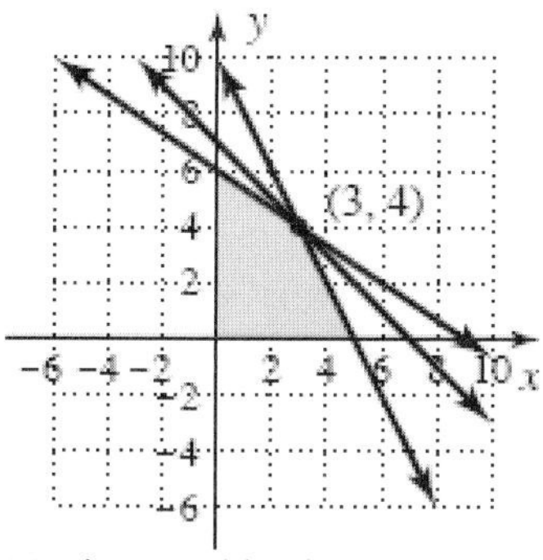

Maximum Objective Function:
$f(x, y) = 30x + 45y$

Corner Point	Objective Function $f(x, y) = 30x + 45y$	Result
(0, 0)	$f(0,0) = 30(0) + 45(0)$	0
(0, 6)	$f(0,6) = 30(0) + 45(6)$	270
(3, 4)	$f(3,4) = 30(3) + 45(4)$	270
(5, 0)	$f(5,0) = 30(5) + 45(0)$	150

Maximum value: (3, 4) and (0, 6)

20. $\begin{cases} 3m + 2e \le 1000 \\ 2m + 1e \le 525 \\ m \ge 0 \\ e \ge 0 \end{cases}$

$\begin{cases} m \le \dfrac{-2e + 1000}{3} \\ m \le \dfrac{-1e + 525}{2} \\ m \ge 0 \\ e \ge 0 \end{cases}$

Corner Point	Maximize $P(e, m) = 85m + 50e$	Result
(0, 0)	$P(e, m) = 85(0) + 50(0)$	0
(0,267.5)	$P(e, m) = 85(267.5) + 50(0)$	22737.50
(425,50)	$P(e, m) = 85(50) + 50(425)$	25,500
(500,0)	$P(e, m) = 85(0) + 50(500)$	25,000

Maximize profits with 50 milk cows, 425 egg-laying chickens.

21. $f(x) = |x - 3|$;
$f(x) = |0 - 3| = 3$
y-intercept: (0,3)
vertex: (3,0)

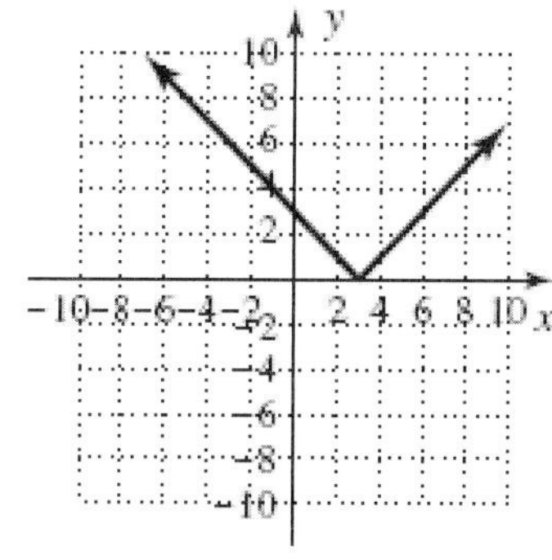

22. $g(x) = 2|x + 2|$;
$g(x) = 2|0 + 2| = 4$
y-intercept: (0,4)
vertex: (-2,0)

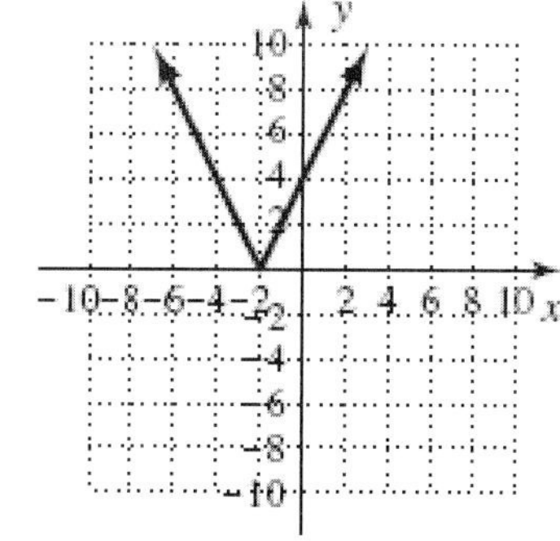

23. $h(x)=|-2x+3|$;
$h(x)=|-2(0)+3|=3$
y-intercept: (0,3)
vertex: (1.5,0)

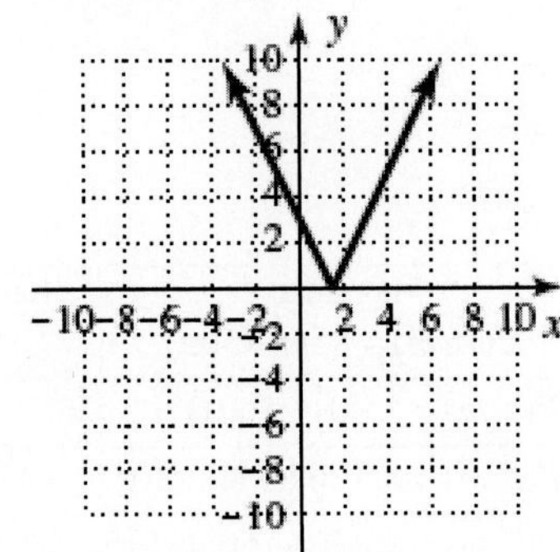

24. $\frac{|2x+5|}{3}+8=9$
$\frac{|2x+5|}{3}=1$
$|2x+5|=3$
$\begin{cases} y_1=|2x+5| \\ y_2=3 \end{cases}$

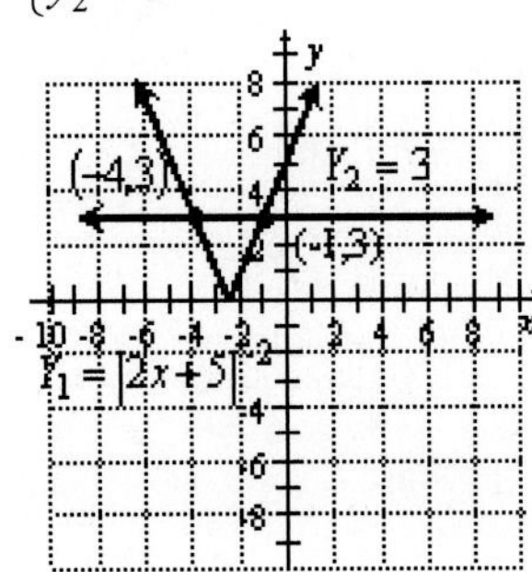

$x=-4;\ x=-1$

25. $-3|x+2|-2<-14$
$-3|x+2|<-12$
$|x+2|>4$
$\begin{cases} y_1=|x+2| \\ y_2=4 \end{cases}$

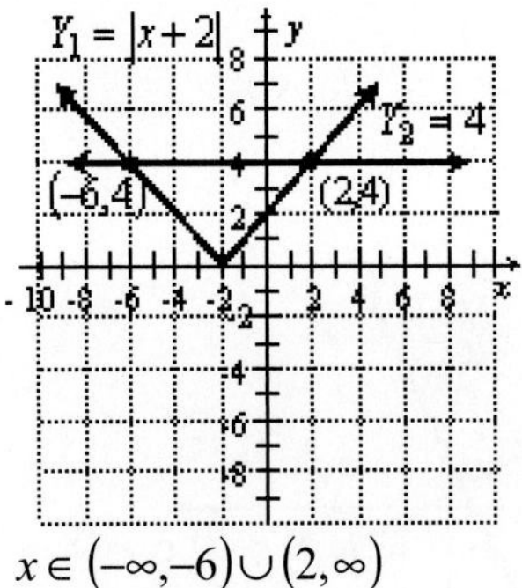

$x\in(-\infty,-6)\cup(2,\infty)$

26. $\left|\frac{x}{2}-9\right|\geq -7$

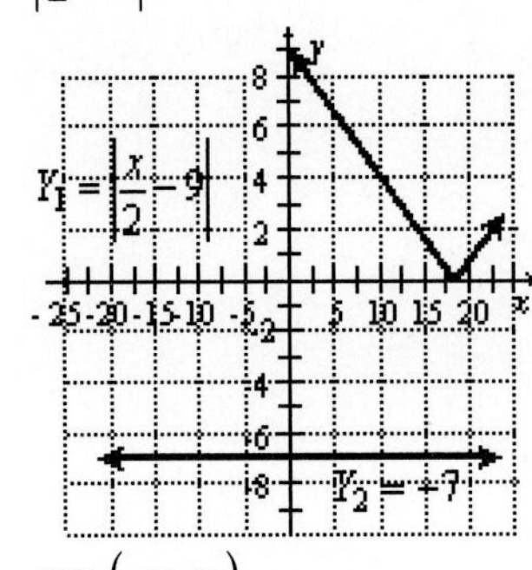

$x\in(-\infty,\infty)$

27. $5|m-2|-12\leq 8$
$5|m-2|\leq 20$
$|m-2|\leq 4$
$m-2\leq 4$ and $m-2\geq -4$
$m\leq 6$ and $m\geq -2$
$x\in[-2,6]$

28. $\frac{|3x-2|}{2}+6=4$
$\frac{|3x-2|}{2}=-2$
$|3x-2|=-4$
No solution

29. $|0.2x-3.1|-1.9<3.2$
$|0.2x-3.1|<5.1$
$0.2x-3.1<5.1$ and $0.2x-3.1>-5.1$
$0.2x<8.2$ and $0.2x>-2$
$x<41$ and $x>-10$
$x\in(-10,41)$

30. $\begin{cases} x-2y=6 \\ 4x-3y=4 \end{cases}$

$$\left[\begin{array}{cc|c} 1 & -2 & 6 \\ 4 & -3 & 4 \end{array}\right] -4R1+R2 \to R2$$

$$\left[\begin{array}{cc|c} 1 & -2 & 6 \\ 0 & 5 & -20 \end{array}\right] \frac{1}{5}R2 \to R2$$

$$\left[\begin{array}{cc|c} 1 & -2 & 6 \\ 0 & 1 & -4 \end{array}\right]$$

$y=-4$;

$$x-2y=6$$
$$x-2(-4)=6$$
$$x+8=6$$
$$x=-2$$

(-2, -4)

31. $\begin{cases} x-2y+2z=7 \\ 2x+2y-z=5 \\ 3x-y+z=6 \end{cases}$

$$\left[\begin{array}{ccc|c} 1 & -2 & 2 & 7 \\ 2 & 2 & -1 & 5 \\ 3 & -1 & 1 & 6 \end{array}\right] -2R1+R2 \to R2$$

$$\left[\begin{array}{ccc|c} 1 & -2 & 2 & 7 \\ 0 & 6 & -5 & -9 \\ 3 & -1 & 1 & 6 \end{array}\right] -3R1+R3 \to R3$$

$$\left[\begin{array}{ccc|c} 1 & -2 & 2 & 7 \\ 0 & 6 & -5 & -9 \\ 0 & 5 & -5 & -15 \end{array}\right] \frac{1}{6}R2 \to R2$$

$$\left[\begin{array}{ccc|c} 1 & -2 & 2 & 7 \\ 0 & 1 & -\frac{5}{6} & -\frac{3}{2} \\ 0 & 5 & -5 & -15 \end{array}\right] -5R2+R3 \to R3$$

$$\left[\begin{array}{ccc|c} 1 & -2 & 2 & 7 \\ 0 & 1 & -\frac{5}{6} & -\frac{3}{2} \\ 0 & 0 & -\frac{5}{6} & -\frac{15}{2} \end{array}\right] -\frac{6}{5}R3 \to R3$$

$$\left[\begin{array}{ccc|c} 1 & -2 & 2 & 7 \\ 0 & 1 & -\frac{5}{6} & -\frac{3}{2} \\ 0 & 0 & 1 & 9 \end{array}\right]$$

$z=9$;

$$y-\frac{5}{6}z=-\frac{3}{2}$$
$$y-\frac{5}{6}(9)=-\frac{3}{2}$$
$$y-\frac{15}{2}=-\frac{3}{2}$$
$$y=-\frac{3}{2}+\frac{15}{2}$$

$y=6$;

$$x-2y+2z=7$$
$$x-2(6)+2(9)=7$$
$$x-12+18=7$$
$$x+6=7$$
$$x=1$$

$(1,6,9)$

32. $\begin{cases} 2w+x+2y-3z=-19 \\ w-2x-y+4z=15 \\ x+2y-z=1 \\ 3w-2x-5z=-60 \end{cases}$

(-2, 7, 1, 8)

33. $A+B$

$$=\begin{bmatrix} \frac{-1}{4} & \frac{-3}{4} \\ \frac{-1}{8} & \frac{-7}{8} \end{bmatrix}+\begin{bmatrix} -7 & 6 \\ 1 & -2 \end{bmatrix}$$

$$=\begin{bmatrix} \frac{-1}{4}+(-7) & \frac{-3}{4}+6 \\ \frac{-1}{8}+1 & \frac{-7}{8}+(-2) \end{bmatrix}$$

$$=\begin{bmatrix} -7.25 & 5.25 \\ 0.875 & -2.875 \end{bmatrix}$$

34. $B-A$

$$=\begin{bmatrix} -7 & 6 \\ 1 & -2 \end{bmatrix}-\begin{bmatrix} \frac{-1}{4} & \frac{-3}{4} \\ \frac{-1}{8} & \frac{-7}{8} \end{bmatrix}$$

$$=\begin{bmatrix} -7-\left(\frac{-1}{4}\right) & 6-\left(\frac{-3}{4}\right) \\ 1-\left(\frac{-1}{8}\right) & -2-\left(\frac{-7}{8}\right) \end{bmatrix}$$

$$=\begin{bmatrix} -6.75 & 6.75 \\ 1.125 & -1.125 \end{bmatrix}$$

35. $C-B$

$$=\begin{bmatrix}-1 & 3 & 4\\ 5 & -2 & 0\\ 6 & -3 & 2\end{bmatrix}-\begin{bmatrix}-7 & 6\\ 1 & -2\end{bmatrix}$$

Not possible

36. $8A$

$$=8\begin{bmatrix}\frac{-1}{4} & \frac{-3}{4}\\ \frac{-1}{8} & \frac{-7}{8}\end{bmatrix}$$

$$=\begin{bmatrix}8\left(\frac{-1}{4}\right) & 8\left(\frac{-3}{4}\right)\\ 8\left(\frac{-1}{8}\right) & 8\left(\frac{-7}{8}\right)\end{bmatrix}$$

$$=\begin{bmatrix}-2 & -6\\ -1 & -7\end{bmatrix}$$

37. BA

$$=\begin{bmatrix}-7 & 6\\ 1 & -2\end{bmatrix}\begin{bmatrix}\frac{-1}{4} & \frac{-3}{4}\\ \frac{-1}{8} & \frac{-7}{8}\end{bmatrix}$$

$$=\begin{bmatrix}-7\left(\frac{-1}{4}\right)+6\left(\frac{-1}{8}\right) & -7\left(\frac{-3}{4}\right)+6\left(\frac{-7}{8}\right)\\ 1\left(\frac{-1}{4}\right)+(-2)\left(\frac{-1}{8}\right) & 1\left(\frac{-3}{4}\right)+(-2)\left(\frac{-7}{8}\right)\end{bmatrix}$$

$$=\begin{bmatrix}1 & 0\\ 0 & 1\end{bmatrix}$$

38. $C+D$

$$=\begin{bmatrix}-1 & 3 & 4\\ 5 & -2 & 0\\ 6 & -3 & 2\end{bmatrix}+\begin{bmatrix}2 & -3 & 0\\ 0.5 & 1 & -1\\ 4 & 0.1 & 5\end{bmatrix}$$

$$=\begin{bmatrix}-1+2 & 3+(-3) & 4+0\\ 5+0.5 & -2+1 & 0+(-1)\\ 6+4 & -3+0.1 & 2+5\end{bmatrix}$$

$$=\begin{bmatrix}1 & 0 & 4\\ 5.5 & -1 & -1\\ 10 & -2.9 & 7\end{bmatrix}$$

39. $D-C$

$$=\begin{bmatrix}2 & -3 & 0\\ 0.5 & 1 & -1\\ 4 & 0.1 & 5\end{bmatrix}-\begin{bmatrix}-1 & 3 & 4\\ 5 & -2 & 0\\ 6 & -3 & 2\end{bmatrix}$$

$$=\begin{bmatrix}2-(-1) & -3-3 & 0-4\\ 0.5-5 & 1-(-2) & -1-0\\ 4-6 & 0.1-(-3) & 5-2\end{bmatrix}$$

$$=\begin{bmatrix}3 & -6 & -4\\ -4.5 & 3 & -1\\ -2 & 3.1 & 3\end{bmatrix}$$

40. BC

$$=\begin{bmatrix}-7 & 6\\ 1 & -2\end{bmatrix}\begin{bmatrix}-1 & 3 & 4\\ 5 & -2 & 0\\ 6 & -3 & 2\end{bmatrix}$$

Not possible

41. $-4D$

$$=-4\begin{bmatrix}2 & -3 & 0\\ 0.5 & 1 & -1\\ 4 & 0.1 & 5\end{bmatrix}$$

$$=\begin{bmatrix}-4(2) & -4(-3) & -4(0)\\ -4(0.5) & -4(1) & -4(-1)\\ -4(4) & -4(0.1) & -4(5)\end{bmatrix}$$

$$=\begin{bmatrix}-8 & 12 & 0\\ -2 & -4 & 4\\ -16 & -0.4 & -20\end{bmatrix}$$

42. CD

$$=\begin{bmatrix}-1 & 3 & 4\\ 5 & -2 & 0\\ 6 & -3 & 2\end{bmatrix}\begin{bmatrix}2 & -3 & 0\\ 0.5 & 1 & -1\\ 4 & 0.1 & 5\end{bmatrix}$$

$$=\begin{bmatrix}-1(2)+3(0.5)+4(4) & -1(-3)+3(1)+4(0.1) & -1(0)+3(-1)+4(5)\\ 5(2)+-2(0.5)+0(4) & 5(-3)+-2(1)+0(0.1) & 5(0)+-2(-1)+0(5)\\ 6(2)+-3(0.5)+2(4) & 6(-3)+-3(1)+2(0.1) & 6(0)+-3(-1)+2(5)\end{bmatrix}$$

$$=\begin{bmatrix}15.5 & 6.4 & 17\\ 9 & -17 & 2\\ 18.5 & -20.8 & 13\end{bmatrix}$$

43. $A=\begin{bmatrix}1 & 0\\ 0 & 1\end{bmatrix}$

$|A|=1(1)-0(0)=1;$

$B=\begin{bmatrix}0.2 & 0.2\\ -0.6 & 0.4\end{bmatrix}$

$|B|=0.2(0.4)-0.2(-0.6)=0.2;$

$$C=\begin{bmatrix}2&-1\\3&1\end{bmatrix}$$
$|C|=2(1)-(-1)(3)=5;$

$$D=\begin{bmatrix}10&-6\\-15&9\end{bmatrix}$$
$|D|=10(9)-(-6)(-15)=0$

Matrix D is singular

44. $AB=\begin{bmatrix}1&0\\0&1\end{bmatrix}\begin{bmatrix}0.2&0.2\\-0.6&0.4\end{bmatrix}=\begin{bmatrix}0.2&0.2\\-0.6&0.4\end{bmatrix}$

$BA=\begin{bmatrix}0.2&0.2\\-0.6&0.4\end{bmatrix}\begin{bmatrix}1&0\\0&1\end{bmatrix}=\begin{bmatrix}0.2&0.2\\-0.6&0.4\end{bmatrix}$

It is an identity matrix.

45. $BC=\begin{bmatrix}0.2&0.2\\-0.6&0.4\end{bmatrix}\begin{bmatrix}2&-1\\3&1\end{bmatrix}=\begin{bmatrix}1&0\\0&1\end{bmatrix}$

$CB=\begin{bmatrix}2&-1\\3&1\end{bmatrix}\begin{bmatrix}0.2&0.2\\-0.6&0.4\end{bmatrix}=\begin{bmatrix}1&0\\0&1\end{bmatrix}$

It is the inverse of B.

46. Matrix E is singular.

$E=\begin{bmatrix}1&-2&3\\-2&1&-5\\-1&-1&-2\end{bmatrix};\ |E|=0$

$F=\begin{bmatrix}1&-1&1\\0&1&0\\-2&1&-1\end{bmatrix};\ |F|=1$

$G=\begin{bmatrix}1&0&0\\0&1&0\\0&0&1\end{bmatrix};\ |G|=1$

$H=\begin{bmatrix}-1&0&-1\\0&1&0\\2&1&1\end{bmatrix};\ |H|=1$

47. $GF=\begin{bmatrix}1&0&0\\0&1&0\\0&0&1\end{bmatrix}\begin{bmatrix}1&-1&1\\0&1&0\\-2&1&-1\end{bmatrix}$

$=\begin{bmatrix}1&-1&1\\0&1&0\\-2&1&-1\end{bmatrix};$

$FG=\begin{bmatrix}1&-1&1\\0&1&0\\-2&1&-1\end{bmatrix}\begin{bmatrix}1&0&0\\0&1&0\\0&0&1\end{bmatrix}$

$=\begin{bmatrix}1&-1&1\\0&1&0\\-2&1&-1\end{bmatrix}$

It is an identity matrix.

48. $FH=\begin{bmatrix}1&-1&1\\0&1&0\\-2&1&-1\end{bmatrix}\begin{bmatrix}-1&0&-1\\0&1&0\\2&1&1\end{bmatrix}$

$=\begin{bmatrix}1&0&0\\0&1&0\\0&0&1\end{bmatrix};$

$HF=\begin{bmatrix}-1&0&-1\\0&1&0\\2&1&1\end{bmatrix}\begin{bmatrix}1&-1&1\\0&1&0\\-2&1&-1\end{bmatrix}$

$=\begin{bmatrix}1&0&0\\0&1&0\\0&0&1\end{bmatrix}$

It is the inverse of F.

49. $EH\neq HE$

$EH=\begin{bmatrix}1&-2&3\\-2&1&-5\\-1&-1&-2\end{bmatrix}\begin{bmatrix}-1&0&-1\\0&1&0\\2&1&1\end{bmatrix}$

$=\begin{bmatrix}5&1&2\\-8&-4&-3\\-3&-3&-1\end{bmatrix}$

$EH=\begin{bmatrix}-1&0&-1\\0&1&0\\2&1&1\end{bmatrix}\begin{bmatrix}1&-2&3\\-2&1&-5\\-1&-1&-2\end{bmatrix}$

$=\begin{bmatrix}0&3&-1\\-2&1&-5\\-1&-4&-1\end{bmatrix}$

$EH\neq HE$; verified

50. $\begin{cases}2x-5y=14\\-3y+4x=-14\end{cases}$

$\begin{cases}2x-5y=14\\4x-3y=-14\end{cases}$

$\left[\begin{array}{cc|c}2&-5&14\\4&-3&-14\end{array}\right]\ \frac{1}{2}R1\to R1$

$$\left[\begin{array}{cc|c} 1 & -\frac{5}{2} & 7 \\ 4 & -3 & -14 \end{array}\right] \quad -4R1+R2 \to R2$$

$$\left[\begin{array}{cc|c} 1 & -\frac{5}{2} & 7 \\ 0 & 7 & -42 \end{array}\right] \quad \frac{1}{7}R2 \to R2$$

$$\left[\begin{array}{cc|c} 1 & -\frac{5}{2} & 7 \\ 0 & 1 & -6 \end{array}\right]$$

$y=-6$

$$x-\frac{5}{2}y=7$$
$$x-\frac{5}{2}(-6)=7$$
$$x+15=7$$
$$x=-8$$

(-8, -6)

51. $\begin{cases} 0.5x-2.2y+3z=-8 \\ -0.6x-y+2z=-7.2 \\ x+1.5y-0.2z=2.6 \end{cases}$

$$\left[\begin{array}{ccc|c} 0.5 & -2.2 & 3 & -8 \\ -0.6 & -1 & 2 & -7.2 \\ 1 & 1.5 & -0.2 & 2.6 \end{array}\right]$$

(2, 0, -3)

52. $\begin{cases} 5x+6y=8 \\ 10x-2y=-9 \end{cases}$

$$D=\begin{vmatrix} 5 & 6 \\ 10 & -2 \end{vmatrix}=-10-60=-70;$$

$$D_x=\begin{vmatrix} 8 & 6 \\ -9 & -2 \end{vmatrix}=-16+54=38;$$

$$D_y=\begin{vmatrix} 5 & 8 \\ 10 & -9 \end{vmatrix}=-45-80=-125;$$

$$x=\frac{D_x}{D}=\frac{38}{-70}=\frac{-19}{35};$$

$$y=\frac{D_y}{D}=\frac{-125}{-70}=\frac{25}{14}$$

$$\left(\frac{-19}{35},\frac{25}{14}\right)$$

53. $\begin{cases} 2x+y=-2 \\ -x+y+5z=12 \\ 3x-2y+z=-8 \end{cases}$

$$D=\begin{vmatrix} 2 & 1 & 0 \\ -1 & 1 & 5 \\ 3 & -2 & 1 \end{vmatrix}=38;$$

$$D_x=\begin{vmatrix} -2 & 1 & 0 \\ 12 & 1 & 5 \\ -8 & -2 & 1 \end{vmatrix}=-74;$$

$$D_y=\begin{vmatrix} 2 & -2 & 0 \\ -1 & 12 & 5 \\ 3 & -8 & 1 \end{vmatrix}=72;$$

$$D_z=\begin{vmatrix} 2 & 1 & -2 \\ -1 & 1 & 12 \\ 3 & -2 & -8 \end{vmatrix}=62;$$

$$x=\frac{D_x}{D}=\frac{-74}{38}=-\frac{37}{19};$$

$$y=\frac{D_y}{D}=\frac{72}{38}=\frac{36}{19};$$

$$z=\frac{D_z}{D}=\frac{62}{38}=\frac{31}{19}$$

$$\left(\frac{-37}{19},\frac{36}{19},\frac{31}{19}\right)$$

54. $\begin{cases} 2x+y-z=-1 \\ x-2y+z=5 \\ 3x-y+2z=8 \end{cases}$

$$D=\begin{vmatrix} 2 & 1 & -1 \\ 1 & -2 & 1 \\ 3 & -1 & 2 \end{vmatrix}=-10;$$

$$D_x=\begin{vmatrix} -1 & 1 & -1 \\ 5 & -2 & 1 \\ 8 & -1 & 2 \end{vmatrix}=-10;$$

$$D_y=\begin{vmatrix} 2 & -1 & -1 \\ 1 & 5 & 1 \\ 3 & 8 & 2 \end{vmatrix}=10;$$

$$D_z=\begin{vmatrix} 2 & 1 & -1 \\ 1 & -2 & 5 \\ 3 & -1 & 8 \end{vmatrix}=-20;$$

$$x=\frac{D_x}{D}=\frac{-10}{-10}=1;$$

$$y=\frac{D_y}{D}=\frac{10}{-10}=-1;$$

$$z = \frac{D_z}{D} = \frac{-20}{-10} = 2$$

(1, -1, 2)

55. (6, 1), (-1, -6) and (-6, 2)

$$A = \frac{\left|\begin{vmatrix} x_1 & y_1 & 1 \\ x_2 & y_2 & 1 \\ x_3 & y_3 & 1 \end{vmatrix}\right|}{2}$$

$$A = \frac{\left|\begin{vmatrix} 6 & 1 & 1 \\ -1 & -6 & 1 \\ -6 & 2 & 1 \end{vmatrix}\right|}{2} = \left|\frac{-91}{2}\right| = \frac{91}{2}\text{ cm}^2$$

56. $\dfrac{7x^2-5x+17}{x^3-2x^2+3x-6} = \dfrac{7x^2-5x+17}{x^2(x-2)+3(x-2)}$

$$= \frac{7x^2-5x+17}{(x^2+3)(x-2)} = \frac{A}{x-2} + \frac{Bx+C}{x^2+3}$$

$$7x^2-5x+17 = A(x^2+3)+(Bx+C)(x-2)$$

$$7x^2-5x+17$$
$$= Ax^2+3A+Bx^2-2Bx+Cx-2C$$

$$7x^2-5x+17$$
$$= Ax^2+Bx^2-2Bx+Cx+3A-2C$$

$$7x^2-5x+17$$
$$= x^2(A+B)+x(-2B+C)+(3A-2C)$$

$$\begin{cases} A+B=7 \\ -2B+C=-5 \\ 3A-2C=17 \end{cases}$$

$$\begin{bmatrix} 1 & 1 & 0 \\ 0 & -2 & 1 \\ 3 & 0 & -2 \end{bmatrix}\begin{bmatrix} A \\ B \\ C \end{bmatrix} = \begin{bmatrix} 7 \\ -5 \\ 17 \end{bmatrix}$$

$$A^{-1} = \begin{bmatrix} \frac{4}{7} & \frac{2}{7} & \frac{1}{7} \\ \frac{3}{7} & \frac{-2}{7} & \frac{-1}{7} \\ \frac{6}{7} & \frac{3}{7} & \frac{-2}{7} \end{bmatrix}$$

$$X = A^{-1}B$$

$$X = \begin{bmatrix} 5 \\ 2 \\ -1 \end{bmatrix}$$

$$\frac{5}{x-2} + \frac{2x-1}{x^2+3}$$

Chapter 6 Mixed Review

1. a. $\begin{cases} -3x+5y=10 \\ 6x+20=10y \end{cases}$

$$\begin{cases} y = \frac{3}{5}x+2 \\ y = \frac{3}{5}x+2 \end{cases}$$

Consistent/dependent

b. $\begin{cases} 4x-3y=9 \\ -2x+5y=-10 \end{cases}$

$$\begin{cases} y = \frac{4}{3}x-3 \\ y = \frac{2}{5}x-2 \end{cases}$$

Consistent/independent

c. $\begin{cases} x-3y=9 \\ -6y+2x=10 \end{cases}$

$$\begin{cases} y = \frac{1}{3}x-3 \\ y = \frac{1}{3}x-\frac{5}{3} \end{cases}$$

Inconsistent

3. $\begin{cases} 2x+3y=5 \\ -x+5y=17 \end{cases}$

$x = 5y-17$

$$2x+3y=5$$
$$2(5y-17)+3y=5$$
$$10y-34+3y=5$$
$$13y=39$$
$$y=3;$$
$$-x+5y=17$$
$$-x+5(3)=17$$
$$-x+15=17$$
$$-x=2$$
$$x=-2$$

(-2, 3)

5. $\begin{cases} x+2y-3z=-4 \\ -3x+4y+z=1 \\ 2x-6y+z=1 \end{cases}$

R1 + 3R2 = Sum

$$\begin{array}{r} x+2y-3z=-4 \\ -9x+12y+3z=3 \\ \hline -8x+14y=-1 \end{array}$$

R1 + 3R3 = Sum

$$\begin{array}{r} x+2y-3z=-4 \\ 6x-18y+3z=3 \\ \hline 7x-16y=-1 \end{array}$$

$$\begin{cases} -8x+14y=-1 \\ 7x-16y=-1 \end{cases}$$

7R1 + 8R2 = Sum

$$\begin{array}{r} -56x+98y=-7 \\ 56x-128y=-8 \\ \hline -30y=-15 \end{array}$$

$$y=\frac{1}{2};$$

$$-8x+14\left(\frac{1}{2}\right)=-1$$

$$-8x+7=-1$$

$$-8x=-8$$

$$x=1;$$

$$x+2y-3z=-4$$

$$1+2\left(\frac{1}{2}\right)-3z=-4$$

$$1+1-3z=-4$$

$$2-3z=-4$$

$$-3z=-6$$

$$z=2$$

$$\left(1,\frac{1}{2},2\right)$$

7. $\begin{cases} \frac{1}{2}x+\frac{2}{3}y=3 \\ \frac{-2}{5}x-\frac{1}{4}y=1 \end{cases}$

$$\left[\begin{array}{cc|c} \frac{1}{2} & \frac{2}{3} & 3 \\ \frac{-2}{5} & \frac{-1}{4} & 1 \end{array}\right] 2R1 \to R1$$

$$\left[\begin{array}{cc|c} 1 & \frac{4}{3} & 6 \\ \frac{-2}{5} & \frac{-1}{4} & 1 \end{array}\right] \frac{2}{5}R1+R2 \to R2$$

$$\left[\begin{array}{cc|c} 1 & \frac{4}{3} & 6 \\ 0 & \frac{17}{60} & \frac{17}{5} \end{array}\right] \frac{60}{17}R2 \to R2$$

$$\left[\begin{array}{cc|c} 1 & \frac{4}{3} & 6 \\ 0 & 1 & 12 \end{array}\right]$$

$$y=12;$$

$$x+\frac{4}{3}y=6$$

$$x+\frac{4}{3}(12)=6$$

$$x+16=6$$

$$x=-10$$

(-10, 12)

9. $-2|x+3|-7\ge -15$

$$-2|x+3|\ge -8$$

$$|x+3|\le 4$$

$x+3\le 4$ and $x+3\ge -4$

$x\le 1$ $\quad$ $x\ge -7$

$$x\in[-7,1]$$

11. a. $-2[A][C]$

$$=-2\begin{bmatrix} 2 & -1 \\ 0 & 3 \end{bmatrix}\begin{bmatrix} 1 & -4 & 2 \\ -2 & 0 & -1 \end{bmatrix}$$

$$=-2\begin{bmatrix} 4 & -8 & 5 \\ -6 & 0 & -3 \end{bmatrix}$$

$$=\begin{bmatrix} -8 & 16 & -10 \\ 12 & 0 & 6 \end{bmatrix}$$

b. $[C][D]$

$$=\begin{bmatrix} 1 & -4 & 2 \\ -2 & 0 & -1 \end{bmatrix}\begin{bmatrix} 3 & 0 & 1 \\ -1 & 2 & 0 \\ 1 & 1 & -4 \end{bmatrix}$$

$$=\begin{bmatrix} 9 & -6 & -7 \\ -7 & -1 & 2 \end{bmatrix}$$

13. $\begin{cases} -x-2z=5 \\ 2y+z=-4 \\ -x+2y=3 \end{cases}$

$$\left[\begin{array}{ccc|c} -1 & 0 & -2 & 5 \\ 0 & 2 & 1 & -4 \\ -1 & 2 & 0 & 3 \end{array}\right] -R1 \to R1$$

$$\left[\begin{array}{ccc|c} 1 & 0 & 2 & -5 \\ 0 & 2 & 1 & -4 \\ -1 & 2 & 0 & 3 \end{array}\right] R1+R3 \to R3$$

$$\left[\begin{array}{ccc|c} 1 & 0 & 2 & -5 \\ 0 & 2 & 1 & -4 \\ 0 & 2 & 2 & -2 \end{array}\right] \frac{1}{2}R2 \to R2$$

$$\left[\begin{array}{ccc|c} 1 & 0 & 2 & -5 \\ 0 & 1 & \frac{1}{2} & -2 \\ 0 & 2 & 2 & -2 \end{array}\right] -2R2+R3 \to R3$$

$$\left[\begin{array}{ccc|c} 1 & 0 & 2 & -5 \\ 0 & 1 & \frac{1}{2} & -2 \\ 0 & 0 & 1 & 2 \end{array}\right] -\frac{1}{2}R3+R2 \to R2$$

$z = 2$;

$$y + \frac{1}{2}z = -2$$
$$y + \frac{1}{2}(2) = -2$$
$$y + 1 = -2$$
$$y = -3;$$

$$x + 2z = -5$$
$$x + 2(2) = -5$$
$$x + 4 = -5$$
$$x = -9$$

(-9, -3, 2)

15. $\begin{cases} -x+5y-2z=1 \\ 2x+3y-z=3 \\ 3x-y+3z=-2 \end{cases}$

$$D = \begin{vmatrix} -1 & 5 & -2 \\ 2 & 3 & -1 \\ 3 & -1 & 3 \end{vmatrix} = -31;$$

$$D_x = \begin{vmatrix} 1 & 5 & -2 \\ 3 & 3 & -1 \\ -2 & -1 & 3 \end{vmatrix} = -33;$$

$$D_y = \begin{vmatrix} -1 & 1 & -2 \\ 2 & 3 & -1 \\ 3 & -2 & 3 \end{vmatrix} = 10;$$

$$D_z = \begin{vmatrix} -1 & 5 & 1 \\ 2 & 3 & 3 \\ 3 & -1 & -2 \end{vmatrix} = 57;$$

$$x = \frac{D_x}{D} = \frac{-33}{-31} = \frac{33}{31};$$

$$y = \frac{D_y}{D} = \frac{10}{-31} = \frac{-10}{31};$$

$$z = \frac{D_z}{D} = \frac{57}{-31} = \frac{-57}{31}$$

$$\left(\frac{33}{31}, \frac{-10}{31}, \frac{-57}{31}\right)$$

17. $\begin{cases} 4x+2y \le 14 \\ 2x+3y \le 15 \\ y \ge 0 \\ x \ge 0 \end{cases}$

$$\begin{cases} y \le -2x+7 \\ y \le -\frac{2}{3}x+5 \\ y \ge 0 \\ x \ge 0 \end{cases}$$

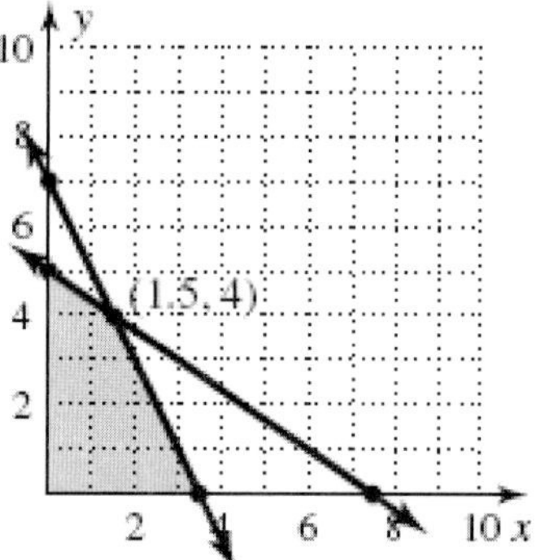

19. Let u represent the number of unicyles.
Let b represent the number of bicycles.
Let t represent the number of tricycles.

$$\begin{cases} u+b+t=21 \\ u+2b+3t=40 \\ b=2t-1 \end{cases}$$

-R1 + R2 = Sum

$$-u-b-t=-21$$
$$\underline{u+2b+3t=40}$$
$$b+2t=19$$

$$\begin{cases} b=2t-1 \\ b+2t=19 \end{cases}$$

$$b+2t=19$$
$$2t-1+2t=19$$
$$4t=20$$
$$t=5;$$

$$b=2t-1$$
$$b=2(5)-1$$
$$b=10-1$$
$$b=9;$$

$u+b+t=21$
$u+9+5=21$
$u+14=21$
$u=7$
7 unicycles
9 bicycles
5 tricycles

Chapter 6 Practice Test

1. $\begin{cases} 3x+2y=12 \\ -x+4y=10 \end{cases}$

$\begin{cases} y=-\dfrac{3}{2}x+6 \\ y=\dfrac{1}{4}x+\dfrac{5}{2} \end{cases}$

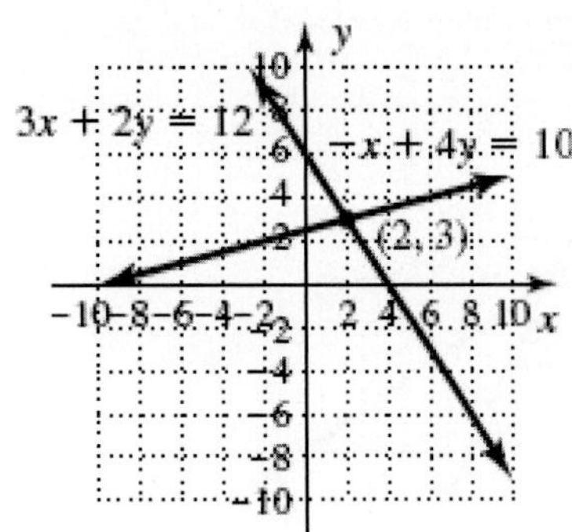

2. $\begin{cases} 3x-y=2 \\ -7x+4y=-6 \end{cases}$

$y=3x-2$
$-7x+4y=-6$
$-7x+4(3x-2)=-6$
$-7x+12x-8=-6$
$5x=2$
$x=\dfrac{2}{5};$
$3x-y=2$
$3\left(\dfrac{2}{5}\right)-y=2$
$\dfrac{6}{5}-y=2$
$-y=\dfrac{4}{5}$
$y=-\dfrac{4}{5}$
$\left(\dfrac{2}{5},-\dfrac{4}{5}\right)$

3. $\begin{cases} 5x+8y=1 \\ 3x+7y=5 \end{cases}$

-3R1 + 5R2 = Sum
$-15x-24y=-3$
$15x+35y=25$
$11y=22$
$y=2;$
$3x+7y=5$
$3x+7(2)=5$
$3x+14=5$
$3x=-9$
$x=-3$
(-3, 2)

4. $\begin{cases} x+2y-z=-4 \\ 2x-3y+5z=27 \\ -5x+y-4z=-27 \end{cases}$

-2R1 + R2 = Sum
$-2x-4y+2z=8$
$2x-3y+5z=27$
$-7y+7z=35$
5R1 + R3 = Sum
$5x+10y-5z=-20$
$-5x+y-4z=-27$
$11y-9z=-47$
$\begin{cases} -7y+7z=35 \\ 11y-9z=-47 \end{cases}$
9R1 + 7R2 = Sum
$-63y+63z=315$
$77y-63z=-329$
$14y=-14$
$y=-1;$
$-7y+7z=35$
$-7(-1)+7z=35$
$7+7z=35$
$7z=28$
$z=4;$
$x+2y-z=-4$
$x+2(-1)-4=-4$
$x-2-4=-4$
$x-6=-4$
$x=2$
(2, -1, 4)

5. a. $A-B$

$$=\begin{bmatrix}-3 & -2\\ 5 & 4\end{bmatrix}-\begin{bmatrix}3 & 3\\ -3 & -5\end{bmatrix}$$

$$=\begin{bmatrix}-3-3 & -2-3\\ 5-(-3) & 4-(-5)\end{bmatrix}$$

$$=\begin{bmatrix}-6 & -5\\ 8 & 9\end{bmatrix}$$

b. $\frac{2}{5}B=\frac{2}{5}\begin{bmatrix}3 & 3\\ -3 & -5\end{bmatrix}=\begin{bmatrix}1.2 & 1.2\\ -1.2 & -2\end{bmatrix}$

c. AB

$$=\begin{bmatrix}-3 & -2\\ 5 & 4\end{bmatrix}\begin{bmatrix}3 & 3\\ -3 & -5\end{bmatrix}$$

$$=\begin{bmatrix}-3 & 1\\ 3 & -5\end{bmatrix}$$

d. A^{-1}

$$=\begin{bmatrix}-2 & -1\\ 2.5 & 1.5\end{bmatrix}$$

e. $|A|=-3(4)-(-2)(5)=-12+10=-2$

6. a. $C-D$

$$=\begin{bmatrix}0.5 & 0 & 0.2\\ 0.4 & -0.5 & 0\\ 0.1 & -0.4 & -0.1\end{bmatrix}-\begin{bmatrix}0.5 & 0.1 & 0.2\\ -0.1 & 0.1 & 0\\ 0.3 & 0.4 & 0.8\end{bmatrix}$$

$$=\begin{bmatrix}0 & -0.1 & 0\\ 0.5 & -0.6 & 0\\ -0.2 & -0.8 & -0.9\end{bmatrix}$$

b. $-0.6D=-0.6\begin{bmatrix}0.5 & 0.1 & 0.2\\ -0.1 & 0.1 & 0\\ 0.3 & 0.4 & 0.8\end{bmatrix}$

$$=\begin{bmatrix}-0.3 & -0.06 & -0.12\\ 0.06 & -0.06 & 0\\ -0.18 & -0.24 & -0.48\end{bmatrix}$$

c. DC

$$=\begin{bmatrix}0.5 & 0.1 & 0.2\\ -0.1 & 0.1 & 0\\ 0.3 & 0.4 & 0.8\end{bmatrix}\begin{bmatrix}0.5 & 0 & 0.2\\ 0.4 & -0.5 & 0\\ 0.1 & -0.4 & -0.1\end{bmatrix}$$

$$=\begin{bmatrix}0.31 & -0.13 & 0.08\\ -0.01 & -0.05 & -0.02\\ 0.39 & -0.52 & -0.2\end{bmatrix}$$

d. $D^{-1}=\begin{bmatrix}\frac{40}{17} & 0 & \frac{-10}{17}\\ \frac{40}{17} & 10 & \frac{-10}{17}\\ \frac{-35}{17} & -5 & \frac{30}{17}\end{bmatrix}$

e. $|D|=0.034$

7. $\begin{cases}4x-5y-6z=5\\ 2x-3y+3z=0\\ x+2y-3z=5\end{cases}$

$$\left[\begin{array}{ccc|c}4 & -5 & -6 & 5\\ 2 & -3 & 3 & 0\\ 1 & 2 & -3 & 5\end{array}\right]\ R1\leftrightarrow R3$$

$$\left[\begin{array}{ccc|c}1 & 2 & -3 & 5\\ 2 & -3 & 3 & 0\\ 4 & -5 & -6 & 5\end{array}\right]\ -2R1+R2\to R2$$

$$\left[\begin{array}{ccc|c}1 & 2 & -3 & 5\\ 0 & -7 & 9 & -10\\ 4 & -5 & -6 & 5\end{array}\right]\ -4R1+R3\to R3$$

$$\left[\begin{array}{ccc|c}1 & 2 & -3 & 5\\ 0 & -7 & 9 & -10\\ 0 & -13 & 6 & -15\end{array}\right]\ -\frac{1}{7}R2\leftrightarrow R2$$

$$\left[\begin{array}{ccc|c}1 & 2 & -3 & 5\\ 0 & 1 & -\frac{9}{7} & \frac{10}{7}\\ 0 & -13 & 6 & -15\end{array}\right]\ 13R2+R3\to R3$$

$$\left[\begin{array}{ccc|c}1 & 2 & -3 & 5\\ 0 & 1 & -\frac{9}{7} & \frac{10}{7}\\ 0 & 0 & -\frac{75}{7} & \frac{25}{7}\end{array}\right]\ -\frac{7}{75}R3\to R3$$

$$\left[\begin{array}{ccc|c}1 & 2 & -3 & 5\\ 0 & 1 & -\frac{9}{7} & \frac{10}{7}\\ 0 & 0 & 1 & -\frac{1}{3}\end{array}\right]$$

$z=-\frac{1}{3}$;

$$y-\frac{9}{7}z=\frac{10}{7}$$
$$y-\frac{9}{7}\left(\frac{-1}{3}\right)=\frac{10}{7}$$
$$y+\frac{3}{7}=\frac{10}{7}$$
$$y=\frac{7}{7}$$
$$y=1;$$
$$x+2y-3z=5$$
$$x+2(1)-3\left(\frac{-1}{3}\right)=5$$
$$x+2+1=5$$
$$x+3=5$$
$$x=2$$

$\left(2,1,-\frac{1}{3}\right)$

8. $\begin{cases}2x+3y+z=3\\ x-2y-z=4\\ x-y-2z=-1\end{cases}$

$$D=\begin{vmatrix}2 & 3 & 1\\ 1 & -2 & -1\\ 1 & -1 & -2\end{vmatrix}=10;$$

$$D_x=\begin{vmatrix}3 & 3 & 1\\ 4 & -2 & -1\\ -1 & -1 & -2\end{vmatrix}=30;$$

$$D_y=\begin{vmatrix}2 & 3 & 1\\ 1 & 4 & -1\\ 1 & -1 & -2\end{vmatrix}=-20;$$

$$D_z=\begin{vmatrix}2 & 3 & 3\\ 1 & -2 & 4\\ 1 & -1 & -1\end{vmatrix}=30;$$

$$x=\frac{D_x}{D}=\frac{30}{10}=3;$$
$$y=\frac{D_y}{D}=\frac{-20}{10}=-2;$$
$$z=\frac{D_z}{D}=\frac{30}{10}=3$$

(3, -2, 3)

9. $\begin{cases}2x-5y=11\\ 4x+7y=4\end{cases}$

$$\left[\begin{array}{cc|c}2 & -5 & 11\\ 4 & 7 & 4\end{array}\right]$$

$$A=\begin{bmatrix}2 & -5\\ 4 & 7\end{bmatrix};\ B=\begin{bmatrix}11\\ 4\end{bmatrix}$$

$$A^{-1}=\begin{bmatrix}\frac{7}{34} & \frac{5}{34}\\ \frac{-2}{17} & \frac{1}{17}\end{bmatrix}$$

$$A^{-1}B=\begin{bmatrix}\frac{97}{34}\\ \frac{-18}{17}\end{bmatrix}$$

$\left(\frac{97}{34},-\frac{18}{17}\right)$

10. $\begin{cases}x-2y+2z=7\\ 2x+2y-z=5\\ 3x-y+z=6\end{cases}$

$$\left[\begin{array}{ccc|c}1 & -2 & 2 & 7\\ 2 & 2 & -1 & 5\\ 3 & -1 & 1 & 6\end{array}\right]$$

$$A=\begin{bmatrix}1 & -2 & 2\\ 2 & 2 & -1\\ 3 & -1 & 1\end{bmatrix};\ B=\begin{bmatrix}7\\ 5\\ 6\end{bmatrix}$$

$$A^{-1}=\begin{bmatrix}-0.2 & 0 & 0.4\\ 1 & 1 & -1\\ 1.6 & 1 & -1.2\end{bmatrix}$$

$$A^{-1}B=\begin{bmatrix}1\\ 6\\ 9\end{bmatrix}$$

(1, 6, 9)

11. Let l represent the length of the paper.
Let w represent the width of the paper.

$\begin{cases}2l+2w=114.3\\ l=2w-7.62\end{cases}$

$$2l+2w=114.3$$
$$2(2w-7.62)+2w=114.3$$
$$4w-15.24+2w=114.3$$
$$6w=129.54$$
$$w=21.59;$$
$$l=2w-7.62$$
$$l=2(21.59)-7.62$$
$$l=43.18-7.62$$
$$l=35.56$$

21.59 cm by 35.56 cm

12. Let h represent the land area of Tahiti.
Let n represent the land area of Tonga.

$$\begin{cases} h+n=692 \\ h=n+112 \end{cases}$$

$$h+n=692$$
$$n+112+n=692$$
$$2n+112=692$$
$$2n=580$$
$$n=290;$$
$$h=290+112$$
$$h=402$$

Tahiti is 402 mi^2
Tonga is 290 mi^2

13. $$\begin{cases} 2C+3B+P=1.39 \\ 3C+2B+2P=1.73 \\ C+4B+3P=1.92 \end{cases}$$

-2R1 + R2 = Sum
$$-4C-6B-2P=-2.78$$
$$\underline{3C+2B+2P=1.73}$$
$$-C-4B=-1.05$$

-3R1 + R3 = Sum
$$-6C-9B-3P=-4.17$$
$$\underline{C+4B+3P=1.92}$$
$$-5C-5B=-2.25$$

$$\begin{cases} -C-4B=-1.05 \\ -5C-5B=-2.25 \end{cases}$$

-5R1 + R2 = Sum
$$5C+20B=5.25$$
$$\underline{-5C-5B=-2.25}$$
$$15B=3$$
$$B=0.20;$$
$$-C-4B=-1.05$$
$$-C-4(0.20)=-1.05$$
$$-C-0.80=-1.05$$
$$-C=-0.25$$
$$C=0.25;$$
$$2C+3B+P=1.39$$
$$2(0.25)+3(0.2)+P=1.39$$
$$0.50+0.60+P=1.39$$
$$1.10+P=1.39$$
$$P=0.29$$

Corn: 25¢
Beans: 20¢
Peas: 29¢

14. Let s represent the amount in the savings account.
Let b represent the amount in the bond account.
Let t represent the amount in the stock account.

$$\begin{cases} s+b+t=30000 \\ 0.05s+0.07b+0.09t=2080 \\ t=b-8000 \end{cases}$$

$$\begin{cases} s+b+t=30000 \\ 5s+7b+9t=208000 \\ -b+t=-8000 \end{cases}$$

-5R1 + R2 = Sum
$$-5s-5b-5t=-150000$$
$$\underline{5s+7b+9t=208000}$$
$$2b+4t=58000$$

$$\begin{cases} t=b-8000 \\ 2b+4t=58000 \end{cases}$$

$$2b+4t=58000$$
$$2b+4(b-8000)=58000$$
$$2b+4b-32000=58000$$
$$6b=90000$$
$$b=15000;$$
$$t=b-8000$$
$$t=15000-8000$$
$$t=7000;$$
$$s+b+t=30000$$
$$s+15000+7000=30000$$
$$s+22000=30000$$
$$s=8000$$

\$8,000 at 5%
\$15,000 at 7%
\$7,000 at 9%

15. $$\frac{-2}{3}|x-2|+7<5$$
$$\frac{-2}{3}|x-2|<-2$$
$$|x-2|>3$$
$$\begin{cases} Y_1=|x-2| \\ Y_2=3 \end{cases}$$

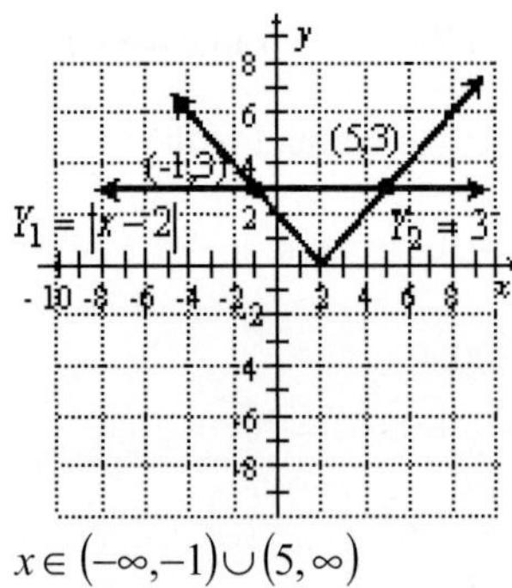

$x \in (-\infty,-1) \cup (5,\infty)$

16. $5|x+3|-12 \geq 8$

$5|x+3| \geq 20$

$|x+3| \geq 4$

$x+3 \geq 4$ or $x+3 \leq -4$

$x \geq 1$ $x \leq -7$

$x \in (-\infty,-7] \cup [1,\infty)$

17. $\begin{cases} x-y \leq 2 \\ x+2y \geq 8 \end{cases}$

$\begin{cases} y \geq x-2 \\ y \geq -\dfrac{1}{2}x+4 \end{cases}$

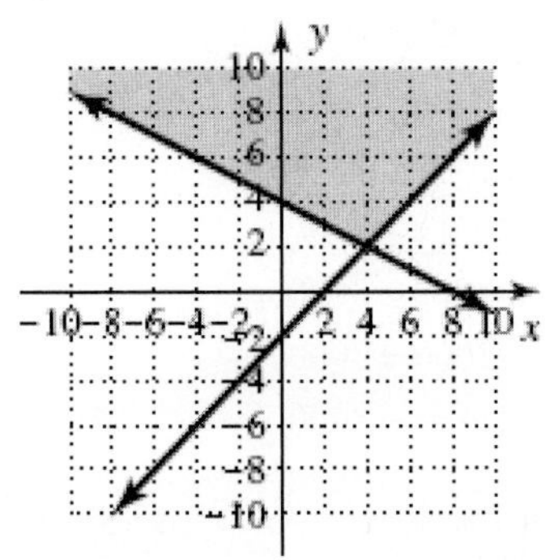

18. $P = 50x-12y$

$\begin{cases} x+2y \leq 8 \\ 8x+5y \geq 40 \\ x, y \geq 0 \end{cases}$

$\begin{cases} y \leq -\dfrac{1}{2}x+4 \\ y \geq -\dfrac{8}{5}x+8 \\ x, y \geq 0 \end{cases}$

Corner Point	Objective Function $P = 50x-12y$	Result
(0, 0)	$P(0,0) = 50(0)-12(0)$	0
(0,4)	$P(0,4) = 50(0)-12(4)$	-48
(5, 0)	$P(5,0) = 50(5)-12(0)$	250
(3.64,2.18)	$P(3.64,2.18)$ $= 50(3.64)-12(2.18)$	155.63

(5,0)

19. $P(x,y) = 4.25x+5y$

$\begin{cases} x+y \leq 50 \\ 2x+3y \leq 120 \end{cases}$

Corner Point	Objective Function $P(x,y) = 4.25x+5y$	Result
(0, 0)	$P(0,0) = 4.25(0)+5(0)$	0
(0,40)	$P(0,40) = 4.25(0)+5(40)$	200
(30, 20)	$P(30,20) = 4.25(30)+5(20)$	227.5
(50, 0)	$P(50,0) = 4.25(50)+5(0)$	212.50

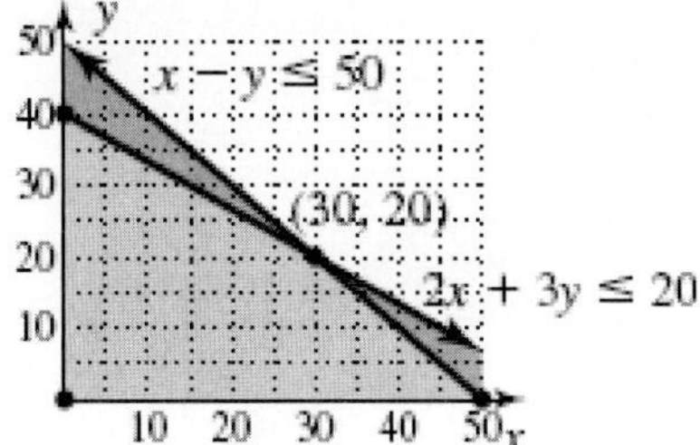

30 plain, 20 deluxe

20. $\dfrac{4x^2-4x+3}{x^3-27} = \dfrac{4x^2-4x+3}{(x-3)(x^2+3x+9)}$

$= \dfrac{A}{x-3} + \dfrac{Bx+C}{x^2+3x+9}$

$4x^2-4x+3 = Ax^2+3Ax+9A+Bx^2-3Bx+Cx-3C$

$= (A+B)x^2+(3A-3B+C)x+(9A-3C)$

$\begin{cases} A+B=4 \\ 3A-3B+C=-4 \\ 9A-3C=3 \end{cases}$

$\begin{bmatrix} 1 & 1 & 0 \\ 3 & -3 & 1 \\ 9 & 0 & -3 \end{bmatrix} \begin{bmatrix} A \\ B \\ C \end{bmatrix} = \begin{bmatrix} 4 \\ -4 \\ 3 \end{bmatrix}$

$A^{-1} = \begin{bmatrix} \frac{1}{3} & \frac{1}{9} & \frac{1}{27} \\ \frac{2}{3} & \frac{-1}{9} & \frac{-1}{27} \\ 1 & \frac{1}{3} & \frac{-2}{9} \end{bmatrix}$

$X = A^{-1}B$

$$X = \begin{bmatrix} 1 \\ 3 \\ 2 \end{bmatrix}$$

$$= \frac{1}{x-3} + \frac{3x+2}{x^2+3x+9}$$

Calculator Exploration and Discovery

Exercise 1: Answers will vary.

Exercise 2:

$$\begin{cases} 8x+3y=30 \\ 5x+4y=23 \\ x+2y=10 \\ x, y \geq 0 \end{cases}$$

(0, 5)

Exercise 3:

$$\begin{cases} 2x+2y=15 \\ x+y=6 \\ x+4y=9 \\ x, y \geq 0 \end{cases}$$

(5, 1)

Strengthening Core Skills

Exercise 1:

$$\begin{cases} 2x+y=-2 \\ -x+3y-2z=-15 \\ 3x-y+2z=9 \end{cases}$$

$$A = \begin{bmatrix} 2 & 1 & 0 \\ -1 & 3 & -2 \\ 3 & -1 & 2 \end{bmatrix};$$

$$A^{-1} = \begin{bmatrix} 1 & -0.5 & -0.5 \\ -1 & 1 & 1 \\ -2 & 1.25 & 1.75 \end{bmatrix};\ B = \begin{bmatrix} -2 \\ -15 \\ 9 \end{bmatrix}$$

$$A^{-1}B = \begin{bmatrix} 1 \\ -4 \\ 1 \end{bmatrix}$$

(1, -4, 1)

Chapter 1-6 Cumulative Review

1. $y = \frac{2}{3}x + 2$

x-intercept:

$$0 = \frac{2}{3}x + 2$$

$$-2 = \frac{2}{3}x$$

$$-3 = x$$

(-3, 0)

y-intercept:

$$y = \frac{2}{3}(0) + 2$$

$$y = 2$$

(0, 2)

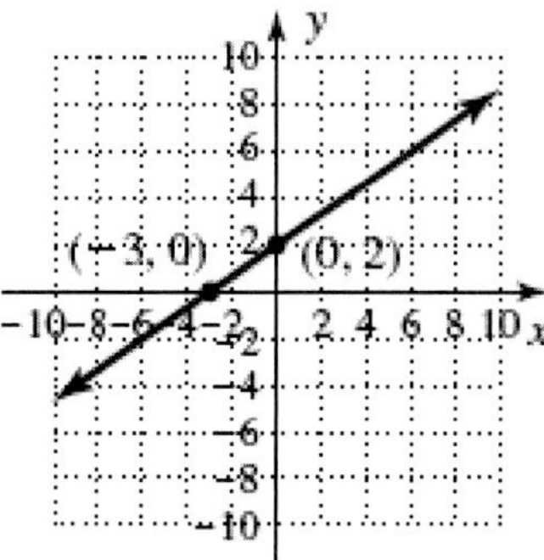

2. $f(x) = |x-2| + 3$

x-intercept: None
y-intercept: (0, 3)
Shifts up 3, right 2

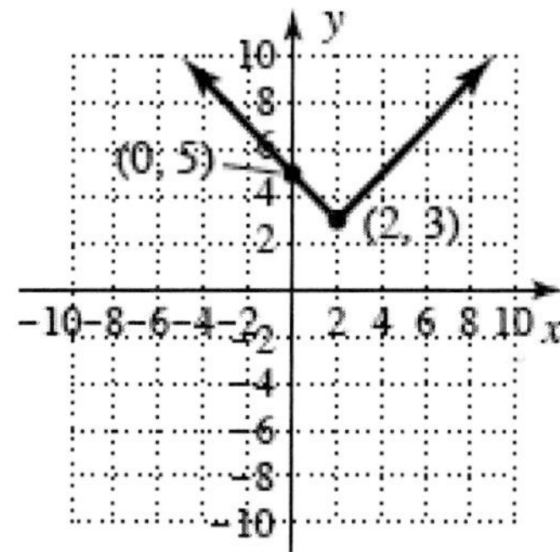

3. $g(x) = \sqrt{x-3} + 1$

x-intercept: None
y-intercept: None
Shifts up 1, right 3

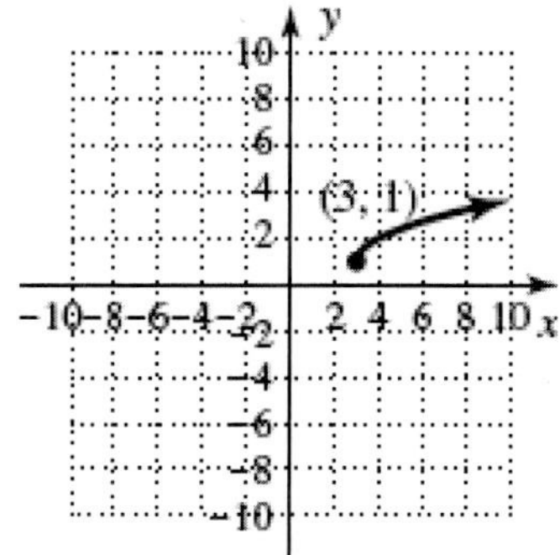

4. $h(x) = \frac{1}{x-1} + 2$

Vertical asymptote: $x = 1$
Horizontal asymptote: $y = 2$
x-intercept: $\left(\frac{1}{2},0\right)$
y-intercept: (0, 1)

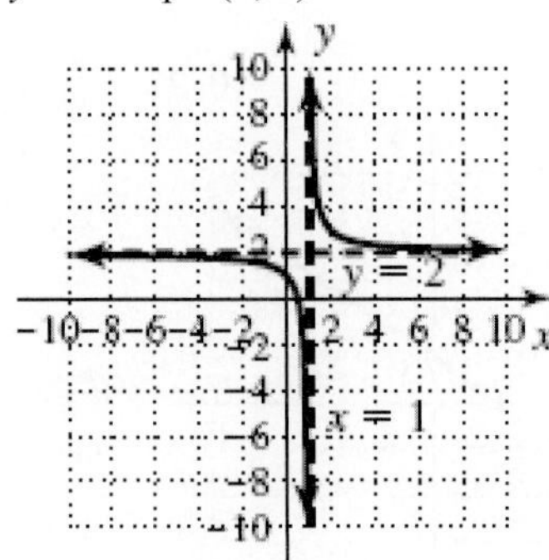

5. $g(x) = (x-3)(x+1)(x+4)$
x-intercepts: (3, 0), (-1, 0), (-4, 0)
y-intercept: (0, -12)

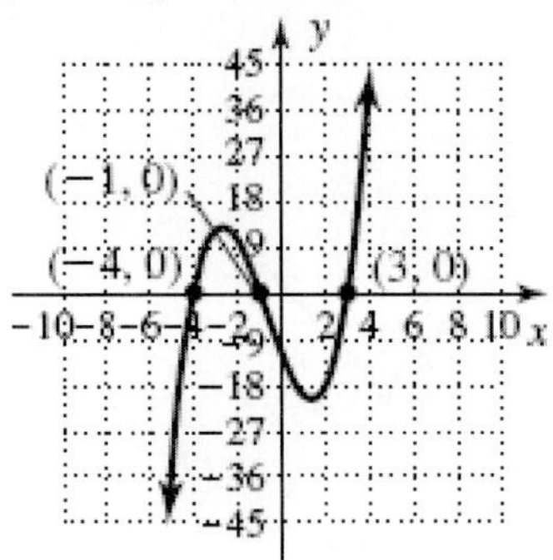

6. $y = 2^x + 3$
x-intercept: None
y-intercept: (0, 4)
$y = 2^0 + 3 = 1 + 3 = 4$

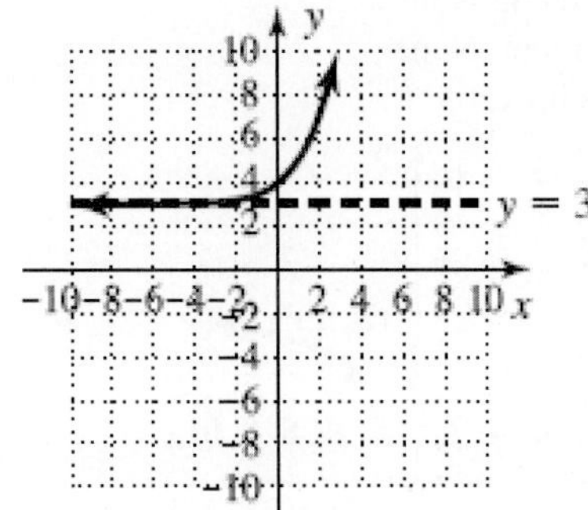

7. a) Domain: $x \in (-\infty,\infty)$
b) Range: $y \in (-\infty,4]$
c) $f(x)\uparrow: (-\infty,-1)$
$f(x)\downarrow: (-1,\infty)$
d) N/A
e) Maximum: (-1, 4)
f) $f(x) > 0: (-4,2)$
$f(x) < 0: (-\infty,-4)\cup(2,\infty)$
g) $\frac{7}{4}$

8. $C(x) = 3x + 10;\ \ R(x) = -x^2 + 123x - 1990$
Profit = Revenue – Cost
$P(x) = -x^2 + 123x - 1990 - 3x - 10$
$P(x) = -x^2 + 120x - 2000$
$x = \frac{-b}{2a} = \frac{-120}{2(-1)} = \frac{-120}{-2} = 60$
$P(60) = -(60)^2 + 120(60) - 2000$
$P(60) = -3600 + 7200 - 2000$
$P(60) = 1600$
6000 balls would produce a maximum profit of \$1600.

9. $g(v) = v^3 - 9v^2 + 2v - 18$
$g(v) = v^2(v-9) + 2(v-9)$
$g(v) = (v-9)(v^2+2);$

$v - 9 = 0$ $\quad$ $v = 9$

$v^2 + 2 = 0$ $\quad$ $v^2 = -2$ $\quad$ $v = \pm\sqrt{2}i$

Zeroes: 9, $\pm\sqrt{2}i$

10. Roots: $x=2$; $x = 2 \pm 3i$
$f(x) = (x-2)(x-(2+3i))(x-(2-3i))$
$f(x) = (x-2)(x-2-3i)(x-2+3i)$
$f(x) = (x-2)$
$(x^2 - 2x + 3ix - 2x + 4 - 6i - 3ix + 6i - 9i^2)$
$f(x) = (x-2)(x^2 - 4x + 13)$
$f(x) = x^3 - 4x^2 + 13x - 2x^2 + 8x - 26$
$f(x) = x^3 - 6x^2 + 21x - 26$

11. $f(x) = 2x - 5;\ \ g(x) = 3x^2 + 2x$
$(g - f)(x) = g(x) - f(x)$
$= 3x^2 + 2x - 2x + 5$
$= 3x^2 + 5$

12. $f(x) = 2x - 5;\ \ g(x) = 3x^2 + 2x$

$$
\begin{aligned}
(fg)(-2) &= f(-2)g(-2) \\
&= (2(-2)-5)(3(-2)^2+2(-2)) \\
&= (-4-5)(3(4)-4) \\
&= -9(12-4) \\
&= -9(8) \\
&= -72
\end{aligned}
$$

13. $f(x)=2x-5;\quad g(x)=3x^2+2x$

$(g \circ f)(2)=g(f(2));$

$f(2)=2(2)-5=4-5=-1;$

$g(-1)=3(-1)^2+2(-1)=3-2=1$

14. $f(x)=x^2-3x$

$$
\begin{aligned}
&\frac{f(x+h)-f(x)}{h} \\
&= \frac{(x+h)^2-3(x+h)-(x^2-3x)}{h} \\
&= \frac{x^2+2xh+h^2-3x-3h-x^2+3x}{h} \\
&= \frac{2xh+h^2-3h}{h} \\
&= \frac{h(2x+h-3)}{h} \\
&= 2x+h-3
\end{aligned}
$$

15. $x^4-6x^3-13x^2+24x+36$

Possible roots:

$$\frac{\{\pm1,\pm36,\pm2,\pm18,\pm3,\pm12,\pm4,\pm9,\pm6\}}{\{\pm1\}}$$

2	1	-6	-13	24	36
		2	-8	-42	-36
	1	-4	-21	-18	0

-2	1	-4	-21	-18
		-2	12	18
	1	-6	-9	0

x^2-6x-9

$a=1, b=-6, c=-9$

$$x=\frac{6\pm\sqrt{(-6)^2-4(1)(-9)}}{2(1)}$$

$$x=\frac{6\pm\sqrt{72}}{2}$$

$$x=\frac{6\pm6\sqrt{2}}{2}$$

$x=3\pm3\sqrt{2}$

$(x-2)(x+2)(x-3-3\sqrt{2})(x-3+3\sqrt{2})$

16. $x^2-3x-10<0$

$(x-5)(x+2)=0$

$x=5;\quad x=-2$

$x\in(-\infty,-2)\cup(5,\infty)$

17. $\dfrac{x-2}{x+3}\le 3$

Vertical asymptote: $x = -3$

$$\frac{x-2}{x+3}-3\le 0$$

$$\frac{x-2-3(x+3)}{x+3}\le 0$$

$$\frac{-2x-11}{x+3}\le 0$$

$x=-\dfrac{11}{2};\quad x=-3$

neg | pos | neg; $-\frac{11}{2}$ (closed), -3 (open)

$$x\in\left(-\infty,-\frac{11}{2}\right]\cup(-3,\infty)$$

18. a. $\sqrt{x}-2=\sqrt{3x+4}$

$$
\begin{aligned}
(\sqrt{x}-2)^2 &= (\sqrt{3x+4})^2 \\
x-4\sqrt{x}+4 &= 3x+4 \\
-4\sqrt{x} &= 2x \\
16x &= 4x^2 \\
0 &= 4x^2-16x \\
0 &= 4x(x-4)
\end{aligned}
$$

$x=4;\quad x=0$ both are extraneous solutions; no solution

b. $x^{\frac{3}{2}}+8=0$

$\sqrt{x^3}=-8$

Not possible, no solution

c. $2|n+4|+3=13$

$2|n+4|=10$

$|n+4|=5$

$n+4=5$ and $n+4=-5$

$n=1 \qquad n=-9$

d. $x^2-6x+13=0$

$$x=\frac{6\pm\sqrt{(-6)^2-4(1)(13)}}{2(1)}$$

$$x=\frac{6\pm\sqrt{36-52}}{2}$$
$$x=\frac{6\pm\sqrt{-16}}{2}$$
$$x=\frac{6\pm 4i}{2}$$
$$x=3\pm 2i$$

e. $x^{-2}-3x^{-1}-40=0$

Let $u=x^{-1}$ and $u^2=x^{-2}$

$$u^2-3u-40=0$$
$$(u-8)(u+5)=0$$

$u=8$ $\quad$ $u=-5$

$x^{-1}=8$ and $x^{-1}=-5$

$x=\frac{1}{8}$ $\quad$ $x=\frac{-1}{5}$

f. $4\cdot 2^{x+1}=\frac{1}{8}$

$$2^{x+1}=\frac{1}{32}$$
$$2^{x+1}=\left(\frac{1}{2}\right)^5$$
$$2^{x+1}=(2)^{-5}$$
$$x+1=-5$$
$$x=-6$$

g. $3^{x-2}=7$

$$(x-2)\ln 3=\ln 7$$
$$x-2=\frac{\ln 7}{\ln 3}$$
$$x=\frac{\ln 7}{\ln 3}+2$$

h. $\log_3 81=x$

$$3^x=81$$
$$3^x=3^4$$
$$x=4$$

i. $\log_3 x+\log_3(x-2)=1$

$$\log_3(x(x-2))=1$$
$$\log_3(x^2-2x)=1$$
$$3=x^2-2x$$
$$0=x^2-2x-3$$
$$0=(x-3)(x+1)$$
$$x=3$$

19. $\begin{cases}4x+3y=13\\-9x+5y=6\end{cases}$

$$\left[\begin{array}{cc|c}4&3&13\\-9&5&6\end{array}\right]\ \frac{1}{4}R1\to R1$$

$$\left[\begin{array}{cc|c}1&\frac{3}{4}&\frac{13}{4}\\-9&5&6\end{array}\right]\ 9R1+R2\to R2$$

$$\left[\begin{array}{cc|c}1&\frac{3}{4}&\frac{13}{4}\\0&\frac{47}{4}&\frac{141}{4}\end{array}\right]\ \frac{4}{47}R2\to R2$$

$$\left[\begin{array}{cc|c}1&\frac{3}{4}&\frac{13}{4}\\0&1&3\end{array}\right]\ -\frac{3}{4}R2+R1\to R1$$

$y=3$;

$$x+\frac{3}{4}y=\frac{13}{4}$$
$$x+\frac{3}{4}(3)=\frac{13}{4}$$
$$x+\frac{9}{4}=\frac{13}{4}$$
$$x=1$$

(1, 3)

20. $\begin{cases}x+2y-z=0\\2x-5y+4z=6\\-x+3y-4z=-5\end{cases}$

$$\left[\begin{array}{ccc|c}1&2&-1&0\\2&-5&4&6\\-1&3&-4&-5\end{array}\right]$$

$$A=\begin{bmatrix}1&2&-1\\2&-5&4\\-1&3&-4\end{bmatrix};\ B=\begin{bmatrix}0\\6\\-5\end{bmatrix}$$

$$A^{-1}=\begin{bmatrix}\frac{8}{15}&\frac{1}{3}&\frac{1}{5}\\\frac{4}{15}&\frac{-1}{3}&\frac{-2}{5}\\\frac{1}{15}&\frac{-1}{3}&\frac{-3}{5}\end{bmatrix}$$

$$A^{-1}B=\begin{bmatrix}1\\0\\1\end{bmatrix}$$

(1, 0, 1)

21. $A=Pe^{rt}$

$$12000 = 5000e^{0.09t}$$
$$2.4 = e^{0.09t}$$
$$\ln 2.4 = \ln e^{0.09t}$$
$$\ln 2.4 = 0.09t$$
$$\frac{\ln 2.4}{0.09} = t$$
$$t \approx 9.7 \text{ years}$$

22. $\begin{cases} |x+2| & -5 \le x \le 0 \\ 2(x-1)^2 & 0 < x \le 5 \end{cases}$

23. $\dfrac{4x-6}{x^2-9} = \dfrac{4x-6}{(x-3)(x+3)}$

$$4x-6 = \frac{A}{x-3} + \frac{B}{x+3}$$
$$4x-6 = A(x+3) + B(x-3)$$
$$4x-6 = Ax + 3A + Bx - 3B$$
$$4x-6 = Ax + Bx + 3A - 3B$$
$$4x-6 = x(A+B) + 3A - 3B$$
$$\begin{cases} A+B = 4 \\ 3A-3B = -6 \end{cases}$$
$$\begin{bmatrix} 1 & 1 \\ 3 & -3 \end{bmatrix}\begin{bmatrix} A \\ B \end{bmatrix} = \begin{bmatrix} 4 \\ -6 \end{bmatrix}$$
$$A^{-1} = \begin{bmatrix} \frac{1}{2} & \frac{1}{6} \\ \frac{1}{2} & \frac{-1}{6} \end{bmatrix}$$
$$X = A^{-1}B$$
$$X = \begin{bmatrix} 1 \\ 3 \end{bmatrix}$$
$$\frac{1}{x-3} + \frac{3}{x+3}$$

24. $h(x) = \dfrac{9-x^2}{x^2-4} = \dfrac{(3-x)(3+x)}{(x-2)(x+2)}$

Vertical asymptotes: $x = -2$, $x = 2$
x-intercepts: (3, 0), (-3, 0)
y-intercept: (0, -2.25)

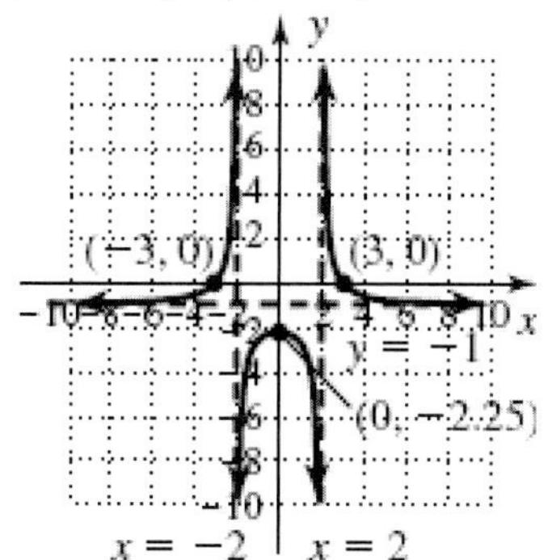

25. a)

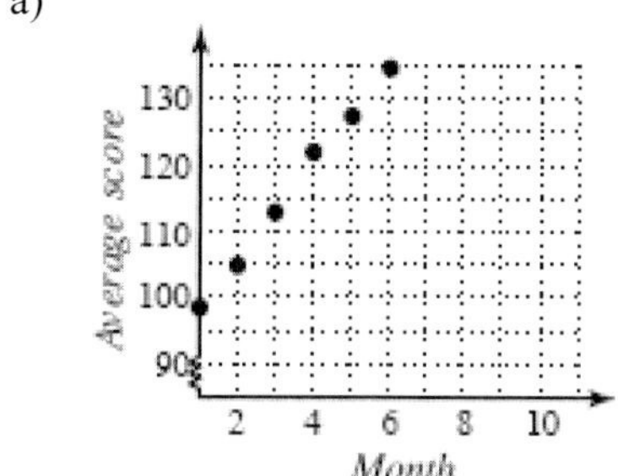

b) $y = 8.14x + 86.67$

c) Increase of 8 points per month.

d) 9th month

7.1 Technology Highlight

1. $y = \pm 4.8$; $y = \pm 3.6$

2. (-3,0), (3,0), (0,6), (0,-6)

7.1 Exercises

1. Radius, center

3. Complete the square, 1

5. Answers will vary.

7. Center (0,0), radius 3
   $x^2 + y^2 = 9$

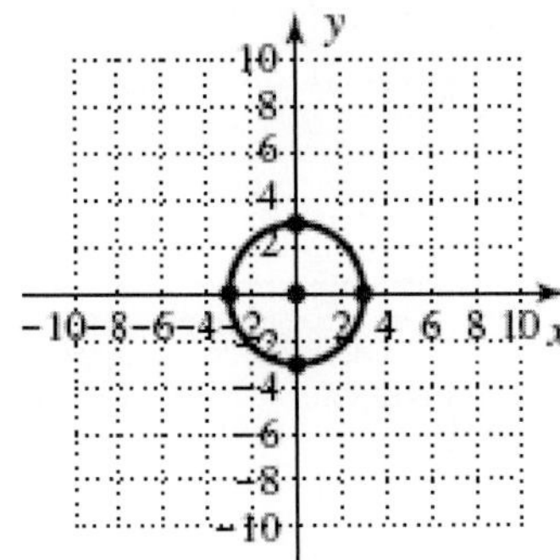

9. Center (5,0), radius $\sqrt{3}$
   $(x-5)^2 + y^2 = 3$

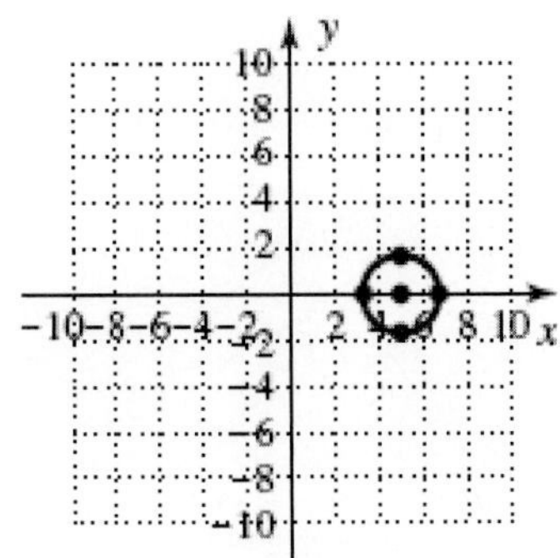

11. Center (4,-3), radius 2
    $(x-4)^2 + (y+3)^2 = 4$

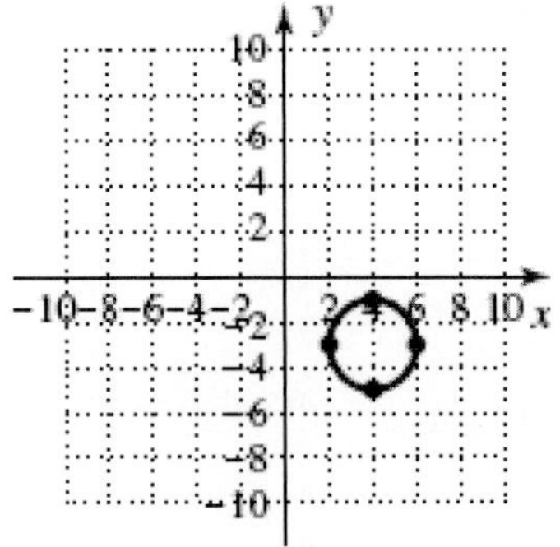

13. Center (-7,-4), radius $\sqrt{7}$
    $(x+7)^2 + (y+4)^2 = 7$

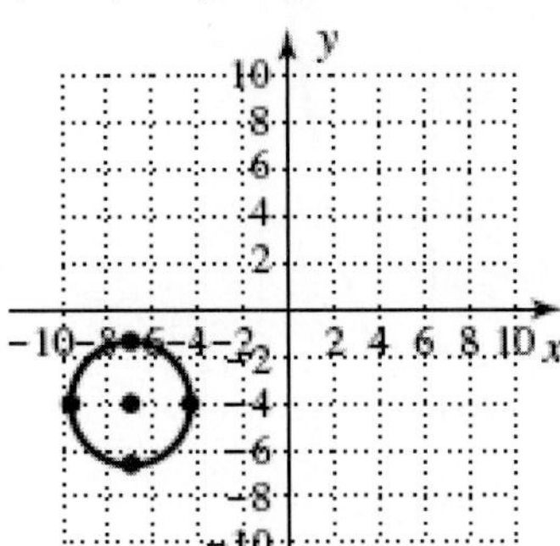

15. Center (1,-2), radius $2\sqrt{3}$
    $(x-1)^2 + (y+2)^2 = 12$

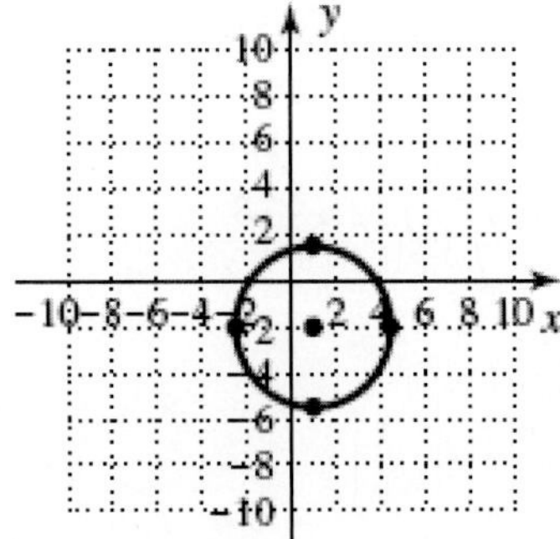

17. Center (4,5), diameter $4\sqrt{3}$

    radius $= \frac{1}{2} \cdot$ diameter

    $r = \frac{1}{2}\left(4\sqrt{3}\right) = 2\sqrt{3}$

    $(x-4)^2 + (y-5)^2 = \left(2\sqrt{3}\right)^2$

    $(x-4)^2 + (y-5)^2 = 12$

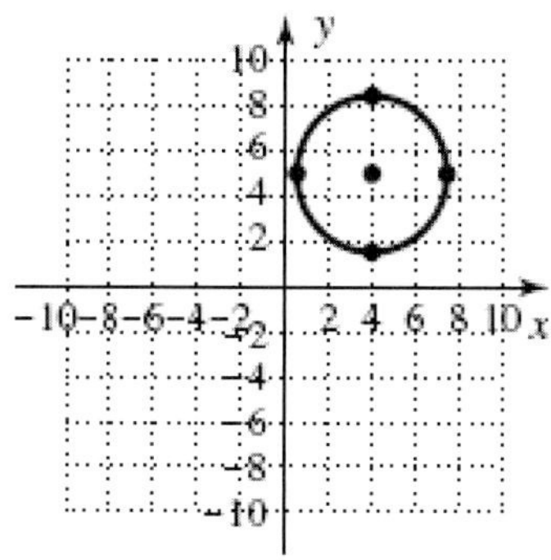

19. Center at (7,1), graph contains the point (1,-7)
$(x-7)^2+(y-1)^2=r^2$;
$(1-7)^2+(-7-1)^2=r^2$
$36+64=r^2$
$100=r^2$;
$(x-7)^2+(y-1)^2=100$

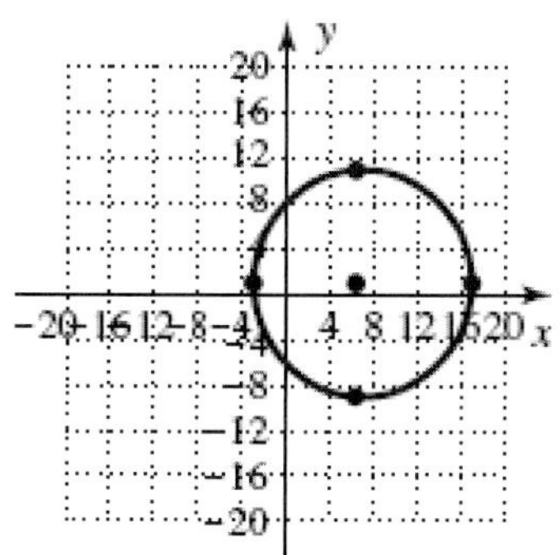

21. Center at (3,4), graph contains the point (7,9)
$(x-3)^2+(y-4)^2=r^2$;
$(7-3)^2+(9-4)^2=r^2$
$16+25=r^2$
$41=r^2$;
$(x-3)^2+(y-4)^2=41$

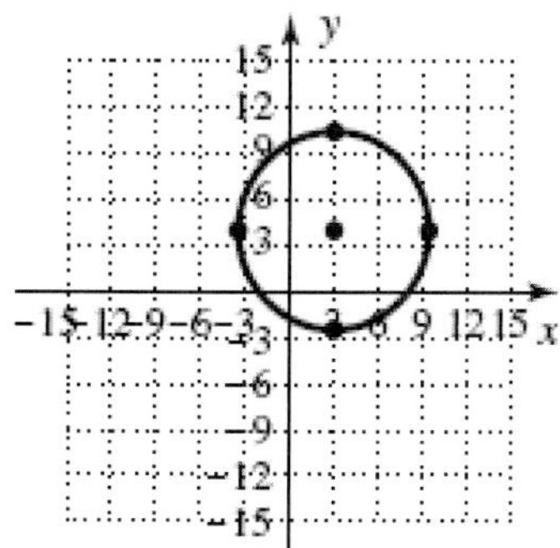

23. Diameter has endpoints (5,1) and (5,7); midpoint of diameter = center of circle
$\left(\frac{5+5}{2},\frac{1+7}{2}\right)=(5,4)$;
radius = distance from center to endpt
$r=\sqrt{(5-5)^2+(1-4)^2}=3$;
$(x-5)^2+(y-4)^2=9$

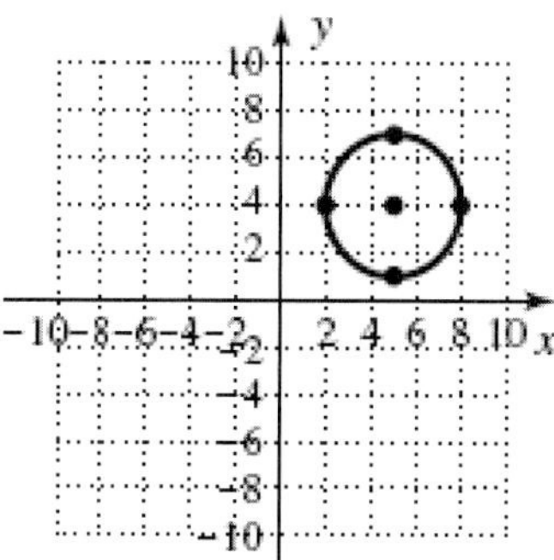

25. Center: $(2,3), r=2$
$D: x\in[0,4]$
$R: y\in[1,5]$

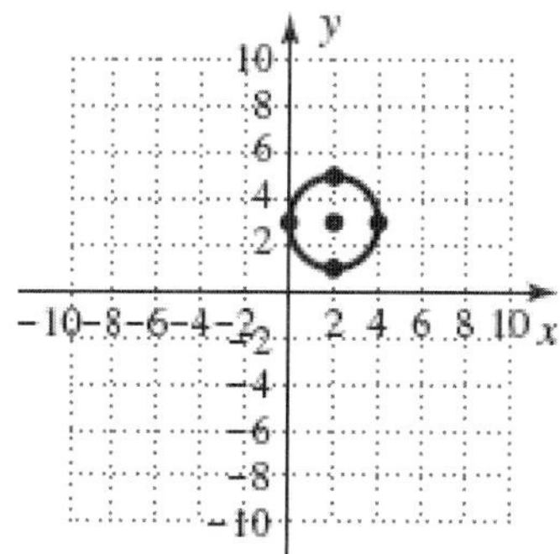

27. Center: $(-1,2), r=2\sqrt{3}$
$D: x\in\left[-1-2\sqrt{3},-1+2\sqrt{3}\right]$
$R: y\in\left[2-2\sqrt{3},2+2\sqrt{3}\right]$

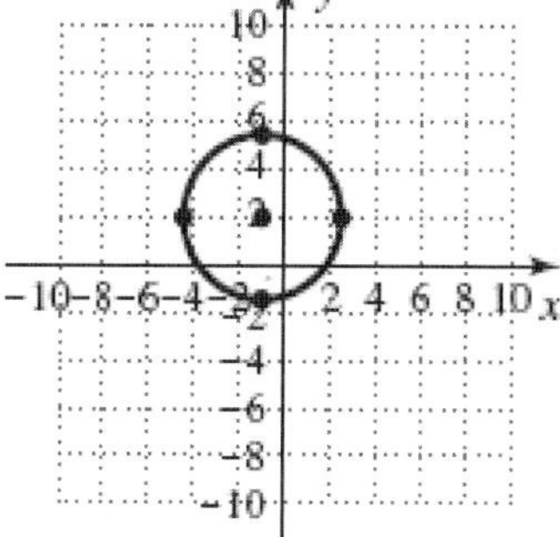

29. Center: $(-4,0), r=9$

$D: x \in [-13,5]$
$R: y \in [-9,9]$

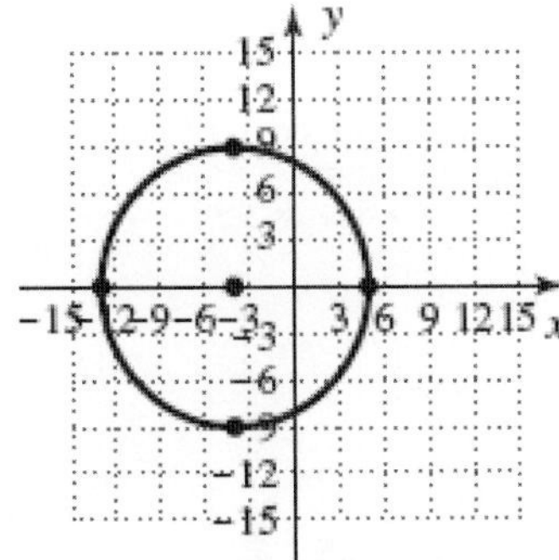

31. $x^2 + y^2 - 10x - 12y + 4 = 0$
$x^2 - 10x + y^2 - 12y = -4$
$x^2 - 10x + 25 + y^2 - 12y + 36 = -4 + 25 + 36$
$(x-5)^2 + (y-6)^2 = 57$
Center: $(5,6)$, Radius: $r = \sqrt{57}$

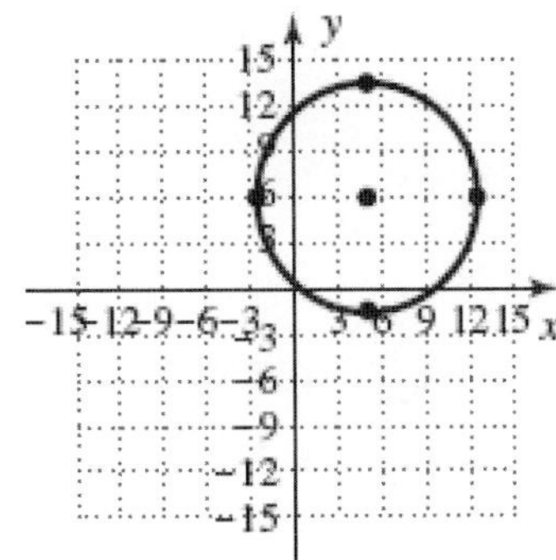

33. $x^2 + y^2 - 10x + 4y + 4 = 0$
$x^2 - 10x + y^2 + 4y = -4$
$x^2 - 10x + 25 + y^2 + 4y + 4 = -4 + 25 + 4$
$(x-5)^2 + (y+2)^2 = 25$
Center: $(5,-2)$, Radius: $r = 5$

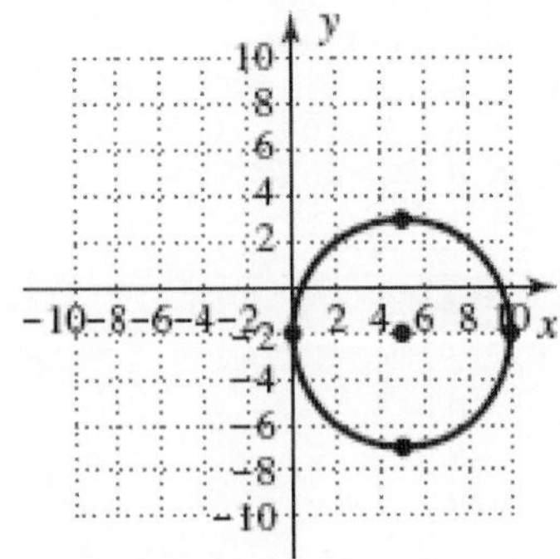

35. $x^2 + y^2 + 6y - 5 = 0$
$x^2 + y^2 + 6y = 5$
$x^2 + y^2 + 6y + 9 = 5 + 9$
$x^2 + (y+3)^2 = 14$
Center: $(0,-3)$, Radius: $r = \sqrt{14}$

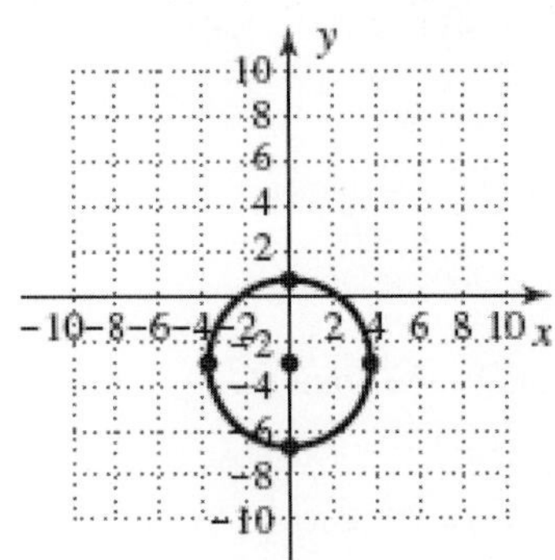

37. $x^2 + y^2 + 4x + 10y + 18 = 0$
$x^2 + 4x + y^2 + 10y = -18$
$x^2 + 4x + 4 + y^2 + 10y + 25 = -18 + 4 + 25$
$(x+2)^2 + (y+5)^2 = 11$
Center: $(-2,-5)$, Radius: $r = \sqrt{11}$

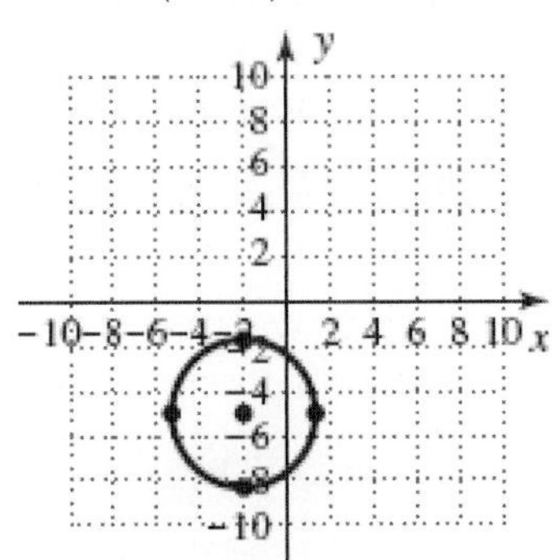

$\frac{(x+2)^2}{11} + \frac{(y+5)^2}{11} = 1$
$a = \sqrt{11}, b = \sqrt{11}, r = \sqrt{11}$
They are equal.

39. $x^2 + y^2 + 14x + 12 = 0$
$x^2 + 14x + y^2 = -12$
$x^2 + 14x + 49 + y^2 = -12 + 49$
$(x+7)^2 + y^2 = 37$
Center: $(-7,0)$, Radius: $r = \sqrt{37}$

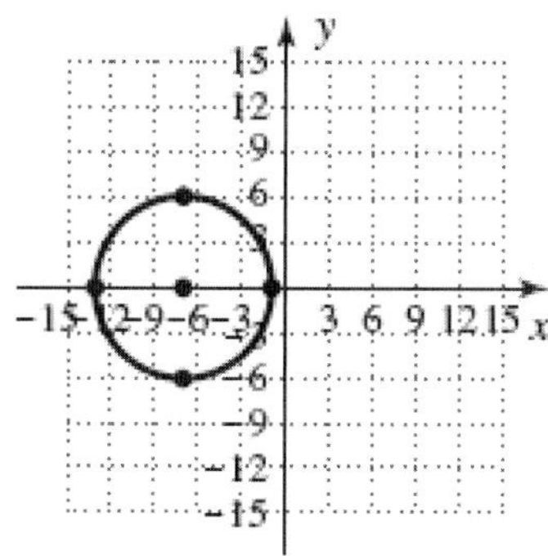

$$\frac{(x+7)^2}{37}+\frac{(y)^2}{37}=1$$

$a=\sqrt{37}, b=\sqrt{37}, r=\sqrt{37}$

They are equal.

41. $2x^2+2y^2-12x+20y+4=0$

$x^2+y^2-6x+10y+2=0$

$x^2-6x+y^2+10y=-2$

$x^2-6x+9+y^2+10y+25=-2+9+25$

$(x-3)^2+(y+5)^2=32$

Center: $(3,-5)$, Radius: $r=4\sqrt{2}$

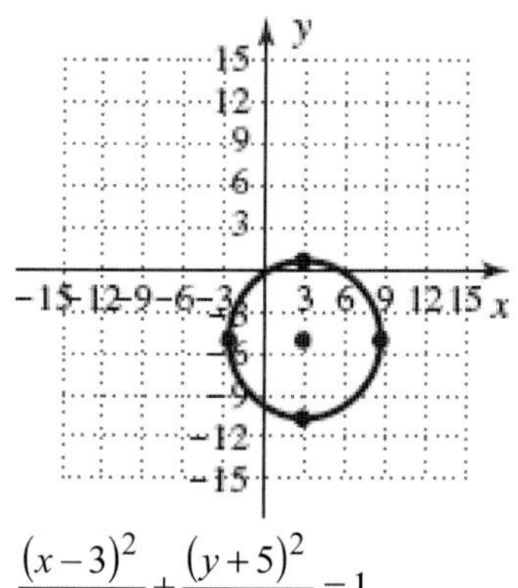

$$\frac{(x-3)^2}{32}+\frac{(y+5)^2}{32}=1$$

$a=4\sqrt{2}, b=4\sqrt{2}, r=4\sqrt{2}$

They are equal.

43. $\frac{(x-1)^2}{9}+\frac{(y-2)^2}{16}=1$

Center: (1,2)

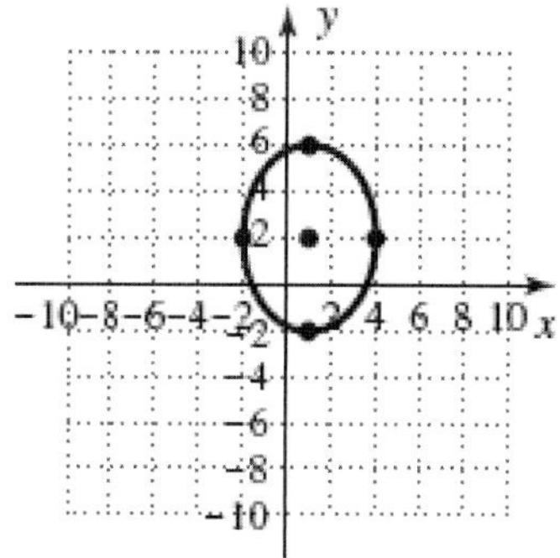

45. $\frac{(x-2)^2}{25}+\frac{(y+3)^2}{4}=1$

Center: (2, -3)

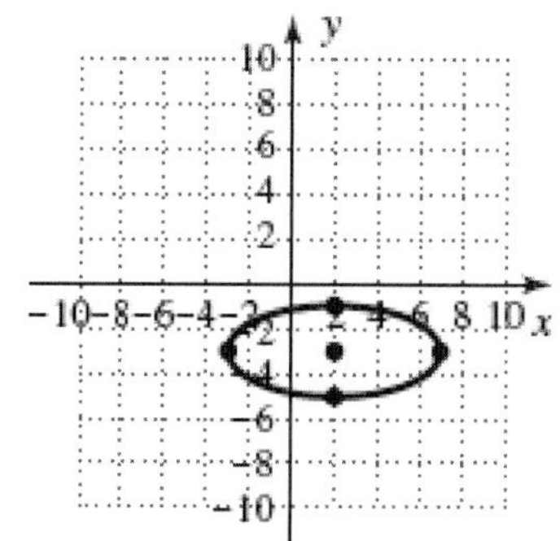

47. $\frac{(x+1)^2}{16}+\frac{(y+2)^2}{9}=1$

Center (-1,-2)

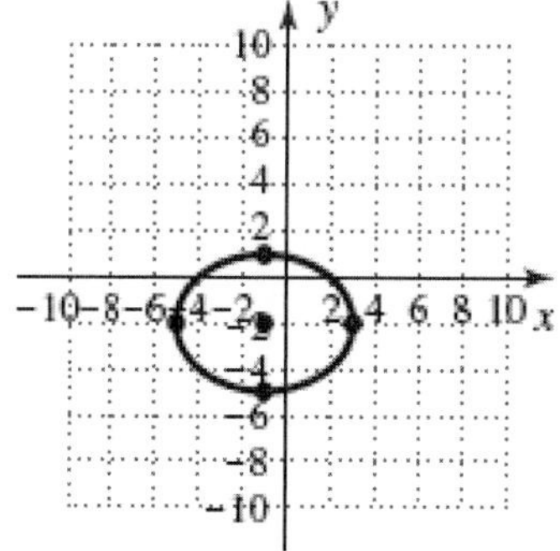

49. $(x+1)^2+4(y-2)^2=16$

$$\frac{(x+1)^2}{16}+\frac{(y-2)^2}{4}=1$$

$$\frac{(x+1)^2}{4^2}+\frac{(y-2)^2}{2^2}=1$$

Ellipse

Center: (-1,2), $a=4, b=2$

Vertices: $(-5,2), (3,2)$

Endpts of minor axis: $(-1,0), (-1,4)$

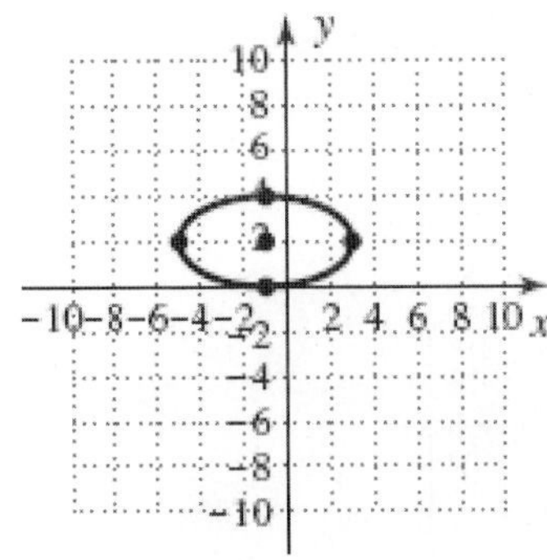

51. $2(x-2)^2+2(y+4)^2=18$

$(x-2)^2+(y+4)^2=9$

Circle

Center: (2,-4), Radius: 3

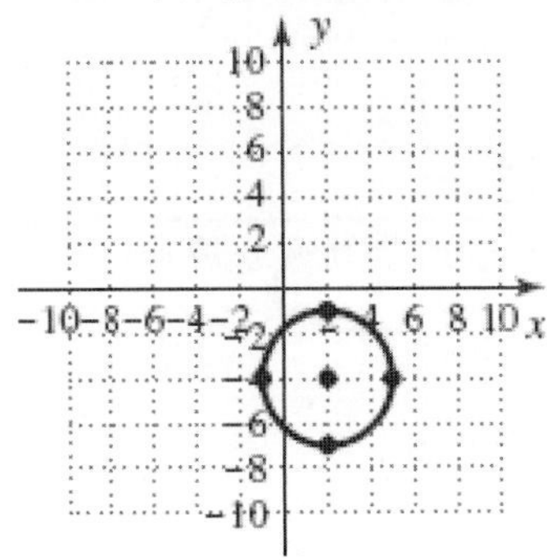

53. $4(x-1)^2+9(y-4)^2=36$

$$\frac{(x-1)^2}{9}+\frac{(y-4)^2}{4}=1$$

$$\frac{(x-1)^2}{3^2}+\frac{(y-4)^2}{2^2}=1$$

Ellipse

Center: (1,4), $a=3, b=2$

Vertices: $(4,4),(-2,4)$

Endpts of minor axis: $(1,2),(1,6)$

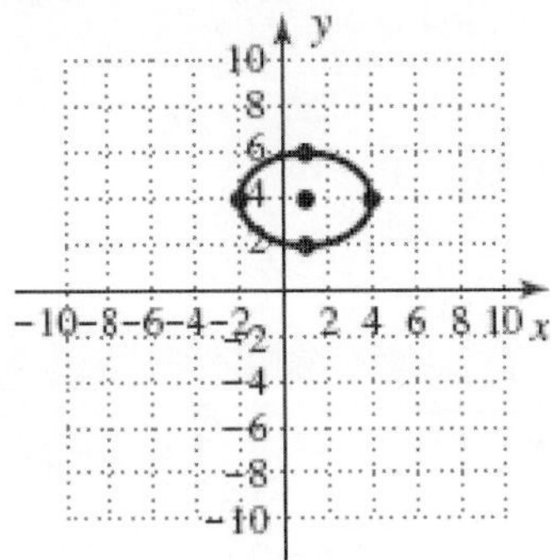

55. $x^2+4y^2=16$

a. $\dfrac{x^2}{16}+\dfrac{y^2}{4}=1$

Center: (0,0), $a=4, b=2$

b. Vertices: $(-4,0),(4,0)$

Endpts of minor axis: $(0,-2),(0,2)$

c.

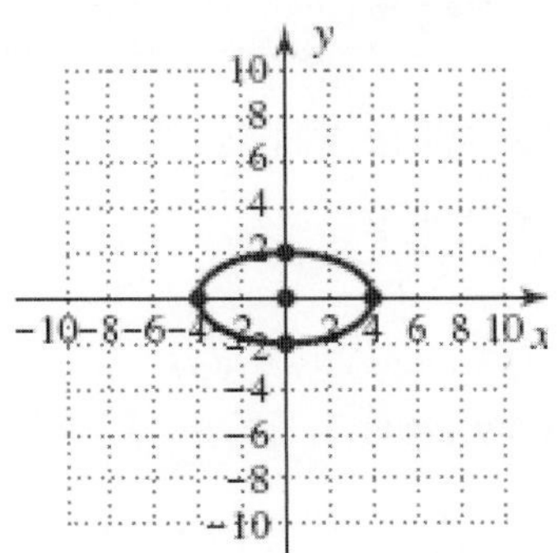

57. $16x^2+9y^2=144$

a. $\dfrac{x^2}{9}+\dfrac{y^2}{16}=1$

Center: (0,0), $a=3, b=4$

b. Vertices: $(0,-4),(0,4)$

Endpts of minor axis: $(-3,0),(3,0)$

c.

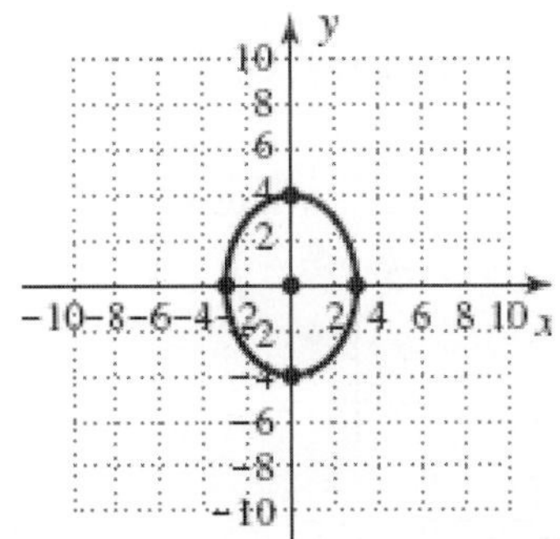

59. $2x^2+5y^2=10$

a. $\dfrac{x^2}{5}+\dfrac{y^2}{2}=1$

Center: (0,0), $a=\sqrt{5}, b=\sqrt{2}$

b. Vertices: $(-\sqrt{5},0),(\sqrt{5},0)$

Endpts of minor axis: $(0,-\sqrt{2}),(0,\sqrt{2})$

c.

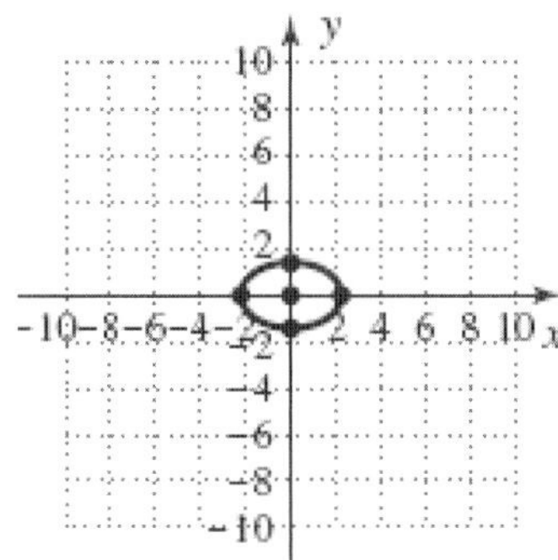

61. $4x^2 + y^2 + 6y + 5 = 0$

$4x^2 + y^2 + 6y = -5$

$4x^2 + y^2 + 6y + 9 = -5 + 9$

$4(x)^2 + (y+3)^2 = 4$

$x^2 + \frac{(y+3)^2}{4} = 1$

Center: $(0,-3)$

Vertices $(0,-3-2),(0,-3+2)$;
$(0,-5),(0,-1)$

Endpts of minor axis $(0-1,-3),(0+1,-3)$;
$(-1,-3),(1,-3)$

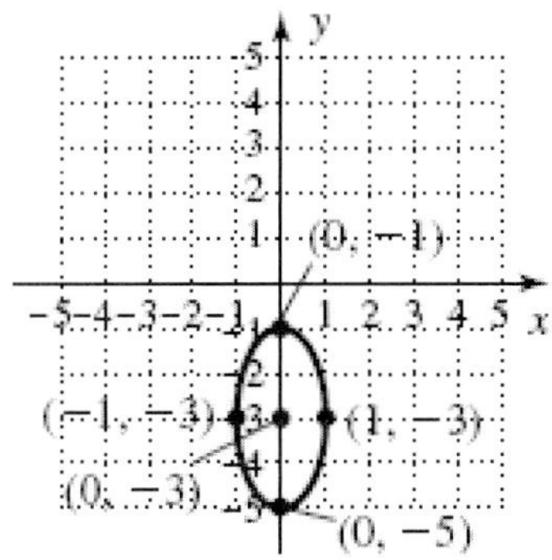

63. $x^2 + 4y^2 - 8y + 4x - 8 = 0$

$x^2 + 4x + 4y^2 - 8y = 8$

$x^2 + 4x + 4(y^2 - 2y) = 8$

$x^2 + 4x + 4 + 4(y^2 - 2y + 1) = 8 + 4 + 4$

$(x+2)^2 + 4(y-1)^2 = 16$

$\frac{(x+2)^2}{16} + \frac{(y-1)^2}{4} = 1$

$a = 4,\ b = 2$

Center: $(-2,1)$

Vertices : $(-2-4,1),(-2+4,1)$;
$(-6,1),(2,1)$

Endpts of minor axis : $(-2,1-2),(-2,1+2)$;

$(-2,-1),(-2,3)$

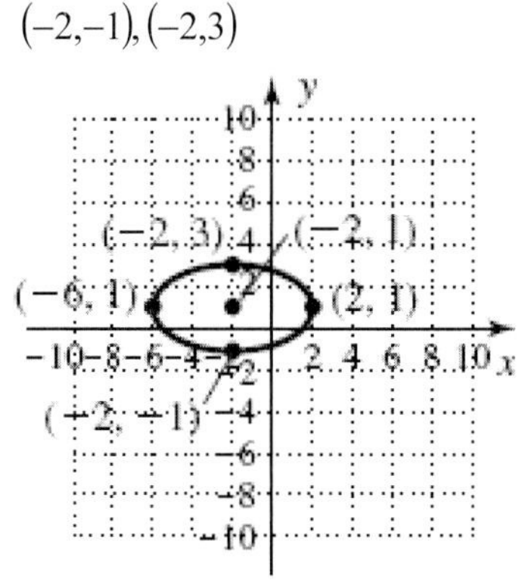

65. $5x^2 + 2y^2 + 20y - 30x + 75 = 0$

$5x^2 - 30x + 2y^2 + 20y = -75$

$5(x^2 - 6x) + 2(y^2 + 10y) = -75$

$5(x^2 - 6x + 9) + 2(y^2 + 10y + 25) = -75 + 45 + 50$

$5(x-3)^2 + 2(y+5)^2 = 20$

$\frac{(x-3)^2}{4} + \frac{(y+5)^2}{10} = 1$

$a = 2,\ b = \sqrt{10}$

Center: $(3,-5)$

Vertices : $(3,-5-\sqrt{10}),(3,-5+\sqrt{10})$

Endpts of minor axis : $(3-2,-5),(3+2,-5)$
$(1,-5),(5,-5)$

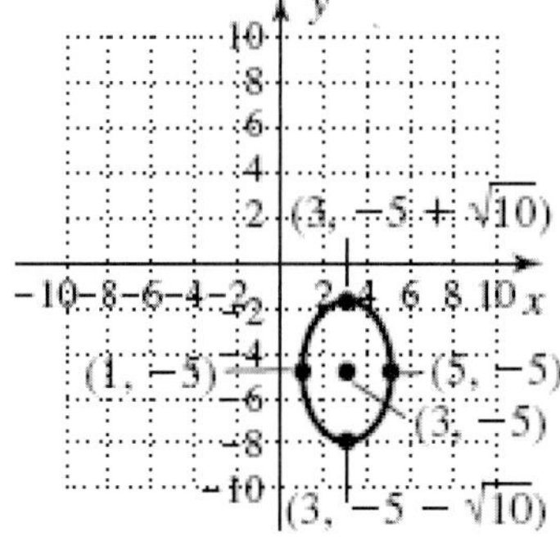

67. $2x^2 + 5y^2 - 12x + 20y - 12 = 0$

$2x^2 - 12x + 5y^2 + 20y = 12$

$2(x^2 - 6x) + 5(y^2 + 4y) = 12$

$2(x^2 - 6x + 9) + 5(y^2 + 4y + 4) = 12 + 18 + 20$

$2(x-3)^2 + 5(y+2)^2 = 50$

$\frac{(x-3)^2}{25} + \frac{(y+2)^2}{10} = 1$

$a = 5,\ b = \sqrt{10}$

Center: $(3,-2)$
Vertices $(3-5,-2),(3+5,-2)$;
$(-2,-2),(8,-2)$
Endpts of minor axis :
$\left(3,-2-\sqrt{10}\right),\left(3,-2+\sqrt{10}\right)$

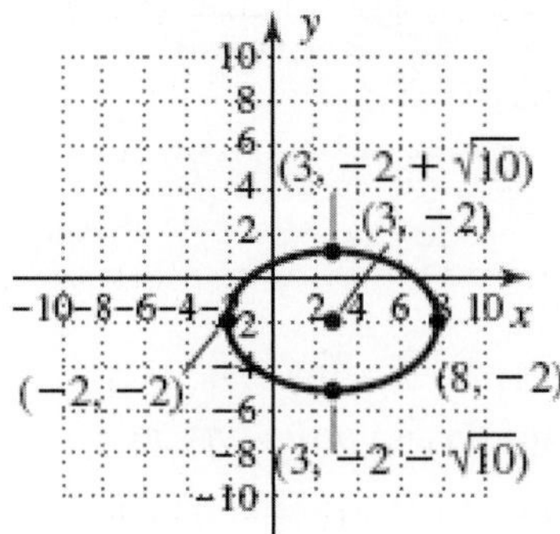

69. $A=2r^2$
$A=2(5)^2=50 \text{ units}^2$

71. $(x-5)^2+(y-12)^2=25^2$;
$(15-5)^2+(36-12)^2 \ ? \ 25^2$
$(10)^2+(24)^2 \ ? \ 25^2$
$676>625$
No

73. Red: $(x-2)^2+(y-2)^2=4$;
Center: (2,2), Radius: 2
Blue: $(x-2)^2+y^2=16$;
Center: (2,0), Radius: 4
Area of blue: $\pi(16)-\pi(4)=12\pi \text{ units}^2$

75. $x^2+y^2+8x-6y=0$
$x^2+8x+y^2-6y=0$
$x^2+8x+16+y^2-6y+9=0+16+9$
$(x+4)^2+(y-3)^2=25$;
$x^2+y^2-10x+4y=0$
$x^2-10x+y^2+4y=0$
$x^2-10x+25+y^2+4y+4=0+25+4$
$(x-5)^2+(y+2)^2=29$;
Distance between centers: (-4,3), (5,-2)
$d=\sqrt{(-4-5)^2+(3-(-2))^2}$
$=\sqrt{81+25}=\sqrt{106}\approx 10.30$;

Sum of the radii: $5+\sqrt{29}\approx 10.39$
No, Distance between the centers is less than the sum of the radii.

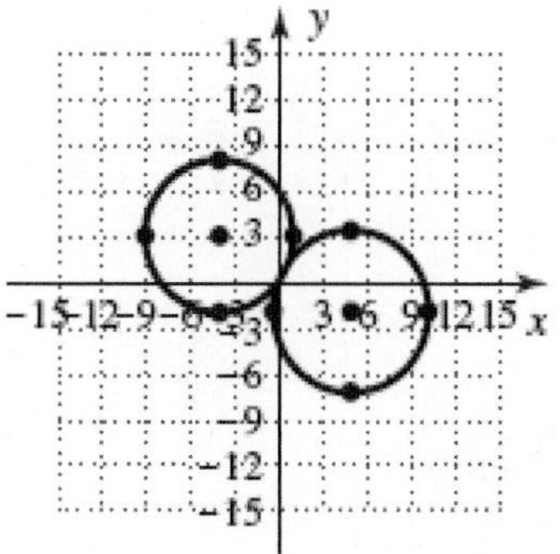

77. $a=\frac{72}{2}=36; b=\frac{70.5}{2}=35.25$
$\frac{x^2}{36^2}+\frac{y^2}{(35.25)^2}=1$

79. $4x^2+9y^2=900$
$\frac{x^2}{15^2}+\frac{y^2}{10^2}=1$;
$9x^2+25y^2=900$
$\frac{x^2}{10^2}+\frac{y^2}{6^2}=1$;
$A=\pi(15)(10)-\pi(10)(6)=90\pi$
$=9{,}000\pi \text{ yd}^2$

81. $\frac{x^2}{15^2}+\frac{y^2}{8^2}=1$
$\frac{9^2}{15^2}+\frac{y^2}{8^2}=1$
$\frac{y^2}{8^2}=0.64$
$y^2=40.96$
$y=6.4 \text{ ft}$

83. Answers will vary.
Aphelion: 155 million miles
Semi-major axis: 142 million miles

85. There are no x-intercepts.
x^2 and y^2 must have the same sign.

87. $\frac{(x-h)^2}{a^2}+\frac{(y-k)^2}{b^2}=1$

$b^2(x-h)^2+a^2(y-k)^2=a^2b^2$;

Eq.1: $b^2(-1-h)^2+a^2(1-k)^2=a^2b^2$;

Eq.2: $b^2(5-h)^2+a^2(1-k)^2=a^2b^2$;

Eq.3: $b^2(2-h)^2+a^2(3-k)^2=a^2b^2$;

Eq.4: $b^2(2-h)^2+a^2(-1-k)^2=a^2b^2$;

Set Eq.1=Eq.2:

$b^2(-1-h)^2+a^2(1-k)^2=b^2(5-h)^2+a^2(1-k)^2$

$b^2(-1-h)^2=b^2(5-h)^2$

$-1-h=\pm(5-h)$

$-1-h\neq 5-h$ or $-1-h=-5+h$

$-1-h\neq 5-h$ or $4=2h$

$-1\neq 5$ or $2=h$;

Set Eq.3=Eq.4:

$b^2(2-h)^2+a^2(3-k)^2=b^2(2-h)^2+a^2(-1-k)^2$

$a^2(3-k)^2=a^2(-1-k)^2$

$3-k=\pm(-1-k)$

$3-k\neq -1-k$ or $3\text{-}k=1+k$

$3-k\neq -1-k$ or $2=2k$

$3\neq -1$ or $1=k$;

eq.3:

$b^2(2-2)^2+a^2(3-1)^2=a^2b^2$

$4=b^2$;

eq.1:

$b^2(-1-2)^2+a^2(1-1)^2=a^2b^2$

$9=a^2$;

$\frac{(x-2)^2}{9}+\frac{(y-1)^2}{4}=1$

89. $f(x)=x^4-4x^3-7x^2+22x+24$

$x=1, x=2, x=-4$ are NOT zeroes.

$x=3$ is a zero.

1	1	-4	-7	22	24
		1	-3	-10	12
	1	-3	-10	12	36

2	1	-4	-7	22	24
		2	-4	-22	0
	1	-2	-11	0	24

3	1	-4	-7	22	24
		3	-3	-30	-24
	1	-1	-10	-8	0

-4	1	-4	-7	22	24
		-4	32	-100	312
	1	-8	25	-78	336

91. $x^3-4x^2+17x-26=0$

$(1+2i\sqrt{3})^3-4(1+2i\sqrt{3})^2+17(1+2i\sqrt{3})-26=0$

$(1+2i\sqrt{3})(-11+4i\sqrt{3})-4(-11+4i\sqrt{3})+17+34i\sqrt{3}-26=0$

$-11-18i\sqrt{3}-24+44-16i\sqrt{3}+34i\sqrt{3}-9=0$

$0=0$ verified;

The conjugate $1-2i\sqrt{3}$ must also be a root.

93. a. $D: x\in(-\infty,\infty), R: y\in(-\infty,4]$

b. $f(x)\geq 0$ for $x\in[-3,-1]\cup[1,5]$

c. $\max: y=2$ at $x=-2$, (-2,2),

$\max: y=4$ at $x=3$, (3,4)

$\min: y=-1$ at $x=0$, (-1,0)

d. $f(x)\uparrow$ for $x\in(-\infty,-2)\cup(0,3)$

$f(x)\downarrow$ for $x\in(-2,0)\cup(3,\infty)$

7.2 Technology Highlight

1. (0,2), (0,-2); $y=\pm 2.5612497$

2. $y=\pm\frac{3}{4}x$, The asymptotes meet at (0,0) because that is the center. The asymptotes will not intersect at the origin anytime the center is not at the origin.

7.2 Exercises

1. transverse

3. midway

5. Answers will vary.

7. $\frac{x^2}{9}-\frac{y^2}{4}=1$

Center: (0,0)

Vertices: (-3,0), (3,0)

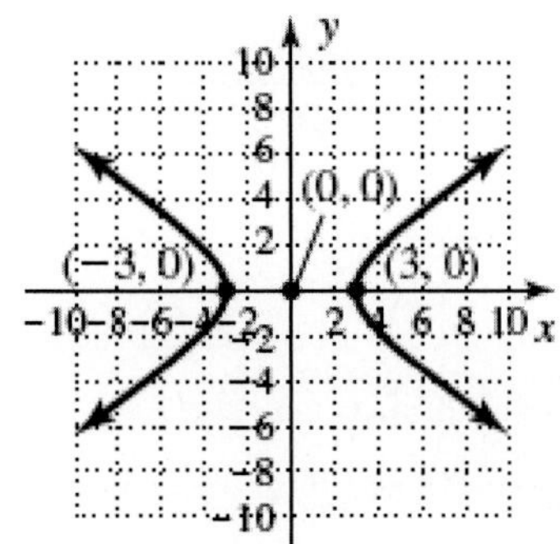

9. 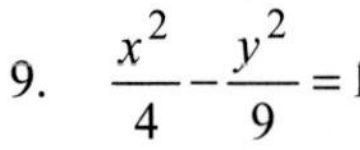 $\dfrac{x^2}{4}-\dfrac{y^2}{9}=1$

Center: (0,0)

Vertices: (-2,0), (2,0)

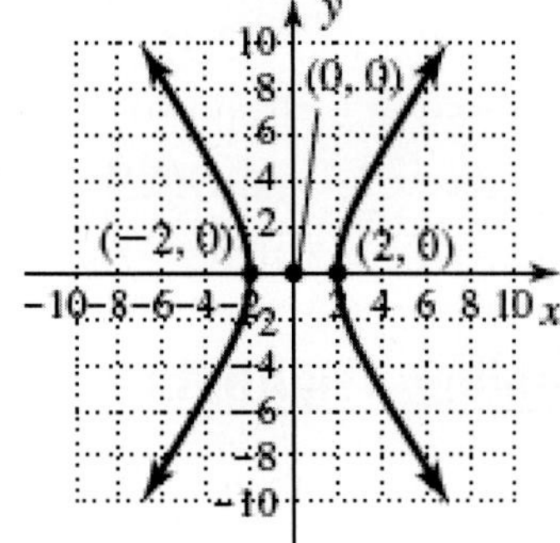

11. $\dfrac{x^2}{49}-\dfrac{y^2}{16}=1$

Center: (0,0)

Vertices: (-7,0), (7,0)

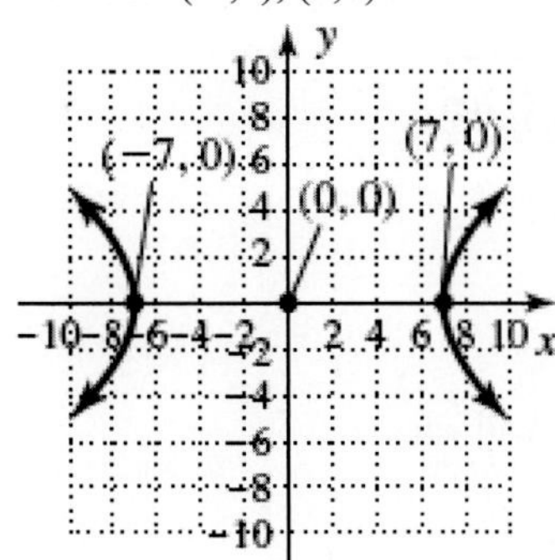

13. $\dfrac{x^2}{36}-\dfrac{y^2}{16}=1$

Center: (0,0)

Vertices: (-6,0), (6,0)

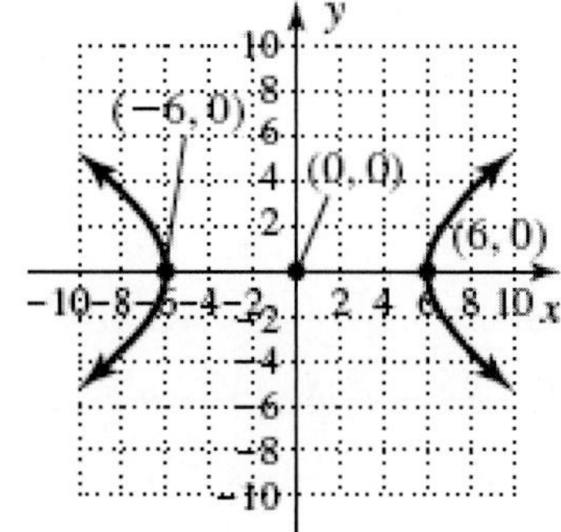

15. $\dfrac{y^2}{9}-\dfrac{x^2}{1}=1$

Center: (0,0)

Vertices: (0,-3), (0,3)

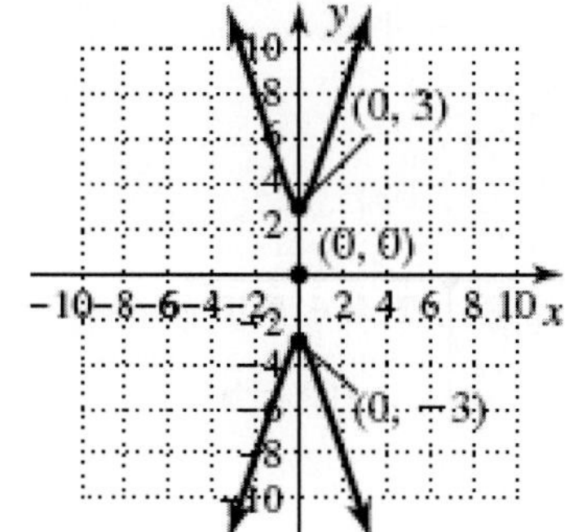

17. $\dfrac{y^2}{12}-\dfrac{x^2}{4}=1$

Center: (0,0)

Vertices: $\left(0,-2\sqrt{3}\right), \left(0,2\sqrt{3}\right)$

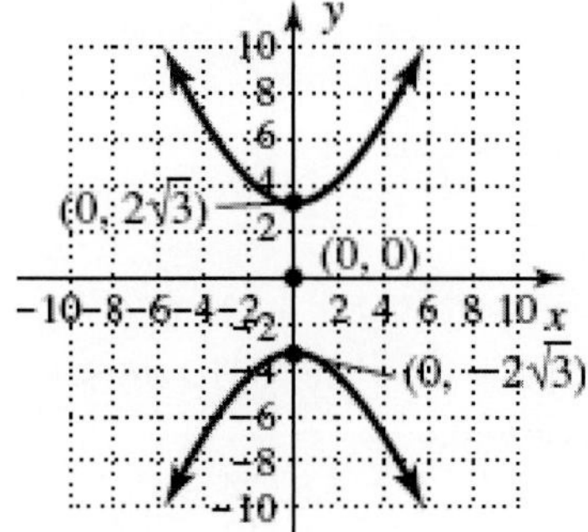

19. $\dfrac{y^2}{9}-\dfrac{x^2}{9}=1$

Center: (0,0)

Vertices: (0,-3), (0,3)

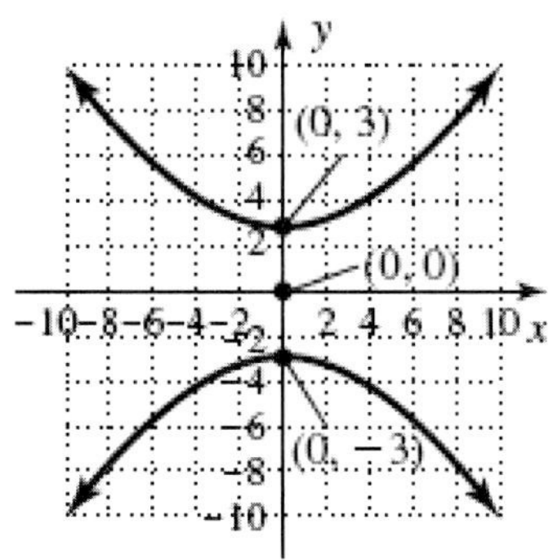

21. $\dfrac{y^2}{36} - \dfrac{x^2}{25} = 1$

Center: (0,0)
Vertices: (0,-6), (0,6)

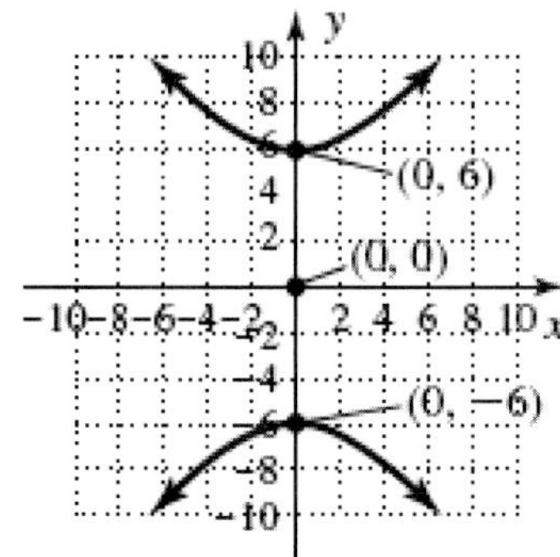

23. Vertices: (-4,-2), (2,-2)
Transverse Axis: $y = -2$

$\dfrac{-4+2}{2} = -1$

Center: (-1,-2)
Conjugate Axis: $x = -1$

25. Vertices: (4,1), (4,-3)
Transverse Axis: $x = 4$

$\dfrac{1+-3}{2} = -1$

Center: (4,-1)
Conjugate Axis: $y = -1$

27. $\dfrac{(y+1)^2}{4} - \dfrac{x^2}{25} = 1$

Center: (0,-1)
$a = 5,\ b = 2$
Vertices: $(0,-1-2), (0,-1+2)$;
$(0,-3), (0,1)$
Transverse axis: $x = 0$
Conjugate axis: $y = -1$

Asymptotes: $y+1 = \pm\dfrac{2}{5}x$

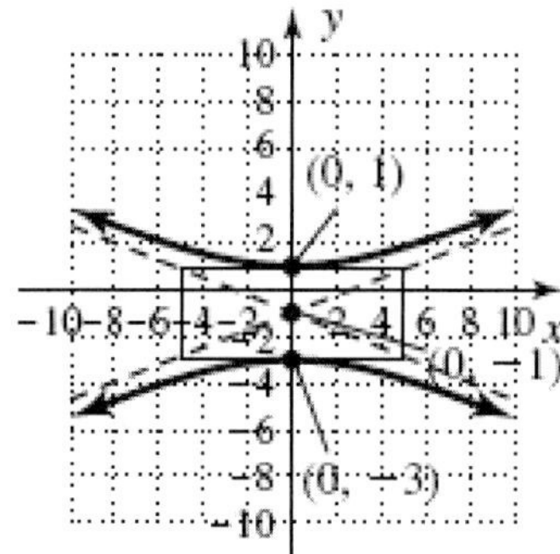

29. $\dfrac{(x-3)^2}{36} - \dfrac{(y+2)^2}{49} = 1$

Center: (3,-2)
$a = 6,\ b = 7$
Vertices: $(3-6,-2), (3+6,-2)$;
$(-3,-2), (9,-2)$
Transverse axis: $y = -2$
Conjugate axis: $x = 3$
Asymptotes: $y+2 = \pm\dfrac{7}{6}(x-3)$

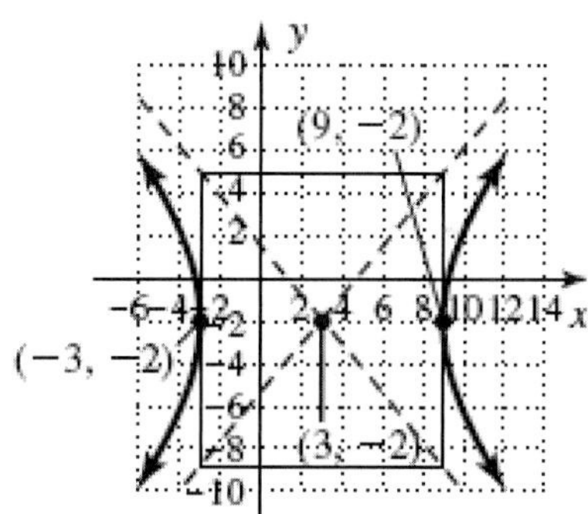

31. $\dfrac{(y+1)^2}{7} - \dfrac{(x+5)^2}{9} = 1$

Center: (-5,-1)
$a = 3,\ b = \sqrt{7}$
Vertices: $\left(-5,-1+\sqrt{7}\right), \left(-5,-1-\sqrt{7}\right)$
Transverse axis: $x = -5$
Conjugate axis: $y = -1$

Asymptotes: $y+1 = \pm\dfrac{\sqrt{7}}{3}(x+5)$

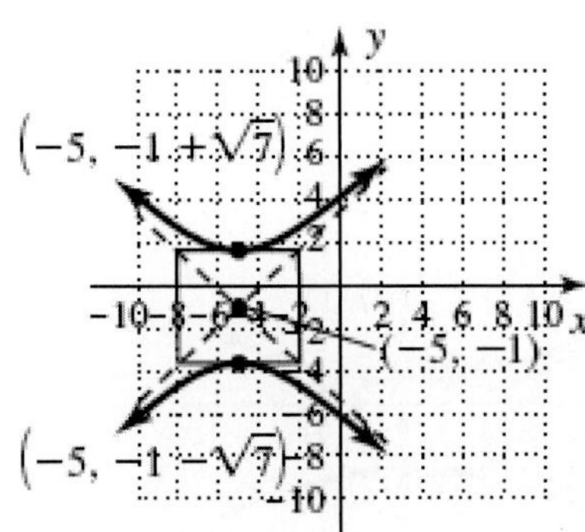

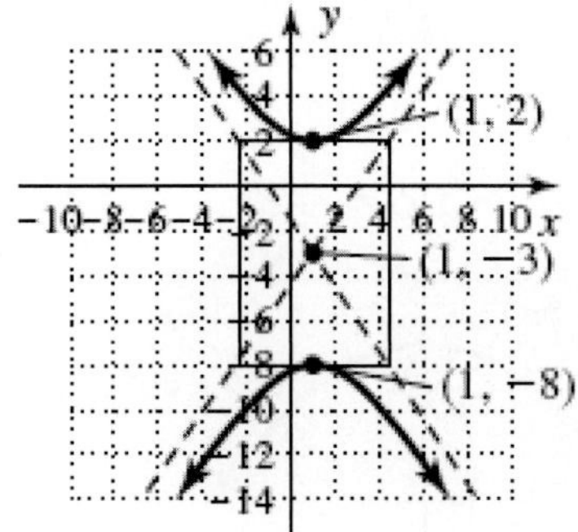

33. $(x-2)^2-4(y+1)^2=16$

$\frac{(x-2)^2}{16}-\frac{(y+1)^2}{4}=1$

Center: (2,-1)

$a=4,\ b=2$

Vertices: $(2-4,-1),(2+4,-1)$;

$(-2,-1),(6,-1)$

Transverse axis: $y=-1$

Conjugate axis: $x=2$

Asymptotes: $y+1=\pm\frac{1}{2}(x-2)$

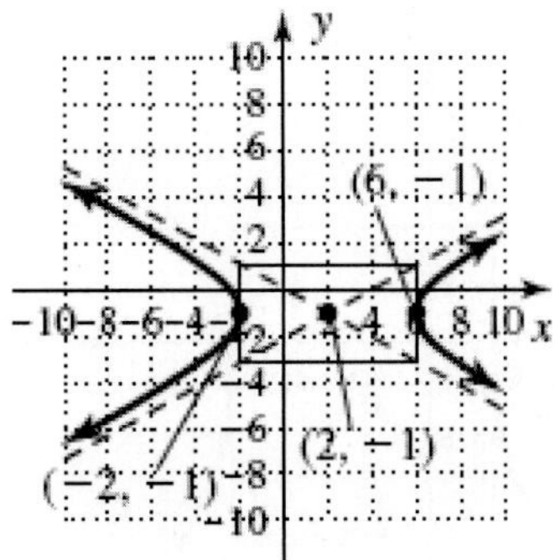

35. $2(y+3)^2-5(x-1)^2=50$

$\frac{(y+3)^2}{25}-\frac{(x-1)^2}{10}=1$

Center: (1,-3)

$a=\sqrt{10},\ b=5$

Vertices: $(1,-3+5),(1,-3-5)$;

$(1,2),(1,-8)$

Transverse axis: $x=1$

Conjugate axis: $y=-3$

Asymptotes: $y+3=\pm\frac{5}{\sqrt{10}}(x-1)$

$y+3=\pm\frac{\sqrt{10}}{2}(x-1)$

37. $12(x-4)^2-5(y-3)^2=60$

$\frac{(x-4)^2}{5}-\frac{(y-3)^2}{12}=1$

Center: (4,3)

$a=\sqrt{5},\ b=\sqrt{12}=2\sqrt{3}$

Vertices: $(4+\sqrt{5},3),(4-\sqrt{5},3)$

Transverse axis: $y=3$

Conjugate axis: $x=4$

Asymptotes: $y-3=\pm\frac{2\sqrt{3}}{\sqrt{5}}(x-4)$

$y-3=\pm\frac{2\sqrt{3}}{\sqrt{5}}\cdot\frac{\sqrt{5}}{\sqrt{5}}(x-4)$

$y-3=\pm\frac{2\sqrt{15}}{5}(x-4)$

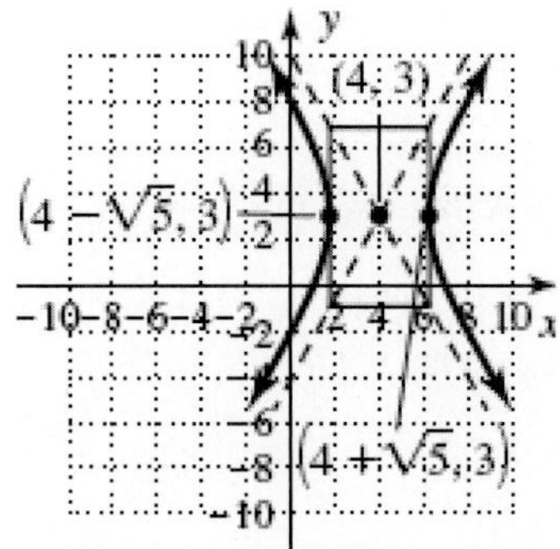

39. $16x^2-9y^2=144$

$\frac{x^2}{9}-\frac{y^2}{16}=1$

Center: (0,0)

$a=3,\ b=4$

Vertices: $(0+3,0),(0-3,0)$;

$(3,0),(-3,0)$

Transverse axis: $y=0$

Conjugate axis: $x=0$

Asymptotes: $y = \pm\frac{4}{3}x$

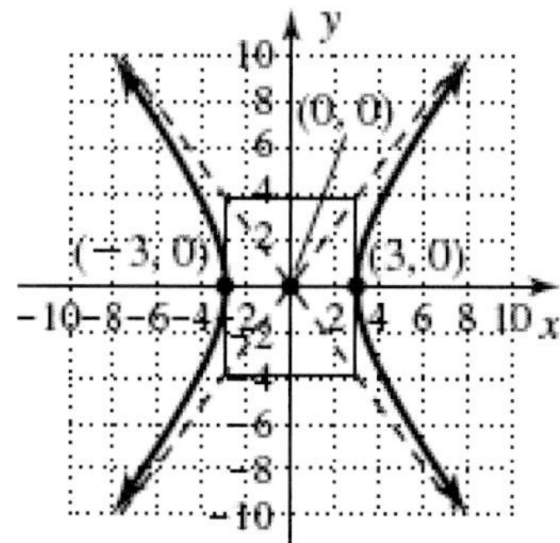

41. $9y^2 - 4x^2 = 36$

$\frac{y^2}{4} - \frac{x^2}{9} = 1$

Center: (0,0)

$a = 3,\ b = 2$

Vertices: $(0,0-2),(0,0+2)$;

$(0,-2),(0,2)$

Transverse axis: $x = 0$

Conjugate axis: $y = 0$

Asymptotes: $y = \pm\frac{2}{3}x$

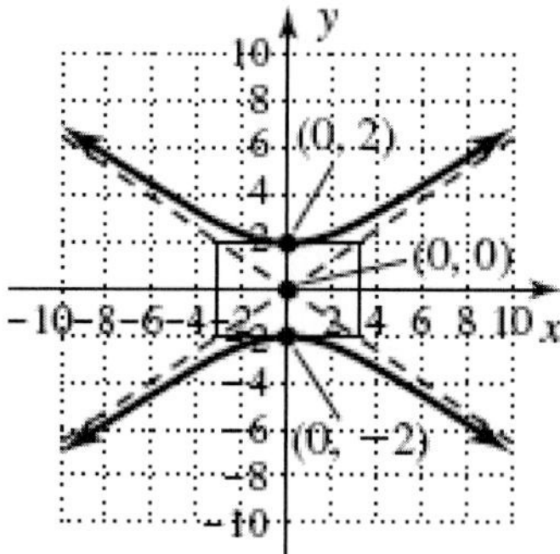

43. $12x^2 - 9y^2 = 72$

$\frac{x^2}{6} - \frac{y^2}{8} = 1$

Center: (0,0)

$a = \sqrt{6},\ b = \sqrt{8} = 2\sqrt{2}$

Vertices: $(\sqrt{6},0),(-\sqrt{6},0)$

Transverse axis: $y = 0$

Conjugate axis: $x = 0$

Asymptotes: $y = \pm\frac{2\sqrt{2}}{\sqrt{6}}x$

$y = \pm\frac{2\sqrt{2}}{\sqrt{6}}\cdot\frac{\sqrt{6}}{\sqrt{6}}x$

$y = \pm\frac{2\sqrt{3}}{3}x$

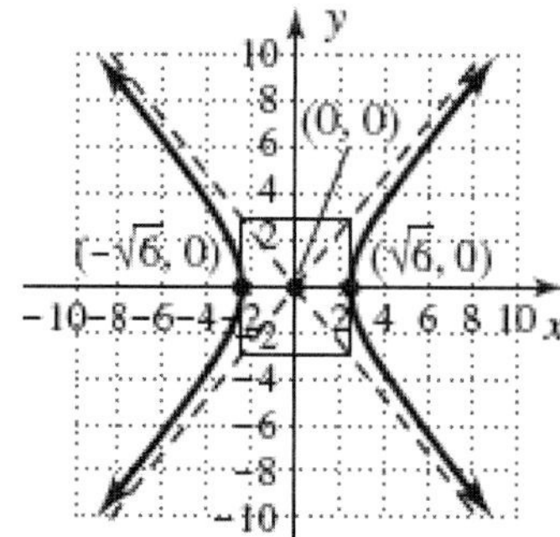

45. $4x^2 - y^2 + 40x - 4y + 60 = 0$

$4x^2 + 40x - y^2 - 4y = -60$

$4(x^2 + 10x) - (y^2 + 4y) = -60$

$4(x^2 + 10x + 25) - (y^2 + 4y + 4) = -60 + 100 - 4$

$4(x+5)^2 - (y+2)^2 = 36$

$\frac{(x+5)^2}{9} - \frac{(y+2)^2}{36} = 1$

Center: (-5,-2)

$a = 3,\ b = 6$

Vertices: $(-5-3,-2),(-5+3,-2)$;

$(-8,-2),(-2,-2)$

Transverse axis: $y = -2$

Conjugate axis: $x = -5$

Asymptotes: $y + 2 = \pm\frac{6}{3}(x+5)$

$y + 2 = \pm 2(x+5)$

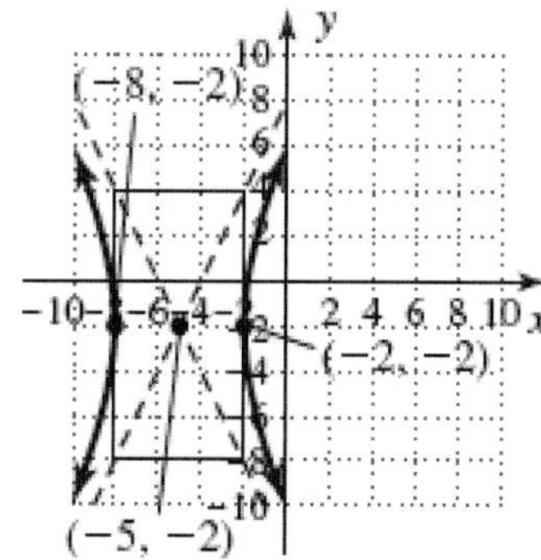

47. $x^2 - 4y^2 - 24y - 4x - 36 = 0$

$x^2 - 4x - 4y^2 - 24y = 36$

$(x^2 - 4x) - 4(y^2 + 6y) = 36$

$\left(x^2-4x+4\right)-4\left(y^2+6y+9\right)=36+4-36$

$(x-2)^2-4(y+3)^2=4$

$\frac{(x-2)^2}{4}-\frac{(y+3)^2}{1}=1$

Center: (2,-3)

$a=2,\ b=1$

Vertices: $(2-2,-3),(2+2,-3);$

$(0,-3),(4,-3)$

Transverse axis: $y=-3$

Conjugate axis: $x=2$

Asymptotes: $y+3=\pm\frac{1}{2}(x-2)$

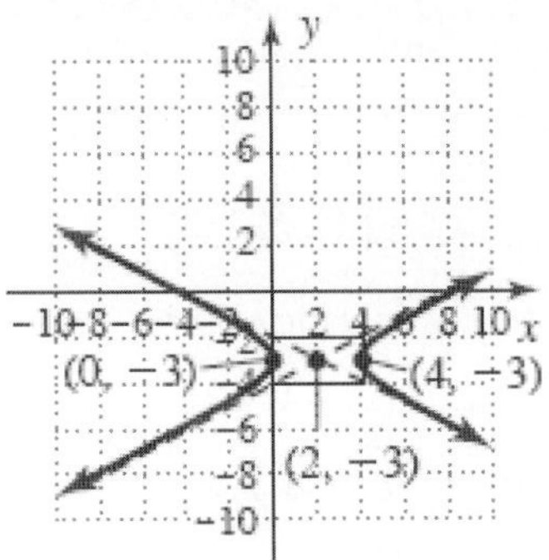

49. $-4x^2-4y^2=-24$

$x^2+y^2=6$

The equation contains a sum of second degree terms with equal coefficients. The equation represents a circle.

51. $x^2+y^2=2x+4y+4$

The equation contains a sum of second degree terms with equal coefficients. The equation represents a circle.

53. $2x^2-4y^2=8$

The equation contains a difference of second degree terms. The equation represents a hyperbola.

55. $x^2+5=2y^2$

$x^2-2y^2=-5$

$2y^2-x^2=5$

The equation contains a difference of second degree terms. The equation represents a hyperbola.

57. $2x^2=-2y^2+x+20$

$2x^2-x+2y^2=20$

The equation contains a sum of second degree terms with equal coefficients. The equation represents a circle.

59. $16x^2+5y^2-3x+4y=538$

The equation contains a sum of second degree terms with unequal coefficients. The equation represents an ellipse.

61. $y=\sqrt{\frac{36-4x^2}{-9}}$

a. $y=\sqrt{\frac{-4\left(-9+x^2\right)}{-9}}$

$y=\sqrt{\frac{4\left(x^2-9\right)}{9}}$

$y=\frac{2}{3}\sqrt{x^2-9}$

b. $x^2-9\geq 0$

$(x+3)(x-3)\geq 0$

pos neg pos

-3 3

$x\in(-\infty,-3]\cup[3,\infty)$

c. $y=-\frac{2}{3}\sqrt{x^2-9}$

63. $25y^2-1600x^2=40000$

$\frac{y^2}{1600}-\frac{x^2}{25}=1$

$\frac{y^2}{40^2}-\frac{x^2}{5^2}=1$

40 yards

65. $1600x^2-400(y-50)^2=640000$

$\frac{x^2}{400}-\frac{(y-50)^2}{1600}=1$

$\frac{x^2}{20^2}-\frac{(y-50)^2}{40^2}=1$

20 + 20 = 40 feet

67. Answers will vary.

69. Answers will vary.

71. $9(x-2)^2-25(y-3)^2=225$

$\frac{(x-2)^2}{25}-\frac{(y-3)^2}{9}=1$

Center: (2,3)

$a=5,\ b=3$

Vertices: (2+5,3), (2-5,3);

(7,3), (-3,3)

Distance between vertices is diameter.

$d=\sqrt{(7--3)^2+(3-3)^2}=10$

Thus, the radius of the circle is 5.

$(x-2)^2+(y-3)^2=25$

72. $9(x-2)^2-25(y-3)^2=225$

$\frac{(x-2)^2}{25}-\frac{(y-3)^2}{9}=1;$

$\frac{(x-2)^2}{25}+\frac{(y-3)^2}{9}=1$

75. $f(x)=\begin{cases}4-x^2 & -2\le x<3\\ 5 & x\ge 3\end{cases}$

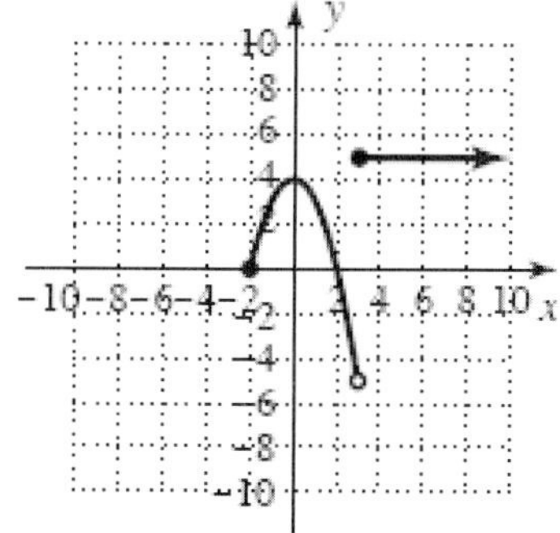

77. 48 inches = 48(2.54)cm=121.92cm

$A=\frac{1}{2}(121.92+200)(54)=8691.84\text{ cm}^2;$

200 - 121.92 = 78.08;

$\frac{78.08}{2}=39.04;$

$39.04^2+54^2=x^2$

$66.63=x$

$P\approx 200+121.92+2(66.63)=455.19\text{ cm}$

79. 39.8tons = 39.8(2000)lbs = 79600 lbs;

$\begin{cases}800x+1000y\le 79600\\ 100x+60y\le 6960\end{cases}$

EQ. 1: $y=-\frac{4}{5}x+79.6$

EQ. 2: $y=-\frac{5}{3}x+116$

$-\frac{4}{5}x+79.6=-\frac{5}{3}x+116$

$\frac{13}{15}x=36.4$

$x=42$

(42,46), (69.6,0) or (0,79.6);

R(x, y) = 300x + 375y;

R(42,46) = 300(42) + 375(46) = $29,850;

R(69.6,0) = 300(69.6) + 375(0) = $20,880;

R(0,79.6) = 300(0) + 375(79.6) = $29,850;

42 solid, 46 liquid produce maximum revenue. The other maximum (0,79.6) would not be used since the question asked for a combination of solid and liquid containers.

Chapter 7 Mid-Chapter

1. $(x-4)^2+(y+3)^2=9$

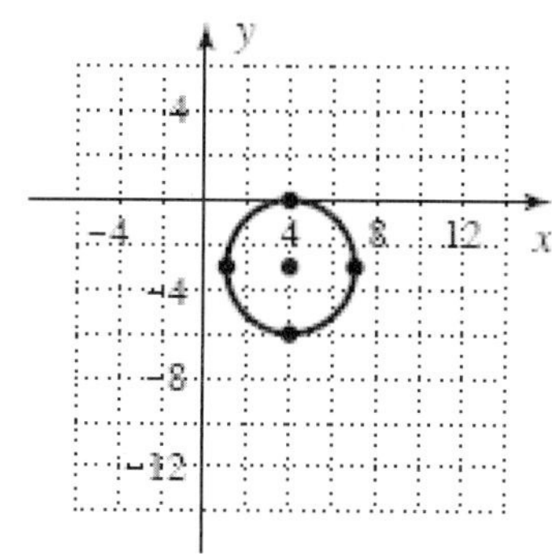

2. $x^2+y^2-10x+4y+4=0$

$x^2-10x+y^2+4y=-4$

$(x^2-10x+25)+(y^2+4y+4)=-4+25+4$

$(x-5)^2+(y+2)^2=25$

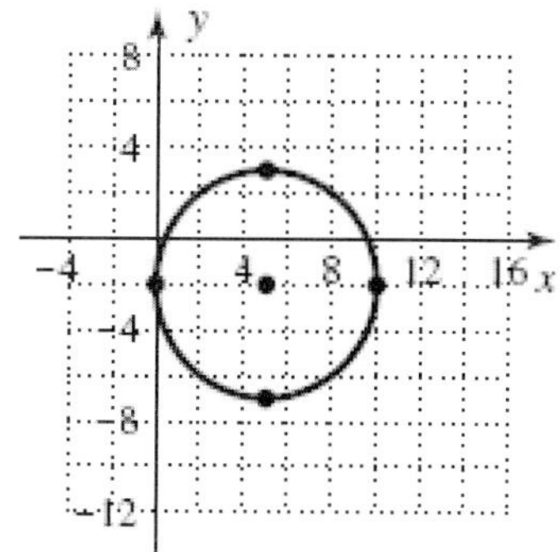

3. $\frac{(x-2)^2}{16}+\frac{(y+3)^2}{1}=1$

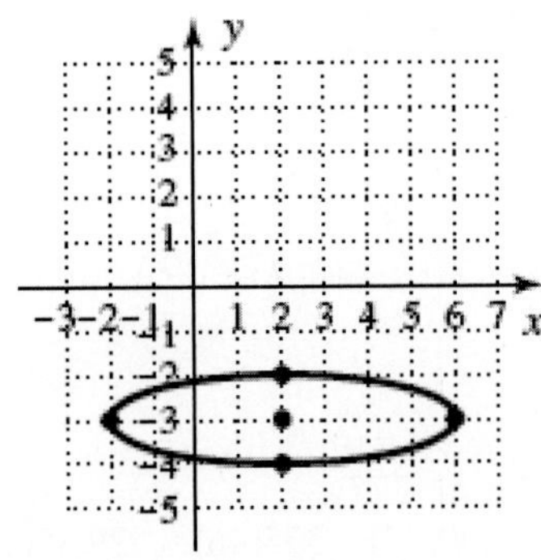

4. $9x^2+4y^2+18x-24y+9=0$

$9x^2+18x+4y^2-24y=-9$

$9\left(x^2+2x\right)+4\left(y^2-6y\right)=-9$

$9\left(x^2+2x+1\right)+4\left(y^2-6y+9\right)=-9+9+36$

$9(x+1)^2+4(y-3)^2=36$

$\frac{(x+1)^2}{4}+\frac{(y-3)^2}{9}=1$

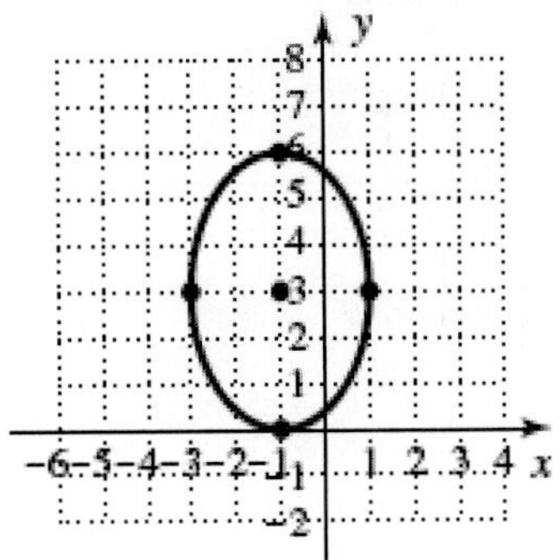

5. $\frac{(x+3)^2}{9}-\frac{(y-4)^2}{4}=1$

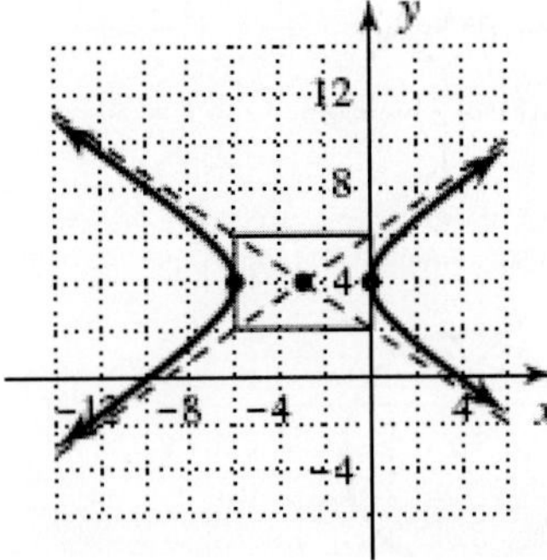

6. $9x^2-4y^2+18x-24y-63=0$

$9x^2+18x-4y^2-24y=63$

$9\left(x^2+2x\right)-4\left(y^2+6y\right)=63$

$9\left(x^2+2x+1\right)-4\left(y^2+6y+9\right)=63+9-36$

$9(x+1)^2-4(y+3)^2=36$

$\frac{(x+1)^2}{4}-\frac{(y+3)^2}{9}=1$

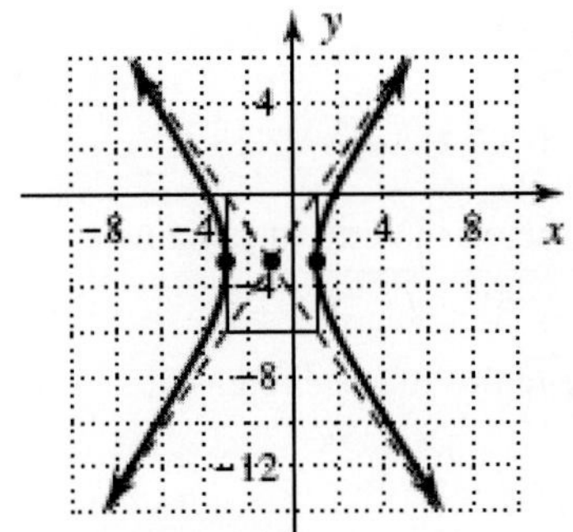

7. a. $\frac{(x+3)^2}{4}+\frac{(y-1)^2}{16}=1$

Center: (-3, 1)

$a=2,\ b=4$

$D: x\in[-3-2,-3+2]$

$D: x\in[-5,-1]$;

$R: y\in[1-4,1+4]$

$R: y\in[-3,5]$

b. $(x-3)^2+(y-2)^2=16$

Center: (3,2), $r=4$

$D: x\in[3-4,3+4]$

$D: x\in[-1,7]$;

$R: y\in[2-4,2+4]$

$R: y\in[-2,6]$

c. $y=(x-3)^2-4$

Vertex: (3, -4)

$D: x\in(-\infty,\infty)$

$R: y\in[-4,\infty)$

8. $(x+2)^2+(y-5)^2=r^2$

$(0+2)^2+(3-5)^2=r^2$

$4+4=r^2$

$8=r^2$

$(x+2)^2+(y-5)^2=8$

9. $\frac{(x)^2}{16}+\frac{(y)^2}{4}=1$

10. $(x-20)^2+(y-30)^2=2500$;

$(10-20)^2+(78-30)^2=2404$

Yes, distance is $\sqrt{2404} \approx 49$ mi

Chapter 7-Reinforcing Basic Concepts

1. $100x^2 - 400x - 18y^2 - 108y + 230 = 0$

$$100\left(x^2 - 4x\right) - 18\left(y^2 + 6y\right) = -230$$
$$100\left(x^2 - 4x + 4\right) - 18\left(y^2 + 6y + 9\right) = -230 + 400 - 162$$
$$100(x-2)^2 - 18(y+3)^2 = 8$$
$$\frac{25(x-2)^2}{2} - \frac{9(y+3)^2}{4} = 1$$

2. $28x^2 - 56x + 48y^2 + 192y + 195 = 0$

$$28\left(x^2 - 2x\right) + 48\left(y^2 + 4y\right) = -195$$
$$28\left(x^2 - 2x + 1\right) + 48\left(y^2 + 4y + 4\right) = -195 + 28 + 192$$
$$28(x-1)^2 + 48(y+2)^2 = 25$$
$$\frac{28(x-1)^2}{25} + \frac{48(y+2)^2}{25} = 1$$

7.3 Exercises

1. a.Circle and Line
3 or 4 not possible

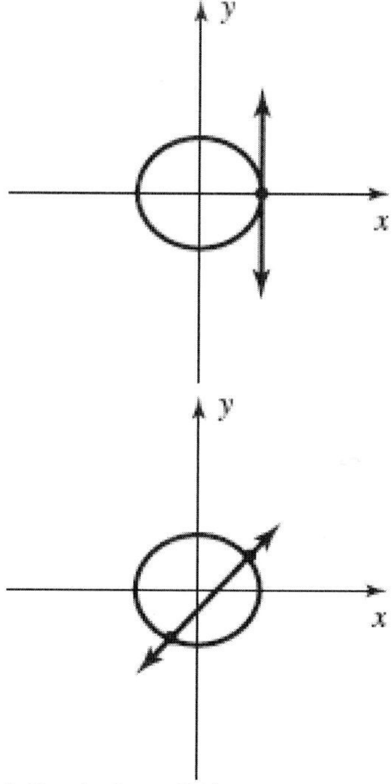

b.Parabola and Line
3 or 4 not possible

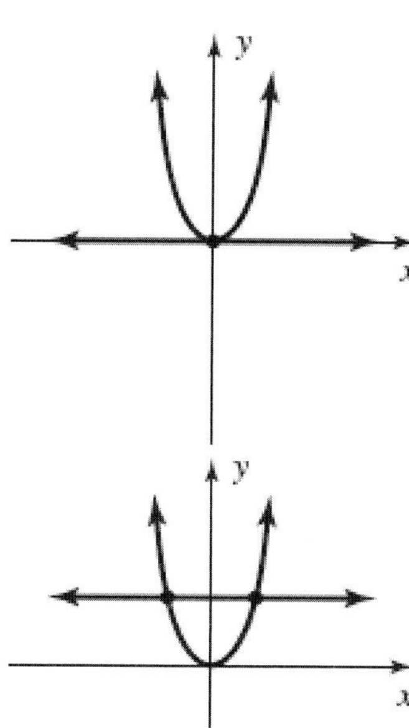

c.Circle and Parabola

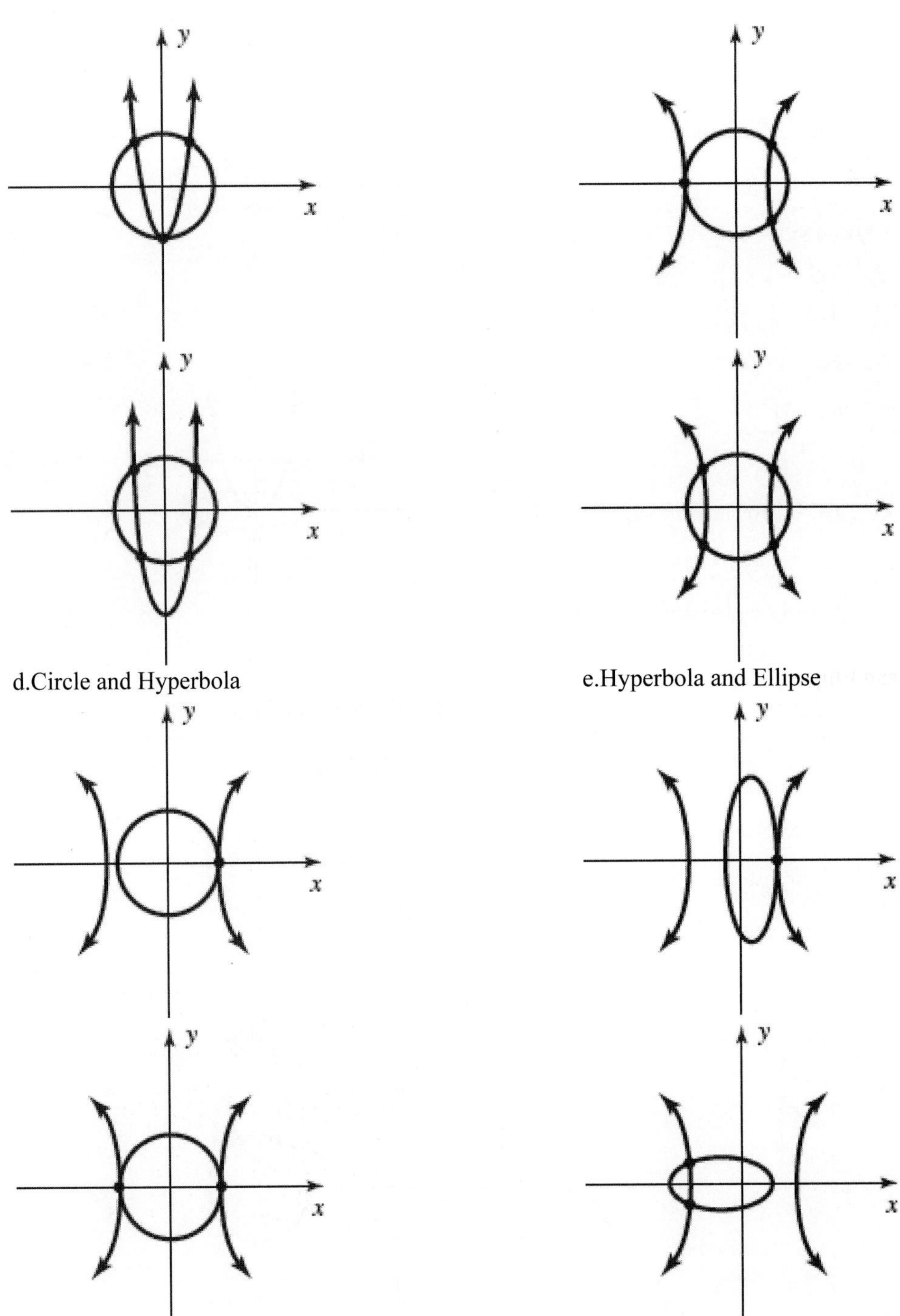
y
x
d.Circle and Hyperbola
e.Hyperbola and Ellipse

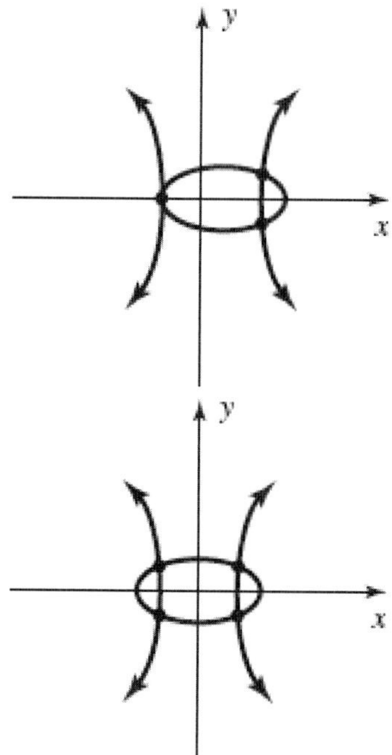

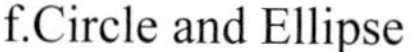
f.Circle and Ellipse

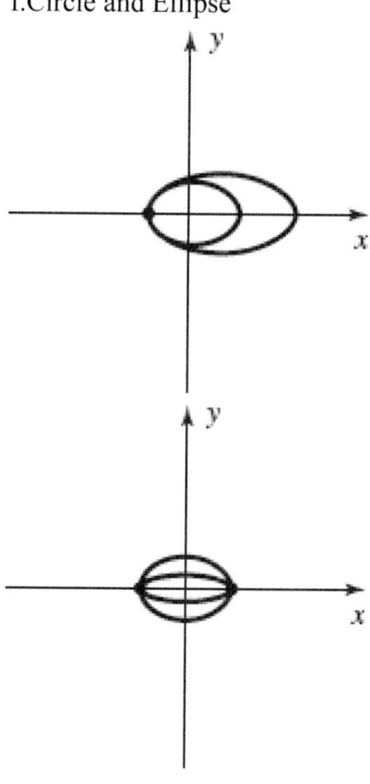

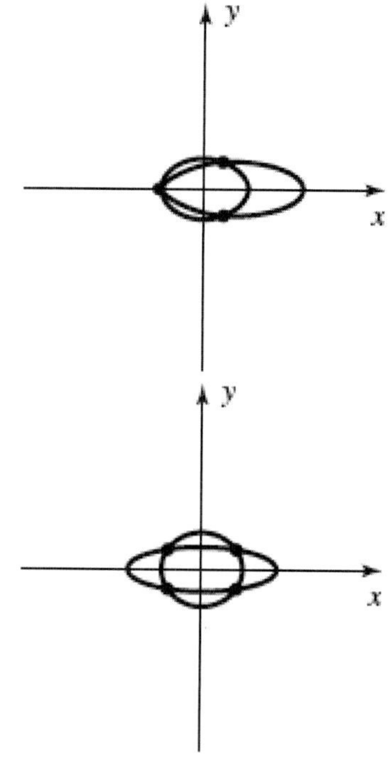

3. Region, solutions

5. Answers will vary.

7. $\begin{cases} x^2 + y = 6 & \text{Parabola} \\ x + y = 4 & \text{Line} \end{cases}$

Solutions: $(-1,5),(2,2)$

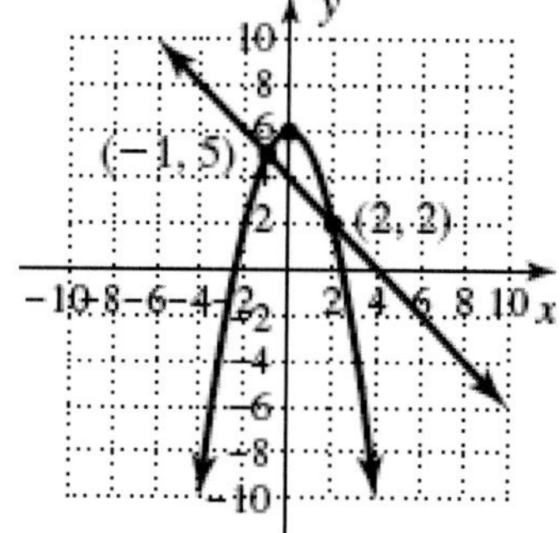

9. $\begin{cases} y - x^2 = -1 & \text{Parabola} \\ 4x^2 + y^2 = 100 & \text{Ellipse} \end{cases}$

Solutions: $(-3,8),(3,8)$

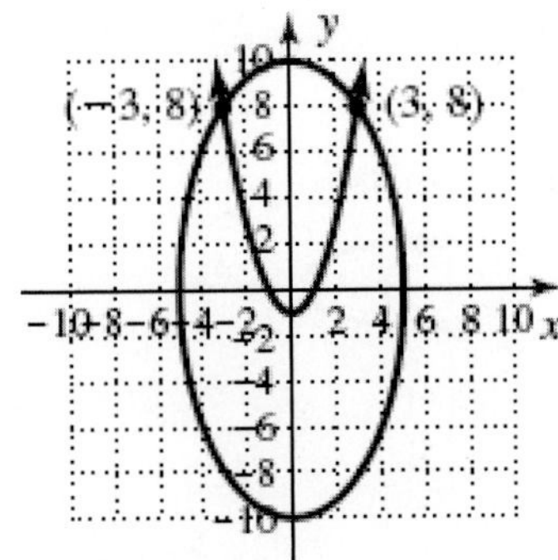

11. $\begin{cases} x^2 - y^2 = 9 & \text{Hyperbola} \\ x^2 + y^2 = 41 & \text{Circle} \end{cases}$

Solutions: $(-5,4),(5,4),(-5,-4),(5,-4)$

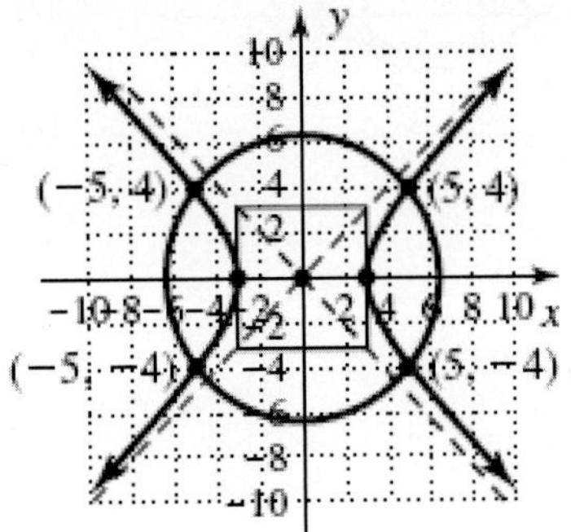

13. $\begin{cases} x^2 + y^2 = 25 \\ y - x = 1 \end{cases}$

2nd equation : $y = x + 1$

$$x^2 + (x+1)^2 = 25$$
$$x^2 + x^2 + 2x + 1 = 25$$
$$2x^2 + 2x - 24 = 0$$
$$x^2 + x - 12 = 0$$
$$(x+4)(x-3) = 0$$

$x = -4, x = 3$;

$y = -4 + 1 = -3$;

$y = 3 + 1 = 4$

Solutions: $(-4,-3),(3,4)$

15. $\begin{cases} x^2 + 4y^2 = 25 \\ x + 2y = 7 \end{cases}$

$x = 7 - 2y$

$$(7-2y)^2 + 4y^2 = 25$$
$$49 - 28y + 4y^2 + 4y^2 = 25$$
$$8y^2 - 28y + 24 = 0$$
$$2y^2 - 7y + 6 = 0$$
$$(2y-3)(y-2) = 0$$

$y = \frac{3}{2}, y = 2$;

$x = 7 - 2\left(\frac{3}{2}\right) = 4$;

$x = 7 - 2(2) = 3$

Solutions: $\left(4, \frac{3}{2}\right), (3,2)$

17. $\begin{cases} x^2 + y = 13 \\ 9x^2 - y^2 = 81 \end{cases}$

$x^2 = 13 - y$

$$9(13-y) - y^2 = 81$$
$$117 - 9y - y^2 = 81$$
$$y^2 + 9y - 36 = 0$$
$$(y-3)(y+12) = 0$$

$y = 3$ or $y = -12$;

$x^2 = 13 - 3$ or $x^2 = 13 - (-12)$

$x^2 = 10$ or $x^2 = 25$

$x = \pm\sqrt{10}$ or $x = \pm 5$

Solutions: $(\sqrt{10},3), (-\sqrt{10},3)$,
$(5,-12), (-5,-12)$

19. $\begin{cases} x^2 + y^2 = 25 \\ 2x^2 - 3y^2 = 5 \end{cases}$

$\begin{cases} 3x^2 + 3y^2 = 75 \\ 2x^2 - 3y^2 = 5 \end{cases}$

$5x^2 = 80$

$x^2 = 16$

$x = \pm 4$;

$(4)^2 + y^2 = 25$

$y^2 = 9$

$y = \pm 3$;

Solutions: $(4,3),(4,-3),(-4,3),(-4,-3)$

21. $\begin{cases} x^2 - y = 4 \\ x^2 - y^2 = 16 \end{cases}$

$\begin{cases} x^2 - y = 4 \\ -x^2 + y^2 = -16 \end{cases}$

$$y^2 - y = -12$$
$$y^2 - y + 12 = 0$$

$a = 1, b = -1, c = 12$

$$y = \frac{-(-1) \pm \sqrt{(-1)^2 - 4(1)(12)}}{2(1)}$$

$y = \frac{1 \pm \sqrt{-47}}{2}$

No real solution.

23. $\begin{cases} 5x^2 - 2y^2 = 75 \\ 2x^2 + 3y^2 = 125 \end{cases}$

$\begin{cases} 15x^2 - 6y^2 = 225 \\ 4x^2 + 6y^2 = 250 \end{cases}$

$19x^2 = 475$
$x^2 = 25$
$x = \pm 5$

$5(\pm 5)^2 - 2y^2 = 75$
$-2y^2 = -50$
$y^2 = 25$
$y = \pm 5$

Solutions: $(5,-5), (5,5), (-5,5), (-5,-5)$

25. $\begin{cases} y = \log x + 5 \\ y = 6 - \log(x-3) \end{cases}$

$\log x + 5 = 6 - \log(x-3)$
$\log x + \log(x-3) = 1$
$\log x(x-3) = 1$
$10 = x^2 - 3x$
$0 = x^2 - 3x - 10$
$0 = (x-5)(x+2)$
$x = 5$ or $x = -2$
$x =$ -2 is extraneous
$y = \log 5 + 5$;
Solution: $(5, \log 5 + 5)$

27. $\begin{cases} y = 4^{x+3} \\ y - 2^{x^2+3x} = 0 \end{cases}$

$4^{x+3} = 2^{x^2+3x}$
$2^{2x+6} = 2^{x^2+3x}$
$2x + 6 = x^2 + 3x$
$0 = x^2 + x - 6$
$(x+3)(x-2) = 0$
$x = -3$ or $x = 2$;
$y = 4^{-3+3} = 1$;
$y = 4^{2+3} = 1024$;
Solutions: $(-3,1), (2,1024)$

29. $\begin{cases} x^3 - y = 2x \\ y - 5x = -6 \end{cases}$

$y = 5x - 6$
$x^3 - 5x + 6 = 2x$
$x^3 - 7x + 6 = 0$

Possible rational roots: $\frac{\pm 1, \pm 6, \pm 2, \pm 3}{\pm 1}$

$\{\pm 1, \pm 6, \pm 2, \pm 3\}$;
$(x+3)(x-2)(x-1) = 0$
$x = -3$ or $x = 2$ or $x = 1$;
$y = 5(-3) - 6 = -21$;
$y = 5(2) - 6 = 4$;
$y = 5(1) - 6 = -1$;
Solutions: $(-3,-21), (2,4), (1,-1)$

31. $\begin{cases} x^2 + 2y^2 = 17 \\ y + x^2 = 11 \end{cases}$

$\begin{cases} x^2 + 2y^2 = 17 \\ x^2 + y = 11 \end{cases}$

$\begin{cases} x^2 + 2y^2 = 17 \\ -x^2 - y = -11 \end{cases}$

$2y^2 - y = 6$
$2y^2 - y - 6 = 0$
$(2y+3)(y-2) = 0$

$y = -\frac{3}{2}$ or $y = 2$;

$-\frac{3}{2} + x^2 = 11$

$x^2 = \frac{25}{2}$

$x = \pm\sqrt{\frac{25}{2}} = \pm\frac{5\sqrt{2}}{2}$;

$2 + x^2 = 11$
$x^2 = 9$
$x = \pm 3$;
Solutions:

$(3,2), (-3,2), \left(\frac{5\sqrt{2}}{2}, \frac{-3}{2}\right), \left(-\frac{5\sqrt{2}}{2}, \frac{-3}{2}\right)$

33. $\begin{cases} 3y^2 - 5x^2 = 7 \\ xy = 6 \end{cases}$

$x = \frac{6}{y}$;

$3y^2 - 5\left(\frac{6}{y}\right)^2 = 7$

$3y^2 - \frac{180}{y^2} = 7$

$3y^4 - 180 = 7y^2$

$3y^4 - 7y^2 - 180 = 0$

$\left(3y^2 + 20\right)\left(y^2 - 9\right) = 0$

$3y^2 + 20 = 0$ or $y^2 = 9$

$y^2 = -\dfrac{20}{3}$ or $y = \pm 3$

not real or $y = \pm 3$;

$x = \dfrac{6}{3} = 2;$

$x = \dfrac{6}{-3} = -2;$

Solutions: $(2,3), (-2,-3)$

35. $\begin{cases} 2y^2 + xy - 7 = x^2 \\ x - 2y = 5 \end{cases}$

$x = 2y + 5;$

$2y^2 + (2y+5)y - 7 = (2y+5)^2$

$2y^2 + 2y^2 + 5y - 7 = 4y^2 + 20y + 25$

$-15y = 32$

$y = \dfrac{-32}{15};$

$x = 2\left(\dfrac{-32}{15}\right) + 5 = \dfrac{11}{15};$

Solution: $\left(\dfrac{11}{15}, \dfrac{-32}{15}\right)$

37. $\begin{cases} 3x^2 + 4y^2 = 35 \\ 5y^2 - x^2 = 1 \end{cases}$

$4y^2 = -3x^2 + 35$

$y^2 = \dfrac{5}{2}x^2 - 15$

$y1 = \pm\sqrt{-\dfrac{3}{4}x^2 + \dfrac{35}{4}};$

$5y^2 = x^2 + 1$

$y1 = \pm\sqrt{\dfrac{5}{2}x^2 - 15}$

$y2 = \pm\sqrt{\dfrac{1}{5}x^2 + \dfrac{1}{5}}$

Solutions:
$(3, 1.41), (-3, 1.41), (3, -1.41), (-3, -1.41)$

39. $\begin{cases} y = 2^x - 3 \\ y + 2x^2 = 9 \end{cases}$

$y1 = 2^x - 3;$

$y2 = -2x^2 + 9;$

Solutions: $(-2.43, -2.81), (2, 1)$

41. $\begin{cases} y = \dfrac{1}{(x-3)^2} + 2 \\ (x-3)^2 + y^2 = 10 \end{cases}$

$y1 = \dfrac{1}{(x-3)^2} + 2;$

$y^2 = -(x-3)^2 + 10$

$y2 = \pm\sqrt{-(x-3)^2 + 10};$

Solutions: $(0.72, 2.19), (2, 3), (4, 3), (5.28, 2.19)$

43. $\begin{cases} y - x^2 \ge 1 \text{ parabola} \\ x + y \le 3 \text{ line} \end{cases}$

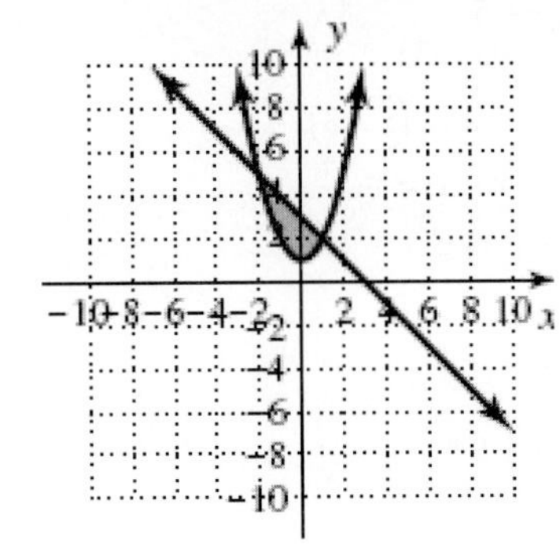

45. $\begin{cases} x^2 + y^2 > 9 \quad \text{circle} \\ 25x^2 + 16y^2 \le 400 \quad \text{ellipse} \end{cases}$

$\dfrac{x^2}{16} + \dfrac{y^2}{25} = 1$

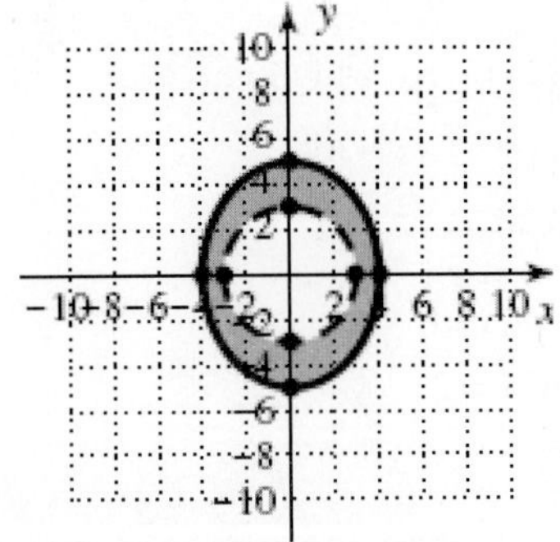

47. $\begin{cases} 4y^2 - x^2 \ge 16 \quad \text{hyperbola} \\ 25x^2 + 16y^2 \le 400 \quad \text{ellipse} \end{cases}$

$$\frac{x^2}{16}+\frac{y^2}{25}=16$$
$$\frac{y^2}{4}-\frac{x^2}{16}=1$$

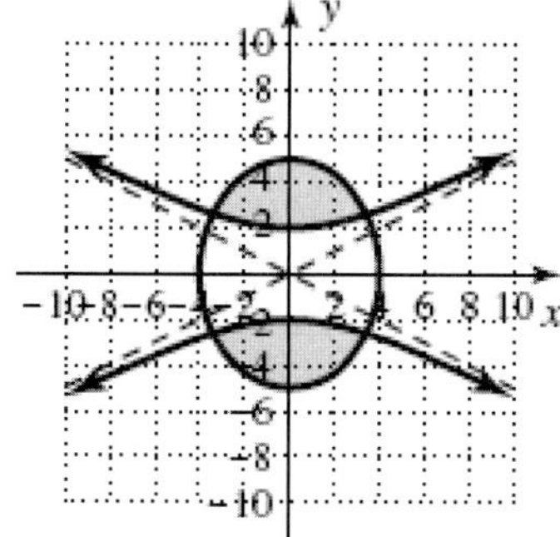

49. $\begin{cases} x^2+y^2 \le 16 \text{ circle} \\ x+2y>10 \text{ line} \end{cases}$

$2y>-x+10$

$y>-\frac{1}{2}x+5$

No solution

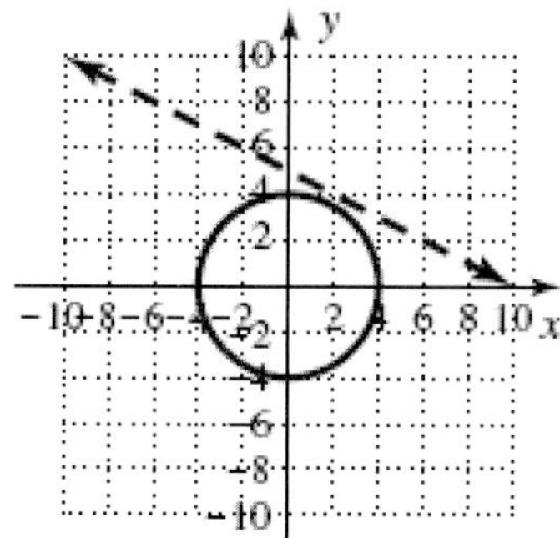

51. $h=b\sqrt{1-\left(\frac{d}{a}\right)^2}$;

$h=30\sqrt{1-\left(\frac{20}{50}\right)^2}$

$h=27.5$ ft.;

$h=30\sqrt{1-\left(\frac{30}{50}\right)^2}$

$h=24$ ft.;

$h=30\sqrt{1-\left(\frac{40}{50}\right)^2}$

$h=18$ ft.

53. $85=lw$

$37=2l+2\text{w}$;

$\frac{85}{l}=\text{w}$

$37=2l+2\left(\frac{85}{l}\right)$

$37l=2l^2+170$

$0=2l^2-37l+170$

$0=(2l-17)(l-10)$

$l=\frac{17}{2}, l=10$;

$w=\frac{85}{\left(\frac{17}{2}\right)}=10$;

$w=\frac{85}{(10)}=8.5$;

$8.5\text{m}\times 10\text{m}$

55. Area : 45km^2

Length : $\sqrt{106}\text{km}$

$45=lw$

$l=\frac{45}{w}$

$l^2+w^2=106$

$\left(\frac{45}{w}\right)^2+w^2=106$

$2025+w^4=106w^2$

$w^4-106w^2+2025=0$

$(w^2-25)(w^2-81)=0$

$w=\pm5,\pm9$

5 km, 9 km

57. Surface Area $=928\text{ ft}^2$

Edges $=164$ ft

$4w+4l+4w=164$

$4l+8w=164$

$4l=164-8w$

$l=41-2w$;

$928=4lw+2w^2$

$928=4(41-2w)w+2w^2$

$928=164w-8w^2+2w^2$

$6w^2-164w+928=0$

$3w^2-82w+464=0$

$(3w-58)(w-8)=0$

$w=\frac{58}{3}$ or $w=8$;

w=8, $l=41-2(8)=25$

8 ft x 8 ft x 25 ft

59. a. $8P^2-8P-4D=12$

$-4D=-8P^2+8P+12$

$D = 2P^2 - 2P - 3$
minimum: \$1.83 (when $D = 0$)

b. $\begin{cases} 10P^2 + 6D = 144 \\ 8P^2 - 8P - 4D = 12 \end{cases}$

$10P^2 + 6D = 144$

$\frac{144 - 10P^2}{6} = D$;

$8P^2 - 8P - 4D = 12$

$8P^2 - 8P - 4\left(\frac{144 - 10P^2}{6}\right) = 12$

$2P^2 - 2P - \left(\frac{144 - 10P^2}{6}\right) = 3$

$12P^2 - 12P - 144 + 10P^2 = 18$
$22P^2 - 12P - 162 = 0$
$11P^2 - 6P - 81 = 0$
$(11P + 27)(P - 3) = 0$
$P = -\frac{27}{11}$ or $P = \$3$;

$10(3)^2 + 6D = 144$
$6D = 54$
$D = 9$
90,000 gallons

61. Answers will vary.

63. Hyperbola $xy = 4$

Ellipse $x^2 + 4y^2 = 20$

$x = \frac{4}{y}$

$\left(\frac{4}{y}\right)^2 + 4y^2 = 20$

$16 + 4y^4 = 20y^2$
$y^4 - 5y^2 + 4 = 0$
$\left(y^2 - 1\right)\left(y^2 - 4\right) = 0$
$y = \pm 1, y = \pm 2$;

$x = \frac{4}{1} = 4$;

$x = \frac{4}{-1} = -4$;

$x = \frac{4}{2} = 2$;

$x = \frac{4}{-2} = -2$;

endpoints of parallelogram:
$(4,1), (-4,-1), (2,2), (-2,-2)$;

$d = \sqrt{(-4-2)^2 + (-1-2)^2} = \sqrt{45}$;

slope of line passing through (-4,-1) and (2,2): $m = \frac{2 - -1}{2 - -4} = \frac{1}{2}$;

equation of line:

$y - 2 = \frac{1}{2}(x - 2)$

$y - 2 = \frac{1}{2}x - 1$

$y = \frac{1}{2}x + 1$;

slope of line perpendicular: -2
equation of line perpendicular:
$y - -2 = -2(x - -2)$
$y + 2 = -2x - 4$
$y = -2x - 6$;

$-2x - 6 = \frac{1}{2}x + 1$

$-\frac{5}{2}x = 7$

$x = -\frac{14}{5}$;

$y = \frac{1}{2}\left(-\frac{14}{5}\right) + 1$

$y = -\frac{2}{5}$;

point of intersection: $\left(-\frac{14}{5}, -\frac{2}{5}\right)$;

$d = \sqrt{\left(-2 - -\frac{14}{5}\right)^2 + \left(-2 - -\frac{2}{5}\right)^2} = \frac{\sqrt{80}}{5}$;

Area: $bh = \sqrt{45}\left(\frac{\sqrt{80}}{5}\right) = \frac{60}{5} = 12$ units^2

65. Height : 18 inches

Area : 4806 in^2

$4806 = lw + 2(18l) + 2(18w)$
$4806 = lw + 36l + 36w$
$4806 - 36l = lw + 36w$
$4806 - 36l = w(l + 36)$
$\frac{4806 - 36l}{l + 36} = w$;
$108(231) = 18(l)w$
$108(231) = 18l\left(\frac{4806 - 36l}{l + 36}\right)$
$24948(l + 36) = 18l(4806 - 36l)$
$24948l + 898128 = 86508l - 648l^2$

$648l^2 - 61560l + 898128 = 0$

$l^2 - 95l + 1386 = 0$

$(l-18)(l-77) = 0$

18 in x 18 in x 77 in.

67. a. $3x^2 + 4x - 12 = 0$

$a = 3, b = 4, c = -12$

$$x = \frac{-4 \pm \sqrt{16 - 4(3)(-12)}}{6}$$

$$x = \frac{-4 \pm \sqrt{160}}{6}$$

$$x = \frac{-2 \pm 2\sqrt{10}}{3}$$

b. $\sqrt{3x+1} - \sqrt{2x} = 1$

$\sqrt{3x+1} = 1 + \sqrt{2x}$

$\left(\sqrt{3x+1}\right)^2 = \left(1 + \sqrt{2x}\right)^2$

$3x + 1 = 1 + 2\sqrt{2x} + 2x$

$x = 2\sqrt{2x}$

$(x)^2 = \left(2\sqrt{2x}\right)^2$

$x^2 = 4(2x)$

$x^2 - 8x = 0$

$x(x-8) = 0$

$x = 0, x = 8$

c. $\dfrac{1}{x+2} + \dfrac{3}{x^2+5x+6} = \dfrac{2}{x+3}$

$\dfrac{1}{x+2} + \dfrac{3}{(x+2)(x+3)} = \dfrac{2}{x+3}$

$(x+2)(x+3)\left[\dfrac{1}{x+2} + \dfrac{3}{(x+2)(x+3)} = \dfrac{2}{x+3}\right]$

$x + 3 + 3 = 2(x+2)$

$x + 6 = 2x + 4$

$2 = x$

$x = 2$

69. a. $y = 2|x+3| - 1$

Vertex: (-4, 3)

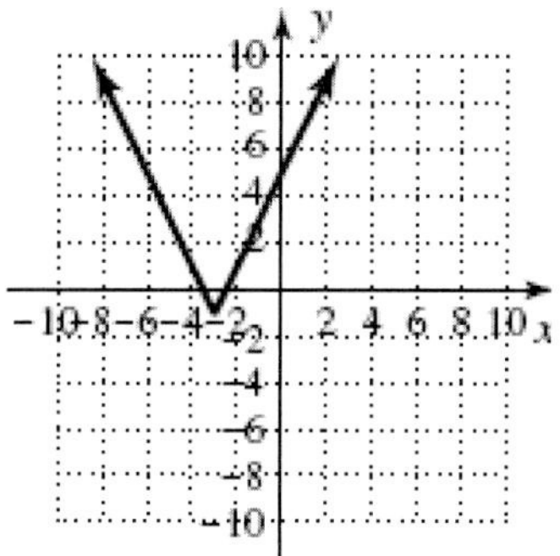

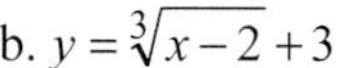

b. $y = \sqrt[3]{x-2} + 3$

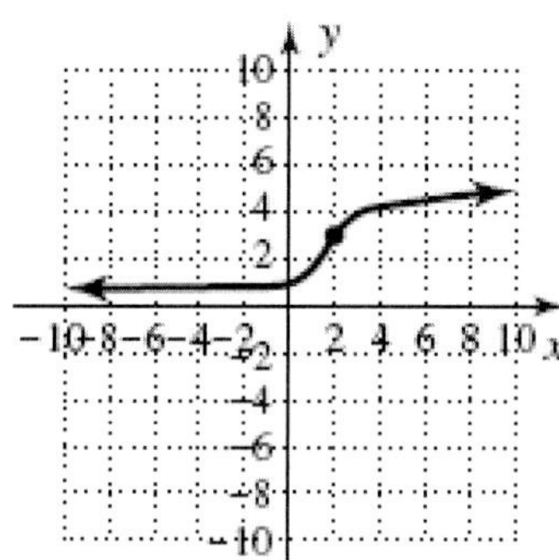

c. $y = -(x+4)^2 + 3$

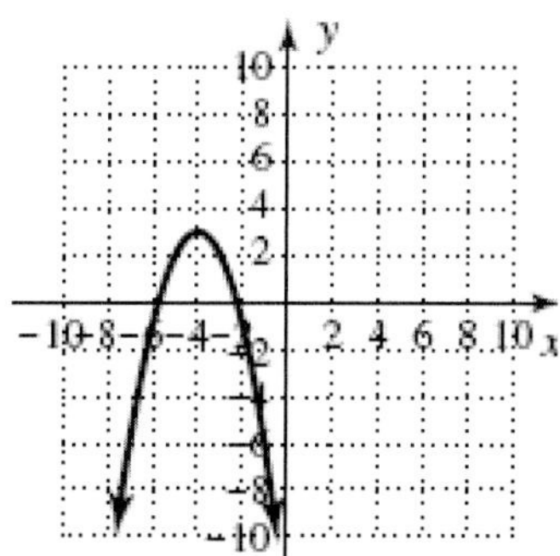

d. $y = 2^{x+1} - 3$

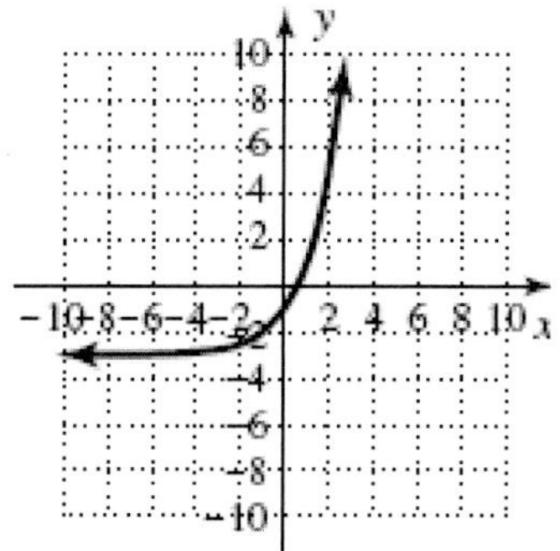

71. a. Let 2001 be year 0.

$$m = \frac{\Delta\text{value}}{\Delta\text{time}} = \frac{4500 - 3300}{0 - 3} = -400$$

The computer depreciates by \$400 a year.

b. $y = -400x + 4500$

c. $y = -400(7) + 4500 = \$1700$

d. $700 = -400x + 4500$

$-3800 = -400x$

$9.5 = x$

9.5 years

7.4 Technology Highlight

1. $\dfrac{x^2}{25} + \dfrac{y^2}{4} = 1$

The sum of corresponding distances is $k = 10$ or $2a = 2(5) = 10$

2. $\dfrac{x^2}{16} - \dfrac{y^2}{9} = 1$; verified

The difference of corresponding distances is $k = 8$ or $2a = 2(4) = 8$

7.4 Exercises

1. $c^2 = \left|a^2 - b^2\right|$

3. $2a, 2b$

5. Answers will vary.

7. $c = 6, b = 8$;

$a^2 - 8^2 = 6^2$

$a^2 = 100$

$a = 10$;

$2a = 20$

9. $c = 8, b = 6$;

$a^2 - 6^2 = 8^2$

$a^2 = 100$

$a = 10$;

$2a = 20$

11. $4x^2 + 25y^2 - 16x - 50y - 59 = 0$

$4(x^2 - 4x + 4) + 25(y^2 - 2y + 1) = 59 + 16 + 25$

$4(x-2)^2 + 25(y-1)^2 = 100$

$\dfrac{(x-2)^2}{25} + \dfrac{(y-1)^2}{4} = 1$;

$a = 5,\ b = 2$

$a^2 - b^2 = c^2$

$25 - 4 = c^2$

$21 = c^2$

$c = \sqrt{21}$;

a. Center: $(2,1)$

b. Vertices: $(2-5,1)$ and $(2+5,1)$

$(-3,1)$ and $(7,1)$

c. Foci: $(2-\sqrt{21},1)$ and $(2+\sqrt{21},1)$

d. Endpoint of minor axis:

$(2,1+2)$ and $(2,1-2)$

$(2,3)$ and $(2,-1)$

e.

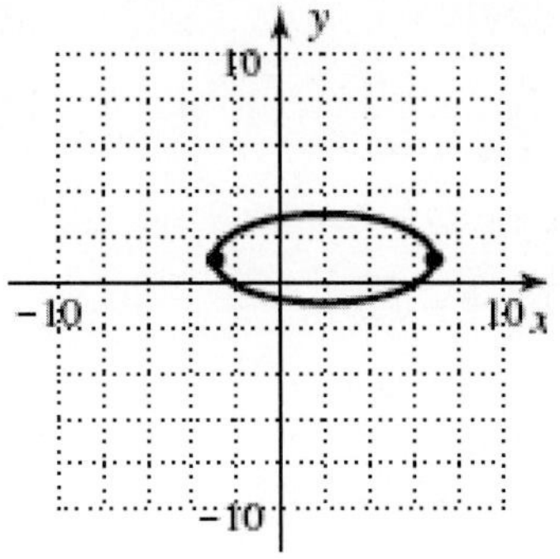

y
10
-10
10 x
-10

13. $25x^2 + 16y^2 - 200x + 96y + 144 = 0$

$25(x^2 - 8x + 16) + 16(y^2 + 6y + 9) = -144 + 400 + 144$

$25(x-4)^2 + 16(y+3)^2 = 400$

$\dfrac{(x-4)^2}{16} + \dfrac{(y+3)^2}{25} = 1$;

$a = 4,\ b = 5$

$c^2 = 25 - 16 = 9$

$c = 3$;

a. Center: $(4,-3)$

b. Vertices: $(4,-3+5)$ and $(4,-3-5)$

$(4,2)$ and $(4,-8)$

c. Foci: $(4,-3+3)$ and $(4,-3-3)$

$(4,0)$ and $(4,-6)$

d. Endpoint of minor axis:

$(4-4,-3)$ and $(4+4,-3)$

$(0,-3)$ and $(8,-3)$

e.

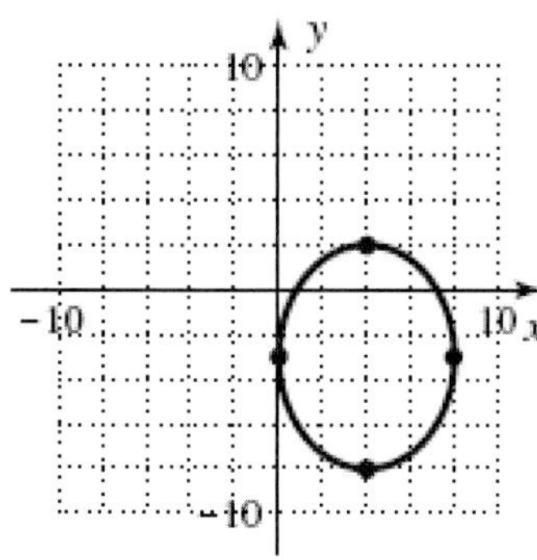

15. $6x^2+24x+9y^2+36y+6=0$

$$6(x^2+4x+4)+9(y^2+4y+4)=-6+24+36$$
$$6(x+2)^2+9(y+2)^2=54$$
$$\frac{(x+2)^2}{9}+\frac{(y+2)^2}{6}=54;$$

$c^2=9-6=3$

$c=\sqrt{3};$

a. Center: $(-2,-2)$

b. Vertices:

$(-2-3,-2)$ and $(-2+3,-2)$

$(-5,-2)$ and $(1,-2)$

c. Foci: $\left(-2-\sqrt{3},-2\right)$ and $\left(-2+\sqrt{3},-2\right)$

d. Endpoint of minor axis:

$\left(-2,-2+\sqrt{6}\right)$ and $\left(-2,-2-\sqrt{6}\right)$

e.

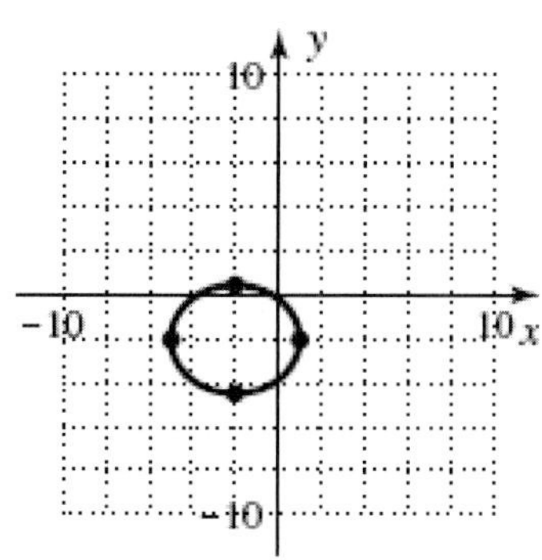

17. Vertices at $(-6,0)$ and $(6,0)$;

Foci at $(-4,0)$ and $(4,0)$, $\leftrightarrow$;

V: $(-6,0),(6,0), a=6$

F: $(-4,0),(4,0), c=4$

$4^2=6^2-b^2$

$b^2=36-16=20$

$b=\sqrt{20};$

Center: $(0,0)$

$$\frac{x^2}{36}+\frac{y^2}{20}=1$$

19. Foci at $(0,-4)$ and $(0,4)$, $\updownarrow$;

Length of minor axis: 6 units, $a=3$

Foci: $(0,-4),(0,4), c=4$

$16=b^2-9$

$b^2=25$

$b=5;$

$$\frac{x^2}{9}+\frac{y^2}{25}=1$$

21. $\sqrt{(-5-5)^2+(0-2.25)^2}$

$-\sqrt{(5-5)^2+(0-2.25)^2}=2a$

$\sqrt{(-5-5)^2+(0-2.25)^2}-2.25=2a$

$\sqrt{100+5.0625}-2.25=2a$

$8=2a$

$a=4, c=5$

$25=4^2+b^2$

$9=b^2$

$3=b;$

$2b=2(3)=6;$

Dimensions: 8 x 6

23.

$$\sqrt{(-0-6)^2+(-10-7.5)^2}-\sqrt{(0-6)^2+(10-7.5)^2}=2b$$
$$\sqrt{36+(17.5)^2}-\sqrt{36+(2.5)^2}=2b$$
$$\sqrt{342.25}-\sqrt{42.25}=2b$$
$$12=2b;$$

$b=6, c=10$

$100=a^2+36$

$a^2=64$

$a=8;$

$2a=2(8)=16;$

Dimensions: 16 x 12

25. $4x^2-9y^2-24x+72y-144=0, \leftrightarrow$

$$4(x^2-6x+9)-9(y^2-8y+16)=144-144+36$$
$$4(x-3)^2-9(y-4)^2=36$$
$$\frac{(x-3)^2}{9}-\frac{(y-4)^2}{4}=1;$$

$a=3, b=2;$

$c^2=9+4=13$

$c=\sqrt{13};$

a. Center: $(3,4)$

b. Vertices: $(3-3,4)$ and $(3+3,4)$

$(0,4)$ and $(6,4)$

c.Foci: $(3-\sqrt{13},4)$ and $(3+\sqrt{13},4)$
d. $2a=6, 2b=4;$
e.Asymptotes: $y-4=\pm\frac{2}{3}(x-3);$

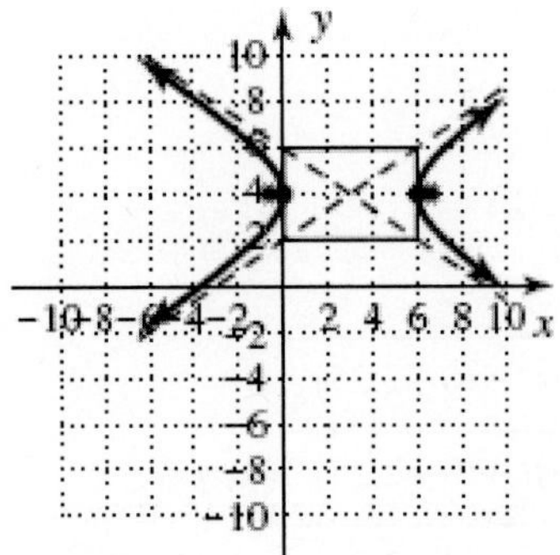

27. $16x^2-4y^2+24y-100=0, \leftrightarrow$
$$16(x^2)-4(y^2-6y+9)=100-36$$
$$16(x^2)-4(y-3)^2=64$$
$$\frac{x^2}{4}-\frac{(y-3)^2}{16}=1;$$
$a=2, b=4;$
$c^2=4+16$
$c^2=20$
$c=2\sqrt{5};$
a. Center: $(0,3)$
b.Vertices: $(0-2,3)$ and $(0+2,3)$
$(-2,3)$ and $(2,3)$
c.Foci: $(-2\sqrt{5},3)$ and $(2\sqrt{5},3)$
d. $2a=4, 2b=8;$
e.Asymptotes: $y-\frac{9}{2}=\pm2(x)$

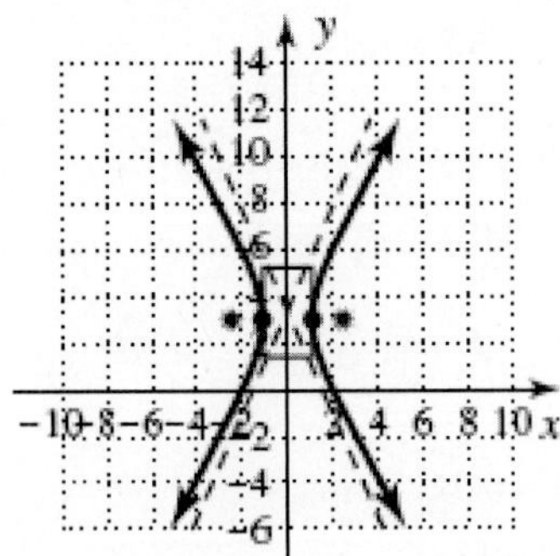

29. $9x^2-3y^2-54x-12y+33=0, \leftrightarrow$
$$9(x^2-6x+9)-3(y^2+4y+4)=-33+81-12$$
$$9(x-3)^2-3(y+2)^2=36$$
$$\frac{(x-3)^2}{4}-\frac{(y+2)^2}{12}=1;$$
$a=2, b=\sqrt{12}=2\sqrt{3}$
$c^2=4+12=16$
$c=4;$
a. Center: $(3,-2)$
b.Vertices: $(3-2,-2)$ and $(3+2,-2)$
$(1,-2)$ and $(5,-2)$
c.Foci: $(-1,-2),(7,-2)$
d. $2a=4, 2b=4\sqrt{3};$
e.Asymptotes: $(y+2)=\pm\sqrt{3}(x-3)$

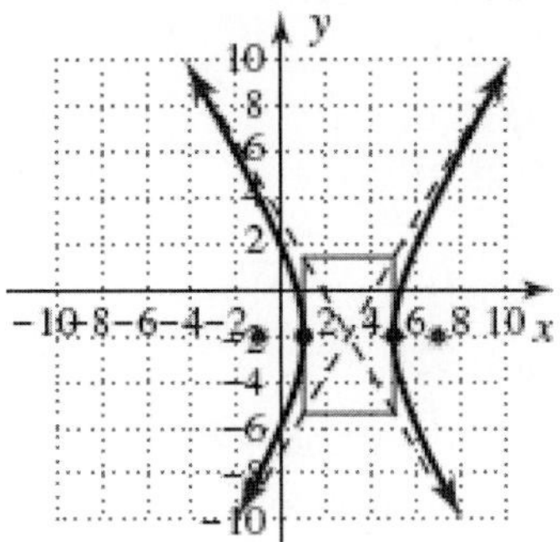

31. Vertices: $(-6,0)$ and $(6,0)$
Foci: $(-8,0)$ and $(8,0)$
$8^2=6^2+b^2$
$64=36+b^2$
$b^2=28;$
$$\frac{x^2}{36}-\frac{y^2}{28}=1$$

33. Foci: $(0,-3\sqrt{2})$ and $(0,3\sqrt{2})$
Length of conjugate axis: 6 units
$2b=6$
$b=3;$
$18=a^2+9$
$a^2=9;$
$$\frac{y^2}{9}-\frac{x^2}{9}=1$$

35. $\frac{x^2}{49}+\frac{y^2}{4}=1, \leftrightarrow$
Ellipse

$c^2 = 49 - 4 = 45$
$c = 3\sqrt{5};$
Foci: $(-3\sqrt{5},0),(3\sqrt{5},0)$

37. $\frac{x^2}{9}+\frac{y^2}{25}=1, \updownarrow$
Ellipse
$c^2 = 25-9=16$
$c=4;$
Foci: $(0,4),(0-4)$

39. $\frac{x^2}{18}+\frac{y^2}{12}=1, \leftrightarrow$
Ellipse
$c^2 = 18-12=6$
$c=\sqrt{6};$
Foci: $(-\sqrt{6},0),(\sqrt{6},0)$

41. $\frac{x^2}{4}-\frac{y^2}{9}=1, \leftrightarrow$
Hyperbola
$c^2 = 4+9=13$
$c=\sqrt{3};$
Foci: $(-\sqrt{13},0),(\sqrt{13},0)$

43. $\frac{y^2}{36}-\frac{x^2}{25}=1, \updownarrow$
Hyperbola
$c^2 = 36+25=61$
$c=\sqrt{61};$
Foci: $(0,\sqrt{61}),(0,-\sqrt{61})$

45. $\frac{x^2}{28}-\frac{y^2}{32}=1, \leftrightarrow$
Hyperbola
$c^2 = 28+32=60$
$c=2\sqrt{15};$
Foci: $(-2\sqrt{15},0),(2\sqrt{15},0)$

47. $e=\frac{c}{a};$
$\frac{x^2}{9}+\frac{y^2}{25}=1;$
$a=5, b=3$
$c^2 = 25-9$
$c^2 = 16$
$c=4;$
$e=\frac{4}{5}=0.8;$
$\frac{x^2}{16}+\frac{y^2}{36}=1$
$a=6, b=4$
$c^2=36-16=20$
$c=2\sqrt{5};$
$e=\frac{2\sqrt{5}}{6}=\frac{\sqrt{5}}{3}=0.75$
The ellipse closest to being circular is
$\frac{x^2}{16}+\frac{y^2}{36}=1$ because $e=0.75$ is closer to zero.

49. $a=4, b=3$
$c^2=a^2-b^2=16-9=7$
$c=\sqrt{7};$
Spines are $\sqrt{7}$ or ≈ 2.65 ft from center
The height of the spine occurs at $(\sqrt{7},y)$ on the hyperbola $\frac{x^2}{16}+\frac{y^2}{9}=1$
$\frac{(\sqrt{7})^2}{16}+\frac{y^2}{9}=1$
$\frac{7}{16}+\frac{y^2}{9}=1$
$\frac{y^2}{9}=1-\frac{7}{16}$
$y^2=9\left(\frac{9}{16}\right)$
$y=\frac{9}{4}=2.25$
Height of the spine: 2.25 ft.

51. $a=12, b=8;$
$c=\sqrt{12^2-8^2}$
$c=\sqrt{80}=4\sqrt{5}=8.9$ ft
8.9 ft from center
$2(4\sqrt{5})\approx 17.9$ ft apart

53. $P=2\pi\sqrt{\frac{a^2+b^2}{2}}$

Aphelion (max)
$c-(-a)=156$ mil miles
Perihelion
$a-c=128$
$\begin{cases} a+c=156 \\ a-c=128 \end{cases}$
$2a=284$
Semi major $a=142$ mil miles;
$142-c=128$
$c=142-128=14;$
$14^2=142^2-b^2$
$b^2=142^2-14^2=19968$
$b\approx 141$ mil miles;
Semi minor $\approx 141;$
$P=2\pi\sqrt{\dfrac{142^2+141^2}{2}}$
$=2\pi\sqrt{20022.5}$ mil miles;
$\dfrac{889.076 \text{ mil miles}}{1.296 \text{ mil miles/day}}\approx 686$ days

55. 0.4 milliseconds to closer;
0.5 milliseconds to further;
300 km/millisecond;
0.5(300) = 150 km; 0.4(300) = 120 km;
$\sqrt{150^2}-\sqrt{120^2}=2a$
$150-120=2a$
$30=2a$
$a=15;$
$c=50;$
$50^2=15^2+b^2$
$2500=225+b^2$
$b^2=2275;$
$\dfrac{x^2}{225}-\dfrac{y^2}{2275}=1;$
$\dfrac{x^2}{225}-\dfrac{(60)^2}{2275}=1$
$\dfrac{x^2}{225}=1-\dfrac{(60)^2}{2275}$
$x^2=\dfrac{52875}{91}$
$x=\pm 24.1;$
$(24.1,60)$ or $(-24.1,60)$

57. $\dfrac{x^2}{4}-\dfrac{y^2}{9}=1$
$a=2, b=3;$
$L=\dfrac{2b^2}{a}$
$L=\dfrac{2(3)^2}{2}=9$ units;
$4+9=c^2$
$c=\sqrt{13};$
$\dfrac{(\sqrt{13})^2}{4}-\dfrac{y^2}{9}=1$
$\dfrac{13}{4}-\dfrac{y^2}{9}=1$
$-\dfrac{y^2}{9}=-\dfrac{9}{4}$
$y^2=\dfrac{81}{4}$
$y=\pm\dfrac{9}{2}$
Verified
$\left(-\sqrt{13},\dfrac{9}{2}\right),\left(\sqrt{13},\dfrac{9}{2}\right)$

59. $9(x-2)^2-25(y-3)^2=225$
$c^2=a^2+b^2=9+25=34$
$c=\sqrt{34}=r$
Circle: $(x-2)^2+(y-3)^2=34$

61. Ellipse $\dfrac{x^2}{a^2}+\dfrac{y^2}{b^2}=1;$
$c^2=a^2-b^2;$
$\dfrac{a^2-b^2}{a^2}+\dfrac{y^2}{b^2}=1$
$b^2a^2-b^2b^2+a^2y^2=a^2b^2$
$a^2y^2-b^2b^2=-a^2b^2+a^2b^2$
$a^2y^2=b^4$
$y^2=\dfrac{b^4}{a^2}$
$y=\dfrac{b^2}{a};$
Focal Chord is $2y$ so $L=\dfrac{2b^2}{a}$.
Verified.

63. $\log_3 20=\dfrac{\log 20}{\log 3}\approx 2.73$

65. a. $f(x)=x^3-7x+6$

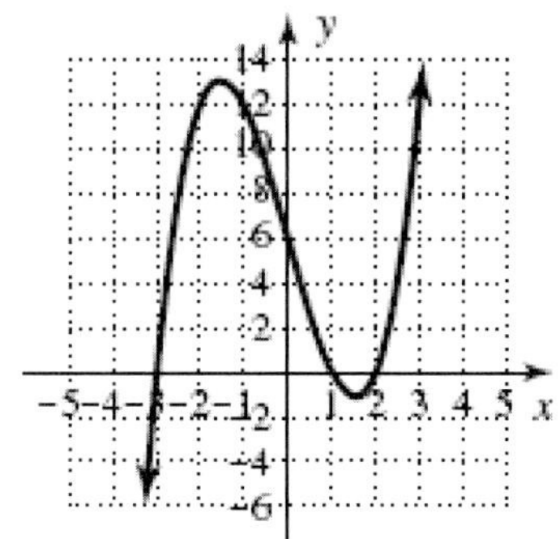

b. $h(x)=\dfrac{x^3}{x^2-4}$

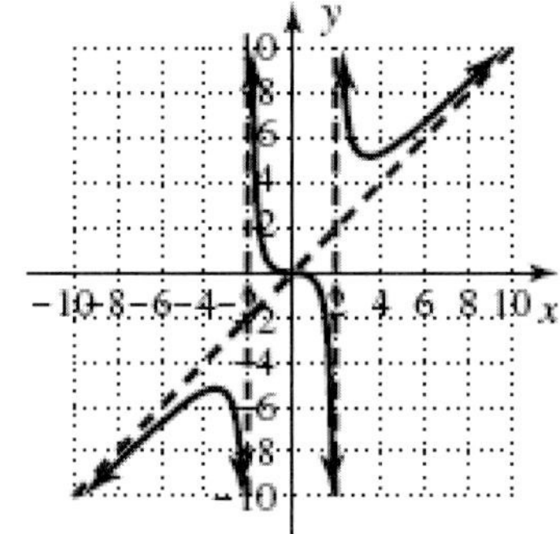

67. a. $-2|x-3|+10>4$

$$-2|x-3|>-6$$
$$|x-3|<3$$
$$Y_1=|x-3|$$
$$Y_2=3$$

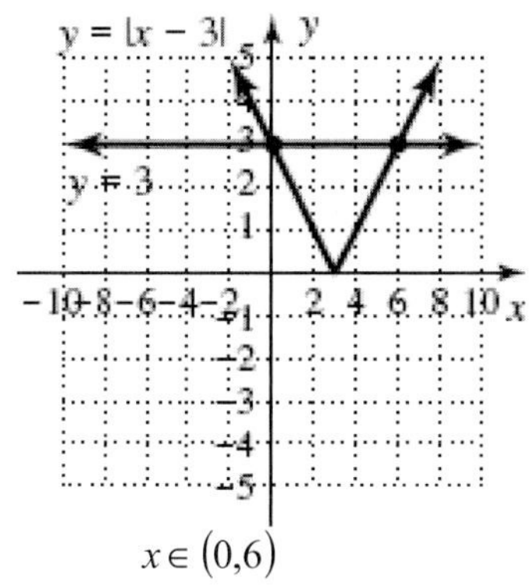

$$x\in(0,6)$$

b. $-2|x-3|+10>4$

$$-2|x-3|>-6$$
$$|x-3|<3$$
$$-3<x-3<3$$
$$0<x<6$$

7.5 Technology Highlight

1. $y=(x+3)^2-2$

$$(x+3)^2=y+2$$
$$4p=1$$
$$p=\frac{1}{4};$$
$\left(-3,-\frac{7}{4}\right)$; verified

2. $y=-2(x-1)^2+5$

$$(x-1)^2=-\frac{1}{2}(y-5)$$
$$4p=-\frac{1}{2}$$
$$p=-\frac{1}{8};$$
$\left(1,\frac{39}{8}\right)$; verified

3. $y=\frac{1}{2}(x-4)^2+2$

$$(x-4)^2=2(y-2)$$
$$4p=2$$
$$p=\frac{1}{2};$$
$\left(4,\frac{5}{2}\right)$; verified

4. $y=x^2-4x+7$

$$(x-2)^2=(y-3)$$
$$4p=1$$
$$p=\frac{1}{4};$$
$\left(2,\frac{13}{4}\right)$; verified

7.5 Exercises

1. Horizontal, right, $a<0$

3. $(p,0)$, $x=-p$

5. Answers will vary.

7. $y=x^2-2x-3$;

$$0=(x-3)(x+1)$$
x-intercepts: $(-1,0),(3,0)$;
$$y=(0)^2-2(0)-3=-3$$

y-intercept: $(0,-3)$;

$x=\frac{-(-2)}{2(1)}=1$

$y=(1)^2-2(1)-3=-4$

Vertex: $(1,-4)$;

Domain: $x\in(-\infty,\infty)$

Range: $y\in[-4,\infty)$

$y=(x^2-2x+1)-3-1$

$y=(x-1)^2-4$

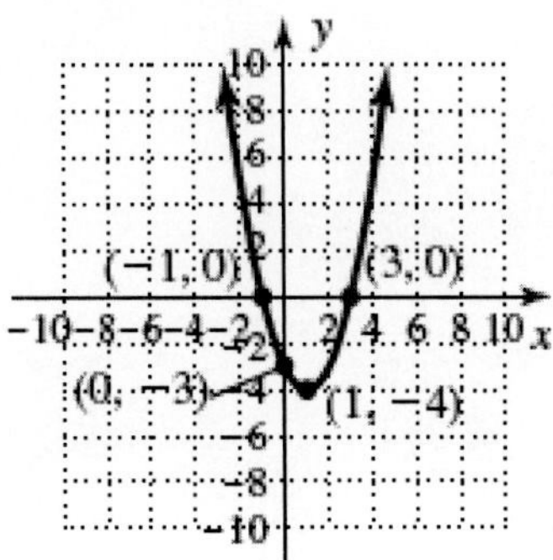

9. $y=2x^2-8x-10$

$0=2(x-5)(x+1)$

x-intercepts: $(-1,0),(5,0)$;

$y=2(0)^2-8(0)-10=-10$

y-intercept: $(0,-10)$;

$x=\frac{-(-8)}{2(2)}=2$

$y=2(2)^2-8(2)-10=-18$

Vertex: $(2,-18)$;

Domain: $x\in(-\infty,\infty)$

Range: $y\in[-18,\infty)$

$y=2(x^2-4x+4)-10-8$

$y=2(x-2)^2-18$

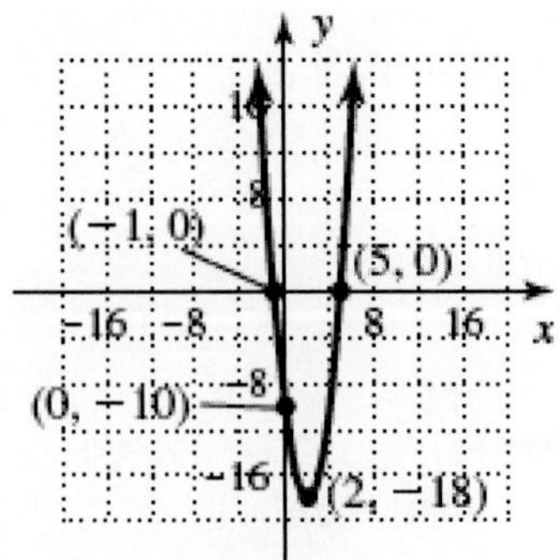

11. $y=2x^2+5x-7$;

$0=(2x+7)(x-1)$

x-intercepts: $(-3.5,0),(1,0)$;

$y=2(0)^2+5(0)-7=-7$

y-intercept: $(0,-7)$;

$x=\frac{-(5)}{2(2)}=-1.25$

$y=2(-1.25)^2+5(-1.25)-7=-10.125$

Vertex: $(-1.25,-10.125)$;

Domain: $x\in(-\infty,\infty)$

Range: $y\in[-10.125,\infty)$

$y=2\left(x^2+\frac{5}{2}x+\frac{25}{16}\right)-7-\frac{25}{8}$

$y=2\left(x+\frac{5}{4}\right)^2-\frac{81}{8}$

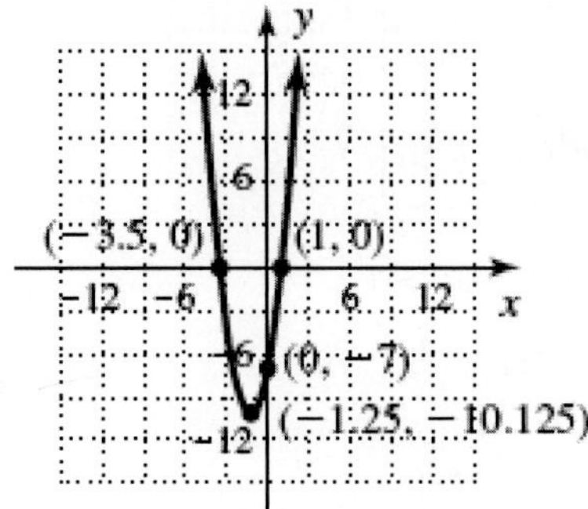

13. $x=y^2-2y-3$

$x=(0)^2-2(0)-3=-3$

x-intercept: $(-3,0)$;

$0=(y-3)(y+1)$

y-intercepts: $(0,3),(0,-1)$;

$y=\frac{-(-2)}{2(1)}=1$

$x=(1)^2-2(1)-3=-4$

Vertex: $(-4,1)$;

Domain: $x\in[-4,\infty)$

Range: $y\in(-\infty,\infty)$

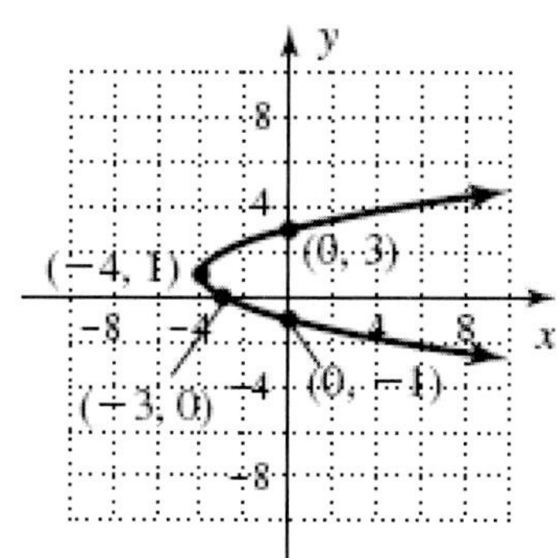

15. $x = -y^2 + 6y + 7$

$x = -(0)^2 + 6(0) + 7 = 7$

x-intercept: $(7,0)$;

$0 = (-y + 7)(y + 1)$

y-intercepts: $(0,7), (0,-1)$;

$y = \dfrac{-(6)}{2(-1)} = 3$

$x = -(3)^2 + 6(3) + 7 = 16$

Vertex: $(16,3)$;

Domain: $x \in (-\infty,16]$

Range: $y \in (-\infty,\infty)$

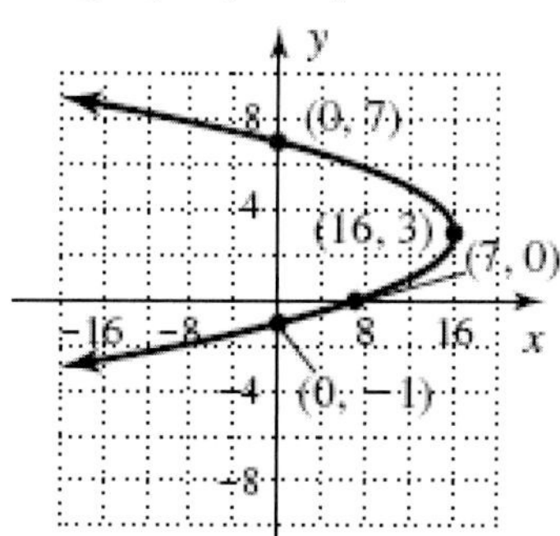

17. $x = -y^2 + 8y - 16$

$x = -(0)^2 + 8(0) - 16 = -16$

x-intercept: $(-16,0)$;

$0 = (-y + 4)(y - 4)$

y-intercept: $(0,4)$;

$y = \dfrac{-(8)}{2(-1)} = 4$

$x = -(4)^2 + 8(4) - 16 = 0$

Vertex: $(0,4)$;

Domain: $x \in (-\infty,0]$

Range: $y \in (-\infty,\infty)$

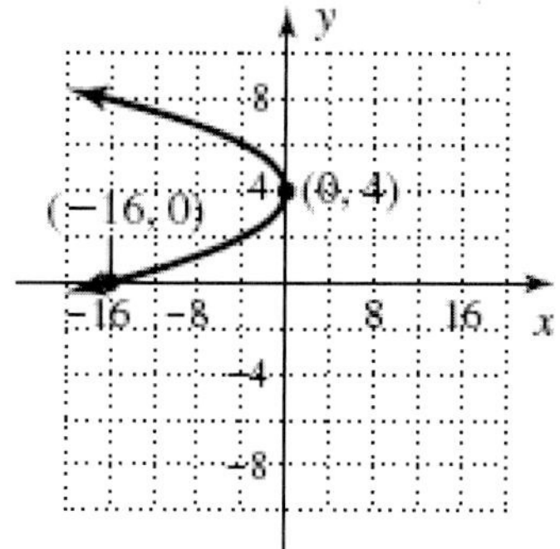

19. $x = y^2 - 6y$

$x = (y^2 - 6y + 9) - 9$

$x = (y - 3)^2 - 9$

Vertex: $(-9,3)$;

$x = (0)^2 - 6(0) = 0$

x-intercept: $(0,0)$;

$0 = y(y - 6)$

y-intercepts: $(0,0), (0,6)$;

Domain: $x \in [-9,\infty)$

Range: $y \in (-\infty,\infty)$

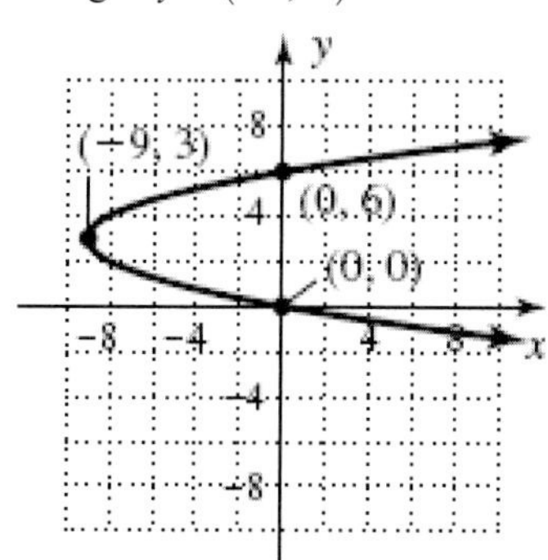

21. $x = y^2 - 4$

Vertex: $(-4,0)$;

$x = (0)^2 - 4 = -4$

x-intercept: $(-4,0)$;

$0 = (y + 2)(y - 2)$

y-intercepts: $(0,2), (0,-2)$;

Domain: $x \in [-4,\infty)$

Range: $y \in (-\infty,\infty)$

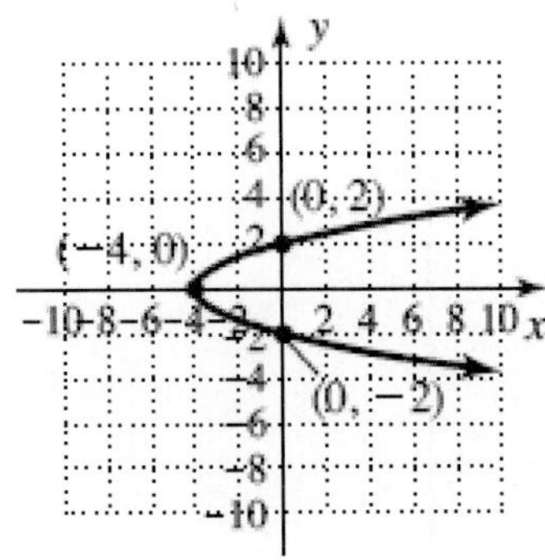

23. $x = -y^2 + 2y - 1$

$x = -\left(y^2 - 2y\right) - 1$

$x = -\left(y^2 - 2y + 1\right) - 1 + 1$

$x = -(y-1)^2 + 0$

Vertex: $(0,1)$;

$x = -(0)^2 + 2(0) - 1 = -1$

x-intercept: $(-1,0)$;

$0 = (-y+1)(y-1)$

y-intercept: $(0,1)$;

Domain: $x \in (-\infty, 0]$

Range: $y \in (-\infty, \infty)$

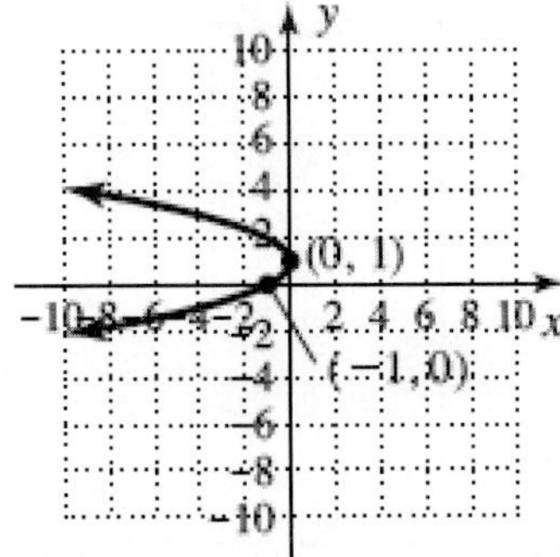

25. $x = y^2 + y - 6$

$x = \left(y^2 + y + \frac{1}{4}\right) - 6 - \frac{1}{4}$

$x = \left(y + \frac{1}{2}\right)^2 - \frac{25}{4}$

Vertex: $(-6.25, -0.5)$;

$x = (0)^2 + (0) - 6 = -6$

x-intercept: $(-6,0)$;

$0 = (y+3)(y-2)$

y-intercepts: $(0,2), (0,-3)$;

Domain: $x \in [-6.25, \infty)$

Range: $y \in (-\infty, \infty)$

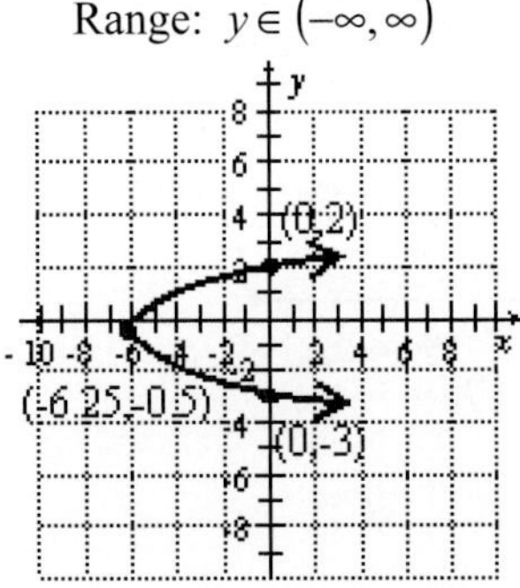

27. $x = y^2 - 10y + 4$

$x = \left(y^2 - 10y + 25\right) + 4 - 25$

$x = (y-5)^2 - 21$

Vertex: $(-21,5)$;

$x = (0)^2 - 10(0) + 4 = 4$

x-intercept: $(4,0)$;

$a = 1, b = -10, c = 4$

$$y = \frac{-(-10) \pm \sqrt{(-10)^2 - 4(1)(4)}}{2(1)}$$

$$y = \frac{10 \pm \sqrt{84}}{2} = \frac{10 \pm 2\sqrt{21}}{2} = 5 \pm \sqrt{21}$$

y-intercepts: $\left(0, 5 - \sqrt{21}\right), \left(0, 5 + \sqrt{21}\right)$;

Domain: $x \in [-21, \infty)$

Range: $y \in (-\infty, \infty)$

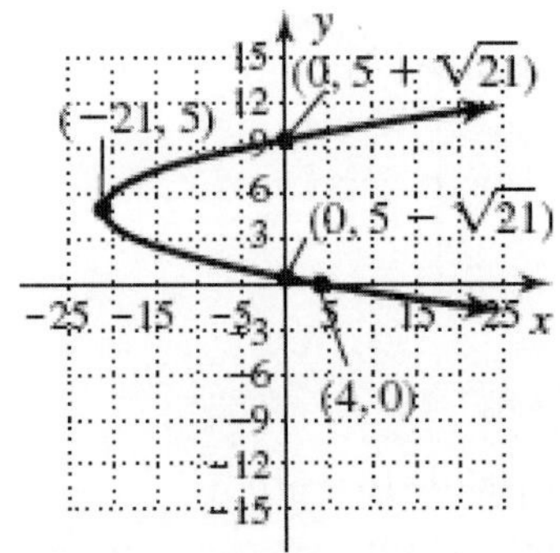

29. $x = 3 - 8y - 2y^2$

$x = -2y^2 - 8y + 3$

$x = -2\left(y^2 + 4y + 4\right) + 3 + 8$

$x = -2(y+2)^2 + 11$

Vertex: $(11,-2)$;

$x = 3 - 8(0) - 2(0)^2 = 3$

x-intercept: $(3,0)$;

$a = -2, b = -8, c = 3$

$$y=\frac{-(-8)\pm\sqrt{(-8)^2-4(-2)(3)}}{2(-2)}$$

$$y=\frac{8\pm\sqrt{88}}{-4}=\frac{8\pm2\sqrt{22}}{-4}=\frac{-4\pm\sqrt{22}}{2}$$

y-intercepts: $\left(0,\frac{-4+\sqrt{22}}{2}\right),\left(0,\frac{-4-\sqrt{22}}{2}\right)$;

Domain: $x\in(-\infty,11]$

Range: $y\in(-\infty,\infty)$

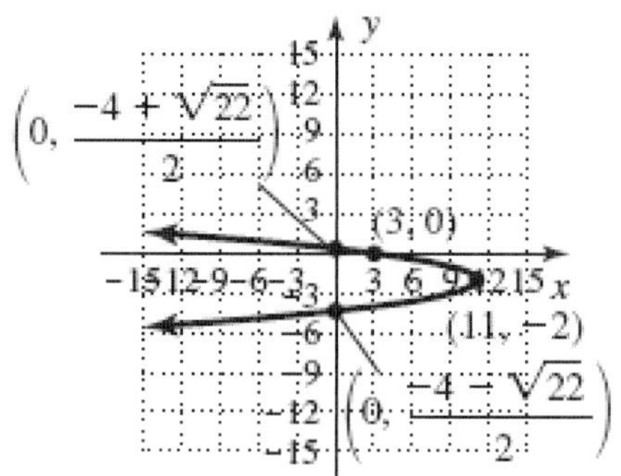

31. $y=(x-2)^2+3$

Vertex: $(2,3)$;

$0\neq(x-2)^2+3$

x-intercept: None;

$y=(0-2)^2+3=7$

y-intercept: $(0,7)$;

Domain: $x\in(-\infty,\infty)$

Range: $y\in[3,\infty)$

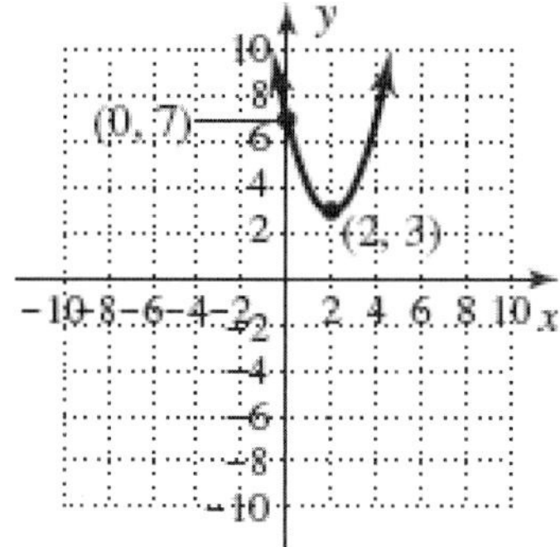

33. $x=(y-3)^2+2$

Vertex: $(2,3)$;

$x=(0-3)^2+2=11$

x-intercept: $(11,0)$;

$0\neq(y-3)^2+2$

y-intercept: None ;

Domain: $x\in[2,\infty)$

Range: $y\in(-\infty,\infty)$

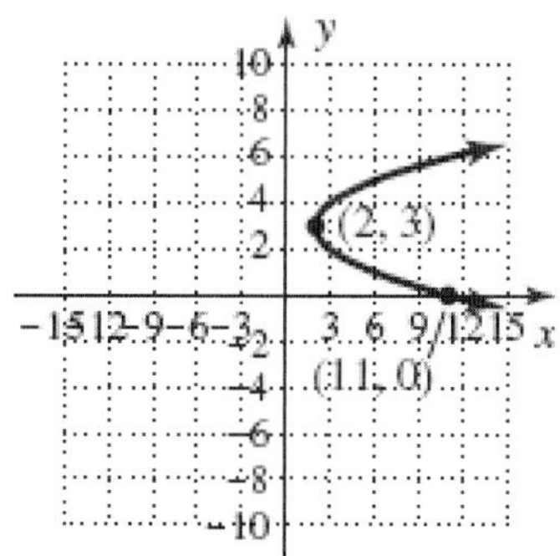

35. $x=2(y-3)^2+1$

Vertex: $(1,3)$;

$x=2(0-3)^2+1=19$

x-intercept: $(19,0)$;

$0\neq2(y-3)^2+1$

y-intercept: None;

Domain: $x\in[1,\infty)$

Range: $y\in(-\infty,\infty)$

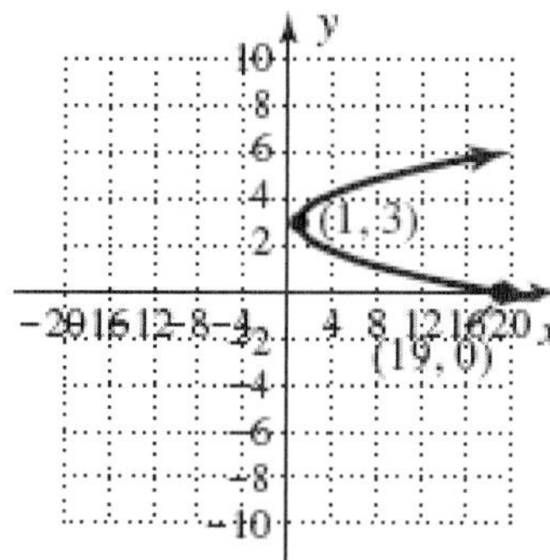

37. $x^2=8y$

Vertex: $(0,0)$

$8=4p$

$2=p$

Focus: $(0,2)$

Length of Focal Chord: 8

Directrix: $y=-2$

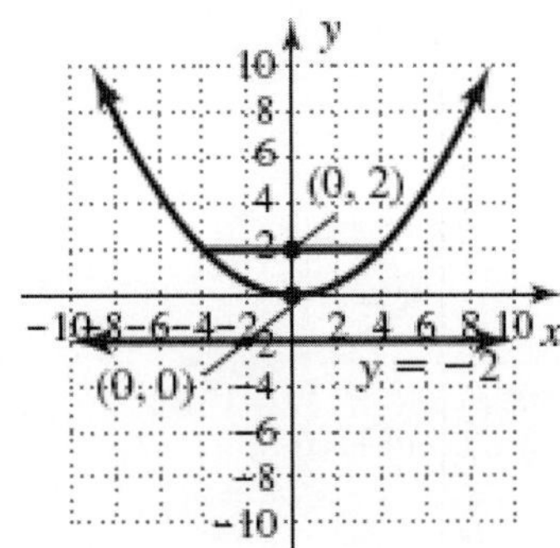

39. $x^2 = -24y$

Vertex: $(0,0)$

$-24 = 4p$

$-6 = p$

Focus: $(0,-6)$

Length of Focal Chord: 24

Directrix: $y = 6$

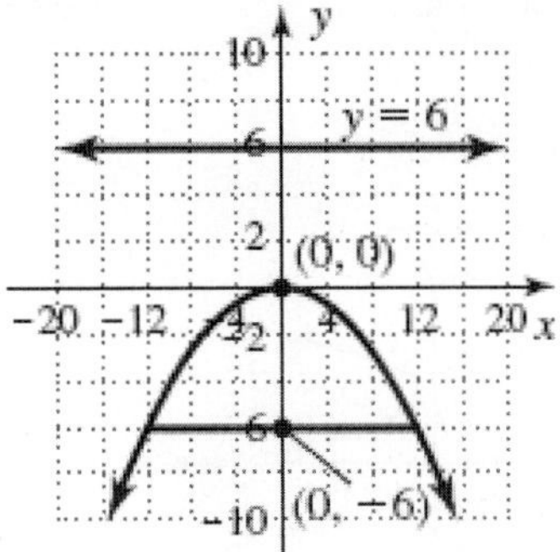

41. $x^2 = 6y$

Vertex: $(0,0)$

$6 = 4p$

$\frac{3}{2} = p$

Focus: $\left(0, \frac{3}{2}\right)$

Length of Focal Chord: 6

Directrix: $y = \frac{-3}{2}$

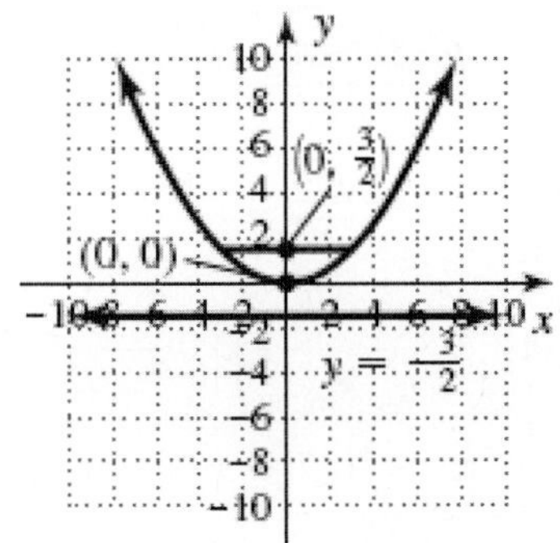

43. $y^2 = -4x$

Vertex: $(0,0)$

$-4 = 4p$

$-1 = p$

Focus: $(-1,0)$

Length of Focal Chord: 4

Directrix: $x = 1$

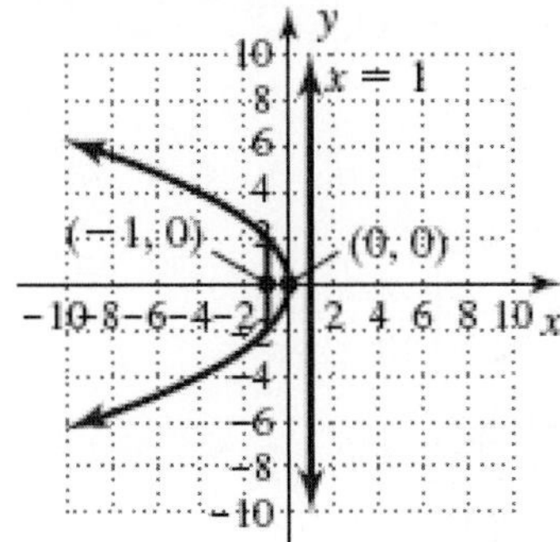

45. $y^2 = 18x$

Vertex: $(0,0)$

$18 = 4p$

$\frac{9}{2} = p$

Focus: $\left(\frac{9}{2}, 0\right)$

Length of Focal Chord: 18

Directrix: $x = -\frac{9}{2}$

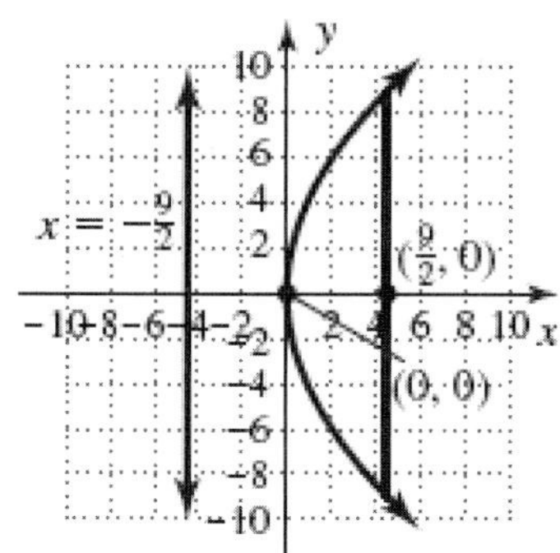

47. $y^2 = -10x$

Vertex: $(0,0)$

$-10 = 4p$

$-\frac{5}{2} = p$

Focus: $\left(-\frac{5}{2}, 0\right)$

Length of Focal Chord: 10

Directrix: $x = \frac{5}{2}$

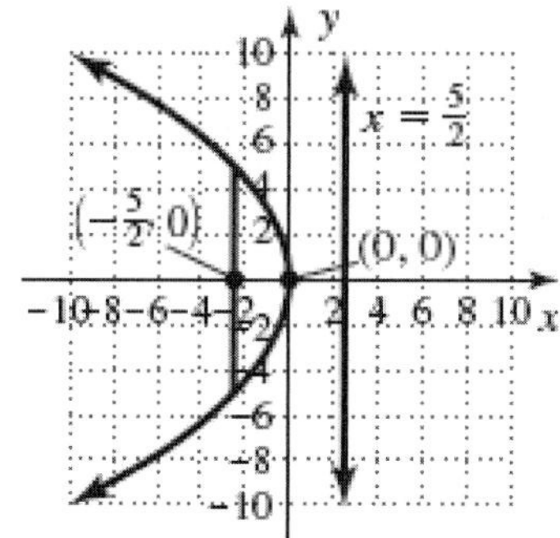

49. $x^2 - 8x - 8y + 16 = 0$

$x^2 - 8x + 16 = 8y$

$(x-4)^2 = 8y$

Vertex: $(4,0)$

$8 = 4p$

$2 = p$

Focus: $(4, 0+2) = (4,2)$

Length of Focal Chord: 8

Directrix: $y = -2$

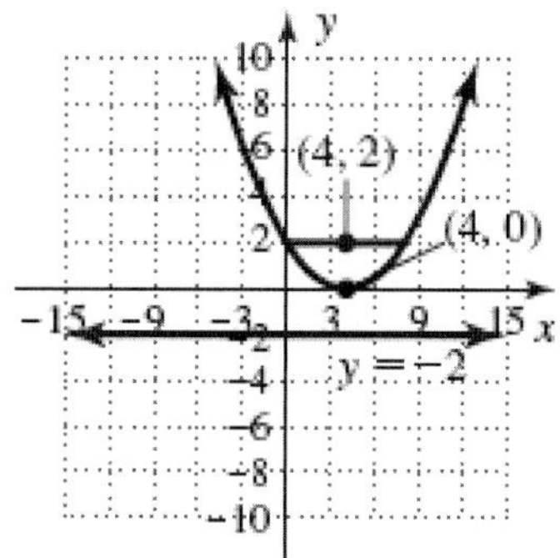

51. $x^2 - 14x - 24y + 1 = 0$

$x^2 - 14x = 24y - 1$

$x^2 - 14x + 49 = 24y - 1 + 49$

$(x-7)^2 = 24y + 48$

$(x-7)^2 = 24(y+2)$

Vertex: $(7,-2)$

$24 = 4p$

$6 = p$

Focus: $(7, -2+6) = (7,4)$

Length of Focal Chord: 24

Directrix: $y = -2 - 6 = -8$

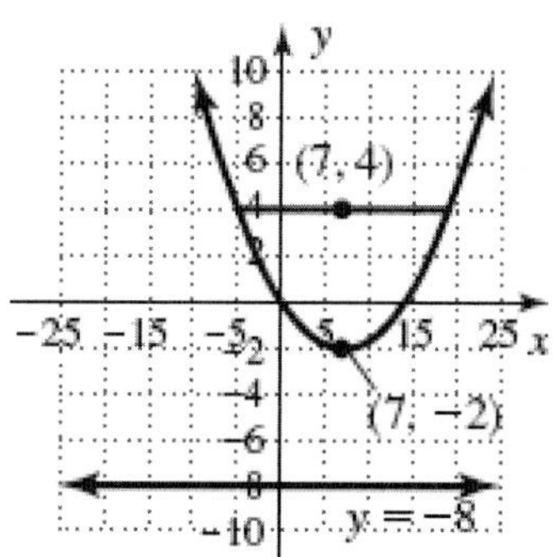

53. $3x^2 - 24x - 12y + 12 = 0$

$3x^2 - 24x = 12y - 12$

$3(x^2 - 8x + 16) = 12y - 12 + 48$

$3(x-4)^2 = 12y + 36$

$3(x-4)^2 = 12(y+3)$

$(x-4)^2 = 4(y+3)$

Vertex: $(4,-3)$

$4 = 4p$

$1 = p$

Focus: $(4, -3+1) = (4,-2)$

Length of Focal Chord: 4

Directrix: $y = -3 - 1 = -4$

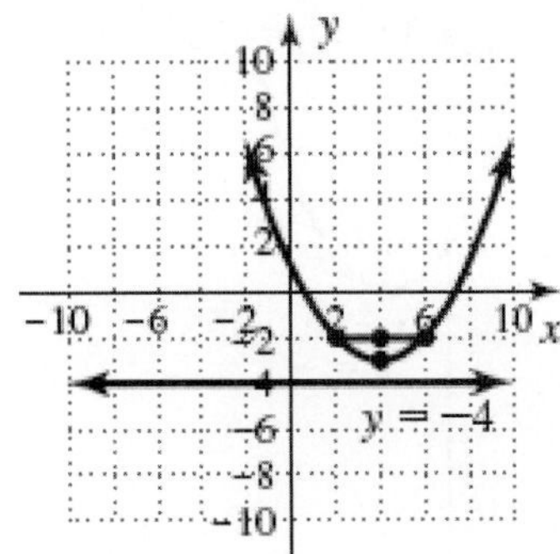

55. $y^2-12y-20x+36=0$

$y^2-12y+36=20x$

$(y-6)^2=20x$

Vertex: $(0,6)$

$4p=20$

$p=5$

Focus: $(0+5,6)=(5,6)$

Length of Focal Chord: 20

Directrix: $x=0-5=-5$

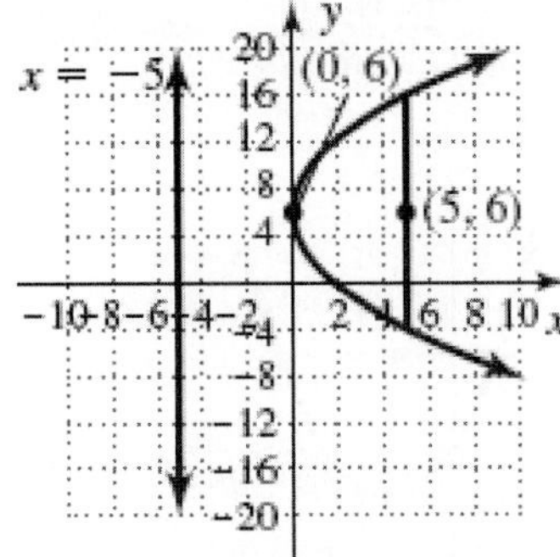

57. $y^2-6y+4x+1=0$

$y^2-6y=-4x-1$

$y^2-6y+9=-4x-1+9$

$(y-3)^2=-4(x-2)$

Vertex: $(2,3)$

$4p=-4$

$p=-1$

Focus: $(2-1,3)=(1,3)$

Length of Focal Chord: 4

Directrix: $x=2-(-1)=3$

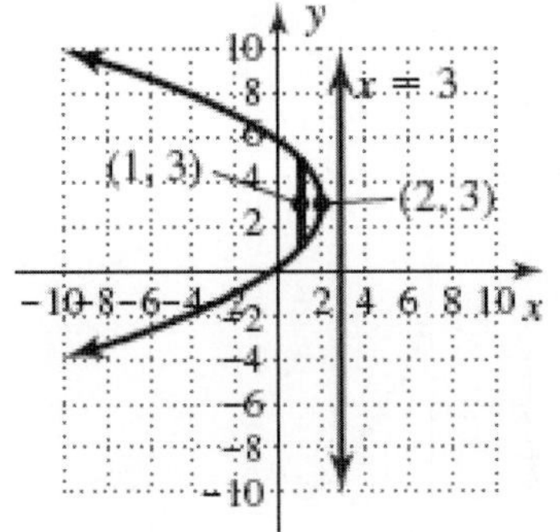

59. $2y^2-20y+8x+2=0$

$y^2-10y+4x+1=0$

$y^2-10y=-4x-1$

$y^2-10y+25=-4x-1+25$

$(y-5)^2=-4(x-6)$

Vertex: $(6,5)$

$4p=-4$

$p=-1$

Focus: $(6-1,5)=(5,5)$

Length of Focal Chord: 4

Directrix: $x=6-(-1)=7$

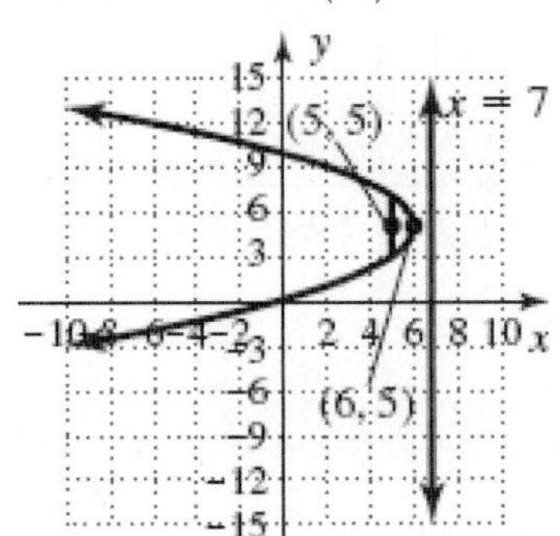

61. $A=\frac{2}{3}ab$

$A=\frac{2}{3}(3)(8)=16\text{ units}^2$

63. $25x=16y^2$

$\frac{25}{16}x=y^2$

Vertex: $(0,0)$

$4p=\frac{25}{16}$

$p=\frac{25}{64}$

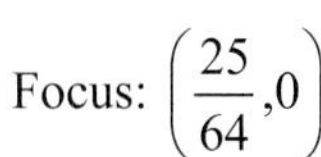

Focus: $\left(\frac{25}{64},0\right)$

Length of Focal Chord: $\frac{25}{16}$

Directrix: $x=-\frac{25}{64}$

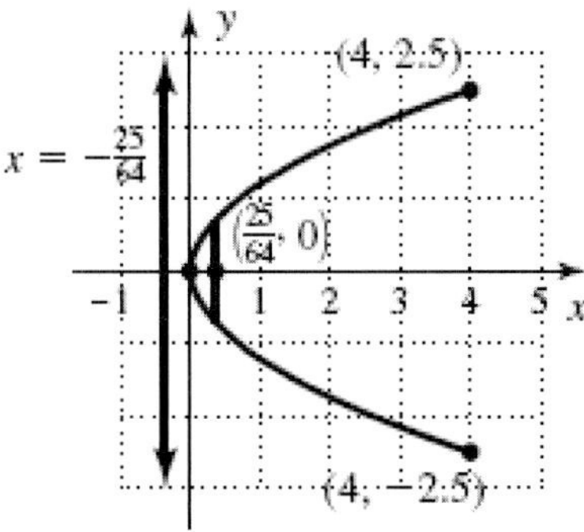

65. $y^2=54x$
$4p=54$
$p=13.5$
Focus: $(13.5,0)$;
36 inch diameter, $y=18$
$(18)^2=54x$
$x=6$
Parabolic receiver: 6 inches

67. $x^2=167y$
$4p=167$
$p=41.75$
Focus: $(0,41.75)$
100 feet diameter, $x=50$
$(50)^2=167y$
$y\approx 14.97$
Parabolic receiver: 14.97 ft

69. $y=2x^2-8x$
$y=2(x^2-4x)$
$y+8=2(x^2-4x+4)$
$y+8=2(x-2)^2$
$(x-2)^2=\frac{1}{2}(y+8)$
Vertex: $(2,-8)$
$4p=\frac{1}{2}$
$p=\frac{1}{8}$

71. b) $x=-(y+3)^2+2$
Answers will vary.

73. $y=-x^2-6x-5$
$y=-(x^2+6x)-5$
$y=-(x^2+6x+9)-5+9$
$y=-(x+3)^2+4$
Vertex: (-3,4);
$(x+3)^2+(y-4)^2=r^2$;
$(-1+3)^2+(0-4)^2=r^2$
$4+16=r^2$
$20=r^2$
$(x+3)^2+(y-4)^2=20$
$A=\pi\left(\sqrt{20}\right)^2=20\pi$ units^2

75. $y=2527.4(1.414)^x$
a. $y=2527.4(1.414)^5\approx 14{,}286$
b. $150000=2527.4(1.414)^x$
$59.34952916=(1.414)^x$
$\ln 59.34952916=\ln(1.414)^x$
$\ln 59.34952916=x\ln(1.414)$
$\frac{\ln 59.34952916}{\ln(1.414)}=x$
$x\approx 11.79$
1990+11=2001
c. $250000=2527.4(1.414)^x$
$98.91588193=(1.414)^x$
$\ln 98.91588193=\ln(1.414)^x$
$\ln 98.91588193=x\ln(1.414)$
$\frac{\ln 98.91588193}{\ln(1.414)}=x$
$x\approx 13.26$
1990+13=2003

77. $f(x)=x^5+2x^4+17x^3+34x^2-18x-36$
Answers will vary.

79. $3-(x+2)+4x=2(x-1)+x+1$
$3-x-2+4x=2x-2+x+1$
$1+3x=3x-1$
$1\neq -1$
No Solution

Chapter 7 Summary and Concept Review

1. $x^2 + y^2 = 16$

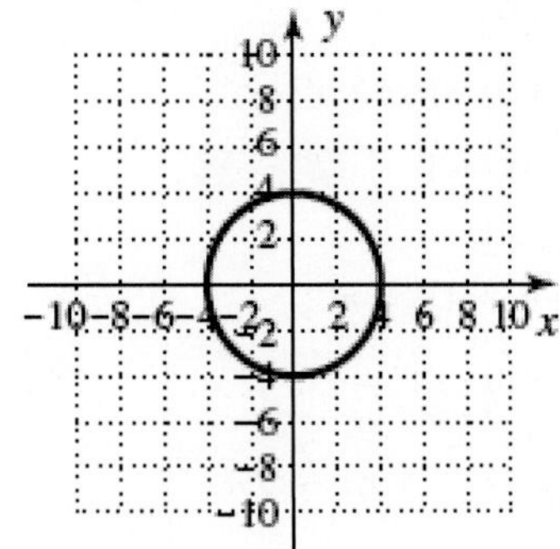

2. $x^2 + 4y^2 = 36$

$\frac{x^2}{36} + \frac{y^2}{9} = 1$

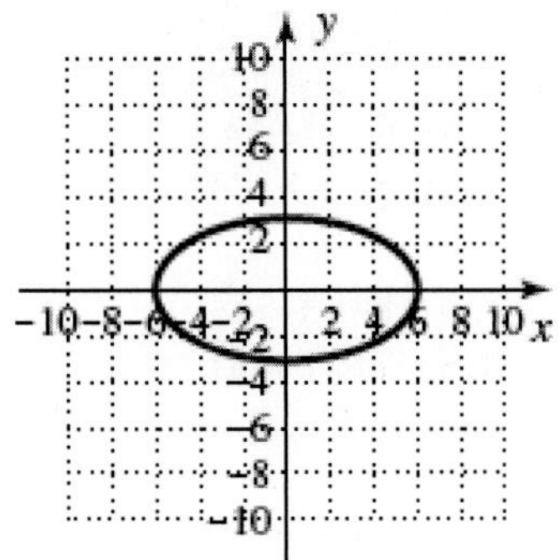

3. $9x^2 + y^2 - 18x - 27 = 0$

$9x^2 - 18x + y^2 = 27$

$9(x^2 - 2x) + y^2 = 27$

$9(x^2 - 2x + 1) + y^2 = 27 + 9$

$9(x-1)^2 + y^2 = 36$

$\frac{(x-1)^2}{4} + \frac{y^2}{36} = 1$

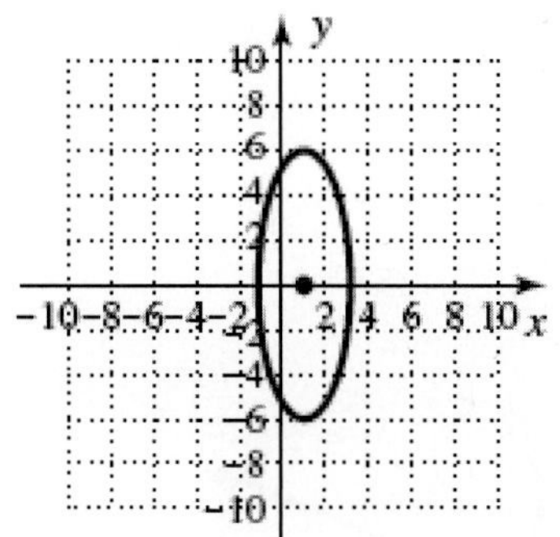

4. $x^2 + y^2 + 6x + 4y + 12 = 0$

$x^2 + 6x + y^2 + 4y = -12$

$(x^2 + 6x + 9) + (y^2 + 4y + 4) = -12 + 9 + 4$

$(x+3)^2 + (y+2)^2 = 1$

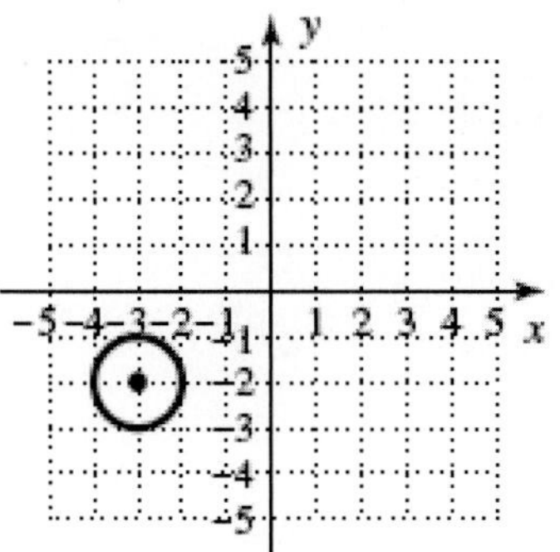

5. $\frac{(x+3)^2}{16} + \frac{(y-2)^2}{9} = 1$

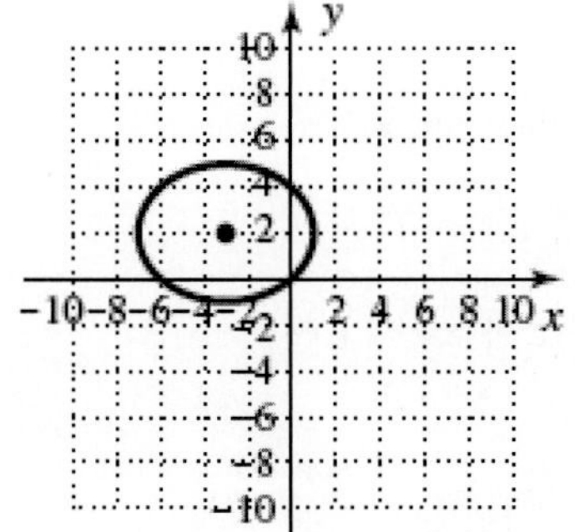

6. Endpoints: $(-4,5)$ and $(2,-3)$

midpoint: $\left(\frac{-4+2}{2}, \frac{5+-3}{2}\right) = (-1,1)$;

$d = \sqrt{(-4--1)^2 + (5-1)^2} = 5$;

$(x+1)^2 + (y-1)^2 = 25$

7. $4y^2 - 25x^2 = 100$

$\frac{y^2}{25}-\frac{x^2}{4}=1$
$a=2,\ b=5$
Hyperbola
Center: $(0,0)$
Vertices: $(0,5),(0,-5)$
$a^2+b^2=c^2$
$25+4=c^2$
$\sqrt{29}=c$
Foci: $\left(0,\sqrt{29}\right),\left(0,-\sqrt{29}\right)$
Asymptotes: $y=\pm\frac{5}{2}x$

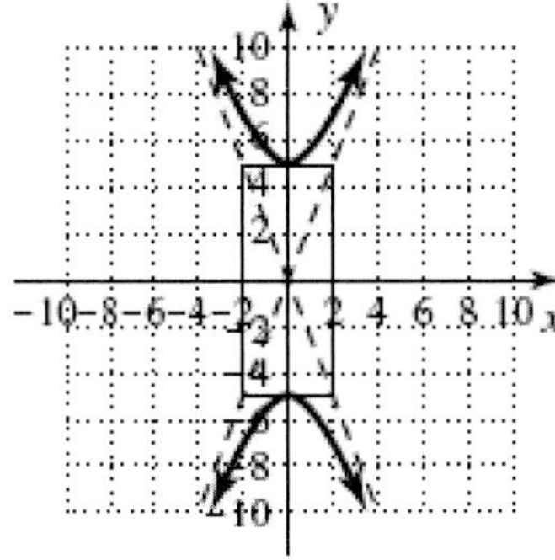

8. $\frac{(y-3)^2}{16}-\frac{(x+2)^2}{9}=1$
Hyperbola
Center: $(-2,3)$
$a=3,\ b=4$
Vertices: $(-2,3+4),(-2,3-4)$
$(-2,7),(-2,-1)$;
$a^2+b^2=c^2$
$16+9=c^2$
$5=c$
Foci: $(-2,3+5),(-2,3-5)$;
$(-2,8),(-2,-2)$
Asymptotes: $y-3=\pm\frac{4}{3}(x+2)$

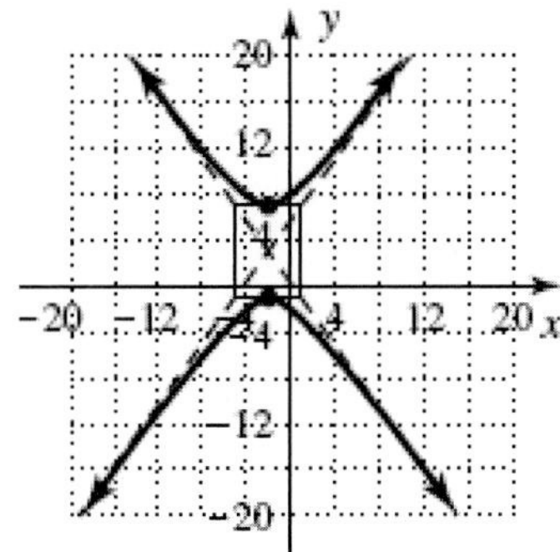

9. $\frac{(x+2)^2}{9}-\frac{(y-1)^2}{4}=1$
Hyperbola
Center: $(-2,1)$
$a=3,\ b=2$
Vertices: $(-2-3,1),(-2+3,1)$
$(-5,1),(1,1)$;
$a^2+b^2=c^2$
$9+4=c^2$
$\sqrt{13}=c$
Foci: $\left(-2-\sqrt{13},1\right),\left(-2+\sqrt{13},1\right)$
Asymptotes: $y-1=\pm\frac{2}{3}(x+2)$

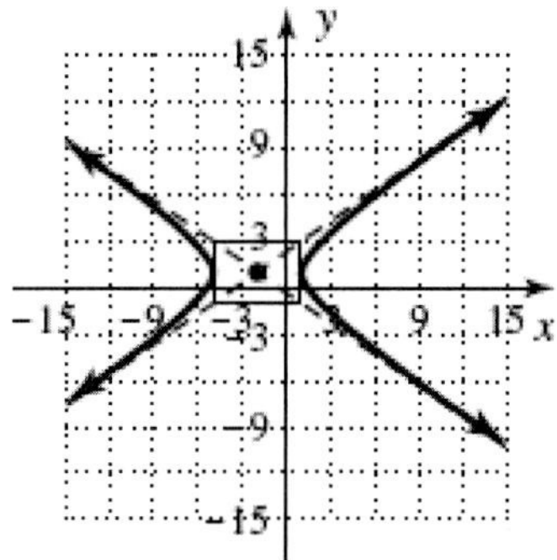

10. $9y^2-x^2-18y-72=0$
$9y^2-18y-x^2=72$
$9\left(y^2-2y\right)-x^2=72$
$9\left(y^2-2y+1\right)-x^2=72+9$
$9(y-1)^2-x^2=81$
$\frac{(y-1)^2}{9}-\frac{x^2}{81}=1$
Hyperbola
Center: $(0,1)$
$a=9,\ b=3$
Vertices: $(0,1+3),(0,1-3)$
$(0,4),(0,-2)$;
$a^2+b^2=c^2$
$9+81=c^2$
$\sqrt{89}=c$
Foci: $\left(0,1+\sqrt{89}\right),\left(0,1-\sqrt{89}\right)$
Asymptotes: $y-1=\pm\frac{1}{3}x$

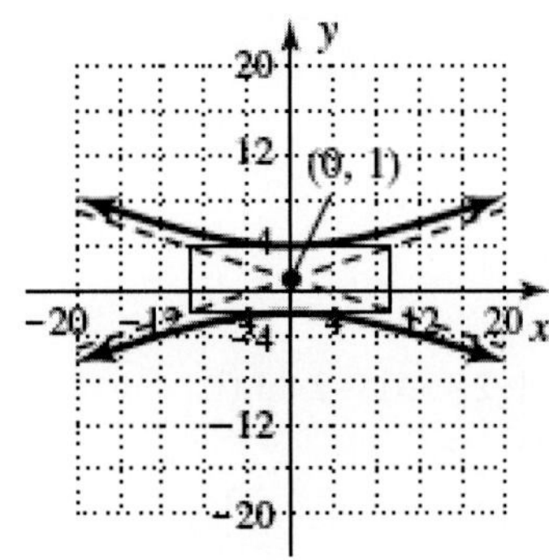

11. $x^2-4y^2-12x-8y+16=0$

$x^2-12x-4y^2-8y=-16$

$(x^2-12x)-4(y^2+2y)=-16$

$(x^2-12x+36)-4(y^2+2y+1)=-16+36-4$

$(x-6)^2-4(y+1)^2=16$

$\dfrac{(x-6)^2}{16}-\dfrac{(y+1)^2}{4}=1$

Hyperbola

Center: $(6,-1)$

$a=4,\ b=2$

Vertices: $(6+4,-1),(6-4,-1)$

$(10,-1),(2,-1)$;

$a^2+b^2=c^2$

$16+4=c^2$

$2\sqrt{5}=c$

Foci: $\left(6+2\sqrt{5},-1\right),\left(6-2\sqrt{5},-1\right)$

Asymptotes: $y+1=\pm\dfrac{1}{2}(x-6)$

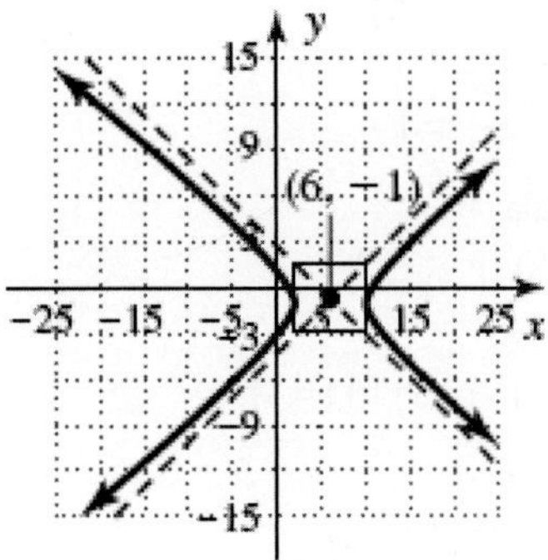

12. Vertices: $(-3,0),(3,0)$

Asymptotes: $y=\pm\dfrac{4}{3}x$;

Center: $(0,0)$

$\dfrac{(x)^2}{9}-\dfrac{(y)^2}{16}=1$

$a^2+b^2=c^2$

$9+16=c^2$

$5=c$

Foci: $(5,0),(-5,0)$

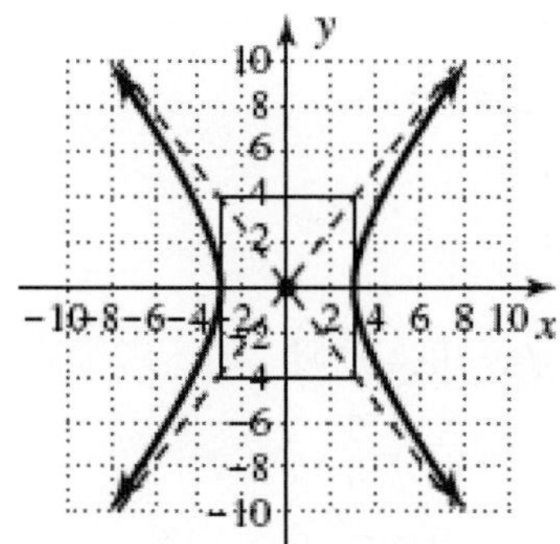

13. $\begin{cases} x^2+y^2=25 & \text{Circle} \\ y-x=-1 & \text{Line} \end{cases}$

2nd equation: $y=x-1$

$x^2+(x-1)^2=25$

$x^2+x^2-2x+1=25$

$2x^2-2x-24=0$

$x^2-x-12=0$

$(x-4)(x+3)=0$

$x=4, x=-3$;

$y=4-1=3$;

$y=-3-1=-4$

Solutions: $(4,3),(-3,-4)$

14. $\begin{cases} x^2-y^2=5 & \text{Hyperbola} \\ x^2+y^2=13 & \text{Circle} \end{cases}$

$2x^2=18$

$x^2=9$

$x=\pm 3$;

$(\pm 3)^2+y^2=13$

$y^2=4$

$y=\pm 2$;

Solutions: $(3,2),(3,-2),(-3,2),(-3,-2)$

15. $\begin{cases} x=y^2-1 & \text{Parabola} \\ x+4y=-5 & \text{Line} \end{cases}$

$y^2-1+4y=-5$

$y^2+4y+4=0$

$(y+2)^2=0$

$y=-2$;

$x=(-2)^2-1=3$

Solution: $(3,-2)$

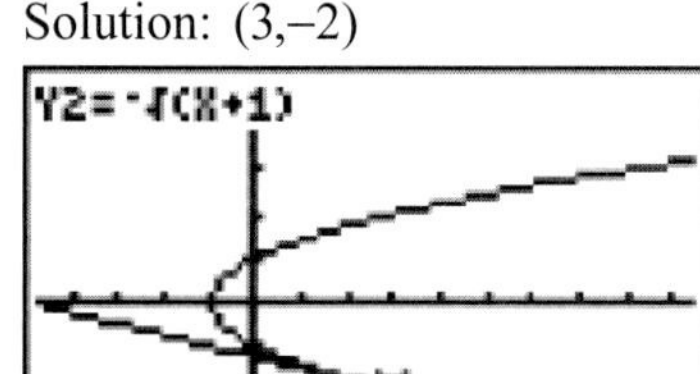

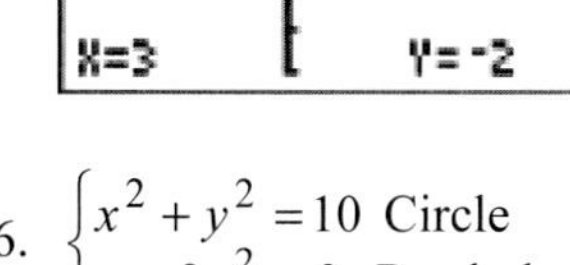

16. $\begin{cases} x^2+y^2=10 & \text{Circle} \\ y-3x^2=0 & \text{Parabola} \end{cases}$

$x^2=\dfrac{y}{3}$

$\dfrac{y}{3}+y^2=10$

$3y^2+y-30=0$

$(3y+10)(y-3)=0$

$3y+10=0$ or $y-3=0$

$y\neq-\dfrac{10}{3}$ or $y=3$

Solutions: $(1,3),(-1,3)$

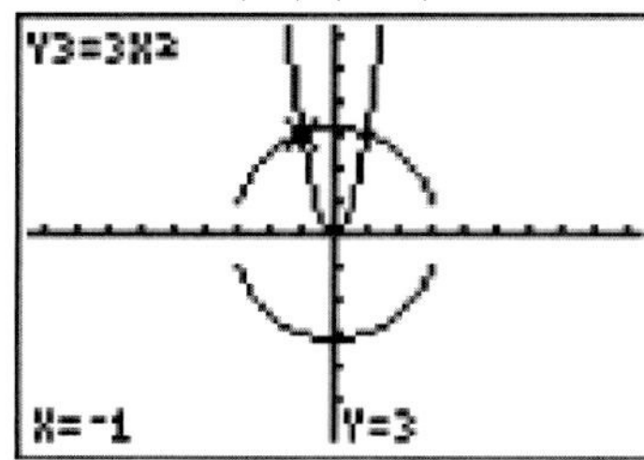

17. $\begin{cases} y\le x^2-2 & \text{Parabola} \\ x^2+4y^2\le 16 & \text{Ellipse} \end{cases}$

Approximate points where conics cross:
$(-1.94,1.75),(1.94,1.75),(0,-2)$

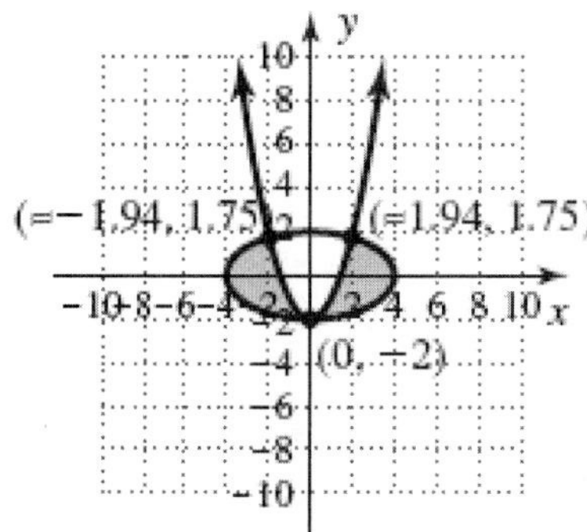

18. $\begin{cases} 4y^2-9x^2\ge 36 & \text{Hyperbola} \\ x^2+y^2\le 48 & \text{Circle} \end{cases}$

$\begin{cases} -9x^2+4y^2=36 \\ x^2+y^2=48 \end{cases}$

$\begin{cases} -9x^2+4y^2=36 \\ 9x^2+9y^2=432 \end{cases}$

$13x^2=468$
$x^2=36$
$x=\pm6;$

$y^2=12$
$y=\pm2\sqrt{3}$

Points where conics cross: $\left(\pm6,\pm2\sqrt{3}\right)$

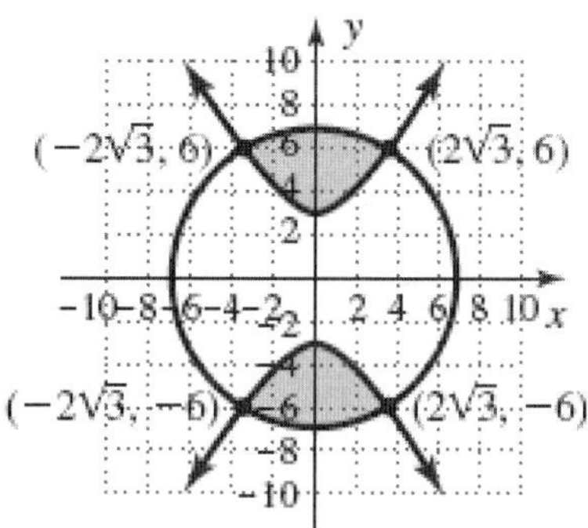

19. $4x^2+25y^2-16x-50y-59=0$

$4x^2-16x+25y^2-50y=59$

$4\left(x^2-4x\right)+25\left(y^2-2y\right)=59$

$4\left(x^2-4x+4\right)+25\left(y^2-2y+1\right)=59+16+25$

$4(x-2)^2+25(y-1)^2=100$

$\dfrac{(x-2)^2}{25}+\dfrac{(y-1)^2}{4}=1$

Ellipse
Center: $(2,1)$
Vertices: $(2-5,1),(2+5,1)$
$(-3,1),(7,1)$;

$a^2=b^2+c^2$
$25=4+c^2$
$\sqrt{21}=c$

Foci: $\left(2+\sqrt{21},1\right),\left(2-\sqrt{21},1\right)$
CA: $(2,1+2),(2,1-2)$
$(2,3),(2,-1)$

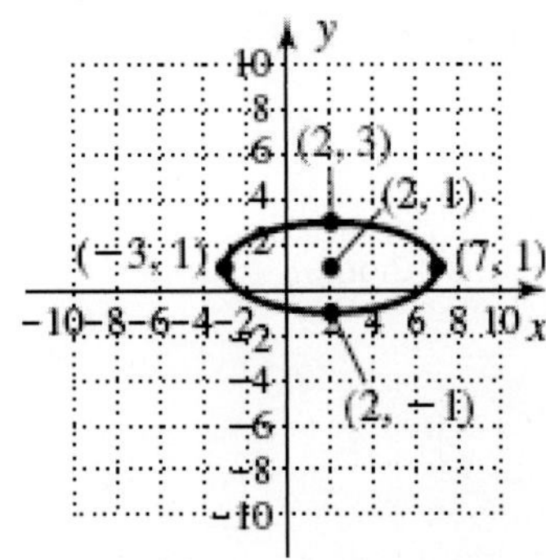

20. $4x^2-36y^2-40x+144y-188=0$

$4x^2-40x-36y^2+144y=188$

$4(x^2-10x)-36(y^2-4y)=188$

$4(x^2-10x+25)-36(y^2-4y+4)=188+100-144$

$4(x-5)^2-36(y-2)^2=144$

$\dfrac{(x-5)^2}{36}-\dfrac{(y-2)^2}{4}=1$

Hyperbola

Center: $(5,2)$

$a=6,\ b=2$

Vertices: $(5-6,2),(5+6,2)$

$(-1,2),(11,2)$;

$a^2+b^2=c^2$

$36+4=c^2$

$2\sqrt{10}=c$

Foci: $(5+2\sqrt{10},2),(5-2\sqrt{10},2)$

CA: $(5,2+2),(5,2-2)$

$(5,4),(5,0)$

Asymptotes: $y-2=\pm\dfrac{1}{3}(x-5)$

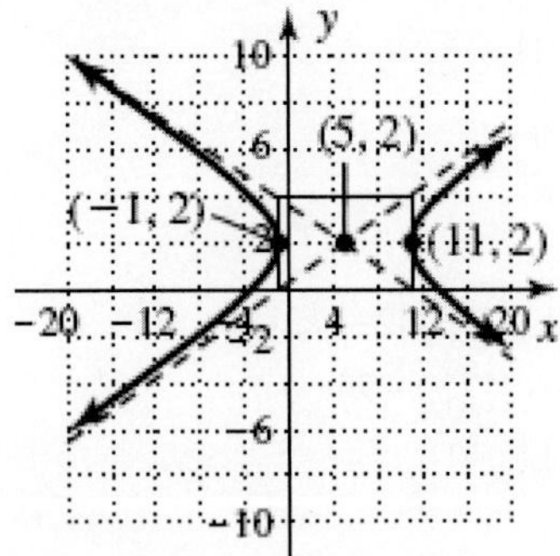

21. a. Vertices: $(-13,0),(13,0)$

Foci: $(-12,0),(12,0)$

$a^2=b^2+c^2$

$169=b^2+144$

$25=b^2$

$\dfrac{x^2}{169}+\dfrac{y^2}{25}=1$

b. Foci: $(0,-16),(0,16)$

Length of major axis: 40 units

$a^2=b^2+c^2$

$400=b^2+256$

$144=b^2$

$\dfrac{x^2}{144}+\dfrac{y^2}{400}=1$

22. a. Vertices: $(-15,0),(15,0)$

Foci: $(-17,0),(17,0)$

$a^2+b^2=c^2$

$225+b^2=289$

$b^2=64$

$\dfrac{x^2}{225}-\dfrac{y^2}{64}=1$

b. Foci: $(0,-5),(0,5)$

Vertical Length of central rectangle: 8 units

$a^2+b^2=c^2$

$4^2+b^2=5^2$

$b^2=9$

$\dfrac{y^2}{16}-\dfrac{x^2}{9}=1$

23. $x=y^2-4$

Parabola

x-intercept: (-4,0)

y-intercepts: (0,2), (0,-2)

Vertex: $(-4,0)$

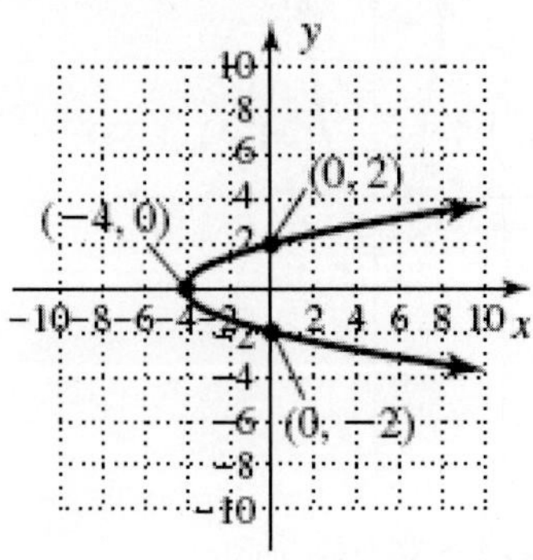

24. $x=y^2+y-6$

$$x = \left(y^2 + y + \frac{1}{4}\right) - 6 - \frac{1}{4}$$

$$x = \left(y + \frac{1}{2}\right)^2 - \frac{25}{4};$$

Parabola

$x = 0^2 + 0 - 6 = -6$

x-intercept: (-6,0)

$0 = y^2 + y - 6$

$0 = (y+3)(y-2)$

$y = -3$ or $y = 2$

y-intercepts: (0,-3), (0,2)

Vertex: $(-6.25, -0.5)$

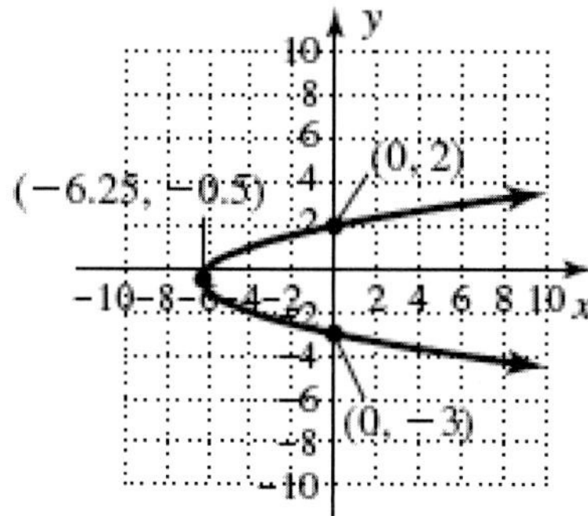

25. $x^2 = -20y$

$4p = -20$

$p = -5$

Parabola

Vertex: $(0,0)$

Focus: $(0,-5)$

Length of Focal Chord: 20

Directrix: $y = 5$

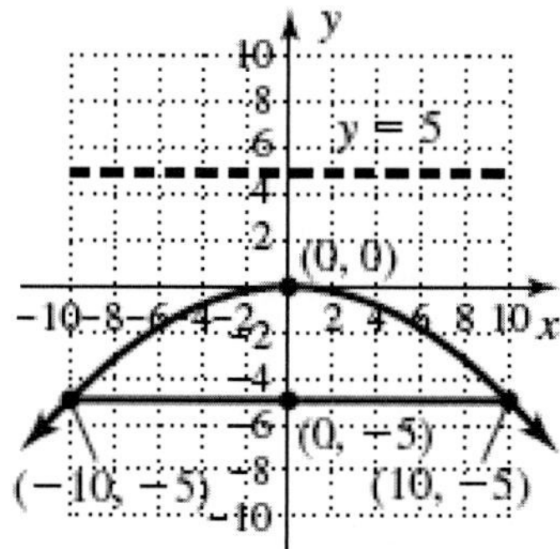

26. $x^2 - 8x - 8y + 16 = 0$

$x^2 - 8x + 16 = 8y$

$(x-4)^2 = 8y$

$4p = 8$

$p = 2$

Parabola

Vertex: $(4,0)$

Focus: $(4, 0+2) = (4,2)$

Length of Focal Chord: 8

Directrix: $y = -2$

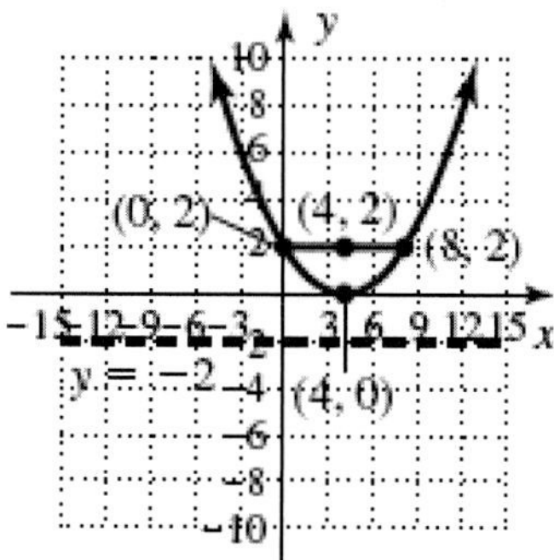

Chapter 7 Mixed Review

1. $9x^2 + 9y^2 = 54$

$x^2 + y^2 = 6$

Circle

Center: $(0,0)$, Radius : $r = \sqrt{6}$

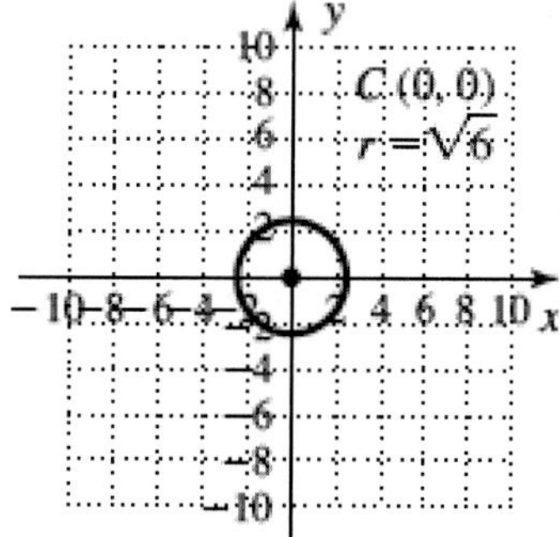

3. $9y^2 - 25x^2 = 225$

$$\frac{y^2}{25} - \frac{x^2}{9} = 1$$

Hyperbola

Center: $(0,0)$

Vertices: $(0,5), (0,-5)$

$a = 3,\ b = 5;$

$a^2 + b^2 = c^2$

$25 + 9 = c^2$

$\sqrt{34} = c$

Foci: $(0, \sqrt{34}), (0, -\sqrt{34})$

CA: $(3,0), (-3,0)$

Asymptotes: $y = \pm\frac{5}{3}x$

$\text{L}=\frac{2\cdot 9}{5} = 3\frac{3}{5} = 3.6$

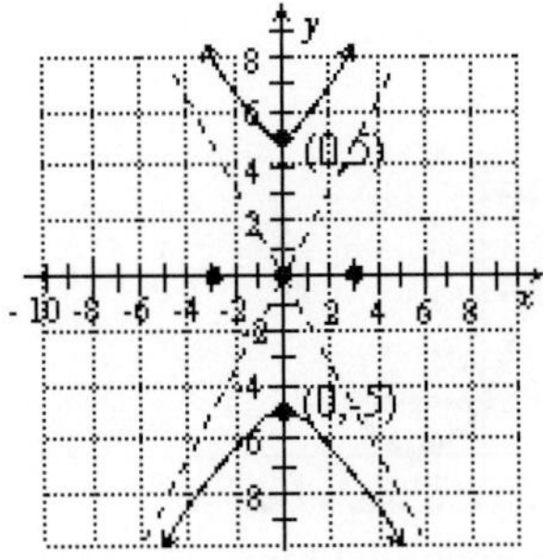

5. $\frac{(y-2)^2}{25} - \frac{(x+3)^2}{16} = 1$

Hyperbola

Center: $(-3,2)$

Vertices: $(-3,7),(-3,-3)$

$a = 4,\ b = 5;$

$a^2 + b^2 = c^2$

$25 + 16 = c^2$

$\sqrt{41} = c$

Foci: $\left(-3,2+\sqrt{41}\right),\left(-3,2-\sqrt{41}\right)$

CA: $(-3-4,2),(-3+4,2)$

$(-7,2),(1,2)$

Asymptotes: $y-2 = \pm\frac{5}{4}(x+3)$

$\text{L}=\frac{2\cdot 16}{5} = 6.4$

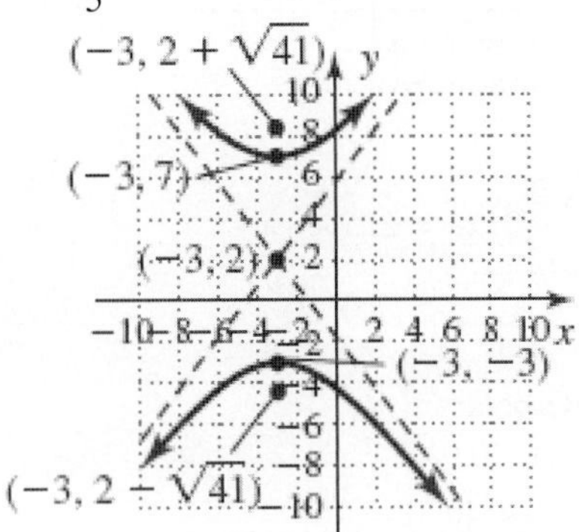

7. $16(x+2)^2 + 4(y-1)^2 = 64$

$\frac{(x+2)^2}{4} + \frac{(y-1)^2}{16} = 1$

Ellipse

Center: $(-2,1)$

Vertices: $(-2,1+4),(-2,1-4)$

$(-2,5),(-2,-3)$

$a^2 = b^2 + c^2$

$16 = 4 + c^2$

$2\sqrt{3} = c$

Foci: $\left(-2,1-2\sqrt{3}\right),\left(-2,1+2\sqrt{3}\right)$

CA: $(-2+2,1),(-2-2,1)$

$(0,1),(-4,1)$

$\text{L}=\frac{2\cdot 4}{4} = 2$

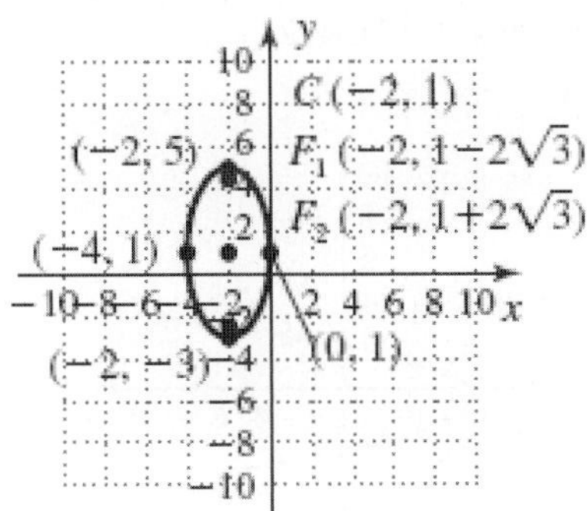

9. $x^2 + y^2 - 8x + 12y + 16 = 0$

$x^2 - 8x + y^2 + 12y = -16$

$\left(x^2 - 8x + 16\right) + \left(y^2 + 12y + 36\right) = -16 + 16 + 36$

$(x-4)^2 + (y+6)^2 = 36$

Circle

Center: $(4,-6)$, Radius : $r = 6$

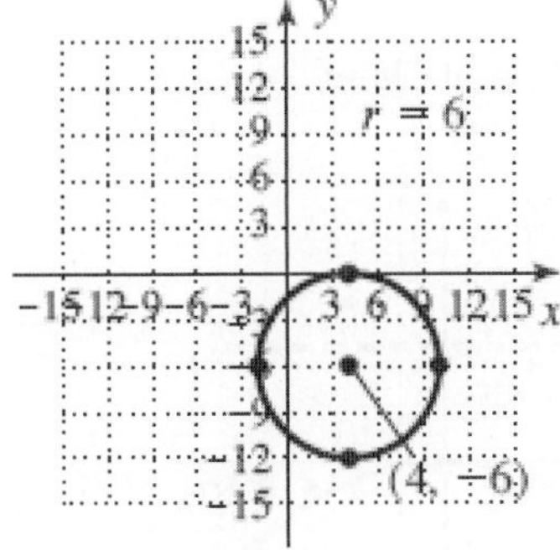

11. $y = 2x^2 - 8x - 10$

$y = 2\left(x^2 - 4x\right) - 10$

$y = 2\left(x^2 - 4x + 4\right) - 10 - 8$

$y = 2(x-2)^2 - 18;$

$y + 18 = 2(x-2)^2$

$\frac{1}{2}(y+18) = (x-2)^2$

$4p = \frac{1}{2}$

$p = \frac{1}{8}$

Parabola

Vertex: $(2,-18)$

Focus: $\left(2,-18+\frac{1}{8}\right)=\left(2,\frac{-143}{8}\right)$

Directrix: $y = -\frac{145}{8}$

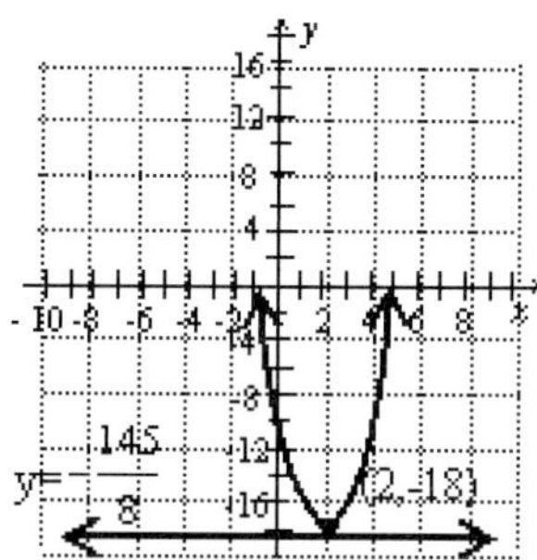

13. $x = -y^2 + 2y + 3$

$x = -(y^2 - 2y) + 3$

$x = -(y^2 - 2y + 1) + 3 + 1$

$x = -(y-1)^2 + 4;$

$x - 4 = -(y-1)^2$

$-(x-4) = (y-1)^2$

$4p = -1$

$p = -\frac{1}{4}$

Parabola

Vertex: $(4,1)$

Focus: $\left(4-\frac{1}{4},1\right)=\left(\frac{15}{4},1\right)$

Directrix: $x = \frac{17}{4}$

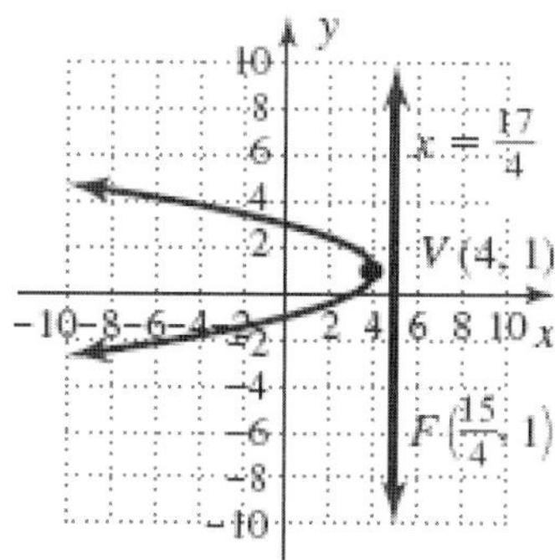

15. $x = y^2 - 9$

$(x+9) = y^2$

$4p = 1$

$p = \frac{1}{4}$

Parabola

Vertex: $(-9,0)$

Focus: $\left(-9+\frac{1}{4},0\right)=\left(-\frac{35}{4},0\right)$

Directrix: $x = -\frac{37}{4}$

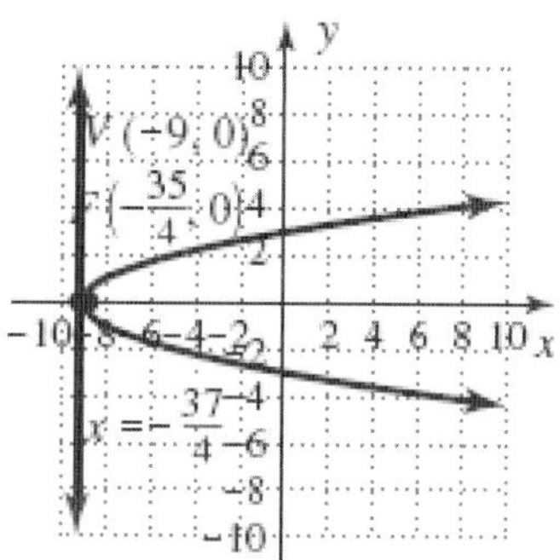

17. $x^2 = -24y$

$4p = -24$

$p = -6$

Parabola

Vertex: $(0,0)$

Focus: $(0,-6)$

Directrix: $y = 6$

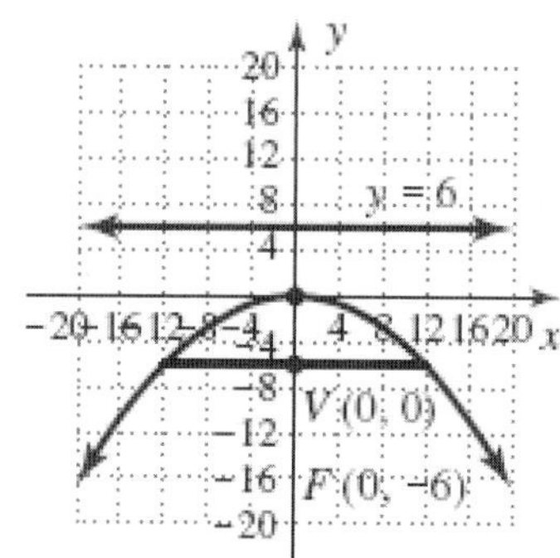

19. $4x^2 + 16y^2 - 12x - 48y - 19 = 0$

$4x^2 - 12x + 16y^2 - 48y = 19$

$4(x^2 - 3x) + 16(y^2 - 3y) = 19$

$4\left(x^2-3x+\frac{9}{4}\right)+16\left(y^2-3y+\frac{9}{4}\right)=19+9+36$

$4\left(x-\frac{3}{2}\right)^2+16\left(y-\frac{3}{2}\right)^2=64$

$\frac{\left(x-\frac{3}{2}\right)^2}{16}+\frac{\left(y-\frac{3}{2}\right)^2}{4}=1$

Ellipse

Center: $\left(\frac{3}{2},\frac{3}{2}\right)$

Vertices: $\left(\frac{3}{2}+4,\frac{3}{2}\right),\left(\frac{3}{2}-4,\frac{3}{2}\right)$

$\left(\frac{11}{2},\frac{3}{2}\right),\left(-\frac{5}{2},\frac{3}{2}\right)$

$a^2=b^2+c^2$
$16=4+c^2$
$2\sqrt{3}=c$

Foci: $\left(\frac{3}{2}-2\sqrt{3},\frac{3}{2}\right),\left(\frac{3}{2}+2\sqrt{3},\frac{3}{2}\right)$

CA: $\left(\frac{3}{2},\frac{3}{2}\right)+2,\left(\frac{3}{2},\frac{3}{2}-2\right)$

$\left(\frac{3}{2},\frac{7}{2}\right),\left(\frac{3}{2},-\frac{1}{2}\right)$

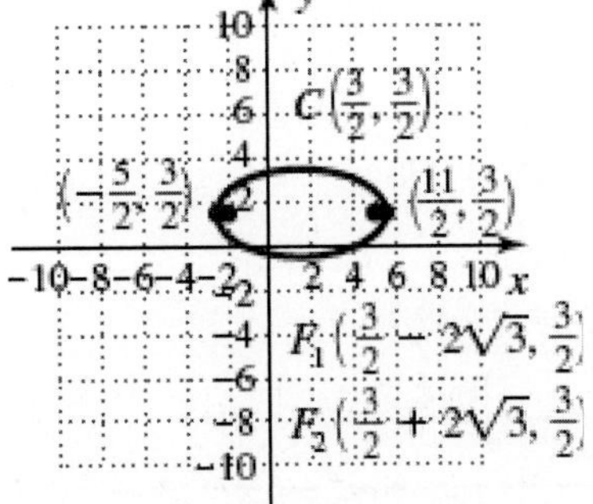

21. $x^2+y^2+8x+12y+2=0$

$x^2+8x+y^2+12y=-2$

$\left(x^2+8x+16\right)+\left(y^2+12y+36\right)=-2+16+36$

$(x+4)^2+(y+6)^2=50$

Circle

Center: $(-4,-6)$, Radius : $r=5\sqrt{2}$

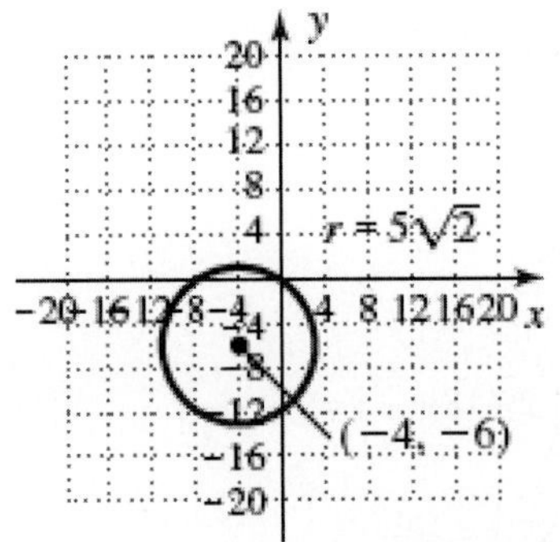

23. $d(P_1P_3)=\sqrt{(-4-(-6))^2+(-9-3)^2}=\sqrt{148}$

$d(P_2P_3)=\sqrt{(6-(-6))^2+(1-3)^2}=\sqrt{148}$;

$d(P_1P_3)=d(P_2P_3)$

25. apehelion(max) $c-(-a)=70$ million ;

perihelion(min) $a-c=46$ million ;

$\begin{cases}a+c=70\\a-c=46\end{cases}$

a. (-46,0)

b. $2a=2(58)=116$
$a=58$ million
$58+c=70$
$c=12$ million ;
$a^2-b^2=c^2$
$58^2-b^2=12^2$
$\sqrt{3220}=b$
$b\approx 56.75$;
$2b\approx 113.50$

a. $\frac{x^2}{3364}+\frac{y^2}{3220}=1$

Chapter 7 Practice Test

1. Circle (C)
2. Ellipse (D)
3. Parabola (A)
4. Hyperbola (B)
5. $(x-4)^2+(y+3)^2=9$
   Center: (4,-3), $r=3$

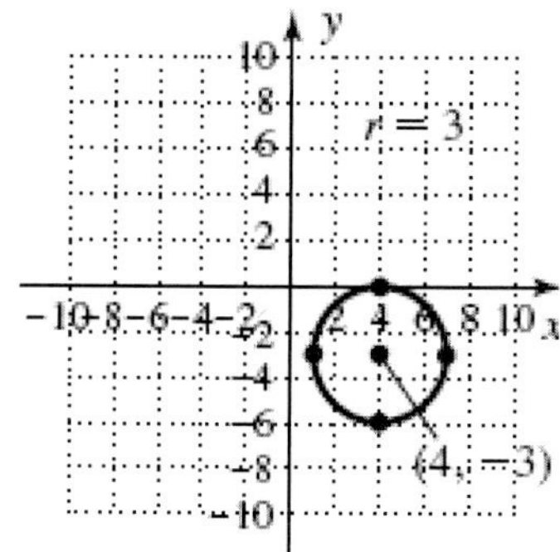

6. $\frac{(x-2)^2}{16}+\frac{(y+3)^2}{1}=1$

Ellipse

Center: $(2,-3)$

Vertices: $(2-4,-3),(2+4,-3)$

$(-2,-3),(6,-3)$

$a^2=b^2+c^2$

$16=1+c^2$

$\sqrt{15}=c$

Foci: $(2-\sqrt{15},-3),(2+\sqrt{15},-3)$

CA: $(2,-3+1),(2,-3-1)$

$(2,-2),(2,-4)$

7. $\frac{(x+3)^2}{9}-\frac{(y-4)^2}{4}=1$

Hyperbola

Center: $(-3,4)$

Vertices: $(-3+3,4),(-3-3,4)$

$(0,4),(-6,4)$

$a^2+b^2=c^2$

$9+4=c^2$

$\sqrt{13}=c$

Foci: $(-3-\sqrt{13},4),(-3+\sqrt{13},4)$

CA: $(-3,4-2),(-3,4+2)$

$(-3,2),(-3,6)$

Asymptotes: $y-4=\pm\frac{2}{3}(x+3)$

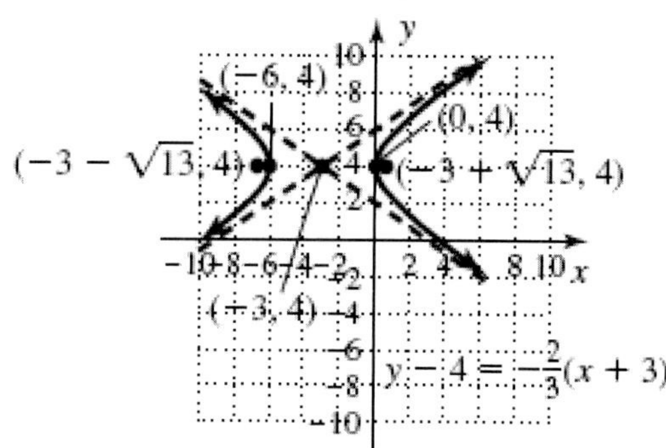

8. $x^2+y^2-10x+4y+4=0$

$x^2-10x+y^2+4y=-4$

$(x^2-10x+25)+(y^2+4y+4)=-4+25+4$

$(x-5)^2+(y+2)^2=25$

Circle

Center: $(5,-2)$, Radius : $r=5$

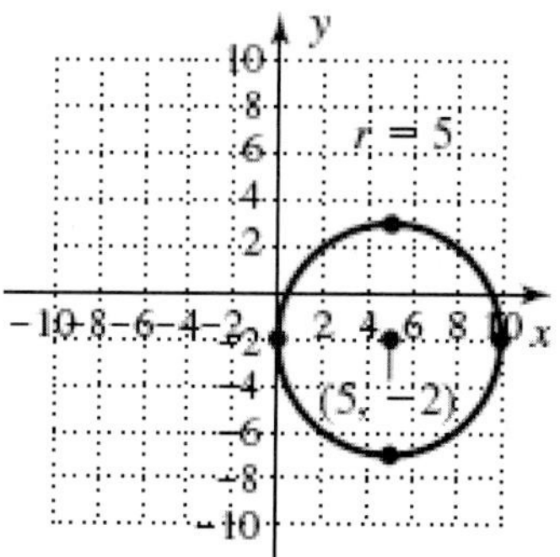

9. $9x^2+4y^2+18x-24y+9=0$

$9x^2+18x+4y^2-24y=-9$

$9(x^2+2x)+4(y^2-6y)=-9$

$9(x^2+2x+1)+4(y^2-6y+9)=-9+9+36$

$9(x+1)^2+4(y-3)^2=36$

$\frac{(x+1)^2}{4}+\frac{(y-3)^2}{9}=1$

Ellipse

Center: $(-1,3)$

Vertices: $(-1,3+3),(-1,3-3)$

$(-1,6),(-1,0)$

$a^2=b^2+c^2$

$9=4+c^2$

$\sqrt{5}=c$

Foci: $(-1,3-\sqrt{5}),(-1,3+\sqrt{5})$

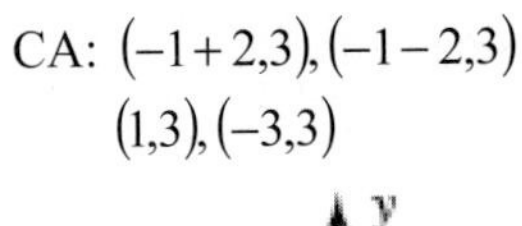

CA: $(-1+2,3),(-1-2,3)$

$(1,3),(-3,3)$

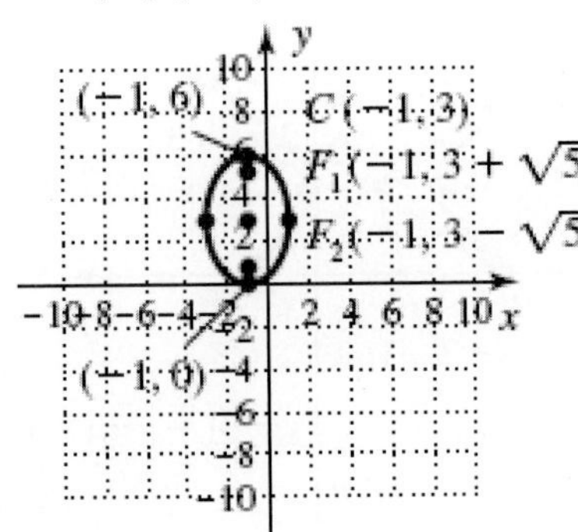

10. $9x^2-4y^2+18x-24y-63=0$

$9x^2+18x-4y^2-24y=63$

$9(x^2+2x)-4(y^2+6y)=63$

$9(x^2+2x+1)-4(y^2+6y+9)=63+9-36$

$9(x+1)^2-4(y+3)^2=36$

$\dfrac{(x+1)^2}{4}-\dfrac{(y+3)^2}{9}=1$

Hyperbola

Center: $(-1,-3)$

Vertices: $(-1-2,-3),(-1+2,-3)$

$(-3,-3),(1,-3)$

$a^2+b^2=c^2$

$4+9=c^2$

$\sqrt{13}=c$

Foci: $(-1-\sqrt{13},-3),(-1+\sqrt{13},-3)$

CA: $(-1,-3+3),(-1,-3-3)$

$(-1,0),(-1,-6)$

Asymptotes: $y+3=\pm\dfrac{3}{2}(x+1)$

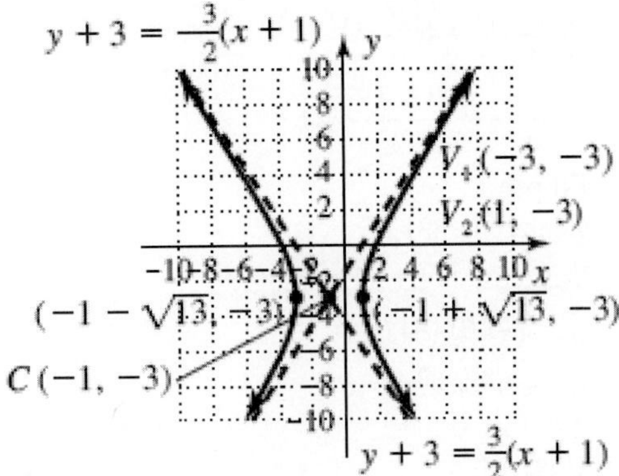

11. $x=(y+3)^2-2$

$(x+2)=(y+3)^2$

$4p=1$

$p=\dfrac{1}{4}$

Parabola

Vertex: $(-2,-3)$

Focus: $\left(-2+\dfrac{1}{4},-3\right)=\left(-\dfrac{7}{4},-3\right)$

Directrix: $x=-\dfrac{9}{4}$

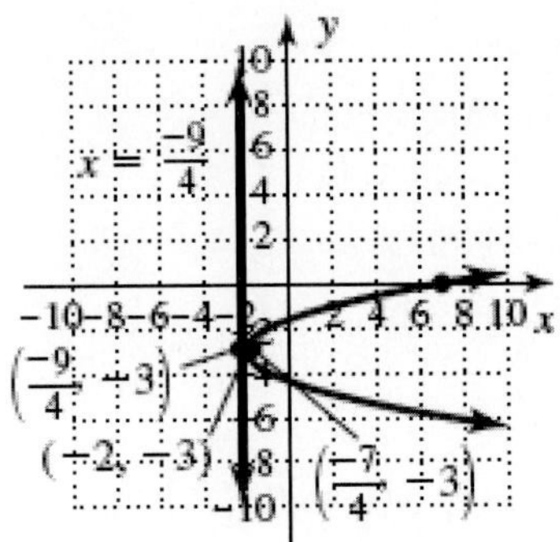

12. $y^2-6y-12x-15=0$

$y^2-6y=12x+15$

$y^2-6y+9=12x+15+9$

$(y-3)^2=12x+24$

$(y-3)^2=12(x+2)$

$4p=12$

$p=3$

Parabola

Vertex: $(-2,3)$

Focus: $(-2+3,3)=(1,3)$

Directrix: $x=-5$

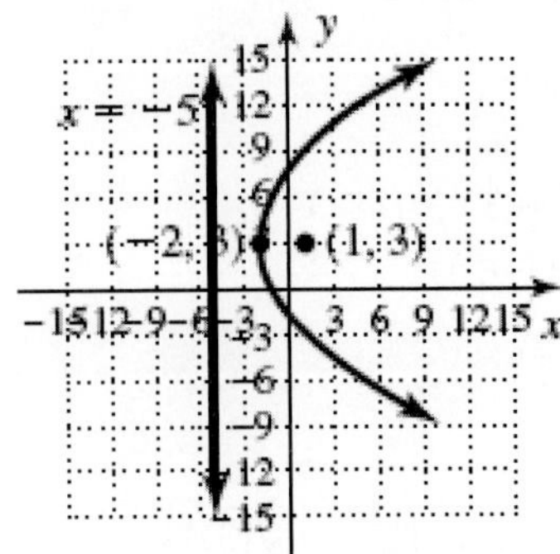

13. a. $\begin{cases}4x^2-y^2=16 & \text{Hyperbola}\\ y-x=2 & \text{Line}\end{cases}$

2nd equation: $y = x + 2$

$$4x^2 - (x+2)^2 = 16$$
$$4x^2 - x^2 - 4x - 4 = 16$$
$$3x^2 - 4x - 20 = 0$$
$$(3x-10)(x+2) = 0$$
$$x = \frac{10}{3}, x = -2;$$
$$y = \frac{10}{3} + 2 = \frac{16}{3};$$
$$y = -2 + 2 = 0$$

Solutions: $\left(\frac{10}{3}, \frac{16}{3}\right), (-2, 0)$

b. $\begin{cases} 4y^2 - x^2 = 4 \text{ Hyperbola} \\ x^2 + y^2 = 4 \text{ Circle} \end{cases}$

$$5y^2 = 8$$
$$y^2 = \frac{8}{5}$$
$$y = \pm\sqrt{\frac{8}{5}} = \pm\frac{2\sqrt{10}}{5};$$
$$4\left(\pm\frac{2\sqrt{10}}{5}\right)^2 - x^2 = 4$$
$$-x^2 = 4 - 4\left(\frac{8}{5}\right)$$
$$x^2 = \frac{12}{5}$$
$$x = \pm\frac{2\sqrt{15}}{5};$$

Solutions: $\left(\frac{2\sqrt{15}}{5}, \frac{2\sqrt{10}}{5}\right), \left(\frac{2\sqrt{15}}{5}, -\frac{2\sqrt{10}}{5}\right),$
$\left(-\frac{2\sqrt{15}}{5}, \frac{2\sqrt{10}}{5}\right), \left(-\frac{2\sqrt{15}}{5}, -\frac{2\sqrt{10}}{5}\right)$

14. $\begin{cases} x^2 + y^2 = 25^2 \\ x + y + 25 = 60 \end{cases}$

$\begin{cases} x^2 + y^2 = 25^2 \\ x + y = 35 \end{cases}$

2nd equation: $y = 35 - x$

$$x^2 + (35 - x)^2 = 625$$
$$x^2 + 1225 - 70x + x^2 = 625$$
$$2x^2 - 70x + 600 = 0$$
$$x^2 - 35x + 300 = 0$$
$$(x-15)(x-20) = 0$$
$$x = 15 \text{ or } x = 20$$

Length of one side = 15feet,

Length of other side= 20 feet

15. $(x+2)^2 + (y-5)^2 = r^2;$

$(0+2)^2 + (3-5)^2 = r^2$

$8 = r^2;$

$(x+2)^2 + (y-5)^2 = 8$

16. $a^2 = b^2 + c^2$

$4^2 = b^2 + 2^2$

$12 = b^2$

$\frac{x^2}{16} + \frac{y^2}{12} = 1$

17. $\frac{x^2}{(141.65)^2} + \frac{y^2}{(141.03)^2} = 1$

$a^2 = b^2 + c^2$

$(141.65)^2 = (141.03)^2 + c^2$

$175.2616 = c^2$

$13.23864041 = c;$

aphelion:

$a + c = 141.65 + 13.24 = 154.89$ million miles

perihelion:

$a - c = 141.65 - 13.24 = 128.41$ million miles

18. $y = (x-1)^2 - 4;$

$D: x \in (-\infty, \infty)$

$R: y \in [-4, \infty)$

Focus: $\left(1, -\frac{15}{4}\right)$

19. $(x-1)^2 + (y-1)^2 = 25$

Center: (1,1), Radius = 5

$D: x \in [-4, 6]$

$R: y \in [-4, 6]$

20. $\frac{(x+3)^2}{9} + \frac{y^2}{36} = 1$

$D: x \in [-6, 0]$

$R: y \in [-6, 6]$

$a^2 = b^2 + c^2$

$6^2 = 3^2 + c^2$

$27 = c^2$

$\sqrt{27} = c$

Foci: $(-3, -3\sqrt{3}), (-3, 3\sqrt{3})$

Chapter 7 Calculator Exploration

1. $e > 1$

2. $e = 1$

Chapter 7 Strengthening Core Skills

1. $100x^2 - 400x - 18y^2 - 108y + 230 = 0$

$$100\left(x^2 - 4x\right) - 18\left(y^2 + 6y\right) = -230$$

$$100\left(x^2 - 4x + 4\right) - 18\left(y^2 + 6y + 9\right) = -230 + 400 - 162$$

$$100(x-2)^2 - 18(y+3)^2 = 8$$

$$\frac{(x-2)^2}{\frac{2}{25}} - \frac{(y+3)^2}{\frac{4}{9}} = 1$$

$$\frac{(x-2)^2}{\left(\frac{\sqrt{2}}{5}\right)^2} - \frac{(y+3)^2}{\left(\frac{2}{3}\right)^2} = 1$$

$$a = \frac{\sqrt{2}}{5}, b = \frac{2}{3}$$

2. $28x^2 - 56x + 48y^2 + 192y + 195 = 0$

$$28\left(x^2 - 2x\right) + 48\left(y^2 + 4y\right) = -195$$

$$28\left(x^2 - 2x + 1\right) + 48\left(y^2 + 4y + 4\right) = -195 + 28 + 192$$

$$28(x-1)^2 + 48(y+2)^2 = 25$$

$$\frac{(x-1)^2}{\frac{25}{28}} + \frac{(y+2)^2}{\frac{25}{48}} = 1$$

$$\frac{(x-1)^2}{\left(\frac{5}{2\sqrt{7}}\right)^2} + \frac{(y+2)^2}{\left(\frac{5}{4\sqrt{3}}\right)^2} = 1$$

$$\frac{(x-1)^2}{\left(\frac{5\sqrt{7}}{14}\right)^2} + \frac{(y+2)^2}{\left(\frac{5\sqrt{3}}{12}\right)^2} = 1$$

$$a = \frac{5\sqrt{7}}{14}, b = \frac{5\sqrt{3}}{12}$$

3. $\frac{4(x+3)^2}{49} + \frac{25(y-1)^2}{36} = 1$

$$\frac{(x+3)^2}{\frac{49}{4}} + \frac{(y-1)^2}{\frac{36}{25}} = 1$$

$$\frac{(x+3)^2}{\left(\frac{7}{2}\right)^2} + \frac{(y-1)^2}{\left(\frac{6}{5}\right)^2} = 1$$

Center: (-3,1)

$$a = \frac{7}{2}, b = \frac{6}{5}$$

Vertices: $\left(-3 + \frac{7}{2}, 1\right), \left(-3 - \frac{7}{2}, 1\right)$

$$\left(\frac{1}{2}, 1\right), \left(-\frac{13}{2}, 1\right)$$

$$(0.5, 1), (-6.5, 1)$$

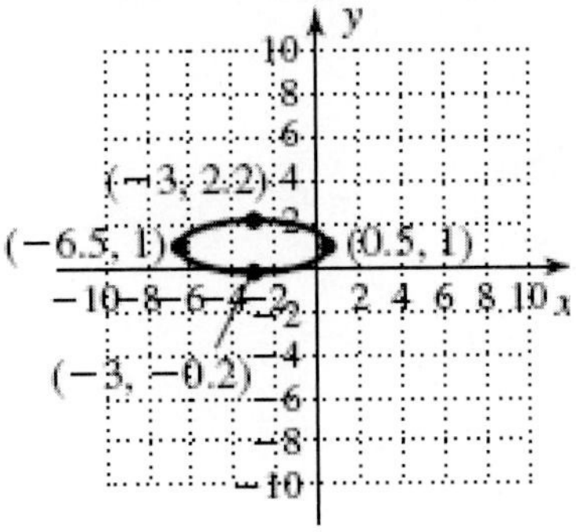

(−3, 2.2)
(−6.5, 1)
(0.5, 1)
(−3, −0.2)

4. $\frac{9(x+3)^2}{80} - \frac{4(y-1)^2}{81} = 1$

$$\frac{(x+3)^2}{\frac{80}{9}} - \frac{(y-1)^2}{\frac{81}{4}} = 1$$

$$\frac{(x+3)^2}{\left(\frac{4\sqrt{5}}{3}\right)^2} - \frac{(y-1)^2}{\left(\frac{9}{2}\right)^2} = 1$$

Center: (-3,1)

$$a = \frac{4\sqrt{5}}{3}, b = \frac{9}{2}$$

Vertices: $\left(-3 + \frac{4\sqrt{5}}{3}, 1\right), \left(-3 + \frac{4\sqrt{5}}{3}, 1\right)$

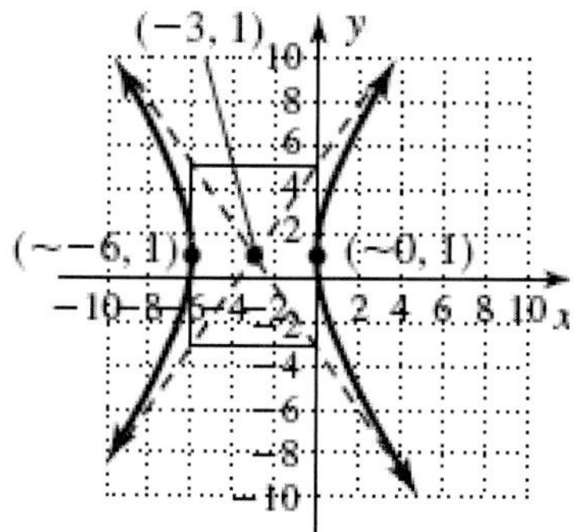

Cumulative Review: Chapters 1-7

1. $x^3 - 2x^2 + 4x - 8 = 0$
$x^2(x-2) + 4(x-2) = 0$
$(x-2)(x^2+4) = 0$
$x - 2 = 0$ or $x^2 + 4 = 0$
$x = 2$ or $x = \pm 2i$

2. $2|n+4| + 3 = 13$
$2|n+4| = 10$
$|n+4| = 5$
$n + 4 = 5$ or $n + 4 = -5$
$n = 1$ or $n = -9$

3. $\sqrt{x+2} - 2 = \sqrt{3x+4}$
$(\sqrt{x+2} - 2)^2 = (\sqrt{3x+4})^2$
$x + 2 - 4\sqrt{x+2} + 4 = 3x + 4$
$-4\sqrt{x+2} = 2x - 2$
$(-4\sqrt{x+2})^2 = (2x-2)^2$
$16(x+2) = 4x^2 - 8x + 4$
$16x + 32 = 4x^2 - 8x + 4$
$0 = 4x^2 - 24x - 28$
$0 = 4(x^2 - 6x - 7)$
$(x-7)(x+1) = 0$
$x - 7 = 0$ or $x + 1 = 0$
$x = 7$ or $x = -1$
Both roots are extraneous. Thus, no solution.

4. $x^{\frac{3}{2}} + 8 = 0$
$x^{\frac{3}{2}} = -8$
No solution

5. $x^2 - 6x + 13 = 0$
$a = 1, b = -6, c = 13$
$$x = \frac{-(-6) \pm \sqrt{(-6)^2 - 4(1)(13)}}{2(1)}$$
$$x = \frac{6 \pm \sqrt{-16}}{2} = \frac{6 \pm 4i}{2} = 3 \pm 2i$$

6. $4 \cdot 2^{x+1} = \frac{1}{8}$
$2^2 \cdot 2^{x+1} = 2^{-3}$
$2^{x+3} = 2^{-3}$
$x + 3 = -3$
$x = -6$

7. $3^{x-2} = 7$
$\ln 3^{x-2} = \ln 7$
$(x-2)\ln 3 = \ln 7$
$x - 2 = \frac{\ln 7}{\ln 3}$
$x = 2 + \frac{\ln 7}{\ln 3}$

8. $\log_3 81 = x$
$3^x = 81$
$3^x = 3^4$
$x = 4$

9. $\log_3 x + \log_3(x-2) = 1$
$\log_3 x(x-2) = 1$
$3^1 = x(x-2)$
$3 = x^2 - 2x$
$0 = x^2 - 2x - 3$
$(x-3)(x+1) = 0$
$x - 3 = 0$ or $x + 1 = 0$
$x = 3$ or $x = -1$
$x = 3, x = -1$ is extraneous.

10. $y = \frac{2}{3}x + 2$
slope = $\frac{2}{3}$, y-intercept: (0,2)

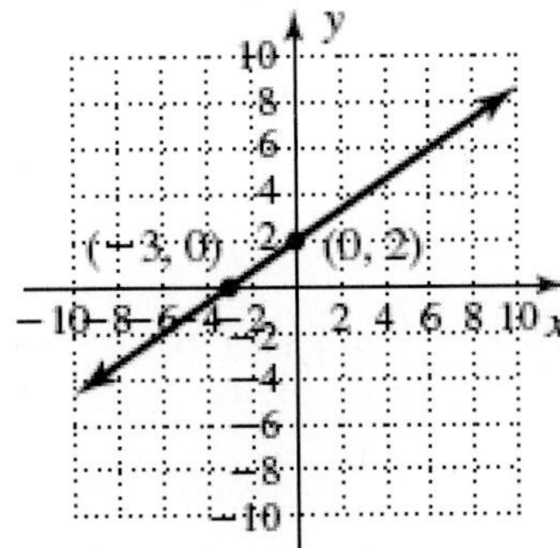

11. $y = |x-2| + 3$

Vertex: (2,3)

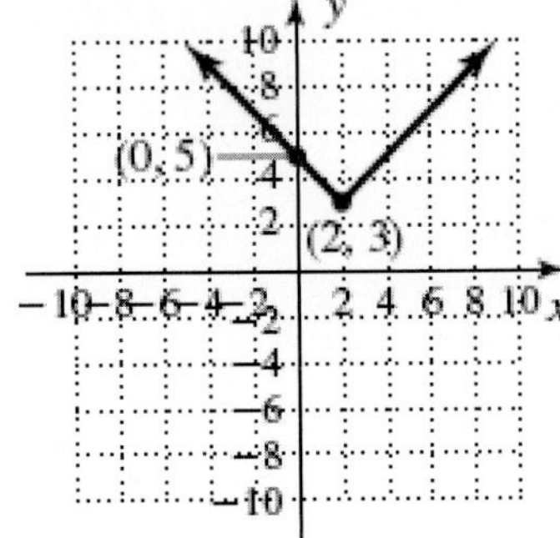

12. $y = \dfrac{1}{x-1} + 2$

$y = \dfrac{1+2x-2}{x-1}$

$y = \dfrac{2x-1}{x-1}$

Horizontal asymptote: $y = 2$

(deg num = deg den)

Vertical asymptote: $x = 1$

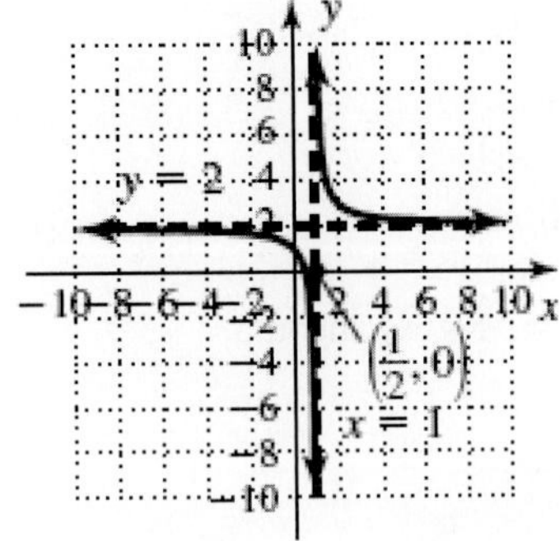

13. $y = \sqrt{x-3} + 1$

Node: (3,1)

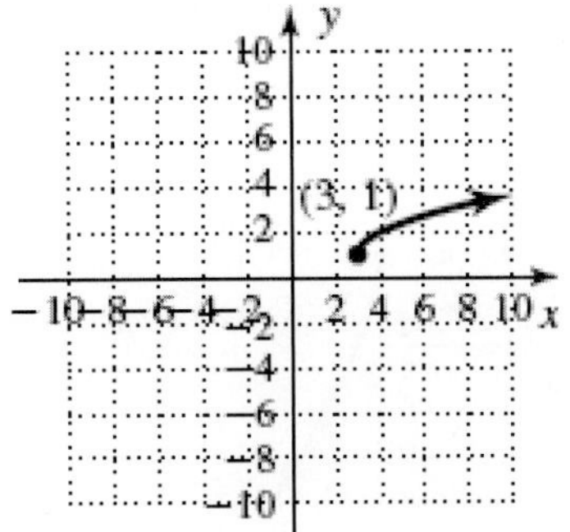

14. a. $g(x) = (x-3)(x+1)(x+4)$

Down/Up

$0 = (x-3)(x+1)(x+4)$

x-intercepts: $(-4,0), (-1,0), (3,0)$;

$g(0) = (0-3)(0+1)(0+4) = -12$

y-intercept: (0,-12)

local maximum: (-2.7,12.60)

local minimum: (1.36,-20.75)

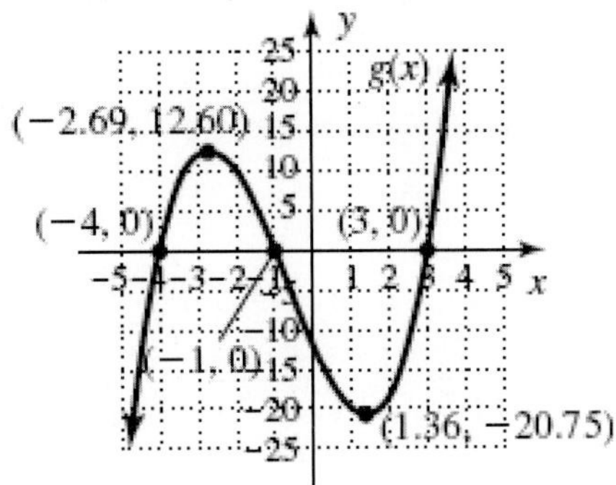

b. $f(x) = x^4 + x^3 - 13x^2 - x + 12$

Possible rational roots:

$\dfrac{\pm1, \pm12, \pm2, \pm6, \pm3, \pm4}{\pm1}$;

$\{\pm1, \pm12, \pm2, \pm6, \pm3, \pm4\}$

$f(x) = (x+4)(x+1)(x-1)(x-3)$

Up/Up

$0 = (x+4)(x+1)(x-1)(x-3)$

x-intercepts: $(-4,0), (-1,0), (1,0), (3,0)$;

$f(0) = 0^4 + 0^3 - 13(0)^2 - 0 + 12 = 12$

y-intercept: (0,12)

local minimums:

(-2.94,-48.13),(2.22,-19.06)

local maximum:

(-0.04,12.02)

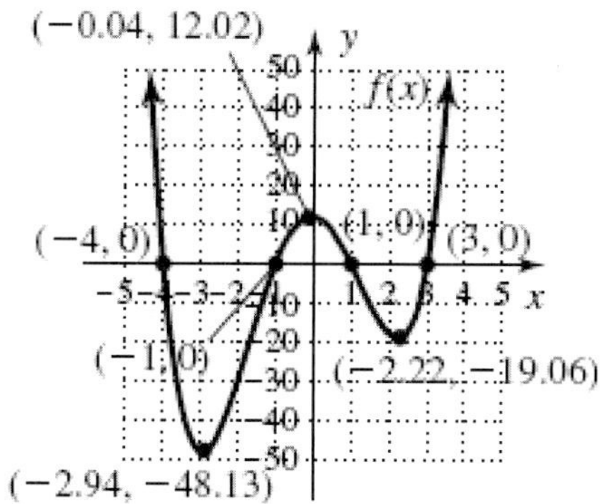

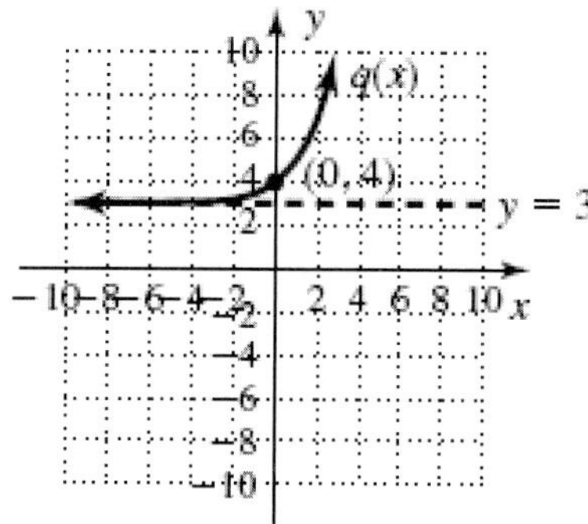

15. $h(x)=\dfrac{x-2}{x^2-9}$

Horizontal asymptote: $y = 0$
(deg num < deg den)

$h(x)=\dfrac{x-2}{(x+3)(x-3)}$

Vertical asymptotes: $x = 3$ or $x = -3$;

$h(0)=\dfrac{0-2}{0^2-9}=\dfrac{2}{9}\approx 0.22$

y-intercept: (0,0.22);

$0=\dfrac{x-2}{x^2-9}$

x-intercept: (2,0);

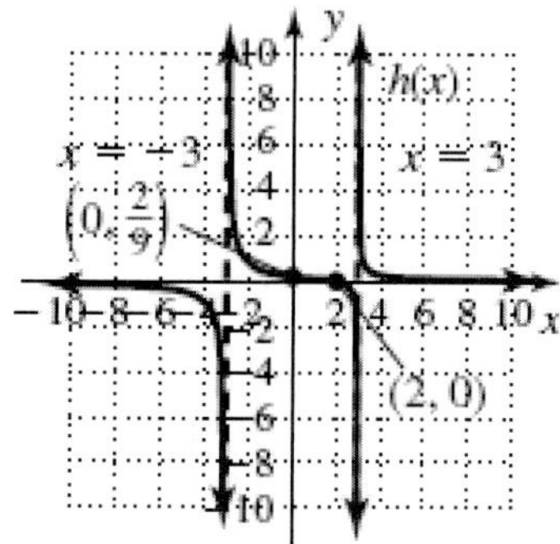

16. $y=2^x+3$

$y=2^0+3=1+3=4$

y-intercept: (0,4);
Horizontal asymptote: $y=3$

17. $f(x)=\log_2(x+1)$

$f(0)=\log_2(0+1)=0$

y-intercept: (0,0);
$x+1=0$
Vertical asymptote: $x=-1$

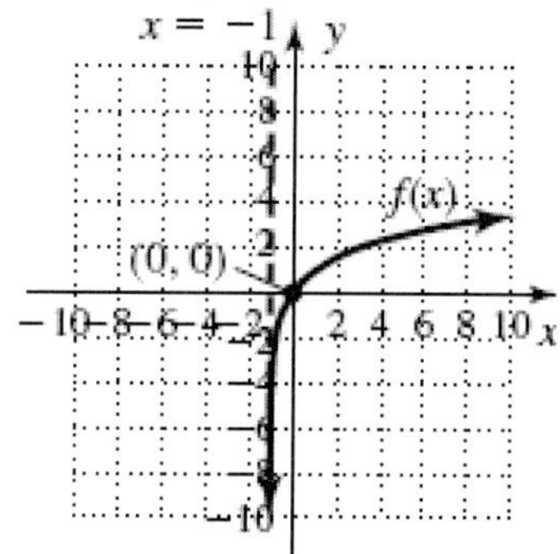

18. $x=y^2+4y+7$

$x-7+4=(y^2+4y+4)$

$x-3=(y+2)^2$

$4p=1$

$p=\dfrac{1}{4}$

Parabola
Vertex: $(3,-2)$

$x=(0)^2+4(0)+7=7$

x-intercept: (7,0);

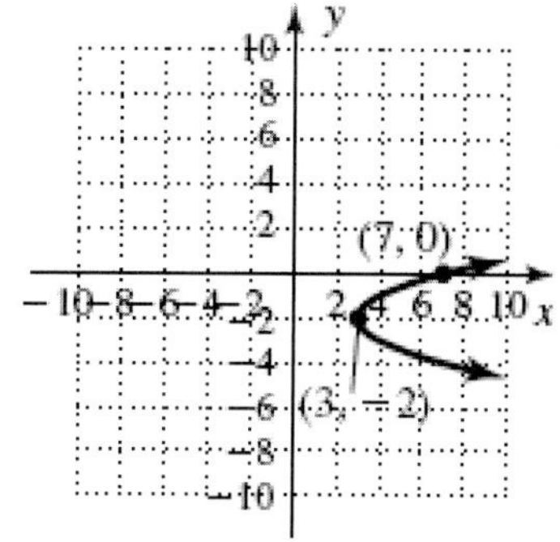

19. $x^2+y^2+10x-4y+20=0$

$x^2+10x+y^2-4y=-20$

$(x^2+10x+25)+(y^2-4y+4)=-20+25+4$

$(x+5)^2+(y-2)^2=9$

Center : (–5,2), Radius = 3

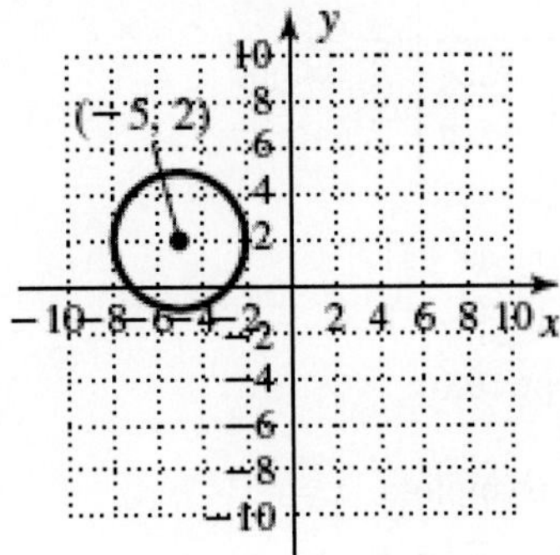

20. $4(x-1)^2-36(y+2)^2=144$

$\frac{(x-1)^2}{36}-\frac{(y+2)^2}{4}=1$

Hyperbola

Center: $(1,-2)$

$a=6,\ b=2$

Vertices: $(1-6,-2),(1+6,-2)$

$(-5,-2),(7,-2)$;

$a^2+b^2=c^2$

$36+4=c^2$

$2\sqrt{10}=c$

Foci: $\left(1-2\sqrt{10},-2\right),\left(1+2\sqrt{10},-2\right)$

CA: $(1,-2+2),(1,-2-2)$

$(1,0),(1,-4)$

Asymptotes: $y+2=\pm\frac{1}{3}(x-1)$

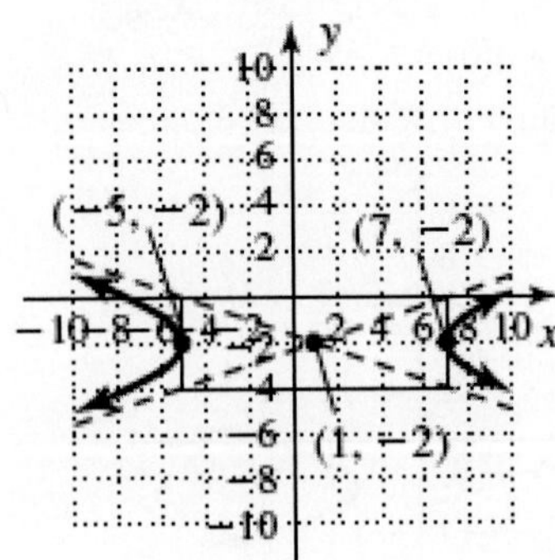

21. (a) $x\in(-\infty,\infty)$

(b) $y\in(-\infty,4]$

(c) $f(x)\uparrow(-\infty,-1), f(x)\downarrow(-1,\infty)$

(d) NA

(e) $\max(-1,4)$

(f) $f(x)>0: x\in(-4,2)$

(g) $f(x)<0: x\in(-\infty,-4)\cup(2,\infty)$

22. $\begin{cases}4x+3y=13\\-9y+5z=19\\x-4z=-4\end{cases}$

eq.1 $12x+9y=39$

eq.2 $\underline{-9y+5z=19}$

$12x+5z=58$

eq.3 $\underline{-12x+48z=48}$

53z=106

z=2;

$x-4(2)=-4$

$x=4$;

$4(4)+3y=13$

$y=-1$;

Solution: (4,-1,2)

23. $\begin{cases}x^2+y^2=25\\64x^2+12y^2=768\end{cases}$

$\begin{cases}-12x^2-12y^2=-300\\64x^2+12y^2=768\end{cases}$

$52x^2=468$

$x^2=9$

$x=\pm3$;

$(\pm3)^2+y^2=25$

$y^2=16$

$y=\pm4$;

$(3,4),(3,-4),(-3,4),(-3,-4)$

24. $12000=5000\left(1+\frac{0.09}{4}\right)^{4t}$

$\ln\frac{12}{5}=\ln\left(1+\frac{0.09}{4}\right)^{4t}$

$\ln\frac{12}{5}=4t\ln\left(1+\frac{0.09}{4}\right)$

$\frac{\ln\frac{12}{5}}{4\ln\left(1+\frac{0.09}{4}\right)}=t$

$t\approx 9.8$ years

25. Let x represent the number of liters of pure antifreeze.
Let y represent the number of liters of 40% antifreeze drained off.

$$\begin{cases} y = x \\ 0.40(10) - 0.40x + y = 0.60(10) \end{cases}$$

$$4 - 0.4x + x = 6$$
$$0.6x = 2$$
$$x = 3\frac{1}{3} \text{ liters}$$

8.1 Technology Highlight

Exercise 1: $a_n = \frac{1}{3^n}$

First 10 terms:
$\frac{1}{3}, \frac{1}{9}, \frac{1}{27}, \frac{1}{81}, \frac{1}{243}, \frac{1}{729}, \frac{1}{2187},$
$\frac{1}{6561}, \frac{1}{19683}, \frac{1}{59049}$
Sum of first 10 terms: sum $\to 0.5$

Exercise 2: $a_n = \frac{2}{n(n+1)}$

First 10 terms:
$1, \frac{1}{3}, \frac{1}{6}, \frac{1}{10}, \frac{1}{15}, \frac{1}{21}, \frac{1}{28}, \frac{1}{36}, \frac{1}{45}, \frac{1}{55}$
Sum of first 10 terms: sum $\to 2$

Exercise 3: $a_n = \frac{1}{(2n-1)(2n+1)}$

First 10 terms:
$\frac{1}{3}, \frac{1}{15}, \frac{1}{35}, \frac{1}{63}, \frac{1}{99}, \frac{1}{143}, \frac{1}{195}, \frac{1}{255}, \frac{1}{323}, \frac{1}{399}$
Sum of first 10 terms: sum $\to 0.5$

8.1 Exercises

1. Pattern; order

3. Increasing

5. Formula defining the sequence uses the preceding term. Answers will vary.

7. $a_n = 2n-1$
$a_1 = 2(1)-1 = 2-1 = 1$;
$a_2 = 2(2)-1 = 4-1 = 3$;
$a_3 = 2(3)-1 = 6-1 = 5$;
$a_4 = 2(4)-1 = 8-1 = 7$;
$a_8 = 2(8)-1 = 16-1 = 15$;
$a_{12} = 2(12)-1 = 24-1 = 23$;
$1, 3, 5, 7; a_8 = 15; a_{12} = 23$

9. $a_n = 3n^2 - 3$
$a_1 = 3(1)^2 - 3 = 3(1) - 3 = 3 - 3 = 0$;
$a_2 = 3(2)^2 - 3 = 3(4) - 3 = 12 - 3 = 9$;
$a_3 = 3(3)^2 - 3 = 3(9) - 3 = 27 - 3 = 24$;
$a_4 = 3(4)^2 - 3 = 3(16) - 3 = 48 - 3 = 45$;
$a_8 = 3(8)^2 - 3 = 3(64) - 3 = 192 - 3 = 189$;
$a_{12} = 3(12)^2 - 3 = 3(144) - 3 = 432 - 3 = 429$;
$0, 9, 24, 45;\ a_8 = 189;\ a_{12} = 429$

11. $a_n = (-1)^n n$
$a_1 = (-1)^1(1) = -1(1) = -1$;
$a_2 = (-1)^2(2) = 1(2) = 2$;
$a_3 = (-1)^3(3) = -1(3) = -3$;
$a_4 = (-1)^4(4) = 1(4) = 4$;
$a_8 = (-1)^8(8) = 1(8) = 8$;
$a_{12} = (-1)^{12}(12) = 1(12) = 12$;
-1, 2, -3, 4; $a_8 = 8$; $a_{12} = 12$

13. $a_n = \frac{n}{n+1}$
$a_1 = \frac{1}{1+1} = \frac{1}{2}$;
$a_2 = \frac{2}{2+1} = \frac{2}{3}$;
$a_3 = \frac{3}{3+1} = \frac{3}{4}$;
$a_4 = \frac{4}{4+1} = \frac{4}{5}$;
$a_8 = \frac{8}{8+1} = \frac{8}{9}$;
$a_{12} = \frac{12}{12+1} = \frac{12}{13}$;
$\frac{1}{2}, \frac{2}{3}, \frac{3}{4}, \frac{4}{5}; a_8 = \frac{8}{9}; a_{12} = \frac{12}{13}$

15. $a_n = \left(\frac{1}{2}\right)^n$
$a_1 = \left(\frac{1}{2}\right)^1 = \frac{1}{2}$;
$a_2 = \left(\frac{1}{2}\right)^2 = \frac{1}{4}$;
$a_3 = \left(\frac{1}{2}\right)^3 = \frac{1}{8}$;
$a_4 = \left(\frac{1}{2}\right)^4 = \frac{1}{16}$;

$a_8 = \left(\frac{1}{2}\right)^8 = \frac{1}{256}$;

$a_{12} = \left(\frac{1}{2}\right)^{12} = \frac{1}{4096}$;

$\frac{1}{2}, \frac{1}{4}, \frac{1}{8}, \frac{1}{16}; a_8 = \frac{1}{256}; a_{12} = \frac{1}{4096}$

17. $a_n = \frac{1}{n}$

$a_1 = \frac{1}{1} = 1$;

$a_2 = \frac{1}{2}$;

$a_3 = \frac{1}{3}$;

$a_4 = \frac{1}{4}$;

$a_8 = \frac{1}{8}$;

$a_{12} = \frac{1}{12}$;

$1, \frac{1}{2}, \frac{1}{3}, \frac{1}{4}; a_8 = \frac{1}{8}; a_{12} = \frac{1}{12}$

19. $a_n = \frac{(-1)^n}{n(n+1)}$

$a_1 = \frac{(-1)^1}{1(1+1)} = \frac{-1}{2}$;

$a_2 = \frac{(-1)^2}{2(2+1)} = \frac{1}{2(3)} = \frac{1}{6}$;

$a_3 = \frac{(-1)^3}{3(3+1)} = \frac{-1}{3(4)} = \frac{-1}{12}$;

$a_4 = \frac{(-1)^4}{4(4+1)} = \frac{1}{4(5)} = \frac{1}{20}$;

$a_8 = \frac{(-1)^8}{8(8+1)} = \frac{1}{8(9)} = \frac{1}{72}$;

$a_{12} = \frac{(-1)^{12}}{12(12+1)} = \frac{1}{12(13)} = \frac{1}{156}$;

$\frac{-1}{2}, \frac{1}{6}, \frac{-1}{12}, \frac{1}{20}; a_8 = \frac{1}{72}; a_{12} = \frac{1}{156}$

21. $a_n = (-1)^n 2^n$

$a_1 = (-1)^1 2^1 = -1(2) = -2$;

$a_2 = (-1)^2 2^2 = 1(4) = 4$;

$a_3 = (-1)^3 2^3 = -1(8) = -8$;

$a_4 = (-1)^4 2^4 = 1(16) = 16$;

$a_8 = (-1)^8 2^8 = 1(256) = 256$;

$a_{12} = (-1)^{12} 2^{12} = 1(4096) = 4096$;

-2, 4, -8, 16; $a_8 = 256$; $a_{12} = 4096$

23. $a_n = n^2 - 2$

$a_9 = (9)^2 - 2 = 81 - 2 = 79$

25. $a_n = \frac{(-1)^{n+1}}{n}$

$a_5 = \frac{(-1)^{5+1}}{5} = \frac{(-1)^6}{5} = \frac{1}{5}$

27. $a_n = 2\left(\frac{1}{2}\right)^{n-1}$

$a_7 = 2\left(\frac{1}{2}\right)^{7-1} = 2\left(\frac{1}{2}\right)^6 = 2\left(\frac{1}{64}\right) = \frac{1}{32}$

29. $a_n = \left(1 + \frac{1}{n}\right)^n$

$a_{10} = \left(1 + \frac{1}{10}\right)^{10} = \left(\frac{11}{10}\right)^{10}$

31. $a_n = \frac{1}{(n)(2n+1)}$

$a_4 = \frac{1}{(4)[2(4)+1]} = \frac{1}{4(9)} = \frac{1}{36}$

33. $\begin{cases} a_1 = 2 \\ a_n = 5a_{n-1} - 3 \end{cases}$

$a_2 = 5a_{2-1} - 3 = 5a_1 - 3$
$= 5(2) - 3 = 10 - 3 = 7$;
$a_3 = 5a_{3-1} - 3 = 5a_2 - 3$
$= 5(7) - 3 = 35 - 3 = 32$;
$a_4 = 5a_{4-1} - 3 = 5a_3 - 3$
$= 5(32) - 3 = 160 - 3 = 157$;
$a_5 = 5a_{5-1} - 3 = 5a_4 - 3$
$= 5(157) - 3 = 785 - 3 = 782$;
2, 7, 32, 157, 782

35. $\begin{cases} a_1 = -1 \\ a_n = (a_{n-1})^2 + 3 \end{cases}$

$a_2 = (a_{2-1})^2 + 3 = (a_1)^2 + 3$
$= (-1)^2 + 3 = 1 + 3 = 4;$
$a_3 = (a_{3-1})^2 + 3 = (a_2)^2 + 3$
$= (4)^2 + 3 = 16 + 3 = 19;$
$a_4 = (a_{4-1})^2 + 3 = (a_3)^2 + 3$
$= (19)^2 + 3 = 361 + 3 = 364;$
$a_5 = (a_{5-1})^2 + 3 = (a_4)^2 + 3$
$= (364)^2 + 3 = 132496 + 3 = 132499;$
$-1, 4, 19, 364, 132499$

37. $\begin{cases} c_1 = 64, c_2 = 32 \\ c_n = \dfrac{c_{n-2} - c_{n-1}}{2} \end{cases}$

$c_3 = \dfrac{c_{3-2} - c_{3-1}}{2} = \dfrac{c_1 - c_2}{2}$
$= \dfrac{64 - 32}{2} = \dfrac{32}{2} = 16;$
$c_4 = \dfrac{c_{4-2} - c_{4-1}}{2} = \dfrac{c_2 - c_3}{2}$
$= \dfrac{32 - 16}{2} = \dfrac{16}{2} = 8;$
$c_5 = \dfrac{c_{5-2} - c_{5-1}}{2} = \dfrac{c_3 - c_4}{2}$
$= \dfrac{16 - 8}{2} = \dfrac{8}{2} = 4;$
$64, 32, 16, 8, 4$

39. $\dfrac{8!}{5!} = \dfrac{8 \cdot 7 \cdot 6 \cdot 5!}{5!} = 8 \cdot 7 \cdot 6 = 336$

41. $\dfrac{9!}{7!2!} = \dfrac{9 \cdot 8 \cdot 7!}{7!2!} = \dfrac{9 \cdot 8}{2 \cdot 1} = \dfrac{72}{2} = 36$

43. $\dfrac{8!}{2!6!} = \dfrac{8 \cdot 7 \cdot 6!}{2!6!} = \dfrac{8 \cdot 7}{2 \cdot 1} = \dfrac{56}{2} = 28$

45. $a_n = \dfrac{n!}{(n+1)!}$

$a_1 = \dfrac{1!}{(1+1)!} = \dfrac{1!}{2!} = \dfrac{1!}{2 \cdot 1!} = \dfrac{1}{2}$
$a_2 = \dfrac{2!}{(2+1)!} = \dfrac{2!}{3!} = \dfrac{2!}{3 \cdot 2!} = \dfrac{1}{3}$
$a_3 = \dfrac{3!}{(3+1)!} = \dfrac{3!}{4!} = \dfrac{3!}{4 \cdot 3!} = \dfrac{1}{4}$
$a_4 = \dfrac{4!}{(4+1)!} = \dfrac{4!}{5!} = \dfrac{4!}{5 \cdot 4!} = \dfrac{1}{5}$
$\dfrac{1}{2}, \dfrac{1}{3}, \dfrac{1}{4}, \dfrac{1}{5}$

47. $a_n = \dfrac{(n+1)!}{(3n)!}$

$a_1 = \dfrac{(1+1)!}{(3 \cdot 1)!} = \dfrac{2!}{3!} = \dfrac{2!}{3 \cdot 2!} = \dfrac{1}{3}$
$a_2 = \dfrac{(2+1)!}{(3 \cdot 2)!} = \dfrac{3!}{6!} = \dfrac{3!}{6 \cdot 5 \cdot 4 \cdot 3!} = \dfrac{1}{120}$
$a_3 = \dfrac{(3+1)!}{(3 \cdot 3)!} = \dfrac{4!}{9!} = \dfrac{4!}{9 \cdot 8 \cdot 7 \cdot 6 \cdot 5 \cdot 4!} = \dfrac{1}{15120}$
$a_4 = \dfrac{(4+1)!}{(3 \cdot 4)!} = \dfrac{5!}{12!}$
$= \dfrac{5!}{12 \cdot 11 \cdot 10 \cdot 9 \cdot 8 \cdot 7 \cdot 6 \cdot 5!} = \dfrac{1}{3991680}$
$\dfrac{1}{3}, \dfrac{1}{120}, \dfrac{1}{15120}, \dfrac{1}{3991680}$

49. $a_n = \dfrac{n^n}{n!}$

$a_1 = \dfrac{1^1}{1!} = \dfrac{1}{1} = 1$
$a_2 = \dfrac{2^2}{2!} = \dfrac{4}{2!} = \dfrac{4}{2 \cdot 1} = \dfrac{4}{2} = 2$
$a_3 = \dfrac{3^3}{3!} = \dfrac{27}{3 \cdot 2 \cdot 1} = \dfrac{27}{6} = \dfrac{9}{2}$
$a_4 = \dfrac{4^4}{4!} = \dfrac{256}{4 \cdot 3 \cdot 2 \cdot 1} = \dfrac{256}{24} = \dfrac{32}{3}$
$1, 2, \dfrac{9}{2}, \dfrac{32}{3}$

51. $2, 4, 8, 16, 32, \ldots$
$2^6 = 64, a_n = 2^n$

53. $1, 4, 9, 16, 25, \ldots$
$6^2 = 36, a_n = n^2$

55. $-3, 4, 11, 18, 25, \ldots$
$7(6) - 10 = 32, a_n = 7n - 10$

57. $\frac{1}{2},\frac{1}{4},\frac{1}{8},\frac{1}{16},\frac{1}{32},...$

$\frac{1}{2^6}=\frac{1}{64}, a_n=\frac{1}{2^n}$

59. $a_n=n$

$S_5=a_1+a_2+a_3+a_4+a_5$

$S_5=1+2+3+4+5$

$S_5=15$

61. $a_n=2n-1$

$S_8=a_1+a_2+a_3+a_4+a_5+a_6+a_7+a_8$

$S_8=1+3+5+7+9+11+13+15$

$S_8=64$

63. $a_n=\frac{1}{n}$

$S_5=a_1+a_2+a_3+a_4+a_5$

$S_5=1+\frac{1}{2}+\frac{1}{3}+\frac{1}{4}+\frac{1}{5}$

$S_5=\frac{137}{60}$

65. $\sum_{i=1}^{4}(3i-5)$

$=(3(1)-5)+(3(2)-5)+(3(3)-5)+(3(4)-5)$

$=-2+1+4+7=10$

67. $\sum_{k=1}^{5}(2k^2-3)$

$=(2(1)^2-3)+(2(2)^2-3)+(2(3)^2-3)+$

$(2(4)^2-3)+(2(5)^2-3)$

$=(2-3)+(8-3)+(18-3)+(32-3)+(50-3)$

$=-1+5+15+29+47=95$

69. $\sum_{k=1}^{7}(-1)^k k$

$=(-1)^1(1)+(-1)^2(2)+(-1)^3(3)+(-1)^4(4)+$

$(-1)^5(5)+(-1)^6(6)+(-1)^7(7)$

$=-1+2-3+4-5+6-7=-4$

71. $\sum_{i=1}^{4}\frac{i^2}{2}$

$=\frac{1^2}{2}+\frac{2^2}{2}+\frac{3^2}{2}+\frac{4^2}{2}$

$=\frac{1}{2}+\frac{4}{2}+\frac{9}{2}+\frac{16}{2}=15$

73. $\sum_{j=3}^{7}2j$

$=2(3)+2(4)+2(5)+2(6)+2(7)$

$=6+8+10+12+14=50$

75. $\sum_{k=3}^{8}\frac{(-1)^k}{k(k-2)}$

$=\frac{(-1)^3}{3(3-2)}+\frac{(-1)^4}{4(4-2)}+\frac{(-1)^5}{5(5-2)}+\frac{(-1)^6}{6(6-2)}+$

$\frac{(-1)^7}{7(7-2)}+\frac{(-1)^8}{8(8-2)}$

$=\frac{-1}{3}+\frac{1}{8}-\frac{1}{15}+\frac{1}{24}-\frac{1}{35}+\frac{1}{48}$

$=\frac{-27}{112}$

77. $4+8+12+16+20$

$\sum_{n=1}^{5}(4n)$

79. $-1+4-9+16-25+36$

$\sum_{n=1}^{6}(-1)^n n^2$

81. $a^n=n+3; S_5$

$\sum_{n=1}^{5}(n+3)$

83. $a^n=\frac{n^2}{3}$; third partial sum

$\sum_{n=1}^{3}\frac{n^2}{3}$

85. $a^n=\frac{n}{2^n}$; sum for $n=3$ to 7

$\sum_{n=3}^{7}\frac{n}{2^n}$

87. $\frac{1}{20}+\frac{1}{10}+\frac{3}{20}+\frac{1}{5}+\frac{1}{4}$

$=\frac{1}{20}(1+2+3+4+5)$

$\frac{1}{20}\sum_{n=1}^{5} n$

89. $\frac{1}{8}+\frac{4}{27}+\frac{9}{64}+\frac{16}{125}+\ldots$

$\sum_{n=1}^{\infty}\frac{n^2}{(n+1)^3}$

91. $a_n = 3n-2 : S_n = \frac{n(3n-1)}{2}$

$a_n = 3n-2 = 1,4,7,10\ldots,(3n-2),\ldots$

$S_5 = \frac{5(3(5)-1)}{2} = \frac{5(14)}{2} = \frac{70}{2} = 35$;

$a_1 = 3(1)-2 = 1$;

$a_2 = 3(2)-2 = 4$;

$a_3 = 3(3)-2 = 7$;

$a_4 = 3(4)-2 = 10$;

$a_5 = 3(5)-2 = 13$;

$1+4+7+10+13 = 35$

93. $a_n = (0.8)^{n-1}(6000)$

$a_1 = (0.8)^{1-1}(6000) = 6000$;

$a2 = (0.8)^{2-1}(6000) = 0.8(6000) = 4800$;

$a_3 = (0.8)^{3-1}(6000) = (0.8)^2(6000) = 3840$;

$a_4 = (0.8)^{4-1}(6000) = (0.8)^3(6000) = 3072$;

$a_5 = (0.8)^{5-1}(6000) = (0.8)^4(6000) = 2457.60$;

\$6000; \$4800; \$3840; \$3072; \$2457.60

95. 5.20, 5.70, 6.20, 6.70, 7.20

$8(7.20)(240) = \$13{,}824$

97. $b_0 = 1500;\ b_n = 1.05b_{n-1} + 100$

$b_1 = 1.05b_{1-1} + 100 = 1.05(1500) + 100 = 1675$;

$b_2 = 1.05b_{2-1} + 100 = 1.05b_1 + 100$
$= 1.05(1675) + 100 = 1858.75$;

$b_3 = 1.05b_{3-1} + 100 = 1.05b_2 + 100$
$= 1.05(1858.75) + 100 = 2051.69$;

$b_4 = 1.05b_{4-1} + 100 = 1.05b_3 + 100$
$= 1.05(2051.69) + 100 = 2254.27$;

$b_5 = 1.05b_{5-1} + 100 = 1.05b_4 + 100$
$= 1.05(2254.27) + 100 = 2466.98$;

$b_6 = 1.05b_{6-1} + 100 = 1.05b_6 + 100$
$= 1.05(2466.98) + 100 = 2690.33$

Approximately 2690

99. $a_n = \begin{cases} \frac{a_{n-1}}{2} & \text{if } a_{n-1} \text{ is even} \\ 3(a_{n-1})+1 & \text{if } a_{n-1} \text{ is odd} \end{cases}$

$a_0 = 5;\ a_0 = 12;\ a_0 = 11$

For $a_0 = 5$

$a_1 = 3(5)+1 = 16$;

$a_2 = \frac{16}{2} = 8$;

$a_3 = \frac{8}{2} = 4$;

$a_4 = \frac{4}{2} = 2$;

$a_5 = \frac{2}{2} = 1$

For $a_0 = 12$

$a_1 = \frac{12}{2} = 6$;

$a_2 = \frac{6}{2} = 3$;

$a_3 = 3(3)+1 = 10$;

$a_4 = \frac{10}{2} = 5$;

$a_5 = 3(5)+1 = 16$;

$a_6 = \frac{16}{2} = 8$;

$a_7 = \frac{8}{2} = 4$;

$a_8 = \frac{4}{2} = 2$;

$a_9 = \frac{2}{2} = 1$

For $a_0 = 11$

$a_1 = 3(11)+1 = 34$;

$a_2 = \frac{34}{2} = 17$;

$a_3 = 3(17)+1 = 52$;

$a_4 = \frac{52}{2} = 26$;

$a_5 = \frac{26}{2} = 13$;

$a_6 = 3(13)+1 = 40$;

$a_7 = \frac{40}{2} = 20$;

$a_8 = \frac{20}{2} = 10$;

$a_9 = \frac{10}{2} = 5$

$a_{10} = 3(5)+1 = 16$;

$a_{11} = \frac{16}{2} = 8$;

$a_{12} = \frac{8}{2} = 4$;

$a_{13} = \frac{4}{2} = 2$;

$a_{14} = \frac{2}{2} = 1$

Answers will vary.

101. $\sum_{k=0}^{n} \frac{1}{k!}$

$n = 4$

$\sum_{k=0}^{4} \frac{1}{k!}$

$S_4 = a_0 + a_1 + a_2 + a_3 + a_4$

$= 1 + 1 + \frac{1}{2} + \frac{1}{6} + \frac{1}{24} = \frac{65}{24}$;

$\sum_{k=0}^{8} \frac{1}{k!}$

$S_8 = S_4 + a_5 + a_6 + a_7 + a_8$

$= \frac{65}{24} + \frac{1}{120} + \frac{1}{720} + \frac{1}{5040} + \frac{1}{40320}$

≈ 2.718;

$\sum_{k=0}^{12} \frac{1}{k!}$

$S_{12} = S_8 + a_9 + a_{10} + a_{11} + a_{12}$

$= 2.718 + \frac{1}{362880} + \frac{1}{3268800} + \frac{1}{39916800}$

$+ \frac{1}{479001600}$

≈ 2.718;

Approaches e

103. $\sum_{k=1}^{n} \frac{1}{2^k}$

$\sum_{k=1}^{4} \frac{1}{2^k}$

$S_4 = a_1 + a_2 + a_3 + a_4$

$= \frac{1}{2} + \frac{1}{4} + \frac{1}{8} + \frac{1}{16} = \frac{15}{16}$;

$\sum_{k=1}^{8} \frac{1}{2^k}$

$S_8 = S_4 + a_5 + a_6 + a_7 + a_8$

$= \frac{15}{16} + \frac{1}{32} + \frac{1}{64} + \frac{1}{128} + \frac{1}{256} = \frac{255}{256}$;

$\sum_{k=1}^{12} \frac{1}{2^k}$

$S_{12} = S_8 + a_9 + a_{10} + a_{11} + a_{12}$

$= \frac{255}{256} + \frac{1}{512} + \frac{1}{1024} + \frac{1}{2048} + \frac{1}{4096}$

≈ 0.9998

Approaches 1

105. $f(x) = \sqrt{x}$

$\frac{\sqrt{x+h} - \sqrt{x}}{h}$

$\frac{\sqrt{x+h} - \sqrt{x}}{h} \cdot \frac{\sqrt{x+h} + \sqrt{x}}{\sqrt{x+h} + \sqrt{x}}$

$= \frac{x+h-x}{h(\sqrt{x+h} + \sqrt{x})}$

$= \frac{h}{h(\sqrt{x+h} + \sqrt{x})}$

$= \frac{1}{\sqrt{x+h} + \sqrt{x}}$

107. $\begin{cases} 25x + y - 2z = -14 \\ 2x - y + z = 40 \\ -7x + 3y - z = -13 \end{cases}$

$$\left[\begin{array}{ccc|c} 25 & 1 & -2 & -14 \\ 2 & -1 & 1 & 40 \\ -7 & 3 & -1 & -13 \end{array}\right]$$

$\frac{1}{25} R1 \Rightarrow R1$

$$\left[\begin{array}{ccc|c} 1 & 0.04 & -0.08 & -0.56 \\ 2 & -1 & 1 & 40 \\ -7 & 3 & -1 & -13 \end{array}\right]$$

-2R1 + R2

$$\left[\begin{array}{ccc|c} 1 & 0.04 & -0.08 & -0.56 \\ 0 & -1.08 & 1.16 & 41.12 \\ -7 & 3 & -1 & -13 \end{array}\right]$$

7R1 + R3

$$\left[\begin{array}{ccc|c} 1 & 0.04 & -0.08 & -0.56 \\ 0 & -1.08 & 1.16 & 41.12 \\ 0 & 3.28 & -1.56 & -16.92 \end{array}\right]$$

$$\frac{-1}{1.08}R2 \Rightarrow R2$$

$$\left[\begin{array}{ccc|c} 1 & 0.04 & -0.08 & -0.56 \\ 0 & 1 & -1.074 & -38.074 \\ 0 & 3.28 & -1.56 & -16.92 \end{array}\right]$$

-3.28R2 + R3

$$\left[\begin{array}{ccc|c} 1 & 0.04 & -0.08 & -0.56 \\ 0 & 1 & -1.074 & -38.074 \\ 0 & 0 & 1.96272 & 107.96272 \end{array}\right]$$

$$\frac{1}{1.96272}R3 \Rightarrow R3$$

$$\left[\begin{array}{ccc|c} 1 & 0.04 & -0.08 & -0.56 \\ 0 & 1 & -1.074 & -38.074 \\ 0 & 0 & 1 & 55 \end{array}\right]$$

$z = 55;$

$$y - 1.074z = -38.074$$
$$y - 1.074(55) = -38.074$$
$$y - 59.07 = -38.074$$
$$y = 21;$$
$$2x - y + z = 40$$
$$2x - 21 + 55 = 40$$
$$2x + 34 = 40$$
$$2x = 6$$
$$x = 3;$$

(3, 21, 55)

109. $(2,-3)$; $r = 7$

$$(x-h)^2 + (y-k)^2 = r^2$$
$$(x-2)^2 + (y+3)^2 = 7^2$$
$$(x-2)^2 + (y+3)^2 = 49$$

8.2 Technology Highlight

Exercise 1: $a_n = 0.5 + (n-1)0.25$

33.75

Exercise 2: $a_n = \frac{2}{3} + (n-1)\frac{5}{6}$

$\frac{13}{2}, \frac{27}{3}, \frac{37}{3}$; 21; 241.5

8.2 Exercises

1. Common difference.

3. $\frac{n(a_1 + a_n)}{2}$; n^{th}

5. Answers will vary.

7. $-5, -2, 1, 4, 7, 10, \ldots$
$-2-(-5) = -2+5 = 3$;
$1-(-2) = 1+2 = 3$;
$4-1 = 3$;
$7-4 = 3$;
$10-7 = 3$;
Arithmetic; $d = 3$

9. $-0.5, 3, 5.5, 8, 10.5, \ldots$
$3-(-0.5) = 3+0.5 = 3.5$;
$5.5-3 = 2.5$;
$8-5.5 = 2.5$;
$10.5-8 = 2.5$;
Arithmetic; $d = 2.5$

11. $2, 3, 5, 7, 11, 13, 17, \ldots$
$3-2 = 1$;
$5-3 = 2$;
Not arithmetic; all prime.

13. $\frac{1}{24}, \frac{1}{12}, \frac{1}{8}, \frac{1}{6}, \frac{5}{24}, \ldots$
$\frac{1}{12} - \frac{1}{24} = \frac{2}{24} - \frac{1}{24} = \frac{1}{24}$;
$\frac{1}{8} - \frac{1}{12} = \frac{3}{24} - \frac{2}{24} = \frac{1}{24}$;
$\frac{1}{6} - \frac{1}{8} = \frac{4}{24} - \frac{3}{24} = \frac{1}{24}$;
$\frac{5}{24} - \frac{1}{6} = \frac{5}{24} - \frac{4}{24} = \frac{1}{24}$;
Arithmetic; $d = \frac{1}{24}$

15. $1, 2, 4, 9, 16, 25, 36, \ldots$
$2-1 = 1$;
$4-2 = 2$;
Not arithmetic; $a_n = n^2$

17. $\pi, \frac{5\pi}{6}, \frac{2\pi}{3}, \frac{\pi}{2}, \frac{\pi}{3}, \frac{\pi}{6}, \ldots$

$\frac{5\pi}{6} - \pi = \frac{5\pi}{6} - \frac{6\pi}{6} = \frac{-\pi}{6}$;

$\frac{2\pi}{3} - \frac{5\pi}{6} = \frac{4\pi}{6} - \frac{5\pi}{6} = \frac{-\pi}{6}$;

$\frac{\pi}{2} - \frac{2\pi}{3} = \frac{3\pi}{6} - \frac{4\pi}{6} = \frac{-\pi}{6}$;

$\frac{\pi}{3} - \frac{\pi}{2} = \frac{2\pi}{6} - \frac{3\pi}{6} = \frac{-\pi}{6}$;

$\frac{\pi}{6} - \frac{\pi}{3} = \frac{\pi}{6} - \frac{2\pi}{6} = \frac{-\pi}{6}$;

Arithmetic; $d = \frac{-\pi}{6}$

19. $a_1 = 2, d = 3$

$2 + 3 = 5$;
$5 + 3 = 8$;
$8 + 3 = 11$;
2, 5, 8, 11

21. $a_1 = 7, d = -2$

$7 - 2 = 5$;
$5 - 2 = 3$;
$3 - 2 = 1$;
7, 5, 3, 1

23. $a_1 = 0.3, d = 0.03$

$0.3 + 0.03 = 0.33$;
$0.33 + 0.03 = 0.36$;
$0.36 + 0.03 = 0.39$;
0.3, 0.33, 0.36, 0.39

25. $a_1 = \frac{3}{2}, d = \frac{1}{2}$

$\frac{3}{2} + \frac{1}{2} = \frac{4}{2} = 2$;

$2 + \frac{1}{2} = \frac{4}{2} + \frac{1}{2} = \frac{5}{2}$;

$\frac{5}{2} + \frac{1}{2} = \frac{6}{2} = 3$;

$\frac{3}{2}, 2, \frac{5}{2}, 3$

27. $a_1 = \frac{3}{4}, d = -\frac{1}{8}$

$\frac{3}{4} - \frac{1}{8} = \frac{6}{8} - \frac{1}{8} = \frac{5}{8}$;

$\frac{5}{8} - \frac{1}{8} = \frac{4}{8} = \frac{1}{2}$;

$\frac{1}{2} - \frac{1}{8} = \frac{4}{8} - \frac{1}{8} = \frac{3}{8}$;

$\frac{3}{4}, \frac{5}{8}, \frac{1}{2}, \frac{3}{8}$

29. $a_1 = -2, d = -3$

$-2 - 3 = -5$;
$-5 - 3 = -8$;
$-8 - 3 = -11$;
-2, -5, -8, -11

31. $2, 7, 12, 17, \ldots$

$a_1 = 2, d = 5$

$a_n = a_1 + (n-1)d$

$a_n = 2 + (n-1)5$

$a_n = 2 + 5n - 5$

$a_n = 5n - 3$;

$a_6 = 5(6) - 3 = 30 - 3 = 27$;

$a_{10} = 5(10) - 3 = 50 - 3 = 47$;

$a_{12} = 5(12) - 3 = 60 - 3 = 57$

33. $\$5.10, \$5.25, \$5.40, \ldots$

$a_1 = 5.10, d = 0.15$

$a_n = a_1 + (n-1)d$

$a_n = 5.10 + (n-1)(0.15)$

$a_n = 5 + 0.15n - 0.15$

$a_n = 0.15n + 4.95$;

$a_6 = 0.15(6) + 4.95 = 0.90 + 4.95 = 5.85$;

$a_{10} = 0.15(10) + 4.95 = 1.50 + 4.95 = 6.45$

$a_{12} = 0.15(12) + 4.95 = 1.80 + 4.95 = 6.75$

\$5.85, \$6.45, \$6.75

35. $\frac{3}{2}, \frac{9}{4}, 3, \frac{15}{4}, \ldots$

$a_1 = \frac{3}{2}, d = \frac{3}{4}$

$a_n = a_1 + (n-1)d$

$a_n = \frac{3}{2} + (n-1)\left(\frac{3}{4}\right)$

$a_n = \frac{3}{2} + \frac{3}{4}n - \frac{3}{4}$

$a_n = \frac{3}{4}n + \frac{3}{4}$;

$a_6 = \frac{3}{4}(6) + \frac{3}{4} = \frac{9}{2} + \frac{3}{4} = \frac{18}{4} + \frac{3}{4} = \frac{21}{4}$;

$a_{10} = \frac{3}{4}(10) + \frac{3}{4} = \frac{15}{2} + \frac{3}{4} = \frac{30}{4} + \frac{3}{4} = \frac{33}{4}$;

$a_{12} = \frac{3}{4}(12) + \frac{3}{4} = 9 + \frac{3}{4} = \frac{36}{4} + \frac{3}{4} = \frac{39}{4}$

37. $a_1 = 5, d = 4$; Find a_{15}
$a_n = a_1 + (n-1)d$
$a_n = 5 + (n-1)4$
$a_n = 5 + 4n - 4$
$a_n = 4n + 1$;
$a_{15} = 4(15) + 1 = 60 + 1 = 61$

39. $a_1 = \frac{3}{2}, d = -\frac{1}{12}$; Find a_7
$a_n = a_1 + (n-1)d$
$a_n = \frac{3}{2} + (n-1)\left(-\frac{1}{12}\right)$
$a_n = \frac{3}{2} - \frac{1}{12}n + \frac{1}{12}$
$a_n = \frac{18}{12} - \frac{1}{12}n + \frac{1}{12}$
$a_n = -\frac{1}{12}n + \frac{19}{12}$;
$a_7 = -\frac{1}{12}(7) + \frac{19}{12} = \frac{-7}{12} + \frac{19}{12} = \frac{12}{12} = 1$

41. $a_1 = -0.025, d = 0.05$; Find a_{50}
$a_n = a_1 + (n-1)d$
$a_n = -0.025 + (n-1)(0.05)$
$a_n = -0.025 + 0.05n - 0.05$
$a_n = 0.05n - 0.075$;
$a_{50} = 0.05(50) - 0.075 = 2.5 - 0.075 = 2.425$

43. $a_1 = 2, a_n = -22, d = -3$
$a_n = a_1 + (n-1)d$
$-22 = 2 + (n-1)(-3)$
$-22 = 2 - 3n + 3$
$-22 = -3n + 5$
$-27 = -3n$
$9 = n$

45. $a_1 = 0.4, a_n = 10.9, d = 0.25$
$a_n = a_1 + (n-1)d$
$10.9 = 0.4 + (n-1)(0.25)$
$10.9 = 0.4 + 0.25n - 0.25$
$10.9 = 0.15 + 0.25n$
$10.75 = 0.25n$
$43 = n$

47. $-3, -0.5, 2, 4.5, 7, \ldots, 47$
$a_1 = -3, a_n = 47, d = 2.5$
$a_n = a_1 + (n-1)d$
$47 = -3 + (n-1)(2.5)$
$47 = -3 + 2.5n - 2.5$
$47 = -5.5 + 2.5n$
$52.5 = 2.5n$
$21 = n$

49. $\frac{1}{12}, \frac{1}{8}, \frac{1}{6}, \frac{5}{24}, \frac{1}{4}, \ldots, \frac{9}{8}$
$a_1 = \frac{1}{12}, a_n = \frac{9}{8}, d = \frac{1}{24}$
$a_n = a_1 + (n-1)d$
$\frac{9}{8} = \frac{1}{12} + (n-1)\left(\frac{1}{24}\right)$
$\frac{9}{8} = \frac{1}{12} + \frac{1}{24}n - \frac{1}{24}$
$\frac{9}{8} = \frac{1}{24} + \frac{1}{24}n$
$\frac{13}{12} = \frac{1}{24}n$
$26 = n$

51. $a_3 = 7, a_7 = 19$
$a_7 = a_3 + 4d$
$19 = 7 + 4d$
$12 = 4d$
$3 = d$;
$a_7 = a_1 + 6d$
$19 = a_1 + 6(3)$
$19 = a_1 + 18$
$1 = a_1$;
$d = 3, a_1 = 1$

53. $a_2 = 1.025, a_{26} = 10.125$
$a_{26} = a_2 + 24d$
$10.125 = 1.025 + 24d$
$9.1 = 24d$
$\frac{91}{240} = d$;

$a_{26} = a_1 + 25d$

$10.125 = a_1 + 25\left(\frac{91}{240}\right)$

$10.125 = a_1 + \frac{455}{48}$

$\frac{31}{48} = a_1;$

$d = \frac{91}{240}, a_1 = \frac{31}{48}$

55. $a_{10} = \frac{13}{18}, a_{24} = \frac{27}{2}$

$a_{24} = a_{10} + 14d$

$\frac{27}{2} = \frac{13}{18} + 14d$

$\frac{115}{9} = 14d$

$\frac{115}{126} = d;$

$a_{24} = a_1 + 23d$

$\frac{27}{2} = a_1 + 23\left(\frac{115}{126}\right)$

$\frac{27}{2} = a_1 + \frac{2645}{126}$

$\frac{-472}{63} = a_1;$

$d = \frac{115}{126}, a_1 = \frac{-472}{63}$

57. $\sum_{n=1}^{30}(3n-4)$

Initial terms: 1, 2, 5, ….

$a_1 = -1; d = 3; n = 30$

$a_{30} = a_1 + 29d$

$a_{30} = -1 + 29(3) = -1 + 87 = 86$;

$S_{30} = n\left(\frac{a_1 + a_n}{2}\right)$

$S_{30} = 30\left(\frac{-1+86}{2}\right)$

$S_{30} = 30\left(\frac{85}{2}\right)$

$S_{30} = 1275$

59. $\sum_{n=1}^{37}\left(\frac{3}{4}n+2\right)$

Initial terms: $\frac{11}{4}, \frac{7}{2}, \frac{17}{4}, \ldots$

$a_1 = \frac{11}{4}; d = \frac{3}{4}; n = 37$

$a_{37} = a_1 + 36d$

$a_{37} = \frac{11}{4} + 36\left(\frac{3}{4}\right) = \frac{11}{4} + 27 = \frac{119}{4};$

$S_{37} = n\left(\frac{a_1 + a_{37}}{2}\right)$

$S_{37} = 37\left(\frac{\frac{11}{4} + \frac{119}{4}}{2}\right)$

$S_{37} = 37\left(\frac{\frac{65}{2}}{2}\right)$

$S_{37} = 601.25$

61. $\sum_{n=4}^{15}(3-5n)$

Initial terms: …,-17, -22, -27, …

$a_4 = -17; d = -5; n = 12$

$a_{15} = a_4 + 11d$

$a_{15} = -17 + 11(-5) = -17 - 55 = -72;$

$S_{12} = n\left(\frac{a_4 + a_{15}}{2}\right)$

$S_{12} = 12\left(\frac{-17+(-72)}{2}\right)$

$S_{12} = 12\left(\frac{-89}{2}\right)$

$S_{12} = -534$

−534

63. $-12 + (-9.5) + (-7) + (-4.5) + \ldots$

Find S_{15}; $a_1 = -12$; $d = 2.5$; $n = 15$

$S_n = \frac{n}{2}[2a_1 + (n-1)d]$

$S_{15} = \frac{15}{2}[2(-12) + (15-1)(2.5)]$

$S_{15} = \frac{15}{2}[-24 + 14(2.5)]$

$S_{15}=\frac{15}{2}[-24+35]$

$S_{15}=\frac{15}{2}[11]$

$S_{15}=82.5$

65. $0.003+0.173+0.343+0.513+\ldots$

Find S_{30}; $a_1=0.003$; $d=0.17$; $n=30$

$S_n=\frac{n}{2}[2a_1+(n-1)d]$

$S_{30}=\frac{30}{2}[2(0.003)+(30-1)(0.17)]$

$S_{30}=15[0.006+(29)(0.17)]$

$S_{30}=15[0.006+4.93]$

$S_{30}=15[4.936]$

$S_{30}=74.04$

67. $\sqrt{2}+2\sqrt{2}+3\sqrt{2}+4\sqrt{2}+\ldots$

Find S_{20}; $a_1=\sqrt{2}$; $d=\sqrt{2}$; $n=20$

$S_n=\frac{n}{2}[2a_1+(n-1)d]$

$S_{20}=\frac{20}{2}\left[2\sqrt{2}+(20-1)\sqrt{2}\right]$

$S_{20}=10\left(2\sqrt{2}+19\sqrt{2}\right)$

$S_{20}=10\left(21\sqrt{2}\right)$

$S_{20}=210\sqrt{2}$

69. $2+5+8+11+\ldots+74$

$a_1=2$; $d=3$; $a_n=74$

$a_n=a_1+(n-1)d$

$74=2+(n-1)(3)$

$74=2+3n-3$

$74=-1+3n$

$75=3n$

$25=n$;

$S_n=\frac{n(a_1+a_n)}{2}$

$S_{25}=\frac{25(2+74)}{2}=\frac{25(76)}{2}=950$;

$n=25, S_{25}=950$

71. $4+9+14+19+\ldots+154$

$a_1=4$; $d=5$; $a_n=154$

$a_n=a_1+(n-1)d$

$154=4+(n-1)(5)$

$154=4+5n-5$

$154=-1+5n$

$155=5n$

$31=n$;

$S_n=\frac{n(a_1+a_n)}{2}$

$S_{31}=\frac{31(4+154)}{2}=\frac{31(158)}{2}=2449$;

$n=31, S_{31}=2449$

73. $7+10+13+16+\ldots+100$

$a_1=7$; $d=3$; $a_n=100$

$a_n=a_1+(n-1)d$

$100=7+(n-1)(3)$

$100=7+3n-3$

$100=4+3n$

$96=3n$

$32=n$;

$S_n=\frac{n(a_1+a_n)}{2}$

$S_{32}=\frac{32(7+100)}{2}=\frac{32(107)}{2}=1712$;

$n=32, S_{32}=1712$

75. $(-13)+(-9)+(-5)+(-1)+\ldots+183$

$a_1=-13$; $d=4$; $a_n=183$

$a_n=a_1+(n-1)d$

$183=-13+(n-1)(4)$

$183=-13+4n-4$

$183=-17+4n$

$200=4n$

$50=n$;

$S_n=\frac{n(a_1+a_n)}{2}$

$S_{50}=\frac{50(-13+183)}{2}=\frac{50(170)}{2}=4250$;

$n=50, S_{50}=4250$

77. $\frac{1}{2}+\frac{3}{4}+1+\frac{5}{4}+\ldots+\frac{13}{4}$

$a_1=\frac{1}{2}$; $d=\frac{1}{4}$; $a_n=\frac{13}{4}$

$a_n=a_1+(n-1)d$

$\frac{13}{4}=\frac{1}{2}+(n-1)\left(\frac{1}{4}\right)$

$\frac{13}{4}=\frac{1}{2}+\frac{1}{4}n-\frac{1}{4}$

$\frac{13}{4}=\frac{1}{4}+\frac{1}{4}n$

$3=\frac{1}{4}n$

$12=n;$

$S_n=\frac{n(a_1+a_n)}{2}$

$S_{12}=\frac{12\left(\frac{1}{2}+\frac{13}{4}\right)}{2}=\frac{12\left(\frac{15}{4}\right)}{2}=\frac{45}{2};$

$n=12, S_{12}=\frac{45}{2}$

79. $S_n=\frac{n(n+1)}{2}$

$1+2+3+4+5+6=21;$

$S_6=\frac{6(6+1)}{2}=\frac{6(7)}{2}=\frac{42}{2}=21;$

$S_{75}=\frac{75(75+1)}{2}=\frac{75(76)}{2}=\frac{5700}{2}=2850$

81. Initial terms: 2, 4, 6, 8, ….

$a_1=2;\ d=2;\ n=30$

$a_n=a_1+(n-1)d$

$a_{30}=2+(30-1)(2)=2+29(2)=2+58=60;$

$S_n=\frac{n(a_1+a_n)}{2}$

$S_{30}=\frac{30(2+60)}{2}=\frac{30(62)}{2}=930$

83. $a_0=2500$ at initial time of 0

$a_1=2500-85=2415;\ d=-85;\ n=19$

$a_n=a_1+(n-1)d$

$a_{19}=2415+(19-1)(-85)$

$a_{19}=2415+18(-85)$

$a_{19}=2415-1530$

$a_{19}=\$885$

85. $a_0=318$ at initial time of 0 hours

$a_1=318-13=305;\ d=-13;\ a_n=6$

$a_n=a_1+(n-1)d$

$6=305+(n-1)(-13)$

$6=305-13n+13$

$6=318-13n$

$-312=-13n$

$24=n$

24 hours

87. (a) $a_1=30;\ d=\frac{-3}{2};\ n=10$

$a_n=a_1+(n-1)d$

$a_{10}=30+(10-1)\left(\frac{-3}{2}\right)$

$a_{10}=30+9\left(\frac{-3}{2}\right)$

$a_{10}=30-\frac{27}{2}$

$a_{10}=16.5;$

16.5 inches

(b) $S_n=\frac{n(a_1+a_n)}{2}$

$S_{10}=\frac{10(30+16.5)}{2}=\frac{10(46.5)}{2}=232.5;$

232.5 inches

89. $a_1=80;\ d=8;\ n=10$

$a_n=a_1+(n-1)d$

$a_{10}=80+(10-1)(8)=80+9(8)=80+72=152;$

152 seats;

$a_{25}=80+(25-1)(8)=80+24(8)$

$=80+192=272;$

$S_n=\frac{n(a_1+a_n)}{2}$

$S_{25}=\frac{25(80+272)}{2}=\frac{25(352)}{2}=4400;$

4400 seats

91. $a_1=100;\ d=20;\ a_n=2500$

$a_n=a_1+(n-1)d$

$a_7=100+(7-1)(20)=100+6(20)=220;$

\$220;

$a_{12}=100+(12-1)(20)=100+11(20)=320;$

$S_n=\frac{n(a_1+a_n)}{2}$

$S_{12}=\frac{12(100+320)}{2}=\frac{12(420)}{2}=2520$

\$2520; yes

93. $f(x)=3+2(x-1)$; $a_n=3+2(n-1)$
Answers will vary.

95. Let n represent the number of sides
$S_n=180(n-2)$;
$S_{10}=180(10-2)=180(8)=1440^\circ$

97.
$$2530=500e^{0.45t}$$
$$5.06=e^{0.45t}$$
$$\ln 5.06=\ln e^{0.45t}$$
$$\ln 5.06=0.45t$$
$$\frac{\ln 5.06}{0.45}=t$$
$$3.6\approx t$$

99. $\begin{cases} 2x-3y=-1 \\ -4x+9y=4 \end{cases}$

$$\left[\begin{array}{cc|c} 2 & -3 & -1 \\ -4 & 9 & 4 \end{array}\right] \quad \frac{1}{2}R1 \Rightarrow R1$$

$$\left[\begin{array}{cc|c} 1 & -\frac{3}{2} & -\frac{1}{2} \\ -4 & 9 & 4 \end{array}\right] \quad 4R1+R2 \Rightarrow R2$$

$$\left[\begin{array}{cc|c} 1 & -\frac{3}{2} & -\frac{1}{2} \\ 0 & 3 & 2 \end{array}\right] \quad \frac{1}{3}R2 \Rightarrow R2$$

$$\left[\begin{array}{cc|c} 1 & -\frac{3}{2} & -\frac{1}{2} \\ 0 & 1 & \frac{2}{3} \end{array}\right]$$

$y=\frac{2}{3}$;
$$2x-3y=-1$$
$$2x-3\left(\frac{2}{3}\right)=-1$$
$$2x-2=-1$$
$$2x=1$$
$$x=\frac{1}{2}$$
$$\left(\frac{1}{2},\frac{2}{3}\right)$$

101. (0, 972), (5, 1217)
$$m=\frac{1217-972}{5-0}=\frac{245}{5}=49$$
$$y-y_1=m(x-x_1)$$
$$y-972=49(x-0)$$
$$y-972=49x$$
$$y=49x+972;$$
$$f(x)=49x+972;$$
$$f(8)=49(8)+972=392+972=1364$$

8.3 Technology Highlight

Exercise 1:

(a) $a_n=3\cdot(0.2)^{n-1}$
3, 0.6, 0.12, 0.024, 0.0048

(b) 3.7988, 3.749952, 3.749999616
$\to 3.75$
$$S_\infty=\frac{a_1}{1-r}=\frac{3}{1-0.2}=\frac{3}{0.8}=3.75$$

Exercise 2:

(a) $a_n=2000\cdot(1.05)^{n-1}$
2000, 2100, 2205, 2315.25, 2431.0125

(b) 2814.200845

(c) 1.05, 1.05; geometric sequence, $r=1.05$

8.3 Exercises

1. Multiplying.

3. a_1r^{n-1}

5. Answers will vary.

7. $4, 8, 16, 32, \ldots$
$\frac{8}{4}=2$; $\frac{16}{8}=2$; $\frac{32}{16}=2$
$r=2$

9. $3, -6, 12, -24, 48, \ldots$
$\frac{-6}{3}=-2$; $\frac{12}{-6}=-2$; $\frac{-24}{12}=-2$; $\frac{48}{-24}=-2$
$r=-2$

11. $2, 5, 10, 17, 26, \ldots$
$\frac{5}{2}$; $\frac{10}{5}=2$; $\frac{17}{10}$; $\frac{26}{17}$; not geometric
$a_n=n^2+1$

13. $3, 0.3, 0.03, 0.003, \ldots$
$\frac{0.3}{3}=0.1$; $\frac{0.03}{0.3}=0.1$; $\frac{0.003}{0.03}=0.1$

$r = 0.1$

15. -1, 3, -12, 60, -360, …
$\frac{3}{-1} = -3;\ \frac{-12}{3} = -4;\ \frac{60}{-12} = -5;\ \frac{-360}{60} = -6$
Not geometric; ratio of terms decreases by 1.

17. $25, 10, 4, \frac{8}{5}, ...$
$\frac{10}{25} = \frac{2}{5};\ \frac{4}{10} = \frac{2}{5};\ \frac{\frac{8}{5}}{4} = \frac{8}{20} = \frac{2}{5}$
$r = \frac{2}{5}$

19. $\frac{1}{2}, \frac{1}{4}, \frac{1}{8}, \frac{1}{16}, ...$
$\frac{\frac{1}{4}}{\frac{1}{2}} = \frac{1}{2};\ \frac{\frac{1}{8}}{\frac{1}{4}} = \frac{1}{2};\ \frac{\frac{1}{16}}{\frac{1}{8}} = \frac{1}{2}$
$r = \frac{1}{2}$

21. $3, \frac{12}{x}, \frac{48}{x^2}, \frac{192}{x^3}, ...$
$\frac{\frac{12}{x}}{3} = \frac{12}{3x} = \frac{4}{x};\ \frac{\frac{48}{x^2}}{\frac{12}{x}} = \frac{48x}{12x^2} = \frac{4}{x};$
$\frac{\frac{192}{x^3}}{\frac{48}{x^2}} = \frac{192x^2}{48x^3} = \frac{4}{x}$
$r = \frac{4}{x}$

23. $240, 120, 40, 10, 2, ...$
$\frac{120}{240} = \frac{1}{2};\ \frac{40}{120} = \frac{1}{3};\ \frac{10}{40} = \frac{1}{4};\ \frac{2}{10} = \frac{1}{5}$
Not geometric $a_n = \frac{240}{n!}$

25. $a_1 = 5, r = 2$
$a_2 = 5 \cdot 2 = 10;$
$a_3 = 10 \cdot 2 = 20;$
$a_4 = 20 \cdot 2 = 40;$
$5, 10, 20, 40$

27. $a_1 = -6, r = -\frac{1}{2}$
$a_2 = -6\left(\frac{-1}{2}\right) = 3;$
$a_3 = 3\left(\frac{-1}{2}\right) = \frac{-3}{2};$
$a_4 = \frac{-3}{2}\left(\frac{-1}{2}\right) = \frac{3}{4};$
$-6, 3, \frac{-3}{2}, \frac{3}{4}$

29. $a_1 = 4, r = \sqrt{3}$
$a_2 = 4\sqrt{3};$
$a_3 = 4\sqrt{3}\left(\sqrt{3}\right) = 12;$
$a_4 = 12\sqrt{3};$
$4, 4\sqrt{3}, 12, 12\sqrt{3}$

31. $a_1 = 0.1, r = 0.1$
$a_2 = 0.1(0.1) = 0.01;$
$a_3 = 0.01(0.1) = 0.001;$
$a_4 = 0.001(0.1) = 0.0001;$
$0.1, 0.01, 0.001, 0.0001$

33. $a_1 = -24,\ r = \frac{1}{2}$; find a_7
$a_n = a_1 r^{n-1}$
$a_7 = -24\left(\frac{1}{2}\right)^{7-1} = -24\left(\frac{1}{2}\right)^6$
$= -24\left(\frac{1}{64}\right) = -\frac{3}{8}$

35. $a_1 = -\frac{1}{20},\ r = -5$; find a_4
$a_n = a_1 r^{n-1}$
$a_4 = -\frac{1}{20}(-5)^{4-1} = -\frac{1}{20}(-5)^3$
$= -\frac{1}{20}(-125) = \frac{25}{4}$

37 $a_1 = 2, r = \sqrt{2}$; find a_7
$a_n = a_1 r^{n-1}$
$a_7 = 2\left(\sqrt{2}\right)^{7-1} = 2\left(\sqrt{2}\right)^6 = 2(8) = 16$

38. $a_1 = \sqrt{3}, \text{r} = \sqrt{3}$; find a_8

39. $\frac{1}{27}, -\frac{1}{9}, \frac{1}{3}, -1, 3, \ldots$

$$r = \frac{\frac{-1}{9}}{\frac{1}{27}} = \frac{-27}{9} = -3$$

$$a_1 = \frac{1}{27};\ r = -3$$

$$a_n = \frac{1}{27}(-3)^{n-1}$$

$$a_6 = \frac{1}{27}(-3)^{6-1} = \frac{1}{27}(-3)^5 = \frac{1}{27}(-243) = -9;$$

$$a_{10} = \frac{1}{27}(-3)^{10-1} = \frac{1}{27}(-3)^9 = \frac{1}{27}(-19683) = -729;$$

$$a_{12} = \frac{1}{27}(-3)^{12-1} = \frac{1}{27}(-3)^{11} = \frac{1}{27}(-177147) = -6561$$

41. $729, 243, 81, 27, 9, \ldots$

$$r = \frac{243}{729} = \frac{1}{3}$$

$$a_1 = 729;\ r = \frac{1}{3}$$

$$a_n = 729\left(\frac{1}{3}\right)^{n-1}$$

$$a_6 = 729\left(\frac{1}{3}\right)^{6-1} = 729\left(\frac{1}{3}\right)^5 = 729\left(\frac{1}{243}\right) = 3;$$

$$a_{10} = 729\left(\frac{1}{3}\right)^{10-1} = 729\left(\frac{1}{3}\right)^9 = 729\left(\frac{1}{19683}\right) = \frac{1}{27};$$

$$a_{12} = 729\left(\frac{1}{3}\right)^{12-1} = 729\left(\frac{1}{3}\right)^{11} = 729\left(\frac{1}{177147}\right) = \frac{1}{243}$$

43. $\frac{1}{2}, \frac{\sqrt{2}}{2}, 1, \sqrt{2}, 2, \ldots$

$$r = \frac{\frac{\sqrt{2}}{2}}{\frac{1}{2}} = \frac{2\sqrt{2}}{2} = \sqrt{2}$$

$$a_1 = \frac{1}{2};\ r = \sqrt{2}$$

$$a_n = \frac{1}{2}\left(\sqrt{2}\right)^{n-1}$$

$$a_6 = \frac{1}{2}\left(\sqrt{2}\right)^{6-1} = \frac{1}{2}\left(\sqrt{2}\right)^5 = \frac{1}{2}\left(4\sqrt{2}\right) = 2\sqrt{2};$$

$$a_{10} = \frac{1}{2}\left(\sqrt{2}\right)^{10-1} = \frac{1}{2}\left(\sqrt{2}\right)^9 = \frac{1}{2}\left(16\sqrt{2}\right) = 8\sqrt{2};$$

$$a_{12} = \frac{1}{2}\left(\sqrt{2}\right)^{12-1} = \frac{1}{2}\left(\sqrt{2}\right)^{11} = \frac{1}{2}\left(32\sqrt{2}\right) = 16\sqrt{2}$$

45. $0.2, 0.08, 0.032, 0.0128, \ldots$

$$r = \frac{0.08}{0.2} = 0.4$$

$$a_1 = 0.2;\ r = 0.4$$

$$a_n = 0.2(0.4)^{n-1}$$

$$a_6 = 0.2(0.4)^{6-1} = 0.2(0.4)^5 = 0.2(0.01024) = 0.002048;$$

$$a_{10} = 0.2(0.4)^{10-1} = 0.2(0.4)^9 = 0.2(0.000261244) = 0.0000524288;$$

$$a_{12} = 0.2(0.4)^{12-1} = 0.2(0.4)^{11} = 0.2(0.00004194304) = 0.000008388608$$

47. $a_1 = 9,\ a_n = 729,\ r = 3$

$$a_n = a_1 r^{n-1}$$

$$729 = 9(3)^{n-1}$$

$$81 = 3^{n-1}$$

$$3^4 = 3^{n-1}$$

$$4 = n - 1$$

$$5 = n$$

49. $a_1 = 16, a_n = \frac{1}{64}, r = \frac{1}{2}$

$$a_n = a_1 r^{n-1}$$

$$\frac{1}{64}=16\left(\frac{1}{2}\right)^{n-1}$$
$$\frac{1}{1024}=\left(\frac{1}{2}\right)^{n-1}$$
$$\left(\frac{1}{2}\right)^{10}=\left(\frac{1}{2}\right)^{n-1}$$
$$10=n-1$$
$$11=n$$

51. $a_1=-1, a_n=-1296,\ r=\sqrt{6}$

$$a_n=a_1r^{n-1}$$
$$-1296=-1\left(\sqrt{6}\right)^{n-1}$$
$$1296=\left(\sqrt{6}\right)^{n-1}$$
$$\left(\sqrt{6}\right)^8=\left(\sqrt{6}\right)^{n-1}$$
$$8=n-1$$
$$9=n$$

53. $2,-6,18,-54,\ldots,-4374$

$$r=\frac{-6}{2}=-3;\ a_1=2;\ a_n=-4374$$
$$a_n=a_1r^{n-1}$$
$$-4374=2(-3)^{n-1}$$
$$-2187=(-3)^{n-1}$$
$$(-3)^7=(-3)^{n-1}$$
$$7=n-1$$
$$8=n$$

55. $64,32\sqrt{2},32,16\sqrt{2},\ldots,1$

$$r=\frac{32\sqrt{2}}{64}=\frac{\sqrt{2}}{2};\ a_1=64;\ a_n=1$$
$$a_n=a_1r^{n-1}$$
$$1=64\left(\frac{\sqrt{2}}{2}\right)^{n-1}$$
$$\frac{1}{64}=\left(\frac{\sqrt{2}}{2}\right)^{n-1}$$
$$\left(\frac{\sqrt{2}}{2}\right)^{12}=\left(\frac{\sqrt{2}}{2}\right)^{n-1}$$
$$12=n-1$$
$$13=n$$

57. $\frac{3}{8},-\frac{3}{4},\frac{3}{2},-3,\ldots,96$

$$r=\frac{\frac{-3}{4}}{\frac{3}{8}}=\frac{-24}{12}=-2;\ a_1=\frac{3}{8};\ a_n=96$$
$$a_n=a_1r^{n-1}$$
$$96=\frac{3}{8}(-2)^{n-1}$$
$$256=(-2)^{n-1}$$
$$(-2)^8=(-2)^{n-1}$$
$$8=n-1$$
$$9=n$$

59. $a_3=324,\ a_7=64$

$$a_7=a_3\cdot r^4$$
$$64=324r^4$$
$$\frac{64}{324}=r^4$$
$$\sqrt[4]{\frac{16}{81}}=r$$
$$\frac{2}{3}=r;$$
$$a_7=a_1\cdot\left(\frac{2}{3}\right)^6$$
$$64=a_1\left(\frac{2}{3}\right)^6$$
$$64=a_1\left(\frac{64}{729}\right)$$
$$729=a_1$$
$$r=\frac{2}{3},\ a_1=729$$

61. $a_4=\frac{4}{9},\ a_8=\frac{9}{4}$

$$a_8=a_4\cdot r^4$$
$$\frac{9}{4}=\frac{4}{9}(r)^4$$
$$\frac{81}{16}=r^4$$
$$\sqrt[4]{\frac{81}{16}}=r$$
$$\frac{3}{2}=r;$$

$$a_8 = a_1\left(\frac{3}{2}\right)^7$$

$$\frac{9}{4} = a_1\left(\frac{3}{2}\right)^7$$

$$\frac{9}{4} = a_1\left(\frac{2187}{128}\right)$$

$$\frac{32}{243} = a_1$$

$$r = \frac{3}{2},\ a_1 = \frac{32}{243}$$

63. $a_4 = \frac{32}{3},\ a_8 = 54$

$$a_8 = a_4 \cdot r^4$$

$$54 = \left(\frac{32}{3}\right)r^4$$

$$\frac{81}{16} = r^4$$

$$\sqrt[4]{\frac{81}{16}} = r$$

$$\frac{3}{2} = r;$$

$$a_8 = a_1\left(\frac{3}{2}\right)^7$$

$$54 = a_1\left(\frac{3}{2}\right)^7$$

$$54 = a_1\left(\frac{2187}{128}\right)$$

$$\frac{256}{81} = a_1$$

$$r = \frac{3}{2},\ a_1 = \frac{256}{81}$$

65. $a_1 = 8, r = -2$, find S_{12}

$$S_n = \frac{a_1(1-r^n)}{1-r}$$

$$S_{12} = \frac{8(1-(-2)^{12})}{1+3} = \frac{8(1-4096)}{1+2}$$

$$= \frac{8(-4095)}{3} = -10{,}920$$

67. $a_1 = 96, r = \frac{1}{3}$, find S_5

$$S_n = \frac{a_1(1-r^n)}{1-r}$$

$$S_5 = \frac{96\left(1-\left(\frac{1}{3}\right)^5\right)}{1-\frac{1}{3}} = \frac{96\left(1-\frac{1}{243}\right)}{\frac{2}{3}}$$

$$= \frac{96\left(\frac{242}{243}\right)}{\frac{2}{3}} = \frac{3872}{27} \approx 143.41$$

69. $a_1 = 8, r = \frac{3}{2}$, find S_7

$$S_n = \frac{a_1(1-r^n)}{1-r}$$

$$S_7 = \frac{8\left(1-\left(\frac{3}{2}\right)^7\right)}{1-\frac{3}{2}} = \frac{8\left(1-\frac{2187}{128}\right)}{-\frac{1}{2}}$$

$$= \frac{8\left(\frac{-2059}{128}\right)}{-\frac{1}{2}} = \frac{2059}{8} = 257.375$$

71. $2+6+18+\ldots$; find S_6

$$a_1 = 2;\ r = \frac{6}{2} = 3$$

$$S_n = \frac{a_1(1-r^n)}{1-r}$$

$$S_6 = \frac{2(1-3^6)}{1-3} = \frac{2(1-729)}{-2} = \frac{2(-728)}{-2} = 728$$

73. $16-8+4-\ldots$; find S_8

$$a_1 = 16;\ r = \frac{-8}{16} = \frac{-1}{2}$$

$$S_n = \frac{a_1(1-r^n)}{1-r}$$

$$S_8 = \frac{16\left(1-\left(\frac{-1}{2}\right)^8\right)}{1-\left(-\frac{1}{2}\right)} = \frac{16\left(1-\frac{1}{256}\right)}{\frac{3}{2}}$$

$$= \frac{16\left(\frac{255}{256}\right)}{\frac{3}{2}} = \frac{85}{8} = 10.625$$

75. $\frac{4}{3} + \frac{2}{9} + \frac{1}{27} + ...$; find S_9

$a_1 = \frac{4}{3};\ r = \frac{1}{6}$

$$S_n = \frac{a_1(1 - r^n)}{1 - r}$$

$$S_9 = \frac{\frac{4}{3}\left(1 - \left(\frac{1}{6}\right)^9\right)}{1 - \frac{1}{6}} = \frac{\frac{4}{3}\left(1 - \frac{1}{10077696}\right)}{\frac{5}{6}}$$

$$= \frac{\frac{4}{3}\left(\frac{10077695}{10077696}\right)}{\frac{5}{6}} \approx 1.60$$

77. $\sum_{j=1}^{5} 4^j$

Initial terms: 4, 16, 64, ….

$a_1 = 4;\ r = \frac{16}{4} = 4;\ n = 5$

$$S_n = \frac{a_1(1 - r^n)}{1 - r}$$

$$S_5 = \frac{4(1 - 4^5)}{1 - 4} = \frac{4(1 - 1024)}{-3}$$

$$= \frac{4(-1023)}{-3} = 1{,}364$$

79. $\sum_{k=1}^{8} 5\left(\frac{2}{3}\right)^{k-1}$

$5\left(\frac{2}{3}\right)^0 = 5$;

$5\left(\frac{2}{3}\right)^1 = \frac{10}{3}$;

$5\left(\frac{2}{3}\right)^2 = \frac{20}{9}$;

$5\left(\frac{2}{3}\right)^3 = \frac{40}{27}$

Initial terms: $5, \frac{10}{3}, \frac{20}{9}, \frac{40}{27}, ...$

$a_1 = 5;\ r = \frac{\frac{20}{9}}{\frac{10}{3}} = \frac{2}{3};\ n = 8$

$$S_n = \frac{a_1(1 - r^n)}{1 - r}$$

$$S_8 = \frac{5\left(1 - \left(\frac{2}{3}\right)^8\right)}{1 - \frac{2}{3}} = \frac{5\left(1 - \frac{256}{6561}\right)}{\frac{1}{3}}$$

$$= \frac{5\left(\frac{6305}{6561}\right)}{\frac{1}{3}} = \frac{31525}{2187} \approx 14.41$$

81. $\sum_{i=4}^{10} 9\left(-\frac{1}{2}\right)^{i-1}$

$9\left(\frac{-1}{2}\right)^3 = \frac{-9}{8}$;

$9\left(\frac{-1}{2}\right)^4 = \frac{9}{16}$;

$9\left(\frac{-1}{2}\right)^5 = \frac{-9}{32}$;

$9\left(\frac{-1}{2}\right)^6 = \frac{9}{64}$

Initial terms: $\frac{-9}{8}, \frac{9}{16}, \frac{-9}{32}, \frac{9}{64}, ...$

$a_1 = \frac{-9}{8};\ r = \frac{\frac{-9}{32}}{\frac{9}{16}} = \frac{-1}{2};\ n = 7$

$$S_n = \frac{a_1(1 - r^n)}{1 - r}$$

$$S_7 = \frac{\frac{-9}{8}\left(1 - \left(\frac{-1}{2}\right)^7\right)}{1 - \left(\frac{-1}{2}\right)} = \frac{\frac{-9}{8}\left(1 + \frac{1}{128}\right)}{\frac{3}{2}}$$

$$= \frac{\frac{-9}{8}\left(\frac{129}{128}\right)}{\frac{3}{2}} = \frac{-387}{512} \approx -0.76$$

83. $a_2 = -5, a_5 = \frac{1}{25}$, find S_5

Find r:

$a_5 = a_2 r^3$

$\frac{1}{25} = -5r^3$

$\frac{-1}{125} = r^3$

$\sqrt[3]{\frac{-1}{125}} = r$

$\frac{-1}{5} = r;$

Find a_1:

$a_5 = a_1 r^4$

$\frac{1}{25} = a_1\left(\frac{-1}{5}\right)^4$

$\frac{1}{25} = a_1\left(\frac{1}{625}\right)$

$25 = a_1;$

$a_1 = 25;\ r = \frac{-1}{5}$

$S_n = \frac{a_1(1-r^n)}{1-r}$

$$S_5 = \frac{25\left(1-\left(\frac{-1}{5}\right)^5\right)}{1-\left(\frac{-1}{5}\right)} = \frac{25\left(1+\frac{1}{3125}\right)}{\frac{6}{5}}$$

$$= \frac{25\left(\frac{3126}{3125}\right)}{\frac{6}{5}} = \frac{521}{25}$$

85. $a_3 = \frac{4}{9}, a_7 = \frac{9}{64}$, find S_6

Find r:

$a_7 = a_3 r^4$

$\frac{9}{64} = \frac{4}{9} r^4$

$\frac{81}{256} = r^4$

$\sqrt[4]{\frac{81}{256}} = r$

$\frac{3}{4} = r;$

Find a_1:

$a_7 = a_1 r^6$

$\frac{9}{64} = a_1\left(\frac{3}{4}\right)^6$

$\frac{9}{64} = a_1\left(\frac{729}{4096}\right)$

$\frac{64}{81} = a_1;$

$a_1 = \frac{64}{81};\ r = \frac{3}{4}$

$S_n = \frac{a_1(1-r^n)}{1-r}$

$$S_6 = \frac{\frac{64}{81}\left(1-\left(\frac{3}{4}\right)^6\right)}{1-\frac{3}{4}} = \frac{\frac{64}{81}\left(1-\frac{729}{4096}\right)}{\frac{1}{4}}$$

$$= \frac{\frac{64}{81}\left(\frac{3367}{4096}\right)}{\frac{1}{4}} = \frac{3367}{1296}$$

87. $a_3 = 2\sqrt{2}, a_6 = 8$, find S_7

Find r:

$a_6 = a_3 r^3$

$8 = 2\sqrt{2}(r)^3$

$\frac{4}{\sqrt{2}} = r^3$

$2\sqrt{2} = r^3$

$2 \cdot 2^{\frac{1}{2}} = r^3$

$2^{\frac{3}{2}} = r^3$

$\left(2^{\frac{3}{2}}\right)^{\frac{1}{3}} = r$

$\sqrt{2} = r;$

Find a_1:

$a_6 = a_1 r^5$

$8 = a_1(\sqrt{2})^5$

$8 = a_1\left(2^{\frac{5}{2}}\right)$

$$\frac{2^3}{2^{\frac{5}{2}}} = a_1$$

$$2^{3-\frac{5}{2}} = a_1$$

$$\sqrt{2} = a_1$$

$$a_1 = \sqrt{2};\ r = \sqrt{2}$$

$$S_n = \frac{a_1\left(1-r^n\right)}{1-r}$$

$$S_7 = \frac{\sqrt{2}\left(1-\left(\sqrt{2}\right)^7\right)}{1-\sqrt{2}} = \frac{\sqrt{2}\left(1-\left(\sqrt{2}\right)^6\left(\sqrt{2}\right)\right)}{1-\sqrt{2}}$$

$$= \frac{\sqrt{2}\left(1-8\left(\sqrt{2}\right)\right)}{1-\sqrt{2}} = \frac{\sqrt{2}-16}{1-\sqrt{2}}$$

$$= \frac{\sqrt{2}-16}{1-\sqrt{2}} \cdot \frac{1+\sqrt{2}}{1+\sqrt{2}}$$

$$= \frac{\sqrt{2}+2-16-16\sqrt{2}}{1-2}$$

$$= \frac{-14-15\sqrt{2}}{-1} = 14+15\sqrt{2}$$

89. $3+6+12+24+\ldots$

$$\frac{6}{3} = 2;\ \text{No}$$

91. $9+3+1+\ldots$

$$\frac{3}{9} = \frac{1}{3};\ \left|\frac{1}{3}\right| < 1$$

$$a_1 = 9;\ r = \frac{1}{3}$$

$$S_\infty = \frac{a_1}{1-r}$$

$$S_\infty = \frac{9}{1-\frac{1}{3}} = \frac{9}{\frac{2}{3}} = \frac{27}{2}$$

93. $25+10+4+\frac{8}{5}+\ldots$

$$\frac{10}{25} = \frac{2}{5};\ \left|\frac{2}{5}\right| < 1$$

$$a_1 = 25;\ r = \frac{2}{5}$$

$$S_\infty = \frac{a_1}{1-r}$$

$$S_\infty = \frac{25}{1-\frac{2}{5}} = \frac{25}{\frac{3}{5}} = \frac{125}{3}$$

95. $6+3+\frac{3}{2}+\frac{3}{4}+\ldots$

$$\frac{3}{6} = \frac{1}{2};\ \left|\frac{1}{2}\right| < 1$$

$$a_1 = 6;\ r = \frac{1}{2}$$

$$S_\infty = \frac{a_1}{1-r}$$

$$S_\infty = \frac{6}{1-\frac{1}{2}} = \frac{6}{\frac{1}{2}} = 12$$

97. $6-3+\frac{3}{2}-\frac{3}{4}+\ldots$

$$\frac{-3}{6} = \frac{-1}{2};\ \left|\frac{-1}{2}\right| < 1$$

$$a_1 = 6;\ r = \frac{-1}{2}$$

$$S_\infty = \frac{a_1}{1-r}$$

$$S_\infty = \frac{6}{1-\frac{-1}{2}} = \frac{6}{\frac{3}{2}} = \frac{12}{3} = 4$$

99. $3+0.3+0.03+0.003+\ldots$

$$\frac{0.3}{3} = \frac{1}{10};\ \left|\frac{1}{10}\right| < 1$$

$$a_1 = 3;\ r = \frac{1}{10}$$

$$S_\infty = \frac{a_1}{1-r}$$

$$S_\infty = \frac{3}{1-\frac{1}{10}} = \frac{3}{\frac{9}{10}} = \frac{30}{9} = \frac{10}{3}$$

101. $\sum_{k=1}^{\infty} \frac{3}{4}\left(\frac{2}{3}\right)^k$

$$\frac{3}{4}\left(\frac{2}{3}\right)^1 = \frac{1}{2};$$

$$\frac{3}{4}\left(\frac{2}{3}\right)^2 = \frac{3}{4}\left(\frac{4}{9}\right) = \frac{1}{3};$$

$\frac{3}{4}\left(\frac{2}{3}\right)^3 = \frac{3}{4}\left(\frac{8}{27}\right) = \frac{2}{9}$;

Initial terms: $\frac{1}{2} + \frac{1}{3} + \frac{2}{9} + \ldots.$

$\frac{\frac{1}{3}}{\frac{1}{2}} = \frac{2}{3};\ \left|\frac{2}{3}\right| < 1$

$a_1 = \frac{1}{2};\ r = \frac{2}{3}$

$S_\infty = \frac{a_1}{1-r}$

$S_\infty = \frac{\frac{1}{2}}{1-\frac{2}{3}} = \frac{\frac{1}{2}}{\frac{1}{3}} = \frac{3}{2}$

103. $\sum_{j=1}^{\infty} 9\left(-\frac{2}{3}\right)^j$

$9\left(\frac{-2}{3}\right)^1 = -6$;

$9\left(\frac{-2}{3}\right)^2 = 9\left(\frac{4}{9}\right) = 4$;

$9\left(\frac{-2}{3}\right)^3 = 9\left(\frac{-8}{27}\right) = \frac{-8}{3}$

Initial terms: $-6 + 4 - \frac{8}{3} + \ldots.$

$\frac{4}{-6} = \frac{2}{-3};\ \left|\frac{-2}{3}\right| < 1$

$a_1 = -6;\ r = \frac{-2}{3}$

$S_\infty = \frac{a_1}{1-r}$

$S_\infty = \frac{-6}{1-\frac{-2}{3}} = \frac{-6}{\frac{5}{3}} = \frac{-18}{5}$

105. $S_n = \frac{n^2(n+1)^2}{4};\ 1^3 + 2^3 + 3^3 + \ldots 8^3$

$S_8 = \frac{8^2(8+1)^2}{4} = \frac{64(9)^2}{4} = \frac{64(81)}{4} = 1296$

$1^3 + 2^3 + 3^3 + 4^3 + 5^3 + 6^3 + 7^3 + 8^3$

$= 1 + 8 + 27 + 64 + 125 + 216 + 343 + 512$

$= 1296$

107. $a_1 = 24;\ r = 0.8;\ n = 7$

Initial terms: 24, 19.2, 15.36, …

$a_n = a_1(r)^{n-1}$

$a_7 = 24(0.8)^6 = 24(0.262144) \approx 6.3$;

$S_\infty = \frac{a_1}{1-r}$

$S_\infty = \frac{24}{1-0.8} = \frac{24}{0.2} = 120$

about 6.3 ft; 120 ft

109. $a_1 = 50000;\ r = 0.03;\ n = 10$

$a_n = a_1(1+r)^n$

$a_{10} = 50000(1+0.03)^{10}$

$a_{10} = 50000(1.03)^{10}$

$a_{10} = 50000(1.343916379)$

$a_{10} = 67195.82$

about \$67,196

111. $a_1 = 26000;\ r = 0.20;\ n = 5$

$a_n = a_1(1-r)^n$

$a_5 = 26000(1-0.2)^5$

$a_5 = 26000(0.8)^5$

$a_5 = 26000(0.32768)$

$a_5 = 8519.68$;

$2791 = 26000(1-0.2)^n$

$0.107346 = 0.8^n$

$\ln 0.107346 = n \ln 0.8$

$\frac{\ln 0.107346}{\ln 0.8} = n$

$10 \approx n$;

about \$8,520; 10 years

113. $a_0 = 160;\ a_1 = 160(0.97) = 155.2;$

$r = 0.03;\ n = 8$

$a_n = a_1(1-r)^{n-1}$

$a_8 = 155.2(1-0.03)^{8-1}$

$a_8 = 155.2(0.97)^7$

$a_8 \approx 125.4$;

$118 = 155.2(0.97)^{n-1}$

$\frac{118}{155.2} = 0.97^{n-1}$

$\ln\frac{118}{155.2} = (n-1)\ln 0.97$

$\frac{\ln\frac{118}{155.2}}{\ln 0.97} = n-1$

$9 \approx n-1$

$10 \approx n$

about 125.4 gpm; about 10 months

115. $a_1 = 277;\ \ r = 0.023;\ \ n = 10$

$a_n = a_1(1+r)^n$

$a_{10} = 277(1+0.023)^{10}$

$a_{10} = 277(1.023)^{10}$

$a_{10} \approx 347.7$

about 347.7 million

117. $a_1 = 50;\ \ r = 2;\ \ n = 10$

10 half-hours in 5 hours

$a_n = a_1(r)^n$

$a_{10} = 50(2)^{10} = 50(1024) = 51200$;

$204800 = 50(2)^n$

$4096 = 2^n$

$(2)^{12} = 2^n$

$12 = n$

51,200 bacteria; 12 half hours later or 6 hours

119. $a_1 = \frac{4}{5}(20) = 16;\ \ r = \frac{4}{5};\ \ n = 7$

$a_n = a_1(r)^{n-1}$

$a_7 = 16\left(\frac{4}{5}\right)^6 = 16\left(\frac{4096}{15625}\right) \approx 4.2$;

$S_\infty = \frac{a_1}{1-r}$

$S_\infty = \frac{16}{1-\frac{4}{5}} = \frac{16}{\frac{1}{5}} = 80$ up

Down: $S = \frac{20}{1-\frac{4}{5}} = \frac{20}{\frac{1}{5}} = 100$

approximately 4.2 m; 180 m

121. $a_1 = 462;\ \ r = \frac{2}{5};\ \ n = 5$

$a_n = a_1(1-r)^n$

$a_5 = 462\left(1-\frac{2}{5}\right)^5$

$a_5 = 462\left(\frac{3}{5}\right)^5$

$a_5 = 462\left(\frac{243}{3125}\right)$

$a_5 = 35.9$

$12.9 = 462\left(\frac{3}{5}\right)^n$

$0.0279 = \left(\frac{3}{5}\right)^n$

$\ln 0.0279 = n\ln 0.6$

$\frac{\ln 0.0279}{\ln 0.6} = n$

35.9 in^3; about 7 strokes

123. $a_0 = 40000;\ \ d = 1750;\ \ r = 0.96$

$40000 + 1750n = 40000(1.04)^n$

$Y_1 = Y_2$

Using a grapher: about 6 years

125. $a_n = 1000(1.05)^{n-1}$

After 6 years, a_7 is needed

$a_7 = 1000(1.05)^{7-1} = 1000(1.05)^6 = 1340.10$

A(6) represents the amount in the account after 6 years; A(7) represents the amount in the account after 7 years.

a_n generates the terms of the sequence before any interest is applied, while A(t) gives the amount in the account after interest has been applied. Here $a_7 = A(6)$.

127. $S_n = \sum_{k=1}^{n} \log(k)$

$S_n = \log n!$

129. $f(x) = x^2 + 5x + 9$

$x = \frac{-5 \pm \sqrt{(5)^2 - 4(1)(9)}}{2(1)}$

$x = \frac{-5 \pm \sqrt{25-36}}{2}$

$x = \frac{-5 \pm \sqrt{-11}}{2}$

$x = \frac{-5}{2} \pm \frac{\sqrt{11}}{2} i$

131. $h(x) = \frac{x^2}{x-1}$

Vertical asymptote: $x = 1$
Horizontal asymptote: none
(deg num > deg den)
Oblique asymptote: $y = x$

$$\begin{array}{r} x \\ x-1 \overline{)\, x^2 \quad} \\ -\underline{(x^2 - x)} \\ x \end{array}$$

y-intercept: (0,0)

$h(0) = \frac{0^2}{0-1} = 0$

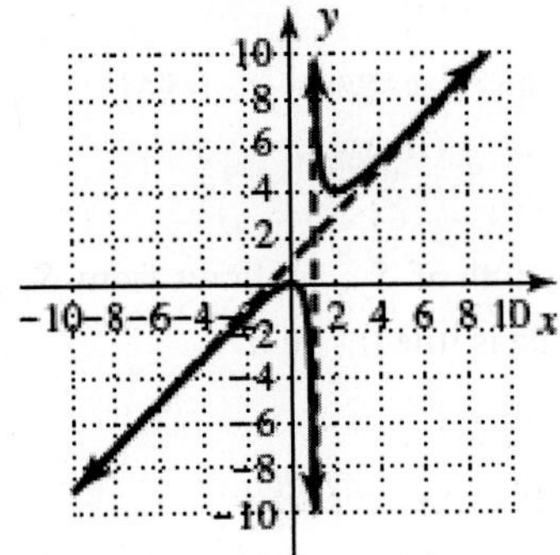

133. $p(t) = \frac{4200}{1+10e^{-0.055t}}$

$p(50) = \frac{4200}{1+10e^{-0.055(50)}} \approx 2562.1$;

$p(75) = \frac{4200}{1+10e^{-0.055(75)}} \approx 3615.6$;

$p(100) = \frac{4200}{1+10e^{-0.055(100)}} \approx 4035.1$;

$p(150) = \frac{4200}{1+10e^{-0.055(150)}} \approx 4189.1$

$X = \begin{bmatrix} -1 \\ -2 \\ 1 \end{bmatrix}$

$\frac{-x-2}{x^2+3} + \frac{1}{x+1}$

8.4 Exercises

1. Finite; universally

3. Induction hypothesis

5. Answers will vary.

7. $a_n = 10n - 6$
$a_4 = 10(4) - 6 = 40 - 6 = 34$;
$a_5 = 10(5) - 6 = 50 - 6 = 44$;
$a_k = 10k - 6$;
$a_{k+1} = 10(k+1) - 6 = 10k + 10 - 6 = 10k + 4$

9. $a_n = n$
$a_4 = 4$;
$a_5 = 5$;
$a_k = k$;
$a_{k+1} = k + 1$

11. $a_n = 2^{n-1}$
$a_4 = 2^{4-1} = 2^3 = 8$;
$a_5 = 2^{5-1} = 2^4 = 16$;
$a_k = 2^{k-1}$;
$a_{k+1} = 2^{k+1-1} = 2^k$

13. $S_n = n(5n-1)$
$S_4 = 4(5(4)-1) = 4(20-1) = 4(19) = 76$;
$S_5 = 5(5(5)-1) = 5(25-1) = 5(24) = 120$;
$S_k = k(5k-1)$;
$S_{k+1} = (k+1)(5(k+1)-1) = (k+1)(5k+5-1)$
$= (k+1)(5k+4)$

15. $S_n = \frac{n(n+1)}{2}$
$S_4 = \frac{4(4+1)}{2} = \frac{4(5)}{2} = 10$;
$S_5 = \frac{5(5+1)}{2} = \frac{5(6)}{2} = 15$;
$S_k = \frac{k(k+1)}{2}$;
$S_{k+1} = \frac{(k+1)(k+1+1)}{2} = \frac{(k+1)(k+2)}{2}$

17. $S_n = 2^n - 1$

$S_4 = 2^4 - 1 = 16 - 1 = 15$;

$S_5 = 2^5 - 1 = 32 - 1 = 31$;

$S_k = 2^k - 1$;

$S_{k+1} = 2^{k+1} - 1$

19. $a_n = 10n - 6$; $S_n = n(5n-1)$
$S_4 = 4(5(4)-1) = 4(20-1) = 4(19) = 76$;
$a_5 = 10(5) - 6 = 50 - 6 = 44$;
$S_5 = 5(5(5)-1) = 5(25-1) = 5(24) = 120$;
$S_4 + a_5 = S_5$
$76 + 44 = 120$
$120 = 120$
Verified

21. $a_n = n$; $S_n = \frac{n(n+1)}{2}$
$S_4 = \frac{4(4+1)}{2} = \frac{4(5)}{2} = 10$;
$a_5 = 5$;
$S_5 = \frac{5(5+1)}{2} = \frac{5(6)}{2} = 15$;
$S_4 + a_5 = S_5$
$10 + 5 = 15$
$15 = 15$
Verified

23. $a_n = 2^{n-1}$; $S_n = 2^n - 1$
$S_4 = 2^4 - 1 = 16 - 1 = 15$;
$a_5 = 2^{5-1} = 2^4 = 16$;
$S_5 = 2^5 - 1 = 32 - 1 = 31$;
$S_4 + a_5 = S_5$
$15 + 16 = 31$
$31 = 31$
Verified

25. $a_n = n^3$; $S_n = (1+2+3+4+...+n)^2$
The needed components are:
$a_n = n^3$; $a_k = k^3$; $a_{k+1} = (k+1)^3$
$S_k = (1+2+3+4+...+k)^2$;
$S_{k+1} = (1+2+3+4+...+k+(k+1))^2$
1. Show S_n is true for $n = 1$.
$S_1 = (1)^2 = 1$
Verified
2. Assume S_k is true:
$1+8+27+\ldots+k^3 = (1+2+3+4+...+k)^2$
and use it to show the truth of S_{k+1} follows. That is:
$1+8+27+\ldots+k^3+(k+1)^3$
$= (1+2+3+...+k+(k+1))^2$
$S_k + a_{k+1} = S_{k+1}$
Working with the left hand side:
$1+8+27+...+k^3+(k+1)^3$
$= 1+8+27+...+k^3+(k+1)^2(k+1)$
$= 1+8+27+...+k^3+k(k+1)^2+(k+1)^2$
$= 1+8+27+...+k^3+k(k+1)(k+1)+(k+1)^2$
$= 1+8+27+...+k^3$
$+2\frac{k(k+1)}{2}(k+1)+(k+1)^2$
$= (1+2+3+...+k)^2$
$+2(1+2+3+...+k)(k+1)+(k+1)^2$
Factoring as a trinomial:
$= ((1+2+3+...+k)+(k+1))^2$
Since the truth of S_{k+1} follows from S_k, the formula is true for all n.

27. $2+4+6+8+10+...+2n$;
The needed components are:
$a_n = 2n$; $a_k = 2k$; $a_{k+1} = 2(k+1)$
$S_n = n(n+1)$; $S_k = k(k+1)$;
$S_{k+1} = (k+1)(k+1+1) = (k+1)(k+2)$
1. Show S_n is true for $n = 1$.
$S_1 = 1(1+1) = 1(2) = 2$
Verified
2. Assume S_k is true:
$2+4+6+8+10+\ldots+2k = k(k+1)$
and use it to show the truth of S_{k+1} follows. That is:
$2+4+6+\ldots+2k+2(k+1) = (k+1)(k+2)$
$S_k + a_{k+1} = S_{k+1}$
Working with the left hand side:
$2+4+6+\ldots+2k+2(k+1)$
$= k(k+1)+2(k+1)$
$= k^2+k+2k+2$
$= k^2+3k+2$
$= (k+1)(k+2)$
$= S_{k+1}$

Since the truth of S_{k+1} follows from S_k, the formula is true for all n.

29. $5+10+15+20+25+...+5n$

The needed components are:

$a_n = 5n$; $a_k = 5k$; $a_{k+1} = 5(k+1)$;

$S_n = \frac{5n(n+1)}{2}$; $S_k = \frac{5k(k+1)}{2}$;

$S_{k+1} = \frac{5(k+1)(k+1+1)}{2} = \frac{5(k+1)(k+2)}{2}$

1. Show S_n is true for $n = 1$.

   $S_1 = \frac{5(1)(1+1)}{2} = \frac{5(2)}{2} = 5$

   Verified

2. Assume S_k is true:

   $5+10+15+...+5k = \frac{5k(k+1)}{2}$

   and use it to show the truth of S_{k+1} follows. That is:

   $5+10+15+...+5k+5(k+1)$
   $= \frac{5(k+1)(k+1+1)}{2}$

   $S_k + a_{k+1} = S_{k+1}$

   Working with the left hand side:

   $5+10+15+...+5k+5(k+1)$

   $= \frac{5k(k+1)}{2} + 5(k+1)$

   $= \frac{5k(k+1)+10(k+1)}{2}$

   $= \frac{(k+1)(5k+10)}{2}$

   $= \frac{5(k+1)(k+2)}{2}$

   $= S_{k+1}$

   Since the truth of S_{k+1} follows from S_k, the formula is true for all n.

31. $5+9+13+17+...+4n+1$

The needed components are:

$a_n = 4n+1$; $a_k = 4k+1$;

$a_{k+1} = 4(k+1)+1 = 4k+4+1 = 4k+5$;

$S_n = n(2n+3)$; $S_k = k(2k+3)$;

$S_{k+1} = (k+1)(2(k+1)+3) = (k+1)(2k+5)$

1. Show S_n is true for $n = 1$.

   $S_1 = 1(2(1)+3) = 5$

   Verified

2. Assume S_k is true:

   $5+9+13+17+...+4k+1 = k(2k+3)$

   and use it to show the truth of S_{k+1} follows. That is:

   $5+9+13+17+...+4k+1+4(k+1)+1$
   $= (k+1)(2(k+1)+3)$

   $S_k + a_{k+1} = S_{k+1}$

   Working with the left hand side:

   $5+9+13+17+...+4k+1+4k+5$

   $= k(2k+3)+4k+5$

   $= 2k^2+3k+4k+5 = 2k^2+7k+5$

   $= (k+1)(2k+5) = S_{k+1}$

   Since the truth of S_{k+1} follows from S_k, the formula is true for all n.

33. $3+9+27+81+243+...+3^n$

The needed components are:

$a_n = 3^n$; $a_k = 3^k$; $a_{k+1} = 3^{k+1}$;

$S_n = \frac{3(3^n-1)}{2}$; $S_k = \frac{3(3^k-1)}{2}$;

$S_{k+1} = \frac{3(3^{k+1}-1)}{2}$

1. Show S_n is true for $n = 1$.

   $S_1 = \frac{3(3^1-1)}{2} = \frac{3(3-1)}{2} = \frac{3(2)}{2} = 3$

   Verified

2. Assume S_k is true:

   $3+9+27+...+3^k = \frac{3(3^k-1)}{2}$

   and use it to show the truth of S_{k+1} follows. That is:

   $3+9+27+...+3^k+3^{k+1} = \frac{3(3^{k+1}-1)}{2}$

   $S_k + a_{k+1} = S_{k+1}$

   Working with the left hand side:

   $3+9+27+...+3^k+3^{k+1}$

   $= \frac{3(3^k-1)}{2} + 3^{k+1}$

   $= \frac{3(3^k-1)+2(3^{k+1})}{2}$

   $= \frac{3^{k+1}-3+2(3^{k+1})}{2}$

   $= \frac{3(3^{k+1})-3}{2}$

$$= \frac{3\left(3^{k+1}-1\right)}{2}$$
$$= S_{k+1}$$

Since the truth of S_{k+1} follows from S_k, the formula is true for all n.

35. $2+4+8+16+32+64+...+2^n$

The needed components are:

$a_n = 2^n$; $a_k = 2^k$; $a_{k+1} = 2^{k+1}$;

$S_n = 2^{n+1}-2$; $S_k = 2^{k+1}-2$;

$S_{k+1} = 2^{k+1+1}-2 = 2^{k+2}-2$

1. Show S_n is true for $n = 1$.

$S_n = 2^{n+1}-2$

$S_1 = 2^{1+1}-2 = 2^2-2 = 4-2 = 2$

Verified

2. Assume S_k is true:

$2+4+8+...+2^k = 2^{k+1}-2$

and use it to show the truth of S_{k+1} follows. That is:

$2+4+8+...+2^k+2^{k+1} = 2^{k+1}-2$

$S_k + a_{k+1} = S_{k+1}$

Working with the left hand side:

$2+4+8+...+2^k+2^{k+1}$

$= 2^{k+1}-2+2^{k+1}$

$= 2\left(2^{k+1}\right)-2$

$= 2^{k+2}-2$

$= S_{k+1}$

Since the truth of S_{k+1} follows from S_k, the formula is true for all n.

37. $\frac{1}{1(3)}+\frac{1}{3(5)}+\frac{1}{5(7)}+...+\frac{1}{(2n-1)(2n+1)}$

The needed components are:

$a_n = \frac{1}{(2n-1)(2n+1)}$; $a_k = \frac{1}{(2k-1)(2k+1)}$;

$a_{k+1} = \frac{1}{(2(k+1)-1)(2(k+1)+1)} = \frac{1}{(2k+1)(2k+3)}$;

$S_n = \frac{n}{2n+1}$; $S_k = \frac{k}{2k+1}$;

$S_{k+1} = \frac{k+1}{2(k+1)+1} = \frac{k+1}{2k+3}$

1. Show S_n is true for $n = 1$.

$S_n = \frac{n}{2n+1}$

$S_1 = \frac{1}{2(1)+1} = \frac{1}{2+1} = \frac{1}{3}$

Verified

2. Assume S_k is true:

$\frac{1}{3}+\frac{1}{15}+\frac{1}{35}+...+\frac{1}{(2k-1)(2k+1)} = \frac{k}{2k+1}$

and use it to show the truth of S_{k+1} follows. That is:

$\frac{1}{3}+\frac{1}{15}+\frac{1}{35}+...+\frac{1}{(2k-1)(2k+1)}$

$+\frac{1}{(2(k+1)-1)(2(k+1)+1)} = \frac{k+1}{2(k+1)+1}$

$S_k + a_{k+1} = S_{k+1}$

Working with the left hand side:

$$\frac{1}{3}+\frac{1}{15}+\frac{1}{35}+...+\frac{1}{(2k-1)(2k+1)}+\frac{1}{(2k+1)(2k+3)}$$

$$= \frac{k}{2k+1}+\frac{1}{(2k+1)(2k+3)}$$

$$= \frac{k(2k+3)+1}{(2k+1)(2k+3)}$$

$$= \frac{2k^2+3k+1}{(2k+1)(2k+3)}$$

$$= \frac{(2k+1)(k+1)}{(2k+1)(2k+3)}$$

$$= \frac{k+1}{2k+3}$$

$$= S_{k+1}$$

Since the truth of S_{k+1} follows from S_k, the formula is true for all n.

39. $S_n : 3^n \geq 2n+1$

$S_k : 3^k \geq 2k+1$

$S_{k+1} : 3^{k+1} \geq 2(k+1)+1$

1. Show S_n is true for $n = 1$.

S_1 :

$3^1 \geq 2(1)+1$

$3 \geq 2+1$

$3 \geq 3$

Verified

2. Assume $S_k : 3^k \geq 2k+1$ is true and use it to show the truth of S_{k+1} follows. That is: $3^{k+1} \geq 2k+3$.

Working with the left hand side:

$$3^{k+1} = 3(3^k) \\ \geq 3(2k+1) \\ \geq 6k+3$$

Since k is a positive integer,
$6k+3 \geq 2k+3$

Showing $S_{k+1}: 3^{k+1} \geq 2k+3$

Verified

41. $S_n: 3 \cdot 4^{n-1} \leq 4^n - 1$

$S_k: 3 \cdot 4^{k-1} \leq 4^k - 1$

$S_{k+1}: 3 \cdot 4^k \leq 4^{k+1} - 1$

1. Show S_n is true for $n = 1$.

S_1:

$3 \cdot 4^{1-1} \leq 4^1 - 1$

$3 \cdot 4^0 \leq 4 - 1$

$3 \cdot 1 \leq 3$

$3 \leq 3$

Verified

2. Assume $S_k: 3 \cdot 4^{k-1} \leq 4^k - 1$ is true. and use it to show the truth of S_{k+1} follows. That is: $3 \cdot 4^k \leq 4^{k+1} - 1$.

Working with the left hand side:

$$3 \cdot 4^k = 3 \cdot 4(4^{k-1}) \\ = 4 \cdot 3(4^{k-1}) \\ \leq 4(4^k - 1) \\ \leq 4^{k+1} - 4$$

Since k is a positive integer,
$4^{k+1} - 4 \leq 4^{k+1} - 1$

Showing that $3 \cdot 4^k \leq 4^{k+1} - 1$

43. $n^2 - 7n$ is divisible by 2

1. Show S_n is true for $n = 1$.

$S_n: n^2 - 7n = 2m$

S_1:

$$(1)^2 - 7(1) = 2m \\ 1 - 7 = 2m \\ -6 = 2m$$

Verified

2. Assume $S_k: k^2 - 7k = 2m$ for $m \in Z$. and use it to show the truth of S_{k+1} follows. That is:

$(k+1)^2 - 7(k+1) = 2p$ for $p \in Z$.

Working with the left hand side:

$$= (k+1)^2 - 7(k+1) \\ = k^2 + 2k + 1 - 7k - 7 \\ = k^2 - 7k + 2k - 6 \\ = 2m + +2k - 6 \\ = 2(m + k - 3)$$

is divisible by 2.

45. $n^3 + 3n^2 + 2n$ is divisible by 3

1. Show S_n is true for $n = 1$.

$S_n: n^3 + 3n^2 + 2n = 3m$

S_1:

$(1)^3 + 3(1)^2 + 2(1) = 3m$

$1 + 3 + 2 = 3m$

$6 = 3m$

$2 = m$

Verified

2. Assume $S_k: k^3 + 3k^2 + 2k = 3m$ for $m \in Z$ and use it to show the truth of S_{k+1} follows.

That is:

$S_{k+1}: (k+1)^3 + 3(k+1)^2 + 2(k+1) = 3p$ for $p \in Z$.

Working with the left hand side:

$(k+1)^3 + 3(k+1)^2 + 2(k+1)$ is true.

$$= k^3 + 3k^2 + 3k + 1 + 3(k^2 + 2k + 1) + 2k + 2 \\ = k^3 + 3k^2 + 2k + 3(k^2 + 2k + 1) + 3k + 3 \\ = k^3 + 3k^2 + 2k + 3(k^2 + 2k + 1) + 3(k+1) \\ = 3m + 3(k^2 + 2k + 1) + 3(k+1)$$

is divisible by 3.

47. $6^n - 1$ is divisible by 5

1. Show S_n is true for $n = 1$.

$S_n: 6^n - 1 = 5m$

S_1:

$6^1 - 1 = 5m$

$6 - 1 = 5m$

$5 = 5m$

$1 = m$

Verified

2. Assume $S_k: 6^k - 1 = 5m$ for $m \in Z$ and use it to show the truth of S_{k+1} follows.

That is: $S_{k+1}: 6^{k+1} - 1 = 5p$ for $p \in Z$.

Working with the left hand side:

$= 6^k - 1$
$= 6(6^k) - 1$
$= 6(5m+1) - 1$
$= 30m + 6 - 1$
$= 30m + 5$
$= 5(6m+1)$
is divisible by 5,
Verified

49. $S_n : 3n^2 - 1 > 4(n-1)^2$

Assume $S_k : 3k^2 - 1 > 4(k-1)^2$

Use it to show $S_{k+1} : 3(k+1)^2 - 1 > 4(k)^2$

$3(8+1)^2 - 1 > 4(8)^2$

$3(9)^2 - 1 > 4(64)$

$3(81) - 1 > 256$

$242 > 256$

The relation cannot be verified for all n;
Using a grapher table with
$y_1 = 3k^2 - 1$ and $y_2 = 4(k-1)^2$,
$n = 8$ is the first natural number for which the statement is false.

51. $\dfrac{x^n - 1}{x - 1} = (1 + x + x^2 + x^3 + ... + x^{n-1})$

The needed components are:

$a_k = x^{k-1}$, $S_k = \dfrac{x^k - 1}{x-1}$, $S_{k+1} = \dfrac{x^{k+1} - 1}{x-1}$

1. Show S_n is true for $n = 1$.

$$S_k = \frac{x^k - 1}{x - 1}$$

$$S_1 = \frac{x^1 - 1}{x - 1} = 1$$

Verified

2. Assume S_k is true:

$$1 + x + x^2 + x^3 + ... + x^{k-1} = \frac{x^k - 1}{x - 1}$$

and use it to show the truth of S_{k+1} follows. That is:

$$1 + x + x^2 + x^3 + ... + x^{k-1} + x^{k+1-1} = \frac{x^{k+1} - 1}{x - 1}$$

$S_k + a_{k+1} = S_{k+1}$

Working with the left hand side:

$1 + x + x^2 + x^3 + ... + x^{k-1} + x^{k+1-1}$

$$= \frac{x^k - 1}{x - 1} + x^k$$

$$= \frac{x^k - 1}{x - 1} + \frac{x^k(x-1)}{x - 1}$$

$$= \frac{x^k - 1 + x^k(x-1)}{x - 1}$$

$$= \frac{x^k - 1 + x^{k+1} - x^k}{x - 1}$$

$$= \frac{x^{k+1} - 1}{x - 1}$$

$= S_{k+1}$

Since the truth of S_{k+1} follows from S_k, the formula is true for all n.

53. $A = \begin{bmatrix} -1 & 2 \\ 3 & 1 \end{bmatrix}$; $B = \begin{bmatrix} 2 & -1 \\ 4 & 3 \end{bmatrix}$

A + B

$$\begin{bmatrix} -1 & 2 \\ 3 & 1 \end{bmatrix} + \begin{bmatrix} 2 & -1 \\ 4 & 3 \end{bmatrix}$$

$$= \begin{bmatrix} -1+2 & 2+(-1) \\ 3+4 & 1+3 \end{bmatrix}$$

$$= \begin{bmatrix} 1 & 1 \\ 7 & 4 \end{bmatrix}$$

A – B

$$\begin{bmatrix} -1 & 2 \\ 3 & 1 \end{bmatrix} - \begin{bmatrix} 2 & -1 \\ 4 & 3 \end{bmatrix}$$

$$= \begin{bmatrix} -1-2 & 2-(-1) \\ 3-4 & 1-3 \end{bmatrix}$$

$$= \begin{bmatrix} -3 & 3 \\ -1 & -2 \end{bmatrix}$$

2A – 3B

$$2\begin{bmatrix} -1 & 2 \\ 3 & 1 \end{bmatrix} - 3\begin{bmatrix} 2 & -1 \\ 4 & 3 \end{bmatrix}$$

$$= \begin{bmatrix} -2 & 4 \\ 6 & 2 \end{bmatrix} - \begin{bmatrix} 6 & -3 \\ 12 & 9 \end{bmatrix}$$

$$= \begin{bmatrix} -2-6 & 4-(-3) \\ 6-12 & 2-9 \end{bmatrix}$$

$$= \begin{bmatrix} -8 & 7 \\ -6 & -7 \end{bmatrix}$$

AB

$$\begin{bmatrix} -1 & 2 \\ 3 & 1 \end{bmatrix}\begin{bmatrix} 2 & -1 \\ 4 & 3 \end{bmatrix}$$

$$=\begin{bmatrix}-1(2)+2(4) & -1(-1)+2(3)\\ 3(2)+1(4) & 3(-1)+1(3)\end{bmatrix}$$
$$=\begin{bmatrix}-2+8 & 1+6\\ 6+4 & -3+3\end{bmatrix}$$
$$=\begin{bmatrix}6 & 7\\ 10 & 0\end{bmatrix}$$

BA

$$\begin{bmatrix}2 & -1\\ 4 & 3\end{bmatrix}\begin{bmatrix}-1 & 2\\ 3 & 1\end{bmatrix}$$
$$=\begin{bmatrix}2(-1)+(-1)(3) & 2(2)+(-1)(1)\\ 4(-1)+3(3) & 4(2)+3(1)\end{bmatrix}$$
$$=\begin{bmatrix}-2-3 & 4-1\\ -4+9 & 8+3\end{bmatrix}$$
$$=\begin{bmatrix}-5 & 3\\ 5 & 11\end{bmatrix}$$

B^{-1}

$$\left[\begin{array}{cc:cc}2 & -1 & 1 & 0\\ 4 & 3 & 0 & 1\end{array}\right] \quad \frac{1}{2}R_1 \Rightarrow R_1$$
$$\left[\begin{array}{cc:cc}1 & -\frac{1}{2} & \frac{1}{2} & 0\\ 4 & 3 & 0 & 1\end{array}\right] \quad -4R_1 + R_2 \Rightarrow R_2$$
$$\left[\begin{array}{cc:cc}1 & -\frac{1}{2} & \frac{1}{2} & 0\\ 0 & 5 & -2 & 1\end{array}\right] \quad \frac{1}{5}R_2 \Rightarrow R_2$$
$$\left[\begin{array}{cc:cc}1 & -\frac{1}{2} & \frac{1}{2} & 0\\ 0 & 1 & -\frac{2}{5} & \frac{1}{5}\end{array}\right] \quad \frac{1}{2}R_2 + R_1 \Rightarrow R_1$$
$$\left[\begin{array}{cc:cc}1 & 0 & \frac{3}{10} & \frac{1}{10}\\ 0 & 1 & -\frac{2}{5} & \frac{1}{5}\end{array}\right]$$
$$\begin{bmatrix}\frac{3}{10} & \frac{1}{10}\\ -\frac{2}{5} & \frac{1}{5}\end{bmatrix}$$
$$\begin{bmatrix}0.3 & 0.1\\ -0.4 & 0.2\end{bmatrix}$$

55. Domain: $(-\infty,\infty)$
Range: $[-2,\infty)$

57. $3e^{(2x-1)}+5=17$
$3e^{(2x-1)}=12$
$e^{(2x-1)}=4$
$\ln e^{(2x-1)}=\ln 4$
$(2x-1)\ln e=\ln 4$
$2x-1=\ln 4$
$2x=1+\ln 4$
$x=\dfrac{1+\ln 4}{2}$

Mid-Chapter Check

1. $a_n=7n-4$
$a_1=7(1)-4=7-4=3$;
$a_2=7(2)-4=14-4=10$;
$a_3=7(3)-4=21-4=17$;
$a_9=7(9)-4=63-4=59$

2. $a_n=n^2+3$
$a_1=1^2+3=1+3=4$;
$a_2=(2)^2+3=4+3=7$;
$a_3=(3)^2+3=9+3=12$;
$a_9=(9)^2+3=81+3=84$

3. $a_n=(-1)^n(2n-1)$
$a_1=(-1)^1(2(1)-1)=-1(2-1)=-1(1)=-1$;
$a_2=(-1)^2(2(2)-1)=1(4-1)=1(3)=3$;
$a_3=(-1)^3(2(3)-1)=-1(6-1)=-1(5)=-5$;
$a_9=(-1)^9(2(9)-1)=-1(18-1)=-17$

4. $\sum_{n=1}^{4}3^{n+1}$
$=3^{1+1}+3^{2+1}+3^{3+1}+3^{4+1}$
$=3^2+3^3+3^4+3^5$
$=9+27+81+243$
$=360$

5. $1+4+7+10+13+16$
$\sum_{n=1}^{5}(3k-2)$

6. $S_n=\dfrac{n(a_1+a_n)}{2}$; d
summation formula for an arithmetic series

7. $a_n=a_1r^{n-1}$; e
nth term formula for a geometric series

8. $S_\infty = \dfrac{a_1}{1-r}$; a
sum of an infinite geometric series

9. $a_n = a_1 + (n-1)d$; b
nth term formula for an arithmetic series

10. $S_n = \dfrac{a_1(1-r^n)}{1-r}$; c
sum of a finite geometric series

11. (a) 2, 5, 8, 11, ...
$a_1 = 2$; $d = 5-2 = 3$
$a_n = 2+(n-1)3$
$a_n = 2+3n-3$
$a_n = 3n-1$

(b) $\frac{3}{2}, \frac{9}{4}, 3, \frac{15}{4}, \ldots$
$a_1 = \frac{3}{2}$; $d = \frac{9}{4} - \frac{3}{2} = \frac{3}{4}$
$a_n = \frac{3}{2} + (n-1)\frac{3}{4}$
$a_n = \frac{3}{2} + \frac{3}{4}n - \frac{3}{4}$
$a_n = \frac{3}{4}n + \frac{3}{4}$

12. 2 + 5 + 8 + 11 + … + 74
$a_1 = 2;\ d = 3$
$74 = 2+(n-1)(3)$
$74 = 2+3n-3$
$74 = 3n-1$
$75 = 3n$
$25 = n;$
$S_{25} = \dfrac{25(2+74)}{2} = \dfrac{25(76)}{2} = 950$

13. $\frac{1}{2} + \frac{3}{2} + \frac{5}{2} + \frac{7}{2} + \ldots + \frac{31}{2}$
$a_1 = \frac{1}{2};\ d = 1$
$\frac{31}{2} = \frac{1}{2} + (n-1)(1)$
$\frac{31}{2} = \frac{1}{2} + n - 1$
$\frac{31}{2} = n - \frac{1}{2}$
$16 = n;$
$S_{16} = \dfrac{16\left(\frac{1}{2}+\frac{31}{2}\right)}{2} = \dfrac{16(16)}{2} = 128$

14. $a_3 = -8;\ a_7 = 4$
$a_7 = a_3 + 4d$
$4 = -8 + 4d$
$12 = 4d$
$3 = d;$
$a_7 = a_1 + 6d$
$4 = a_1 + 6(3)$
$4 = a_1 + 18$
$-14 = a_1;$
$a_{10} = -14 + (10-1)(3) = 13$
$S_{10} = \dfrac{10(13-14)}{2} = -5$

15. $a_3 = -81;\ a_7 = -1$
$a_7 = a_3 \cdot r^4$
$-1 = -81r^4$
$\frac{1}{81} = r^4$
$\frac{1}{3} = r;$
$a_7 = a_1 r^6$
$-1 = a_1\left(\frac{1}{3}\right)^6$
$-1 = \frac{1}{729}a_1$
$-729 = a_1;$
$S_{10} = \dfrac{-729\left(1-\left(\frac{1}{3}\right)^{10}\right)}{1-\frac{1}{3}}$
$= \dfrac{-729\left(1-\frac{1}{59049}\right)}{\frac{2}{3}}$

$$= \frac{-729\left(\frac{59048}{59049}\right)}{\frac{2}{3}}$$

$$= \frac{-29524}{27}$$

16. (a) 2, 6, 18, 54, …

$a_1 = 2; \; r = \frac{6}{2} = 3;$

$a_n = 2(3)^{n-1}$

(b) $\frac{1}{2}, \frac{1}{4}, \frac{1}{8}, \frac{1}{16}, \ldots$

$a_1 = \frac{1}{2}; \; r = \frac{\frac{1}{4}}{\frac{1}{2}} = \frac{1}{2};$

$a_n = \frac{1}{2}\left(\frac{1}{2}\right)^{n-1} = \left(\frac{1}{2}\right)^n$

17. $\frac{1}{54} + \frac{1}{18} + \frac{1}{6} + \ldots + \frac{81}{2}$

$a_1 = \frac{1}{54}; \; r = \frac{\frac{1}{18}}{\frac{1}{54}} = 3$

$\frac{81}{2} = \frac{1}{54}(3)^{n-1}$
$2187 = 3^{(n-1)}$
$3^7 = 3^{(n-1)}$
$7 = n - 1$
$n = 8;$

$$S_8 = \frac{\frac{1}{54}(1-3^8)}{1-3} = \frac{\frac{1}{54}(1-6561)}{-2}$$

$$= \frac{\frac{1}{54}(-6560)}{-2} = \frac{6560}{108} = \frac{1640}{27}$$

18. $-49 + (-7) + (-1) + \left(-\frac{1}{7}\right) + \ldots$

$a_1 = -49;$

$r = \frac{-7}{-49} = \frac{1}{7}$ Since $r < 1,$

$$S_\infty = \frac{-49}{1-\frac{1}{7}} = \frac{-49}{\frac{6}{7}} = -49 \cdot \frac{7}{6} = \frac{-343}{6}$$

19. 60, 59, 58,…, 10

$a_1 = 60; \; d = -1$
$10 = 60 + (n-1)(-1)$
$10 = 60 - n + 1$
$10 = 61 - n$
$-51 = -n$
$51 = n;$

$$S_{51} = \frac{51(60+10)}{2} = \frac{51(70)}{2} = 1785$$

20. $a_1 = 8; \; r = 0.96$

$a_{15} = 8(0.96)^{15-1} = 8(0.96)^{14} \approx 4.5$ ft

$$S_{25} = \frac{8(1-0.96^{25})}{1-0.96}$$

$$= \frac{8(0.6396032831)}{0.04} \approx 127.9 \text{ ft}$$

8.5 Technology Highlight

Exercise 1:

(a) ${}_9C_2 = 36$

(b) ${}_9C_3 = 84$

(c) ${}_9C_4 = 126$

(d) ${}_9C_5 = 126$

Exercise 2: ${}_{45}C_5 = 1{,}221{,}759$

Exercise 3: ${}_6C_3 = 20$

8.5 Exercises

1. Experiment, well defined.

3. Distinguishable.

5. Answers will vary.

7. (a)

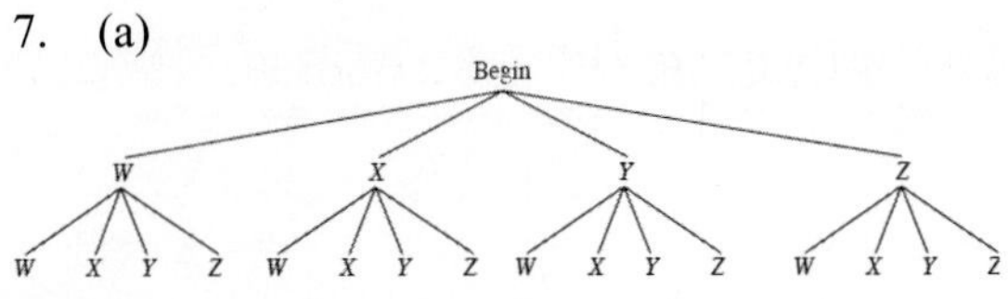

(b) $WW, WX, WY, WZ,$

$XW, XX, XY, XZ,$
$YW, YX, YY, YZ,$
ZW, ZX, ZY, ZZ

9. 32

11. $25^3 = 15{,}625$

13. $26 \cdot 26 \cdot 4 \cdot 10 \cdot 10 \cdot 10 = 2{,}704{,}000$

15. (a) $9^5 = 59{,}049$
    (b) $9 \cdot 8 \cdot 7 \cdot 6 \cdot 5 = 15{,}120$

17. $4 \cdot 6 \cdot 5 \cdot 3 = 360$
    360 if double vegetables are not allowed,
    $4 \cdot 6 \cdot 6 \cdot 3 = 432$
    432 if double vegetables are allowed.

19. (a) $5 \cdot 4 \cdot 3 \cdot 2 = 120$
    (b) $5^4 = 625$
    (c) $2 \cdot 3 \cdot 2 \cdot 1 = 12$

21. $4 \cdot 3 \cdot 2 \cdot 1 = 24$

23. $2 \cdot 2 \cdot 1 \cdot 1 = 4$

25. $5 \cdot 4 \cdot 3 \cdot 2 \cdot 1 = 120$

27. $1 \cdot 3 \cdot 9 \cdot 8 = 216$

29. ${}_{10}P_3 = 10 \cdot 9 \cdot 8 = 720$;

$${}_nP_r = \frac{n!}{(n-r)!}$$

$${}_{10}P_3 = \frac{10!}{(10-3)!} = \frac{10 \cdot 9 \cdot 8 \cdot 7!}{7!} = 720$$

31 ${}_9P_4 = 9 \cdot 8 \cdot 7 \cdot 6 = 3024$;

$${}_nP_r = \frac{n!}{(n-r)!}$$

$${}_9P_4 = \frac{9!}{(9-4)!} = \frac{9 \cdot 8 \cdot 7 \cdot 6 \cdot 5!}{5!} = 3024$$

33. ${}_8P_7 = 8 \cdot 7 \cdot 6 \cdot 5 \cdot 4 \cdot 3 \cdot 2 = 40320$

$${}_nP_r = \frac{n!}{(n-r)!}$$

$${}_8P_7 = \frac{8!}{(8-7)!}$$

$$= \frac{8 \cdot 7 \cdot 6 \cdot 5 \cdot 4 \cdot 3 \cdot 2 \cdot 1!}{1!} = 40320$$

35. T, R and A
    ${}_3P_3 = 3 \cdot 2 \cdot 1 = 6$
    TRA, TAR, RTA, RAT, ART, ATR
    3 actual words

37. ${}_{10}P_2 = 10 \cdot 9 = 90$

39. ${}_8P_3 = 8 \cdot 7 \cdot 6 = 336$

41. (a) ${}_6P_6 = 6 \cdot 5 \cdot 4 \cdot 3 \cdot 2 \cdot 1 = 720$
    (b) ${}_6P_3 = 6 \cdot 5 \cdot 4 = 120$
    (c) ${}_4P_4 = \cdot 4 \cdot 3 \cdot 2 \cdot 1 = 24$

43. $\frac{{}_nP_n}{p!} = \frac{{}_6P_6}{2!} = \frac{6 \cdot 5 \cdot 4 \cdot 3 \cdot 2 \cdot 1}{2 \cdot 1} = 360$

45. $\frac{{}_nP_n}{p!q!} = \frac{{}_6P_6}{2!3!} = \frac{6 \cdot 5 \cdot 4 \cdot 3 \cdot 2 \cdot 1}{2 \cdot 1 \cdot 3 \cdot 2 \cdot 1} = \frac{120}{2} = 60$

47. $\frac{{}_nP_n}{p!q!} = \frac{{}_6P_6}{2!3!} = \frac{6 \cdot 5 \cdot 4 \cdot 3 \cdot 2 \cdot 1}{2 \cdot 1 \cdot 3 \cdot 2 \cdot 1} = \frac{120}{2} = 60$

49. Logic
    $\frac{{}_nP_n}{p!} = \frac{{}_5P_5}{1!} = \frac{5 \cdot 4 \cdot 3 \cdot 2 \cdot 1}{1} = 120$

51. Lotto
    $\frac{{}_nP_n}{p!q!} = \frac{{}_5P_5}{2!2!} = \frac{5 \cdot 4 \cdot 3 \cdot 2 \cdot 1}{2 \cdot 1 \cdot 2 \cdot 1} = \frac{60}{2} = 30$

53. A, A, A, N, N, B
    $\frac{{}_nP_n}{p!q!} = \frac{{}_6P_6}{3!2!} = \frac{6 \cdot 5 \cdot 4 \cdot 3 \cdot 2 \cdot 1}{3 \cdot 2 \cdot 1 \cdot 2 \cdot 1} = \frac{120}{2} = 60$
    BANANA

55. ${}_9C_4$

    (a) ${}_nC_r = \frac{{}_nP_r}{r!}$

    $${}_9C_4 = \frac{{}_9P_4}{4!} = \frac{9 \cdot 8 \cdot 7 \cdot 6}{4 \cdot 3 \cdot 2 \cdot 1} = \frac{3024}{24} = 126;$$

    (b) ${}_nC_r = \frac{n!}{r!(n-r)!}$

$${}_9C_4 = \frac{9!}{4!(9-4)!} = \frac{9 \cdot 8 \cdot 7 \cdot 6 \cdot 5!}{4 \cdot 3 \cdot 2 \cdot 1 \cdot 5!}$$
$$= \frac{3024}{24} = 126$$

57. ${}_8C_5$

(a) ${}_nC_r = \frac{{}_nP_r}{r!}$

$${}_8C_5 = \frac{{}_8P_5}{5!} = \frac{8 \cdot 7 \cdot 6 \cdot 5 \cdot 4}{5 \cdot 4 \cdot 3 \cdot 2 \cdot 1}$$
$$= \frac{6720}{120} = 56;$$

(b) ${}_nC_r = \frac{n!}{r!(n-r)!}$

$${}_8C_5 = \frac{8!}{5!(8-5)!} = \frac{8 \cdot 7 \cdot 6 \cdot 5!}{5! \cdot 3 \cdot 2 \cdot 1}$$
$$= \frac{336}{6} = 56$$

59. ${}_6C_6$

(a) ${}_nC_r = \frac{{}_nP_r}{r!}$

$${}_6C_6 = \frac{{}_6P_6}{6!} = \frac{6 \cdot 5 \cdot 4 \cdot 3 \cdot 2 \cdot 1}{6 \cdot 5 \cdot 4 \cdot 3 \cdot 2 \cdot 1} = 1;$$

(b) ${}_nC_r = \frac{n!}{r!(n-r)!}$

$${}_6C_6 = \frac{6!}{6!(6-6)!} = \frac{6!}{6!} = 1$$

61. ${}_9C_4, {}_9C_5$

$${}_9C_4 = \frac{9!}{4!(9-4)!} = \frac{9!}{4!5!} = \frac{9 \cdot 8 \cdot 7 \cdot 6 \cdot 5!}{4 \cdot 3 \cdot 2 \cdot 1 \cdot 5!}$$
$${}_9C_4 = \frac{3024}{24} = 126;$$
$${}_9C_5 = \frac{9!}{5!(9-5)!} = \frac{9!}{5!4!} = \frac{9 \cdot 8 \cdot 7 \cdot 6 \cdot 5!}{5! \cdot 4 \cdot 3 \cdot 2 \cdot 1}$$
$${}_9C_5 = \frac{3024}{24} = 126$$

Verified

63. ${}_8C_5, {}_8C_3$

$${}_8C_5 = \frac{8!}{5!(8-5)!} = \frac{8!}{5!3!}$$
$$= \frac{8 \cdot 7 \cdot 6 \cdot 5!}{5! \cdot 3 \cdot 2 \cdot 1} = \frac{336}{6} = 56;$$
$${}_8C_3 = \frac{8!}{3!(8-3)!} = \frac{8!}{3!5!}$$
$$= \frac{8 \cdot 7 \cdot 6 \cdot 5!}{3 \cdot 2 \cdot 1 \cdot 5!} = \frac{336}{6} = 56$$

Verified

65. $${}_{12}C_4 = \frac{12!}{4!(12-4)!} = \frac{12!}{12!8!} = \frac{12 \cdot 11 \cdot 10 \cdot 9 \cdot 8!}{4 \cdot 3 \cdot 2 \cdot 1 \cdot 8!}$$
$$= \frac{11880}{24} = 495$$

67. $${}_{14}C_3 = \frac{14!}{3!(14-3)!} = \frac{14!}{3!11!} = \frac{14 \cdot 13 \cdot 12 \cdot 11!}{3 \cdot 2 \cdot 1 \cdot 11!}$$
$$= \frac{2184}{6} = 364$$

69. $${}_{10}C_5 = \frac{10!}{5!(10-5)!} = \frac{10!}{5!5!} = \frac{10 \cdot 9 \cdot 8 \cdot 7 \cdot 6 \cdot 5!}{5 \cdot 4 \cdot 3 \cdot 2 \cdot 1 \cdot 5!}$$
$$= \frac{30240}{120} = 252$$

71. $8! = 40{,}320$

73. $${}_nP_r = \frac{n!}{(n-r)!}$$
$${}_8P_3 = \frac{8!}{(8-3)!} = \frac{8!}{5!} = \frac{8 \cdot 7 \cdot 6 \cdot 5!}{5!} = 336$$

75. $${}_{20}C_5 = \frac{20!}{5!(20-5)!} = \frac{20!}{5!15!}$$
$$= \frac{20 \cdot 19 \cdot 18 \cdot 17 \cdot 16 \cdot 15!}{5 \cdot 4 \cdot 3 \cdot 2 \cdot 1 \cdot 15!} = \frac{1860480}{120} = 15{,}504$$

77. $${}_8C_4 = \frac{8!}{4!(8-4)!} = \frac{8 \cdot 7 \cdot 6 \cdot 5 \cdot 4!}{4 \cdot 3 \cdot 2 \cdot 1 \cdot 4!}$$
$$= \frac{1680}{24} = 70$$

79. (a) $7! = 5{,}040;$

$$n! \approx \sqrt{2\pi} \cdot \left(n^{n+0.5}\right) \cdot e^{-n}$$
$$7! \approx \sqrt{2\pi} \cdot \left(7^{7+0.5}\right) \cdot e^{-7}$$
$$7! \approx \sqrt{2\pi} \cdot \left(7^{7.5}\right)\left(e^{-7}\right)$$
$$7! \approx 4980.395832;$$
$$\frac{5040 - 4980}{5040} = 0.0119 \approx 1.2\%$$

(b) $10! = 3{,}628{,}800;$

$$10! \approx \sqrt{2\pi} \cdot \left(10^{10+0.5}\right) \cdot e^{-10}$$
$$10! \approx \sqrt{2\pi} \cdot \left(10^{10.5}\right) \cdot e^{-10}$$
$$10! \approx 3598695.619;$$

81. $6^5 = 7776$

83. $9 \cdot 6 \cdot 6 = 324$

85. ${}_{15}P_4 = \dfrac{15!}{(15-4)!}$

$${}_{15}P_4 = \frac{15 \cdot 14 \cdot 13 \cdot 12 \cdot 11!}{11!} = 32{,}760$$

87. ${}_{15}P_5 = \dfrac{15 \cdot 14 \cdot 13 \cdot 12 \cdot 11 \cdot 10!}{10!} = 360{,}360$;

360360(4) = 1,441,440;

$${}_{15}P_4 = \frac{15 \cdot 14 \cdot 13 \cdot 12 \cdot 11!}{11!} = 32{,}760;$$

1,441,440 + 32,760 = 1,474,200

89. $8 \cdot 10 \cdot 10 = 800$

91. Exchanges: $8 \cdot 10 \cdot 10 = 800$
Area Codes: $8 \cdot 10 \cdot 10 = 800 - 16 = 784$
Final digits: $10 \cdot 10 \cdot 10 \cdot 10 = 10000$
$784 \cdot 800 \cdot 10000 = 6{,}272{,}000{,}000$

93. $9 \cdot 10 \cdot 10 \cdot 24 \cdot 24 = 518{,}400$

95. $9 \cdot 9 \cdot 8 \cdot 24 \cdot 23 = 357{,}696$

97. Five

$${}_8P_5 = \frac{8!}{(8-5)!} = \frac{8 \cdot 7 \cdot 6 \cdot 5 \cdot 4 \cdot 3!}{3!} = 6{,}720$$

99. One

$${}_8P_1 = \frac{8!}{(8-1)!} = \frac{8 \cdot 7!}{7!} = 8$$

101. $7 \cdot 2 \cdot 6! = 10{,}080$
7 ways they can sit side by side;
2 ways they can sit together, teacher 1 on the left and teacher 2 on the right, or teacher 2 on the left and teacher 1 on the right;
the students can be seated randomly.

103. $1 \cdot 7! = 1 \cdot 7 \cdot 6 \cdot 5 \cdot 4 \cdot 3 \cdot 2 \cdot 1 = 5{,}040$

105. $2 \cdot 2 \cdot 6 \cdot 5 \cdot 4 \cdot 3 \cdot 2 \cdot 1 = 2880$

107. ${}_{15}C_6 = \dfrac{15!}{6!(15-6)!}$

$$= \frac{15 \cdot 14 \cdot 13 \cdot 12 \cdot 11 \cdot 10 \cdot 9!}{6 \cdot 5 \cdot 4 \cdot 3 \cdot 2 \cdot 1 \cdot 9!}$$

$${}_{15}C_6 = \frac{3603600}{720} = 5005$$

109. ${}_{10}P_3 = \dfrac{10 \cdot 9 \cdot 8 \cdot 7!}{7!} = 720$

111. $26 \cdot 25 \cdot 9 \cdot 9 = 52{,}650$; no

113. $n! \approx \sqrt{2\pi} \cdot \left(n^{n+0.5}\right) \cdot e^{-n}$
According to the grapher, the approximation gets better as (n) gets larger.

115. A 5-move win: "X" wins
8 winning positions for X (two diagonals, three rows, three columns).
Choose two of the remaining squares for O's.
$8\left({}_6C_2\right) = 120$
Take into account the orders for reaching these positions
for X: ${}_3P_2 = 6$
for O: ${}_2P_2 = 2$
Thus, $120(6)(2) = 1440$

A 6-move win: "O" wins
There are two winning positions in which the O's are on the diagonals. That leaves 6 squares for the 3 X's (none of which allow X to win first). The orders in which the O's could be chosen is 3! and similarly for the X's. Thus, the number of paths to a 'diagonal' winning position is
$2\left({}_6C_3\right)(3!)(3!) = 1440$
The six winning positions of O's in the columns and rows are different - for each case one must exclude the two positions where there are 3 X's in a remaining row or column. (Since, in that case, X would have won first.) Thus, for the six winning positions in which the O's are in a row or column, the number of paths is:
$6\left({}_6C_3 - 2\right)(3!)(3!) = 3888$
Thus, 1440+3888=5328.

117. $h(x) = \dfrac{x^3 - x}{x^2 - 4}$

$$= \frac{x\left(x^2 - 1\right)}{(x+2)(x-2)} = \frac{x(x+1)(x-1)}{(x+2)(x-2)}$$

$x^3 - x = 0$
$x(x^2 - 1) = 0$
$x(x-1)(x+1) = 0$

$x = 0;\ \ x = 1;\ \ x = -1$
x-intercepts: (0, 0), (1, 0), (-1, 0)
$h(0) = \frac{0^3 - 0}{0^2 - 4} = \frac{0}{4} = 0$
y-intercept: (0, 0)
$x^2 - 4 = 0$
$(x-2)(x+2) = 0$
$x = 2;\ \ x = -2$
Vertical Asymptotes: $x = 2; x = -2,\ \ y = x$
No Horizontal Asymptotes
Oblique Asymptote: $y = x$

$$\begin{array}{r} x \\ x^2 - 4 \overline{)\, x^3 - x} \\ -(x^3 - 4x) \\ \hline 3x \end{array}$$

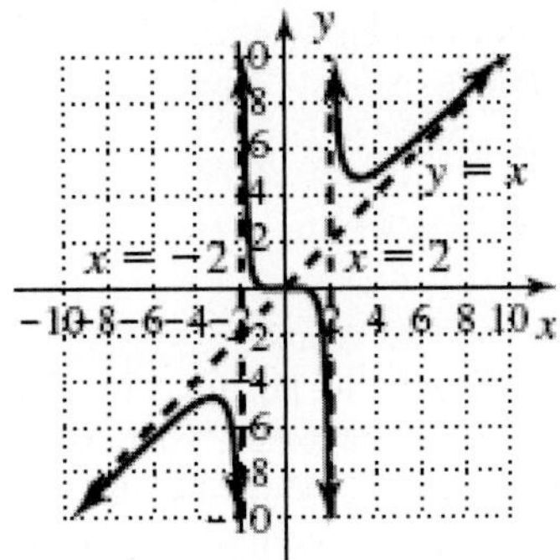

119. $1 + 5 + 9 + 13 + ... + 197$

$a_1 = 1;\ \ r = 5 - 1 = 4;\ \ a_n = 197$
$a_n = 1 + (n-1)4$
$a_{35} = 1 + (35-1)4 = 1 + 34(4) = 1 + 136 = 137;$
$S_n = \frac{n(a_1 + a_n)}{2}$
$S_{35} = \frac{35(1+137)}{2} = \frac{35(138)}{2} = 2415$

121. $\frac{(x-2)^2}{4} - \frac{(y+3)^2}{9} = 1$

Center: (2, -3)
$a = 2, b = 3$
Vertices: (2-2, -3), (2+2, -3)
(0,-3), (4,-3)

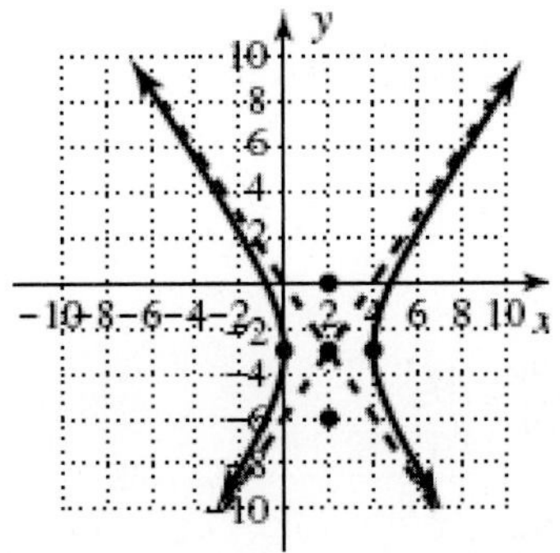

8.6 Technology Highlight

Exercise 1: The "middle values" are repeated for n =7 and n =5 but not n = 6 because of the symmetry of ${}_nC_r$.

Exercise 2: $({}_{10}C_4)({}_8C_3) = 11{,}760$

8.6 Exercises

1. $P(E) = \frac{n(E)}{n(S)}$

3. $0 \le P(E) \le 1$
   $P(S) = 1$, and $P(\sim S) = 0$

5. Answers will vary.

7. $S = \{HH, HT, TH, TT\}$; $\frac{1}{4}$

9. S ={coach of Patriots, Cougars, Angels, Sharks, Eagles, Stars}; $\frac{1}{6}$

11. S = {nine index cards 1-9}
    $P(E) = \frac{4}{9}$

13. S ={52 cards}
    a. drawing a Jack: $\frac{4}{52} = \frac{1}{13}$
    b. drawing a spade: $\frac{13}{52} = \frac{1}{4}$
    c. drawing a black card: $\frac{26}{52} = \frac{1}{2}$
    d. drawing a red three: $\frac{2}{52} = \frac{1}{26}$

15. $S=\{\text{three males, five females}\}$

$P(E_1)=\frac{1}{8}$;

$P(E_2)=\frac{5}{8}$;

$P(E_3)=\frac{6}{8}=\frac{3}{4}$

17. $S=\{\text{Spinner 1-4}\}$

a. P(green): $\frac{3}{4}$

b. P(less than 5): $\frac{4}{4}=1$

c. P(2): $\frac{1}{4}$

d. P(prime number): $\frac{2}{4}=\frac{1}{2}$

(1 is not considered a prime number)

19. $P(\sim C)=1-P(C)=1-\frac{13}{52}=\frac{39}{52}=\frac{3}{4}$

21. $10!=3{,}628{,}800$

$P(\sim 2)=1-P(2)=1-\frac{1}{7}=\frac{6}{7}$

23. $P(\sim F)=1-P(F)=1-0.009=0.991$

25. a. $P(\text{Sum}<4)=\frac{3}{36}=\frac{1}{12}$

b. $P(\sim \text{Sum}<11)=1-P(\text{Sum}>10)=1-\frac{3}{36}$

$=\frac{33}{36}=\frac{11}{12}$

c. $P(\sim \text{Sum}9)=1-P(9)=1-\frac{4}{36}=\frac{32}{36}=\frac{8}{9}$

d. $P(\sim D)=1-P(D)=1-\frac{6}{36}=\frac{30}{36}=\frac{5}{6}$

27. $n(E)={}_6C_3\cdot{}_4C_2;\, n(S)={}_{10}C_5$

$P(E)=\frac{{}_6C_3\cdot{}_4C_2}{{}_{10}C_5}=\frac{120}{252}=\frac{10}{21}$

29. $n(E)={}_9C_6\cdot{}_5C_3;\, n(S)={}_{14}C_9$

$P(E)=\frac{{}_9C_6\cdot{}_5C_3}{{}_{14}C_9}=\frac{840}{2002}=\frac{60}{143}$

31. a. $P(\text{all red})=\frac{{}_{26}C_5}{{}_{52}C_5}$

$=\frac{65780}{2598960}=\frac{253}{9996}\approx 0.025$

b. $P(\text{all numbered})=\frac{{}_{36}C_5}{{}_{52}C_5}$

$=\frac{376992}{2598960}=\frac{66}{455}\approx 0.145$

b; $0.145-0.025=0.12$

about 12 %

33. a. exactly two vegetarians

$P(E)=\frac{{}_9C_2\cdot{}_{15}C_4}{{}_{24}C_6}=\frac{49140}{134596}=0.3651$

b. exactly four non-vegetarians

$P(E)=\frac{{}_{15}C_4\cdot{}_9C_2}{{}_{24}C_6}=\frac{49140}{134596}=0.3651$

c. at least three vegetarians

$1-(P(0\text{veg})+P(1\text{veg})+P(2\text{veg}))$

$=1-\left(\frac{{}_9C_0\cdot{}_{15}C_6+{}_9C_1\cdot{}_{15}C_5+{}_9C_2\cdot{}_{15}C_4}{{}_{24}C_6}\right)$

$=1-\left(\frac{81172}{134596}\right)=0.3969$

35. a. E_1 = boxcars; E_2 = snake eyes

$P(E_1\cup E_2)=P(E_1)+P(E_2)$

$P(E_1\cup E_2)=\frac{1}{36}+\frac{1}{36}=\frac{2}{36}=\frac{1}{18}$

b. E_1 = sum of 7; E_2 = sum of 11

$P(E_1\cup E_2)=P(E_1)+P(E_2)$

$P(E_1\cup E_2)=\frac{6}{36}+\frac{2}{36}=\frac{8}{36}=\frac{2}{9}$

c. E_1 = even numbered sum;

E_2 = prime sum

$P(E_1\cup E_2)$

$=P(E_1)+P(E_2)-P(E_1\cap E_2)$

$P(E_1\cup E_2)=\frac{18}{36}+\frac{15}{36}-\frac{1}{36}=\frac{32}{36}=\frac{8}{9}$

d. E_1 = odd numbered sum;

E_2 = multiple of four

$P(E_1\cup E_2)=P(E_1)+P(E_2)$

$P(E_1\cup E_2)=\frac{18}{36}+\frac{9}{36}=\frac{27}{36}=\frac{3}{4}$

e. E_1 = a sum of 15; E_2=multiple of 12

$P(E_1\cup E_2)=P(E_1)+P(E_2)$

$P(E_1\cup E_2)=\frac{0}{36}+\frac{1}{36}=\frac{1}{36}$

f. E = prime number sum

$P(E)=\frac{15}{36}=\frac{5}{12}$

37. $P(E_1)=0.7; P(E_2)=0.5; P(E_1\cap E_2)=0.3$

$P(E_1\cup E_2)=P(E_1)+P(E_2)-P(E_1\cap E_2)$

$P(E_1\cup E_2)=0.7+0.5-0.3=0.9$

39. $P(E_1)=\frac{3}{8}; P(E_2)=\frac{3}{4}; P(E_1\cup E_2)=\frac{15}{18}$

$P(E_1\cup E_2)=P(E_1)+P(E_2)-P(E_1\cap E_2)$

$\frac{15}{18}=\frac{3}{8}+\frac{3}{4}-P(E_1\cap E_2)$

$\frac{15}{18}=\frac{9}{8}-P(E_1\cap E_2)$

$-\frac{7}{24}=-P(E_1\cap E_2)$

$\frac{7}{24}=P(E_1\cap E_2)$

41. $P(E_1\cup E_2)=0.72; P(E_2)=0.56;$

$P(E_1\cap E_2)=0.43$

$P(E_1\cup E_2)=P(E_1)+P(E_2)-P(E_1\cap E_2)$

$0.72=P(E_1)+0.56-0.43$

$0.72=P(E_1)+0.13$

$0.59=P(E_1)$

43. a. $P(\text{multiple of 3 and odd})=\frac{6}{36}=\frac{1}{6}$

b. $P(\text{sum}>5\text{ and a }3)=\frac{7}{36}$

c. $P(\text{even and}>9)=\frac{4}{36}=\frac{1}{9}$

d. $P(\text{odd and}<10)=\frac{16}{36}=\frac{4}{9}$

45. a. $P(\text{woman and sergeant})=\frac{4}{50}=\frac{2}{25}$

b. $P(\text{man and private})=\frac{9}{50}$

c. $P(\text{private and sergeant})=\frac{0}{50}=0$

d. $P(\text{woman and officer})=\frac{4}{50}=\frac{2}{25}$

e. $P(\text{person in military})=\frac{50}{50}=1$

47. $\frac{9\cdot5\cdot10+9\cdot5\cdot10-9\cdot5\cdot5}{9\cdot10\cdot10}=\frac{3}{4}$

49. A number greater than 4000 can have digits 4, 5, 6, 7, 8, or 9 in the first position.
(6 choices)
Multiples of 5 must end in 0 or 5.
(2 choices)

$\frac{6\cdot10\cdot10\cdot10}{9\cdot10\cdot10\cdot10}+\frac{9\cdot10\cdot10\cdot2}{9\cdot10\cdot10\cdot10}-\frac{6\cdot10\cdot10\cdot2}{9\cdot10\cdot10\cdot10}$

$=\frac{6}{9}+\frac{1}{5}-\frac{6}{45}=\frac{11}{15}$

51. $P(n)=\left(\frac{1}{4}\right)^n$

a. $P(\text{spins a 2})=\left(\frac{1}{4}\right)^1=\frac{1}{4}$

b. $P(\text{all 4 spin a 2})=\left(\frac{1}{4}\right)^4=\frac{1}{256}$

c. Answers will vary.

53. a. $P(x\geq 2)=0.25+0.08=0.33$

b. $P(x<2)=0.07+0.28+0.32=0.67$

c. $P(x\leq 4)$

$=0.08+0.25+0.32+0.28+0.07=1$

d. $P(x>4)=0$

e. $P(x<2\text{ or x}>4)=0.67+0=0.67$

f. $P(x\geq 3)=0.08$

55. Total = 200

a. $P(\text{Isosceles})=\frac{\frac{1}{2}(200)}{200}=\frac{1}{2}$

b. $P(\text{Right triangle})=\frac{\frac{1}{2}(200)}{200}=\frac{1}{2}$

c. $5^2+h^2=10^2$

$h^2=75$

$h=5\sqrt{3};$

$P(\text{Equilateral})=\frac{\frac{1}{2}(10)5\sqrt{3}}{200}=\frac{\sqrt{3}}{8}$

≈ 0.2165

57. a. $P(x \ge 4) = \frac{\pi(6)^2}{\pi(8)^2} = \frac{36}{64} = \frac{9}{16}$

b. $P(x \ge 6) = \frac{\pi(4)^2}{\pi(8)^2} = \frac{16}{64} = \frac{1}{4}$

c. $P(\text{exactly } 8) = \frac{\pi(2)^2}{\pi(8)^2} = \frac{4}{64} = \frac{1}{16}$

d. $P(x = 4) = \frac{\pi(6)^2 - \pi(4)^2}{\pi(8)^2} = \frac{20\pi}{64\pi} = \frac{5}{16}$

59. n = 13, 3R, 6B, 4W

a. $P(\text{red, blue}) = \frac{3}{13} \cdot \frac{6}{12} = \frac{3}{26}$

b. $P(\text{blue, red}) = \frac{6}{13} \cdot \frac{3}{12} = \frac{3}{26}$

c. $P(\text{white, white}) = \frac{4}{13} \cdot \frac{3}{12} = \frac{1}{13}$

d. $P(\text{blue, not red}) = \frac{6}{13} \cdot \frac{9}{12} = \frac{9}{26}$

e. $P(\text{white, not blue}) = \frac{4}{13} \cdot \frac{6}{12} = \frac{2}{13}$

f. $P(\text{not red, not blue})$

$= \frac{6}{13} \cdot \frac{3}{12} + \frac{6}{13} \cdot \frac{4}{12} + \frac{4}{13} \cdot \frac{3}{12} + \frac{4}{13} \cdot \frac{3}{12}$

$= \frac{3}{26} + \frac{4}{26} + \frac{2}{26} + \frac{2}{26} = \frac{11}{26}$

61. Let C represent correct.
Let W represent wrong.

a. $P(\text{Grade} \ge 80\%) = \left(\frac{1}{2}\right)^3 = \frac{1}{8}$

With 3 questions, the only grade greater than or equal to 80% would be a 100% since 2 out of three questions correct only give 67%.
Possible outcomes: CCC, CCW, CWW, CWC, WCC, WCW, WWC, WWW

b. $P(\text{Grade} \ge 80\%) = \left(\frac{1}{2}\right)^4 = \frac{1}{16}$

With 4 questions, the only grade greater than or equal to 80% would be a 100% since 3 out of four questions correct only give 75%.Possible outcomes: CCCC, CCCW, CCWC,CCWW, CWCC, CWCW, CWWC, CWWW, WCCW, WCCC, WCWC, WCWW, WWCC, WWCW, WWWC , WWWW.

c. $P(\text{Grade} \ge 80\%) = \frac{1}{32} + \frac{5}{32} = \frac{6}{32} = \frac{3}{16}$

With 5 questions, the only grade greater than 80% would be a 100% but 4 out of five questions would give 80%. Thus, we need 5 or 4 correct answers.
Possible outcomes: CCCCC, CCCCW, CCCWC,CCWW, CCWCC, CCWCW, CCWWC, CCWWW, CWCCC, CWCCW, CWCWC, CWCWW, CWWCC, CWWCW, CWWWC, CWWWW, WCCCC, WCCCW, WCCWC, WCWW, WCWCC, WCWCW, WCWWC, WCWWW, WWCCC, WWCCW, WWCWC, WWCWW, WWWCC, WWWCW, WWWWC, WWWWW.

63. a. $P(\text{career and opposed}) = \frac{47}{100}$

b. $P(\text{medical and supported}) = \frac{8}{100} = \frac{2}{25}$

c. $P(\text{military and opposed}) = \frac{3}{100}$

d. $P(\text{legal or business and opposed})$

$= \frac{18}{100} = \frac{9}{50}$

e. $P(\text{academic or medical and supported})$

$= \frac{11}{100}$

65. a. $\frac{{}_6C_4 \cdot {}_5C_4}{{}_{15}C_8} = \frac{5}{429}$

b. $\frac{{}_4C_3 \cdot {}_6C_5}{{}_{15}C_8} = \frac{8}{2145}$

67. $\frac{8!}{2!3!} = \frac{8 \cdot 7 \cdot 6 \cdot 5 \cdot 4 \cdot 3!}{2 \cdot 3!} = 3360$;

$P(\text{parallel}) = \frac{1}{3360}$

69. $\left(\frac{1}{2}\right)^x$ where x = number of flips

$P(\text{exactly 20 heads}) = \left(\frac{1}{2}\right)^{20} = \frac{1}{1048576}$

P(winning the lottery) = will vary; but P(exactly 20 heads) > P(winning the lottery).

71. $P(E_1)=P(E_2)=\frac{100}{288}$;

$P(E_1 \cap E_2)=\frac{30}{288}$;

$P(E_1 \cup E_2)=P(E_1)+P(E_2)-P(E_1 \cap E_2)$

$=\frac{100}{288}+\frac{100}{288}-\frac{30}{288}=\frac{170}{288}$;

$\frac{\text{Area of small}}{\text{Area of large}}=\frac{17\text{x}10}{12\text{x}24}=\frac{170}{288}$

73. $\begin{cases} x-2y+3z=10 \\ 2x+y-z=18 \\ 3x-2y+z=26 \end{cases}$

$\left[\begin{array}{ccc|c} 1 & -2 & 3 & 10 \\ 2 & 1 & -1 & 18 \\ 3 & -2 & 1 & 26 \end{array}\right] \quad -2R_1+R_2 \Rightarrow R_2$

$\left[\begin{array}{ccc|c} 1 & -2 & 3 & 10 \\ 0 & 5 & -7 & -2 \\ 3 & -2 & 1 & 26 \end{array}\right] \quad -3R_1+R_3 \Rightarrow R_3$

$\left[\begin{array}{ccc|c} 1 & -2 & 3 & 10 \\ 0 & 5 & -7 & -2 \\ 0 & 4 & -8 & -4 \end{array}\right] \quad \frac{1}{5}R_2 \Rightarrow R_2$

$\left[\begin{array}{ccc|c} 1 & -2 & 3 & 10 \\ 0 & 1 & -\frac{7}{5} & -\frac{2}{5} \\ 0 & 4 & -8 & -4 \end{array}\right] \quad -4R_2+R_3 \Rightarrow R_3$

$\left[\begin{array}{ccc|c} 1 & -2 & 3 & 10 \\ 0 & 1 & -\frac{7}{5} & -\frac{2}{5} \\ 0 & 0 & -\frac{12}{5} & -\frac{12}{5} \end{array}\right] \quad -\frac{5}{12}R_3 \Rightarrow R_3$

$\left[\begin{array}{ccc|c} 1 & -2 & 3 & 10 \\ 0 & 1 & -\frac{7}{5} & -\frac{2}{5} \\ 0 & 0 & 1 & 1 \end{array}\right]$

$z=1$;

$y-\frac{7}{5}z=-\frac{2}{5}$

$y-\frac{7}{5}(1)=-\frac{2}{5}$

$y=-\frac{2}{5}+\frac{7}{5}$

$y=\frac{5}{5}$

$y=1$;

$x-2y+3z=10$

$x-2(1)+3(1)=10$

$x+1=10$

$x=9$;

$(9,1,1)$

75. $\frac{(x-5)^2}{4}+\frac{(y-3)^3}{25}=1$

Center: (5,3)

Vertices: (5,3-5), (5,3+5);

(5,-2), (5,8)

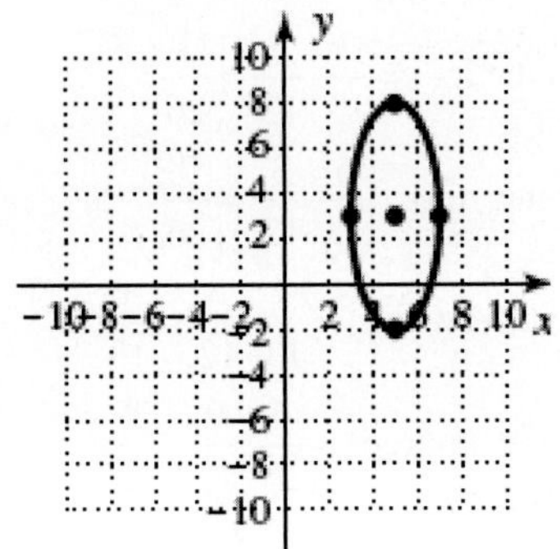

77. $\frac{x^2-1}{x} \geq 0$

$\frac{(x+1)(x-1)}{x} \geq 0$

neg pos neg pos

-1 0 1

$x \in [-1,0) \cup [1,\infty)$

8.7 Technology Highlight

Exercise 1: {1 6 15 20 15 6 1}

$(a+b)^6$

$=a^6+6a^5b+15a^4b^2+20a^3b^3$

$+15a^2b^4+6ab^5+b^6$

Exercise 2: {1, 9, 36, …}

$(a+2b)^9$

$=a^9+9a^8(2b)+36a^7(2b)^2$

$=a^9+18a^8b+144a^7b^2$

8.7 Exercises

1. One

3. $(a+(-2b))^5$

5. Answers will vary.

7. $(x+y)^5$

$x^5+5x^4y+10x^3y^2+10x^2y^3+5xy^4+y^5$

9. $(2x+3)^4$

$=(2x)^4+4(2x)^3(3)+6(2x)^2(3)^2$
$+4(2x)(3)^3+3^4$
$=16x^4+4(24x^3)+6(36x^2)+4(54x)+81$
$=16x^4+96x^3+216x^2+216x+81$

11. $(p-q)^7$

$=p^7-7p^6q+21p^5q^2-35p^4q^3$
$+35p^3q^4-21p^2q^5+7pq^6-q^7$

13. $\binom{n}{r}=\frac{n!}{r!(n-r)!}$

$\binom{7}{4}=\frac{7!}{4!(7-4)!}=\frac{7\cdot 6\cdot 5\cdot 4!}{4!\cdot 3\cdot 2\cdot 1}=\frac{210}{6}=35$

15. $\binom{n}{r}=\frac{n!}{r!(n-r)!}$

$\binom{5}{3}=\frac{5!}{3!(5-3)!}=\frac{5\cdot 4\cdot 3!}{3!2\cdot 1}=\frac{20}{2}=10$

17. $\binom{n}{r}=\frac{n!}{r!(n-r)!}$

$\binom{20}{17}=\frac{20!}{17!(20-17)!}=\frac{20\cdot 19\cdot 18\cdot 17!}{17!3\cdot 2\cdot 1}$

$=\frac{6840}{6}=1140$

19. $\binom{n}{r}=\frac{n!}{r!(n-r)!}$

$\binom{40}{3}=\frac{40!}{3!(40-3)!}=\frac{40\cdot 39\cdot 38\cdot 37!}{3\cdot 2\cdot 1\cdot 37!}$

$=\frac{59280}{6}=9880$

21. $\binom{n}{r}=\frac{n!}{r!(n-r)!}$

$\binom{6}{0}=\frac{6!}{0!(6-0)!}=\frac{6!}{6!}=1$

23. $\binom{n}{r}=\frac{n!}{r!(n-r)!}$

$\binom{15}{15}=\frac{15!}{15!(15-15)!}=\frac{15!}{15!(0)!}=1$

25. $(c+d)^5$; $a=c$; $b=d$; $n=5$

$(c+d)^5$

$=\binom{5}{0}c^5d^0+\binom{5}{1}c^4d+\binom{5}{2}c^3d^2+\binom{5}{3}c^2d^3$
$+\binom{5}{4}cd^4+\binom{5}{5}c^0d^5$

$=1c^5+5c^4d+\frac{5!}{2!3!}c^3d^2+\frac{5!}{3!2!}c^2d^3$
$+\frac{5!}{4!1!}cd^4+1d^5$

$=c^5+5c^4d+10c^3d^2+10c^2d^3+5cd^4+d^5$

27. $(a-b)^6$; $a=a$; $b=b$; $n=6$

$(a-b)^6$

$=\binom{6}{0}a^6b^0-\binom{6}{1}a^5b^1+\binom{6}{2}a^4b^2-\binom{6}{3}a^3b^3$
$+\binom{6}{4}a^2b^4-\binom{6}{5}ab^5+\binom{6}{6}a^0b^6$

$=1a^6b^0-6a^5b+\frac{6!}{2!4!}a^4b^2-\frac{6!}{3!3!}a^3b^3$
$+\frac{6!}{2!4!}a^2b^4-\frac{6!}{1!5!}ab^5+1a^0b^6$

$=a^6-6a^5b+15a^4b^2-20a^3b^3$
$+15a^2b^4-6ab^5+b^6$

29. $(2x-3)^4$; $a=2x$; $b=3$; $n=4$

$=\binom{4}{0}(2x)^4(3)^0-\binom{4}{1}(2x)^3(3)^1+\binom{4}{2}(2x)^2(3)^2$
$-\binom{4}{3}(2x)^1(3)^3+\binom{4}{4}(2x)^0(3)^4$

$=1(16x^4)-4(8x^3)(3)+\frac{4!}{2!2!}(4x^2)(9)$
$-\frac{4!}{3!1!}(2x)(27)+1(1)(81)$

$=16x^4-96x^3+6(36x^2)-4(54x)+81$
$=16x^4-96x^3+216x^2-216x+81$

31. $(1-2i)^3$; $a=1$; $b=2i$; $n=3$

$(1-2i)^3$
$=\binom{3}{0}1^3(2i)^0-\binom{3}{1}1^2(2i)+\binom{3}{2}1(2i)^2-\binom{3}{3}1^0(2i)^3$
$=1(1)-3(1)(2i)+\frac{3!}{2!1!}(4i^2)-(8i^3)$
$=1-6i+3(-4)+8i$
$=1-6i-12+8i$
$=-11+2i$

33. $(x+2y)^9$; $a=x$; $b=2y$; $n=9$
$=\binom{9}{0}x^9(2y)^0+\binom{9}{1}x^8(2y)^1+\binom{9}{2}x^7(2y)^2$
$=1(x^9)(1)+9x^8(2y)+\frac{9!}{2!7!}x^7(4y^2)$
$=x^9+18x^8y+36x^7(4y^2)$
$=x^9+18x^8y+144x^7y^2$

35. $\left(v^2-\frac{1}{2}w\right)^{12}$; $a=v^2$; $b=\frac{1}{2}w$; $n=12$
$=\binom{12}{0}(v^2)^{12}\left(\frac{1}{2}w\right)^0-\binom{12}{1}(v^2)^{11}\left(\frac{1}{2}w\right)^1$
$+\binom{12}{2}(v^2)^{10}\left(\frac{1}{2}w\right)^2$
$=v^{24}-12(v^{22})\left(\frac{1}{2}w\right)+\frac{12!}{2!10!}v^{20}\left(\frac{1}{4}w^2\right)$
$=v^{24}-6v^{22}w+66v^{20}\left(\frac{1}{4}w^2\right)$
$=v^{24}-6v^{22}w+\frac{33}{2}v^{20}w^2$

37. $(x+y)^7$; 4th term
$a=x$; $b=y$; $n=7$; $r=3$
$\binom{n}{r}a^{n-r}b^r$
$\binom{7}{3}(x)^{7-3}y^3=\frac{7!}{3!4!}x^4y^3=35x^4y^3$

39. $(p-2)^8$; 7th term
$a=p$; $b=2$; $n=8$; $r=6$
$\binom{n}{r}a^{n-r}b^r$
$\binom{8}{6}p^{8-6}(2)^6=\frac{8!}{6!2!}p^2(64)$
$=28p^2(64)=1792p^2$

41. $(2x+y)^{12}$; 11th term
$a=2x$; $b=y$; $n=12$; $r=10$
$\binom{n}{r}a^{n-r}b^r$
$\binom{12}{10}(2x)^{12-10}y^{10}=\frac{12!}{10!2!}(2x)^2y^{10}$
$=66(4x^2)y^{10}=264x^2y^{10}$

43. a. $P(\text{exactly } 3)$
$=\binom{5}{3}(1-0.347)^{5-3}(0.347)^3$
$=10(0.653)^2(0.347)^3\approx 0.178$; 17.8%

b. $P(\text{at least } 3)=P(3)+P(4)+P(5)$
$=\binom{5}{3}(0.653)^2(0.347)^3+\binom{5}{4}(0.653)^1(0.347)^4$
$+\binom{5}{5}(0.653)^0(0.347)^4$
$=0.1782+0.0473+0.01449\approx 0.2399$;
23.99%

45. a. $P(\text{exactly } 5)$
$=\binom{8}{5}(1-0.94)^{8-5}(0.94)^5$
$=56(0.06)^3(0.94)^5\approx 0.0088$; 0.88%

b. $P(\text{exactly } 6)$
$=\binom{8}{6}(1-0.94)^{8-6}(0.94)^6$
$=28(0.06)^2(0.94)^6\approx 0.0695$; 6.95%

c. $P(\text{at least } 6)=P(6)+P(7)+P(8)$
$=\binom{8}{6}(0.06)^2(0.94)^6+\binom{8}{7}(0.06)(0.94)^7$
$+\binom{8}{8}(0.06)^0(0.94)^8$
$=0.0695+0.3113+0.6096$
$=0.9904\approx 0.99$; 99%

d. $P(\text{none})=P(\text{all on time})$
$=\binom{8}{8}(0.06)^0(0.94)^8\approx 0.6096$; 60.96%

47. $(5-2x)^4$
(a) $(5-2x)^2(5-2x)^2$

$$= (25-20x+4x^2)(25-20x+4x^2)$$
$$= 625-500x+100x^2-500x+400x^2$$
$$-80x^3+100x^2-80x^3+16x^4$$
$$=16x^4-160x^3+600x^2-1000x+625$$

(b) $(5-2x)^4$

$$=(5)^4(2x)^0-4(5)^3(2x)^1+6(5)^2(2x)^2 -4(5)^1(2x)^3+(5)^0(2x)^4$$
$$=625-8x(125)+6(25)(4x^2) -20(8x^3)+16x^4$$
$$=16x^4-160x^3+600x^2-1000x+625$$

(c) $(5-2x)^4$; $a=5$; $b=2x$; $n=4$

$$(5-2x)^4$$
$$=\binom{4}{0}5^4-\binom{4}{1}5^3(2x)+\binom{4}{2}5^2(2x)^2 -\binom{4}{3}5(2x)^3+\binom{4}{4}(2x)^4$$
$$=625-4(125)(2x)+\frac{4!}{2!2!}(25)(4x^2) -\frac{4!}{3!1!}(5)(8x^3)+16x^4$$
$$=625-1000x+6(100x^2) -4(40x^3)+16x^4$$
$$=16x^4-160x^3+600x^2-1000x+625$$

49. $P(k)=\binom{n}{k}\left(\frac{1}{2}\right)^k\left(\frac{1}{2}\right)^{n-k}$

$$P(5)=\binom{10}{5}\left(\frac{1}{2}\right)^5\left(\frac{1}{2}\right)^{10-5}$$
$$=\frac{10!}{5!5!}\left(\frac{1}{32}\right)\left(\frac{1}{2}\right)^5$$
$$=252\left(\frac{1}{32}\right)\left(\frac{1}{32}\right)$$
$$=\frac{252}{1024}$$

≈ 0.25; Answers will vary.

51. (a) $\%\text{error}=\dfrac{\text{approximate value}}{\text{actual value}}$

$$=\frac{1.476}{(1.02)^{20}}=0.9933=99.33\%$$

(b) $\%\text{error}=\dfrac{\text{approximate value}}{\text{actual value}}$

$$\frac{1.4}{(1.02)^{20}}=0.94216=94.22\%$$

53. Answers will vary.

55. $(3-2i)^4=-119-120i$

$$3^4-4(3)^3(2i)+6(3)^2(2i)^2-4(3)(2i)^3+(2i)^4$$
$$=81-8i(27)+6(9)(-4)-12(-8i)+16$$
$$=81-216i-216+96i+16$$
$$=-119-120i$$

57. $f(x)=\begin{cases}x+2 & x\le 2\\ (x-4)^2 & x>2\end{cases}$

$$f(3)=(3-4)^2=(-1)^2=1$$

59. $g(x)=x^3-x^2-6x$

$$x^3-x^2-6x=0$$
$$x(x^2-x-6)=0$$
$$x(x-3)(x+2)=0$$

$x=0$; $x=3$; $x=-2$

x-intercepts: (0, 0), (3, 0), (-2, 0)

y-intercept: (0, 0)

$g(x)\uparrow: (-\infty,-1)\cup(2,\infty)$

$g(x)\downarrow: (-1,2)$

$g(x)>0: (-2,0)\cup(3,\infty)$

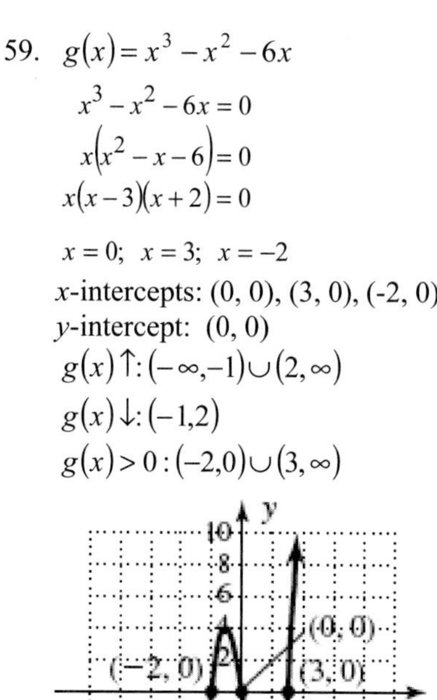

61. a.

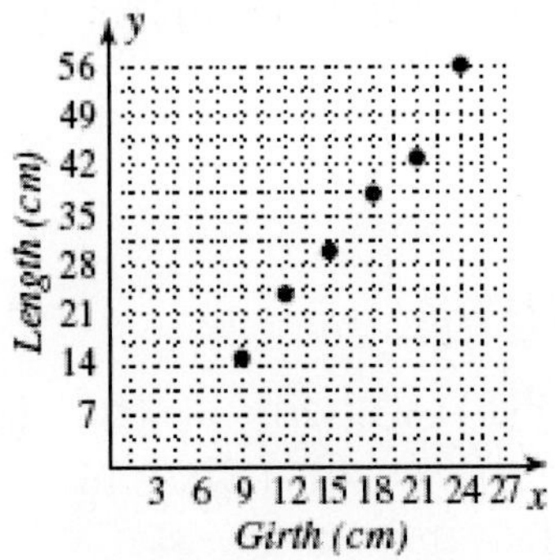

b. Yes, appears to be linear.

c. From grapher: $l(g)=2.57g-8.10$

$l(35)=2.57(35)-8.10=81.85\approx 82$ cm

Chapter 8 Summary Exercises

1. $a_n=5n-4$

$a_1=5(1)-4=5-4=1;$

$a_2=5(2)-4=10-4=6;$

$a_3=5(3)-4=15-4=11;$

$a_4=5(4)-4=20-4=16;$

$a_{10}=5(10)-4=50-4=46$

$1,6,11,16; a_{10}=46$

2. $a_n=\dfrac{n+1}{n^2-1}$

$a_1=\dfrac{(1)+1}{(1)^2-1}=\dfrac{2}{0}$ Not defined;

$a_2=\dfrac{(2)+1}{(2)^2-1}=\dfrac{3}{3}=1;$

$a_3=\dfrac{(3)+1}{(3)^2-1}=\dfrac{4}{8}=\dfrac{1}{2};$

$a_4=\dfrac{(4)+1}{(4)^2-1}=\dfrac{5}{15}=\dfrac{1}{3};$

$a_4=\dfrac{(5)+1}{(5)^2-1}=\dfrac{6}{24}=\dfrac{1}{4};$

$a_{10}=\dfrac{(10)+1}{(10)^2-1}=\dfrac{11}{99}=\dfrac{1}{9}$

Not defined, $1,\dfrac{1}{2},\dfrac{1}{3},\dfrac{1}{4}; a_{10}=\dfrac{1}{9}$

3. $1,16,81,256,\ldots$

$a_n=n^4$

$a_6=6^4=1296$

4. $-17,-14,-11,-8,\ldots$

$a_n=-17+(n-1)(3)$

$a_6=-17+(6-1)(3)$

$=-17+5(3)$

$=-17+15=-2$

5. $\dfrac{1}{2},\dfrac{1}{4},\dfrac{1}{8},\ldots$

$a_n=\dfrac{1}{2^n}$

$S_8=a_1+a_2+a_3+a_4+a_5+a_6+a_7+a_8$

$S_8=\dfrac{1}{2}+\dfrac{1}{4}+\dfrac{1}{8}+\dfrac{1}{16}+\dfrac{1}{32}+\dfrac{1}{64}+\dfrac{1}{128}+\dfrac{1}{256}$

$S_8=\dfrac{255}{256}$

6. $-21,-19,-17,\ldots$

$S_8=a_1+a_2+a_3+a_4+a_5+a_6+a_7+a_8$

$S_8=-21-19-17-15-13-11-9-7$

$=-112$

7. $\sum_{n=1}^{7} n^2$

$=1^2+2^2+3^2+4^2+5^2+6^2+7^2$

$=1+4+9+16+25+36+49$

$=140$

8. $\sum_{n=1}^{5}(3n-2)$

$=(3(1)-2)+(3(2)-2)+(3(3)-2)$

$+(3(4)-2)+(3(5)-2)$

$=(3-2)+(6-2)+(9-2)$

$+(12-2)+(15-2)$

$=1+4+7+10+13$

$=35$

9. $a_n=\dfrac{n!}{(n-2)!}$

$a_2=\dfrac{2!}{(2-2)!}=\dfrac{2!}{0!}=\dfrac{2\cdot 1}{1}=2;$

$a_3=\dfrac{3!}{(3-2)!}=\dfrac{3!}{1!}=\dfrac{3\cdot 2}{1}=6;$

$a_4=\dfrac{4!}{(4-2)!}=\dfrac{4\cdot 3\cdot 2!}{2!}=12;$

$$a_5 = \frac{5!}{(5-2)!} = \frac{5\cdot 4\cdot 3!}{3!} = 20\,;$$

$$a_6 = \frac{6!}{(6-2)!} = \frac{6\cdot 5\cdot 4!}{4!} = 30$$

$2, 6, 12, 20, 30$

10. $\begin{cases} a_1 = \dfrac{1}{2} \\ a_{n+1} = 2a_n - \dfrac{1}{4} \end{cases}$

$$a_{1+1} = 2a_1 - \frac{1}{4}$$

$$a_2 = 2a_1 - \frac{1}{4}$$

$$a_2 = 2\left(\frac{1}{2}\right) - \frac{1}{4} = 1 - \frac{1}{4} = \frac{3}{4}\,;$$

$$a_{2+1} = 2a_2 - \frac{1}{4}$$

$$a_3 = 2a_2 - \frac{1}{4}$$

$$a_3 = 2\left(\frac{3}{4}\right) - \frac{1}{4} = \frac{3}{2} - \frac{1}{4} = \frac{5}{4}\,;$$

$$a_{3+1} = 2a_3 - \frac{1}{4}$$

$$a_4 = 2a_3 - \frac{1}{4}$$

$$a_4 = 2\left(\frac{5}{4}\right) - \frac{1}{4} = \frac{5}{2} - \frac{1}{4} = \frac{9}{4}\,;$$

$$a_{4+1} = 2a_4 - \frac{1}{4}$$

$$a_5 = 2a_4 - \frac{1}{4}$$

$$a_5 = 2\left(\frac{9}{4}\right) - \frac{1}{4} = \frac{9}{2} - \frac{1}{4} = \frac{17}{4}\,;$$

$$\frac{1}{2}, \frac{3}{4}, \frac{5}{4}, \frac{9}{4}, \frac{17}{4}$$

11. $\displaystyle\sum_{n=1}^{7} n^2 + \sum_{n=1}^{7} (3n-2)$

$$\sum_{n=1}^{7} \left(n^2 + 3n - 2\right)$$

$$\begin{aligned}
&= \left(1^2 + 3(1) - 2\right) + \left(2^2 + 3(2) - 2\right) \\
&\quad + \left(3^2 + 3(3) - 2\right) + \left(4^2 + 3(4) - 2\right) \\
&\quad + \left(5^2 + 3(5) - 2\right) + \left(6^2 + 3(6) - 2\right) \\
&\quad + \left(7^2 + 3(7) - 2\right)
\end{aligned}$$

$$\begin{aligned}
&= (1+3-2) + (4+6-2) + (9+9-2) \\
&\quad + (16+12-2) + (25+15-2) + (36+18-2) \\
&\quad + (49+21-2)
\end{aligned}$$

$$= 2 + 8 + 16 + 26 + 38 + 52 + 68$$

$$= 210$$

12. $2, 5, 8, 11, \ldots$; find a_{40}

$$a_1 = 2;\quad d = 3$$

$$a_n = a_1 + (n-1)d$$

$$a_n = 2 + (n-1)3$$

$$a_{40} = 2 + (40-1)(3) = 2 + 39(3)$$

$$= 2 + 117 = 119$$

13. $3, 1, \text{-}1, \text{-}3\ldots$; find a_{35}

$$a_1 = 3;\quad d = -2$$

$$a_n = a_1 + (n-1)d$$

$$a_n = 3 + (n-1)(-2)$$

$$a_{35} = 3 + (35-1)(-2) = 3 + 34(-2)$$

$$= 3 - 68 = -65$$

14. $\text{-}1 + 3 + 7 + 11 + \ldots + 75$

$$a_n = a_1 + (n-1)d$$

$$75 = -1 + (n-1)(4)$$

$$75 = -1 + 4n - 4$$

$$75 = -5 + 4n$$

$$80 = 4n$$

$$20 = n;$$

$$a_1 = -1;\quad d = 4;\quad a_n = 75;\quad n = 20\,;$$

$$S_n = \frac{n(a_1 + a_n)}{2}$$

$$S_{20} = \frac{20(-1+75)}{2} = \frac{20(74)}{2} = 740$$

15. $1 + 4 + 7 + 10 + \ldots + 88$

$$a_n = a_1 + (n-1)d$$

$$88 = 1 + (n-1)(3)$$

$$88 = 1 + 3n - 3$$

$$88 = -2 + 3n$$

$$90 = 3n$$

$$30 = n;$$

$$a_1 = 1;\quad d = 3;\quad a_n = 88;\quad n = 30\,;$$

$$S_n = \frac{n(a_1 + a_n)}{2}$$

$$S_{30} = \frac{30(1+88)}{2} = \frac{30(89)}{2} = 1335$$

16. $3+6+9+12+\ldots; S_{20}$

$a_n = a_1 + (n-1)d$

$a_{20} = 3+(20-1)(3) = 3+19(3) = 60$;

$a_1 = 3;\ d = 3;\ a_{20} = 60;\ n = 20$;

$S_n = \dfrac{n(a_1 + a_n)}{2}$

$S_{20} = \dfrac{20(3+60)}{2} = \dfrac{20(63)}{2} = 630$

17. $1+\frac{3}{4}+\frac{1}{2}+\frac{1}{4}+\ldots; S_{15}$

$a_n = a_1 + (n-1)d$

$a_{15} = 1+(15-1)\left(\dfrac{-1}{4}\right) = 1+14\left(\dfrac{-1}{4}\right) = \dfrac{-5}{2}$;

$a_1 = 1;\ d = \dfrac{-1}{4};\ a_{15} = \dfrac{-5}{2};\ n = 15$

$S_n = \dfrac{n(a_1 + a_n)}{2}$

$S_{15} = \dfrac{15\left(1-\frac{5}{2}\right)}{2} = \dfrac{15\left(\frac{-3}{2}\right)}{2} = -11.25$

18. $\sum_{n=1}^{25}(3n-4)$

$a_1 = 3(1)-4 = 3-4 = -1$;

$a_2 = 3(2)-4 = 6-4 = 2$;

$a_{25} = 3(25)-4 = 75-4 = 71$;

$a_1 = -1;\ d = 3;\ a_{25} = 71;\ n = 25$;

$S_n = \dfrac{n(a_1 + a_n)}{2}$

$S_{25} = \dfrac{25(-1+71)}{2} = \dfrac{25(70)}{2} = 875$

19. $\sum_{n=1}^{40}(4n-1)$

$a_1 = 4(1)-1 = 4-1 = 3$;

$a_2 = 4(2)-1 = 8-1 = 7$;

$a_{40} = 4(40)-1 = 160-1 = 159$;

$a_1 = 3;\ d = 4;\ a_{40} = 159;\ n = 40$;

$S_n = \dfrac{n(a_1 + a_n)}{2}$

$S_{40} = \dfrac{40(3+159)}{2} = \dfrac{40(162)}{2} = 3240$

20. $a_1 = 5, r = 3$; find a_7

$a_n = a_1 r^{n-1}$

$a_7 = 5(3)^{7-1} = 5(3)^6 = 5(729) = 3645$

21. $a_1 = 4, r = \sqrt{2}$; find a_7

$a_n = a_1 r^{n-1}$

$a_7 = 4\left(\sqrt{2}\right)^{7-1} = 4\left(\sqrt{2}\right)^6 = 4(8) = 32$

22. $a_1 = \sqrt{7}, r = \sqrt{7}$; find a_8

$a_n = a_1 r^{n-1}$

$a_8 = \sqrt{7}\left(\sqrt{7}\right)^{8-1} = \sqrt{7}\left(\sqrt{7}\right)^7 = 2401$

23. $16-8+4-\ldots$find S_7

$a_1 = 16;\ r = \dfrac{-1}{2}$;

$S_n = \dfrac{a_1\left(1-r^n\right)}{1-r}$

$S_7 = \dfrac{16\left(1-\left(\frac{-1}{2}\right)^7\right)}{1-\left(\frac{-1}{2}\right)} = \dfrac{16\left(1-\left(\frac{-1}{128}\right)\right)}{\frac{3}{2}}$

$= \dfrac{16\left(\frac{129}{128}\right)}{\frac{3}{2}} = 10.75$

24. $2+6+18+\ldots$find S_8

$a_1 = 2;\ r = 3$;

$S_n = \dfrac{a_1\left(1-r^n\right)}{1-r}$

$S_8 = \dfrac{2\left(1-3^8\right)}{1-3} = \dfrac{2(1-6561)}{-2}$

$= \dfrac{2(-6560)}{-2} = 6560$

25. $\frac{4}{5}+\frac{2}{5}+\frac{1}{5}+\frac{1}{10}+\ldots$find S_{12}

$a_1 = \dfrac{4}{5};\ r = \dfrac{\frac{2}{5}}{\frac{4}{5}} = \dfrac{1}{2}$;

$S_n = \dfrac{a_1\left(1-r^n\right)}{1-r}$

$$S_{12}=\frac{\frac{4}{5}\left(1-\left(\frac{1}{2}\right)^{12}\right)}{1-\frac{1}{2}}=\frac{\frac{4}{5}\left(1-\frac{1}{4096}\right)}{\frac{1}{2}}$$

$$=\frac{\frac{4}{5}\left(\frac{4095}{4096}\right)}{\frac{1}{2}}=\frac{819}{512}$$

26. $4+8+12+24+...$

$\frac{8}{4}=2;\ \frac{12}{8}=\frac{3}{2}$

Does not exist

27. $5+0.5+0.05+0.005+...$

$a_1=5;\ r=\frac{0.5}{5}=\frac{1}{10};$

$$S_\infty=\frac{a_1}{1-r}$$

$$S_\infty=\frac{5}{1-\frac{1}{10}}=\frac{5}{\frac{9}{10}}=\frac{50}{9}$$

28. $6-3+\frac{3}{2}-\frac{3}{4}+...$

$a_1=6;\ r=\frac{-3}{6}=\frac{-1}{2};$

$$S_\infty=\frac{a_1}{1-r}$$

$$S_\infty=\frac{6}{1-\left(\frac{-1}{2}\right)}=\frac{6}{\frac{3}{2}}=4$$

29. $\sum_{n=1}^{8}5\left(\frac{2}{3}\right)^n$

$a_1=5\left(\frac{2}{3}\right)=\frac{10}{3};$

$a_2=5\left(\frac{2}{3}\right)^2=5\left(\frac{4}{9}\right)=\frac{20}{9};$

$r=\frac{\frac{20}{9}}{\frac{10}{3}}=\frac{2}{3};$

$a_1=\frac{10}{3};\ r=\frac{2}{3};$

$$S_n=\frac{a_1\left(1-r^n\right)}{1-r}$$

$$S_8=\frac{\frac{10}{3}\left(1-\left(\frac{2}{3}\right)^8\right)}{1-\frac{2}{3}}=\frac{\frac{10}{3}\left(1-\frac{256}{6561}\right)}{\frac{1}{3}}$$

$$=\frac{\frac{10}{3}\left(\frac{6305}{6561}\right)}{\frac{1}{3}}=\frac{63050}{6561}$$

30. $\sum_{n=1}^{\infty}12\left(\frac{4}{3}\right)^n$

$a_1=12\left(\frac{4}{3}\right)=16;$

$a_2=12\left(\frac{4}{3}\right)^2=12\left(\frac{16}{9}\right)=\frac{64}{3};$

$r=\frac{\frac{64}{3}}{16}=\frac{4}{3}$

Does not exist; r > 1.

31. $\sum_{n=1}^{\infty}5\left(\frac{1}{2}\right)^n$

$a_1=5\left(\frac{1}{2}\right)=\frac{5}{2};$

$a_2=5\left(\frac{1}{2}\right)^2=5\left(\frac{1}{4}\right)=\frac{5}{4};$

$r=\frac{\frac{5}{4}}{\frac{5}{2}}=\frac{1}{2};$

$a_1=\frac{5}{2};\ r=\frac{1}{2};$

$$S_\infty=\frac{a_1}{1-r}$$

$$S_\infty=\frac{\frac{5}{2}}{1-\frac{1}{2}}=\frac{\frac{5}{2}}{\frac{1}{2}}=5$$

32. $a_1=26000;\ d=1220;\ n=10$

$a_n=a_1+(n-1)d$

$a_9=26000+(10-1)(1220)$

$= 26000 + 9(1220) = 26000 + 10980 = 36980$

$S_n = \dfrac{n(a_1 + a_n)}{2}$

$S_{10} = \dfrac{10(26000 + 36980)}{2}$

$= \dfrac{10(62980)}{2} = 314900$

Salary after nine years: \$36,980
Total earnings: \$314,900

33. $a_1 = 121500;\ \ r = \dfrac{2}{3}$

$a_n = a_1 r^n$

$a_7 = 121500\left(\dfrac{2}{3}\right)^7$

$= 121500\left(\dfrac{128}{2187}\right) \approx 7111.1 ft^3$

34. $a_1 = 1225;\ \ r = 1.07$

$a_n = a_1 r^{n-1}$

$a_9 = 1225(1.07)^{9-1} = 1225(1.07)^8$

$= 2104.77807;$

$a_9 \approx 2105$ credit hours

$S_n = \dfrac{a_1(1 - r^n)}{1 - r}$

$S_9 = \dfrac{1225(1 - 1.07^9)}{1 - 1.07} = 14673.03622;$

$S_9 \approx 14{,}673$ credit hours

35. $1 + 2 + 3 + 4 + 5 + ... + n;$

The needed components are:

$a_n = n;\ a_k = k;\ a_{k+1} = k + 1;$

$S_n = \dfrac{n(n+1)}{2};\ S_k = \dfrac{k(k+1)}{2};$

$S_{k+1} = \dfrac{(k+1)(k+2)}{2}$

1. Show S_n is true for $n = 1$.

   $S_1 = \dfrac{1(1+1)}{2} = \dfrac{2}{2} = 1$

   Verified

2. Assume S_k is true:

   $1 + 2 + 3 + ... + k = \dfrac{k(k+1)}{2}$

   and use it to show the truth of S_{k+1} follows. That is:

   $1 + 2 + 3 + ... + k + (k+1)$

   $= \dfrac{(k+1)(k+1+1)}{2}$

   $S_k + a_{k+1} = S_{k+1}$

   Working with the left hand side:

   $1 + 2 + 3 + ... + k + (k+1)$

   $= \dfrac{k(k+1)}{2} + k + 1$

   $= \dfrac{k(k+1) + 2(k+1)}{2}$

   $= \dfrac{(k+1)(k+2)}{2}$

   $= S_{k+1}$

   Since the truth of S_{k+1} follows from S_k, the formula is true for all *n*.

36. $1 + 4 + 9 + 16 + 25 + 36 + ... + n^2$

The needed components are:

$a_n = n^2;\ a_k = k^2;\ a_{k+1} = (k+1)^2$

$S_n = \dfrac{n(n+1)(2n+1)}{6};\ S_k = \dfrac{k(k+1)(2k+1)}{6};$

$S_{k+1} = \dfrac{(k+1)(k+2)(2(k+1)+1)}{6}$

$= \dfrac{(k+1)(k+2)(2k+3)}{6}$

1. Show S_n is true for $n = 1$.

   $S_1 = \dfrac{1(1+1)(2(1)+1)}{6} = \dfrac{2(3)}{6} = \dfrac{6}{6} = 1$

   Verified

2. Assume S_k is true:

   $1 + 4 + 9 + ... + k^2 = \dfrac{k(k+1)(2k+1)}{6}$

   and use it to show the truth of S_{k+1} follows. That is:

   $1 + 4 + 9 + ... + k^2 + (k+1)^2$

   $= \dfrac{(k+1)(k+1+1)(2(k+1)+1)}{6}$

   $S_k + a_{k+1} = S_{k+1}$

   Working with the left hand side:

   $1 + 4 + 9 + ... + k^2 + (k+1)^2$

   $= \dfrac{k(k+1)(2k+1)}{6} + (k+1)^2$

$= \dfrac{k(k+1)(2k+1)+6(k+1)^2}{6}$

$= \dfrac{(k+1)(k(2k+1)+6(k+1))}{6}$

$= \dfrac{(k+1)(2k^2+k+6k+6)}{6}$

$= \dfrac{(k+1)(2k^2+7k+6)}{6}$

$= \dfrac{(k+1)(k+2)(2k+3)}{6}$

$= S_{k+1}$

Since the truth of S_{k+1} follows from S_k, the formula is true for all n.

37. $S_n : 3^n \geq 2n+1$

$S_k : 3^k \geq 2k+1$

$S_{k+1} : 3^{k+1} \geq 2(k+1)+1$

$S_{k+1} : 3^{k+1} \geq 2k+3$

1. Show S_n is true for $n = 1$.

   S_1 :

   $3^1 \geq 2(1)+1$

   $3 \geq 3$

2. Assume $S_k : 3^k \geq 2k+1$ is true and use it to show the truth of S_{k+1} follows.

   That is: $3^{k+1} \geq 2k+3$.

   Working with the left hand side:

   $3^{k+1} = 3(3^k)$

   $= 3(2k+1)$

   $= 6k+3$

   $\geq 2k+3$

   Since k is a positive integer, $6k+3 \geq 2k+3$ showing $3^{k+1} \geq 2k+3$.

   Verified.

38. $S_n : 6 \cdot 7^{n-1} \leq 7^n - 1$

$S_k : 6 \cdot 7^{k-1} \leq 7^k - 1$

$S_{k+1} : 6 \cdot 7^k \leq 7^{k+1} - 1$

1. Show S_n is true for $n = 1$.

   S_1 :

   S_1 :

   $6 \cdot 7^{1-1} \leq 7^1 - 1$

   $6 \cdot 1 \leq 7 - 1$

   $6 \leq 6$

   Verified

2. Assume $S_k : 6 \cdot 7^{k-1} \leq 7^k - 1$ is true and use it to show the truth of S_{k+1} follows. That is: $6 \cdot 7^k \leq 7^{k+1} - 1$.

   Working with the left hand side:

   $6 \cdot 7^k = 6 \cdot 7(7^{k-1})$

   $= 7 \cdot 6(7^{k-1})$

   $= 7(7^k - 1)$

   $= 7^{k+1} - 7$

   $\leq 7^{k+1} - 1$

   Since k is a positive integer, $7^{k+1} - 7 \leq 7^{k+1} - 1$ showing $6 \cdot 7^k \leq 7^{k+1} - 1$.

39. $3^n - 1$ is divisible by 2

1. Show S_n is true for $n = 1$.

   $S_n : 3^n - 1 = 2m$

   S_1 :

   $3^1 - 1 = 2m$

   $2 = 2m$

2. Assume $S_k : 3^k - 1 = 2m$ for $m \in Z$. and use it to show the truth of S_{k+1} follows.

   That is: $S_{k+1} : 3^{k+1} - 1 = 2p$ for $p \in Z$.

   Working with the left hand side:

   $= 3^{k+1} - 1$

   $= 3(3^k) - 1$

   $= 3(2m+1) - 1$

   $= 6m + 3 - 1$

   $= 6m + 2$

   $= 2(3m+1)$

   $3^{k+1} - 1$ is divisible by 2, which is $= S_{k+1}$.

   Verified

40.

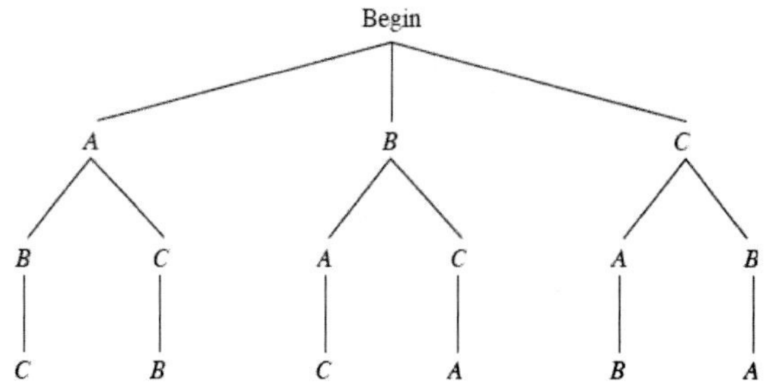

```
                 Begin
        /          |          \
       A           B           C
      / \         / \         / \
     B   C       A   C       A   B
     |   |       |   |       |   |
     C   B       C   A       B   A
```

41. (a) $10 \cdot 9 \cdot 8 = 720$
(b) $10 \cdot 10 \cdot 10 = 1000$

42. $3 \cdot 4 \cdot 2 = 24$

43. $_{12}C_3 = \frac{_{12}P_3}{3!} = \frac{1320}{6} = 220$

44. $_5C_5 + _5C_4 + _5C_3 + _5C_2 + _5C_1 + _5C_0$
$= \frac{5!}{5!0!} + \frac{5!}{4!1!} + \frac{5!}{3!2!} + \frac{5!}{2!3!} + \frac{5!}{1!4!} + \frac{5!}{0!5!}$
$= 1 + 5 + 10 + 10 + 5 + 1$
$= 32$

45. a. $7! = 7 \cdot 6 \cdot 5 \cdot 4 \cdot 3 \cdot 2 \cdot 1 = 5040$
b. $_7P_4 = \frac{7!}{(7-4)!} = \frac{7 \cdot 6 \cdot 5 \cdot 4 \cdot 3!}{3!} = 840$
c. $_7C_4 = \frac{_7P_4}{4!} = \frac{840}{4 \cdot 3 \cdot 2 \cdot 1} = \frac{840}{24} = 35$

46. a. $6 \cdot 5 \cdot 4 \cdot 3 \cdot 2 \cdot 1 = 720$
b. $6 \cdot 5 \cdot 4 = 120$
c. $1 \cdot 4 \cdot 3 \cdot 2 \cdot 1 \cdot 1 = 24$

47. $\frac{_8P_8}{3!2} = \frac{8!}{3!2!} = \frac{8 \cdot 7 \cdot 6 \cdot 5 \cdot 4 \cdot 3!}{3!2 \cdot 1} = \frac{6720}{2} = 3360$

48 a. $_{12}C_3 = \frac{_{12}P_3}{3!} = \frac{1320}{6} = 220$
b. $_{12}P_3 = 12 \cdot 11 \cdot 10 = 1320$

49. $P(\text{ten or face card}) = \frac{4}{52} + \frac{12}{52} = \frac{16}{52} = \frac{4}{13}$

50. $P(\text{queen or face card})$
$\frac{4}{52} + \frac{12}{52} - \frac{4}{52} = \frac{12}{52} = \frac{3}{13}$

51. $P(\sim 3) = 1 - P(3) = 1 - \frac{1}{6} = \frac{5}{6}$

52. $P(E_1) = \frac{3}{8}, P(E_2) = \frac{3}{4}, P(E_1 \cup E_2) = \frac{5}{6}$
$P(E_1 \cap E_2) = \frac{3}{8} + \frac{3}{4} - \frac{5}{6} = \frac{7}{24}$

53. $n(E) = {_7C_4} \cdot {_5C_3};\ n(S) = {_{12}C_7}$
$P(E) = \frac{n(E)}{n(S)} = \frac{_7C_4 \cdot {_5C_3}}{_{12}C_7} = \frac{350}{792} = \frac{175}{396}$

54. a. $0.43 + 0.178 = 0.608$
b. $0.002 + 0.07 + 0.32 = 0.392$
c. 1
d. 0
e. $1 - 0.002 - 0.07 = 0.928$
f. 0.178

55. a. $\binom{7}{5} = \frac{7!}{5!2!} = \frac{7 \cdot 6 \cdot 5!}{5! \cdot 2 \cdot 1} = \frac{42}{2} = 21$
b. $\binom{8}{3} = \frac{8!}{3!5!} = \frac{8 \cdot 7 \cdot 6 \cdot 5!}{3 \cdot 2 \cdot 1 \cdot 5!} = \frac{336}{6} = 56$

56. a. $(x-y)^4$
$= x^4 - 4x^3y + 6x^2y^2 - 4xy^3 + y^4$
b. $(1+2i)^5$
$= 1^5 + 5(1^4)(2i) + 10(1^3)(2i)^2$
$+ 10(1^2)(2i)^3 + 5(1)(2i)^4 + (2i)^5$
$= 1 + 10i + 10(4i^2) + 10(8i^3)$
$+ 5(16i^4) + 32i^5$
$= 1 + 10i - 40 - 80i + 80 + 32i$
$= 41 - 38i$

57. a. $(a+\sqrt{3})^8$
$\binom{8}{0}a^8 + \binom{8}{1}a^7(\sqrt{3})$
$+ \binom{8}{2}a^6(\sqrt{3})^2 + \binom{8}{3}a^5(\sqrt{3})^3$
$= a^8 + 8a^7\sqrt{3} + \frac{8!}{2!6!}a^6(3)$
$+ \frac{8!}{3!5!}a^5(3\sqrt{3})$
$= a^8 + 8\sqrt{3}a^7 + 28a^6(3) + 56a^5(3\sqrt{3})$
$= a^8 + 8\sqrt{3}a^7 + 84a^6 + 168\sqrt{3}a^5$
b. $(5a+2b)^7$
$\binom{7}{0}(5a)^7 + \binom{7}{1}(5a)^6(2b)$
$+ \binom{7}{2}(5a)^5(2b)^2 + \binom{7}{3}(5a)^4(2b)^3$
$= 78125a^7 + 7(15625a^6)(2b)$
$+ \frac{7!}{2!5!}(3125a^5)(4b^2) + \frac{7!}{3!4!}(625a^4)(8b^3)$
$= 78125a^7 + 218750a^6b$
$+ 21(12500a^5b^2) + 35(5000a^4b^3)$
$= 78125a^7 + 218750a^6b$
$+ 262500a^5b^2 + 175000a^4b^3$

58. a. $(x+2y)^7$; 4th

$\binom{n}{r}a^{n-r}b^r$

$\binom{7}{3}x^4(2y)^3 = \frac{7!}{3!4!}x^4(8y^3) = 280x^4y^3$

b. $(2a-b)^{14}$; 10th

$\binom{n}{r}a^{n-r}b^r$

$\binom{14}{9}(2a)^5(-b)^9 = -\frac{14!}{9!5!}32a^5b^9$

$= -64064a^5b^9$

Chapter 8 Mixed Review

1. a. 120,163,206,249,...
   Arithmetic

   b. 4,4,4,4,4,4,....
   $a_n = 4$

   c. 1,2,6,24,120,720,5040,...
   $a_n = n!$

   d. 2.00,1.95,1.90,1.85,...
   Arithmetic

   e. $\frac{5}{8}, \frac{5}{64}, \frac{5}{512}, \frac{5}{4096}, \ldots$
   Geometric

   f. –5.5,6.05,–6.655,7.3205,...
   Geometric

   g. $0.\overline{1}, 0.\overline{2}, 0.\overline{3}, 0.\overline{4}$
   Arithmetic

   h. 525,551.25,578.8125,...
   Geometric

   i. $\frac{1}{2}, \frac{1}{4}, \frac{1}{6}, \frac{1}{8}, \ldots$
   $a_n = \frac{1}{2n}$

3. $2 \cdot 25 \cdot 24 \cdot 23 = 27600$

5. $a_1 = 0.1, r = 5$

$a_2 = 5(0.1) = 0.5$;

$a_3 = 5(0.5) = 2.5$;

$a_4 = 5(2.5) = 12.5$;

$a_5 = 5(12.5) = 62.5$;

$a_n = a_1 r^{n-1}$

$a_{20} = 0.1(5)^{19} = 1907348632812$

$0.1, 0.5, 2.5, 12.5, 62.5; a_{20} = 1907348632812$

7. $P(\sim \text{doubles}) = 1 - P(D) = 1 - \frac{6}{36} = \frac{30}{36} = \frac{5}{6}$

9. a. $\sum_{n=1}^{\infty}\left(\frac{2}{3}\right)^n$

$a_1 = \frac{2}{3};\ a_2 = \left(\frac{2}{3}\right)^2 = \frac{4}{9};\ r = \frac{\frac{4}{9}}{\frac{2}{3}} = \frac{2}{3}$

$S_\infty = \frac{a_1}{1-r}$

$S_\infty = \frac{\frac{2}{3}}{1-\frac{2}{3}} = \frac{\frac{2}{3}}{\frac{1}{3}} = 2$

b. $\sum_{n=1}^{10}(9+2n)$

$a_1 = 9 + 2(1) = 9 + 2 = 11$;

$a_{10} = 9 + 2(10) = 9 + 20 = 29$;

$S_n = \frac{n(a_1 + a_n)}{2}$

$S_{10} = \frac{10(11+29)}{2} = \frac{10(40)}{2} = 200$

c. $\sum_{n=1}^{5}12n + \sum_{n=1}^{5}(-5) + \sum_{n=1}^{5}n^2$

$\sum_{n=1}^{5}12n$

$a_1 = 12;\ a_5 = 12(5) = 60$;

$S_5 = \frac{5(12+60)}{2} = \frac{5(72)}{2} = 180$;

$\sum_{n=1}^{5}(-5)$

$a_1 = -5;\ a_5 = -5$;

$S_5 = \frac{5(-5-5)}{2} = \frac{5(-10)}{2} = -25$;

$\sum_{n=1}^{5}n^2 = 1 + 4 + 9 + 16 + 25 = 55$;

$\sum_{n=1}^{5}12n + \sum_{n=1}^{5}(-5) + \sum_{n=1}^{5}n^2$

$= 180 - 25 + 55 = 210$

11. $(a+b)^n$

a. first 3 terms for $n = 20$

$$\binom{20}{0}a^{20} + \binom{20}{1}a^{19}b + \binom{20}{2}a^{18}b^2$$

$$= a^{20} + 20a^{19}b + \frac{20!}{2!18!}a^{18}b^2$$

$$= a^{20} + 20a^{19}b + 190a^{18}b^2$$

b. last 3 terms for $n = 20$

$$\binom{20}{18}a^2b^{18} + \binom{20}{19}ab^{19} + \binom{20}{20}b^{20}$$

$$= \frac{20!}{18!2!}a^2b^{18} + 20ab^{19} + b^{20}$$

$$= 190a^2b^{18} + 20ab^{19} + b^{20}$$

c. fifth term where $n = 35$

$a = a;\ \ b = b;\ \ n = 35;\ \ r = 4$

$$\binom{n}{r}a^{n-r}b^r$$

$$\binom{35}{4}a^{35-4}b^4 = \frac{35!}{4!31!}a^{31}b^4$$

$$= 52360a^{31}b^4$$

d. fifth term where $n = 35, p = 0.2, q = 0.8$

$$\binom{n}{r}a^{n-r}b^r$$

$$\binom{35}{4}(0.2)^{35-4}(0.8)^4$$

$$= \frac{35!}{4!31!}(0.2)^{31}(0.8)^4$$

$$= 52360(0.2)^{31}(0.8)^4$$

$$= 4.6\text{x}10^{-18}$$

13. $3 + 6 + 9 + ... + 3n = \dfrac{3n(n+1)}{2}$

The needed components are:

$a_n = 3n$; $a_k = 3k$; $a_{k+1} = 3(k+1)$;

$S_n = \dfrac{3n(n+1)}{2}$; $S_k = \dfrac{3k(k+1)}{2}$;

$S_{k+1} = \dfrac{3(k+1)(k+2)}{2}$

1. Show S_n is true for $n = 1$.

$$S_1 = \frac{3(1)(1+1)}{2} = \frac{3(2)}{2} = 3$$

Verified

2. Assume S_k is true:

$$3 + 6 + 9 + ... + 3k = \frac{3k(k+1)}{2}$$

and use it to show the truth of S_{k+1} follows. That is:

$$3 + 6 + 9 + ... + 3k + 3(k+1)$$

$$= \frac{3(k+1)(k+1+1)}{2}$$

$S_k + a_{k+1} = S_{k+1}$

Working with the left hand side:

$$3 + 6 + 9 + ... + 3k + 3(k+1)$$

$$= \frac{3k(k+1)}{2} + 3(k+1)$$

$$= \frac{3k(k+1) + 6(k+1)}{2}$$

$$= \frac{(k+1)(3k+6)}{2}$$

$$= \frac{3(k+1)(k+2)}{2}$$

$$= S_{k+1}$$

Since the truth of S_{k+1} follows from S_k, the formula is true for all *n*.

15. $P(E_1 \text{ or } E_2) = \dfrac{15}{2000} + \dfrac{5}{550} \approx 0.01659$

17. $0.36 + 0.0036 + 0.00036 + 0.00000036 + ...$

$a_1 = 0.36;\ \ r = 0.01$

$$S_\infty = \frac{a_1}{1-r}$$

$$S_\infty = \frac{0.36}{1-0.01} = \frac{0.36}{0.99} = \frac{4}{11}$$

19. $\begin{cases} a_1 = 10 \\ a_{n+1} = a_n\left(\frac{1}{5}\right) \end{cases}$

$$a_{1+1} = a_1\left(\frac{1}{5}\right)$$

$$a_2 = 10\left(\frac{1}{5}\right) = 2;$$

$$a_{2+1} = a_2\left(\frac{1}{5}\right)$$

$$a_3 = 2\left(\frac{1}{5}\right) = \frac{2}{5};$$

$a_{3+1} = a_3\left(\frac{1}{5}\right)$

$a_4 = \frac{2}{5}\left(\frac{1}{5}\right) = \frac{2}{25};$

$a_{4+1} = a_4\left(\frac{1}{5}\right)$

$a_5 = \frac{2}{25}\left(\frac{1}{5}\right) = \frac{2}{125};$

$10, 2, \frac{2}{5}, \frac{2}{25}, \frac{2}{125}$

Chapter 8 Practice Test

1. a. $a_n = \frac{2n}{n+3}$

$a_1 = \frac{2(1)}{1+3} = \frac{2}{4} = \frac{1}{2};$

$a_2 = \frac{2(2)}{2+3} = \frac{4}{5};$

$a_3 = \frac{2(3)}{3+3} = \frac{6}{6} = 1;$

$a_4 = \frac{2(4)}{4+3} = \frac{8}{7};$

$a_8 = \frac{2(8)}{8+3} = \frac{16}{11};$

$a_{12} = \frac{2(12)}{12+3} = \frac{24}{15} = \frac{8}{5}$

$\frac{1}{2}, \frac{4}{5}, 1, \frac{8}{7};\ a_8 = \frac{16}{11}, a_{12} = \frac{8}{5}$

b. $a_n = \frac{(n+2)!}{n!}$

$a_1 = \frac{(1+2)!}{1!} = \frac{3!}{1!} = 3\cdot 2\cdot 1 = 6;$

$a_2 = \frac{(2+2)!}{2!} = \frac{4!}{2!} = \frac{4\cdot 3\cdot 2!}{2!} = 12;$

$a_3 = \frac{(3+2)!}{3!} = \frac{5!}{3!} = \frac{5\cdot 4\cdot 3!}{3!} = 20;$

$a_4 = \frac{(4+2)!}{4!} = \frac{6!}{4!} = \frac{6\cdot 5\cdot 4!}{4!} = 30;$

$a_8 = \frac{(8+2)!}{8!} = \frac{10!}{8!} = \frac{10\cdot 9\cdot 8!}{8!} = 90;$

$a_{12} = \frac{(12+2)!}{12!} = \frac{14!}{12!} = \frac{14\cdot 13\cdot 12!}{12!} = 182;$

$6, 12, 20, 30;\ a_8 = 90, a_{12} = 182$

c. $a_n = \begin{cases} a_1 = 3 \\ a_{n+1} = \sqrt{(a_n)^2 - 1} \end{cases}$

$a_{1+1} = \sqrt{(a_1)^2 - 1}$

$a_2 = \sqrt{(3)^2 - 1} = \sqrt{9-1} = \sqrt{8} = 2\sqrt{2};$

$a_{2+1} = \sqrt{(a_2)^2 - 1}$

$a_3 = \sqrt{\left(2\sqrt{2}\right)^2 - 1} = \sqrt{8-1} = \sqrt{7};$

$a_{3+1} = \sqrt{(a_3)^2 - 1}$

$a_4 = \sqrt{\left(\sqrt{7}\right)^2 - 1} = \sqrt{7-1} = \sqrt{6};$

$a_{7+1} = \sqrt{(a_7)^2 - 1}$

$a_8 = \sqrt{\left(\sqrt{3}\right)^2 - 1} = \sqrt{2};$

$a_{11+1} = \sqrt{(a_{11})^2 - 1}$

$a_{12} = \sqrt{i^2 - 1} = \sqrt{-1-1} = \sqrt{-2} = i\sqrt{2};$

$3, 2\sqrt{2}, \sqrt{7}, \sqrt{6};\ a_8 = \sqrt{2}, a_{12} = i\sqrt{2}$

2. a. $\sum_{k=2}^{6}\left(2k^2 - 3\right)$

$= \left(2(2)^2 - 3\right) + \left(2(3)^2 - 3\right) + \left(2(4)^2 - 3\right)$
$+ \left(2(5)^2 - 3\right) + \left(2(6)^2 - 3\right)$

$= (2(4) - 3) + (2(9) - 3) + (2(16) - 3)$
$+ (2(25) - 3) + (2(36) - 3)$

$= 5 + 15 + 29 + 47 + 69$

$= 165$

b. $\sum_{j=2}^{6}(-1)^j\left(\frac{j}{j+1}\right)$

$= (-1)^2\left(\frac{2}{2+1}\right) + (-1)^3\left(\frac{3}{3+1}\right)$
$+ (-1)^4\left(\frac{4}{4+1}\right) + (-1)^5\left(\frac{5}{5+1}\right)$
$+ (-1)^6\left(\frac{6}{6+1}\right)$

$= \frac{2}{3} - \frac{3}{4} + \frac{4}{5} - \frac{5}{6} + \frac{6}{7}$

$= \frac{311}{420}$

c. $\sum_{j=1}^{5}(-2)\left(\frac{3}{4}\right)^j$

$$= -2\left(\frac{3}{4}\right) - 2\left(\frac{3}{4}\right)^2 - 2\left(\frac{3}{4}\right)^3$$
$$-2\left(\frac{3}{4}\right)^4 - 2\left(\frac{3}{4}\right)^5$$
$$= -\frac{3}{2} - 2\left(\frac{9}{16}\right) - 2\left(\frac{27}{64}\right) - 2\left(\frac{81}{256}\right)$$
$$-2\left(\frac{243}{1024}\right)$$
$$= -\frac{3}{2} - \frac{9}{8} - \frac{27}{32} - \frac{81}{128} - \frac{243}{512}$$
$$= -\frac{2343}{512}$$

d. $\sum_{k=1}^{\infty} 7\left(\frac{1}{2}\right)^k$

$$a_1 = 7\left(\frac{1}{2}\right) = \frac{7}{2};$$
$$a_2 = 7\left(\frac{1}{2}\right)^2 = 7\left(\frac{1}{4}\right) = \frac{7}{4};$$
$$r = \frac{\frac{7}{4}}{\frac{7}{2}} = \frac{1}{2}$$
$$S_\infty = \frac{a_1}{1-r}$$
$$S_\infty = \frac{\frac{7}{2}}{1-\frac{1}{2}} = \frac{\frac{7}{2}}{\frac{1}{2}} = 7$$

3. a. $7, 4, 1, -2, \ldots$

$a_1 = 7, d = -3, a_n = 10 - 3n$

b. $-8, -6, -4, -2, \ldots$

$a_1 = -8, d = 2, a_n = 2n - 10$

c. $4, -8, 16, -32, \ldots$

$a_1 = 4, r = -2, a_n = 4(-2)^{n-1}$

d. $10, 4, \frac{8}{5}, \frac{16}{25}, \ldots$

$a_1 = 10, r = \frac{2}{5}, a_n = 10\left(\frac{2}{5}\right)^{n-1}$

4. a. $a_1 = 4, d = 5$; find a_{40}

$$a_n = a_1 + (n-1)d$$
$$a_{40} = 4 + (40-1)(5) = 4 + 39(5)$$
$$= 4 + 195 = 199$$

b. $a_1 = 2, a_n = -22, d = -3$; Find n

$$a_n = a_1 + (n-1)d$$
$$-22 = 2 + (n-1)(-3)$$
$$-22 = 2 - 3n + 3$$
$$-22 = 5 - 3n$$
$$-27 = -3n$$
$$9 = n$$

c. $a_1 = 24, r = \frac{1}{2}$; Find a_6

$$a_n = a_1 r^{n-1}$$
$$a_6 = 24\left(\frac{1}{2}\right)^5 = 24\left(\frac{1}{32}\right) = \frac{3}{4}$$

d. $a_1 = -2, a_n = 486, r = -3$; Find n

$$a_n = a_1 r^{n-1}$$
$$486 = -2(-3)^{n-1}$$
$$-243 = (-3)^{n-1}$$
$$(-3)^5 = (-3)^{n-1}$$
$$5 = n-1$$
$$6 = n$$

5. a. $7 + 10 + 13 + \ldots + 100$

$$a_1 = 7; \quad a_n = 100; \quad d = 3; \quad n = 32$$
$$a_n = a_1 + (n-1)d$$
$$100 = 7 + (n-1)(3)$$
$$100 = 7 + 3n - 3$$
$$100 = 4 + 3n$$
$$96 = 3n$$
$$32 = n;$$
$$S_n = \frac{n(a_1 + a_n)}{2}$$
$$S_{32} = \frac{32(7+100)}{2} = \frac{32(107)}{2} = 1712$$

b. $\sum_{k=1}^{37} (3k+2)$

$$a_1 = 3(1) + 2 = 3 + 2 = 5;$$
$$a_2 = 3(2) + 2 = 6 + 2 = 8;$$
$$a_{37} = 3(37) + 2 = 111 + 2 = 113;$$
$$a_1 = 5; \quad a_{37} = 113; \quad d = 3; \quad n = 37;$$
$$S_n = \frac{n(a_1 + a_n)}{2}$$

$$S_{37}=\frac{37(5+113)}{2}=\frac{37(118)}{2}=2183$$

c. $4-12+36-108+\ldots$ Find S_7

$a_1=4;\ r=\frac{-12}{4}=-3;$

$$S_n=\frac{a_1\left(1-r^n\right)}{1-r}$$

$$S_7=\frac{4\left(1-(-3)^7\right)}{1-(-3)}=\frac{4(1-(-2187))}{4}$$

$=2188$

d. $6+3+\frac{3}{2}+\frac{3}{4}+\ldots$

$a_1=6;\ r=\frac{1}{2}$

$$S_\infty=\frac{a_1}{1-r}$$

$$S_\infty=\frac{6}{1-\frac{1}{2}}=\frac{6}{\frac{1}{2}}=12$$

6. a. $a_1=12;\ r=0.95$

$a_n=a_1r^{n-1}$

$a_7=12(0.95)^6\approx 8.82\,ft$

b. $S_7=\frac{12\left(1-0.95^7\right)}{1-0.95}\approx 72.4\,ft$

7. $a_1=3000;\ r=1.07;\ n=12$

$a_n=a_1r^n$

$a_{12}=3000(1.07)^{12}=6756.57$

\$6756.57

8. $a_1=50000;\ r=0.85;\ n=5$

$a_n=a_1r^n$

$a_5=50000(0.85)^5=22185.27$

\$22,185.27

9. $a_n=5n-3;\ a_k=5k-3;$

The needed components are:

$a_{k+1}=5(k+1)-3=5k+2;$

$S_n=\frac{5n^2-n}{2};\ S_k=\frac{5k^2-k}{2};$

$$S_{k+1}=\frac{5(k+1)^2-(k+1)}{2}$$

1. Show S_n is true for $n=1$.

$a_1=5(1)-3=5-3=2$

$$S_1=\frac{5(1)^2-1}{2}=\frac{5-1}{2}=\frac{4}{2}=2$$

Verified

2. Assume S_k is true:

$$2+7+12+\ldots+5k-3=\frac{5k^2-k}{2}$$

and use it to show the truth of S_{k+1} follows. That is:

$$2+7+12+\ldots+5k-3+(5(k+1)-3)=\frac{5(k+1)^2-(k+1)}{2}$$

$S_k+a_{k+1}=S_{k+1}$

Working with the left hand side:

$2+7+12+\ldots+5k-3+(5k+2)$

$$=\frac{5k^2-k}{2}+5k+2$$

$$=\frac{5k^2-k+2(5k+2)}{2}$$

$$=\frac{5k^2+10k-k+4}{2}$$

$$=\frac{5\left(k^2+2k\right)-k+4}{2}$$

$$=\frac{5(k+1)^2-k+4-5}{2}$$

$$=\frac{5(k+1)^2-k-1}{2}$$

$$=\frac{5(k+1)^2-(k+1)}{2}$$

Since the truth of S_{k+1} follows from S_k, the formula is true for all n.

10. $S_n: 2\cdot 3^{n-1}\le 3^n-1$

$S_k: 2\cdot 3^{k-1}\le 3^k-1$

$S_{k+1}: 2\cdot 3^k\le 3^{k+1}-1$

1. Show S_n is true for $n=1$.

S_1 :

$2\cdot 3^{1-1}\le 3^1-1$

$2\cdot 1\le 2$

$2\le 2$

2. Assume $S_k: 2\cdot 3^{k-1}\le 3^k-1$ is true. and use it to show the truth of S_{k+1} follows. That is: $2\cdot 3^k\le 3^{k+1}-1$. Working with the left hand side:

$$\begin{aligned}2\cdot 3^k &= 2\cdot 3(3^{k-1})\\ &= 3\cdot 2(3^{k-1})\\ &\le 3(3^k-1)\\ &\le 3^{k+1}-3\\ &\le 3^{k+1}-1\end{aligned}$$

Since k is a positive integer,

$3^{k+1}-3 \le 3^{k+1}-1$ showing

$2\cdot 3^k \le 3^{k+1}-1$

Verified.

11. a.

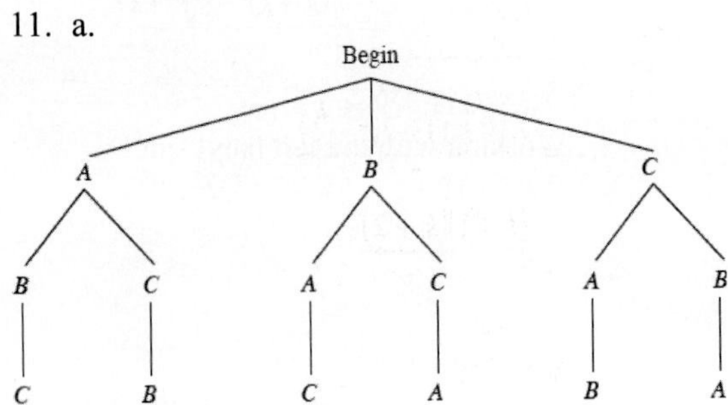

b. ABC, ACB, BAC, BCA, CAB, CBA

12. $9\cdot 8\cdot 7\cdot 25\cdot 24 = 302400$

13. ${}_6C_6 + {}_6C_5 + {}_6C_4 + {}_6C_3 + {}_6C_2 + {}_6C_1 + {}_6C_0$

$$= \frac{6!}{6!0!} + \frac{6!}{5!1!} + \frac{6!}{4!2!} + \frac{6!}{3!3!} + \frac{6!}{2!4!} + \frac{6!}{1!5!} + \frac{6!}{0!6!}$$

$= 1+6+15+20+15+6+1$

$= 64$

14. a. $6! = 6\cdot 5\cdot 4\cdot 3\cdot 2\cdot 1 = 720$

b. ${}_6P_3 = \frac{6!}{(6-3)!} = \frac{6!}{3!} = \frac{6\cdot 5\cdot 4\cdot 3!}{3!} = 120$

c. ${}_6C_3 = \frac{{}_6P_3}{3!} = \frac{120}{3\cdot 2\cdot 1} = 20$

15. $\frac{{}_{13}P_{13}}{2!3!4!4!} = \frac{6227020800}{6912} = 900900$

16. $7\cdot 6\cdot 6\cdot 5\cdot 5\cdot 4\cdot 4\cdot 3 = 302400$

17. a. $(x-2y)^4$

$$\begin{aligned}&= x^4 - 4x^3(2y) + 6x^2(2y)^2\\ &\quad - 4x(2y)^3 + (2y)^4\\ &= x^4 - 8x^3y + 6x^2(4y^2)\\ &\quad - 4x(8y^3) + 16y^4\\ &= x^4 - 8x^3y + 24x^2y^2 - 32xy^3 + 16y^4\end{aligned}$$

b. $(1+i)^4$

$$\begin{aligned}&= 1^4 + 4(1)^3 i + 6(1)^2 i^2 + 4(1)i^3 + i^4\\ &= 1 + 4i - 6 - 4i + 1\\ &= -4\end{aligned}$$

18. First three terms of:

(a) $(x+\sqrt{2})^{10}$

$$\begin{aligned}&\binom{10}{0}x^{10} + \binom{10}{1}x^9\sqrt{2} + \binom{10}{2}x^8(\sqrt{2})^2\\ &= x^{10} + 10x^9\sqrt{2} + \frac{10!}{2!8!}x^8(2)\\ &= x^{10} + 10\sqrt{2}x^9 + 45x^8(2)\\ &= x^{10} + 10\sqrt{2}x^9 + 90x^8\end{aligned}$$

(b) $(a-2b^3)^8$

$$\begin{aligned}&\binom{8}{0}a^8 - \binom{8}{1}a^7(2b^3) + \binom{8}{2}a^6(2b^3)^2\\ &= a^8 - 8a^7(2b^3) + \frac{8!}{2!6!}a^6(4b^6)\\ &= a^8 - 16a^7b^3 + 28a^6(4b^6)\\ &= a^8 - 16a^7b^3 + 112a^6b^6\end{aligned}$$

19. $1 - 0.011 = 0.989$

20. a. P(striped wedge) = $\frac{3}{12} = \frac{1}{4}$

b. P(shaded wedge) = $\frac{5}{12}$

c. P(clear wedge) = $\frac{4}{12} = \frac{1}{3}$

d. P(even number) = $\frac{6}{12} = \frac{1}{2}$

e. P(2 or odd number) = $\frac{1}{12} + \frac{6}{12} = \frac{7}{12}$

f. P(number >9) = $\frac{3}{12} = \frac{1}{4}$

g. P(shade or number > 12) = $\frac{5}{12}$

h. P(shade and number > 12) = $\frac{5}{12}\cdot 0 = 0$

21. a. $0.05 + 0.03 = 0.08$

b. $0.02 + 0.30 + 0.60 = 0.92$

c. $0.02 + 0.30 + 0.60 + 0.05 + 0.03 = 1$

d. 0

e. $0.02 + 0.30 + 0.60 + 0.03 = 0.95$

f. 0.03

22. a. Length of triangle is 48, height is half the diameter of the circle which is 48, therefore the height is 24.

$A = \frac{48(24)}{2} = 576$;

P(triangle) $= \frac{576}{3072} = 0.1875$

b. $A = \pi(24)^2 = 1810$;

P(circle) $= \frac{1810}{3072} = 0.589$

c. P(circle outside triangle) $= 0.589 - 0.1875 = 0.4015$

d. P(lower half of circle) $= \frac{1}{2}(0.589) = 0.2945$

e. P(rectangle, but outside circle)

$64(48) - \pi(24)^2 = 1262$;

$\frac{1262}{3072} = 0.4108$

f. P(lower half-rectangle, outside circle)

$\frac{1}{2}(3072) - \frac{1}{2}1810 = 631$;

$\frac{631}{3072} = 0.2055$

23. a. P(woman or craftsman)

$= \frac{50}{100} + \frac{18}{100} - \frac{9}{100} = \frac{59}{100}$

b. P(man or contractor)

$= \frac{50}{100} + \frac{7}{100} - \frac{4}{100} = \frac{53}{100}$

c. P(man and technician) $= \frac{13}{100}$

d. P(journeyman or apprentice)

$= \frac{13}{100} + \frac{34}{100} = \frac{47}{100}$

24. a. P(both turtles live another 8 yrs) $= 0.85(0.95) = 0.8075$

b. P(neither turtle lives another 8 yrs) $= (1-0.85)(1-0.95) = 0.0075$

c. P(at least one lives another 8 yrs) $1 - 0.0075 = 0.9925$

25. $1 + 2 + 3 + ... + n = \frac{n(n+1)}{2}$

$a_n = n$; $a_k = k$; $a_{k+1} = k+1$

$S_n = \frac{n(n+1)}{2}$; $S_k = \frac{k(k+1)}{2}$;

$S_{k+1} = \frac{(k+1)(k+2)}{2}$

1. Show S_n is true for $n = 1$.

$S_1 = \frac{1(1+1)}{2} = \frac{2}{2} = 1$

Verified

2. Assume S_k is true, show $S_k + a_{k+1} = S_{k+1}$.

$1 + 2 + 3 + ... + k + (k+1)$

$= \frac{k(k+1)}{2} + k + 1$

$= \frac{k(k+1) + 2(k+1)}{2}$

$= \frac{(k+1)(k+2)}{2}$

$= S_{k+1}$

Verified

Calculator Exploration and Discovery

1. $a_1 = \frac{1}{3}$ and $r = \frac{1}{3}$

Using a grapher: $\frac{1}{2}$

2. $a_1 = 0.2$ and $r = 0.2$

Using a grapher: $\frac{1}{4}$

3. $a_n = \frac{1}{(n-1)!}$

Using a grapher: e

4. $a_n = \frac{2}{n(n+1)}$

Using a grapher: 1.9607,1.9801,1.9867,1.9900,1.9920

Strengthening Core Skills

Exercise 1: P(four-of-a-kind)

$\frac{_{13}C_1 \cdot {_4C_1}}{_{52}C_5} = 0.00024$

Exercise 2: P(full house)

$$\frac{{}_{13}C_1 \cdot {}_4C_3 \cdot {}_{12}C_1 \cdot {}_4C_2}{{}_{52}C_5} = 0.001441$$

Exercise 3: P(three-of-a-kind)

$$\frac{{}_{13}C_1 \cdot {}_4C_3 \cdot {}_{12}C_2 \cdot {}_4C_1 \cdot {}_4C_1}{{}_{52}C_5} = 0.021128$$

Exercise 4: P(four of same suit and one other card)

$$\frac{{}_4C_1 \cdot {}_{13}C_4 \cdot {}_3C_1 \cdot {}_{13}C_1}{{}_{52}C_5} = 0.04292$$

Exercise 5: P(straight)

$$\frac{{}_{10}C_1 \cdot {}_4C_1 \cdot {}_4C_1 \cdot {}_4C_1 \cdot {}_4C_1 \cdot {}_4C_1 - 40}{{}_{52}C_5}$$

$= 0.003925$

Chapters 1-8 Cumulative Review

1. a. (9, 52) (11, 98)

$$m = \frac{98-52}{11-9} = \frac{46}{2} = 23$$

23 cards are assembled each hour

b. $y - y_1 = m(x - x_1)$

$y - 52 = 23(x-9)$

$y - 52 = 23x - 207$

$y = 23x - 155$

c. $23(8) = 184$

184 cards

d. $\frac{52}{23} = 2.26$

Approximately 2.25 hours before 9am $\approx$ 6:45am

2. $x^2 - 4x + 5 = 0;\ x = 2 + i$

$0 = (2+i)^2 - 4(2+i) + 5$

$0 = 4 + 4i + i^2 - 8 - 4i + 5$

$0 = 4 - 1 - 8 + 5$

$0 = 0$

Verified

3. $3x^2 + 5x - 7 = 0$

$$x = \frac{-5 \pm \sqrt{(5)^2 - 4(3)(-7)}}{2(3)}$$

$$x = \frac{-5 \pm \sqrt{25+84}}{6}$$

$$x = \frac{-5 \pm \sqrt{109}}{6}$$

$x \approx 0.91; x \approx -2.57$

4. $y = \sqrt{x+4} - 3$

Left 4, down 3

$0 = \sqrt{x+4} - 3$

$3 = \sqrt{x+4}$

$9 = x + 4$

$5 = x$

x-intercept: (5, 0)

$y = \sqrt{0+4} - 3 = -1$

y-intercept: (0, -1)

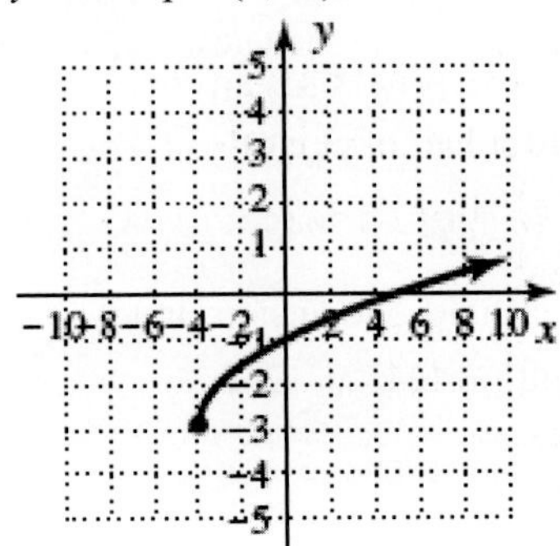

5. $Y = \frac{kVW}{X}$

$10 = \frac{k(5)(12)}{9}$

$90 = 60k$

$\frac{3}{2} = k;$

$y = \frac{3VW}{2X}$

6. $y = \begin{cases} -2 & -3 \le x \le -1 \\ x & -1 < x < 2 \\ x^2 & 2 \le x \le 3 \end{cases}$

Domain: $[-3,3]$

Range: $[-2,-2] \cup (1,2) \cup [4,9]$

7. $f(x)=x^3-5;\ \ g(x)=\sqrt[3]{x+5}$

$$y=x^3-5$$
$$x=y^3-5$$
$$x+5=y^3$$
$$\sqrt[3]{x+5}=y$$

Verified

8. a. $g(x)=0$
   $x=0$

b. $g(x)<0:\ x\in(-1,0)$

c. $g(x)>0:\ x\in(-\infty,-1)\cup(0,\infty)$

d. $g(x)\uparrow:\ x\in(-\infty,-1)\cup(-1,1)$

e. $g(x\downarrow):\ (1,\infty)$

f. Local Max
   $y=3$ at $x=1$

g. Local Min
   None

h. $g(x)=2$
   $x\approx-2.3, x\approx 0.4, x\approx 2$

i. $g(4)\approx\frac{1}{4}$

j. $g(-1)$ undefined; does not exist

k. as $x\to-1^+, g(x)\to-\infty$

l. as $x\to\infty, g(x)\to 0$

m. The domain of $g(x)$: $(-\infty,-1)\cup(-1,\infty)$

9. a. $f(x)=2x^2-3x$

$$\frac{f(x+h)-f(x)}{h}$$
$$=\frac{\left(2(x+h)^2-3(x+h)\right)-\left(2x^2-3x\right)}{h}$$
$$=\frac{\left(2\left(x^2+2xh+h^2\right)-3x-3h\right)-\left(2x^2-3x\right)}{h}$$
$$=\frac{\left(2x^2+4xh+2h^2-3x-3h\right)-\left(2x^2-3x\right)}{h}$$
$$=\frac{2x^2-4xh-2h^2+3x+3h-2x^2+3x}{h}$$
$$=\frac{4xh+2h^2-3h}{h}$$
$$=\frac{h(4x+2h-3)}{h}$$
$$=4x+2h-3$$

b. $h(x)=\frac{1}{x-2}$

$$\frac{f(x+h)-f(x)}{h}$$
$$=\frac{\frac{1}{x+h-2}-\frac{1}{x-2}}{h}$$
$$=\frac{\frac{x-2-(x+h-2)}{(x-2)(x+h-2)}}{h}$$
$$=\frac{\frac{x-2-x-h-2}{(x-2)(x+h-2)}}{h}$$
$$=\frac{\frac{-h}{(x-2)(x+h-2)}}{h}$$
$$=\frac{-1}{(x+h-2)(x-2)}$$

10. $f(x)=x^3+x^2-4x-4$

$$0=x^3+x^2-4x-4$$
$$0=x^2(x+1)-4(x+1)$$
$$0=(x+1)\left(x^2-4\right)$$

$x+1=0\qquad x^2-4=0$

$x=-1\qquad x^2=4$

$x=\pm2$

$x=-1;\ \ x=\pm2$

x-intercepts: (-1, 0), (2, 0), (-2, 0)

$f(0)=0^3+0^2-4(0)-4=-4$

y-intercept: (0, -4)

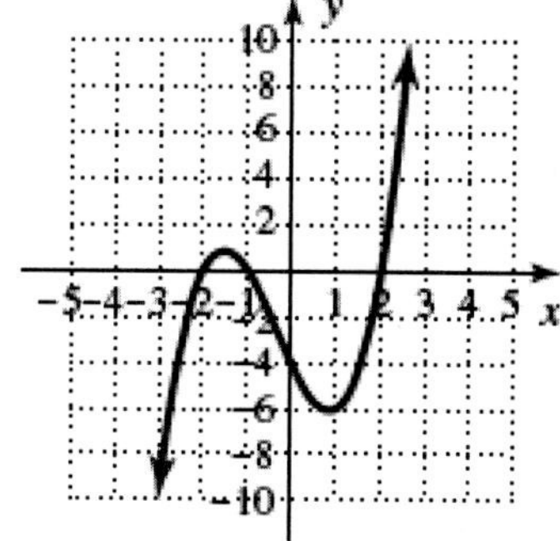

11. $h(x)=\frac{2x^2-8}{x^2-1}$

$$0=2x^2-8$$
$$0=2\left(x^2-4\right)$$
$$x^2-4=0$$
$$x^2=4$$

$x = \pm 2$

x-intercepts: (2, 0), (-2, 0)

$h(0) = \dfrac{2(0)^2 - 8}{0^2 - 1} = 8$

y-intercept: (0, 8)

$0 = x^2 - 1$

$x = \pm 1$

Vertical asymptotes: $x = \pm 1$

Horizontal asymptote: $y = 2$

(deg num = deg den)

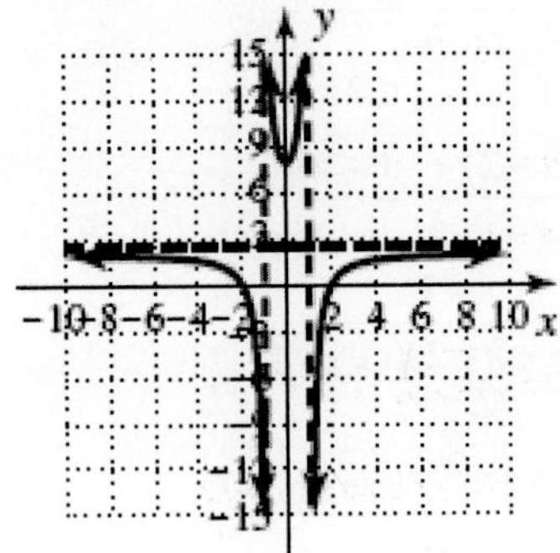

12. a. $x = 10^y$

$\log_{10} x = y$

b. $\dfrac{1}{81} = 3^{-4}$

$\log_3 \dfrac{1}{81} = -4$

13. a. $3 = \log_x(125)$

$x^3 = 125$

b. $\ln(2x-1) = 5$

$e^5 = 2x - 1$

14. $4000 = 2000e^{10r}$

$2 = e^{10r}$

$\ln 2 = 10r$

$r = \dfrac{\ln 2}{10}$

$r = 0.693$

6.93%

15. a. $e^{2x-1} = 217$

$(2x-1)\ln e = \ln 217$

$2x - 1 = \ln 217$

$2x = \ln 217 + 1$

$x = \dfrac{\ln 217 + 1}{2}$

$x \approx 3.19$

b. $\log(3x-2) + 1 = 4$

$\log(3x-2) = 3$

$10^3 = 3x - 2$

$1000 = 3x - 2$

$1002 = 3x$

$334 = x$

$x \approx 334$

16. $\begin{cases} 2a + 3b - 6c = 15 \\ 4a - 6b + 5c = 35 \\ 3a + 2b - 5c = 24 \end{cases}$

$$\left[\begin{array}{ccc|c} 2 & 3 & -6 & 15 \\ 4 & -6 & 5 & 35 \\ 3 & 2 & -5 & 24 \end{array}\right] \quad \frac{1}{2}R_1 \Rightarrow R_1$$

$$\left[\begin{array}{ccc|c} 1 & \frac{3}{2} & -3 & \frac{15}{2} \\ 4 & -6 & 5 & 35 \\ 3 & 2 & -5 & 24 \end{array}\right] \quad -4R_1 + R_2 \Rightarrow R_2$$

$$\left[\begin{array}{ccc|c} 1 & \frac{3}{2} & -3 & \frac{15}{2} \\ 0 & -12 & 17 & 5 \\ 3 & 2 & -5 & 24 \end{array}\right] \quad -3R_1 + R_3 \Rightarrow R_3$$

$$\left[\begin{array}{ccc|c} 1 & \frac{3}{2} & -3 & \frac{15}{2} \\ 0 & -12 & 17 & 5 \\ 0 & \frac{-5}{2} & 4 & \frac{3}{2} \end{array}\right] \quad \frac{-1}{12}R_2 \Rightarrow R_2$$

$$\left[\begin{array}{ccc|c} 1 & \frac{3}{2} & -3 & \frac{15}{2} \\ 0 & 1 & \frac{-17}{12} & \frac{-5}{12} \\ 0 & \frac{-5}{2} & 4 & \frac{3}{2} \end{array}\right] \quad \frac{5}{2}R_2 + R_3 \Rightarrow R_3$$

$$\left[\begin{array}{ccc|c} 1 & \frac{3}{2} & -3 & \frac{15}{2} \\ 0 & 1 & \frac{-17}{12} & \frac{-5}{12} \\ 0 & 0 & \frac{11}{24} & \frac{11}{24} \end{array}\right] \quad \frac{24}{11}R_3 \Rightarrow R_3$$

$$\left[\begin{array}{ccc|c} 1 & \frac{3}{2} & -3 & \frac{15}{2} \\ 0 & 1 & \frac{-17}{12} & \frac{-5}{12} \\ 0 & 0 & 1 & 1 \end{array}\right] \quad \frac{17}{12}R_3 + R_2 \Rightarrow R_2$$

$z = 1;$

$$y - \frac{17}{12}z = -\frac{5}{12}$$

$$y - \frac{17}{12}(1) = -\frac{5}{12}$$

$$y = -\frac{5}{12} + \frac{17}{12}$$

$$y = 1;$$

$$x + \frac{3}{2}y - 3z = \frac{15}{2}$$

$$x + \frac{3}{2}(1) - 3(1) = \frac{15}{2}$$

$$x - \frac{3}{2} = \frac{15}{2}$$

$$x = \frac{18}{2}$$

$$x = 9$$

$(9, 1, 1)$

17. $\begin{cases} 0.7x + 1.2y - 3.2z = -32.5 \\ 1.5x - 2.7y + 0.8z = -7.5 \\ 2.8x + 1.9y - 2.1z = 1.5 \end{cases}$

$$A = \begin{bmatrix} 0.7 & 1.2 & -3.2 \\ 1.5 & -2.7 & 0.8 \\ 2.8 & 1.9 & -2.1 \end{bmatrix} \quad B = \begin{bmatrix} -32.5 \\ -7.5 \\ 1.5 \end{bmatrix}$$

$$X = A^{-1}B$$

$$X = \begin{bmatrix} 5 \\ 10 \\ 15 \end{bmatrix}$$

$(5, 10, 15)$

18. Foci: $(-6,0), (6,0)$

Vertices: $(-4,0), (4,0)$

Center: (0, 0)

$a = 4,\ c = 6$

$c^2 - a^2 = b^2$

$36 - 16 = b^2$

$20 = b^2$

$2\sqrt{5} = b$

$$\frac{x^2}{16} - \frac{y^2}{20} = 1$$

19. $x^2 + 4y^2 - 24y + 6x + 29 = 0$

$$\left(x^2 + 6x\right) + 4\left(y^2 - 6y\right) = -29$$

$$(x+3)^2 + 4(y-3)^2 = -29 + 9 + 36$$

$$(x+3)^2 + 4(y-3)^2 = 16$$

$$\frac{(x+3)^2}{16} + \frac{(y-3)^2}{4} = 1$$

Center: (-3, 3)

Vertices: (-7, 3), (1, 3)

$a = 4,\ b = 2$

$c^2 = 16 - 4 = 12$

$c = 2\sqrt{3}$

Foci: $\left(-3 - 2\sqrt{3}, 3\right)\left(-3 + 2\sqrt{3}, 3\right)$

20. a. 262144,65536,16384,4096...

$a_1 = 262144;\quad r = 0.25$

$$a_n = a_1 r^{n-1}$$

$$a_{20} = 262144(0.25)^{19} = \frac{1}{1048576}$$

$$\approx 0.00000095;$$

$$S_n = \frac{a_1\left(1 - r^n\right)}{1 - r}$$

$$S_{20} = \frac{262144\left(1 - 0.25^{20}\right)}{1 - 0.25}$$

$$= \frac{262144(1)}{0.75} = 349525.\bar{3}$$

b. $\frac{7}{8}, \frac{27}{40}, \frac{19}{40}, \frac{11}{40}, \ldots$

$a_1 = \frac{7}{8};\quad d = -0.2$

$$a_n = a_1 + (n-1)d$$

$$a_{20} = \frac{7}{8} + (20-1)(-0.2)$$

$$= \frac{7}{8} + 19(-0.2) = \frac{7}{8} - 3.8 = -\frac{117}{40};$$

$$S_n = \frac{n(a_1 + a_n)}{2}$$

$$S_{20} = \frac{20\left(\frac{7}{8} - \frac{117}{40}\right)}{2}$$

$$= \frac{20\left(\frac{-41}{20}\right)}{2} = -20.5$$

21. $a_1 = 52;\; a_n = 10;\; d = -1$

$$a_n = a_1 + (n-1)d$$
$$10 = 52 + (n-1)(-1)$$
$$10 = 52 - n + 1$$
$$10 = 53 - n$$
$$-43 = -n$$
$$43 = n;$$
$$S_n = \frac{n(a_1 + a_n)}{2}$$
$$S_{43} = \frac{43(52+10)}{2} = \frac{43(62)}{2} = 1333$$

22. (a) P($20 and $10)

$$= \frac{3}{13}\cdot\frac{6}{12} = \frac{18}{156} = \frac{3}{26}$$

(b) P($10 and $20)

$$= \frac{6}{13}\cdot\frac{3}{12} = \frac{18}{156} = \frac{3}{26}$$

(c) P(both $5)

$$= \frac{4}{13}\cdot\frac{3}{12} = \frac{12}{156} = \frac{1}{13}$$

(d) P($5 and not $20)

$$= \frac{4}{13}\cdot\frac{9}{12} = \frac{36}{156} = \frac{3}{13}$$

(e) P($5 and not $10)

$$= \frac{4}{13}\cdot\frac{6}{12} = \frac{24}{156} = \frac{2}{13}$$

(f) P(not $20 and $20)

$$= \frac{10}{13}\cdot\frac{3}{12} = \frac{30}{156} = \frac{5}{26}$$

23. $P(\text{late}) = 0.04,\ P(\text{on time}) = 1 - 0.04 = 0.96$

(a) $\text{P}(10) = \binom{12}{10}(0.04)^2(0.96)^{10}$

$$\approx 0.07 = 7\%$$

(b) $P(x \ge 11) = \binom{12}{11}(0.04)^1(0.96)^{11}$

$$+\binom{12}{12}(0.04)^0(0.96)^{12} \approx 0.919 = 91.9\%$$

(c) $P(x \ge 10) = \binom{12}{10}(0.04)^2(0.96)^{10}$

$$+\binom{12}{11}(0.04)^1(0.96)^{11}$$

$$+\binom{12}{12}(0.04)^0(0.96)^{12} \approx 0.989 = 98.9\%$$

(d) P(x= 10)

$$= \binom{12}{0}(0.04)^{12}(0.96)^0 \approx 1.7\text{x}10^{-17}$$

virtually nil

24. $3 + 7 + 11 + 15 + \ldots + (4n-1) = n(2n+1)$

The needed components are:

$a_n = 4n - 1;\; a_k = 4k - 1;$

$a_{k+1} = 4(k+1) - 1 = 4k + 3;$

$S_n = n(2n+1);\; S_k = k(2k+1);$

$S_{k+1} = (k+1)(2(k+1)+1) = (k+1)(2k+3)$

1. Show S_n is true for $n = 1$.

   $S_1 = 1(2(1)+1) = 3$

   Verified

2. Assume S_k is true:

   $3 + 7 + 11 + \ldots + 4k - 1 = k(2k+1)$

   and use it to show the truth of S_{k+1} follows. That is:

   $3 + 7 + 11 + \ldots + 4k - 1 + 4(k+1) - 1$
   $= (k+1)(2(k+1)+1)$

   $S_k + a_{k+1} = S_{k+1}$

   Working with the left hand side:

   $3 + 7 + 11 + \ldots + 4k - 1 + 4(k+1) - 1$
   $= k(2k+1) + 4k + 3$
   $= 2k^2 + k + 4k + 3$
   $= 2k^2 + 5k + 3$
   $= (k+1)(2k+3)$
   $= S_{k+1}$

   Since the truth of S_{k+1} follows from S_k, the formula is true for all n.

25. (3, 12), (5, 114), (7, 135), (9, 81)

Using grapher:

$y = -9.75x^2 + 128.4x - 285.15$

(a) July

(b) $f(7) = -9.75(7)^2 + 128.4(7) - 285.15$
≈ 136

(c) $f(6) = -9.75(6)^2 + 128.4(6) - 285.15$
≈ 134

(d) early October to late February